International Conference on Clean Energy and Low Carbon Technologies (CELCT 2023)

E3S Web of Conferences Volume 441 (2023)

Wuhan, China
15-17 September 2023

Editors:

A.-S. Nizami
M.J.K. Bashir
H. Ganjidoust

ISBN: 978-1-7138-8440-8

Printed from e-media with permission by:

Curran Associates, Inc.
57 Morehouse Lane
Red Hook, NY 12571

Some format issues inherent in the e-media version may also appear in this print version.

For additional information, please contact EDP Sciences – Web of Conferences
at the address below.

EDP Sciences – Web of Conferences
17, Avenue du Hoggar
Parc d'Activité de Courtabœuf
BP 112
F-91944 Les Ulis Cedex A
France

Phone: +33 (0) 1 69 18 75 75
Fax: +33 (0) 1 69 28 84 91

contact-edps@webofconferences.org

Additional copies of this publication are available from:

Curran Associates, Inc.
57 Morehouse Lane
Red Hook, NY 12571 USA
Phone: 845-758-0400
Fax: 845-758-2633
Email: curran@proceedings.com
Web: www.proceedings.com

TABLE OF CONTENTS

CLEAN ENERGY UTILIZATION AND ENERGY STORAGE TECHNOLOGY

POLLUTION CONTROL AND LOW-CARBON ENERGY SAVING DEVELOPMENT

Preface

From 15th to 17th September 2023, 2023 International Conference on Clean Energy and Low Carbon Technologies (CELCT 2023) was held in Wuhan, China. The Conference was attended by about 100 researchers, scientists, engineers, and students working in the fields of clean energy and low carbon technologies.

The general objectives of CELCT 2023 were:

To create a forum for sharing, communication and research at the international level, so that participants can be informed of the latest research directions, contents and results of clean energy and low carbon technologies.

To provide approaches to the main problems facing clean energy and low carbon technologies, creating spaces that encourage discussion and joint development of research, technology development, and innovation.

Various excellent papers were approved, after revision, for presentation in the Conference Proceedings and publication in E3S Web of Conferences. The Conference was organized with many speeches about different research areas, various oral and poster presentations, as well as an academic investigation. There was an enabling environment in the Conference for discussion and exchange of ideas, and heated discussion took place during the entire Conference.

Both conference and presentation sessions were attended by many conference attendees, allowing closer relationships, both academic, as well as scientific and personal. The Conference was sponsored by major universities and academic institutions related to clean energy and low-carbon technologies.

CELCT 2023 was actually a space for integration of undisputed knowledge and sharing of peer experiences within the specific areas of clean energy and low-carbon technologies. The cordiality and good treatment of the attendees promoted a climate of friendship, which we could enjoy during all the development of the conference agenda.

We would like to extend an enthusiastic round of thanks to all our conference authors and speakers for their excellent contributions. We are indebted to the Technical Program Committee for their efforts in making and pushing the conference program. Special acknowledgement to the editors and other members of E3S Web of Conferences for their patience and endeavor while waiting for the final versions of the manuscripts.

The Committee of CELCT 2023

Committee Member

Conference Chair
Sunita Singh, IES College of Technology, India
Abdul-Sattar Nizami, Government College University Lahore, Pakistan

Technical Program Committee Chair
Mohammed J K Bashir, Universiti Tunku Abdul Rahman, Malaysia
Hossein Ganjidoust, Tarbiat Modares University, Iran

Technical Program Committee Member
Sonali Paunikar, IES College of Technology, Bhopal, India
Muhammad Hassan, Southwest Jiaotong University, China
Nicola Cantasano, National Research Council, Italy
Jean Pierre Bavumiragira, Tongji University, China
Malek Hassanpour, Osmania University, India

Organizing Committee Chair
Ali Ayati, Quchan University of Technology, Iran / ITMO university, Russia
Afshin Ebrahimi, Isfahan University of Medical Sciences, Iran

Organizing Committee Member
Abhishek Kumar Awasthi, Nanjing University, China
Amin Mojiri, Hiroshima University, Japan
Amir Parnian, National Salinity Research Center, Iran

Statement of Peer review

In submitting conference proceedings to *Web of Conferences*, the editors of the proceedings certify to the Publisher that

1. They adhere to its **Policy on Publishing Integrity** in order to safeguard good scientific practice in publishing.
2. All articles have been subjected to peer review administered by the proceedings editors.
3. Reviews have been conducted by expert referees, who have been requested to provide unbiased and constructive comments aimed, whenever possible, at improving the work.
4. Proceedings editors have taken all reasonable steps to ensure the quality of the materials they publish and their decision to accept or reject a paper for publication has been based only on the merits of the work and the relevance to the journal.

Title, date and place of the conference

2023 International Conference on Clean Energy and Low Carbon Technologies (CELCT 2023)

September 15-17, 2023
Wuhan, China

Proceedings editor(s):

Prof. Abdul-Sattar Nizami, Government College University Lahore, Pakistan
Prof. Mohammed J K Bashir, Jalan Universiti, Malaysia
Prof. Hossein Ganjidoust, Tarbiat Modares University, Iran

Date and editor's signature

Abdul-Sattar Nizami
Mohammed J K Bashir
Hossein Ganjidoust

17, avenue du Hoggar - PA de Courtabœuf – BP 112 - 91944 Les Ulis Cedex A (France)
Tél. : 33 (0)1 69 18 75 75 - Fax : 33(0)1 69 07 45 17 – www.edpsciences.org

E3S Web of Conferences **441**, 01001 (2023) https://doi.org/10.1051/e3sconf/202344101001
CELCT 2023

Application of McKinsey Curve in Building Emission Reduction Technology

Yubo Fan[1,a], Yingying Zhao[1,b], Yue Tian[1,c], Zhaojie Liu[1,d], Wenrong Si[1,e], Xuemei Chen[2,f*], Jingwu Jia[2,g]

[1]State Grid Shanghai Electric Power Research Institute, China East China Electrical Power Test & Research Institute, Shanghai, China
[2]Shanghai Twenty-first Century Energy Conservation Technology Co.,LTD., Shanghai, China

Abstract: In order to promote the application of McKinsey energy reduction cost curves in buildings, This paper introduces the concept of McKinsey abatement cost curves, the current application status of building abatement measures and abatement cost data, makes an analysis for the economic characteristics of different energy efficiency measures in retrofit projects, and presents the retrofit measures, annual energy savings and investment costs used in retrofit programs by taking the example of energy efficiency retrofits in office and commercial buildings in the past 2 years, discloses the data changes in abatement costs for some energy-saving retrofit technologies contrastingly, and suggests that dates should be iterated based on McKinsey abatement cost curves. The McKinsey abatement cost curve is useful for guiding the selection of building energy efficiency measures. The curve is economical and visualized compatibly and is a good reference for making abatement decisions through timely updating the data of retrofit technologies and influencing factors.

1. Introduction

In September 2020, Chinese government proposed to the United Nations General Assembly the "double carbon" goal of "striving to reach peak of CO_2 emissions by 2030 and achieving carbon neutrality by 2060". To achieve these goals, the Ministry of Housing and Urban-Rural Development has proposed in the 14th Five-Year Plan for the Development of Building Energy Efficiency and Green Buildings to strengthen the energy-saving and green renovation level of existing buildings and improve the energy-saving level of new buildings, and to complete the energy-saving renovation of existing buildings with an area of more than 350 million square meters by 2025, and to construct ultra-low energy consumption and near-zero energy consumption buildings with an area of more than 50 million square meters by 2022[1]. In November 2022, the State Administration, the National Development and Reform Commission, the Ministry of Finance, and the Ministry of Ecology and Environment issued the "Implementation Plan for conducting Green and Low-Carbon Leading Actions for Public Institutions in-depth to Promote Carbon Peaking", which clearly proposes that by 2025, the energy use structure of public institutions nationwide will be continuously optimized, energy efficiency will be continuously improved, energy consumption per unit floor space will be reduced by 5% and carbon emissions will be reduced by 7% on the basis of the 2020 level, areas with the necessary conditions will reach public institutions carbon peak by 2025, and the total carbon emissions of public institutions nationwide will reach the peak as early as 2030. In the context of "double carbon", the coordinated development of economy and environment, the economics, feasibility and implementation cycle of energy-saving renovation technology in the process of building emission reduction have become the issues that need to be considered in order to steadily improve building energy use efficiency, gradually optimize building energy use structure and reduce building energy consumption[2].

2. McKinsey Abatement Curve and the original intention of drawing it

When conducting a regional analysis of potential capacity for energy saving and emission reduction, we are facing with dozens choices of energy saving and emission reduction technologies .If there's a set of clear and intuitive expressions can be used to enable decision makers in energy saving and emission reduction planning to scientifically select the available technologies and understand the corresponding implementation costs so as to propose practical energy saving and emission reduction targets, which will become the main use of drawing energy saving and emission reduction cost curves.

As shown in Figure 1, in 2010, the Shanghai Energy Conservation Supervision Center was commissioned by the Changning District Government, under the guidance of WB experts, to conduct extensive research and statistics on a large number of individual retrofitting of large public buildings in the central area of Shanghai, for including adopted technology's frequency, distribution

[a] 351634373@qq.com, [b] zhaoyy@sh.sgcc.com.cn, [c] tianyue@sh.sgcc.com.cn, [d] liushaojie@sh.sgcc.com.cn,
[e] siwr@sh.sgcc.com.cn, [f*] xmchen_tt@163.com, [g] jjwl08@163.com

characteristics, project cycle, energy savings, economic parameters, ease of implementation, ease of maintenance, potential risks , etc. to establish a comprehensive database[3]. Using the McKinsey Abatement Curve Drawing Tool of AHP analysis method and score weighting combined multi-index as the main tool, we successfully compiled the energy saving and emission reduction cost curves for the Changning Hongqiao area[4], which provided an important reference basis for the district government to set the low-carbon development goals.

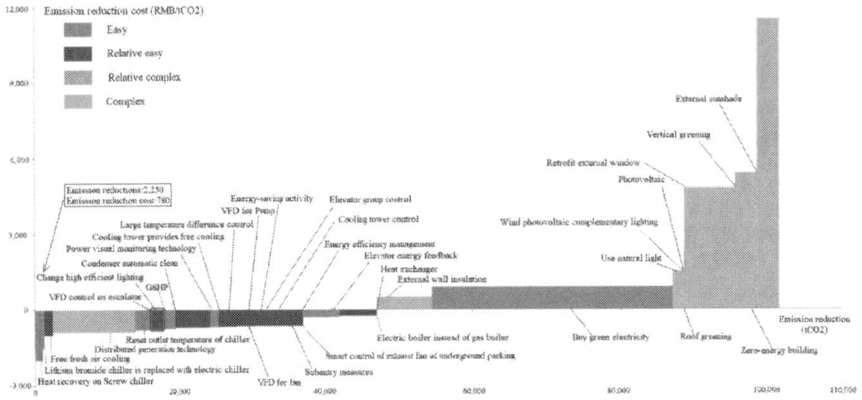

Figure 1 McKinsey Abatement Curve

3. Interpreting the McKinsey Abatement Curve

The cost curve not only covers spatial characteristics, but also implies a temporal characteristic, which refers to the expected year of potential capacity for energy saving and emission reduction. In different expected years, The technologies included in the corresponding curves, the amount of emission reduction for each technology will vary, the cost of abatement for each technology also changes somewhat, Figure 1 shows the data of expected year 2020.

The most important elements in the cost curve are the bar boxes, each of which represents a project (corresponding to a technology or product). The horizontal (X-axis) width of the bar indicates the size of the project's emission reductions, the height of the bar (Y-axis direction) indicates the amount of the unit abatement cost of the project (using NPV dynamic costing). Emission reductions can be measured in tons of carbon dioxide or tons of carbon dioxide equivalent (in the presence of other GHG emissions), and unit abatement costs can be measured in ($/ton of carbon dioxide) or ($/ton of carbon dioxide equivalent). There are usually two calibers of emission reductions: one is the annual emission reductions for the expected year, and the other is the sum of annual emission reductions for the expected year. The first caliber is process independent and the second caliber is process related, the earlier the technology is implemented, the greater the total amount of emission reductions, but for some technologies the unit abatement cost may be higher. Figure 1 shows the caliber X, All bar boxes are arranged from lowest to highest cost (from left to right) by according to per unit abatement. The tops of the bars form a horizontal S shape, which is the reason for the "curve" name. The cost curve provides key information for decision makers, including:

(1) The applicable abatement technologies in the region and the magnitude of the abatement capacity of each technology.

(2) Maximized abatement capacity (potential technical capacity), which would be the limit of the abatement target, and different low-carbon scenario analysis based on regional energy savings or potential abatement capacity curves can be used to analyze energy savings and low-carbon targets under different models.

(3) For Verifying the likelihood of implementation of emission reduction targets set through top-down approaches, generally the emission reduction achieved through technological progress can be calculated based on the emission reduction target. If the emission reduction target is on the left side of the zero point of the unit abatement cost, the target is likely to be achieved through market mechanisms. If on the right side, it indicates that the target requires additional push from the government to be achieved, the abatement cost of the technology corresponding to that target can be considered as the cost to the government in the additional push.

(4) The difficulty of implementing the project gives more information for decision making at the time of implementation. Some energy efficiency and emission reduction technologies have some important but unquantifiable benefits as a side effect, but the decision from cost can cause some bias. The difficulty of implementation will provide important information for scientific decision making, and can be analyzed for energy efficiency measures implementation decisions. Special attention can be paid to techniques that are not too difficult to implement above the X-axis of the curve.

(5) Corresponding curves have a corresponding a comprehensive database involving a large amount of

information, concerning basic data such as investment cost, operation cost, lifetime, energy saving rate and energy efficiency changes related to each technology, and more valuable information can be obtained through in-depth analysis of this comprehensive database[5].

(6) As shown in Figure 1,as for the column box of "More efficient lighting"[6], which indicates that the emission reduction of this technology is 2250 tons of carbon dioxide, and the emission reduction cost is 780 yuan/ton of carbon dioxide.

Ninety-three projects using the same technology were used as the basic data to calculate the economy of the energy-saving transformation technology, and the period was set for 10 years. Calculate the total cost and total emission reduction of the projects to obtain the emission reduction cost curve data.

4. Application in energy-saving renovation examples

Table 1 shows the technical and economic analysis of the energy-saving renovation of Project A. Project A was put into operation in 2007 with a building area of 35,000 square meters. It is business type, and the following renovation was completed in 2021.

Table 1 Technical and economic analysis of project A energy-saving renovation

Project-A	Emission reduction cost (RMB/tCO2)	Emission reduction (tCO2)	Energy saving cost (RMB /tce)	Energy saving (tce)
Lighting system energy saving	129	50	278	11.56
Air conditioning refrigeration and cooling pump replacement	142	51	492	14.62
Cooling tower replacement	622	52	13263	2.43
Fresh air volume control (automatic control)	270	53	363	39.26
Refrigeration Station Group Control	102	54	2023	2.72
Air-cooled heat pumps replace electric boilers	3102	55	1183	143.65

Table 2 shows the technical and economic analysis of the energy-saving renovation of Project B. Project B was put into operation in 2005 with a building area of 25,000 m^2. It is business type, and the following renovation was completed in 2012.

Table 2 Technical and economic analysis of project B energy-saving renovation

Project-B	Emission reduction cost (RMB/tCO2)	Annual Emission Reduction (tCO2)	Energy saving cost (RMB/tce)	Annual Festival Energy (tce)
Window Film	686	50	960	36
Air-cooled heat pump with wind-pulling cover	3527	5	4938	3
Roof exterior windows to aluminum louvers	496	3	694	2
Air-cooled heat pump condensing fin spray	947	15	1325	11
Air-cooled heat pump to water-cooled screw	3237	42	4532	30
Energy saving in public area lighting systems	1341	47	1877	33
Energy saving in tenant area lighting system	2571	63	3600	45
Fresh air unit inverter	893	3	1250	2
Operation Management	551	38	772	27

Compared with the McKinsey reduction curve, adopting the same energy-saving technology[7]:

(1) Annual energy savings are mainly related to the base energy consumption of the project and the scale of the renovation.

(2) Annual emission reductions are mainly related to the project's base energy consumption, the scale of the renovation, and the emission reduction factors.

(3) The annual cost of energy saving and emission reduction, in addition to the above factors, is also related to the cost of equipment and operating life cycle.

A single item common to the McKinsey abatement cost curve in the retrofit example of Project A and Project B was selected for analysis, as shown in Table 3, What can be seen:

(1) For the same technology, the annual emission reduction is mainly related to the basic energy consumption of the building and the scale of renovation. The larger the individual energy consumption corresponding to the technical measures and the larger the scale of renovation, the larger the annual emission reduction .

(2) The abatement cost of energy-saving renovation technology directly affects the frequency of adoption of this technology. Comparing the energy-saving cost curve table with the economic data of renovation cost of A and B projects, it can be seen that in terms of building renovation technology, the economic trend of energy-saving technical renovation measures has not changed, and lighting renovation and inverter renovation of pumps and fans are still the measures with lower abatement cost and are the preferred measures for energy-saving renovation projects.

(3) The abatement costs of measures such as group control of refrigeration stations and operation and management are beginning to fall, and their energy-saving effects will become increasingly evident during the digital construction of buildings in the context of the 3060 double carbon.

(4) Exterior windows and other envelope retrofit measures continue to be poorly economized.

(5) Lighting energy-saving retrofit measures, in the case of lighting product updates and upgrades iterations faster, the price of the product itself decreases, coupled with the new product life extension so that the frequency of product replacement decreases, the cost of emission reduction of this technology significantly decreased.

(6) Water pump and fan frequency conversion transformation measures[8], single unit transformation effect varies greatly because in the building energy-saving transformation measures, the price of the equipment itself does not change significantly, the basic energy consumption and the difficulty of transformation caused by the cost and emission reduction fluctuations of the influencing factors, so the regional emission reduction cost of the technology may be relatively lower than the emission reduction cost of a single building, but for the building itself energy-saving technology selection, energy-saving potential still occupies an indispensable advantage.

(7) Energy-saving renovation measures such as envelope structure, the amount of renovation monomer is small, the renovation target and the products used vary greatly, and the referenceability of the economic cost curve is low.

Table 3 Benchmark analysis between energy saving retrofit example and McKinsey curve

Project (McKinsey)	Emission reduction costs (RMB/tCO_2)	Project-A	Emission reduction costs (RMB/tCO_2)	Project-B	Emission reduction costs ($yuan/tCO_2$)
Lighting system energy saving	780	Lighting system energy saving	129		
Refrigeration Station Group Control	670	Refrigeration Station Group Control	102		
Air conditioning pump inverter	648			Air conditioning pump inverter	1031
Fan Frequency Conversion	650			Fresh air unit inverter	893
Public building exterior window renovation	4800			Exterior window film	686
Energy saving operation management	645			Operation Management	551

5. Development and suggestions

McKinsey reduction cost curve data are based on energy-saving renovation cases in hot summer and cold winter low areas.The data comes from "Changning Energy Consumption Monitoring and carbon emission Comprehensive Management Platform"[9].The curve has guiding significance for the selection of energy saving measures for large public buildings over 20,000 square meters.At present, the case data does not relate to residential buildings.

In the dual carbon strategy studies of applying abatement cost curves [10], the plotting of cost curves contains two types of parameters, one belonging to the set parameters and the other belonging to the input parameters. The set parameters involved, such as: life cycle of a single technology, carbon emission factor, standard coal conversion factor, subsidy cost, electricity tariff, etc., need to be updated timely according to the latest documents; the input parameters involved, such as: single project cost, discount rate, etc., are the factors that mainly affect the cost of carbon emission.

When new products, new technologies and new specifications appear, the relevant impact factors should be iteratively upgraded and managed. Generally speaking, it is necessary to iterate the cost of products in 3~5 years, iterate the innovative application of technology in 5~10 years, and iterate the parameters such as emission reduction factors and discount rates every year. The project database needs to be supplemented and increased in a timely manner so that the cost data can be iterated in a timely manner and the latest parameter data can be obtained to further improve the accuracy of abatement costs and provide a reliable basis for making decisions on abatement strategies.

References

1. Zhang Kai,LU Yumei,LU Haishu,Countermeasures for High Quality Development of Green Buildings in China under the Background of "Double Carbon" Goal, CONSTRUCTION ECONOMY, 2022, Vol. 43, No.3, pp.14-20.

2. WU P, SONG Y Z, ZHU J B. et al. Analyzing the Influence Factors of the Carbon Emissions from China's Building and Construction Industry from 2000 to 2015 [J], Journal of Cleaner Production, 2019, 221: 552-566.
3. Application of Cost Reduction Curve Method in the Low Carbon Strategy of Changning District Shanghai
4. Investment scale estimation of low-carbon development projects in Changning, Shanghai
5. Wei Tian, A Review of Sensitivity Analysis Methods in Building Energy Analysis[J].Renewable and Sustainable Energy Reviews,2013,20(4) :411- 419
6. Shan Yu,Ying Zhong,Guohua Wang,Gongming Zhang,"Energy Saving Performance of LED Lighting Products in Lighting Retrofitting of Public Institutions", Building Energy efficiency, No.12 in 2021, Vol. 49, No.370, pp.160-164.
7. McKinsey&Company. China's Green Revolution: Technology Choice for Energy and Environment Sustainable Development [R/OL]. 2009 [2010-08-18].
8. Ding Yong, Wei Jia,Energy Efficient Analysis of Chilled Water Pump Frequency Renovation, Building Energy efficiency, No.9 in 2015, Vol. 43, No.295, pp.1-7.
9. Li Huimin, Qi Ye,Comparison on China's Carbon Emission Scenarios in 2050,Advances In Climate Change Research, 2011, Vol. 7, No.4, pp.271-280.
10. Zhou Nan, Fridley D, Mcneil M, et al. China's energy and carbon emissions outlook to 2050 [R]. CA: Lawrence Berkeley National Laboratory, 2010

Interfacial Properties of Graphene Oxide/Epoxy Resin Coatings in Corrosive Environments

Ji-Qiang Li[a], Jun-Jun Shang[b*], Qing-Sheng Yang[c*]

Department of Engineering Mechanics, Faculty of Materials and Manufacturing, Beijing University of Technology, Beijing 100124, China

Abstract: Graphene oxide/epoxy composite coatings have attracted much attention from researchers due to their excellent anti-corrosion properties. Graphene oxide as a two-dimensional nano-filler can enhance the overall properties of epoxy resin coatings. In the paper, models of graphene oxide/epoxy composite coatings with different permeation degrees are constructed respectively. The interfacial interaction between GO and epoxy resin and the adsorption properties of epoxy coating with metal are analyzed from an atomic perspective by molecular dynamics simulations. The results indicate that the interfacial adsorption effect of the epoxy coating on the protected metal is stronger than the interfacial bonding of GO on the epoxy resin matrix. Simulations of models with different degrees of permeation suggest that the permeation of the NaCl solution occupies the interfacial space and reduces the adhesion between the two interfaces.

1. Introduction

Graphene oxide/epoxy composite (GO/EP) coatings have excellent anti-corrosive properties [1], but seawater permeation is inevitable. The permeation of seawater can lead to severe degradation of the performance of the polymer composite coating [2]. Previously, the anticorrosion property of the epoxy coating has been investigated in experiments [3,4]. Researchers have also analyzed the diffusion properties of water in polymers through molecular dynamics (MD) simulations to investigate the bonding state of water and epoxy resin (EP) and their effects on the properties of the composites [5]. However, these simulation efforts have paid little attention to the interaction with GO-EP under the influence of water and the adsorption of the epoxy coating to the metal interface.

Especially in the interface of GO-EP, water permeation affects the bond strength between them [6]. The experimental characterization of the interactions amongst water, GO, and EP is a challenging task. MD simulations have received attention in recent years as a complement to experiments to accomplish atomic characterization problems. The dynamic process of the corrosive medium in the coating and its diffusion behavior can be obtained by MD simulation [7,8].

Epoxy coating and metal substrate interface adsorption performance is also a significant indicator of the coating corrosion resistance property. The diffusion of NaCl solution between the epoxy coating and the metal interface will lead to further aggravation of corrosion and, eventually, the exfoliation of the epoxy coating. The adsorption properties of polymer coatings also play an essential role in corrosion resistance [9]. Graphene oxide can improve the anticorrosion properties of EP coatings and enhance the adsorption of the coatings to metals. However, there is still a certain gap in the understanding of the interface adsorption property between the epoxy coating and metal substrate in a corrosive environment. Inspired by these problems, MD simulation is carried out in this paper to provide new insights into the interfacial and adsorption properties of NaCl solutions on GO/EP coatings.

2. Molecular dynamics modeling and simulation methods

To investigate the influence of NaCl solution on the GO-EP interaction and the adsorption of epoxy coating with the metal interface, a GO/EP coating permeation model is developed, and MD simulations are performed using LAMMPS. The simulation considers only the permeation of the solution. No electrochemical reactions occur. The permeation medium is a 3.5% mass fraction NaCl solution, mostly water molecules.

[a]lijq@emails.bjut.edu.cn, [b*]shangjunjun@bjut.edu.cn, [c*]qsyang@bjut.edu.cn

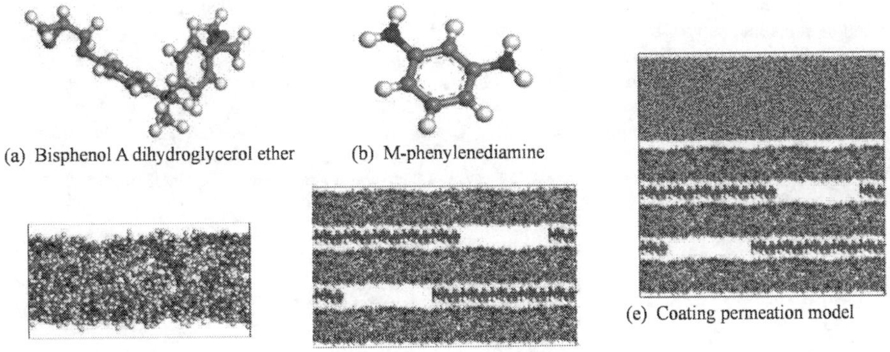

(a) Bisphenol A dihydroglycerol ether (b) M-phenylenediamine (c) EP monomer (d) GO/EP composite coating model (e) Coating permeation model

Figure 1. Coating model

As indicated in Figure 1, the GO/EP coating uses a laminated structure in the z-direction due to the large size of the coating model. Three EP layers and two lapped GO pieces are laminated together to form the coating, where the single EP layer is made of EP unit cell supercell. As is indicated in Figure 1(c), the EP monomer is modeled by bisphenol A diglycidyl ether (DGEBA) and m-phenylenediamine (MPD) curing agent at a molar ratio of 1:1. The initial density of the model is set to 1.0 g/cm^3, the model size is 36×38×60 Å, and the crosslinking degree is controlled to 90%. The cross-linked model is 3×3 arrayed in *x* and *y* directions to obtain the requested monolayer EP layer, and the model size is 110×115×20 Å. The GO/EP coating permeation model is a NaCl solution with a thickness of 36.8 Å over GO/EP coating. The simulations are performed with periodic boundary conditions, and the GO/EP coating is relaxed for 100 ps each under the canonical ensemble, isothermal-isobaric ensemble, and canonical ensemble (NVT, NPT, NVT). The EP molecular chains can fully wrap the nanomaterials and finally reach the equilibrium state. After the relaxation of the GO/EP coating, it is compounded with the NaCl solution. The permeation simulation is performed after the relaxation of the NVT ensemble for 200 ps.

3. Results

3.1 Analysis of permeation results

As shown in Figure 2, the NaCl has permeated to the bottom of the coating at 2000 ps, and the GO has been deformed and displaced. With increased permeation time, GO deformation recovers partly after more downward diffusion of NaCl solution at 3000 ps, which has a strong impact on the study of the interaction of GO with EP and the adsorption between coating and metal interface. Permeation is a continuous process, and GO deformation could not recover before, thus losing its barrier effect.

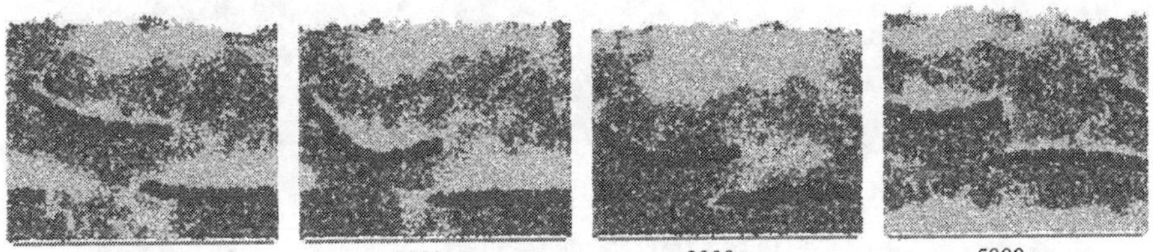

1000 ps 2000 ps 3000 ps 5000 ps

Figure 2. GO/EP composite coating permeation simulation results

3.2 Overlap modeling analysis

The results show that the NaCl solution of this thickness could not achieve a continuous permeation process. The GO/EP coating contains less NaCl solution in the surface layer at the late stage of penetration, and most of the atoms reach the lower layer, which has a great impact on the testing of interfacial properties. Multiple attempts using different thicknesses of NaCl solution and different permeation rates also yield the same results, so the following solution is selected.

Based on the above findings, the overlapping modeling approach is used to directly model GO/EP coatings with different degrees of permeability. Two models of GO/EP coating and NaCl solution are overlapped, and then the model is set to reach a reasonable configuration by energy minimization and relaxation. The NaCl solution is added to the free volume of GO/EP coating.

The GO/EP composite coating model shown in Figure 3(a) is divided into three parts from top to bottom according to the black line, which is defined as the first part, the second part, and the third part. According to the results in Figure 2, the NaCl solution amount is 55.78% of the total solution in the first part at 1000 ps, 38.80% in the second part at 2000 ps, and 30.68% in the third part at 5000 ps. The NaCl solution is compounded with EP according to the above three ratios. The post-permeation model with gradient distribution is built directly after energy minimization and relaxation.

The NaCl solution is compounded with the GO/EP coating according to the above three ratios and GO/EP-Fe-2 in Figure 3(c) is established. GO/EP-Fe-1 is an impermeable model and GO/EP-Fe-3 has a 3.9 Å thick NaCl solution added at the coating-metal interface of GO/EP-Fe-2. GO/EP-Fe-3 represents the further accumulation of permeable media at the interface. An iron substrate is placed underneath the coating, and relaxing the coating and the substrate will fully interact. In the figure, the red atom is the metal substrate, the yellow atom is GO, the white atom is the NaCl solution, and the other atoms are epoxy resin.

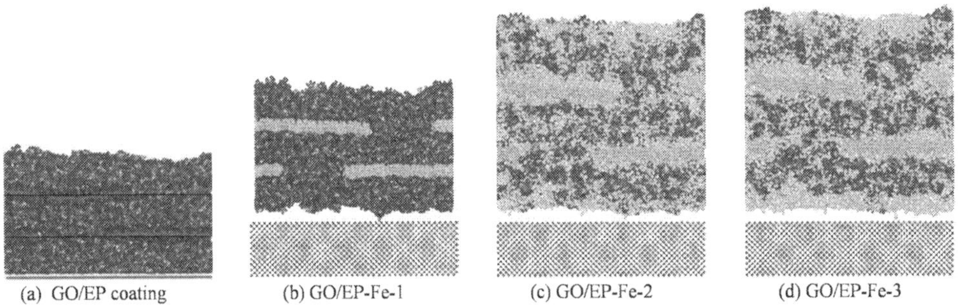

Figure 3. Overlap modeling schematic

The stretching simulation of the coating is performed under the NVT system ensemble by fixing the top and bottom part of the atoms and applying a velocity of 0.001 Å/fs to the top fixed atoms and for 100 ps. As shown in Figure 4, the results show that GO/EP-Fe-1 is pulled apart at the interface between GO and epoxy resin, but some molecular chains are adsorbed on the undersurface of GO.

The interaction potential energy in Figure 5 also shows that the coating-metal interaction potential energy is higher than that between GO and the epoxy, so it is not pulled away from the coating-metal interface. The NaCl solution in GO/EP-Fe-2 occupies the surface of GO and reduces the interaction of GO and epoxy. However, most of the NaCl solution is water molecules. The water molecules, GO, and epoxy will form hydrogen bonds to weaken the damage caused by the NaCl solution [10]. Thus, there is still the adsorption of molecular chains at the interface. GO/EP-Fe-3 is pulled away from the coating-metal interface, and the accumulation with NaCl solution lowers the adsorption capacity of the coating.

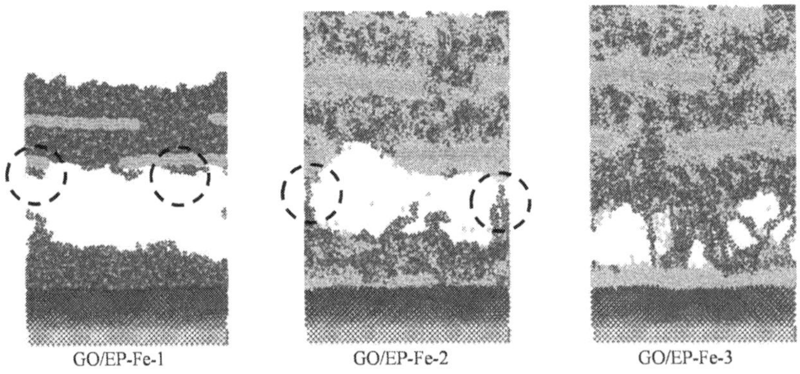

Figure 4. Graph of coating stretching results

The interaction potential energy can be used to characterize the changes in the interfacial properties during the simulation. The low interaction energy means the strong interfacial bonding of the coating. The potential energy of GO/EP-Fe-1 in Figure 5(a) is lower than that of the other models during initiation and stretching. The results indicate that the GO/EP coating has better interfacial properties when not penetrated. The permeation of the NaCl solution destroys the GO-epoxy interface, resulting in higher potential energy values for GO/EP-Fe-2 and GO/EP-Fe-3.

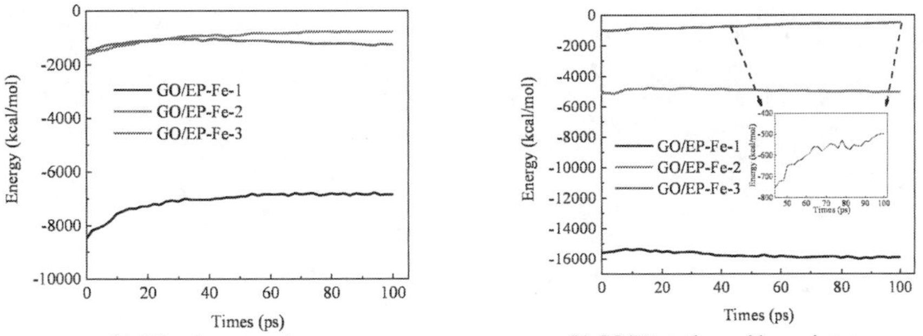

(a) GO and epoxy resin (b) GO/EP coating and iron substrate

Figure 5. Interaction potential energy

The potential energy of GO/EP-Fe-1 in Figure 5(b) is lower than that of the other models during the onset and stretching. The results indicate that the GO/EP coating has better interfacial properties with the iron substrate when not penetrated. The permeation of NaCl solution leads to the destruction of the interface and the decrease of the adsorption capacity of the coating. Thus, the interaction potential energy of GO/EP-Fe-2 and GO/EP-Fe-3 also becomes larger with the deepening of the permeation. The interface between the coating and the iron substrate is pulled apart in GO/EP-Fe-3, corresponding to the gradual increase of the interaction potential energy in the enlarged figure.

4. Conclusions

In this paper, the interfacial interaction between GO and EP and the interfacial adsorption of epoxy coating with the metal are investigated by MD simulation. The results indicate that the interfacial adsorption of the epoxy coating to the metal is stronger than the interfacial bonding of GO to the epoxy matrix, which also indicates the excellent adsorption properties of the epoxy resin. Simulation results for different degrees of permeation of NaCl solution indicate that the permeation of NaCl solution occupies the interfacial space and weakens the bond between the two interfaces.

Interfacial properties play an essential role in the performance of fiber-reinforced polymers, and the fiber-polymer interface undertakes the important task of stress transfer. Epoxy coating and metal substrate interface adsorption performance is also an important guarantee that the coating of anti-corrosion properties gives full play. The results of this effort enhanced our comprehension of the properties of two interfaces of epoxy coatings in corrosive environments. In addition, the adhesion of the epoxy coating-metal is not a purely physical connection, and hydrogen bonds may be formed at the interface, which requires further analysis and research.

Acknowledgments

This work was funded by the National Natural Science Foundation of China (Project No. 11902010,11932002), which is gratefully acknowledged.

References

1. Rajabi, M., Rashed, G.R., Zaarei, D. (2015) Assessment of graphene oxide/epoxy nanocomposite as corrosion resistance coating on carbon steel. Corrosion engineering, science, and technology, 50:509-516.
2. Verma, C., Olasunkanmi, L.O., Akpan, E.D., Quraishi, M.A., Dagdag, O., El Gouri, M., Sherif, E.M., Ebenso, E.E. (2020) Epoxy resins as anticorrosive polymeric materials: a review. Reactive and Functional Polymers, 156:104741.
3. Chen, L., Liu, H., Liu, Z., Song, Q. (2021) Thermal conductivity and anti-corrosion of epoxy resin based composite coatings doped with graphene and graphene oxide. Composites Part C: Open Access, 5:100124.
4. Zhu, X., Yan, Q., Cheng, L., Wu, H., Zhao, H., Wang, L. (2020) Self-alignment of cationic graphene oxide nanosheets for anticorrosive reinforcement of epoxy coatings. Chem. Eng. J., 389:124435.
5. Li, W., Chai, W., Zhang, L., Guo, Y., Wang, W., Chen, S. (2022) Atomic insight into the influences of moisture ingress on the structures and dynamics of graphene-epoxy interfaces. Compos. Sci. Technol., 219:109222.
6. Li, W., Zhang, L., Zhang, M., Chen, S. (2020) Structures of graphene-reinforced epoxy coatings and the dynamic diffusion of guest water: a molecular dynamics study. Ind. Eng. Chem. Res., 59:20749-20756.
7. Bahtiri, B., Arash, B., Rolfes, R. (2022) Elucidating atomistic mechanisms underlying water diffusion in amorphous polymers: an autonomous basin climbing-based simulation method. Comput. Mater. Sci., 212:111565.
8. Yang, Y., Zhao, H. (2022) Water-induced polymer swelling and its application in soft electronics. Appl. Surf. Sci., 577:151895.
9. Wang, S., Zhu, J., Rao, Y., Li, B., Zhao, S., Bai, H., Cui, J. (2019) Polydopamine modified graphene oxide-tio2 nanofiller for reinforcing physical

 https://doi.org/10.1051/e3sconf/202344101002

properties and anticorrosion performance of waterborne epoxy coatings. Applied Sciences, 9:3760.

10. Tam, L., Lau, D. (2015) Moisture effect on the mechanical and interfacial properties of epoxy-bonded material system: an atomistic and experimental investigation. Polymer, 57:132-142.

Modeling Analysis and Study of Unbalanced Funds Impact on Power Spot Market with High Share of Clean Energy

Ruiqin Duan[1a*], Jinding He[1b], Yan Jiang[1c], Shuangquan Liu[1d], Qizhuan Shao[1e], Chuan Zhao[1f], Beiyu Gan[2g], Jianbin Lin[2h]

[1]Power Dispatch Center of Yunnan Power Grid Co. Ltd., Kunming 650011, China
[2]Beijing Tsintergy Technology Co. Ltd., Beijing 100080, China

Abstract-Under the background of "double carbon" and large-scale grid connection of clean energy, the uncertainty of clean energy output and the market reforms to deal with this uncertainty will affect the interests of market participants through settlement and unbalanced funds. Focusing on the background of the power spot market for clean energy, this paper first analyzes the unbalanced funds items specific to Chinese power spot market, then conducts a qualitative analysis of the impact of high proportions of clean energy on unbalanced funds. Finally, from the perspective of quantitative analysis, three prediction models for the impact of unbalanced funds are proposed to provide technical means for prediction and evaluation of the changes in the interests of power market participants under the background of China's energy transition in the future.

1. Introduction

With the construction of the electricity spot market on a large scale nationwide, each province and region are gradually carrying out the construction of the electricity spot market, and both the contract for differences (CFD) and the electricity spot market are settled on the floor. From the perspective of the central counterparty, if the total settlement funds of all counterparties are not zero, the difference is called "unbalanced funds", which means the unbalanced funds are the total market revenue minus the total market expenses. The reason why the electricity market generates cash flow is that the supplier provides a certain type of electricity derivative commodity and the demand side pays the cost. Then the market settlement needs to set up independent caption of account between the supplier and the demand side based on the counterparty relationships sorted out for all the electricity and electricity derivative commodity items to achieve the budget balance. Therefore, in a market with well-established settlement mechanism, the unbalance funds should be 0 regardless of the rounding of bills, measurement errors, etc. When the imbalance is not 0, it is inevitably due to the inequality between the costs incurred by the demand side and the costs received by the suppliers in the electricity and derivatives markets.

At present, China's provincial and regional electricity spot markets are in the preliminary construction stage, usually focusing more on the design of the electricity energy spot market and paying less attention to the settlement of electricity derivative commodities, such as: the need to encourage suppliers to meet the demand for capacity, peaking and other electricity derivative commodities through a unilateral market or cost compensation in order to guarantee the stable operation of the market and the grid; the congestion surplus caused by transmission congestion; and other costs like the China-specific dual-track deviation cost. These costs are subject to unclear settlement items and unclear division of authority and responsibility, resulting in the consolidation of multiple items that should be settled independently under one account of "unbalanced funds" for diversion. Therefore, the unbalanced funds in this paper refers to such broad unbalanced funds, and the study objects are the settlement items that are often combined in the unbalanced funds accounts in the electricity spot market of each province and region in China.

Existing studies in domestic electricity market mainly focus on classifying unbalanced funds and designing allocation mechanism, literature [1-3] classified unbalanced funds from different perspectives, literature [4] carried out the design of unbalanced funds allocation mechanism, literature [5] analyzed the impact brought by different unbalanced funds allocation mechanisms based on ABM model.

Under the policy goal of reaching carbon peak by 2030 and achieving carbon neutrality by 2060[6], clean energy will gradually become the main source of power supply. The uncertainty of clean energy and the market reforms to deal with this uncertainty will affect the interests of market participants through the settlement and unbalanced funds, and the impact of clean energy on unbalanced funds has not been analyzed yet.

Focusing on the background of the power spot market for clean energy, this paper first analyzes the unbalanced

[a*]duanruiqincsg@hotmail.com, [b]hejinding@yn.csg.cn, [c]jiangyan@yn.csg.cn, [d]liushuangquan@gmail.com, [e]shaoqizhuan@yn.csg.cn, [f]zhaochuan@yn.csg.cn, [g]bybeatricegan@126.com, [h]linjb@tsintergy.com

funds items specific to China's power spot market, then conducts a qualitative analysis of the impact caused by high share of clean energy on unbalanced funds, and finally, from the perspective of quantitative analysis, proposes three predictive models of unbalanced funds impact: a prediction model based on simulation clearing, a prediction model based on copula function and fluctuation contagion chain, and a prediction model based on ES and quantile regression. It also puts forward suggestions on the application scenarios of the forecast model for the impact of unbalanced funds, and provides technical means for predicting and evaluating the changes in the interests of power market participants in the context of China's energy transition in the future.

2. Unbalanced Funds Specific to China's Power Spot Market

2.1. Dual-track deviation

Chinese electricity industry is transitioning from planning-based mechanism to market-based mechanism. In the power spot market clearing model, there is no division between units and loads as to whether they are market participants, and the clearing model is the optimization of the output of units and loads on the whole network, so when there is no line congestion and the units and loads on the whole network are settled according to the locational marginal prices (LMP), the market expenditure and revenue are balanced.

$$\sum_i P_{\text{u-m},i}\, Q_{\text{u-m},i} + \sum_j P_{\text{u-p},j}\, Q_{\text{u-p},j} = \sum_\beta P_{\text{l-m},\beta}\, Q_{\text{l-m},\beta} + \sum_\varepsilon P_{\text{l-p},\varepsilon}\, Q_{\text{u-p},\varepsilon} \quad (1)$$

Where, $Q_{\text{u-m},i}$, $Q_{\text{u-p},j}$, $Q_{\text{l-m},\beta}$ and $Q_{\text{u-p},\varepsilon}$ respectively denote the spot power market clearing power of market unit i, planned unit j, market load β, and planned load ε, and $P_{\text{u-m},i}$, $P_{\text{u-p},j}$, $P_{\text{l-m},\beta}$, $P_{\text{l-p},\varepsilon}$ denote the clearing prices of market unit i, planned unit j, market load β, and planned load ε, respectively.

However, the current market settlement requires separating the settlement of units and loads that are still in the planning-based system and those that have become market participants. The revenue and expenditure of units and loads in the market are likely to be unbalanced, i.e. $\sum_i P_{\text{u-m},i}\, Q_{\text{u-m},i} \neq \sum_\beta P_{\text{l-m},\beta}\, Q_{\text{l-m},\beta}$, and the unequal part of these two is the dual-track deviation.

Depending on the dual-track mode adopted by each province, Guangdong province adopts the decoupling mode[7] , in which the unit output is re-classified as "whether or not it is market output" at the settlement stage, so that the market unit output is equal to the market load, and the deviation of the dual-track system is only caused by the price mismatch between market power generation and market consumption, i.e. $P_{\text{u-m},i} \neq P_{\text{u-p},j}$; Shanxi, Shandong and other provinces adopt the non-decoupling mode in which power will not be adjust at the settlement stage[7][8], and the deviation of the dual-track system is brought by the mismatch of both quantities and prices between the market power generation and market consumption.

2.2. CfD basis amount

(1) Basis cost between the medium- and long-term market and the day-ahead market

At present, all provinces and regions have adopted the "day-ahead energy market + real-time energy market" electricity spot market design structure, that is, the market structure adopted by PJM, CAISO and other electricity markets. With this market structure, the real-time power market is the real "spot market", and the day-ahead power market is financial in nature and can be regarded as an ultra-short-term contract for difference in the real-time power market, so PJM market has introduced non-actual electricity generating and producing institutions such as financial institutions as participants in virtual transactions in the day-ahead electricity energy market in order to activate day-ahead energy market transactions and discover market prices. In addition, the futures on financial exchanges and the current medium- and long-term CfDs in China are CfDs for longer time periods. Thus, under such a market structure, the market settlement would be calculated as follows.

$$R_i = \sum_c Q_{c,i} \cdot (P_{c,i} - P_{\text{base},c,i}) + Q_{\text{da},i} \cdot (P_{\text{da},i} - P_{\text{rt},i}) + Q_{\text{rt},i} \cdot P_{\text{rt},i} \quad (2)$$

where R_i represents the revenue of unit i, $Q_{c,i}$, $P_{c,i}$ and $P_{\text{base},c,i}$ denotes the contract volume, contract price and settlement point price of the cth CfD or futures of market participant i, $Q_{\text{da},i}$ and $P_{\text{da},i}$ denotes the clearing volume and price of market participant i in the day-ahead spot market for electricity, , $Q_{\text{rt},i}$ and $P_{\text{rt},i}$ denotes the clearing volume and price of market participant i in the real-time spot market for electricity, respectively.

Under the current settlement mechanism of power market in China's provinces and regions, the revenue $R_{\text{set},i}$ of the unit i can be presented at in the following form.

$$R_{\text{set},i} = \sum_c Q_{c,i} \cdot P_{c,i} + (Q_{\text{da},i} - Q_{c,i}) \cdot P_{\text{da},i} + (Q_{\text{rt},i} - Q_{\text{da},i}) \cdot P_{\text{rt},i} \quad (3)$$

Let $A_{\text{con}} = R_i - R_{\text{set},i}$, then it can be found that equation (3) ignores the basis difference between the CfD settlement point price and the day-ahead market LMP compared to equation (2), resulting in a deviation charge $A_{\text{CfD2da},i}$.

$$A_{\text{CfD2da},i} = \sum_c Q_{c,i} \cdot (P_{\text{da},i} - P_{\text{base},c,i}) \quad (4)$$

Similarly, the expenses C_j , $C_{\text{set},j}$ for load j, and the deviation charges $A_{\text{CfD2da},j}$ can be formulated as in equation (2) (3) (4). Since the CfD and the electricity energy spot are both settled on the floor, the sum A_{CfD2da} of the deviation expense is calculated on both sides of the generation and consumption.

$$A_{\mathrm{CfD2da}} = \sum_{c} Q_c \cdot (P_{\mathrm{da},i} - P_{\mathrm{da},j}) \tag{5}$$

where Q_c represents the contract volume of the cth CfD, and unit i and load j are the counterparties to this CfD, respectively.

(2) Basis cost between the day-ahead market and the real-time market

Based on $\sum_i R_i - \sum_j C_j$,the deviation cost between generation and consumption costs can be calculated as follows.

$$\sum_i R_i - \sum_j C_j = \sum_i Q_{\mathrm{da},i} \cdot P_{\mathrm{rt},i} - \sum_j Q_{\mathrm{da},j} \cdot P_{\mathrm{rt},j} - A_{\mathrm{da,block}} - A_{\mathrm{rt,block}} - A_{\mathrm{CfD2da}} \tag{6}$$

where $A_{\mathrm{da,block}}$, $A_{\mathrm{rt,block}}$ are the congestion surpluses in the day-ahead and real-time markets, respectively. It can be found that there is a deviation cost A_{da2rt} between the revenue on the generation side and the expenditure on the consumption side.

$$A_{\mathrm{da2rt}} = \sum_i Q_{\mathrm{da},i} \cdot P_{\mathrm{rt},i} - \sum_j Q_{\mathrm{da},j} \cdot P_{\mathrm{rt},j} \tag{7}$$

Comparing equations (5) and (7), it can be found that the form of the two deviation cost calculation formulas is the same, for the medium- and long-term CfDs and the day-ahead market for two different time scales of CfDs, whose incomplete hedging causes a price basis difference and thus accumulating the market deviation cost.

3. Qualitative Analysis of the Impact of Clean Energy on the Unbalanced Funds in the Electricity Spot Market

3.1. Impact of high proportion of clean energy on existing unbalanced capital account

According to the unbalanced cash items of electricity in various provinces and regions, the possible impacts of high proportion of clean energy on them are summarized as follows:

(1) Double-track mechanism deviation

Affected by the double-track market mechanism, it is difficult to form a matching balance relationship between priority generation and power purchase on the curve, while after large-scale clean energy integration, its power generation uncertainty leads to an increase in the deviation of power generation and purchase curve, thus expanding the unbalanced funds.

(2) Energy compensation cost for frequency regulation

In the frequency regulation ancillary service market, the clean energy stations that bidding during each trading session, participating in real-time frequency regulation, lead to the deviation between the actual power generation and the planned clearing of the real-time spot market, which results in unbalanced funds.

(3) Capacity compensation cost

The clean energy stations participate in the capacity market, resulting in a decrease in the utilization hours of the units. It is necessary to compensate the clean energy units to recover the fixed costs. The compensation is the market imbalance fund.

3.2. New unbalanced funds that may be brought about by high proportion of clean energy

When the generation of power grid is mainly turned into clean energy, the uncertainty of its output brings about the demand for new power derivatives. Facing with these demands, in order to motivate service providers, the preliminary stage may be in the form of compensation. Through the unbalanced fund account, the service recipient shall bear the cost

(1) Flexibility service

Under the environment of high proportion clean energy electricity market, in order to enhance the flexibility of power system flexible regulation, flexible resources are needed to provide flexibility value, such as system climbing service and system sliding service [9-12].

(2) Adequacy service

In order to ensure the reliability of system power supply, it is necessary to have resources to provide adequacy value, such as the service cost of reserve capacity generated by the electric energy capacity market service, and the peak shaving service cost generated by the flexible resource capacity market [13].

(3) Grid stability service

The uncertainty of clean energy output and the inherent physical characteristics of clean energy power generation units will impact the stability of the power system after the clean energy is integrated to the grid. Therefore, it is also necessary to add the service cost of the moment of inertia [14], the service cost of the stability control machine cutting, and the service cost of the fast load cutting.

4. Prediction method of unbalanced funds impacts on clean energy power spot market

4.1. Unbalanced funds impact prediction model

In the above analysis, it can be known that after the high percentage of clean energy enters the power spot market, it may not only cause new unbalanced funds terms, but also cause fluctuations in the original unbalanced funds terms. Therefore, it is necessary to forecast the possible impact of unbalanced funds in the high share of clean energy power spot environment, as shown in figure 1.

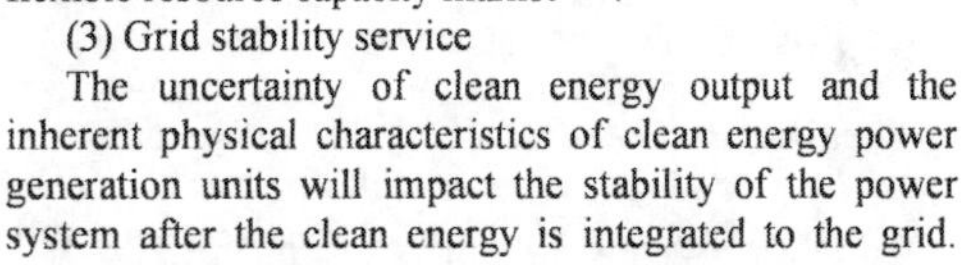

Figure.1 Model framework for forecasting the impact of unbalanced funds

(1) Predict variables based on historical data of power spot market variables

(2) Based on historical data, describe the relationship between power spot market variables and unbalanced funds by modeling, and the resulting model will be used to predict unbalanced funds.

(3) Obtain the prediction of unbalanced funds impacts by setting the variable prediction of the power spot market as the input of the model.

The core of the prediction of the unbalanced funds impact is mainly composed of two parts:1) Prediction of power spot market variables; 2) Modeling of the impact of power spot market variables on the unbalanced funds.

For the prediction of power spot market variables, there already are many mature methods for different variables, such as literature [15] describes different types of load forecasting methods, and the literature [16] reveals wind turbine output forecasting methods for different technical routes. This paper mainly focuses on the second point, i.e., the modeling of the impact of power spot market variables on unbalanced funds.

4.2. Unbalanced funds impact model based on simulated clearing

In the ideal case where the basic data are sufficient and the technical methodology is mature, a prediction model based on power spot simulation clearing can be used. All variables of the high percentage of clean energy power spot market are forecasted and put into the simulation system for clearing, so that the clearing results of the electricity energy and electricity derivative commodity market can be obtained [17]. Based on the market results, the amount of each unbalanced funds term is further calculated for each time period, and finally, the unbalanced funds are forecasted for a certain period in the future. This method is essentially a process of simulating the clearing and settlement of the power spot market for a certain period in the future.

The advantage of this model is that it can obtain the exact amount of unbalanced funds for each future time period. However, the drawbacks of this method are: 1) High forecasting accuracy is required for all boundary conditions of the simulation system, and the market clearing model itself is a nonlinear optimization model, which may magnify the further error of the method boundary conditions forecasting, and the convergence of the obtained unbalanced funds forecasting results cannot be guaranteed. 2) The application scenarios of unbalanced funds impact analysis are usually long-period analysis with annual and monthly cycles, and after time scale lengthening, the deviation of the simulation clearing model may also be cumulatively enlarged.

4.3. Unbalanced funds impact model based on copula function and fluctuation contagion chain

The impact model of power spot variables on unbalanced funds is a nonlinear model, which can be subdivided into two models according to the different types of unbalanced funds:

4.3.1 The amount of unbalanced funds is a function of the volume and price of power spot clearing

Unbalanced funds such as dual-track deviation costs, congestion surplus, and medium- and long-term congestion costs all fall into this category and can be abstracted and expressed as the following equation.

$$A = f(P_i, Q_i) \tag{8}$$

where A denotes the unbalanced funds costs, and P_i, Q_i denote the ith clearing price and clearing volume of the electric energy spot, respectively.

The dual-track deviation can be expressed as follows.

$$A_{t,\mathrm{bi}} = (p_{t,s} - p_{t,l}) \cdot (q_{t,s} - q_{t,l}),\ s \in S, l \in L \tag{9}$$

where $A_{t,\mathrm{bi}}$ represents the dual-track deviation cost at time t, $p_{t,s}$ and $q_{t,s}$ respectively denote the LMP and clearing power of unit s at time t, $p_{t,l}$ and $q_{t,l}$ denote the LMP and load level of load l at time t, and S and L denote the set of market units and market loads, respectively.

The congestion surplus can be calculated as follows with the market portion of generation and consumption balanced.

$$A_{t,\mathrm{block}} = \sum_{s \in S} p_{t,s} \cdot q_{t,s} - \sum_{l \in L} p_{t,l} \cdot q_{t,l} \tag{10}$$

where $A_{t,\mathrm{block}}$ represents the cost of unbalanced funds at time t.

If the power generation and consumption of the market portion are unbalanced, the calculation is as follows.

$$A_{t,\mathrm{block}} = \sum_{s \in S} p_{t,s} \cdot q_{t,s} - \sum_{l \in L} p_{t,l} \cdot q_{t,l} - A_{t,\mathrm{bi}} \tag{11}$$

The medium- and long-term congestion costs can be expressed as:

$$A_{t,s,c} = q_{t,s,c}(p_{t,s} - p_{t,s,c}), c \in C \tag{12}$$

where $A_{t,s,c}$ represents the medium- and long-term congestion costs incurred by unit s in contract c at time t, $p_{t,s}$ and $p_{t,s,c}$ respectively denotes the spot market price of unit s at time t and the agreed price of contract c. When $q_{t,s,c}$ and $p_{t,s,c}$ are the volume and price of the medium- and long-term CfD contracts, respectively, and $p_{t,s}$ is the day-ahead spot price of electricity, then $A_{t,s,c}$ represents the unbalanced funds from the basis difference between the medium- and long-term CfD contracts and the day-ahead spot market of electricity; when $q_{t,s,c}$ and $p_{t,s,c}$ are the clearing volume and price of the day-ahead spot market, respectively, and $p_{t,s}$ is the price of real-time market, $A_{t,s,c}$ represents the unbalanced funds from the basis difference between the day-ahead spot market and the real-time spot market.

4.3.2 The amount of unbalanced funds is a function of the demand for a certain type of electricity commodity

This type of unbalanced funds is the cost of meeting the needs of a certain type of market, that is, the cost of electricity derivative commodities. This type of market demand is often met through cost compensation, which usually includes: capacity, ramping and other service costs; variable costs, start-up and shutdown costs, and other compensation costs. The amount of such unbalanced funds can be expressed as follows.

$$A_{t,d} = q_{t,d} \cdot p_{t,d} \quad (13)$$

where $A_{t,d}$ represents the unbalanced funds generated by d demand at time t, $q_{t,d}$ and $p_{t,d}$ denote the supply and unit price of d demand at time t, respectively.

For (A) type unbalance funds, the effect of the power spot variable on the unbalanced funds can be described as figure 2.

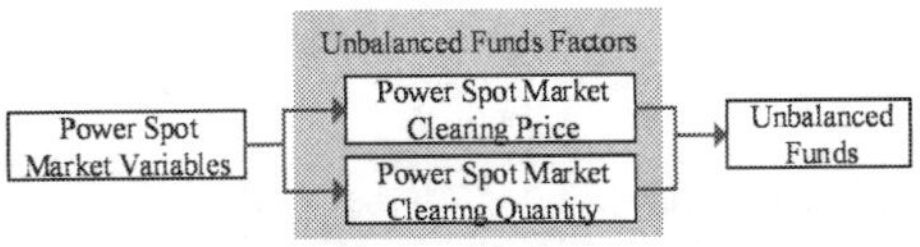

Figure.2 Influence of spot power variable on (A) type unbalanced funds

For (B) type unbalance funds, the effect of the power spot variable on the unbalanced funds can be described as figure 3.

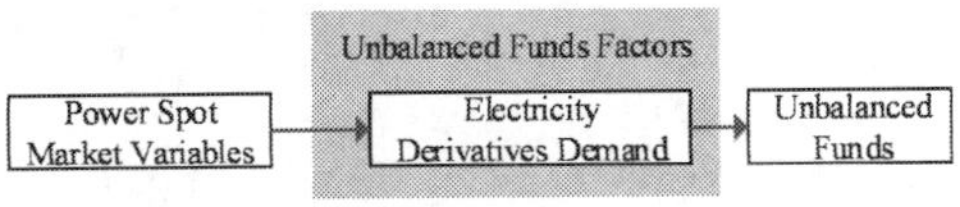

Figure.3 Influence of spot power variable on (B) type unbalanced funds

For (A) type unbalance funds, the effect of the power spot variable on the unbalanced funds can be described as follows.

It can be found that the mapping relationship from power spot market clearing price to unbalanced funds and the mapping relationship from electricity derivative commodity to unbalanced funds can be described by analytical equations, but the mapping relationship from power spot market variables to power spot clearing price and electricity derivatives demand (hereafter collectively referred to as unbalanced funds impact factor) is obviously a nonlinear system that cannot be described by analytical equations, so the Copula function is used to describe the relationship between these two relationship.

Copula function was proposed by Sklar in 1959[18] for separating randomness and coupling between multiple variables, randomness is measured by the respective marginal distributions of the variables and coupling is measured by the Copula function.

Suppose $X_1, X_2, \cdots, X_N$ is the N market factors in the power market with their respective edge distributions $F_1(x_1), F_2(x_2), \cdots, F_N(x_N)$, then there exists a Copula function such that:

$$H(x_1, x_2, \cdots, x_N) = C(F_1(x_1), F_2(x_2), \cdots, F_N(x_N)) \quad (14)$$

The Copula model is obtained by estimating the parameters of the Copula function between the actual market data of a single power spot variable and a single unbalanced funds impact factor; when the unbalanced funds impact is actually predicting, the output of the unbalanced funds impact factor is obtained by using the power spot variable as the input of this model.

If focus on multiple power spot market variables, the correlation between the multiple variables and the unbalanced funds impact factor can be described by a vine-copula [19].

4.4. Unbalanced funds impact model based on ES and quantile regression

The application scenario of unbalanced funds prediction is usually market operation risk assessment, which can assess risk based on the small granularity results of "hourly electricity price and unbalanced funds volume", but if the small granularity results cannot be obtained with guaranteed accuracy, we can also refer to the experience of financial risk field and focus only on the overall level and tail effect of unbalanced funds and thus construct the impact model of unbalanced funds.

ES index is commonly used in the field of risk quantification to measure tail risk, and in the context of unbalanced funds impact analysis, it means the maximum amount of expected value of unbalanced funds that could occur in a given future period under a certain probability level.

$$A_{\mathrm{ES},i,\alpha} = E(A_i \mid A_i \geq A_{\mathrm{VaR},i,\alpha}) \quad (15)$$

where $A_{\mathrm{ES},i,\alpha}$ represents the expected value of the most possible amount of unbalanced funds i under the probability of α, $E(\cdot)$ represents the expected value, A_i represents the amount of unbalanced funds i, and $A_{\mathrm{VaR},i,\alpha}$ represents the maximum possible amount of unbalanced funds i under the probability of α, satisfying the following equation.

$$P(A_i \leq A_{\mathrm{VaR},i,\alpha}) = \alpha \quad (16)$$

where $P(\cdot)$ represents the probability.

When measuring not a single term of unbalanced funds, but comprehensively considering the impact of different unbalanced funds, the indicator SES (systemic expected shortfall) can be used [20], calculated as follows.

$$A_{\mathrm{SES},\alpha} = \sum_i k_i E(A_i \mid A_i \geq A_{\mathrm{VaR},i,\alpha}) \quad (17)$$

where k_i denotes the contribution weight caused by the unbalanced funds i to the total impact A of the unbalanced funds, satisfying $A=\sum_i k_i A_i$.

At this point, $A_{\mathrm{ES},i,\alpha}$ is the MES (marginal expected shortfall) of the imbalanced funds i.

To evaluate the SES, the idea of quantile regression [21] can be used to construct a regression model with the power spot variable as the explanatory variable. With different α as the explanatory variable, the regression model is constructed as follows.

$$A_{\mathrm{VaR},i,\alpha} = g(\mathbf{X}) \tag{18}$$

where $g(\cdot)$ represents the regression model and $\mathbf{X}$ denotes the power spot market variable, which is a vector with the number of variables as the dimension.

The technical tools used for regression can be either econometric linear regression models or intelligent algorithms such as SVR [22], LSTM [23].

5. Conclusion

This paper studies the background and types of unbalanced funds in China's power spot market, conducts a qualitative analysis of the possible impact of the clean energy spot market environment on unbalanced funds, and designs three kinds of prediction models for the impact of unbalanced funds in the clean energy power spot market, including the unbalanced fund impact model based on simulated clearing, the unbalanced fund impact model based on copula function and fluctuation contagion chain, and the unbalanced funds impact model based on ES and quantile regression.

For the impact model of unbalanced funds proposed in this paper, its application scenarios include:

(1) Settlement risk warning

The impact of unbalanced funds on different time scales is assessed to support the work of early warning of settlement risk. The cost of unbalanced funds will eventually be transmitted to customers through the unbalance funds allocation mechanism or spontaneous market adjustment. Therefore, the early warning of unbalanced funds risk is an early warning of customers' electricity cost. Short-term prediction mainly assesses the tail risk of unbalance funds and issues early warnings under weekly and monthly time scales; long-term prediction mainly assesses the fluctuation trend of unbalance funds and evaluates the improvement of customers' electricity costs under multi-month and yearly time scales.

(2) Improving the allocation methods of unbalanced funds

In addition to the need for impact assessment of the total amount of unbalanced funds, it is also necessary to focus on the amount of unbalanced funds caused by different types of market participants under the existing allocation methods. Since the allocation methods can be expressed analytically, the impact analysis results can be obtained based on the forecast results of total unbalanced funds. For market participants who may bear large amounts of unbalanced funds, the fairness and incentive signals of the existing allocation methods need to be considered.

On the other hand, for the unbalanced fund costs incurred to meet the demand for a certain type of electricity derivative commodity, in the context of an energy mix with a high percentage of clean energy, the total costs associated with such demand, such as the cost of capacity compensation, need to be fully evaluated to make a decision on whether to make that power derivative commodity a separate market and encourage resource participation through competitive bidding to fully optimize the resources.

Acknowledgments

This work was supported by Yunnan Power Grid Technology Project (Research on Unbalanced Funds in Spot Market of High Proportion Clean Energy Power, YNKJXM20210124).

References

1. Gu Feng, He Aimin. See clearly the "unbalanced funds" [J]. China Power Enterprise Management, 2020 (25): 22-26.
2. Ye Ze, Chen Nianbin, Xie Qing. Formation Mechanism and Management Countermeasures of Unbalanced Funds in Power Market [J]. China Power Enterprise Management, 2021 (19): 36-41.
3. Summary of Unbalanced Funds in Power Market: Causes, Countermeasures and Prospects [J/OL]. Power System Technology:1-15[2023-02-24].https://doi.org/10.13335/j.1000-3673.pst.2022.2033.
4. Ding Weibin, Tan Zhongfu. Study on the settlement mechanism of the power market considering the treatment of unbalanced funds [J]. Power Construction, 2022,43 (07): 13-23.
5. Wu Zhaoyuan, Zhou Ming, Kou Yinggang, et al. Research on unbalanced fund allocation mechanism based on ABM model [J]. Power Grid Technology, 2021,45 (09): 3408-3416.
6. Xinhua News Agency. Xi Jinping delivers important speech at 75th UN General Assembly general debate [EB/OL] . (2020-09-22) [2021-11-14] . http://www.gov.cn/xinwen/2020-09/22/content_5546168.htm.
7. Southern Energy Regulatory Bureau, Notice on Matters Related to the Supervision of the 2022 Settlement and Trial Operation Market of the Southern (starting from Guangdong) Power Spot Market [Z] 2022.01.17.
8. Notice of the People's Government of Shanxi Province on Printing and Distributing the Administrative Measures for the Operation of Shanxi Electric Power Market (Jin Zheng Ban Fa [2022] No. 87) [Z] 2022.10.26.

9. Shandong Development and Reform Commission of the Shandong Regulatory Office of the National Energy Administration of Shandong Energy Administration, Notice on Doing a Good Job in the Settlement and Trial Operation of Shandong Electric Power Spot Market in 2022 (LJNSH [2022] No. 8) [Z] 2022.09.28.
10. CAISO. Flexible ramping product: revised draft final proposal [R]. California: CAISO, 2015.
11. MISO. Ramp capability for load following in MISO markets white paper[R]. Indiana: MISO, 2016.
12. NYISO . Market assessment design concepts-flexible ramping product [R]. New York: NYISO, 2018.
13. Liu YQ, Zhang HP, Li Qun et al. Design and practice of power peaking auxiliary service market in Northeast Power Grid[J]. Power System Automation,2017,41(10):148-154.
14. JIANG Han, YUE Chengyan, YAN Xingyu, et al. Influence of system inertia on flexibility resource analysis for an interconnection system with a high proportion of intermittent renewable energy[J]. Power System Protection and Control, 2021, 49(18): 44-51(in Chinese).
15. Kang, Chongqing, Xia, Qing, Zhang, Bo-Ming. A review of power system load forecasting research and development direction[J]. Power System Automation,2004(17):1-11.
16. Gu Xingkai,Fan Gaofeng,Wang Xiaorong et al. A review of wind power forecasting techniques[J]. Power Grid Technology,2007(S2):335-338.
17. Zhou Baorong, Li Xiaolin, Zhao Wenmeng, et al. Functional design of the southern regional power market simulation operation system[J]. Southern Power Grid Technology,2020,14(05):65-73.DOI:10.13648/j.cnki.issn1674-0629.2020.05.010.
18. Sklar A. Fonctions de repartition an dimensions et leurs marges[J]. Publ. Inst. Statist. Univ. Paris, 1959(8): 229-231.
19. Bedford Tim, Cooke Roger. Vines: A new graphical model for dependent random variables[J]. The Annals of Statistics, 2002, 30(4): 1031-1068.
20. Acharya V,Pedersen L H,Philippon T,et al.Measuring systemic risk[J].Review of Financial Studies,2017,30(1):2-47.
21. Bassett G W, Koenker R. A Quantile Regression Memoir[M]//Handbook of Quantile Regression. Chapman and Hall/CRC, 2017: 3-5.
22. Cortes C, Vapnik V. Support vector machine[J]. Machine learning, 1995, 20(3): 273-297.
23. Hochreiter S, Schmidhuber J. Long Short — term Memory[J]. Neural Computation, 1997(8) : 1735—1780.

Energy Management of PV-based Parking Lots Considering Utility Satisfaction

Youmna Elsayed[1,a], Mohamed R. Hamouda[2,b], Maher Mohamed[1,c], Mohamed EL-Shimy[1,d]

[1]Electrical power and Machines Department, Ain Shams University, Cairo, Egypt
[2]Electrical and Electronics Engineering, Coventry University branch in Egypt, Cairo, Egypt

Abstract— The urgent need to reduce greenhouse gas emissions has prompted the exploration of cleaner energy options such as renewable energy resources (RES) and electric vehicles (EVs). With the increasing use of RES and EVs, there are new opportunities and challenges for the stakeholders involved, and traditional centralized power system management methods are insufficient to handle the complexity. The energy management of photovoltaic (PV)-based parking lots is one of the challenges tackled in recent research. The main obstacle in applying previous research is the elimination of the utility objectives from the energy management models. This paper proposes an energy management framework that enables the integration of PV-based parking lots in a smart grid environment, considering the satisfaction of the utilities as a key player. The proposed framework is modeled using an optimization software package (GAMS), and the results are compared with two different cases to ensure its feasibility and effectiveness. The proposed model succeeded in taking the utilities satisfaction into account by managing the charging and discharging of the EVs.

1. Introduction

During the last decades, environmental concerns about global warming have arisen. Transportation and burning fossil fuels for electricity are the largest sources of greenhouse gas emissions, and the electricity and transportation sector shares around 25% and 27% of 2020 greenhouse gas emissions, respectively [1]. Therefore, moving to cleaner energy, renewable energy, and electric vehicles (EV) can provide the optimal pathway to reduce CO_2 emissions and help mitigate the impacts of global warming. The future of energy production is looking increasingly bright, thanks to the growth of renewable energy sources. The International Renewable Energy Agency (IRENA) prepared a global roadmap showing that renewables will reach a share of the grid of around 60% in 2050. In China, the electric energy generated from renewable energy sources (RES) came to 7% in 2015 and is expected to grow to 67% in 2050. Meanwhile, in Europe, the share of renewable energy could grow from about 17% to over 70% by 2050 [2]. Among different types of distributed energy resources (DER) technologies, solar-photovoltaic (PV) energy generation is the most popular and widely used[3]. This is due to the low cost of PV raw materials and installation costs because of large-scale production [4].

On the other hand, EVs has been flourished in recent years and have a substantial global market. By 2050, studies show that there will be more than 1 billion EVs on the road, which is proof of the popularity and practicality of this exciting technology. China and the United States are the largest markets for EVs, with 2.6 million and 1.1 million, respectively, in 2019. Although EVs is a green transportation technology, they can also play a key role in supporting renewable energy sources and improving the overall reliability of our energy systems. In 2050, around 14 terawatt-hours (TWh) of EV batteries will be available to provide grid services, compared to 9 TWh of stationary batteries [5]. As the use of EVs increases, there is a growing demand for EV parking lots to be installed in various locations. Solar-based parking lots are proving to be a more advantageous option than grid-based charging stations due to their lower costs and higher environmental friendliness compared to the grid or fossil fuels [6]. Solar-powered charging stations are being established by private hospitals, offices, and industries to offer EV charging facilities to their employees in parking lots.

The massive and widespread integration of RES and EVs into power systems creates new opportunities and challenges for all stakeholders involved, but at the same time, the electric distribution system management has become more complex than before. Traditional centralized concepts of power system management methods are insufficient to handle this complexity and should be replaced by a smart grid. As Smart grid technology provides a decentralized approach to managing the distribution network, optimizing the integration of RES and EVs while maintaining grid viability.

One of the key features of the smart grid is the prosumer's participation in the electricity market. Prosumers are both energy producers and consumers in the electricity market. They can trade electricity with each

[a] 2102202@eng.asu.edu.eg, [b] Ad9748@coventry.ac.uk, [c] Maher_eldessoki@eng.asu.edu.eg, [d] shimymb@gmail.com

other directly to achieve a win-win situation for all players with reasonable tariffs, and the buyers can save on costs while sellers gain more profit [4]. The same is true for EV owners, who can sell or buy energy to and from the grid or other prosumers, creating a more dynamic and flexible as well as resilient energy market.

Numerous research have been conducted in recent years for the sake of building parking lots and handling EV demand and charging. In [7], the author only discusses charging stations powered with PV stations, which are undependable without a connection to the electrical grid. The author illustrates optimal EV scheduling for a solar-powered charging station. Authors in [8] propose a mixture between the grid, PV, and diesel generator for a charging station. Despite the system's dependability, the diesel generator still causes environmental problems. The author examines the optimum method for integrating PV with ESS [9]; however, the proposed approach is not grid-connected. Furthermore, the study in [10] describes a research study that aimed to optimize the sizing of PV and energy storage system (ESS) for EV charging. The study found that both the PV system and the ESS configuration is the most effective and economically feasible option. Other studies provide general overviews of EVs and charging station layouts [11]. Additionally,[12] proposes a charging station equipped with off-grid PV for EVs and hydrogen vehicles, but this work is only viable when used with hydrogen vehicles alongside EVs. A comprehensive scheme for properly deploying charging stations with PV systems that has an improved voltage profile, decreased power loss, and low costs within a distribution network [13]. A collaborative operating approach that combines a ESS, a charging station, and group of buildings is proposed in [14] to reduce operating expenses in a community microgrid.

This paper introduces an energy management framework that examines potential stakeholders' business models. The work presented in this paper has the following contributions:

1) Propose an energy management framework enabling energy transactions between EV-PV and the grid.

2) Develop a unique utility function for the grid to participate efficiently in the framework.

2. Architecture of Proposed System

The proposed architecture for the energy management framework represents an innovative energy marketplace to introduce a more flexible marketplace that benefits both prosumers and utilities. It mainly consists of prosumers (EVs), PV systems, and utilities. In contrast to the conventional scheme shown in Fig.1, where producers and consumers are forced to trade at a fixed price set by the utility, the proposed scheme introduces a more flexible and dynamic approach.

As illustrated in Fig.2, the proposed scheme allows prosumers to trade with each other or with the utility based on a mutual price agreement. This approach creates a more competitive and efficient market by enabling the participant to negotiate prices that benefit all players [12].

The system is designed for commercial buildings with parking lots where EVs are parked from 9 am to 5 pm. Therefore, an aggregator is proposed to manage the transitions between all players effectively. In addition, the aggregator will also handle EV charging and discharging decisions based on various factors such as the state of charge, parking duration, and cost. Table 1 presents the characteristics of the EV [15].

As the goal is to achieve satisfaction for all players involved through an optimization scheme, a new comprehensive approach is required. This approach enables all parties to achieve greater satisfaction.

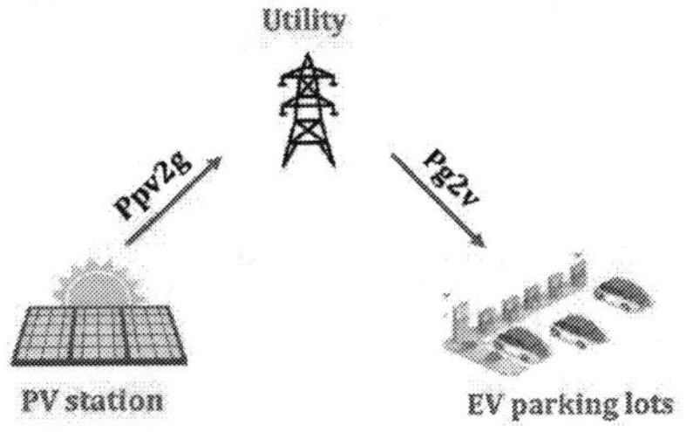

Figure 1. Conventional Regulation Scheme

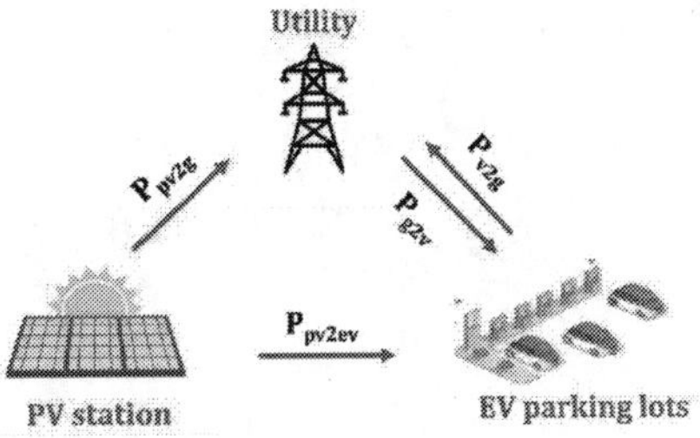

Figure 2. Proposed Scheme

Table 1 EV Characteristic Parameters

	Battery capacity [kWh]	**Max charging power [KW]**
Mitsubishi Outlander	13.8	3.7
Hyundai Ioniq	38.3	7.2
Tesla Model 3	55	7.4
Toyota Prius	8.8	3.3
Nissan Leaf	40	3.6

3. Proplem Formulation

The objective functions of the proposed scheme are defined as follows:

1. The main objective function:

Equation (1) is a multi-objective function that includes two sub-objectives. The first objective represents the parking lots' total running cost, including PV, and EV. The second objective represents utility satisfaction, which will be discussed in detail. That being said, the first term represents the cost of the proposed system, and the second term represents the grid satisfaction cost.

$$Min\ (C) = C_{tot} + C_g. \qquad (1)$$

2. The proposed system Cost of buying electricity:

The total parking lot cost is modeled using equation (2), which involves EVs and PV systems. The first term represents the price of purchasing electricity from the utility. The second term represents the cost of battery degradation. The third term in which describes the PV operation cost. The fourth term, the cost of power exported to the grid and the cost of power supplied to EV from a PV system, represents the profit for the system

$$\mathrm{Min}(C_{tot}) = \sum_{p=1}^{P}\sum_{t=1}^{T}\sum_{i=1}^{N}[P_{i,t,p}^{g2v} \times C_{elec_t} + P_{i,t,p}^{v2g} \times C_{deg} + (P_{i,t,p}^{pv2ev} + P_{t,p}^{pv2g}) \times C_{pv} - (P_{t,p}^{pv2g} + P_{i,t,p}^{v2g} + P_{i,t,p}^{pv2ev}) \times C_{Fit}]. \quad (2)$$

3. The Grid satisfaction objective:

Equation (3) represents the objective function to minimize the utility cost, which is divided into two equations to model different objectives. Equation (4) aims to maximize the power during the on-peak hours in order to encourage the grid to buy more power during these peak hours and reduce the leakage of power during this period, and the utility has op. However, equation (5) aims to minimize grid cost, with the first term representing the power cost purchased by the grid, which is a cost incurred by the utility, and the second term representing the purchasing price of electricity from the utility, which is a profit earned by the utility

$$Min\,(C_g) = Cg1_{i,t,p} + Cg2_{i,t,p} \quad (3)$$

$$Cg1_{i,t,p} = -\sum_{p=1}^{P}\sum_{t=1}^{T}\sum_{i=1}^{N}[(P_{i,t,p}^{v2g} + P_{t,p}^{pv2g}) \times C_{Fit} \quad 7 \le t < 11 \text{ and } 17 \le t < 19 \quad (4)$$

$$Cg2_{i,t,p} = \sum_{p=1}^{P}\sum_{t=1}^{T}\sum_{i=1}^{N}[(P_{i,t,p}^{v2g} + P_{t,p}^{pv2g}) \times C_{Fit} - (P_{i,t,p}^{g2v} \times C_{elec_t})]. \quad (5)$$

The constraints of the proposed scheme are defined as follows:

1) The following charging and discharging power limits for each EV are defined by specific boundaries that cannot be exceeded to prevent damage to the EV's battery and ensure optimal performance:

$$P_{i,t,p}^{g2v} + P_{i,t,p}^{pv2ev} \le P_{i,t}^{Max\,ch} \quad (6)$$

$$P_{i,t,p}^{v2g} \le P_{i,t}^{Max\,dsch}. \quad (7)$$

2) Equations (8) and (9) introduce additional constraints on the charging behavior of the EV. Equation (8) ensures that the EV can either charge or discharge at any specific time, but not both simultaneously, to prevent damage to EV's battery. Furthermore, equation (9) restricts the charging process of the EV to either the PV system or the grid, but not both sources at the same time as follow:

$$(P_{i,t,p}^{g2v} + P_{i,t,p}^{pv2ev}) * P_{i,t,p}^{v2g} = 0 \quad (8)$$

$$P_{i,t,p}^{g2v} * P_{i,t,p}^{pv2ev} = 0 \quad (9)$$

3) When EV is charged or discharged, some power is lost due to system inefficiencies. So, the charging and discharging process is not 100% efficient, with a charging/discharging efficiency of 90% [15]. The EV power equation is as follows:

$$P_{i,t,p}^{EV} = \eta_{ch} \times (P_{i,t,p}^{g2v} + P_{i,t,p}^{pv2ev}) - (P_{i,t,p}^{v2g}/\eta_{ch}) \quad (10)$$

4) The State of Charge (SoC) of EVs can be calculated using equation (11). The arrival SoC of EVs with varying levels, but they must determine their desired SoC at the time of departure based on their needs:

$$SOC_{i,t,p} = \begin{cases} 0 & t < T_{i,p}^{Arrv} \\ SOC_{i,p}^{Arrv} & t = T_{i,p}^{Arrv} \\ SOC_{i,t-1,p} + \frac{P_{i,t,p}^{Ev}}{BC_t} \times \Delta t & T_{i,p}^{Arrv} < t < T_{i,p}^{Dept} \\ SOC_{i,p}^{dept}. & t > T_{i,p}^{Dept} \end{cases} \quad (11)$$

5) Equation (12) defines a constraint that is essential for ensuring optimal performance and a longer lifespan of the battery by preventing deep discharges and overcharges:

$$SOC_{Min} \le SOC_{i,t,p} \le SOC_{Max}. \quad (12)$$

6) When EVs are not connected to the parking lot charging system:

$$\begin{cases} P_{i,t,p}^{v2g} = 0 \\ P_{i,t,p}^{g2v} = 0 \quad t < T_{i,p}^{Arrv} \quad or \quad t > T_{i,p}^{Dept} \\ P_{i,t,p}^{pv2ev} = 0 \end{cases} \quad (13)$$

7) To prevent the PV system from generating excess power beyond its rated capacity with a converter and inverter efficiency of 98% [15]:

$$\frac{\sum_{i=1}^{N} P_{i,t,p}^{pv2ev}}{\eta_{dc-dc}} + \frac{P_{t,p}^{pv2g}}{\eta_{dc-dc} \times \eta_{inv}} < P_t^{pv} \le P_{pv}^{Max} \quad (14)$$

4. Case Study

This section presents a numerical analysis of the feasibility and effectiveness of the proposed energy management framework for EV, PV, and utility over 24 hours. The optimization problem is solved using GAMS solver as a nonlinear problem. In order to evaluate the effectiveness of the proposed scheme in this study, three different case studies are performed. In the first case study, the objective will be to minimize the utility cost only. In the second case, the objective will be to mitigate the proposed system's total cost only. While in the third case, a multi-objective optimization problem is solved, which includes both the utility and the system.

This work investigates one EV parking lots with 25 EVs and a 40 kW PV system. The battery degradation cost is 0.056 C$/kWh. The LCOE for PV is 0.065 c$/kWh [16]. Time-of-use (TOU) pricing is categorized into off-peak, mid-peak, and on-peak hours, with corresponding prices of 0.074 c$/kWh, 0.102 c$/kWh, and 0.151 c$/kWh, respectively, as stated in reference [17]. The feed-in tariff (FIT) value chosen for the study is 0.0713 c$/kWh, as reported in reference [18].

https://doi.org/10.1051/e3sconf/202344101004

4.1. Case study 1: The main objective is to minimize the total cost of the parking lots (Ctot) without considering the Grid satisfaction cost

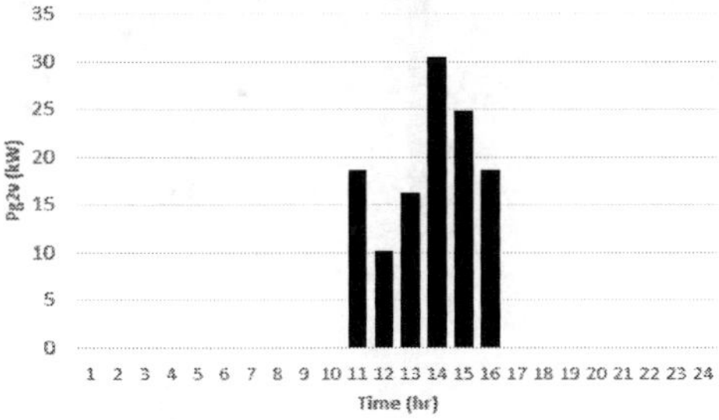

Figure 1 G2V power in case 1.

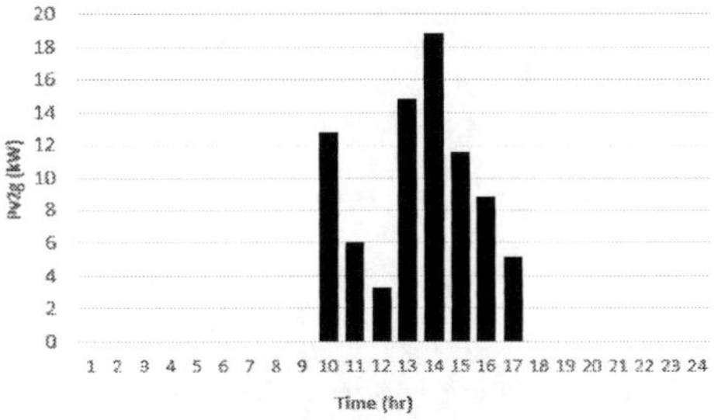

Figure 2 V2G power in case 1.

The proposed case study aims to minimize the total cost of the system while ensuring optimal utilization of distributed energy sources. Fig. 3 shows the power purchased by the EVs from the grid (G2V power).

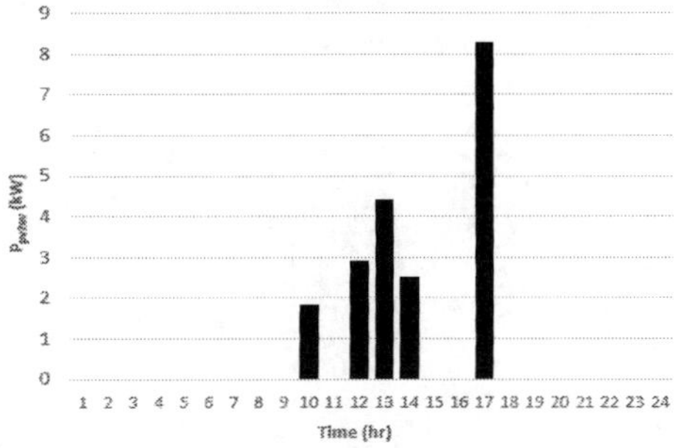

Figure 3 PV2EV power in case 1.

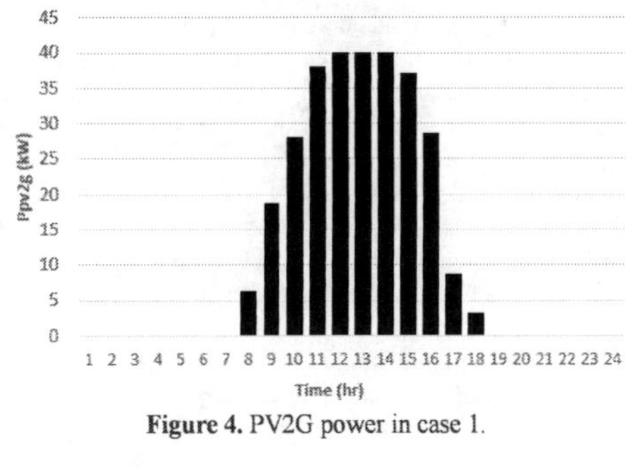

Figure 4. PV2G power in case 1.

It is observed from the results of Fig.4 and Fig.6 that the EVs and PV sell most of the power to the grid, which ensures a profit for both parties. However, the utility is forced to purchase most of the power generated by the EVs and PV, which results in financial losses to the utility. Fig.5 illustrates the quantity of power being drawn from the PV source and consumed by EVs.

4.2. Case study 2: The main objective is to minimize the grid satisfaction cost.

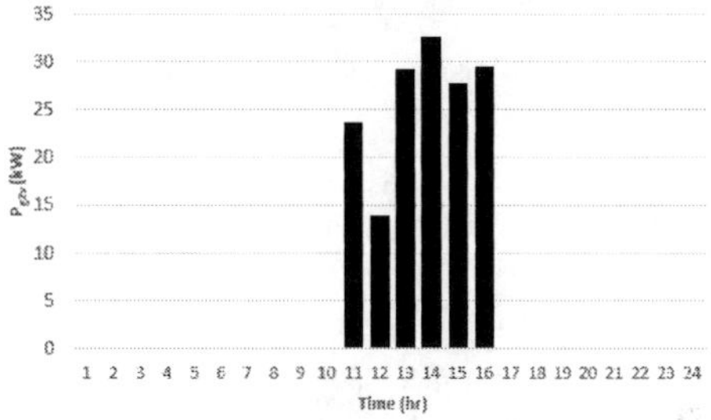

Figure 5 G2V power in case 2

In this section, we present a case study of the proposed scheme with the objective of minimizing grid costs, maximizing power in on-peak hours, and maximizing profits. In Fig.7, demonstrates that a significant amount of power is drawn from the grid to charge EVs compared to other scenarios.

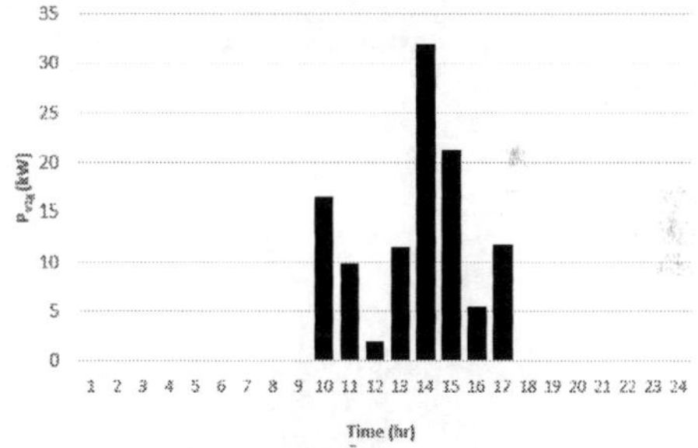

Figure 6 V2G power in case 2.

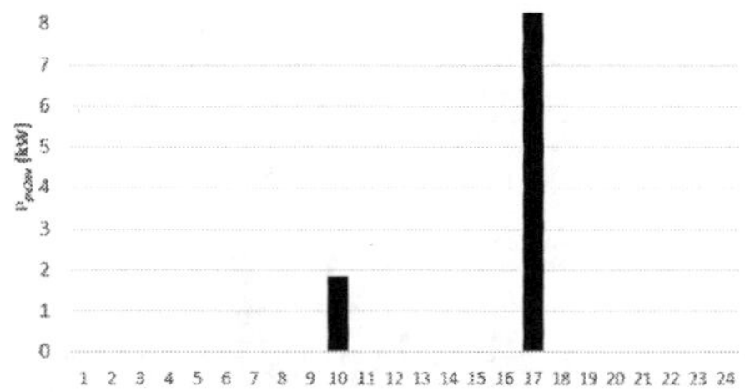

Figure 7. PV2EV power in case 2.

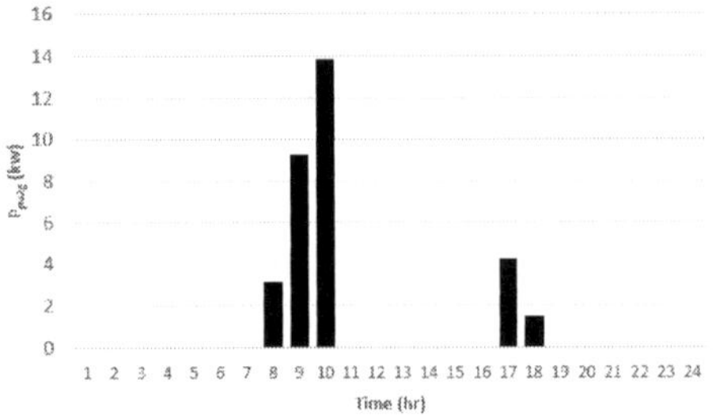

Figure 8 PV2G power in case 2.

During on-peak hours, the grid does not inject power to EVs.

Fig. 8 shows that the EVs sold a significant amount of power to the grid, which contradicts the objective of minimizing grid cost. However, due to the State of Charge (SOC) restrictions, the EVs were compelled to sell their excess power, which included the purchased power to the grid. The results from Fig.7 and Fig.8 show that this approach resulted in financial losses for EVs. Fig 9. the PV system supplies power to the EVs during on-peak hours only, as the EVs are forced to buy most of their power from the grid in order to generate a profit for the grid. In Fig. 10, the PV system sells only a small amount of power to the grid. Fig.9 and Fig. 10 imply that renewable energy sources were not utilized to their full potential in this case.

4.3. Case study 3: The main objective is to minimize the main multiobjective function to account for the total parking cost and the grid satisfaction cost

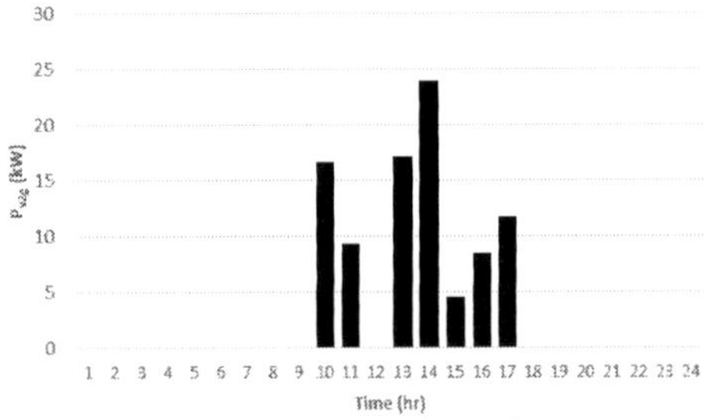

Figure 9 V2G power in case 3.

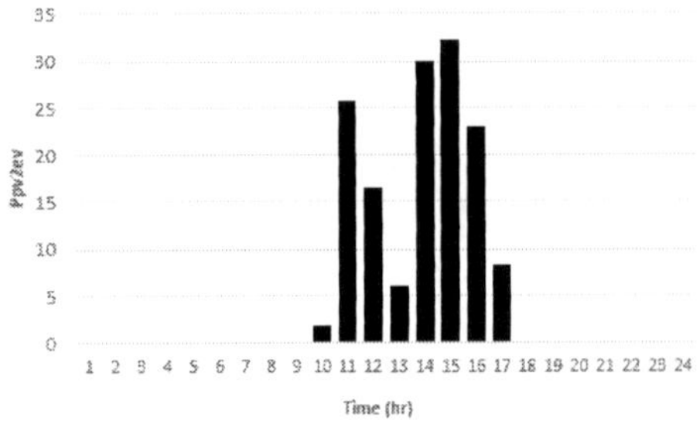

Figure 10 PV2EV power in case 3.

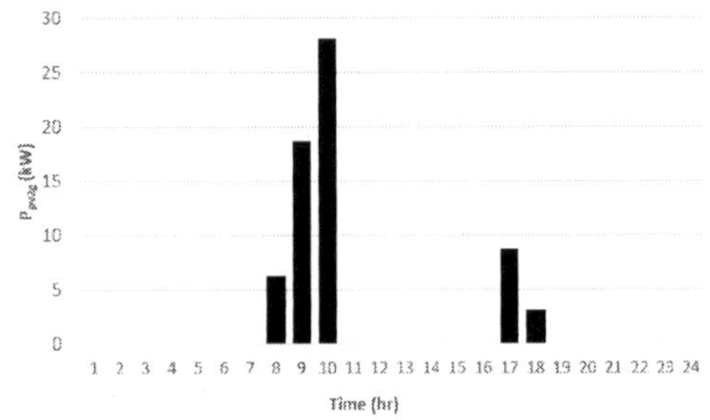

Figure 11 PV2G power in case 3.

This case study presents a multi-objective optimization problem that aims to address the complex energy management challenges faced by EV, PV, and utility systems while ensuring profitable operations for all parties involved. It is essential to highlight that despite the utility not being able to achieve its objective of selling power to EVs, it was still able to fulfill one of its key goals, which was to maximize power in on-peak hours.

Also, the results demonstrate the effectiveness of this case in ensuring the optimal utilization of renewable energy sources, thus promoting sustainability while providing a significant economic advantage to both the EV and PV systems.

Fig.11 and Fig.13 depict the considerable contribution of EV and PV systems in selling power to the grid, especially during on-peak hours. Furthermore, as illustrated in Fig.12, the EV system effectively purchases all its power requirements from the PV system, taking advantage of the lower tariff prices.

Table 2 presents a comparison of the three cases, indicating that case 3 provides the optimal solution, resulting in a fair profit for both case 1 and case 2. This approach ensures that all parties involved are able to benefit economically.

Table 2 Comparison of All Cases' Costs

	CASE 1	CASE 2	CASE 3
Ct (C$)	-175.27	-33.225	-83.288
Cg (C$)	449.558	-9.29	-6.477
C (C$)	274.288	-42.430	-91.661

5. Conclusions

This article proposes an energy management framework for EV, PV, and utility systems that aims to ensure optimal utilization of distributed energy sources while maximizing profits and promoting sustainability. The proposed framework is evaluated through three case studies, which focus on minimizing the total cost of the proposed system, minimizing the utility cost, and addressing a multi-objective optimization problem that includes both the utility and the proposed framework. The results demonstrated that the multi-objective approach provided the optimal solution, ensuring a fair profit for all parties involved.

The findings of this study highlight the importance of considering multiple objectives when designing energy management frameworks for distributed energy systems.

This approach ensures the optimal utilization of renewable energy sources, promotes sustainability and provides a significant economic advantage to both the EV and PV systems. Overall, the proposed framework offers a promising solution for future energy management systems that prioritize sustainability and profitability for all stakeholders involved.

NOMENCLATURE

Sets and Indices:	
t	Time period.
i	Numbers of Electric vehicles.
p	Number of parking lots.
Parameters:	
C_{elec_t}	Utility selling price of electricity [C$/kWh].
C_{deg}	Cost of Battery degradation [C$/kWh].
C_{pv}	Levelized Cost of Energy (LCOE) [C$/kWh].
C_{Fit}	Cost of feed-in tariff [C$/kWh].
$SOC_{i,p}^{arr}$	EV's SoC Arrival [%].
$SOC_{i,p}^{dept}$	EV's SoC Departure [%].
$T_{i,p}^{Arrv}$	Arrival time of EV [hr].
$T_{i,p}^{Dept}$	Departure time of EV [hr].
SOC_{Min}	Minimum State of Charge of battery [%].
SOC_{Max}	Maximum State of Charge of battery [%].
BCi	Battery capacity of EV [kWh].
η_{dc-dc}	DC-DC converter efficiency [%].
η_{inv}	DC-AC inverter efficiency [%].
P_t^{pv}	PV output power [kW].
P_{pv}^{Max}	Maximum PV power [kW].
$P_{i,t}^{Max\,ch}$	EV's maximum charging power [kW].
$P_{i,t}^{Max\,dsch}$	EV's maximum discharging power [kW].
η_{ch}	EV charging efficiency [%].
Variables:	
C	Total cost which includes the proposed system and grid satisfaction [C$].
C_{tot}	Total cost for PV and parking lots [C$].
C_g	Grid satisfaction cost [C$/.
$P_{i,t,p}^{g2v}$	Power transfer from grid to EV [kW].
$P_{i,t,p}^{pv2ev}$	Power transfer from PV to EV [kW].
$P_{i,t,p}^{v2g}$	Power transfer to grid from EV [kW].
$P_{t,p}^{pv2g}$	Power transfer from PV to grid [kW].
$SOC_{i,t,p}$	SoC of EV [%].
$P_{i,t,p}^{EV}$	EV 's total power transfer [kW].

References

1. J. Soussan, The Mitigation of Climate Change. 2017. doi: 10.4324/9781351297967-12.
2. M. Larsson, Global energy transformation: Four necessary steps to make clean energy the next success story. 2009. doi: 10.1057/9780230244092.
3. A. ur Rehman, M. Ali, S. Iqbal, A. Shafiq, N. Ullah, and S. Al Otaibi, "Artificial Intelligence-Based Control and Coordination of Multiple PV Inverters for Reactive Power/Voltage Control of Power Distribution Networks," Energies, vol. 15, no. 17, 2022, doi: 10.3390/en15176297.
4. S. Aznavi, P. Fajri, M. B. Shadmand, and A. Khoshkbar-Sadigh, "Peer-to-Peer Operation Strategy of PV Equipped Office Buildings and Charging Stations Considering Electric Vehicle Energy Pricing," IEEE Trans. Ind. Appl., vol. 56, no. 5, pp. 5848–5857, 2020, doi: 10.1109/TIA.2020.2990585.
5. IRENA, "Electric-Vehicle Smart Charging: Innovation Landscape Brief," Innov. Landsc. Br. Electr. smart Charg. Int. Renew. Energy Agency, Abu Dhabi., pp. 1–24, 2019, [Online]. Available: www.irena.org
6. Q. Cheng et al., "A Smart Charging Algorithm Based on a Fast Charging Station Without Energy Storage System," vol. 7, no. 4, pp. 850–861, 2021, doi: 10.17775/CSEEJPES.2020.00350.
7. M. E. Kabir, S. Member, and C. Assi, "Optimal Scheduling of EV Charging at a Solar Power-Based Charging Station," pp. 1–11, 2020.
8. A. Verma and B. Singh, "Multimode Operation of Solar PV Array , Grid , Battery and Diesel Generator Set Based EV Charging Station," vol. 9994, no. c, 2020, doi: 10.1109/TIA.2020.3001268.
9. Y. Li and J. Wu, "Optimum Integration of Solar Energy With Battery Energy Storage Systems," pp. 1–11, 2020.
10. C. Leone, C. Peretti, A. Paris, and M. Longo, "Photovoltaic and battery systems sizing optimization for ultra-fast charging station integration," J. Energy Storage, vol. 52, no. PC, p. 104995, 2022, doi: 10.1016/j.est.2022.104995.
11. R. Patnaik and S. Mopidevi, "A technological overview & design considerations for developing electric vehicle charging stations," J. Energy Storage, vol. 43, no. September, p. 103225, 2021, doi: 10.1016/j.est.2021.103225.
12. Q. Guo, H. Zhou, W. Lin, and S. Nojavan, "Risk-based design of hydrogen storage-based charging station for hydrogen and electric vehicles using downside risk constraint approach," J. Energy Storage, vol. 48, no. August 2021, p. 103973, 2022, doi: 10.1016/j.est.2022.103973.
13. F. Ahmad, M. Khalid, and B. Ketan, "An enhanced approach to optimally place the solar powered electric vehicle charging station in distribution network," J. Energy Storage, vol. 42, no. August, p. 103090, 2021, doi: 10.1016/j.est.2021.103090.
14. X. Zhong, W. Zhong, Y. Liu, C. Yang, and S. Xie, "Cooperative operation of battery swapping stations and charging stations with electricity and carbon trading," energy, vol. 254, p. 124208, 2022, doi: 10.1016/j.energy.2022.124208.
15. A. Mohammad, R. Zamora, and T. T. Lie, "Transactive Energy Management of PV-Based EV Integrated Parking Lots," pp. 1–9, 2020, doi: 10.1109/JSYST.2020.3043327.

16. E. Summary, "RENEWABLE POWER GENERATION COSTS IN 2021," pp. 1–4, 2021.
17. ONTARIO ENERGY BOARD, https://www.oeb.ca/consumer-information-and-protection/electricity-rates
18. solar electricity handbook, http://www.solarelectricityhandbook.com/canada-feed-in-tariff.html

Risk Assessment and Application of Port Liquefied Natural Gas Trucks Bunkering Based on Hazard Identification and Computational Fluid Dynamics

Kun Li [a], Chang Liu *

China Waterborne Transport Research Institute, Beijing, China

Abstract—In order to safely and orderly promote the application of LNG as a clean energy for inland river ships, the paper takes a port truck to bunker a 2000 tons LNG power bulk carrier for example, carrying out a qualitative risk assessment based on HAZID method, putting forward risk prevention and control measures, and using FLACS software to get CFD accident consequence quantitative calculation. The analysis shows that the overall risk is controllable under certain measures. It is the first time that this method has been applied to the risk assessment of Trucks Bunkering.

1. Introduction

Promoting the use of liquefied natural gas (LNG) for clean energy on ships is an important measure to thoroughly implement General Secretary Xi Jinping's thought on ecological civilization, resolutely win the battle against blue sky and fight the battle against pollution. In order to accelerate the construction of ecological civilization and pollution prevention and control in China[1], the CPC Central Committee and the State Council has formulated the "about strengthening ecological environment protection Resolutely play pollution prevention and control of the opinions of the battle ", "win the sky battle action plan for three years and a series of policy, clearly put forward" active and control ship pollution and ship transformation and upgrading, and promote the use of new energy and clean energy such as natural gas ships, Strengthen the demonstration and application of green ships "recently. Up to now, there are more than 310 LNG powered ships in China, and Guangdong plans to build or renovate 350 ships by the end of this year. The demand for LNG bunkering is increasing day by day.

The construction of LNG bunkering system is the core link to promote the application of LNG in ships. In order to ensure that new and rebuilt ships have gas to fill, it is an important measure to ensure reliable bunkering of LNG by tank truck before shore-based and barging bunkering stations are built to meet the gas needs of ships. LNG is flammable and explosive dangerous goods, and bunker safety is an important basis to promote the sustainable development of ships using LNG clean energy. This project is in accordance with the Notice issued by Maritime Safety Administration of the People's Republic of China on the revision and issuance of Safety Supervision and Management Measures for Floating LIiquefied Natural Gas Bunkering Operation (Haiweifang (2021) No. 148) and Technical Requirements for Bunkering Operation of Natural Gas Fuel Powered Ship Tank Truck (JT/T) On the basis of relevant laws and regulations and standards, carry out qualitative and quantitative analysis of tank car bunkering risk, so as to ensure the safety of LNG fuel bunkering operation of LNG trucks at port[2][3][4].

2. Evaluation Methods

Hazard Identification (HAZID) method is used for qualitative analysis. HAZID identifies, analyzes and evaluates potential safety hazards and their consequences in the process of operation by using the brainstorming method combined with the corresponding checklist. According to the analysis results, measures are proposed to control the risk level and failure scenario input is provided for Quantitative risk assessment (QRA)[5-6]. The specific work flow of HAZID is shown in Fig.1.

The Computational Fluid Dynamics (CFD) software FLACS (Flame Accelaratition Simulatation) was used for quantitative analysis. The software was developed with funding and guidance from the relevant oil companies BP, Conocophillips, ExxonMobil and Statoil, as well as three national legislative bodies: the Health and Safety Executive (HSE) of the United Kingdom, the Norwegian Petroleum Agency (NPD) and the German Ministry of Research and

[a] likun@wti.ac.cn, * liuchang@wti.ac.cn

Technology (BMFT). FLACS is the designated explosion risk assessment tool for major oil and gas companies. FLACS can meet the risk assessment requirements of relevant international standards in the oil and gas industry (such as ISO13702, NorSOKZ-013, ISO/DIS 19901-3). For LNG, FLACS is licensed by the U.S. department of transportation's (DOT) pipeline and dangerous goods safety administration (PHMSA) to simulate the diffusion of LNG vapor clouds under federal regulation 49 CFR193.2059.

2.1. Bunkering Object

This project takes 2000 tons of pure natural aerodynamic bulk cargo as the object, as shown in Fig.2(a), and the main parameters are shown in Table 1. The ship's LNG Marine fuel tank is 30m³, and the single bunkering capacity is about 80% of the tank capacity, i.e. 24m³.

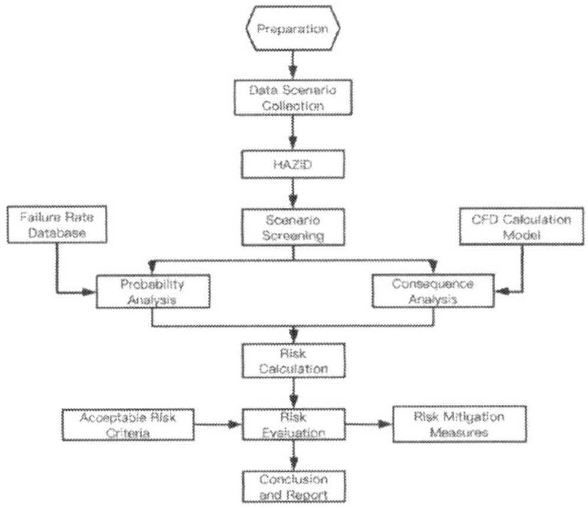

Figure 1. Risk Evaluation Process

3. Application Case

Table 1. 2000 tons LNG powered ship parameters

Ship Type		Total Length	Width	Depth	Navigation Zone	LNG Fuel Tank			Bunkering Liquid Pipe Flange Spesification	Bunkering Gas Pipe Flange Spesification
						Volume	Working Pressure	Design Pressure		
Bulk Cargo Ship	2000 Ton	59.9m	15.8m	4.5m	Inland River A Level	$30m^3$	0.7～0.9MPa	1.2 MPa	DN50	DN50

3.1. LNG Bunkering Facility

The 55m³ capacity truck used in the LNG bunkering facility of this project, as shown in Fig. 2(b), realizes the LNG fuel bunkering of ships through the port integrated skid.

Figure 2. (a)LNG Powered ship; (b)LNG Truck

3.2. Bunkering Medium

The bunkering medium of this project is LNG fuel, and its physical characteristics are shown in Table 2.

Table 2. LNG Physical Property Parameters

Average Molecular Weight	17.3
Main Ingredients	93.6% methane (CH4) approximately, 4.12% ethane (C2H6) approximately
Gasification Temperature	-163℃（Ordinary Pressure）
Liquid Density	447 Kg/m approximately
Gas Density	0.772 Kg/Nm3 （15.5℃）
Gas/Liquid Volume Ratio	612.5：1（15.5℃）
Fire Point	650℃
Average Molecular Weight	17.3

3.3. Bunkering Operation Area Layout

The truck of this project carries out LNG bunkering operation of purely natural gas-powered ships in a port, as shown in Fig.3.

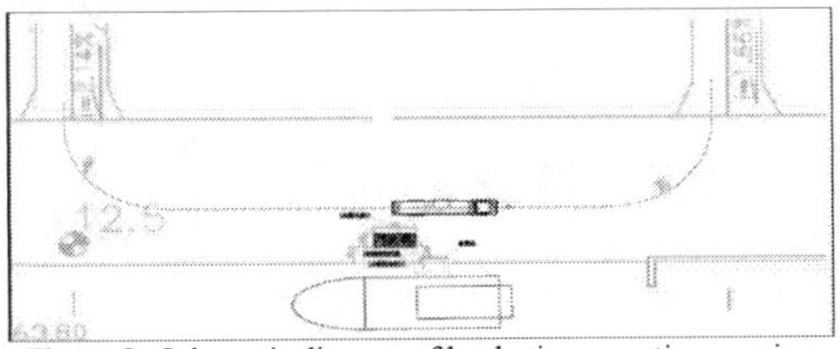

Figure 3. Schematic diagram of bunkering operation area in a port

4. Risk Analysis

4.1. Qualitative Risk

4.1.1 Risk Assessment Node Division

This project adopts truck to bunker LNG fuel for natural gas-powered ships. HAZID risk analysis is divided into three and two nodes according to the operation process. Node 1: truck- integrated skid and truck hose joint, mainly for truck, integrated skid and its connection hose joint risk identification; Node 2: Integrated skid and truck hose joint - ship bunkering station interface, mainly for integrated skid, LNG bunkering station and its connection hose joint risk identification.

4.1.2 Risk Matrix

The risk matrix is a way of combining qualitative or semi-quantitative grading of consequences with the likelihood of producing a certain level of risk or risk level. The risk matrix can be customized or graded according to relevant standards. This report refers to the recommended guidance document for risk assessment as required by IACS REC.146 IGF regulation and ISO 17776:2000 for the oil and gas industry. Offshore mining installations - Risk matrices in the methodological and technical guidelines for hazard identification and risk assessment as a basis for grading identified risk sources. The risk matrix is shown in Fig. 4.

Consequences	**C3.Multiple Deaths**	Medium	Medium	Medium	High	High
	C2.Single Death or Multiple Injuries	Low	Medium	Medium	Medium	High
	C1.Multiple Injuries	Low	Low	Low	Medium	Medium
		L1 <10E-6/y< 1/40000	**L2** 10E-6～10E-5/y, 1/40000～1/4000	**L3** 10E-5～10E-4/y, 1/4000～1/400	**L4** 10E-4～10E-3/y, 1/400～1/40)	**L5** >10E-3/ >1/40
		Probability (probability of occurrence per year, 1 occurrence for N ships)				

Figure 4. Risk Matrix Chart

4.1.3 Analysis Result

HAZID analysis method was used to conduct qualitative risk assessment for the two nodes divided, and some analysis results are shown in Table 3.

Table 3. HAZID Risk Analysis results of LNG Bunkering by Truck for Natural Gas-Powered Ship (Part)

Order and Categar y	Hazard Source Guide Word	Reason	Consequence	Existing Safety Measures	C	L	R	Suggestive Measures	Respon sible Party
1.1 Truck and Unloadi ng Hose	1.1.1 Layout and Environ ment	1. The truck moved unexpectedly during bunkering.	1. Hose damage, resulting in LNG/NG leakage, low temperature injury or fire; 2. Collisions between trucks damage vehicles and pipelines.	1. Before unloading the liquid, make sure the truck engine is off, the car key is unplugged and the handbrake is pulled up; place a triangle of wood at both ends of the wheels. 2. In the process of unloading liquid, forbid idle people to stay and watch, and forbid the driver to turn on the vehicle power.	C 2	L 3	Medi um	1. Set truck limit device, or equivalent fixed device.	Bunker service party
		2. Interference by other irrelevant	1. Accidental collision leads to damage of	1. Set up waiting area for trucks, and set up a warning area around the waiting	C 2	L 1	Low	1. Set warning signs at the entrance and exit of the fence to remind	Bunker service party

Order and Categary	Hazard Source Guide Word	Reason	Consequence	Existing Safety Measures	C	L	R	Suggestive Measures	Responsible Party
		vehicles and personnel.	unloading hose, resulting in LNG/NG leakage, low temperature injury or fire; 2. Open flame may cause fire.	area, non-ruck operators are forbidden to enter the warning area. 2. Set up safety warnings and warning signs in the LNG unloading area.				other vehicles and personnel that LNG bunkering operations are in progress during refueling. 2. Set warning signs on the fence.	
		3. Fire outside the dump area.	1. Fire in unloading area.	1. The unloading area has been equipped with fire-fighting equipment in accordance with the Design Code for Inland LNG Bunkering Terminal (JTS 196-11-2016).	C 2	L 1	Low	1. No piles of flammable and explosive goods or items within the appropriate range, and quantify the size of the range.	Bunker service party
		4. Electrostatic sparks are generated in the unloading area by personnel and trucks.	1. A fire occurs when flammable gas is present.	1. LNG tankers are connected to electrostatic grounding wires. 2. All personnel entering the guarded area wear anti-static clothes and shoes.	C 1	L 2	Low	1. Set up personnel elimination static device at the entrance of the fence.	Bunker service party
2.1 Loading Skid and Bunkering Hose	2.1.1 Layout and Environment	1. The combustible gas probe is improperly positioned.	1. No leak detected.	Not seen.	C 2	L 3	Medium	1. Clearly define the location of the combustible gas probe of the loading sled, and suggest to arrange the combustible gas probe near the joint.	Bunker service party
	2.1.2 Attatchment	5. In the process of bunkering, weather changes (e.g. extreme weather).	1. Affect bunkering operation.	1. Stop bunkering operations immediately during high intensity lightning and frequent lightning strikes. 2. Stop bunkering operation immediately during high intensity lightning, frequent lightning strikes and unexpected events inside or outside the refueling station	C 1	L 3	Low	1. Define emergency procedures for extreme weather in the bunkering program.	Bunker service party

4.2. Quantitative Analysis

4.2.1 Calculation Parameter

In this project, the diameter of the liquid and gas phase pipes of the bunkering hose is DN50, the bunkering rate is 7m/s, the bunkering pressure is 0.7mpa, and the leakage time is 60s, including 30s leakage detection time and 30s equipment shutdown time.

According to the meteorological observation data of the port from 1994 to 2013, the annual average wind speed is 2.1m/s, the annual average temperature is 22.64℃, and the average relative humidity is 75. The average annual dominant wind is NE wind, with wind frequency accounting for 16.3%; the secondary dominant wind is NNE wind, with wind frequency accounting for 11.1%; the average annual static wind frequency in this region is 11.9%, as shown in Table 4.

Table 4. Annual Wind Frequency of Port Location (%)

Wind Direction	N	NNE	NE	ENE	E	ESE	SE	SSE	S	SSW	SW	WSW	W	WNW	NW	NNW	C	Dominant Wind Direction
Wind Frequency	4.8	11.1	16.3	9.8	6.9	5.6	4.6	4.4	4.5	3.5	3.0	2.2	2.2	2.4	3.7	2.9	11.9	NE

4.2.2 Acceptance Criteria

According to GB/T 20368 "Liquefied Natural Gas (LNG) Production, Storage and Shipment", "Guideline for Quantitative Risk Assessment of Oil and Gas" and

"Guideline for LNG Bunkering Operation", the volume concentration of combustible gas in air is 50% of the lower limit of flammability (2.5% for methane) as the diffusion boundary.

4.2.3 FLACS Model Establishment

The overall model of port bunkering area, tank and bunkering skid model, 2000 tons LNG-powered ship model and LNG hose boom model constructed by FLACS software are shown in Fig.5 (a) (b) , Fig. 6 (a) (b) respectively.

Figure 5. (a) CFD Calculation Model; (b)Truck and bunkering Skid Model

Figure 6. (a)2000 tons LNG Storage Ship Model; (b)Hose Boom Model

4.3. Scenario Analysis and Results

Calculate and consider three scenarios: full hole leakage, half hole leakage and 1/10-hole leakage.

4.3.1 Scenario 1: 1/10-hole leakage

According to the simulation calculation, the leakage result of 1/10 hole of the bunkering hose is about 4m in the range of combustible concentration (0.025-0.15), as shown in Fig.7(a).

4.3.2 Scenario 2: half hole leakage

According to the simulation calculation, the half hole leakage result of the bunkering hose is about 30m in the range of combustible concentration (0.025-0.15), as shown in Fig. 7(b).

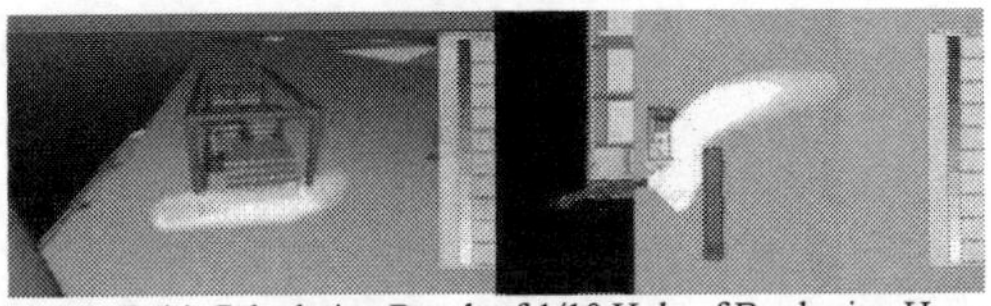

Figure 7. (a) Calculation Result of 1/10 Hole of Bunkering Hose; (b)Calculation Result of Half Hole of Bunkering Hose

4.3.3 Scenario 3: full hole leakage

According to the simulation calculation, the leakage result of the full hole of the bunkering hose is about 80m in the range of combustible concentration (0.025-0.15), as shown in Fig.8.

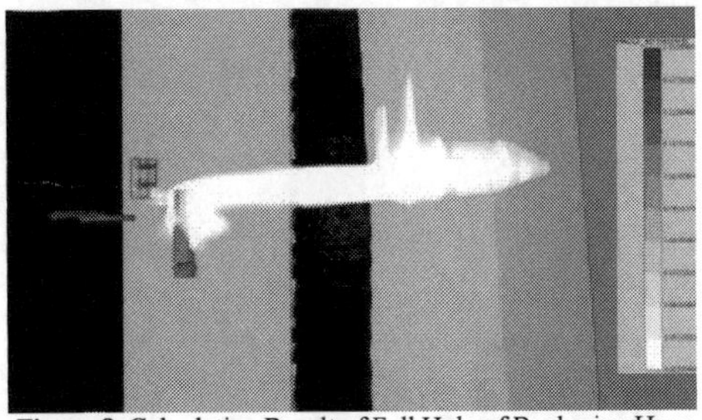

Figure 8. Calculation Result of Full Hole of Bunkering Hose

5. Conclusions

According to the results of qualitative and quantitative analysis, the evaluation conclusions are as follows:

1) Truck- integrated pry with truck hose joint. The main risk lies in the bunkering process: first, the truck moves unexpectedly, which will lead to hose damage, and then lead to LNG/NG leakage, low temperature injury or fire. It is suggested to control the risk by confirming that the truck engine flays off before discharging liquid, pulling up the key and pulling up the handbrake, placing triangular wood at both ends of the wheel and setting the truck limit device or equivalent fixing device, etc. Secondly, the joint between the unloading hose and the truck or the unloading pry leaks, resulting in low temperature injury, and may cause fire in case of open fire. It is suggested to strictly require operators to wear antifreeze gloves, set up restricted areas according to JT1319-2020 "Refueling Requirements for Natural Gas Fuel Powered Ship Truck" and combined with quantitative analysis results, and strictly prohibit ignition sources within the restricted areas to control risks.

2) Integrated pry and truck hose joint - ship bunkering station interface. The main risk lies in the bunkering process: firstly, due to the rolling of the ship, the joint between the ship's bunkering hose and the loading skid fails, and LNG leaks occur. It is suggested to control the risk by strengthening the ship's mooring design and paying more attention to the ship's mooring status during bunkering. The second risk occurs at valve parts and joint leakage, and open fire may cause a fire. It is recommended to check the regular inspection and test records of valves and joints before bunkering to ensure that the valves and joints are in good working condition to control risks.

3) The leakage results of 1/10 hole of the filling hose show that the distance of vapor cloud diffusion along the wharf is about 4m within the range of combustible concentration. In the case of half-hole and full-hole leakage,

the distance of combustible vapor cloud diffusion along the wharf formed in the port operation area is about 30m and 80m respectively. Considering the hose length and number of valves, and the bunkering facility is equipped with a valve, hose and half diameter and the diameter of burst probability is very low, according to the accident database and the relevant international risk assessment practice, the project implementation process of risk prevention and control measures, and set up around the mouth bunkering is not less than 25 m range of restricted areas, A warning area of not less than 50 meters can control the risk within an acceptable range.

References

1. Lu Xiaoyan, Liao Guoxiang. Fulfill the Important Idea of Ocean Community with a Shared Future, Strengthen Marine Ecology and Environment Protection and Integrated Governance[J]. Environment Protection, 2019,47(15):29-31.
2. Han Chao. Study on the Traffic Safety Management Mode of the Water in the Perspective of Multiple Cogovernance Tanking nantiong Marine Bureau as an Example[D]. Shanghai Jiao Tong University, 2016.
3. Duan Bin, Song Wei, Liu Heng, etc. Development Trend of LNG Bunkering Vessel [J] . Marine Engineering, 2020, 50 (3) : 1-3.
4. Wu Shunping, Luo Xiaofeng, Fan Hongjun, etc. Quantitative Risk Assessment of LNG Truck Bunkering [J]. Natural Gas Industry, 2015,35 (12) :111-116.
5. Anonymous. Ammonia-Fueled Ship Design Principle Approved [J]. Sea Technology. 2020,61: 37-37.
6. Choo Boon Leong, Go Yun Ii. Energy storage for large scale/utility renewable energy system - An enhanced safety model and risk assessment[J]. Renewable Energy Focus. 2022, 42: 79-96.

E3S Web of Conferences **441**, 01006 (2023)
CELCT 2023
https://doi.org/10.1051/e3sconf/202344101006

The Preliminary Application of Biochar in Geotechnical Engineering

Ying Zheng [1st *]

Civil Engineering, National Taiwan University, Taipei, Taiwan, China

Abstract—Biochar is a carbonaceous material produced by the pyrolysis of biomass, which has a porous structure and a high specific surface area. In recent years, with a focus on environmental protection and sustainable development, biochar, a new type of environmentally friendly material, has received extensive attention and research. In geotechnical engineering, biochar can be used in soil improvement, contaminated soil treatment, and other fields but also in slope protection, landfill engineering, and other aspects. This review aims to look back at the current research status of biochar application in geotechnical engineering and look forward to future development.

1. Introduction

Biochar, a charcoal produced through biomass pyrolysis, has been recognized as a potential tool to address environmental challenges related to soil fertility, carbon sequestration, and waste management [2] [3] [4] [13]. In recent years, the application of biochar in geotechnical engineering has gained increasing attention due to its unique properties for enhancing soil quality and stability.

Biochar is a well-known substance that is valued for its high surface area. This kind of characteristic is due to its porous structure and adsorption properties. Depending on the feedstock and pyrolysis conditions, the surface area of biochar can range from 50 to 500 m^2/g. Because of its high surface area, biochar can effectively adsorb and retain nutrients, water, and pollutants in the soil. This results in enhanced soil fertility and the prevention of groundwater contamination. Numerous studies have shown that biochar can be used as a soil conditioner to improve the fertility of acidic soils. This method is primarily based on biochar, which has a porous structure and acts as a carrier [10]. Heavy metals in the soil can be solidified through the addition of microorganisms, as shown in Figure 1, which presents biochar application. Biochar can fix heavy metals in the pores by directly or indirectly changing the soil properties to prevent them from spreading to other areas through groundwater [11] [12].

Furthermore, biochar can alter the physical and chemical properties of soil, such as pH, cation exchange capacity (CEC), and water holding capacity (WHC) [3].

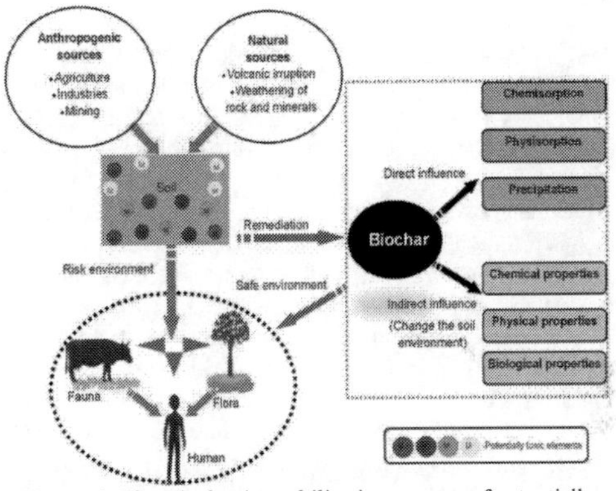

Figure 1. The Biochar immobilization process of potentially toxic substances (PTEs) in the environment.

Due to the presence of mineral components, biochar typically has an alkaline pH, which can neutralize acidic soils and increase the nutrient supply for plant growth [5]. Biochar also has a high CEC, which can promote the retention and release of nutrients in the soil [6]. Additionally, biochar can improve the WHC of soil, reducing the risks of soil erosion and drought stress [4].

In geotechnical engineering, biochar has been applied in various soil improvements projects, such as landfill cover systems, slope stabilization, and road construction [7] [8] [9]. Adding biochar to landfill covers can enhance gas diffusion, reduce leachate production, and improve the overall stability and safety of the landfill [7]. Using biochar in slope stabilization can increase soil shear strength, reduce soil erosion, and prevent landslides and soil settlement [9]. In road construction, biochar can be added as a stabilizer to the soil to improve the strength and

*metall961118@gmail.com

stability of the subgrade and reduce pavement deformation and wear [8].

This branch of civil engineering deals with engineering problems involving geology, soil, rock, etc. It is an essential part of civil engineering. In order to ensure the stability and safety of engineering structures, this project studies the physical and mechanical properties of soil and rock and their impact on engineering buildings. A series of excellent biochar properties can improve soil quality and enhance stability. In view of the fact that such reviews are relatively rare, this study only reviews research on the influences of biochar production and its applications in soil improvement and geotechnical engineering to suggest problems and helpful suggestions.

2. Soil Improvement with Biochar

In the realm of geotechnical engineering, implementing the use of biochar in soil can bring about a multitude of benefits. These advantages include amplified compressive strength in the soil, diminished expansion and shrinkage, heightened stability, and reduced deformation and settlement. Geotechnical engineering benefits from it.

Notability, biochar increases soil's porosity and permeability and improves the facilitating drainage and anti-seepage. Those benefits mentioned above all lead to enhance soil quality. When biochar is used as a soil amendment, it can improve the soil's physical structure, which makes the soil looser and more conducive to plant growth.

Biochar shows the potential of improving soil fertility, which in turn allows plants to be planted on slopes, increasing the impact of slope anti-scalability. Temperatures of pyrolysis and raw material sources affect the functional performance of biochar. The following figure (Figure 2) illustrates the effects of different pyrolysis temperatures on the final biochar product's pH value. The pH value of the final product is alkaline except for the wood ash that is thermally cracked at a low temperature (400 °C).

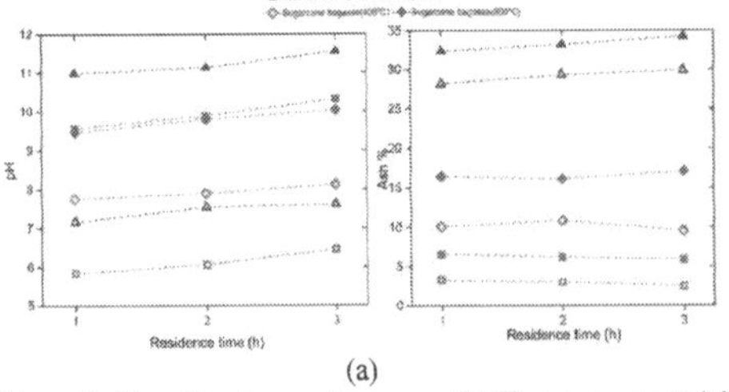

(a)

Figure 2. The pH value, surface area of different raw material biochars after pyrolysis at different temperatures. [1]

As part of their research, Gupta et al. performed mechanical tests on biochar samples formed under various curing conditions with or without biochar samples and water-available porosity to compare the strength of single compression, flexural strength, splitting tensile strength, water permeability, and absorption [26]. This study used wood saw powder to pyrolyze under different temperature conditions (300 °C and 500 °C) to produce two different biochars (Figure 3).

Figure 3. (a) Wood saw powder used in the preparation of biochar; (b) Preparation results of biochar. [26]

Figure 4 shows the results of the compression test. Figure 4a gives a sense of the sample's compressive strength under wet curing, and Figure 4b displays the sample's compressive strength under dry curing. It can be found that the strength of the sample after wet curing is higher when compared with the two figures.

There was a significant difference in compressive strength between samples that were added to biochar that had been pyrolyzed at different temperatures and samples that had not been added. Saturated biochar increases the strength of the modified sample compared to unsaturated biochar and lower temperatures. This phenomenon is because the pores of the pre-impregnated biochar are filled with more water, which can make the hydration reaction of the concrete curing process longer and improve the development of its strength.

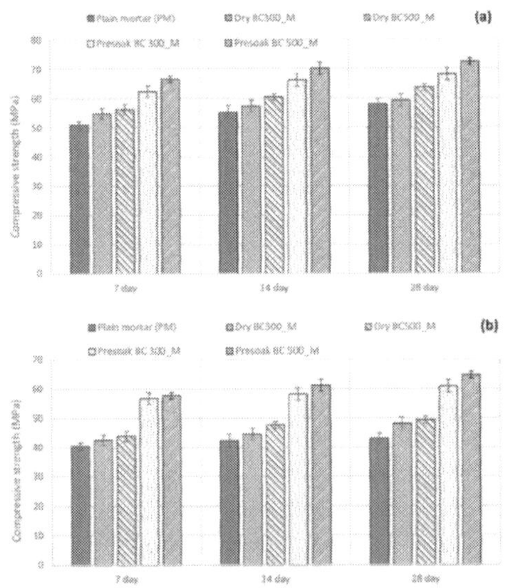

Figure 4. Results of compressive strength of different modified specimens. [26]

In a six-year field study of hillside farmland in the Sichuan Basin, Tuyishimire et al. (2022) investigated the effects of honeysuckle planting and biochar application on soil structure and hydraulic properties [14].

A variety of parameters were used to determine soil structure (soil organic matter ,so called the SOM; aggregate soil's stability, bulk density) and permeability. It was demonstrated that planting honeysuckle alone increased the SOM content, and planting honeysuckle with biochar improved the SOM content to a greater extent, as well as a significant increase in SOM content in deep soil (20 cm - 30 cm and 30 cm - 40 cm). Biochar increased the SOM concentration in the top 20 cm of soil. Combining honeysuckle planting and biochar amendment can improve the stability of soil aggregates. Additionally, the biochar modifier increased K-s directly or indirectly by increasing the SOM. Furthermore, honeysuckle planting and biochar amendments increased saturated soil water content more than saturated soil hydraulic conductivity.

It should be noted, however, that the value of the SOM in bare land was significantly lower, suggesting that both honeysuckle planting and biochar may improve the soil quality by increasing the amount of the SOM in it.

3. Biochar Used in Landfill Projects

Wani et al. discuss using biochar as a soil amendment in geoengineering. Biochar has many advantages and is widely used in agronomy, environmental management and geoengineering [15]. A total of 180 studies were randomly selected from 1996 to 2020 to assess biochar's role in geoengineering. Research papers and parameters were collected using the snowball approach. The main objective was to analyze critical gaps in the research on biochar for soil stabilization in geoengineering applications. Despite this, the production and application of biochar are still confined to the laboratory and require larger-scale research. Several studies have demonstrated conflicting results regarding the effect of biochar application; for instance, some studies suggest that biochar increases soil water retention capacity, while others suggest the opposite.

Further considerations must be given to the properties and cost-effectiveness of biochar, such as its negative impacts, its in-situ application in compacted engineered soils, and its cost-effectiveness in preparation and application. It will be necessary to develop preliminary guidelines on appropriate and optimal biochar percentages for use in green engineering infrastructure in the future. In addition, the study needed more information regarding the availability and properties of biochar feedstock, its cost-effectiveness compared to other soil stabilization options, and its useful life.

Sun W. and Li M. presented the use of biochar to improve the traditional clay cover of landfills to reduce methane emissions into the atmosphere. They also discussed the effects of particle size and biochar content on the air permeability characteristics of the improved soil [27]. It is an ideal material for soil improvement as it has a porous structure, high specific surface area, and strong adsorption properties (Figure 5). Adding biochar can change the pore structure of the soil, thereby improving the mobility of gas in the soil. Although the mechanism affecting landfill overburden's gas flow characteristics remains unclear, environmental factors still play a role in that field.

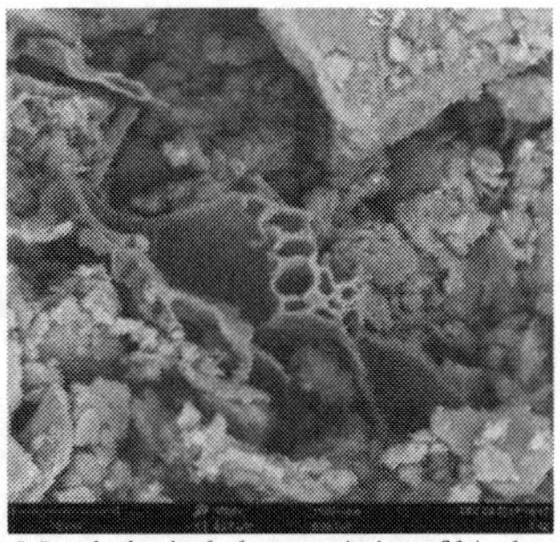

Figure 5. Morphological characteristics of biochar and clay. (From the Internet)

A flexible wall gas phase permeation test device was used to evaluate the air permeability characteristics of biochar-clay mixed soil in this study (Figure 6). By combining nuclear magnetic resonance technology, the gas permeability coefficients of mixed soils and the change laws of soil pore size distributions were analyzed.

Results of the study indicate that for a biochar-clay mixed soil whose dry density has not fallen below 90% of clay's maximum dry density, it contains the optimal amount of biochar, and its gas permeability coefficient reaches the lowest level. The gas permeability coefficient decreases with the increase of biochar when the amount of biochar is less than the optimal value; the increase in biochar causes the gas permeability coefficient to decrease gradually with the increase in biochar.

By comparing the gas permeability coefficients of biochar and clay mixed soil in non-intersecting particle size groups. It is easily to observe a gradual decrease as the biochar particle size is reduced. This study provides guidelines for selecting biochar content and particle size for the design and construction of biochar-modified clay final covers in landfills.

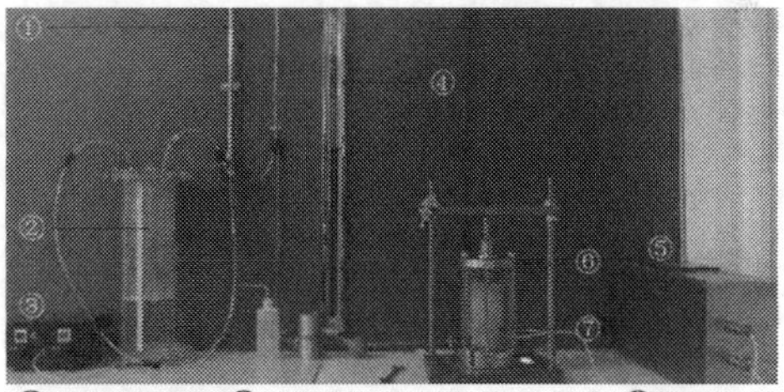

①—Air source; ②—Aqueous gas storage tank; ③—Digital fine-tuning regulator; ④—U-shaped tube (accuracy: 10 Pa); ⑤—water pressure controller;⑥—pressure chamber;⑦ sample

Figure 6. Flexible wall gas phase permeation test setup

Matovic compared Canada's carbon emissions from 1990 to 2008 and biomass energy production in a year(see figure 7), which was strongly suggested by the researcher that Canada could generate electricity by burning biomass to reduce carbon emissions [17]. For emissions, this study only briefly describes the issues related to whether biochar can be used for carbon storage in Canada. It does not mention detailed storage technologies or related methods.

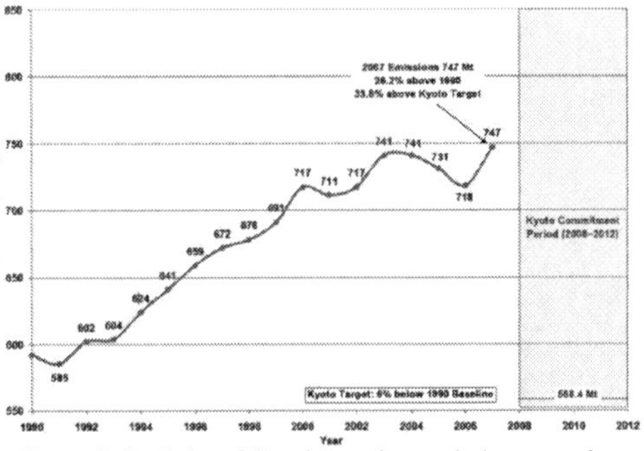

Figure 7. Statistics of Canada's carbon emissions over the years. [17]

Woolf et al. believed biochar could mitigate greenhouse gases in the future [18]. This is mainly through the thermal cracking of biological waste generated to produce biochar and its by-products such as bio-oil, natural gas, and process. The heat generated can be applied as an energy source that humans can use again. Biochar can be utilized as a soil improver to improve land with poor fertility due to its characteristics.

Composing it into the world's carbon cycle through nature is complex, so that it can be applied to carbon sequestration (Figure 8). This study uses three different models to simulate carbon emissions estimation under different scenarios. It compares carbon emissions reduced by biochar for analysis.

This study used sensitivity analysis and the Monte Carlo method to analyze biochar production. The variables analyzed included raw materials from different biomass sources, as shown in Figure 9.The left side of Figure 9 displays the amount of carbon that can be avoided and the amount of carbon emitted by eight raw materials used as biochar. The avoided emissions are positive, and the generated emissions are harmful. The right side of Figure 9 shows the combined results of biochar and biomass combustion. The results show that turning existing biomass into biochar can reduce greenhouse gas production and emissions.

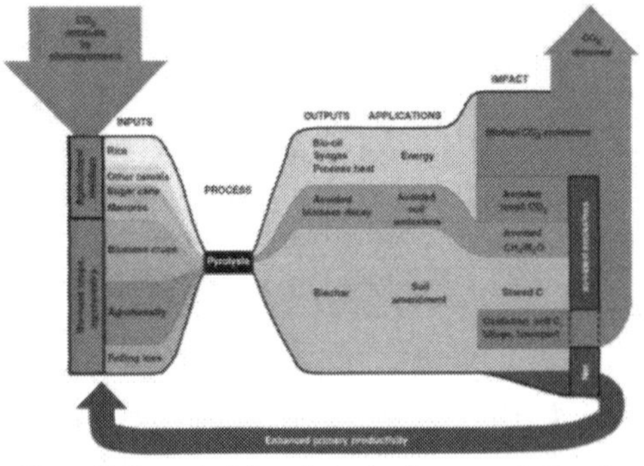

Figure 8. Conceptual flow chart of carbon sequestration via biochar. [18]

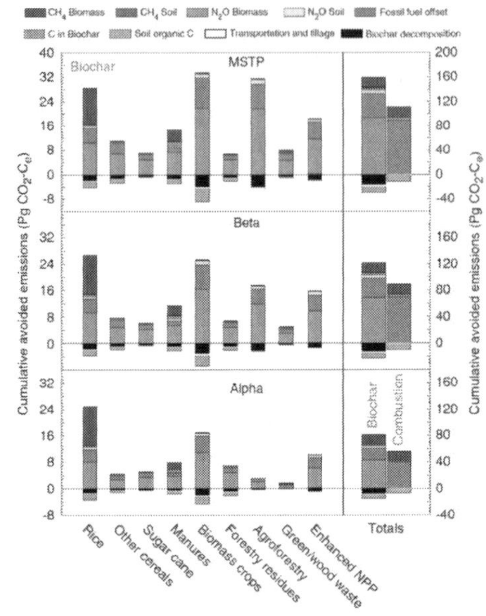

Figure 9. The results of predicting the greenhouse gas emissions [19]

Fuss et al. and others sorted out seven technologies currently related to harmful emissions, including carbon capture, bioenergy, afforestation and reforestation, direct air capture and storage, enhanced weathering, ocean fertilization, biochar, and soil carbon sequestration [19].

Other seven technologies are reviewed. Figure 10 calculates the amount of carbon emission reduction to quantify the potential index of this technology and its cost. This is the basis for this review. Although biochar is quite effective in reducing emissions, it is too expensive for its high price and cannot be used. Large-scale biochar production and sequestration require long-term underground storage, and the impact on local microbial systems cannot be estimated. Therefore, more research is still needed on this technology.

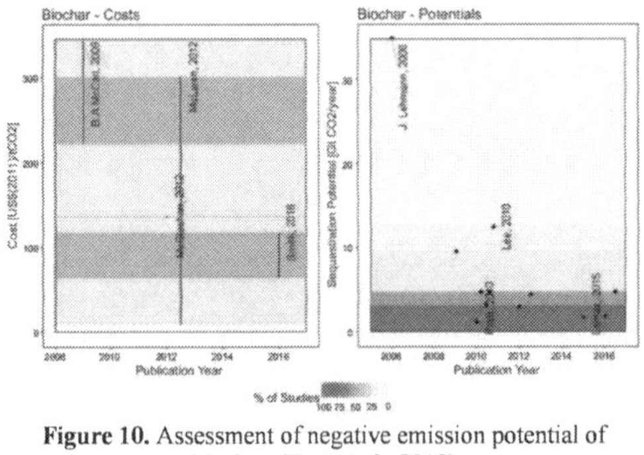

Figure 10. Assessment of negative emission potential of biochar. (Fuss et al., 2018)

4. Conclusions

This review aims to provide an overview of how biochar technology, which has emerged in recent years, can contribute to geotechnical engineering. Additionally, it reduces carbon emissions. Biochar improves the physical and mechanical properties of the soil as well as permeability and benefits plant growth and slope protection [10] [11] [21] [22] [23] [24] [25]. Biochar can also be used as a soil improvement for landfill projects and as a concrete amendment. Biochar improvement can strengthen soil structure and increase soil water retention, which can be utilized as a basis for future green development.

As an environmentally friendly material, biochar has broad application prospects and is of tremendous research importance in geotechnical engineering. It is imperative to study biochar's physical and chemical properties and application effects in the future. Exploring more innovative application methods should also be pursued, as well as promoting their development and application in geotechnical engineering.

References

1. Bandara, T., Franks, A., Xu, J., Bolan, N., Wang, H., & Tang, C. (2019). Chemical and biological immobilization mechanisms of potentially toxic elements in biochar-amended soils. Critical Reviews in Environmental Science and Technology, 1–76.
2. Glaser, B., Lehmann, J., & Zech, W. (2002). Ameliorating physical and chemical properties of highly weathered soils in the tropics with charcoal—a review. Biology and fertility of soils, 35(4), 219-230.
3. Lehmann, J., Gaunt, J., & Rondon, M. (2006). Bio-char sequestration in terrestrial ecosystems–a review. Mitigation and adaptation strategies for global change, 11(2), 395-419.
4. Spokas, K. A., & Reicosky, D. C. (2009). Impacts of sixteen different biochars on soil greenhouse gas production. Annals of Environmental Science, 3, 179-193.
5. Laird, D. A., Fleming, P., Davis, D. D., Horton, R., Wang, B., Karlen, D. L., ... & Brown, R. C. (2010). Impact of biochar amendments on the quality of a typical Midwestern agricultural soil. Geoderma, 158(3-4), 443-449.
6. Joseph, S. D., Camps-Arbestain, M., Lin, Y., Munroe, P., Chia, C. H., Hook, J., ... & Graber, E. R. (2010). An investigation into the reactions of biochar in soil. Soil Research, 48(7), 501-515.
7. Beesley, L., Moreno-Jiménez, E., Gomez-Eyles, J. L., Harris, E., Robinson, B., & Sizmur, T. (2011). A review of biochars' potential role in the remediation, revegetation and restoration of contaminated soils. Environmental pollution, 159(12), 3269-3282.
8. El-Nakhlawy, F. S., El-Shafie, A., & El-Shorbagy, A. (2016). Using Biochar to Improve Soil Physical and Mechanical Properties for Road Subgrade. Journal of Geotechnical and Geological Engineering, 34(1), 19-29.
9. Xu, W., Zhao, J., Liu, X., Liu, W., & Xiong, C. (2016). Effects of biochar amendment on soil physical properties and soil erosion in the loess plateau, China. CATENA, 147, 98-106.
10. Mohan, D., Pittman Jr, C. U., & Steele, P. H. (2006). Pyrolysis of wood/biomass for bio-oil: a critical review. Energy & fuels, 20(3), 848-889.
11. Chen, D., Chen, Y., Zhang, Q., Yin, D., & Liu, J. (2017). Synthesis of iron-modified biochar and its application in heavy metals removal. Journal of hazardous materials, 330, 94-102.
12. Beiyuan, J., Rui, L., Liwei, W., Zhaoliang, Y., Xingtao, Z., Xuewei, P., Xiaoping, X., & Mingxiao, L. (2017). Experimental study on the mechanical properties of biochar reinforced soil. Journal of Earthquake Engineering and Engineering Vibration, 16(2), 409-418.
13. Karaosmanoğlu, F., Işıgıgür-Ergüdenler, A., & Sever, A. (2000). Biochar from the straw-stalk of rapeseed plant. Energy & Fuels, 14(2), 336-339.
14. Tuyishimire, E., Cui, JF., Cheng, J.H. Interactive Effects of Honeysuckle Planting and Biochar Amendment on Soil Structure and Hydraulic Properties of Hillslope Farmland. Agriculture-Basel. 2022, 12(3). DOI10.3390/agriculture12030414.
15. Wani, I., Ramola, S., Garg, A., Kushvaha, V. Critical review of biochar applications in geoengineering infrastructure: moving beyond agricultural and environmental perspectives. Biomass Conversion and Biorefinery, online, 2021, DOI10.1007/s13399-021-01346-8.
16. Sun W. and Li M. Study on air permeability characteristics of overburden soil modified by biochar. Chinese Journal of Rock Mechanics and Engineering. 2022,41(S2). 3543-3550.
17. Matovic, D. (2011). Biochar as a viable carbon sequestration option: Global and Canadian perspective. Energy, 36(4).
18. Woolf, D., Amonette, J. E., Street-Perrott, F. A., Lehmann, J., & Joseph, S. (2010). Sustainable biochar to mitigate global climate change. Nature communications, 1(1), 56.
19. Fuss, S., Lamb, W. F., Callaghan, M. W., Hilaire, J., Creutzig, F., Amann, T., ... & Minx, J. C. (2018). Negative emissions - Part 2: Costs, potentials and side effects. Environmental Research Letters, 13(6), 063002.
20. Lehmann, J., & Joseph, S. (2015). Biochar for environmental management: science, technology and implementation. Routledge.
21. Shackley, S., Carter, S., Knowles, T., Middelink, E., Haefele, S., Haszeldine, S., ... & Sohi, S. (2012). Sustainable gasification–biochar systems? A case-study of rice-husk gasification in Cambodia, part I: Context, chemical properties, environmental and health and safety issues. Energy Policy, 49, 422-433.

22. Sohi, S. P., Krull, E., Lopez-Capel, E., & Bol, R. (2010). A review of biochar and its use and function in soil. Advances in agronomy, 105, 47-82.
23. Cao, X., & Harris, W. (2010). Properties of dairy-manure-derived biochar pertinent to its potential use in remediation. Bioresource technology, 101(14), 5222-5228.
24. Shen, Q., Ren, L., Zhong, H., & Yan, L. (2017). Effect of pyrolysis temperature on the physicochemical properties and heavy metal adsorption of biochar derived from municipal sewage sludge. Bioresource technology, 244, 355-362.
25. Beiyuan, J., Rui, L., Liwei, W., Zhaoliang, Y., Xingtao, Z., Xuewei, P., ... & Mingxiao, L. (2020). Effects of pyrolysis temperature on the properties and cadmium adsorption capacity of biochar derived from tobacco stalk. Journal of environmental management, 271, 110996.
26. Liu, X., Zhang, A., Ji, C., Joseph, S., Bian, R., Li, L., ... & Pan, G. (2013). Biochar's effect on crop productivity and the dependence on experimental conditions—a meta-analysis of literature data. Plant and soil, 373(1-2), 583-594.

CCER Development Based on Scenery, Fire and Storage Integration Project

Ning Qiao*, Chao Zhang, Jisheng Zhang, Haidong Chen, Jing Zhang, Hongjie Tian

State Grid Ningxia Electric Power Company, Yinchuan, Ningxia, China

Abstract: Since 2012, China has established a national voluntary greenhouse gas emission reduction trading mechanism, which realizes the filing of voluntary emission reduction projects such as carbon sinks and the issuance of certified voluntary emission reduction (CCER). Due to the challenges in the development, operation, trading, and other stages of carbon sequestration projects, and the suspension of voluntary emission reduction-related applications by the competent authorities in 2017, the development of CCER carbon sequestration projects in China is relatively insufficient, and the number of registered projects accounts for only 2%. Therefore, this article, based on the CCER methodology development perspective of the Wind-Solar-Storage Integration Project, investigates and analyzes the structure of the project, the baseline methodology, and procedures for CCERs, economic evaluation, and additional challenges of the project, and verifies the feasibility of CCER development for the Wind-Solar-Storage Integration Project in the Ningxia region based on actual conditions. The article also proposes policy recommendations for the development of new energy carbon sequestration projects.

1 Introduction

According to the sixth assessment report (AR6) issued by IPCC [1], the global temperature rise reached 1.2°C in 2020, and the climate change situation is becoming more and more severe. China has made a solemn commitment to peak carbon dioxide emissions and carbon neutrality to the international community, and carbon emission trading is an important means for China to realize this commitment. China's carbon emission trading market is mainly divided into two types: carbon quota trading based on total amount control and voluntary emission reduction trading based on projects. The former deals with carbon emission quotas allocated by the state to various emission control industries based on controlling the total amount of carbon emissions. The trading object of the latter is the national certified voluntary emission reduction (CCER) obtained by the industry through the implementation of voluntary carbon emission reduction projects and the quantitative verification of the emission reduction effect of the projects by relevant departments. CCER is a certified emission reduction with China characteristics based on the extension of the clean development mechanism (CDM) model, which can be used to offset carbon quotas. It is trading, and the offsetting mechanism is an important supplement to carbon quota trading.

At present, in the research related to emission reduction certification, from the perspective of the research object, due to the late start of CCER trading in China and poor data availability [2] (CCER trading was first realized in China in 2015, but the state suspended the filing and approval of CCER projects in 2017), most of the existing literature achievements take CDM as the research object [3-5], and the achievements taking CCER as the research object are limited [6-9]. From the perspective of research content, the discussion on the environmental benefits of CDM or CCER projects in the existing literature results is relatively concentrated [10-12]. Since renewable energy projects are an important source of certified emission reduction, whether CDM or CCER project development can promote the development of renewable energy is also a hot issue for scholars, and the research on these types of projects has affirmed their promotion to the development of renewable energy [13-14]. Therefore, this paper will start with the integration project of scenery, fire, and storage and analyze the feasibility and economy of CCER development with standardized CCER methodology.

2 Integrated wind-Photovoltaic-fire-storage project structure

The integration project of scenery, fire, and storage will give priority to the use of clean energy such as wind power and Photovoltaic, give play to the regulation function of coal and electricity, appropriately allocate energy storage facilities, and coordinate the development and scientific allocation of various resources, which is conducive to giving full play to the advantages of new energy-rich areas,

*2492618731@qq.com

realizing the large-scale consumption of clean electricity, and improving the flexibility and stability of regional power systems. As shown in Figure 1 below, the project's physical structure includes a Photovoltaic power station, a wind power station, a traditional thermal power station with flexibility transformation, a supporting energy storage power station, and all power plants connected to the power grid.

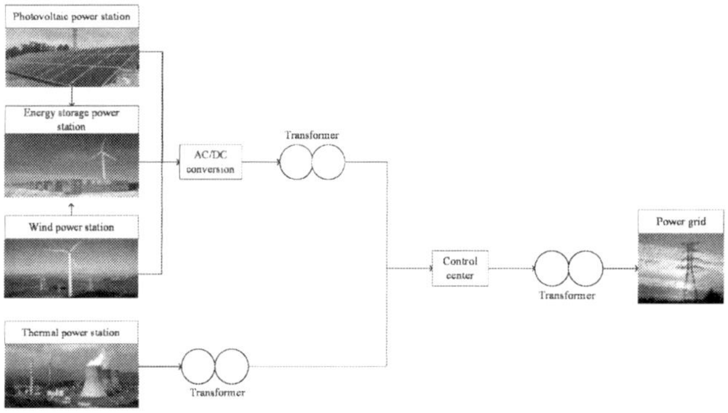

Figure 1. Integrated wind-Photovoltaic-fire-storage project structure

3 CCER baseline methodology program

3.1 Baseline identification and additionality demonstration

The baseline scenario is a scenario that produces the same result (e.g., the same amount of electricity generated) without building the existing project or continuing the historical situation.

It is important to implement integrated wind-fire-storage power generation projects, assess emission reductions, and carry out emissions reduction trading of carbon reduction products from integrated wind-fire-storage power generation projects. First, through the scenery fire storage integrated power generation project carbon reduction products emission reduction development and trading, it can help open voluntary carbon neutral, carbon offset and other related practices, and promote the implementation of national and provincial carbon peak carbon neutral major strategic decisions to further promote regional green low-carbon development. Secondly, the development of wind-fire-storage integrated power generation project carbon reduction product emissions trading can increase economic income and improve people's livelihood for project operators, developers, workers, and other project developers. It can alleviate the difficulties of insufficient subsidies and poor benefits of new energy projects. Thirdly, giving full play to the advantages of regional new energy and promoting the realization of ecological values are conducive to promoting the development of a new energy power generation industry.

3.2 Baseline scenario carbon emissions

Baseline emission (BE_y) includes the CO_2 emissions from the generation of electricity from fossil fuel-fired power plants in the grid by the integrated wind-fired storage project, and are the product of the base arene emission factor ($EF_{CO_2,\text{gridy}}$) and the effective power (EG_y) supply of the project, and its calculation is as follows:

$$BE_y = EG_y \times EF_{CO_2,\text{gridy}} \quad (1)$$

where BE_y represents the baseline emissions (tCO_2) in year y; EG_y represents the net feed-in tariff (MWh) in year y; $EF_{CO_2,\text{gridy}}$ represents the combined marginal CO_2 emission factor (tCO_2/MWh) of grid-connected power generation calculated by the Power System Emission Factor Calculation Tool in year y.

$$EF_{CO_2,\text{gridy}} = EF_{grid,OM,y} \times W_{OM} + EF_{grid,BM,y} \times W_{BM} \quad (2)$$

where $EF_{grid,OM,y}$ represents the marginal emission factor of electricity (tCO_2/MWh) in year y, using the latest marginal emission factor of electricity of the northwest power grid published by the Ministry of Ecology and Environment. $EF_{grid,BM,y}$ represents the capacity marginal emission factor (tCO_2/MWh) in year y, using the latest capacity marginal emission factor of the Northwest Power Grid published by the Ministry of Ecology and Environment. W_{OM} represents the weight of the electricity marginal emission factor, with W_{OM}=0.75 in the first and subsequent crediting periods. W_{BM} represents the weight of the capacity marginal emission factor, with w_OM=0.75 in the first and subsequent crediting periods. W_{OM}=0.25 is for the first and subsequent crediting periods.

3.3 Carbon emissions of the project

In the project scenario, thermal and energy storage can provide auxiliary services to wind power and PV units. Wind power and PV have great randomness, and with the fluctuation of scenery generation, thermal power units will climb up and down as a regulating class resource, and the coal consumption rate per unit of electricity will show a large increase compared to the baseline scenario.

The feed-in tariff of the integrated wind-Photovoltaic-fire-storage project is composed of the following four parts: (1) The electricity of PV units connected to the grid EG_{PV}; (2) The electricity of wind turbines connected to the grid EG_{WD}; (3) The energy storage units are connected to the grid EG_{SE}, which is composed of the electricity that the PV units fail to connect to the grid during the wind and Photovoltaic abandonment hours EG_{DPV} and the electricity that the wind units fail to connect to the grid EG_{DWD}; (4) The electricity that is connected to the grid when the thermal units provide auxiliary services during the peak load hours EG_{TP}. The order of consumption of the four types of power sources is PV, wind power, energy storage and thermal power.

$$EG_y=EG_{PV}+EG_{WD}+EG_{SE}+EG_{TP} \quad (3)$$

$$EG_{SE}=EG_{DPV}+EG_{DWD} \quad (4)$$

For most renewable energy power generation project activities, =0. However, some project activities may produce significant emissions, that is, project emissions, which are calculated by the following formula:

$$PE_y=PE_{PV,y}+PE_{WD,y}+PE_{SE,y}+PE_{TP,y} \quad (5)$$

where PE_y represents project emissions in year y (tCO_2); $PE_{PV,y}$ represents project emissions from PV units in year y (tCO_2); $PE_{WD,y}$ represents project emissions from wind turbines in year y (tCO_2); $PE_{SE,y}$ represents project emissions from energy storage units in year y (tCO_2); $PE_{TP,y}$ represents project emissions from thermal peaking units in year y (tCO_2).

The calculation process of project emissions from each emission source is as follows.

Emissions from Photovoltaic units ($PE_{PV,y}$): For Photovoltaic units, they will consume some electricity from fossil fuels during operation, but this self-consumption electricity can be neglected, then $PE_{PV,y}$ can be set to 0.

Wind turbine emission ($PE_{WD,y}$): For a wind turbine, it will consume some electricity from fossil fuel during operation, but this self-consumption electricity can be ignored, then $PE_{WD,y}$ can be set to 0.

Emissions from energy storage units ($PE_{SE,y}$): For energy storage units, the electricity stored in the operation process comes from the over-generated electricity of new energy units during the time of wind and Photovoltaic abandonment, which can be regarded as a clean power source, and its carbon emission can be ignored.

Thermal power unit emissions ($PE_{TP,y}$): For thermal power units that mainly undertake the task of peaking, they need to carry out flexibility transformation in the process of project operation. With the flexibility transformation of thermal power units, traditional coal power units have faster variable load rates, higher accuracy of load regulation and better primary frequency regulation performance. The emission $PE_{TP,y}$ of thermal power units is calculated as follows:

$$PE_{TP,y} = EG_{TP} \times EF_{CO_2,\mathrm{TP}} \quad (6)$$

where EG_{TP} represents the generation capacity of thermal power units, i.e., the net load curve minus the output curve of PV units, wind power units and energy storage units.

$EF_{CO_2,\mathrm{TP}}$ represents the kilowatt-hour carbon emission factor of thermal power units after flexibility retrofit. The carbon emission factor of thermal units with flexibility modification is calculated as follows:

$$EF_{CO_2,\mathrm{TP}}=COEF_{TP,y} \times F_{TP,y}/EG_{TP} \quad (7)$$

where $F_{TP,y}$ represents the coal consumption of the thermal power unit after flexibility retrofit in year y(t); $COEF_{TP,y}$ represents the CO_2 emission factor factor of coal consumed by the thermal power unit after flexibility retrofit (gCO_2/t).

4 Economic evaluation model

According to the "Guidelines for Engineering Economic Evaluation", the benchmark rate of return for power generation projects is 8%. When the financial internal rate of return of the proposed project is equal to or higher than the benchmark rate of return, the project is financially feasible.

The life cycle cost of the integrated project consists of total investment cost, operation and maintenance cost and decommissioning disposal cost. The total investment cost, or one-time input cost, refers to the total cost of one-time input during the project construction period. Operation and maintenance cost refers to all the expenses required for the operation, maintenance and overhaul of power generation equipment after it is put into operation. Scrap cost refers to the total cost of cleaning or destroying the equipment after its service life reaches retirement age or its life cycle is interrupted for some reason.

$$\mathrm{LCC}=\mathrm{CI}+\sum_{t=1}^{N}\left(\frac{1}{1+r}\right)^t * COM_t + \left(\frac{1}{1+r}\right)^N * CD \quad (8)$$

where LCC denotes the whole life cycle cost of the water-landscape complementary power generation system. CI denotes the initial investment cost; COM_t denotes the first-year operation and maintenance cost; CD denotes the decommissioning disposal cost. r denotes the discount rate. N denotes the life year.

5 Basic data of an example

Taking Ningxia as an example, the project plans to develop wind, Photovoltaic, fire and storage to meet the local load needs, improve the reliability and flexibility of power system operation, and tap the consumption space of renewable energy. It is assumed that the configuration of each unit of the wind, Photovoltaic, fire and storage integration project is as follows: 500 kW Photovoltaic unit, 500 kW wind turbine unit and 500 kW thermal power unit (the maximum peak-shaving capacity of the thermal power unit is changed from 60% to 30% after flexible transformation), and the energy storage is configured according to 10% of the installed capacity of new energy. The output power curve and load curve of each unit in the wind, Photovoltaic, fire and storage integration project scenario and baseline scenario are shown in the figure 2.

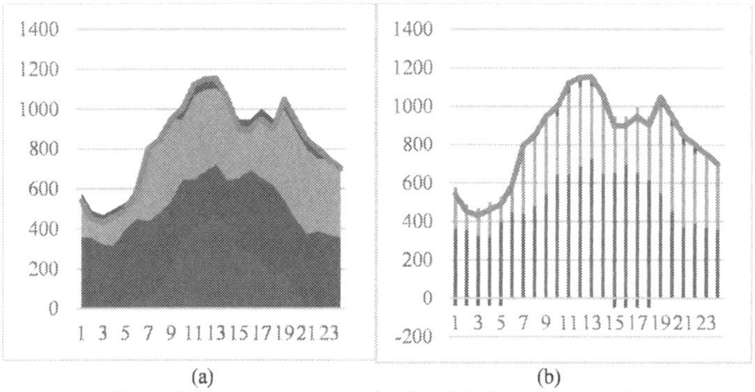

Figure 2. Output power curve of each unit in the project scenario

According to the regional grid baseline emission factors for emission reduction projects in 2019 released by the China Voluntary Emission Reduction Trading Information Platform, the power marginal emission factor $EF_{grid,OM,y}$ in Northwest China is 0.8922 tCO_2/MWh, and the capacity marginal emission factor $EF_{grid,BM,y}$ is 0.4407 tCO_2/MWh, so the combined marginal CO_2 emission factor of grid-connected power generation in Northwest China is 0.7793 tCO_2/MWh. The emission factor $EF_{CO_2,\mathrm{TP}}$ of the thermal peaking unit after flexibility modification is 0.9523 tCO_2/MWh. The unit feed-in tariff is set at RMB 0.35/kWh and the CCER carbon price is RMB 40/tCO_2. The relevant parameters for emission reduction and economic calculation are as shown in table 1 below:

Table 1. Parameter Table of Emission Reduction and Economic Calculation

Parameter	Numerical value	Unit	Parameter	Numerical value	Unit
The unit price of Photovoltaic constructionP_{sr}	3440	Yuan /kW	System residual rateQ_{cz}	Five	%
The unit price of wind power constructionP_{wd}	6708	Yuan /kW	Flexible thermal power carbon emission factor$EF_{CO_2,\mathrm{TP}}$	0.9523	t/MWhCO_2
The unit price of energy storage constructionP_{se}	3600	Yuan /kW	Capacity marginal emission factor$EF_{grid,BM,y}$	0.4407	t/MWhCO_2

6 Demonstration of project additionality

Based on the above parameters, the CCER economic calculation of the integrated wind, Photovoltaic, fire and storage project on the power supply side is carried out. The economic effect evaluation indexes of the project are shown in Table 2.

Table 2. Evaluation index of the economic effect of the CCER project

	Considering CCER	Regardless of CCER
Internal rate of return	8.93%	7.28%
Payback period of investment	9.16 years	10.23 years

The benchmark rate of return used in power generation projects is 8%, and the internal rate, without considering CCER, is less than the benchmark rate of return, which does not achieve an ideal economy. After considering the benefits of CCER, the internal rate of return of the integrated project is higher than the benchmark rate of return, and the project is additional.

7 Conclusion

This paper provides experience in the development of CCER methodology for new energy projects in the context of CCER development for integrated wind-fire-storage projects. The results demonstrate that the CCER methodology for integrated wind-fire-storage projects is feasible and brings considerable economic results.

The integrated wind-fire-storage project uses wind power and Photovoltaic power generation as the main power supply source, supplemented by modified thermal power and energy storage equipment to mitigate the instability of wind and Photovoltaic power generation on the one hand and to regulate the changes in customer demand on the other. The project improves the stability of the power supply in the region, promotes the consumption of new energy, and improves carbon emission reduction.

The basic research on monitoring and accounting data applicable to the integrated wind-fire-storage project, the establishment of the integrated wind-fire-storage carbon sink accounting method system, the formation of the integrated wind-fire-storage methodology, and the implementation of the application of the methodology to develop CCER projects, opens up a new channel for new energy development, helps to carry out international cooperation in the field of new energy and obtains economic income.

Acknowledgments

State Grid Science and Technology Project (Research on the mechanism and analysis model of carbon emission reduction market with the cooperation of source, network, load and storage under the bridging of electricity and carbon market, SGNX0000JYJS2200216).

References

1. IPCC. Climate change 2021: the physical science basis [R]. Geneva: IPCC, 2021
2. Li L, Ye F, Li Y, et al. How will the Chinese Certified Emission Reduction scheme save cost for the national carbon trading system? [J]. Journal of Environmental Management, 2019, 244: 99-109
3. Lim X L, Lam W H. Review on Clean Development Mechanism (CDM) implementation in Malaysia [J]. Renewable and Sustainable Energy Reviews, 2014, 29: 276-285
4. Ye F, Xiong X Y, Li L X, et al. Measuring the effectiveness of the Chinese Certified Emission Reduction scheme in mitigating CO_2 emissions: a system dynamics approach [J]. Journal of Cleaner Production, 2021, 294: 125355
5. Zhao Z Y, Li Z W, Xia B. The impact of the CDM (clean development mechanism on the cost price of wind power electricity: a China study [J]. Energy, 2014, 69: 179-185
6. Zhang Ning, Pang Jun. Study on the economic impact of introducing CCER trading and offsetting mechanism into the national carbon market [J]. Progress in climate change research, 2022, 18(05): 622-636.
7. Liu Haiyan, Zheng Shuang, Sun Yijia, etc. Problems and countermeasures of forestry carbon sink market development based on Saihanba CCER project [J/OL]. Progress in climate change research: 1-9[2023-04-10].
8. Guo Shengwei, Men Xiujie, Sun Haiping, etc. Comparative study on the status quo and trend of green electricity, green certificate and CCER policies in China [J]. China Energy, 2022, 44(03): 75-80.
9. Xie Qiang, Yang Wai, Zou Fangrui. The methodology of carbon trading based on CCER and its application at home and abroad [J]. Construction Science and Technology, 2022, No.447(Z1): 85-89.
10. Deng Qiang. Comparative analysis of emission reduction by anaerobic digestion and incineration of municipal solid waste based on CCER methodology [J]. Resource regeneration, 2022, No.235(02): 30-31.
11. He Ke, Li Fanlve, Chang Huayi. Building a low-carbon community: local consensus and participation of large-scale pig farmers in agricultural carbon trading-taking rural biogas CCER carbon trading project as an example [J]. china rural survey, 2021(05): 71-91.
12. Huan M ,Qinghan S ,Lei C , et al. Cogeneration transition for energy system decarbonization: From basic to flexible and complementary multi-energy sources[J]. Renewable and Sustainable Energy Reviews,2023,187.
13. Shuaishuai F ,Weida C ,Junfei D , et al. Is carbon asset pledge financing contributing to the operation of emission-dependent engineering machinery remanufacturing with emission abatement?[J]. International Journal of Production Economics,2023,266.
14. Meng Y ,Yisheng L . Research on multi-energy collaborative operation optimization of integrated energy system considering carbon trading and demand response[J]. Energy,2023,283.

Analytical Research on the Exploitation of Flexible Resources and Power Grid Regulation Technologies Driven by Industrial Internet under Industrial Load

Yucheng Ding[1,*], Guanglong Xie[1], Xinyang Han[1], Yongyu Fan[2] and Yue Cai[2]

[1]State Grid Energy Research Institute, Beijing 102209, China
[2]State Grid Suzhou Electric Power Company, Suzhou 215004, China

Abstract. With the release and gradual implementation of the "carbon peaking and carbon neutrality" policy, renewable energy is integrated into power grids on a large scale, and its uncertain and intermittent supply challenges the stability of the power system. The new power system integrated with renewable energy is faced with a new challenge in terms of frequency security, while the extreme climate in recent years has made it more difficult to achieve power balance in the power system. As an integral part of the load side, the flexible power regulation of industrial enterprises is now attracting attention. With focus on the exploitation of flexible power regulation resources under industrial load, this paper builds various industrial load models for industrial enterprises, promotes the deep integration of Industrial Internet and Energy Internet, conducts research on the exploitation of flexible resources and power grid regulation technologies driven by Industrial Internet under industrial load, and analyzes the direct and indirect benefits of the research. The research results are of great significance to the safe and stable operation of the power system, the consumption of new energy, and the green and low-carbon transformation of the industry.

1 Introduction

With increasing global climate change, General Secretary Xi Jinping pledged to the world at the UN General Assembly in 2020 that China will actively promote the sustainable transformation of its energy system and aim to have carbon emissions peak before 2030 and achieve carbon neutrality before 2060. In this context, China has seen dramatic growth in installed new energy capacity, with total installed wind and solar capacity expected to exceed 1.2GW by 2030. The uncertainty of new energy sources such as wind and PV power can dramatically reduce the supply-side regulation capacity of the power system when they are integrated into the power grids on a large scale. As this new power system has little ability to load fluctuations like a traditional power system, its security, stability and economic viability decline, and its ability to maintain balanced grids is compromised. In addition, the frequency of extreme weather conditions, such as prolonged high temperatures and droughts in recent years, has caused the maximum electricity load on power grids to surge to new highs, while insufficient rainfall has led to reduced hydropower generation capacity. The combination of these many factors has increased the difficulty of maintaining balanced grids. In Jiangsu, for example, the electricity load of the whole province reached 131 million kW on August 5, 2022, and the rapidly growing electricity consumption led to a gap between supply and demand from that day onward, which worsened until August 20. The largest gap occurred on August 12, standing at 10.68 million kW. Given the current structure and operation mode of power supply in China, the lack of flexible power regulation resources will be a key issue affecting the consumption of renewable energy, which has a high share, and the safe and stable operation of power grids. Therefore, the exploitation of flexible power regulation resources to support balanced grids is in line with the national strategy for carbon peaking and carbon neutrality in the long term and provides a powerful method to address local power shortages and protect domestic production and life for the time being.

At present, the flexible grid regulation resources in China are mainly thermal power units. On the one hand, thermal power units have a lower ramp rate than other power regulation resources, and on the other hand, the share of installed coal power unit capacity will continue to decline in the future. Therefore, the use of thermal power units as the main regulation resource will not be sustainable. There are currently two possible solutions. One is to regulate the power grids with energy storage devices, but the technical indicators of energy storage devices need to be further improved at this stage, installation is costly, and the safety specifications of some forms of energy storage are not yet complete. Without major technological breakthroughs, it would be difficult to promote and deploy grid-side energy storage

* Corresponding author: 675281777@qq.com

on a large scale. The other solution is to make full use of the flexible resources on the load side and to achieve a balanced system through the interaction of power, grid, load and storage. On the load side, the industrial energy consumption in China accounts for about 65% of the total social energy consumption, and the industrial load capacity has a large base. In typical industries such as metal smelting, cement, paper making, and glass, some technologies are not strict with real-time balance of electrical energy, and the existence of extensive transferable electricity load has good power regulation potential and can be ideally transformed into cheap flexible resources. In the metallurgical industry alone, electrolytic and electric arc furnace loads account for 16-18% of the total social electricity consumption, and their participation in power regulation can significantly increase the flexibility of the grids. However, to participate in grid demand response, some of the industrial load is often removed directly. Load outages can seriously disrupt normal social and industrial operations, causing huge economic losses and even more potential negative impacts. Therefore, industrial enterprises are not willing enough to participate in power regulation in this context. It is a key and urgent task to realize outrage-free flexible active power control and to exploit the flexible resources of the industrial load without affecting enterprise production as much as possible.

At the same time, as a new infrastructure for the deep integration of new-generation information and communication technology with the industrial economy, the Industrial Internet can build a new manufacturing and service system that covers the entire industry chain and value chain through the close-loop analysis of the manpower, machine, material, method and environmental data of industrial enterprises. It is an important method to perfectly integrate industrial enterprises with the power system through digital means. At present, the global Industrial Internet structure still needs to be improved, and the related industries in China are facing a window of opportunity in the wave of great new-generation digital industry development. Since 2015, China has successively introduced a number of plans and guidelines, including the Guidance on Promoting "Internet+" Smart Energy Development, in order to promote the deep integration of energy with new technologies such as information technology and to build Energy Internet that promotes the coordinated development, integration and complementarity of "power, grid, load and storage". Under this trend, grid enterprises have participated in the construction of some Industrial Internet platforms by introducing the "power supply + energy efficiency" business in recent years, which accumulates massive industrial energy consumption data, builds databases for grid enterprises, and sets the stage for analyzing the energy consumption characteristics of typical industrial enterprises and exploiting the flexible power regulation resources under industrial load. However, most of the current Industrial Internet projects can only collate and display data on the production, management, energy and other aspects of enterprises by deploying sensors and building platforms to solve observable system problems. They are far from being "smart". The massive amount of data collected does not create added value for enterprises, and various high-tech models and algorithms are not effective in suitable application scenarios. Further research is therefore needed to extend the value chain of Industrial Internet data and to combine the practical requirements of the power system with digital technologies.

To address the issues above, this paper focuses on the exploitation of flexible power regulation resources under industrial load, studies and analyzes the exploitation of flexible resources and power grid regulation technologies driven by Industrial Internet under industrial load, discusses domestic and foreign research on the planning of flexible grid resources, the participation of industrial load in grid regulation, and demonstration projects for the participation of controllable industrial load in grid regulation. This paper also demonstrates the direct and indirect benefits of the research and models various types of typical industrial load in industrial enterprises to tap the power regulation potential of industrial enterprises and industrial parks, accelerate the deep participation of industrial enterprises in grid interaction, create the added value of Industrial Internet platforms, and promote flexible and accurate outrage-free power regulation under industrial load.

2 Domestic and Foreign Research Status

2.1 Research on Modeling Methods and Adjustable Features of Typical Industrial Load

In terms of load power regulation, Reference [1] is the earliest research to control the load power of energy-intensive electrolytic aluminum to participate in primary frequency modulation based on grid frequency deviation. However, this control method is still similar to the principle of graded load shedding, which instructs the saturable reactor to adapt to the maximum pressure drop when the grid frequency deviation exceeds the threshold. This ignores the fact that the load is continuously adjustable and can cause frequency oscillations. A U.S. electrolytic aluminum company proposed a method of electrolytic aluminum load power regulation based on the control of the transformer tap on the load side, but it was also noted that continuous and frequent adjustment of the transformer can lead to mechanical wear and affect the service life of the equipment[2]. Based on the detailed technology process of energy-intensive load production, Reference [3] analyzed in detail the regulation potential of energy-intensive load from the technical aspect of power regulation and built a corresponding economic cost model that combined the technical and economic aspects to provide a power regulation method that industrial enterprises would be willing to accept. Based on isolated grids, Reference [4] proposed a control method of providing grid frequency deviation feedback for the saturated reactor of electrolytic aluminum, regardless of the voltage fluctuations of the electrolytic aluminum high-voltage

https://doi.org/10.1051/e3sconf/202344101008

bus. Grid frequency variations can trigger changes in the equivalent inductance and thus fast power response of the saturated reactor, which can effectively prevent frequency oscillations.

In the macro background of policy support for the development of new energy, China vigorously develops new energy power generation and makes full use of new energy sources such as PV and wind. However, the rapid growth of electricity load and the volatile and intermittent nature of new energy result in difficulties in different regions, mainly including insufficient regional new energy consumption capacity, lack of grid peak and frequency regulation, and immaturity of the grid auxiliary service market. Although the industrial load in China has a large base, high interaction potential, and good economics that can bring favorable benefits when it participates in new energy consumption, its participation in grid regulation is still beset with difficulties due to multiple factors such as process and security.

The existing research on the modeling and adjustable feature of industrial load is as follows: Reference [5] adopted a measurement-based modeling approach, built a comprehensive load static admittance model and a suitable comprehensive load mathematical model of distribution networks in industrial parks, and selected better-performing chaos particle swarm optimization for complete parameter identification of the model. Reference [6] built models of continuous load, strongly correlated load, front load and synchronous load in process flow by analyzing the process flow of the production equipment load in industrial load. In addition, it built an energy storage load model by analyzing the power relationships and constraints of energy storage under industrial load between different time periods. Using the prediction method of the ensemble of optimal input-pruned neural networks based on TRUST-TECH (ELITE), the research predicted the electricity consumption of industrial load by category for cycle optimization.

For the specific industrial load, according to the power impact characteristics of load, Reference [7] proposed a new mathematical model that considered the characteristics of electrolytic aluminum impact load based on a simplified physical model of electrolytic aluminum load; it also proposed a parameter identification method of electrolytic aluminum impact load based on the gray wolf algorithm and the conditions of bus voltage and power variables of the load. Reference [8] provided the first comprehensive analysis of the poly-silicon production process and an equivalent model of the key production steps from the electrical point of view. It finally established an equivalent electrical circuit that could describe the characteristics of poly-silicon production, completed online parameter identification using the least square method, and provided the active power regulation range, response time scale and regulation duration of poly-silicon while taking into account non-electrical quantity state constraints such as temperature. Reference [9] further proposed a modeling method for the continuous power regulation characteristics of different types of load, such as submerged arc furnaces, and provided the corresponding control implementation means.

2.2 Research on the Technologies for Load Participation in Grid Interaction

Using load-side adjustment as a grid peak and frequency regulation method is mainly achieved through demand-side response based on price incentive mechanisms in the more mature European and U.S. electricity markets. The United States is the country that first implemented demand-side response and now has the most mature mechanism, and its independent dispatching organization ISO accepts offers from demand response suppliers in the reserve market and makes unit mix plans with consideration to such offers as input to prior generation plans[10]. The target users of demand-side response in the United States are mainly residential end users, and there are two regulation methods: One is to encourage users to actively reduce electricity consumption during peak load periods by implementing policies such as time-of-use pricing and critical peak pricing[11]; the other method is to reach an agreement with the users who agree to participate in demand response so that the users allow power companies to cut peak load by using methods such as temperature controllers to exert cycle control of the users' air conditions[12]. Reference [13] considered packaging industrial load into industrial VPP and activating load response by adjusting the start-stop interruption of the process. When its power curve was adjusted, the role of the load could be interrupted and the load could participate in the economical and optimal operation of the industrial VPP. This approach was compared with other response modes in order to seek the most economical response strategy for users.

Clearly, the response-ability of industrial load to the power system has been universally recognized by domestic and foreign scholars, and the application of load response in grid optimization and dispatching has also attracted wide attention in the industry.

In China, the promotion of demand-side response in the industry is mainly restricted by the lack of a mature power market. With the gradual advancement of the domestic electricity market reform, Chinese scholars show strong interest in the participation of adjustable load in load response. Reference [14] proposed the participation of energy-intensive enterprises in the local consumption of wind power in microgrids and built a model for the interaction between these enterprises and wind farms. A pricing mechanism was designed, in conjunction with the peak-valley load constraints of power grids, to guide the adjustment of daily electrical load of energy-intensive enterprises based on the wind power output volatility and the peak-valley load size of power grids. With a cement plant as an example, Reference [15] introduced energy storage to compensate for the disadvantage of the load unit switch that could simply enable discrete power changes and proposed a kind of auxiliary service demand response that connected industrial load with energy storage. It should be noted that the ability of industrial load like electrolytic aluminum to meet the key process constraints in the control process is the most important prerequisite for it to participate in grid interaction control, and it is thus necessary to conduct detailed modeling of the impact of power regulation on core production parameters, so as to minimize the impact on the production of electrolytic aluminum during the power support process. Regarding the demand-side response process, Reference [16] was the only research to analyze the impact of power regulation on the electrolyzer temperature based on the principle of energy conservation. Reference [17] explained that the cost for energy-intensive electrolytic aluminum load to participate in the FM auxiliary service was twofold: On the one hand, the loss of energy in the regulation process could result in losses of product yield and profit; on the other hand, the service life of equipment could be affected by the overload regulation of energy-intensive electrolytic aluminum. In both cases, the regulation boundary of electrolytic aluminum load was not effectively constrained to ensure the production safety of electrolytic aluminum.

2.3 Research on Demonstration Participation of Typical Industrial Enterprises in Grid Power Regulation

Although the participation of industrial load in grid interaction has been a clear development direction, the analysis of existing research results and the investigation of several industrial loads find that the methods and systems for designing flexible interaction between industrial load and the market supply and demand regulation and for operation optimization still need to be improved, and most demonstration projects fail to achieve the expected goals in practical operation. The main reasons are as follows: the theoretical research on the behavioral characteristics and flexibility assessment methods of industrial load is inadequate; there is a lack of detailed models for the operational and technical characteristics of multiple types of industrial load, which results in a lack of planning tools; the assessment and analysis of the behavioral characteristics of industrial load in electricity generation and consumption is inadequate and there is a lack of online monitoring technology and analysis of relevant external data, which results in unclear flexibility enhancement targets; the operation method of load-side response to grid frequency regulation is under-researched.

Based on isolated grids and because of their geographical proximity to an aluminum plant, Reference [18] finally realized load regulation in the aluminum plant by adjusting the excitation voltage control and changing the terminal voltage of generators to modify the voltage of the high voltage bus. Reference [19] analyzed in detail the calculation method of capacity curve in the virtual power plant. By combining sampling with OPF and considering the operation constraints of distributed power supply and the safety constraints of distribution networks, the research aggregated the irregular and non-convex active-reactive coupling output range of the virtual power plant. Reference [20] generated a capacity curve of the distribution system based on the assumption that the aggregation model can optimize the active and reactive power support of the distribution system to the grid. It then built a maximum dynamic ellipsoidal model for the feasible range of power consumption of the distribution system by solving a semi-definite programming problem. Finally, the AC-OPF for optimal dispatching of power transmission grids was solved with the aggregated distribution model. Reference [21] proposed a novel two-step projection calculation method to complete the evaluation of the output curve of a virtual power plant based on convex-hull approximation of the critical point. This method could be used to characterize the allowable active and reactive power output range of the virtual power plant and incorporate the operational constraints of the plant into transmission-level operation and market clearing.

Due to the increasing share of renewable energy sources, the power system's demand for reserve capacity participating in grid operation and control is growing, and the regulation potential of demand-side response to participate in auxiliary services in the power market is attracting more and more attention. At present, auxiliary services in the domestic power market are developing rapidly, and many provincial power companies in Hebei, Zhejiang and Anhui have recognized the principal market position of demand-side response resources. Instead of simply being used for peak-load shifting, power demand response is priced and can be traded in the auxiliary service market and the spot market. The market-oriented choice of electricity consumption rests in users themselves.

At present, 13 provinces (regions) have put into operation demand response markets, mostly pursuing the basic goal of building a demand-side flexible peak regulation capacity equal to 3-5% of the maximum directly dispatched electricity load in the previous year. In Shanghai, Jiangsu, Jiangxi, Shanxi, and North China, which are early starters with mature conditions, many attempts have been made to exploit the response resources and start to explore the participation of electric vehicles, virtual power plants, and residential users. Shandong, Xinjiang and many other provinces have successively issued rules for the operation of electricity auxiliary services, sending a clear signal that the participation of demand-side response resources in the auxiliary services of the electricity market has become an inevitable trend.

In addition, due to different load structure characteristics and industrial layout plans in different regions, the focus of response resources exploitation varies, such as commercial buildings in Shanghai and Tianjin, electric vehicles in Shaanxi, Shanxi and Zhejiang, and residential users in Jiangxi and Tianjin. The load regulation potential of industrial enterprises is first exploited in the northern region, where industrial users are concentrated. In 2021, State Grid Jibei Electric Power Company Limited planned to implement the annual key work requirement of "actively exploring load resources that can be interrupted and adjustable to enhance demand-side response capacity" and build the demand response capacity based on the peak-shaving auxiliary service market in North China. In March 2021, the pilot industrial enterprise demand response project of the company completed commissioning for 168 hours, and the response load participated in the peak-shaving auxiliary service market in North China. During the commissioning, the maximum valley-filling load in a single day reached 42,000kW and the accumulated response power exceeded 1.60 million kWh. The grid-load interaction between the electricity load of industrial users and the peak-shaving auxiliary service market in North China was achieved.

3 Industrial Load Model

Different from the flexible and independent residential load, industrial load operates in line with the production target and is strongly coupled and sequentially correlated. The change of one load may trigger the change in the process flow. Therefore, the modeling of industrial load is particularly important.

3.1 Load of Production Equipment

Production equipment follows strict process steps and involves in the process flow of the industry. Therefore, they must strictly follow the process steps in Fig 1.

U is the set of production equipment load in the demand management of industrial load, $U=\{1,\dots,N\}$, and the production equipment load includes N load units. One optimization cycle T of demand management of industrial load is divided equally into K periods. Demand management is conducted in each of these periods.

As shown in Fig 1, each load unit, such as *d1* and *d2*, represents one set of production equipment, each of which has its specific power and operation task. The production equipment is marked with and without an asterisk. The production equipment with an asterisk, such as *d3**, represents a continuous load that cannot be interrupted before the operation task is completed. The coupling between two units of production equipment is expressed by directed edges *e1* and *e2*, representing the sequences of production and processing. For example, *e2* represents that *d2* is the front load of *d3**, which means that only when the production of *d2* is completed, will *d3** be started. The directed edges include 0 and 1. 0 represents that the rear load can be started after an interval or as soon as the front load is completed, and the rear load can be translated horizontally; 1 represents that the rear load is strongly correlated to the front load and must be started as soon as the production of the front load is completed.

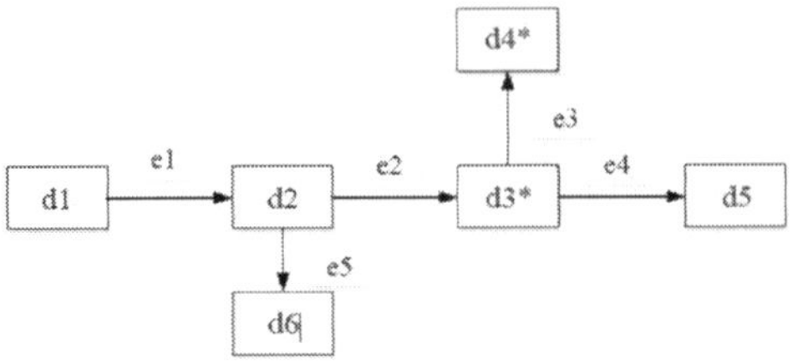

Fig. 1. Flow Chart of Production Equipment Load.

3.2 Load of Air Conditioning

In addition to production and operation equipment, industrial enterprises also use air conditioners, heat pumps, etc. Different from the load of operation equipment, air conditioning load is independent of other loads. It is used in relatively fixed time periods, and its power can be adjusted through temperature control. However, the temperature control and adjustment of the air conditioner will affect the satisfaction and comfort of workers. Therefore, the air conditioning load is modeled with a dual objective.

$$
\begin{gathered}
0 \le P_t^{AC} \le P_{\max}^{AC} \\
(T_{set}^{in} - \Delta C) \le T_t^{in} \le (T_{set}^{in} + \Delta C) \\
P^{AC}(t) = \sum_{t=1}^{T} \omega_t^{AC} \Delta t P_t^{AC} \\
T_{t+1}^{in} = \varepsilon T_t^{in} + (1-\varepsilon)(T_t^{out} \mp \frac{COP * P_t^{AC}}{A})
\end{gathered} \tag{1}
$$

In which: P_t^{AC} and T_t^{in} are the air conditioner power and the indoor temperature at t; $P_{\max}^{AC}$ is the maximum operating power of the air conditioner; T_{set}^{in} is the set temperature of the air conditioner; ΔC is the maximum temperature offset; + and - are heating and cooling modes of the air conditioner, respectively; ω_t^{AC} is the variable of 0-1, representing the cooling of the air conditioner at t: 1 represents cooling while 0 represents no cooling; $P^{AC}(t)$ is the total operating power of the air conditioner; ε is the inertia coefficient of indoor temperature change; T_t^{out} is the outdoor temperature at t; COP is the performance coefficient; A is the coefficient of thermal conductivity.

The air conditioning temperature offset must be controlled within a certain range. Although an excessive offset can reduce the operating power of the air conditioner and the electricity demand of industrial load to achieve energy saving and loss reduction and save considerable electricity costs, it will affect the comfort and satisfaction of workers, reduce their efficiency, affect the production, and cause unpredictable losses. Therefore, the relationship between the indoor temperature controlled by air conditioning and the comfort and satisfaction of workers is modeled and expressed by a quadratic function.

$$\lambda^{AC} = -\sum_{t=1}^{T} \gamma_t^{AC} (T_t^{in} - T_{set}^{in})^2 + 100 \quad (2)$$

In which: λ^{AC} is the comfort and satisfaction of workers under the temperature; T_{set}^{in} is the set indoor temperature; T_t^{in} is the indoor temperature at *t*; γ_t^{AC} is the impact of air conditioning temperature adjustment on the satisfaction of workers at *t*, which can be set as 0 when the workers are out of work or when the plant is empty, in order to avoid unnecessary electricity consumption. The comfort and satisfaction of workers under the temperature is expressed as percentages.

The dual-objective model can effectively control the temperature of the air conditioner to reduce electricity consumption without affecting the comfort and satisfaction of workers. Thus, the industrial enterprise can reduce the electricity consumption of air conditioners based on the electricity demand and also reduce the electricity demand to achieve energy saving and loss reduction and save considerable electricity costs without affecting the comfort and satisfaction of workers at work.

3.3 Load of Energy Storage

With the progress of science and technology, energy storage technology becomes increasingly mature and is ready for vigorous promotion. Industrial enterprises equipped with mature energy storage systems can reduce instantaneous power and electricity consumption while improving stability and reducing the volatility of electricity consumption, achieving energy saving and loss reduction, and saving considerable electricity costs. Coordinating and optimizing the production process of industrial enterprises with the supporting energy storage can not only reduce load volatility and achieve peak-load shifting (by controlling the rate, valley-to-peak difference and variance of industrial load) but also improve the electricity consumption stability of the industrial load.

This paper selects the battery to build an energy storage system model, in which the state of charge characterizes the stored capacity of the battery, reflecting the percentage of the remaining capacity to the total capacity. Meanwhile, the stored capacity of the battery at *t* is a function of the remaining capacity at *t*-1 and the charging and discharging capacity at *t*. The formula is expressed as.

$$E(t) = E(t-1)(1-\delta_E) + (P_E^{cha}(t)\eta_E^{cha} - \frac{P_E^{dis}(t)}{\eta_E^{dis}})\Delta t \quad (3)$$

In which: $E(t)$ and $E(t-1)$ are the stored capacity of the battery at *t* and *t*-1; $P_E^{cha}(t)$ and $P_E^{dis}(t)$ are the charging and discharging power of the battery at *t*, respectively; η_E^{cha} and η_E^{dis} are the charging and discharging efficiency of the battery; δ_E is the self-discharge rate of the battery.

The life of the battery is related to not only the number of charges but also the charging and discharging depths, which will cause battery losses and further affect the life of the battery. Therefore, the battery should not be overcharged or over-discharged. The stored capacity constraint of the battery is expressed as:

$$E_{min} \leq E(t) \leq E_{max} \quad (4)$$

In which: E_{min} and E_{max} are the allowable maximum and minimum stored capacity of the battery, respectively.

Since the capacity of the battery must be balanced within the optimization cycle *T* and should not disrupt the use in the next cycle, its constraint is expressed as:

$$E(0) = E(t) \quad (5)$$

In which: $E(0)$ and $E(t)$ are the stored capacity of the battery at the beginning and end of the optimization cycle, respectively.

The maximum charging and discharging power of the battery is also constrained, which is expressed as:

$$0 \leq P_E^{cha}(t) \leq E^{nom}\alpha_E^{cha}$$
$$0 \leq P_E^{dis}(t) \leq E^{nom}\alpha_E^{dis} \quad (6)$$

In which: E^{nom} is the rated power of the battery; α_E^{cha} and α_E^{dis} are the allowable maximum charging and discharging rates of the battery, respectively.

The battery cannot be under the charging mode and the discharging mode at the same time, and the constraint is expressed as:

$$P_E^{cha}(t)P_E^{dis}(t) = 0 \quad (7)$$

An energy storage system can reduce the instantaneous maximum electricity demand of industrial load and electricity consumption of industrial enterprises while improving stability and reducing the volatility of electricity consumption, achieving energy saving and loss reduction, and saving considerable electricity costs. Coordinating and optimizing the production process of industrial enterprises with the supporting energy storage can fully leverage the advantages of energy storage, increase the space for scheduling of production process, minimize the electricity demand of industrial load, improve the economics, and reduce energy pollution in the production process.

4 Benefits Analysis

4.1 Direct Benefits

The research on the exploitation of flexible resources driven by Industrial Internet under industrial load can primarily contribute to the regulated, market-oriented, commercial and normalized application of multiple types of typical and controllable industrial load in friendly grid interaction, which is in line with the national strategies to ensure the safe and stable operation of power grids and to pursue low-carbon green development and adapts to the industrial application prospects of controllable industrial load resources participating in the interactive regulation of power grids in the context of massive access of new energy and intensified pressure of grid regulation. Specifically, the flexible control of typical controllable industrial load power can avoid the economic loss caused by direct removal of industrial load during load peak hours, minimize the impact of power outages on the electricity consumption of users, and improve the production efficiency of enterprises. In addition, the participation of controllable industrial load in interactive grid control can address the problem of large-scale new energy consumption, effectively reduce the "abandoned wind and PV" energy that could have been used for power generation, and reduce the energy storage configuration requirements and input costs of the power system.

4.2 Indirect Benefits

With the support of the Industrial Internet platform, the regulation potential of controllable industrial load, once fully exploited, will enable the participation of industrial load in interactive grid control, which can improve the regulation ability and adaptability of power grids while reducing the production cost and net carbon emissions of enterprises and facilitating industrial enterprises to help achieve the carbon peaking and carbon neutrality goals.

The participation of industrial load in interactive grid control can help power grid enterprises develop new load-side regulation resources, effectively alleviating the reduced flexible regulation capacity on the power supply side caused by large-scale integration of new energy into power grids. In addition, it can alleviate the operational pressure of power grids while reducing the inefficient output on the power generation side during load peak hours and preventing excessive investment and construction of power generation, transmission and distribution resources due to the short-term peak load of power grids.

By participating in interactive grid control such as peak regulation and new energy consumption, industrial users can reduce the electricity costs and net carbon emissions of production, improve the overall economic and environmental benefits, promote new energy consumption in the context of extensive integration of new energy into power grids, achieve energy saving and emissions reduction, and promote green energy development.

In addition, consistent scientific research practice and data accumulation basis can be provided for the participation of typical controllable industrial load in friendly interactive grid control. When they learn from demonstration applications in this regard, industrial users can be motivated to participate in interactive grid control, promote the coordinated control of industrial adjustable load resources, promote the integrated development of Industrial Internet and Energy Internet in China, and contribute to the realization of the "3060" carbon peaking and carbon neutrality goal.

5 Conclusion

In the existing electricity auxiliary service market environment, the research on the participation of controllable industrial load in the regulation and control of power grids to achieve friendly grid interaction can effectively address problems such as "the extent, duration and method" of grid peak and frequency regulation and other many types of auxiliary services by the participation of industrial load. The research can be important to promoting the active participation of controllable industrial load in grid interaction, which can help power grid companies to address new energy consumption and power supply-demand balance and to achieve the safety, stability and economic operation of the power system. Focusing on the exploitation of flexible resources and power grid regulation technologies driven by Industrial Internet under industrial load, this paper discusses domestic and foreign research on the planning of flexible grid resources, the participation of industrial load in grid regulation, and demonstration projects for the participation of controllable industrial load in grid regulation. It goes on to build industrial load models by category and analyze its direct and indirect benefits, and proposes to realize information interconnection between the energy system and industrial enterprises through Industrial Internet, promote the transformation of industrial load into flexible grid resources, and form mutually beneficial mechanisms to support new energy consumption and the realization of low-carbon goals in society.

Acknowledgments

This research has been funded by the Project Research and application of key technologies for load flexibility resource mining and power grid regulation driven by industrial Internet (SGJSSZ00KJJS2310831) supported by State Grid Corporation of China.

References

1. Aguero J L, Beroqui M, Achilles S. Aluminum plant load modeling for stability studies[C]. IEEE Power Engineering Society Summer Meeting, Alberta, Canada, 1999: 1-6.
2. Xu J, Liao S, Sun Y, et al. An isolated industrial power system driven by wind-coal power for aluminum productions: a case study of frequency control[J]. Power Systems, IEEE Transactions on, 2015, 30(1): 471-483.
3. Paulus M, Borggrefe F. The potential of demand-side management in energy-intensive industries for electricity markets in Germany[J]. Applied Energy, 2011, 88(2): 432-441.
4. Jiang H, Lin J, Song Y. Demand side frequency control scheme in an isolated wind power system for industrial aluminum smelting production[J]. IEEE Transactions on Power Systems, 2014, 29(2): 844-853.
5. Li Duo. "Study on Integrated Load Modeling of Distribution Networks in Industrial Parks with

Distributed Generation from Multiple Energy Sources" [D]. North China University of Water Resources and Electric Power, 2018.

6. Song Yuexin. "Research on the Key Technology for Precise Management of Electricity Demand under Industrial Load" [D]. University of Jinan, 2020.
7. Guo Cheng, Zhu Runlin, Meng Xian, Xie Hao, He Peng, Ye Zhuang, and Yang Lei. "Modeling Method of Electrolytic Aluminum Impact Load Based on Gray Wolf Optimization Algorithm" [J]. Electric Engineering, 2021(11): 65-70.
8. Chen Yuanfeng. "Research on Frequency Control Strategy of Isolated Grids with High-Permeability Wind Power Based on Poly-Silicon Load Control" [D]. Wuhan: Wuhan University, 2018.
9. Tu Xiazhe. "Research on Frequency Control of Local Area Network in the Iron and Steel Industry Based on Load Regulation of Submerged Arc Furnace" [D]. Wuhan University, 2018.
10. Ruan Wenjun, Liu Sha, and Li Yang. "A Review of Demand Response in the United States" [J]. Power Demand Side Management, 2013 (2): 61-64.
11. Wang Haibo and Zhang Li. "Supply-Demand Interactive Scheduling Model Considering Short Production Process of Steel Enterprises" [J]. Automation of Electric Power Systems, 2021, 45(15): 64-76. DOI:10.7500/AEPS20201002004.
12. Liao S, Xu J, Sun Y, et al. Control of Energy-Intensive Load for Power Smoothing in Wind Power Plants[J].Power Systems IEEE Transactions on, 2018,33(6):6142-6154.
13. Nosratabadi S M, Hooshmand R A, Gholipour E. Stochastic profit-based scheduling of industrial virtual power plant using the best demand response strategy[J]. Applied Energy, 2016, 164:590-606.
14. Ji Jin. "Research on the Chance Constrained Model on the Participation of Energy-Intensive Enterprises in Wind Power Consumption" [D]. Lanzhou University of Technology, 2014.
15. X. Zhang, G. Hug, J. Z. Kolter and I. Harjunkoski, "Demand Response of Ancillary Service From Industrial Loads Coordinated With Energy Storage", in IEEE Transactions on Power Systems, vol. 33, no. 1, pp. 951-961, Jan. 2018, Doi: 10.1109/TPWRS.2017.2704524.
16. Liao S, Xu J, Sun Y, Bao Y. Control of Energy-Intensive Load for Power Smoothing in Wind Power Plants. IEEE Transactions on Power Systems 2018; 33(6): 6142-54.
17. Bao Yi. "Research on the Participation of Energy-Intensive Electrolytic Aluminum Load in the Frequency Regulation of the Power System and Auxiliary Service Strategy" [D]. Wuhan University, 2019.
18. Xu J, Liao S, Sun Y. An isolated industrial power system driven by wind-coal power for aluminum production: a case study of frequency control[J]. IEEE Transactions on Power Systems, 2015, 30(1): 471-483.
19. Huang C, Yue D, Xie J. Economic dispatch of power systems with virtual power plant based interval optimization method[J]. CSEE Journal of Power and Energy Systems, 2016, 2(1): 74-80.
20. Samimi A, Nikzad M, Kazemi A. Coupled active and reactive market in smart distribution system considering renewable distributed energy resources and demand response programs[J]. International Transactions on Electrical Energy Systems, 2017, 27(4): 2268-2280.
21. Muller F L, Szabo J, Sundstrom O. Aggregation and disaggregation of energetic flexibility from distributed energy resources[J]. IEEE Transactions on Smart Grid, 2019, 10(2): 1205-1214.

https://doi.org/10.1051/e3sconf/202344101009

Structural Optimization of Ring Type Oxygen Injector for Waste Incineration Research

Ke Wang[a]

Shanghai Institute of Special Equipment Inspection and Technical, Shanghai 200062, China

Abstract: Waste incineration is one of the most effective means to achieve harmless, reduction and resource recovery. As environmental problems become more and more prominent, increasingly stringent environmental standards will greatly limit the development of conventional waste incineration technology, therefore, the development of a new clean and efficient waste incineration technology will be the inevitable development trend of the waste incineration industry. Based on the computational fluid dynamics (CFD) technology, the structural optimization of the ring oxygen injector used in the oxygen-enriched incineration technology of the waste grate incinerator is investigated. The results show that the oxygen injection ring diameter is 0.7071D, the air duct length is 1.4D, and the oxygen injection ring outer hole is 0.28d and inner hole is 0.4d.

1. Introduction

Oxyfuel incineration is the use of oxygen-rich air with a higher oxygen content than air for combustion, which is an energy-saving and environmentally friendly yet efficient incineration technology[1-5] . The combustion of oxygen-enriched incineration can be divided into oxygen premixed and oxygen non-premixed. Oxygen premixing is more widely used in oxyfuel combustion in order to make the flow field reasonably organized and to ensure the stability of combustion. Oxygen premixing means that the oxygen is first mixed evenly with the air/flue gas in the duct and then passed into the furnace to mix with the combustible material for combustion. The oxygen injector is the device that mixes the oxygen with the air/fume.

The uniformity of oxygen and air/fume premixing also has a great impact on combustion, so the oxygen injector must ensure that oxygen and air/fume are mixed quickly and uniformly, avoiding the danger in the pipeline due to insufficient mixing of oxygen and air/fume and unstable combustion due to uneven mixing[6-10] .

In this paper, a ring type oxygen injector is given, and the numerical simulation technology is combined with the optimization study of the ring type oxygen injector ring diameter, air pipe length and oxygen injector ring internal and external spray hole diameter to determine a kind of oxygen injector that can meet the safety, rapid mixing uniformity, structural simplicity, operability and intelligent control to meet the actual needs of engineering.

2. Study Object

In this paper, the structure of the ring type oxygen injector designed according to the theory of fluid dynamics is shown in Figure 1:

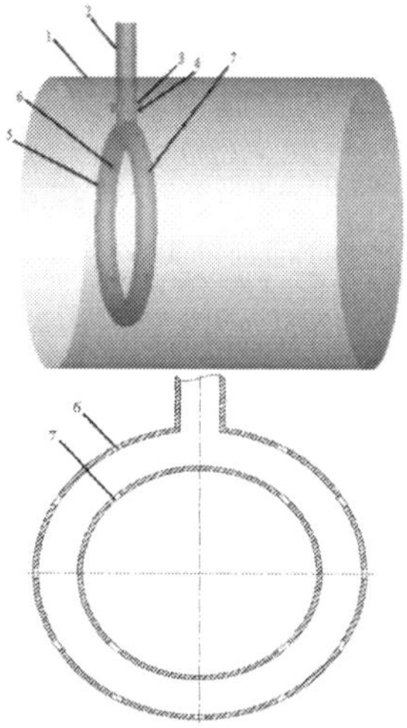

Fig.1 The structure of the ring oxygen injector

Where: air or flue gas pipe 1, oxygen DC pipe 2, first flange 3, second flange 4, oxygen injection ring pipe 5, oxygen injection ring inner spray hole 6, oxygen injection ring outer spray hole 7.

[a] e-mail: 724354507@qq.com

The oxygen injection ring tube 5 is in the form of a ring, and the oxygen inlet end is installed in the direction of the extension of the ring diameter of the oxygen injection ring tube 5, and the oxygen inlet end is in the form of a tube, and a second flange 4 is installed on the port of the oxygen inlet end, and the oxygen injection ring tube has a plurality of oxygen injection ring inner spray holes 6 on the inner wall of the ring, and a plurality of oxygen injection ring outer spray holes 7 on the outer wall of the ring. One port of the oxygen DC tube 2 is located inside the air or flue gas pipe and a first flange 3 is installed at the port, and the other end is connected to an external oxygen source. The size of the oxygen direct current tube is the same as the oxygen inlet end of the oxygen injection ring tube 5, and the first flange 3 and the second flange 4 are of the same type, when the oxygen inlet end of the oxygen injection ring tube is connected to the oxygen direct current tube 2 through the second flange 4 and the first flange 3, on the one hand, the oxygen inlet end of the oxygen injection ring tube 5 and the oxygen direct current tube 3 are coaxial, on the other hand, the length of the oxygen inlet end and the length of the oxygen direct current tube are matched to ensure that the oxygen inlet end of the oxygen injection ring tube 5 and the oxygen direct current tube 3 are coaxial. On the other hand, by matching the length of the oxygen inlet end and the length of the oxygen DC tube to ensure that the centerline of the ring of the oxygen injection ring tube 5 coincides with the centerline of the air or flue gas pipe, to create conditions for the mixing of oxygen and flue gas or air. The normal vectors at the center of all the holes of the inner oxygen injection ring 6 and the outer oxygen injection ring 7 intersect with the centerline of the air or flue gas pipe. When outside oxygen enters the oxygen injection ring tube 5 through the oxygen DC tube 2, it can be ejected through the oxygen injection ring inner spray hole 6 and outer spray hole 7 in a direction perpendicular to the inner wall of the air or flue gas pipe, i.e., along the radius of the air or flue gas pipe.

3. Research content

This research on the structural optimization design of oxygen injector focuses on the key part parameters of oxygen injector, with emphasis on controllability. For the local structure optimization design of the ring oxygen injector is mainly based on the original design of the oxygen injector to optimize the diameter of the oxygen injection ring (ring outer diameter), the length of the air pipe, and the internal and external apertures of the ring. The oxygen injection ring diameter and air pipe length were analyzed by taking the air pipe diameter D as the base and using different multiples to compare one by one; the inner and outer ring apertures were analyzed by taking the oxygen pipe diameter d as the base and taking different multiples to analyze one by one. The coefficient of variation C.V is quoted in the analysis of the study results. it is defined as the ratio of standard deviation to mean, i.e., C.V = (SD/MN) × 100%. The optimal results were obtained by comparing the coefficients of variation of velocity (V), temperature (T) and oxygen concentration (C) for different operating conditions.

4. Analysis of research results

4.1. Effect of oxygen injection ring diameter

The size of the oxygen injection ring diameter (ring outer diameter) has a serious impact on the diffusion of oxygen in the air duct. In order to get the best oxygen injection ring diameter, five different working conditions are given below for numerical calculation. The average turbulent kinetic energy distribution and the coefficient of variation of velocity, temperature and concentration for the five operating conditions are shown in Fig. 2 and Fig. 3.

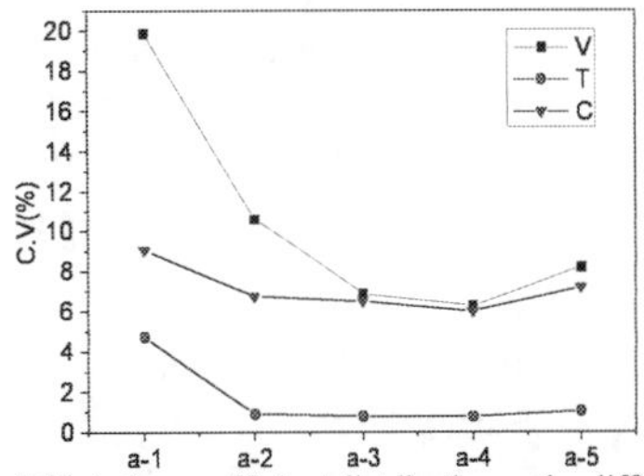

Fig.2 Variation coefficient distribution under different conditions

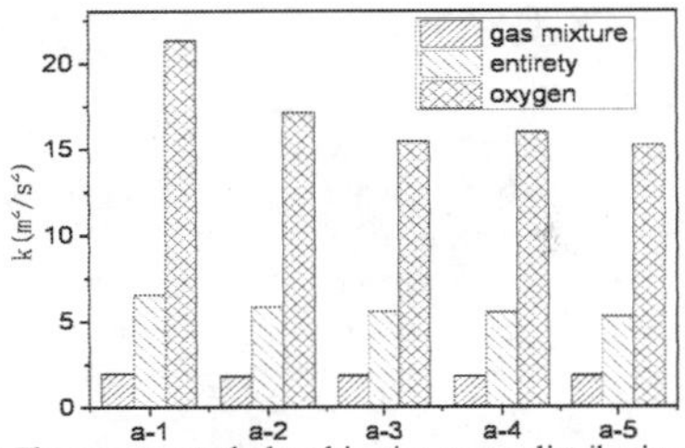

Fig.3 The average turbulent kinetic energy distribution under different conditions

As shown in Fig. 2, the coefficients of variation of velocity, temperature and concentration all tend to decrease and then increase with the increase of oxygen injection ring diameter. The coefficients of variation of temperature distribution are much lower than those of velocity and concentration under the same conditions, which indicates that the temperature of the oxygen and air mixes with large temperature difference and reaches uniformity quickly through strong convective heat exchange. The coefficient of variation of the velocity distribution between working condition a-1 and working condition a-2 is most obviously affected by the diameter of the oxygen injection ring, combined with the analysis of Figure 3, it can be seen that in working condition a-1 to a-2, the oxygen export turbulence intensity changes most drastically, the local disturbance is serious, oxygen can not quickly and better diffuse around the air pipe, so that the oxygen and air at the oxygen injector outlet does not get a good mixture and still maintain a large speed difference. In addition, the gradient of temperature and concentration coefficient of variation can also be proved. The coefficient of variation of the temperature distribution does not change significantly after the diameter of the oxygen injection ring reaches 0.6D. The coefficients of variation of velocity, temperature and concentration reach their minimum values at working condition a-4, i.e. D4=0.7071D, indicating the best mixing effect. At this time, oxygen and air are mixed most uniformly under the same conditions.

4.2. Effect of air duct length

The length of the oxygen injector air pipe has an absolute effect on the residence time of oxygen and air in the oxygen injector, which is also the mixing time of oxygen and air. In order to standardize the measurement of the mixing effect of the oxygen injector, it is hereby stipulated that the coefficient of variation C.V of the speed, temperature and molar concentration of oxygen in the exit section of the mixed gas is less than or equal to 10%. After the previous optimization of the oxygen injection ring diameter, the best oxygen injection ring diameter is selected as 0.7071D, as a premise, the following is the numerical simulation analysis of the five working conditions of different air pipe length, set the air pipe length as L, air pipe diameter as D, then: working condition b-1: L1 = 1.1D; working condition b-2: L2 = 1.2D; working condition b-3: L3 = 1.3D; working condition b -The distribution of the coefficient of variation of velocity, temperature and concentration at the outlet section of the pipe for the five working conditions is shown in Fig. 4.

As can be seen from Figure 4, the coefficients of variation of velocity distribution, temperature distribution, and oxygen concentration distribution of the oxygen injector air duct outlet section under five different operating conditions all decrease with the increase of the length of the oxygen injector air duct. The coefficient of variation of velocity distribution has the largest variation gradient, followed by concentration distribution and the smallest temperature distribution. It is obvious that the coefficient of variation of the temperature distribution is much lower than 10% and therefore need not be considered, while the coefficients of variation of the velocity and concentration distributions are different. The coefficient of variation of the velocity distribution is 15.37% at b-1 and decreases to 10.1% at b-3, but it is still higher than 10% and only below 10% at b-4, which is acceptable. The coefficient of variation of oxygen molar concentration changes more slowly, 11.23% in b-1 working condition, and has been reduced to 9.73% in b-3 working condition to reach qualification. Combined with the above analysis, it can be seen that the coefficient of variation of velocity distribution, temperature distribution and oxygen molar concentration distribution of the outlet section of the oxygen injector air duct under b-4 working condition are qualified, so L4=1.4D can meet the needs.

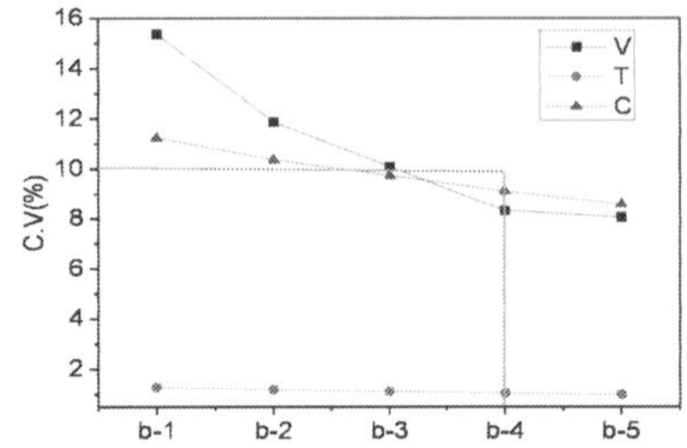

Fig.4 Variation coefficient distribution under different conditions

4.3. Influence of the aperture size of the spray hole inside and outside the oxygen injection ring

The oxygen injection ring in the oxygen injector designed in this paper includes inner and outer spray holes. The size of the inner and outer spray hole apertures is closely related to the momentum of oxygen when injected into the air pipe, and therefore also affects the diffusion of oxygen in the air. In order to get the best inner and outer spray hole apertures, this section optimizes the inner and outer spray hole apertures one by one. Let the diameter of the oxygen injection ring tube be d. Firstly, the aperture diameter of the inner oxygen injection ring is specified as a constant value, and the following five different working conditions are given to determine the best outer aperture diameter: working condition c-1: d1=0.26d; working condition c-2: d2=0.28d; working condition c-3: d3=0.30d; working condition c-4: d4=0.32d; working condition c-5:d5=0.34d. The simulation results The coefficient of variation distributions of velocity, temperature and concentration, and the mean turbulent kinetic energy distributions for the five operating conditions from the data analysis are shown in Fig. 5 and Fig. 6.

As shown in Figure 5, the coefficient of variation of the velocity distribution, temperature distribution, and oxygen molar concentration distribution of the oxygen injector air duct outlet cross-section tends to decrease and then increase with the increase of the outer jet orifice diameter, and reaches the minimum in the c-2 working condition, which can be explained by Figure 6. In the c-1 condition, the oxygen momentum is larger, and the average turbulent kinetic energy in the orifice section, the oxygen injector as a whole and the mixed gas outlet section all reach the maximum, which are higher than the other conditions, which leads to a strong mixed gas disturbance and does not achieve sufficient mixing within a limited distance; in the c-3 condition, the average turbulent kinetic energy of the oxygen injector as a whole is equal to the c-1 condition, and the turbulent kinetic energy of the mixed gas outlet section is also Therefore, the distribution of various coefficients of variation in Fig. 5 is close to that of the c-1 case, and the mixed gas is strongly disturbed, and the mixing effect is not sufficient; in the c-4 and c-5 cases, it can be seen from Fig. 6 that the average turbulent kinetic energy of the nozzle section, the oxygen injector as a whole and the mixed gas outlet section in both cases are significantly lower than those of the first three cases, and the mixing intensity is not sufficient, so the distribution of various coefficients of variation in Fig. 5 is not sufficient. In the same conditions, c-3 working condition to take the oxygen injection ring outside the orifice diameter of 0.28d, turbulent kinetic energy to achieve can make the gas mixture mix fully without disturbing too strong or too weak state, so choose the outside of the orifice diameter of 0.28d for the best.

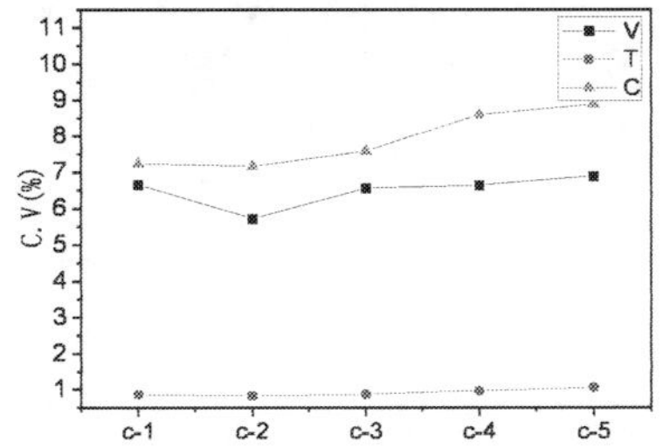

Fig.5 Variation coefficient distribution under different working conditions.

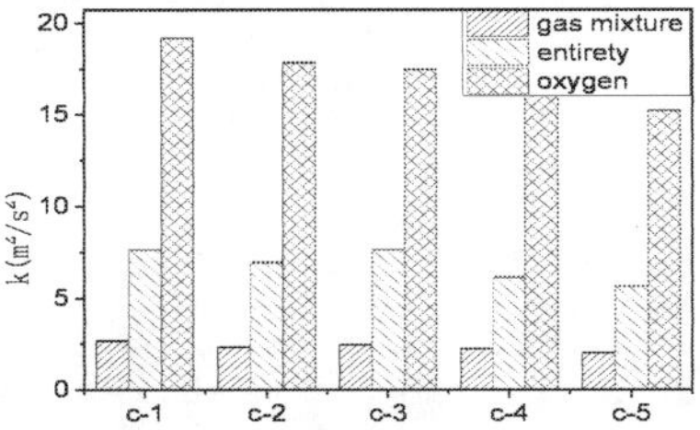

Fig.6 The average turbulent kinetic energy distribution under different conditions

The inner hole is optimized on the basis of choosing the outer spray hole diameter of 0.28d, and setting the diameter of the oxygen injection ring tube as d. The following five different working conditions are given to determine the optimal outer hole diameter: working condition d-1: d1=0.31d; working condition d-2: d2=0.34d; working condition d-3: d3=0.37d; working condition d-4: d4=0.40d; working condition d-5:d5=0.43d. The average turbulent kinetic energy distribution of the coefficient of variation of velocity, temperature and concentration for the five conditions from the data analysis of the simulation results are shown in Fig. 7 and Fig. 8.

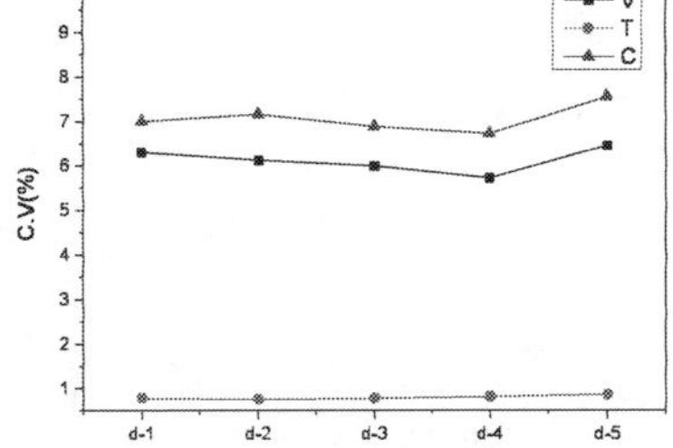

Fig.7 variation coefficient distribution under different working conditions.

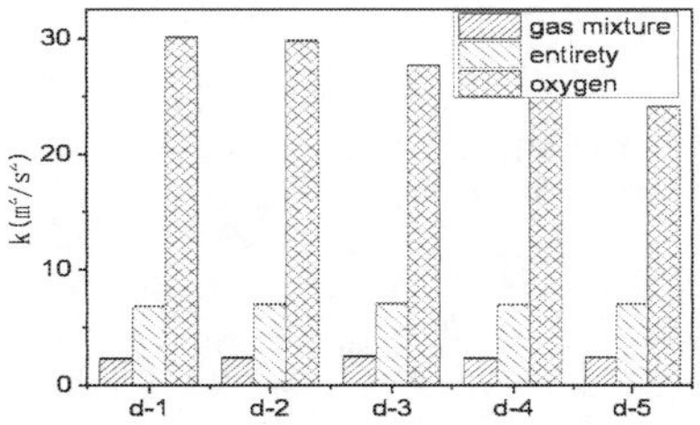

Fig.8 The average turbulent kinetic energy distribution under different conditions

As can be seen from Figure 7, with the increase of the inner jet aperture diameter, the variation trend of the coefficient of variation of each parameter of the oxygenator air duct outlet section is different: the coefficient of variation of the velocity distribution is decreasing and then increasing. The coefficient of variation of temperature distribution does not change obviously, but the overall trend is to increase; the coefficient of variation of concentration distribution is more complicated, first increasing to d-2 condition, then decreasing to the minimum to d-4 condition, and finally continuing to increase. From Figure 8, it can be seen that the average turbulent kinetic energy of the oxygen outlet section gradually decreases as the aperture diameter of the inner jet increases, while the average turbulent kinetic energy of the oxygen injector as a whole and the mixed gas outlet section does not change significantly, but the change trend is the same. A reasonable explanation for the variation trend of each parameter in Figure 7 cannot be found from Figure 8. However, it can be found that although the coefficient of variation of temperature distribution is not the smallest in d-4 working condition, the coefficient of variation of velocity distribution and concentration distribution reaches the smallest value, and it is known that the oxygen injection effect is optimal in d-4 working condition, i.e., when the inner orifice diameter is 0.4d.

5. Conclusion

In order to get the oxygen injector that meets the actual application of engineering, the local optimization of the ring oxygen injector was carried out, and the influence of the diameter of the oxygen injection ring, the influence of the length of the air pipe and the influence of the aperture diameter inside and outside the oxygen injection ring were analyzed. The research results show that the oxygen injection ring diameter of 0.7071D has the best mixing effect between oxygen and air under the same conditions; the length of the air duct reaches 1.4D to achieve the requirement that the coefficient of variation of velocity distribution, concentration distribution and temperature distribution of the oxygen injector exit section is less than 10%; the oxygen injection effect is optimized when the outer hole of the oxygen injection ring is 0.28d and the inner hole is 0.4d.

References

1. Jiang A P. (2002)Study on Combustion Characteristics of typical MSW components and emission characteristics. thesis .
2. Liu G H, Ma XQ, Yu Zhaosheng.(2009) Analysis of municipal solid waste oxyfuel combustion characteristics by CFD [J]. Journal of Engineering for Themal Energy and Power,, 24 (2): 247-251.
3. Giuseppe Liuzzo, Nicola Verdone *, Marco Bravi.(2007) The benefits of flue gas recirculation in waste incineration[J].Waste Management 27:106–116.
4. Han J H, Jeong K, Choi J H, et al. (1997) A hot-flow model analysis of the MSW incinerator[J]. International journal of energy research,21(10): 899-910.
5. Shin D, Ryu C K, Choi S. Computational fluid dynamics evaluation of good combustion performance in waste incinerators[J]. Journal of the Air & Waste Management Association, 1998, 48(4): 345-351.
6. Baukal Jr, C. E. (2013) Oxygen-enhanced combustion. crc press, .
7. Gohlke, O.,Busch, M.(2001) Reduction of combustion by-products in WTE plants: O2 enrichment of underfire air in the Martin Syncom process[J]. Chemosphere, 42 (5): 545-550.
8. Melo, G.,Lacava, P.,Carvalho Jr, J. (1998) A case study of air enrichment in rotary kiln incineration[J]. International communications in heat and mass transfer, 25 (5): 681-692.
9. Müller-Roosen, M.,Ringel, H.(2000) Studies on the High-Temperature Incineration of Polymer-Containing Special Waste with Oxygen Enrichment[J]. Chemical engineering & technology, 23 (12): 1069-1072.
10. Liuzzo, G.,Verdone, N.,Bravi, M.(2007) The benefits of flue gas recirculation in waste incineration[J]. Waste management, 27 (1): 106-116.

 https://doi.org/10.1051/e3sconf/202344101010

Study on scale inhibition properties of chitosan derivatives

Yuxuan Zheng [a*], Yuhua Gao [b], Lihui Zhang [c]

Institute of Energy Sources, Hebei Academy of Sciences, Shijiazhuang Hebei, China

Abstract—The chitosan derivatives scale inhibitor CTS-AMPS was synthesized by using chitosan and 2-acrylamido-2-methylpropane sulfonic acid as raw materials and ammonium persulphate as initiator, and the structure of the product was determined by ATR-FTIR. The scale inhibition performance of CTS-AMPS for municipal water was tested. The results showed that when the concentration factor of municipal wastewater is 2.5 times and the dosage is 30 mg/L, the scale inhibition reached 76.6%. The results of SEM show that the scale inhibition performance of CTS-AMPS is mainly due to lattice distortion.

1. Introduction

At present, the recharge water used in industrial circulating cooling water is still dominated by groundwater and surface water, while the municipal wastewater (MWW) was rarely used[1]. The scale and corrosion inhibitors used in the simultaneous treatment system mostly contain phosphorus, which not only easily lead to water eutrophication, but also difficult to biodegradation[2]. With the shortage of water resources and the requirements of sustainable development strategy, the discharge of phosphorus containing wastewater at home and abroad is becoming increasingly strict. Therefore, the development of environmentally friendly water treatment agents has become the focus of research at home and abroad[3]. Chitosan, with its unique biological structure, is a natural environmental recyclable and degradable point, which is in line with green. In recent years, many scholars at home and abroad have studied different kinds of chitosan derivatives used as scale inhibitors and corrosion inhibitors, but there are common shortcomings such as single performance and poor biodegradable[4,5]. Chitosan derivatives with both scale inhibition, dispersion and corrosion inhibition properties are rarely reported.

In this paper, CTS-AMPS was prepared from CTS and 2-acrylamido-2-methylpropane sulfonic acid (AMPS). The effects of the raw material ratio, initiator dosage on the scale resistance of municipal wastewater were studied. The optimal synthesis conditions are as follows: mass ratio of CTS: AA is 3.00: 1.93; the amount of initiator is 10%; the reaction temperature is 70 °C, and the reaction time is 4 h. The resistance of $CaCO_3$ in municipal wastewater could reach 76.6%, when the concentration factor of MWW is 2.5 times and the dosage is 30 mg/L.

[a*]15533957007@163.com, [b]gaoyuhua77@163.com, [c]zlhkxy@sohu.com

2. Materials and Methods

2.1. Regents

Chitosan, analytical pure, was purchased from Shanghai Aladdin Biochemical Technology Co., LTD; China. Ammonium persulphate, analytical purity, was got from Guangfu Fine Chemical Research Institute; China. AMPS, analytical pure, was purchased from Tianjin Fuchen chemical reagent factory, China. Acetic acid, analytical pure, was purchased from Shanghai Aladdin Biochemical Technology Co., LTD; China.

2.2. Regents Synthesis of CTS-AMPS

In a 250 mL four-mouth flask equipped with a thermometer, reflux condenser tube and drip funnel, 3.00 g of chitosan and 150 mL of glacial acetic acid aqueous solution with a volume fraction of 2% were added, and after N_2 was introduced, magnetic stirring made it completely dissolved, at this time, the aqueous solution had a certain viscosity and was colorless and transparent. At a stirring speed of 500 r/min, a certain amount of APS aqueous solution with a mass fraction of 1% was slowly added to the solution dropwise, the reaction temperature was controlled below 70 °C, the reaction was 0.5 h, and the solution gradually turned light yellow during the reaction. When the temperature rises to 70 °C, start to slowly add 3.84 g AMPS dropwise, after dropping, maintain the reaction temperature of 70 °C, and carry out grafting and copolymerization for 4 h to obtain the light-yellow product CTS-AMPS. The schematic diagram of the composition is shown in Figure 1.

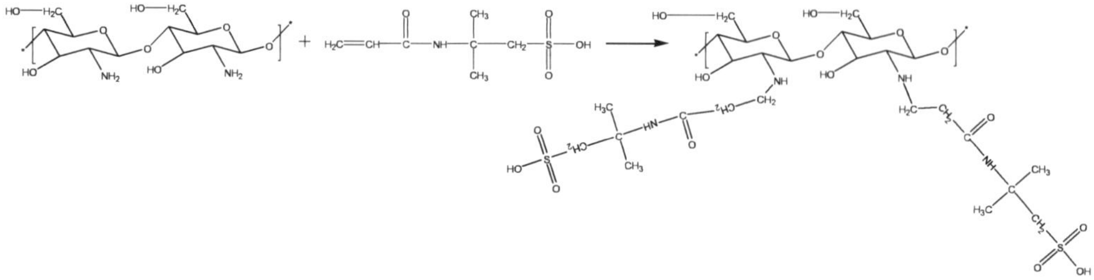

Figure 1. Schematic diagram of CTS-AMPS synthesis

2.3. Characterization of CTS-AMPS

Infrared spectrum of CTS-AMPS was measured via a Frontier Fourier-transform infrared (FTIR) spectrometer (PerkinElmer, USA.) with the attenuated total reflection (ATR) attachment. The wave number range was 4000-650 cm^{-1}.

2.4. Methods for analysis of scale inhibition mechanism

SEM images were observed on an Inspect S50 scanning microscope (FEI, USA). The acceleration voltage was 10 kV and the magnification was 2000 times. The appearances of the CaCO3 crystals with and without CTS-AA-PSI were observed by SEM (FEI Inspect S50).

2.5. Scale inhibition experiment

The experimental water was municipal wastewater (Shijiazhuang city). The parameters are as follows: pH=8.30, The electrical conductivity is 1149 $\mu S/cm^2$, the total alkalinity is 160.13 mg/L($CaCO_3$), The calcium hardness is 342.97 mg/L($CaCO_3$), TP=7.57 mg/L, Cl^- = 40.12 mg/L.

Add 500 mL of municipal reclaimed water to a 1 L beaker, add a certain amount of agent to be tested (the dosage of the agent is calculated according to the solid content of the agent to be tested), stir well, put the beaker into a water bath with a temperature of (80±1 °C) for evaporation concentration, and start timing when the temperature reaches the set temperature. When the solution has evaporated to 200 mL, remove and cool to room temperature before transferring to a 200 mL volumetric flask. If the experimental solution is less than 200 mL, deionized water can be replenished to the scale and shaken well (note: the replenishment water should not exceed 10 mL), then cover and continue to place in the (80±1 °C) water bath and continue to time for 10 h. After the timing is over, remove the volumetric flask and cool to room temperature, then pipette 25 mL of the supernatant, add 20 mL of deionized water, and sequentially add 1 mL of sodium hydroxide solution with a mass fraction of 20% and an appropriate amount of calcium carboxylate indicator, titrated with EDTA standard solution. In the experiment, the blank experiment without adding the agent was used for comparison, and the rest of the experimental steps were the same as above except without the agent to be tested. The scale inhibition rate ($\eta\%$) was calculated by the following formulas[6].

$$\eta\% = \frac{\rho_1-\rho_0}{1.5\rho_2-\rho_0} \quad (1)$$

In the formulas, ρ_0 is the mass concentration of Ca^{2+} in the supernatant after blank heating and concentration, mg/L; ρ_1 is the mass concentration of Ca^{2+} in the supernatant after heating and concentration with scale inhibitor, mg/L; ρ_2 is the mass concentration of Ca^{2+} in municipal water, mg/L.

3. Results and discussion

3.1. Characterization of CTS-AMPS ATR-FTIR analysis.

By comparing Figure2 (a) and (b), it can be found that the skeleton vibration peak of CTS appears at 890 cm^{-1}, while in CTS-AMPS, the skeleton vibration peak strength weakens and the position red shift is at 896 cm^{-1}. Moreover, the peak of N-H stretching vibration appeared at 3320 cm^{-1}, which was corresponded to the I band of amide group, while the peak at 1540 cm^{-1} (Figure 2 a) and 1591 cm^{-1} (Figure 2 b) was that of C=O vibration corresponding to the II band of amide group. There was a sharp absorption peak at 1659 cm^{-1}, which was caused by the –COOH[7]. The peak at 1060 cm^{-1} was the stretching vibration absorption peak of sulfonic group[8]. The appearance of these new peaks indicates that amide and sulfonic groups are introduced into the structure of CTS-AMPS.

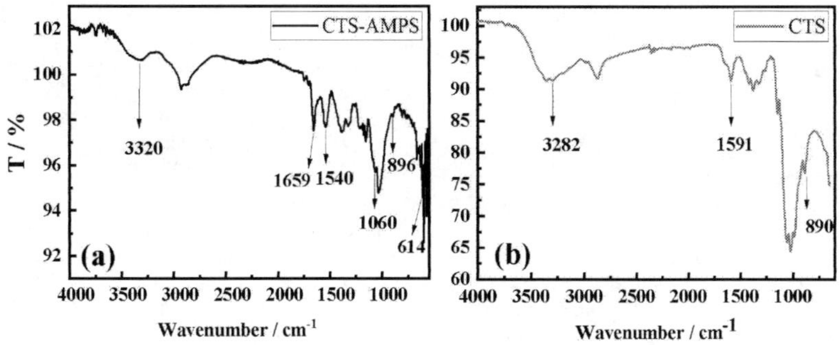

Figure 2. (a) Infrared spectrum of CTS-AMPS (b) CTS

3.2. Scale inhibition performance and mechanism

The scale inhibition rate of CTS-AMPS under different dosage is shown in Figure 3. With the dosage increases, the scale inhibition rate gradually increases, reaching 76.6% when the dosage is 30 mg/L, and 78.2% when the dosage is 40 mg/L. Considering the cost increase caused by excessive dosage, the optimal dosage is determined to be 30 mg/L. Because the composition of municipal water is complex, it is difficult to study its scale inhibition mechanism, so we use the calcium carbonate deposition method to prepare scale samples. Before the scale inhibition experiment began, a slide with the size of 0.5cm2 was placed at the bottom of the conical bottle.

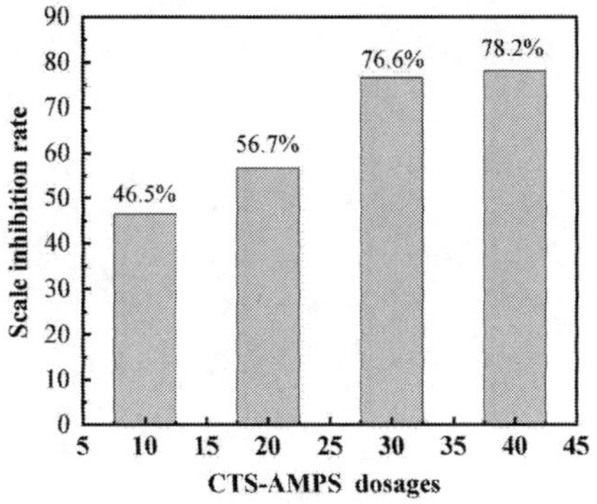

Figure 3. Scale inhibition rate of CTS-AMPS with different dosages

After the experiment, the slide with scale was taken out after drying, the slides were glued to the conductive adhesive and sprayed with gold, and then observed by scanning electron microscope and XRD. Then the corresponding contents of calcite and vaterite are calculated using the following formula, the results are shown in Figure 4[9].

$$I_V^{110}/I_C^{104}=7.690\times X_C/X_V \quad (2)$$

$$X_C+X_V=1 \quad (3)$$

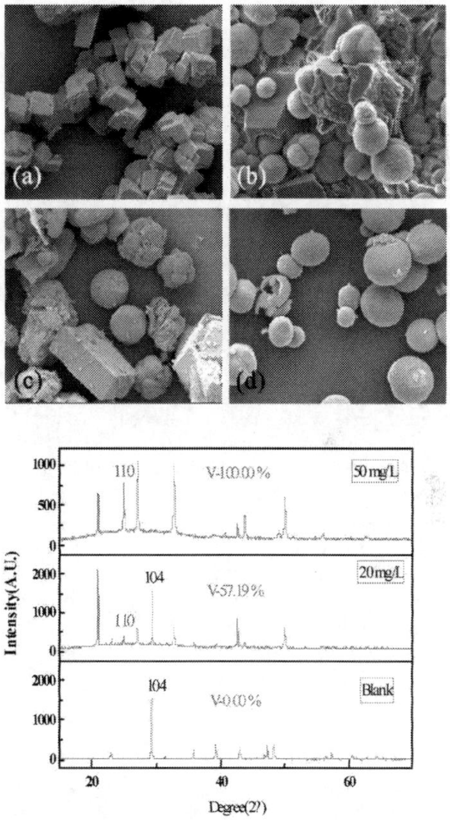

Figure 4. SEM and XRD of $CaCO_3$ samples with CTS-AMPS at different dosages (a,0 b,10 c,20 and d,30 mg/L)

It can be seen from Figure 4 that all the blank samples are angular calcite blocks and the percentage of vaterite is 0%. With the gradual increase of the amount of CTS-AMPS, irregular spheroids begin to appear on the crystal surface of calcite. [10] At a dosage of 20 mg/L, the vaterite content was 57.19%, and the surface becomes rough and irregular until the amount of CTS-AMPS is 30 mg/L, and all the square blocks become spherical (V-100%), which proves that the scale inhibition performance of CTS-AMPS is due to lattice distortion.

4. Conclusions

The chitosan derivatives scale inhibitor CTS-AMPS was synthesized by using chitosan and 2-acrylamido-2-methylpropane sulfonic acid as raw materials and ammonium persulphate as initiator. The scale inhibition performance of CTS-AMPS for municipal water was tested. The results showed that when the concentration factor of municipal wastewater is 2.5 times and the dosage is 30 mg/L, the scale inhibition reached 76.6%. The results of SEM show that the scale inhibition performance of CTS-AMPS is mainly due to lattice distortion.The results show that the chemically modified chitosan derivatives have excellent scale inhibition properties on municipal wastewater, and more kinds of chitosan derivatives can be developed for scale inhibition in the future

Acknowledgments

Funded by Hebei Academy of Sciences High-level Talent Training and Funding Project (2023G04).

References

1. Pramanik B K, Shu L, Jegatheesan V, et al. (2019) Effect of the coagulation/persulfate pre-treatment to mitigate organic fouling in the forward osmosis of municipal wastewater treatment[J]. Journal of environmental management, 249: 109394. https://doi.org/10.1016/j. jenvman.2019.109394.
2. Zhai J, Ren D, Xiao Y, et al. (2023) Synthesis of phosphorus-free terpolymer and its inhibitory performance on scale formation[J]. Journal of Polymer Research, 30(6): 232. https://link.springer.com/article/10.1007/s10965-023-03610-z.
3. Xiao Y, Ren D, Zhai J, et al. (2023) Synthesis and properties of a new polycarboxylic acid multi-polymer scale inhibitor suitable for industrial circulating water[J]. Journal of Applied Polymer Science, e54270. https://doi.org/10.1002/app.54270.
4. Ma W, Yang L, Wu Y, et al. (2023) Synthesis, characterization and properties of a novel environmentally friendly ternary hydrophilic copolymer[J]. RSC advances, 13(17): 11685-11696. https://doi.org/10.1002/jccs.201200441.
5. Mady M F, Abdel-Azeim S, Kelland M A. (2021) Investigation of the antiscaling performance of phosphonated chitosan for upstream petroleum industry application[J]. ACS Sustainable Chemistry & Engineering, 9(48): 16494-16505. https://doi.org/10.1021/acssusc hemeng.1c06786.
6. Sinopec Group. Cooling water analysis and test methods[M]. (1989) Beijing: Petroleum Press, 71-374.
7. Zhou M, Gu Y, Yi R, et al. Synthesis and property study of ter-copolymer P (MA-AMPS-HPA) scale inhibitor[J]. (2020) Journal of Polymer Research, 27: 1-12. https://link.springer. com/article /10.1007/s10965-020-02270-7.
8. Cui C, Zhang S. (2019) Synthesis, characterization and performance evaluation of an environmentally benign scale inhibitor IA/AMPS co-polymer[J]. New Journal of Chemistry, 43(24): 9472-9482. https://doi.org/ 10.1039/C9NJ01355E.
9. Zheng Y, Gao Y, Li H, et al. (2021) Chitosan-acrylic acid-polysuccinimide terpolymer as environmentally friendly scale and corrosion inhibitor in artificial seawater[J]. Desalination, 520: 115367.https://doi.org/10.1016/j.desal.2021.115367.
10. Ji Y, Chen Y, Le J, et al. (2017) Highly effective scale inhibition performance of amino trimethylenephosphonic acid on calcium carbonate[J]. Desalination, 422: 165-173. https://doi.org/10.1016/j.desal.2017.08.027

Optimal Scheduling Strategy for Photovoltaic Charging Stations Participating in the Electricity Market based on Transactive Energy Mechanism

Bao Wang [1,*], Li Zhang [1], Zhumeng Song [1] and Fei Wu [2]

[1] State Grid Anhui Electric Power Co., Ltd. Economic and Technical Research Institute, Hefei, China;
[2] State Grid Anhui Electric Power Co., Ltd., Hefei, China.

Abstract. Regarding the optimization and management of energy in photovoltaic charging stations, existing research mostly focuses on the power grid, with little consideration given to its characteristics as a producer and consumer. This paper proposes an optimized scheduling strategy for photovoltaic charging stations that considers transactive energy mechanisms. Firstly, under the premise of ensuring the safe operation constraints of the distribution network, a distributed transaction model for photovoltaic charging stations was established. Secondly, the optimization model is solved using the Lagrangian decomposition principle and Subgradient method to obtain the scheduling plan of the photovoltaic charging station. The simulation results indicate that the scheduling strategy proposed in this article can promote the on-site consumption of renewable energy and improve the economy of photovoltaic charging stations.

1. Introduction

Environmental pollution and energy issues have become increasingly serious. In the current strategic environment of promoting the development of a low-carbon economy, electric vehicles (EVs) have also received significant support in the development of the automotive industry due to their energy-saving and emission reduction characteristics [1-2]. Charging facilities are the hub for achieving electric vehicle energy supply and interaction [3]. The random charging and load characteristics of large-scale electric vehicles will inevitably have an impact on the stable operation of the power grid. Reasonably configuring a certain capacity of distributed power sources with charging stations can not only suppress the randomness of charging loads but also promote the nearby consumption of renewable energy generation and the low-carbon development of electric vehicles [4-6]. Therefore, it is necessary to study the optimization scheduling method for EV charging stations (EVCS) that include renewable energy.

The existing research on optimization scheduling methods for charging stations mainly includes orderly charging control methods [7-9] aimed at regulating and controlling demand on the grid side, charging station power distribution methods [10,11], etc. Reference [7] proposes a charging and discharging optimization model for EVCSs to participate in the carbon emission trading market and peak shaving auxiliary service market to minimize the grid load fluctuation, which can reduce the load fluctuation while optimizing the operating costs of charging stations. Reference [8] established a charging and discharging optimization strategy for EVs based on the double-layer dynamic game theory with the objective function of minimizing the voltage deviation of the power grid. In reference [9], an optimization scheduling strategy based on the dynamic electricity pricing mechanism was designed to guide EVs charging and discharging, taking into account various interests and achieving multi-objective optimization. As the number of electric vehicles and charging stations increases, centralized control in the above research would significantly raise the computational load and construction control cost of the dispatch center, while neglecting the use of renewable energy. References [10,11] proposed energy management strategies for renewable energy charging stations, which improved the service efficiency of charging stations and reduced the impact of charging loads on the power grid. However, the transaction mechanisms designed in the research are optimized based on the power grid, with less consideration given to the characteristics of the producers and consumers of charging stations. When charging stations reach a certain scale, they can participate in electricity market transactions.

The transactive energy mechanism (TE) has the advantages of distributed regulation and giving economic benefits to participating entities. By integrating economic and power grid control methods, "value" is used as a coordination tool to manage and optimize energy dispatch for producers and consumers [12]. Reference [13] proposes a two-layer hybrid robust stochastic framework based on an transactive energy mechanism,

*Correspondence: lus4403@163.com

 https://doi.org/10.1051/e3sconf/202344101011

which verifies the feasibility of energy trading for high-penetration electric vehicle populations in networked microgrids. Reference [14] proposes a distributed EV charging scheduling strategy based on an transactive energy mechanism, with the optimization goal of peak shaving and valley filling. This strategy not only achieves grid regulation requirements but also improves the economic benefits of electric vehicle charging. Reference [15] proposes an economic scheduling method based on an transactive energy mechanism for the economic scheduling problem of multi-user and consumer clusters.

The above research mainly focuses on the optimization of power grid objectives and establishes energy management and optimization scheduling strategies for producers and consumers, with less consideration given to the characteristics and autonomy of photovoltaic charging stations. Therefore, based on the interactive energy mechanism, this article takes the power grid and photovoltaic charging stations (PVCS) as research objects, considers the uncertainty of photovoltaic output and the charging load characteristics of each charging station, and proposes an optimized scheduling strategy for photovoltaic charging stations.

2. Distributed Trading Strategy

2.1. Structure of PVCS

The research objects are fast charging stations (FCS) and photovoltaic charging stations, Fig.1 and Fig.2 are structural diagrams. The CS includes AC charging stations and DC charging stations, which can meet the charging needs of different types of EV users. The PVCS includes photovoltaic power generation systems, AC/DC converters, and other equipment, with energy sources coming from the power grid and photovoltaic system.

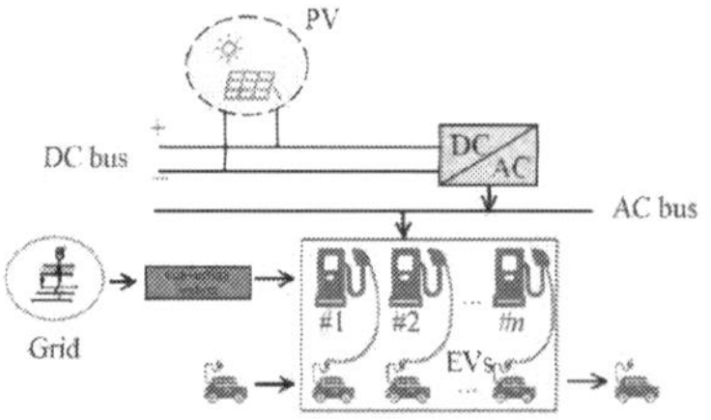

Fig. 1. Structure of photovoltaic charging station for EV

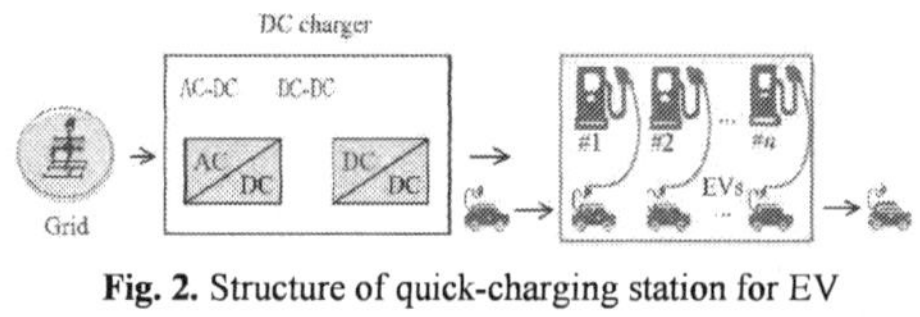

Fig. 2. Structure of quick-charging station for EV

2.2. Distributed Trading Framework for CSs

Considering the information exchange and energy trading between the power grid, trading platforms, CSs, and EV users, a charging station energy trading mechanism framework is proposed as shown in Fig. 3.

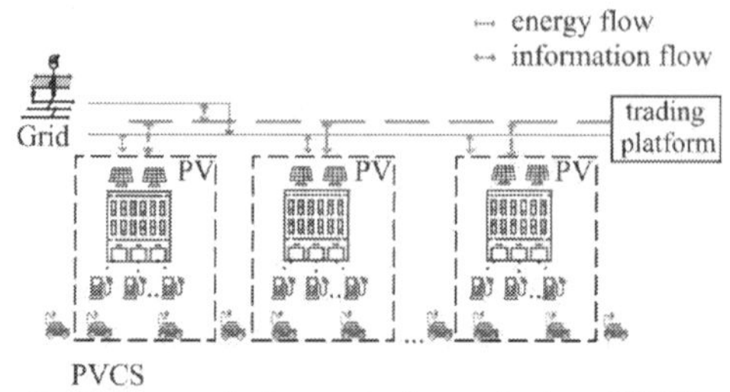

Fig. 3. Framework of energy trading mechanism of EVCS

Considering the randomness of the number of electric vehicles with charging needs at each moment in the charging station, Monte Carlo simulation is used to obtain their quantity. The energy sources of photovoltaic charging stations are photovoltaic power generation systems, trading platforms, and power grids. When the photovoltaic output in the charging station meets the charging needs of electric vehicle users and there is still a surplus, the remaining electricity can be sold to the trading platform; When the photovoltaic output in the charging station cannot meet the charging needs of electric vehicle users, the charging station first purchases electricity from the trading platform, and finally considers the power grid. The distributed trading framework for electric vehicle charging stations established in this paper adopts an interactive energy mechanism, which weakens the dominant position of the power grid in the electric vehicle charging market. It transforms one-way transactions between charging stations and the power grid into multilateral transactions dominated by transactions between charging stations, enhancing the flexibility of charging stations participating in power market transactions, avoiding long-distance transmission, and improving economic efficiency. It also promotes the on-site consumption of distributed power sources.

3. Mathematical model

3.1. PV model

Set the number of EVCS as N_{evcs}, and the PV model of the charging station n is

$$0 \le P^{pv}_{n,t,s} \le P^{pv,\max}_{n,t,s} \tag{1}$$

where $P^{pv}_{n,t,s}$ represents the PV power generation within n of the CS in the s-th scenario; $P^{pv,\max}_{n,t,s}$ is the maximum value of PV output.

The cost function of PV power generation system is

$$C^{pv} = \frac{C'^{pv}}{P^{pv,ave}T^{pv}} \tag{2}$$

where C^{pv} is the unit price of PV power generation cost; $C^{'pv}$ is the cost of PV equipment; $P^{pv,ave}$ is the average annual power generation of PV equipment; T^{pv} is the service life of the PV equipment.

3.2. PV uncertainty model

This paper uses the Monte Carlo method to generate scenarios for PV power generation systems. The multi-scenario method solves the uncertainty problem of PV output. The synchronous backpropagation reduction method is used to cluster and generate typical scenarios [16]. Assuming the initial number of scenes is Ns, the sample distances of scenes $Q_i(t)$ and $Q_j(t)$ is $\|Q_i(t)-Q_j(t)\|_2$, and the clustered scenes need to meet the probability distance constraint condition $D_{di}=\pi_i \min\|Q_i(t)-Q_j(t)\|_2$ to obtain the final number of clustered scenes.

3.3. EV model

The charging demand of EVs is influenced by factors such as the state of charge (SOC) of the battery at the time of grid connection, the SOC at the time of grid disconnection, and the charging time. The charging process model is

$$\begin{cases} P_{i,t}^{Ev,\min} \leqslant P_{i,t}^{Ev,ch} \leqslant P_{i,t}^{Ev,\max} \\ Q_i S_{i,t}^{SOC} = Q_i S_{i,t-1}^{SOC} + P_{i,t}^{Ev,ch} \Delta t \\ M = \begin{cases} 1, t \in \left[T_{\text{start}}, T_{\text{depart}}\right] \\ 0, t \in \left[0, T_{\text{start}}\right] \cup \left[T_{\text{depart}}, T_{\text{end}}\right] \end{cases} \\ S_{i,\min}^{SOC} \leqslant S_{i,t}^{SOC} \leqslant S_{i,\max}^{SOC} \end{cases} \tag{3}$$

where $P_{i,t}^{Ev,ch}$ is the average charging power; $P_{i,t}^{Ev,\max}$ and $P_{i,t}^{Ev,\min}$ are the upper and lower limits of the charging power; $S_{i,t}^{SOC}$ represents the state of charge of the EV, $S_{i,\min}^{SOC}$ and $S_{i,\max}^{SOC}$ represent the upper and lower limits of SOC for the EV; *M* represents the grid connection status of EVs during period *t*.

The EV cluster charging model is

$$P_{n,t}^{EV,ch} = \sum_{i=1}^{N_{ev}} P_{i,t}^{ch} \tag{4}$$

4. Electric vehicle charging station transaction model supported by interactive energy mechanism

4.1. Transaction model

Based on the TE mechanism, PVCS purchase energy through their PV power generation system, trading platform, and power grid to meet the charging needs of EV users. The PVCS prioritizes the use of electricity generated by the PV system. When there is excess PV output, the remaining electricity is sold to the trading platform for profit.

The purchasing power of CS is $P_{n,t,s}^{buy}$, the selling power is $P_{n,t,s}^{sell}$. Among them:

$$P_{n,t,s}^{buy} = P_{n,t,s}^{bfp} + P_{n,t,s}^{bfg} \tag{5}$$

where $P_{n,t,s}^{bfg}$ and $P_{n,t,s}^{bfp}$ are the electricity purchased by the CS from the power grid and trading platform.

The time-of-use electricity price model of the power grid is

$$y_t = c_t + \beta_t P_t^{bfg} \tag{6}$$

where c_t is the initial electricity price issued by the platform at time *t* before the day; β_t is the electricity price adjustment coefficient; P_t^{bfg} represents the electricity purchased by all CSs on the grid.

Distributed PV output has strong uncertainty. This paper uses the multi scenario method to cluster and reduce the initial scenarios, ultimately retaining N_{pv} typical scenarios. Divide the previous 24 hours into N_T time intervals, and each CS adopts the electricity price model represented by equation (6). Establish an objective function to maximize the revenue of all charging stations

$$F = \max \sum_{s=1}^{N_{pv}} \pi_s \sum_{t=1}^{T} \sum_{n=1}^{N_{evcs}} \begin{Bmatrix} [-c_t P_{n,t,s}^{bfg} - \beta_t \left(P_{n,t,s}^{bfg}\right)^2] - \\ g_h \left(P_{n,t,s}^{bfp}\right)^2 + [c_t E_{n,t,s}^{b} / \eta - C_n^{pv}] \end{Bmatrix} \tag{7}$$

where π_s is the probability of scenario *s* occurring, N_{evcs} represents the number of EVCSs, η is the charging efficiency of the CS, $E_{n,t,s}^{b}$ is the electric energy charged by the EV in the CS during time t in scenario s; g_h is the electrical energy loss parameter. The objective function includes the cost of purchasing electricity from the power grid for the CS, network loss costs, EVs charging costs, and PV power generation costs

To ensure the safe operation of the power grid and CSs, the following constraints need to be met

$$\begin{cases} 0 \leqslant P_{n,t,s}^{bfp} \leqslant P_{n,t}^{bfp,\max} \\ 0 \leqslant P_{n,t,s}^{stp} \leqslant P_{n,t}^{stp,\max} \\ 0 \leqslant P_{n,t,s}^{bfg} \\ 0 < P_{ev,t,s} < P_{ev,t,\max} \\ S_{i,\min}^{SOC} \leqslant S_i^{SOC} \leqslant S_{i,\max}^{SOC} \\ P_{n,t}^{bfg} + P_{n,t}^{bfp} + P_{load} \leqslant P_m \end{cases} \tag{8}$$

where *s* represents the scene number; $P_{n,t}^{bfp,\max}$ is the upper limit for CS *n* to purchase electricity from the platform during time *t*; $P_{n,t}^{stp,\max}$ is the upper limit for CS *n* to sell electricity on the platform during time t; P_m is the rated capacity of the transformer

In order to meet the charging demands of conventional loads and EVs, the CS obtains energy from its PV power generation system, trading platform, and power grid. When there is excess PV output, the CS sells the remaining electricity to the trading platform for profit. The power balance constraint of the CS is

$$P_{n,t,s}^{\text{stp}} + P_{n,t,s}^{\text{load}} = P_{n,t,s}^{\text{bfp}} + P_{n,t,s}^{\text{bfg}} + P_{n,t,s}^{\text{pv}} \tag{9}$$

The global constraint for this problem is

$$\sum_{n}^{N_{\text{evcs}}} P_{n,t,s}^{\text{bfp}} = \sum_{n}^{N_{\text{evcs}}} P_{n,t,s}^{\text{stp}} \tag{10}$$

Lagrange multiplier η_t is introduced, and equation (7) is rewritten as

$$F = \max \sum_{s=1}^{N_{\text{pv}}} \pi_s \sum_{t=1}^{T} \sum_{n=1}^{N_{\text{evcs}}} \begin{Bmatrix} \beta_t \left(P_{n,t}^{\text{bfg}}\right)^2 - g_{\text{h}} \left(P_{n,t}^{\text{bfp}}\right)^2 - \\ C_{\text{n}}^{\text{pv}} + c_t E_{n,t}^{\text{b}} / \eta - \eta_t P_{n,t}^{\text{bfp}} + \eta_t P_{n,t}^{\text{stp}} \end{Bmatrix} \tag{11}$$

According to the Lagrangian dual decomposition principle, the objective function is decoupled to obtain

$$F = \max \sum_{s=1}^{N_{\text{pv}}} \pi_s \sum_{t=1}^{T} \begin{Bmatrix} -c_t P_{n,t}^{\text{bfg}} - \beta_t \left(P_{n,t}^{\text{bfg}}\right)^2 - g_{\text{h}} \left(P_{n,t}^{\text{bfp}}\right)^2 + \\ c_t E_{n,t}^{\text{b}} / \eta - C_{\text{n}}^{\text{pv}} + \eta_t P_{n,t}^{\text{bfp}} + \eta_t P_{n,t}^{\text{stp}} \end{Bmatrix} \tag{12}$$

In order to obtain the optimal solution, it is necessary to solve all sub-problems of the CS, and each sub-problem must meet the constraints in equation (9) during the solving process. In this paper, the Subgradient method [15] is used to solve equation (12). The Lagrange multiplier A is updated with equation (13).

$$\eta_t[\mu+1] = \eta_t[\mu] + \sigma_h \left(\sum_{u=1}^{N_{\text{evcs}}} P_{n,t}^{\text{bfp}}[\mu] - \sum_{u=1}^{N_{\text{evcs}}} P_{n,t}^{\text{stp}}[\mu] \right) \tag{13}$$

where μ is the number of iterations in the solving process; σ_h is a constant step coefficient.

The iteration end condition is

$$\left|\eta_t[\mu+1] - \eta_t[\mu]\right| \leqslant \varepsilon_{\text{h}} \tag{14}$$

where ε_{h} is the iterative convergence criterion.

4.2. EVCS Energy Trading Process

Fig. 4 shows the flow chart of electric energy trading for CSs. The optimization scheduling strategy for CSs considering interactive energy mechanisms proposed in this paper requires information exchange between power and price information to achieve multiple iterations and ultimately achieve system balance.

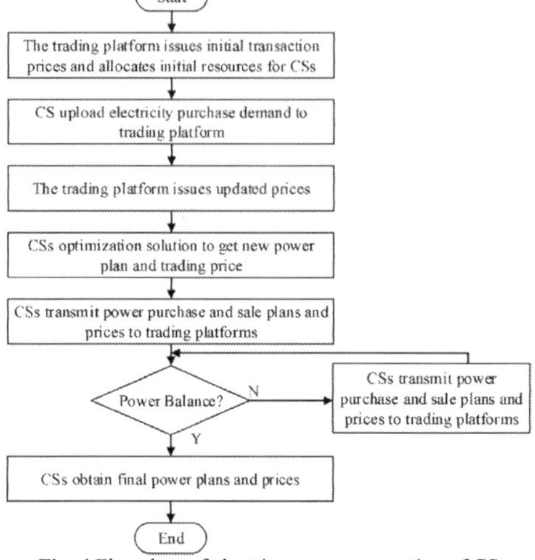

Fig. 4 Flowchart of electric energy transaction of CS

5. Result and Discussion

5.1. Case Analysis

This paper sets a trading cycle of 30 minutes and takes 3 fast charging stations and 2 slow charging stations with PV systems as the research objects. Through simulation analysis, the effectiveness of the proposed optimization scheduling strategy for CSs is verified. Fig. 5 shows the clustering reduction scenarios for each PV output within a day. The parameter settings of the model are shown in Tab.1.

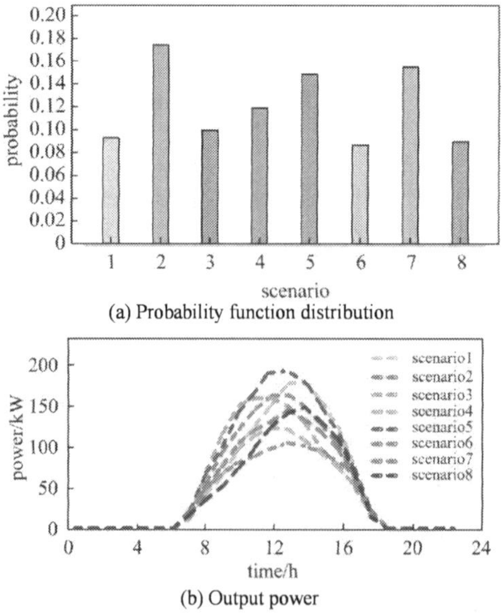

(a) Probability function distribution

(b) Output power

Fig. 5. Situation of each PV station output scenario

 https://doi.org/10.1051/e3sconf/202344101011

Tab. 1 Setting of model parameters

Parameter	Value
$P_{n,t}^{\mathrm{stp,max}}$ /kW	2000
$P_{n,t}^{\mathrm{bfp,max}}$ /kW	2000
P_{m} /kW	10000
β_t	0.001
σ_h	0.0005
ε_{h}	0.000001
$P^{Ev,\mathrm{max}}$ /kW	50

5.2. Transaction situation

Fig. 6 shows the electricity purchase and PV output of each CS from the platform and power grid. CSs 2 and 3 are slow charging stations, while charging stations 4 and 5 are fast charging stations. Due to the relatively high charging power of fast charging stations and the large number of electric vehicle users they serve, the electricity and power purchased from trading platforms and the power grid are higher than those of slow charging stations.

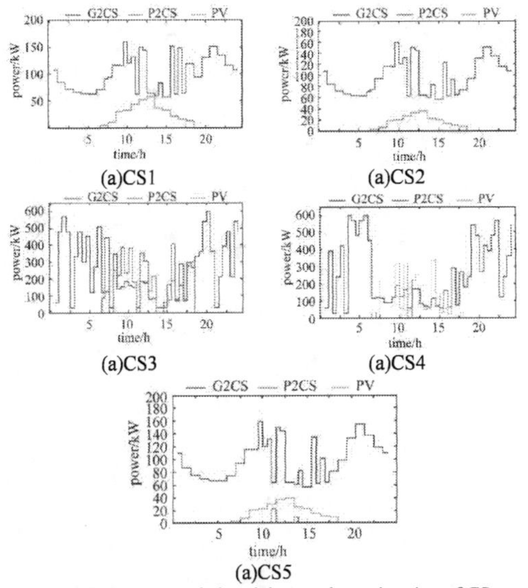

Fig. 6. PV output and electricity purchase situation of CSs

CSs optimize their charging and discharging plans to maximize their interests. Fig. 7 shows the optimization results of different charging stations. Slow charging stations 1, 2, and 3 equipment PV system, and in order to reduce the cost of purchasing electricity, priority is given to using PV output. When the PV output cannot meet the charging needs and basic load of EV users, the CS first purchases electricity from the trading platform, and finally considers the power grid. When the PV output in the CS meets the user's charging demands and there is still a surplus, the remaining electricity can be sold to the trading platform to obtain more profits. Fast charging stations 4 and 5 do not have PV systems and can only purchase electricity from trading platforms and the power grid to meet users' charging demands.

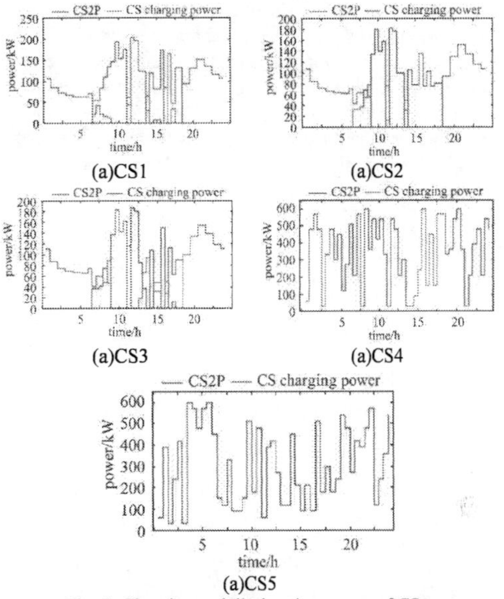

Fig. 7. Charging and discharging power of CSs.

The simulation results show that the proposed optimization scheduling strategy for CSs can improve the utilization rate of renewable energy and increase the operating revenue of CSs. Under the control of this strategy, energy trading is carried out between CSs and platforms, CSs and the power grid in the same area, achieving regional autonomy. Compared with centralized control by the power grid control center, it greatly reduces operating costs and control difficulties and avoids energy loss caused by long-distance transmission of electricity. EVCSs play a market and system-level control role, mobilizing electrical energy to achieve energy balance in the system.

Tab. 2 shows the comparison of benefits between the CS optimization model proposed and the traditional model. From the table, it can be seen that under the charging station transaction model proposed in this paper, the revenue of CSs has significantly increased. The charging station fully utilizes its distributed PV technology, promotes the consumption of renewable resources, and also promotes electricity market transactions, improving economic efficiency.

Tab. 2 Comparison of revenues of each charging station

CS	Revenue/yuan	
	The TE mechanism	The traditional model
1	693.63	517.08
2	657.63	479.76
3	663.77	492.05
4	547.01	365.72
5	616.11	461.69

https://doi.org/10.1051/e3sconf/202344101011

6. Conclusion

This paper proposes an optimized scheduling strategy for PVCSs that considers interactive energy mechanisms, fully utilizing the "distributed" characteristics of EVCSs to achieve information and energy exchange between the power grid, trading platforms, and CSs. Compared with traditional scheduling models, this strategy can enhance the enthusiasm and flexibility of CSs to participate in the market, increase the utilization rate of renewable energy, and improve the economy of charging stations.

References

1. Gössling S, Dolnicar S. A review of air travel behavior and climate change[J]. Wiley Interdisciplinary Reviews: Climate Change, 2023, 14(1): 802.
2. Dall-Orsoletta A, Ferreira P, Dranka G G. Low-carbon technologies and just energy transition: prospects for electric vehicles[J]. Energy Conversion and Management: X, 2022, 16: 100271.
3. Yuan Q, Ye Y, Tang Y, et al. Low Carbon Electric Vehicle Charging Coordination in Coupled Transportation and Power Networks[J]. IEEE Transactions on Industry Applications, 2022.
4. Hasan M K, Habib A K M A, Islam S, et al. Smart Grid Communication Networks for Electric Vehicles Empowering Distributed Energy Generation: Constraints, Challenges, and Recommendations[J]. Energies, 2023, 16(3): 1140.
5. Qi C, Song C, Xiao F, et al. Generalization ability of hybrid electric vehicle energy management strategy based on reinforcement learning method[J]. Energy, 2022, 250: 123826.
6. Makeen P, Ghali H A, Memon S, et al. Smart techno-economic operation of electric vehicle charging station in Egypt[J]. Energy, 2023, 264: 126151.
7. Envelope S X Y A B , Envelope X F W B P , Envelope W Q N B ,et al.An optimization model for charging and discharging battery-exchange buses: Consider carbon emission quota and peak-shaving auxiliary service market[J].Sustainable Cities and Society, 2021, 68.
8. Ma K, Hu X, Yue Z, et al. Voltage regulation with electric taxi based on dynamic game strategy[J]. IEEE Transactions on Vehicular Technology, 2022, 71(3): 2413-2426.
9. Yin W J, Wen T, Zhang C. Cooperative optimal scheduling strategy of electric vehicles based on dynamic electricity price mechanism[J]. Energy, 2023, 263: 125627.
10. ZHANG J, YUAN R, YAN D, et al A non-cooperative game based charging power dispatch in electric vehicle charging station and charging effect analysis[C]//2018 2nd IEEE Conference on Energy Internet and Energy System Integration, Beijing, China, 2018: 1–6.
11. Ullah Z, Wang S, Wu G, et al. Optimal scheduling and techno-economic analysis of electric vehicles by implementing solar-based grid-tied charging station[J]. Energy, 2023, 267: 126560.
12. Liu W, Zhan J, Chung C Y. A novel transactive energy control mechanism for collaborative networked microgrids[J]. IEEE Transactions on Power Systems, 2018, 34(3): 2048-2060.
13. Nasiri N, Zeynali S, Ravadanegh S N. A tactical transactive energy scheduling for the electric vehicle-integrated networked microgrids[J]. Sustainable Cities and Society, 2022, 83: 103943.
14. Wu Z, Chen B. Distributed electric vehicle charging scheduling with transactive energy management[J]. Energies, 2022, 15(1): 163.
15. LIANG Y, ZHU Y, ZHOU R, et al An economic dispatch method for multi-users based on the transactive energy mechanism[J]. Electrotechnical Application, 2021, 40（2）: 18–23.
16. WANG G, ZHAO J, WEN F, et al Stochastic optimization dispatching of plug-in hybrid electric vehicles in coordination with renewable generation in distribution systems[J]. Automation of Electric Power Systems, 2012, 36（19）: 22–29.

Analysis of Sea Ice Area Variation in the Arctic Ocean

Rong Yi[1,*], Minchen Jin[1], Wenfeng Bian[2]

[1]School of Communication Engineering, Hanzhou Dianzi University, No.1158, Ave.2, Xiasha Higher Education Park District, Hangzhou, Zhejiang Province, China
[2]School of Computer Science, Hanzhou Dianzi University, No.215, Ave.6, Qiantang District, Hangzhou, Zhejiang Province, China

Abstract. Against the backdrop of global warming, the changing area of Arctic sea ice is constantly shrinking, and the fluctuation of the Arctic sea ice area is the outcome of multiple factors. This paper mainly studies the volatility, periodicity, trend and related climate factors of Arctic sea ice area variation. It is found that the period of Arctic sea ice area variation T is equal to 3.468 years by establishing a sea ice area change model, and there is a strong negative correlation between the time series of the deseasonalized temperature and the sea ice area through the correlation analysis, with a Pearson coefficient of -0.711. In addition, the Bruun rule leads to the conclusion that for every 1cm increase in sea level, the coastline will retreat by 0.5-1m, which indirectly indicates that without considering other conditions, the continental area will decrease as the sea ice area decreases. Finally, the conclusion obtained is that the CO_2 level has the highest correlation degree by establishing a grey correlation degree model for each factor. Based on this, various measures are proposed to find solutions to the relevant factors and change existing policies to minimize the harm of the warming climate.

1 Introduction

The fluctuation of the Arctic sea ice area can reflect climate change. As various activities on Earth have significantly increased carbon dioxide, the temperature is going up. Studying the area fluctuation data of Arctic sea ice not only reveals the driving forces of climate change, but also explores the greenhouse effect and quantitatively studies the harm of adverse factors of climate change.Nowadays,There are two main methods to study the area of sea ice cover in the Arctic Ocean:satellite observations and numerical simulations.Satellite observations are passive monitoring methods that utilize images captured by satellites to acquire the coverage and distribution of sea ice. Besides,Numerical simulations are active monitoring methods that use numerical simulation software to model the flow and movement of sea ice.

Using this two methods,many researchers have carried out different aspects of research in the study on the area of the Arctic Ocean covered by sea ice. In satellite observations field,aiming at getting correlation between sea ice area and temperature, Ren et al.[1] ,found that there is consistency in the impact of temperature changes on sea ice and snow cover, but there are differences in the response time of snow cover and sea ice to temperature, indicating spatial variability.As for the climate change,Xue et al.[2] found that the increase in CO_2 concentration in the atmosphere affects climate change factors, including temperature, leading to the retreat of Arctic sea ice.Additionally,to sum up the seasonal rule,Liu[3] analyzed the spatial variation trends of long-term ice and seasonal sea ice by performing map algebra operations on the density data within the edge line range of these two types of ice.In numerical simulations field, Liu et al.[4-5] pointed out a relationship between sea ice area variation and ice density in the North and South Poles. Wang et al.[6-8] believed that the reduction of sea ice cover significantly increases the impact of atmospheric heat transfer, while Han et al.[9-10] put forward other perspectives for analyzing changes in the Arctic Ocean sea ice area.H.Yu Models[11] changes in coastal erosion and sedimentation under wave action to search for the effect of the reduction of sea ice area. Additionally,Aldenhoff[12-13] offered other efficient methods to analysis the variation of sea ice area.

Although there has been a lot of research,there is still a vacany in the study of sea ice area characteristics and related quantitative analysis.Thus,this paper mainly studies the volatility, periodicity, trend and related climate factors of changes in Arctic sea ice area as well as the interrelationship between sea ice area variation and climate change, and presents the impact of Arctic sea ice change on the regional area of the Earth. A mathematical model is established through given indicators to minimize the harm of climate change by changing existing policies. The main contributions are as follows:

1. By establishing a model of sea ice area variation, it is determined that the period of sea ice area variation in the Arctic Ocean T is equal to 3.468 years.

* Corresponding author: 1048319067@qq.com

2. There is a strong negative correlation between the time series of the deseasonalized temperature and the sea ice area through the correlation analysis, with a Pearson coefficient of -0.711.Besides,the area change of Arctic sea ice lags behind the temperature from 1 to 2 months.

3. According to Bruun's conclusion that for every 1cm increase in sea level, the coastline will retreat by 0.5-1m, which indirectly indicates that without considering other conditions, the continental area will decrease as the sea ice area decreases.

4. The conclusion obtained is that the CO_2 level has the highest correlation degree by establishing a grey correlation degree model for each factor, so various measures are proposed to find solutions to the relevant factors and change existing policies to minimize the harm of the warming climate.

For the organizational structure below, the third paragraph mainly focuses on the model of sea ice area change and mainly analyzes the correlation between the time series of the deseasonalized temperature and the sea ice area, the fourth paragraph mainly introduces the impact of Arctic sea ice variation on the regional area of the Earth, and the fifth paragraph mainly discusses the correlation coefficients and relevant measures between various data and temperature values, while the conclusion and discussion are narrated in the sixth paragraph.

2 Methods

2.1. Sea Ice Area Change Model

2.1.1 Construction of Sea Ice Change Model:

The years of changes in sea ice reflect the characteristics of the climate changes in the Arctic Ocean over the years. At present, research on the mechanism of sea level changes in the Arctic Ocean is not systematic. Essentially, sea level change is the sum of changes in the specific volume and mass of seawater. The change in specific volume comes from changes in temperature and salinity of seawater, which is actually the rise and fall of sea level caused by changes in seawater density and volume. In this section, a sea ice area change model from volatility, periodicity and trend will be built based on the sea ice area data provided by the National Snow and Ice Data Center of the United States from 1979 to 2020.

After standardizing and normalizing the data for each month, the standard deviation is used as the volatility value to measure each region. Table 1 shows the volatility analysis of representative regions:

Table 1: Fluctuation of Sea Ice Area in the St. Lawrence Area

Corresponding Year	Fluctuation Value (Standard Deviation)
1980	0.316
1990	0.878
2000	0.961
2010	1.235
2020	1.457

The sea ice area in the Saint Lawrence area is relatively low and will drop to 0 in August and September. Although the Arctic center, Eastern Siberia and the Bering Strait boast the largest sea ice area, the sea ice area continues to decrease over time due to global temperature rise. Overall, the sea ice area in the Arctic region reaches its maximum in March and its minimum in September, and the fluctuation of temperature changes increases with age.

2.1.2 Periodic Trend of Sea Ice Area Over Time in Various Regions:

Although the sea ice area varies in different regions due to different dimensions and geographical conditions, the periodicity of sea ice area variation can still be predicted. As shown in Table 2, the periodic statistical data of sea ice area variation in representative Arctic regions:

Table 2: Periodic Variation of Sea Ice Area on Baffin Island

Peak Year	The Period Recent Peak
1981	2 years
1987	5 years
1991	4 years
1993	2 years
1997	4 years
2000	3 years
2005	5 years
2010	5 years
2014	4 years
2017	3 years
2019	2 years

After analysis, regions with higher dimensions have a slower variation period. The common periodic sequence at the North Pole reaches 7 years, while for places with lower dimensions, such as St. Lawrence, the average periodic sequence is only 3 years. By analyzing the data in the above table and averaging the peak period values of 14 regions, the average sea ice area variation period for the 14 regions T equals 3.468 years.

Based on the analysis of the variation trend of sea ice area in the Arctic regions, the average variation trend in Baffin Island, Barents Sea, Eastern Siberia, Greenland, Hudson Bay, Laptev Sea, Sea of Okhotsk and other places is -3,246 square kilometers per year, while the trend of the Arctic midpoint is 22,780 square kilometers per year. On top of that, the sea ice area growth in other Arctic regions is also negative. In general, the sea ice area in various regions of the Arctic is decreasing year by year.

The following factors may affect changes in sea ice area:

1. The dimension; 2. The cold and warm currents; 3. The greenhouse gases; 4. The seawater salinity; 5. The direction of the sea breeze; 6. Other factors.

The variation trends and reasons in some regions of the Arctic are specifically explained below:

▶Although Greenland is affected by warm currents, the trend of change is relatively small due to its higher dimensions.

▶ Although the dimension of the Barents Sea is higher than that of the St. Lawrence Sea, the Barents Sea is affected by the North Atlantic warm current, while St. Lawrence is affected by the cold current, so their sea ice area changes are different.

▶ Although Eastern Siberia is affected by the cold current and its dimension is not low, the sea ice is still falling fast due to the excessive greenhouse gas emissions from production.

2.2 The Correlation Analysis on the Time Series of the Deseasonalized temperature and the Sea Ice Area

The changes in sea ice areas are mainly directly related to temperature. This section mainly analyzes the reasons for sea ice area variation from three aspects: the characteristics of changes within the year, deseasonalized analysis as well as the relationship between sea ice area and temperature changes.

2.2.1 Characteristics of Changes Within the Year:

The Arctic sea ice area and temperature for years are averaged by month, and both average monthly values can be obtained.

2.2.2 Deseasonalized Analysis

Owning to the significant seasonal changes in the Arctic sea ice area, in order to remove the impact of seasonal fluctuations, it is necessary to deseasonalize the time series of the Arctic sea ice area to remove periodicity. Assume a time series X_{ijk} to represent the Arctic sea ice temperature on the kth day of the jth month in the ith year. Consider a time series X_{ij} to represent the Arctic sea ice area in the jth month of the ith year, and the following formula can be obtained from the average Arctic sea ice area in the jth month of the ith year:

$$X_{ij} = \frac{\sum_k X_{ij}}{m} \quad (1)$$

The average temperature for years is:

$$\overline{X_{ij}} = \frac{\sum_k X_{ij}}{m} (i = 1,2,3 \ldots n; j = 1,2 \ldots 12) \quad (2)$$

The time series excluding seasonal fluctuations is:

$$Y_{ij} = X_{ij} - \overline{X_i} \quad (3)$$

By performing deseasonalized analysis on the Arctic temperature, Z_{ij}, Z_{ij} can be obtained that represents the Arctic temperature in the jth month of the ith year without seasonal fluctuations.

2.2.3 Relationship Between Sea Ice Area and Temperature Changes:

With the improvement of production capacity in each country, greenhouse gas emissions have increased, leading to global warming and a significant increase in global temperatures.

The correlation analysis between Arctic sea ice area variation and climate change can indicate the extent of their inherent interrelationships, which can be achieved by calculating correlation coefficients. Calculation of correlation coefficients between the Arctic sea ice area and Arctic temperature without seasonal fluctuations:

$$r_{xy} = \frac{\sum_{i=1}^{n} (x_i - \overline{x})(y_i - \overline{y})}{\sqrt{\sum_{i=1}^{n} (x_i - \overline{x})^2 (y_i - \overline{y})^2}} \quad (4)$$

Among them, x_i is the Arctic sea ice area and y_i is the corresponding temperature in the Arctic. The magnitude of correlation coefficients and positive or negative correlation coefficients can explain the relationship between the Arctic sea ice area variation and temperature. When r is larger than 0, the Arctic sea ice area is positively correlated with temperature. When r is less than 0, the Arctic sea ice area is negatively correlated with temperature. The closer r is to 1 or -1, the closer the relationship between the Arctic sea ice area and temperature is. If r tends to be 0, the smaller the relationship between the Arctic sea ice area and temperature is.

The least square method is used to fit the sea ice area and temperature, and the relationship between the sea ice area and temperature is obtained as follows:

$$\text{area} = -0.3346t + 12.22 \quad (5)$$

There is a significant negative correlation between the average temperature and the corresponding sea ice area in terms of variation. Based on the variation pattern of the sea ice area during a year, it is divided into two parts, namely the sea ice melting period from March to August and the sea ice accumulation period from September to February of the following year, with 6 months of the melting period and accumulation period respectively. There are two different relationships between the sea ice area and temperature at different periods. The slope of the accumulation period is greater than that of the melting period, indicating that when sea ice is in the accumulation period, the decrease in the sea ice area caused by temperature rise is greater than that of the melting period, which shows that the relationship between the sea ice area and temperature is a typical loop curve.

The lag correlation analysis on the time series of the deseasonalized temperature and Arctic sea ice area is performed to get the correlation coefficient between the two at different times:

 https://doi.org/10.1051/e3sconf/202344101012

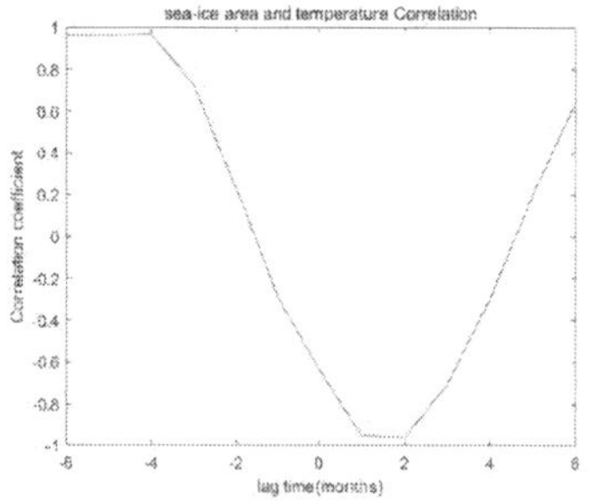

Figure 1: Seasonal Variation Curve of Temperature-Sea Ice Area

From the information in Figure 1, it can be seen that when the lag time is between January and February, the correlation coefficient between the sea ice area and the temperature reaches its maximum negative correlation, suggesting that the area change of Arctic sea ice lags behind the temperature from 1 to 2 months.

2.3 The Sea Ice Area-Land Area Model

It is difficult to obtain data on changes in the regional area of the Earth, so an indirect search for relationships is used. Firstly, the correlation coefficient model of the sea ice area and sea level rise is constructed, and then the relational model of the sea level rise and coastline is built to obtain the relational model of the sea ice area and coastline variation. Next, the influence of natural and human factors on this model is considered, so as to comprehensively analyze and discuss the impact of sea ice area changes on the land area.

2.3.1 Construct a Correlation Coefficient Model Between the Sea Ice Area and Sea Level Rise:

$$r_{xs} = \frac{\sum_{i=1}^{n}(x_i-\bar{x})(s_i-\bar{s})}{\sqrt{\sum_{i=1}^{n}(x_i-\bar{x})^2(s_i-s)^2}} \tag{6}$$

Among them, x_i represents the Arctic sea ice area, and x_i the rising value of sea level. The relationship between the Arctic sea ice area variation and the rising value of sea level can be explained by the magnitude of correlation coefficients and positive or negative correlation coefficients. When r is larger than 0, the Arctic sea ice area is positively correlated with the rising value of sea level. When r is less than 0, the Arctic sea ice area is negatively correlated with the rising value of sea level. The closer r is to 1 or -1, the closer the relationship between the Arctic sea ice area and the rising value of sea level is. If r tends to be 0, the smaller the relationship between the Arctic sea ice area and the rising value of sea level is.

2.3.2 Build a Relational Model Between the Sea Level Rise and Coastline:

According to the Bruun rule in the United States,

$$R = SL_*(B + h_*)^{-1} \tag{7}$$

In the formula, R is the distance of shoreline erosion, S is the rising value of sea level, L∗ and h∗ are the width and thickness of beach sand movement in the wave field; B is the vertical height of eroded sand dunes.

By simplifying the above equation, R=50-100S can be obtained, which is usually used as a rough value for the shoreline retreat caused by sea level rise, i.e. if the sea level rises by 1cm, the shoreline will retreat by 0.5-1m.

Due to the difficulty of directly finding land area, indirect analysis of the rising value of sea level and coastline retreat values is used to analyze the impact of changes in sea ice area on land area variation.

As rising sea levels will inevitably lead to the retreat of continental coastlines, there is a strong positive correlation between the two. However, the retreat of the coastline will result in a decrease in the land area, so the land area will also decrease with the decreasing sea ice area.

Changes will happen in surface water and other factors as the sea ice area change, with green and blue indicating areas where surface water changes have occurred in the past thirty years. Green pixels represent the location where surface water becomes land (absorption, land reclamation and drought), while blue pixels display the location where land becomes surface water (erosion and waterfront construction). In the past thirty years, 173,000 km^2 of land and 115,000 km^2 of water resources have been increased on the Earth's surface.

From this, it can be seen that the land area is constantly increasing under the influence of natural factors. Furthermore, human factors such as sea reclamation, reclamation of usable sea areas and reclamation of lakes and fields have also led to an increase in land area. Therefore, in the current situation, although the reduction of the Arctic sea ice area will lead to a decrease in land area, more natural and human factors have brought about an increase in the total land area instead of a decrease.

2.4 Establishment of the Grey Correlation Degree Model for Various Factors

For the fluctuation of the Arctic sea ice area mentioned above, a grey correlation analysis model is used to construct indicators and propose corresponding suggestions in this section.

2.4.1 Grey Correlation Analysis Model:

1. Determine the analysis sequence:

Parent sequence (parent indicator): A data sequence that reflects the behavioral characteristics of a system and is similar to the dependent variable, which refers to the temperature value here. Subsequence (sub-indicators): data series composed of factors influencing system behavior, which is similar to the independent variable and refers to the factor set U=(forest area, CO_2 emissions, CO_2 level, sea ice area and world population) here.

2. Definition of correlation matrix:

Assume there is a set of reference and comparison sequences as follows:

$$X_j = (X_j(1), X_j(2), X_j(3)\ldots\ldots X_j(n))\ j = 1,2,3\ldots s(8)$$

$$X_i = (X_i(1), X_i(2), X_i(3)\ldots\ldots X_i(n))\ i = 1,2,3\ldots t \quad (9)$$

Based on the above two sequences, the correlation matrix is defined as follows:

$$\delta_{ij}(k) = \frac{minmin\,|x_j(k)-x_i(k)|+\rho\, maxmax\,|x_j(k)-x_i(k)|}{|x_j(k)-x_i(k)|+\rho\, maxmax\,|x_j(k)-x_i(k)|} \quad (10)$$

Among them, δji (k) represents the correlation coefficient between the ith comparative sequence and the kth sample of the jth reference sequence, and $minmin\,|\,x_j(k) - x_i(k)|$ as well as $max\ max\,|\,x_j(k) - x_i(k)|$ show the minimum and maximum values after the difference between the reference sequence matrix and the comparison sequence matrix values.

The equation $|x_j(k) - x_i(k)|$ is called the Hamming distance, and its reciprocal is called the inverse reciprocal distance. The essence of the grey correlation degree is to determine the degree of correlation by the size of the inverse reciprocal.

3. Define sequence correlation

$$z^{(1)}(k) = \frac{x^{(1)}(k)+x^{(1)}(k-1)}{2} \quad (11)$$

Then the new sequence $z^{(1)} = (z^{(1)}(2)z^{(1)}(3)\ldots z^{(1)}(n))$ is the nearest neighbor mean sequence of $x^{(1)}$.

4. Define the grey differential equation of GM (1,1) δji (k), which needs to be integrated, can only reflect the correlation between points and is scattered in the correlation information, making it difficult to characterize the correlation between sequences, thus the correlation is defined here:

$$r_{ij} = \frac{\sum_{k=1}^{n} q_{ij}(k)}{n} \quad (12)$$

Based on this matrix, we can clearly conclude that the effect of the comparison sequence can be seen from the column, and the degree of influence of the reference sequence can be seen from the row, while which sequence plays the main role in the comparison sequence matrix can be analyzed on the basis of the numerical size of the matrix. For example, if the numerical value of a certain column is significantly larger than that of other columns, such a sequence is called the advantage sub-factor, and vice versa is the disadvantage sub-factor; if the value of a certain row is significantly larger than that of other rows, it is called the dominant parent factor, which is relatively sensitive and easily influenced by the driving force of sub factors.

2.4.2 Result Analysis:

The Figure 2 shows the correlation coefficients of the five factors:

Step 1: Perform dimensionless processing on the data.

Step 2: Solve the grey correlation coefficient value between the parent sequence and the feature sequence.

Step 3: Solve the grey correlation value.

Step 4: Sort the grey correlation value and draw a conclusion.

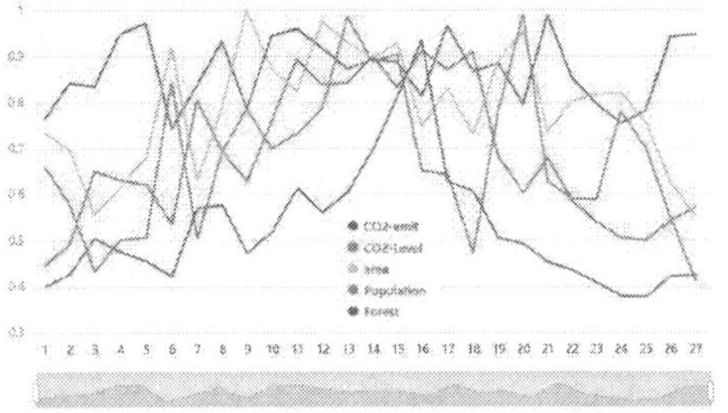

Figure 2: Correlation Coefficient Diagram of Five Factors

The correlation analysis of five-factor sets on temperature are as follows in Table 3:

Table 3. five-factor sets on temperature

Evaluative items	Correlation	Ranking
CO_2 level	0.871	1
Forest area	0.789	2
population	0.684	3
Sea ice area	0.678	4
CO_2 emission	0.525	5

Correlation values are obtained after combining the above correlation coefficients for weighted processing, and they are used to evaluate and sort the five factors. The larger the correlation value, the stronger the correlation between the two. From the above table, it can be seen that for evaluation items in this analysis, the CO_2 level correlation is the highest at 0.871, followed by forest area at 0.789.

3 Conclusion and Discussion

3.1. Conclusion:

This paper studies the overall trend of the Arctic sea ice area variation and provides a spatial analysis of the changes in long-term and seasonal ice, reflecting the changing characteristics of the Arctic sea ice. At the same time, the paper also discusses the relationship between the sea ice area variation and climate change, providing the impact of the Arctic sea ice variation on the regional area of the Earth and establishing mathematical models by providing indicators to minimize the harm of climate change by changing existing policies.

3.2. Discussion:

To clarify the effect Sea Ice Area Variation and obtain the effective scheme,we performed the Correlation model.Various suggestions based on the strength of model correlation are proposed:

1. Reduce the air value level of CO_2, regulate and formulate relevant laws to control the massive production of CO_2 and pursue the idea of carbon neutrality.2. Make efforts to increase afforestation projects and construct projects to protect the natural environment, such as the Three-North Shelterbelt Program Project, which will indirectly affect the air concentration of CO_2, so as to slow down the process of global warming.3. Reasonably control the World population.4. Reduce CO_2 emissions by reducing high-polluting enterprises to weaken the greenhouse effect.

References

1. Ren Yanqun, Liu Suxia. Differential analysis of the correlation between snow/sea ice area and temperature in the Northern Hemisphere [J]. Geographic Research, 2020, 37 (05): 870-882.
2. Xue Yangguang, Guan Hao, Dong Zhaojun, Chen Fei. Analysis of the characteristics of Arctic ice range changes in the last 40 years [J]. Ocean Forecast, 2014, 31 (04) : 85-91.
3. T. Liu, M. Wang, Z. Wang, R. Feng, C. Zhou and L. Zhang, "Joint Total Variation With Nonnegative Constrained Least Square for Sea Ice Concentration Estimation in Low Concentration Areas of Antarctica," in IEEE Geoscience and Remote Sensing Letters, vol. 19, pp. 1-5, 2022, Art no. 2000505, doi: 10.1109/LGRS.2021.3090395.
4. Liu Yanxia, Wang Zemin, Liu Tingting. Analysis of the characteristics of sea ice changes in the North and South Poles from 1979 to 2014 [J]. Remote Sensing Information, 2016, 31 (02) : 24-29.
5. R. R and S. J, "Iceberg Detection in Aerial Images Using Deep Learning Models," 2022 Third International Conference on Smart Technologies in Computing, Electrical and Electronics (ICSTCEE), Bengaluru, India, 2022, pp. 1-8, doi: 10.1109/ICSTCEE56972.2022.10099806.Serreze M C, Barrett a p, Stroeve J C, et al. The emergence of surface-based Arctic amplification [J]. The Cryosphere, 2009, 3 (1) : 11-19.
6. Comiso J C. Large decadal decline of the Arctic multiyear ice cover [J]. Journal of Climate, 2012, 25 (4) : 1176-1193.
7. Wang Qiang, Wekerle C, Danilov S, et al. Recent sea ice decline did not significantly increase the total liquid fresh water content of the Arctic Ocean [J]. Journal of Climate, 2019, 32 (1) : 15-32.
8. Legeais J F, Ablain M, Zawadzki L, et al. An improved and homogeneous altimeter sea level record from the ESA Climate Change Initiative [J]. Earth System Science Data, 2018, 10 : 281-301.
9. Henry O, Prandi P, Llovel W, et al. Tide gauge-based sea level variations since 1950 along the Norwegian and Russian coasts of the Arctic Ocean : Contribution of the steric and mass components [J]. Journal of Geophysical Rsearch : Oceans, 2012, 117 : c06023.
10. K. V. Prakash and P. Shanmugam, "Artificial Neural Network Model for Estimating Ocean Heat Content in the Sea Ice-Covered Arctic Regions Using Satellite Data," in IEEE Access, vol. 10, pp. 109544109557,2022,doi:10.1109/ACCESS.2022.3213942.H. Yu, "Numerical Study of Coastal Erosion Based on Sea Level Rise," 2022 8th International Conference on Hydraulic and Civil Engineering: Deep Space Intelligent Development and Utilization Forum (ICHCE), Xi'an, China, 2022, pp. 495-498, doi: 10.1109/ICHCE57331.2022.10042607.
11. H. Yu, "Numerical Study of Coastal Erosion Based on Sea Level Rise," 2022 8th International Conference on Hydraulic and Civil Engineering: Deep Space Intelligent Development and Utilization Forum (ICHCE), Xi' an, China, 2022, pp. 495-498, doi: 10.1109/ICHCE57331.2022.10042607.
12. W. Aldenhoff, L. E. B. Eriksson, Y. Ye and C. Heuzé, "First-Year and Multiyear Sea Ice Incidence Angle Normalization of Dual-Polarized Sentinel-1 SAR Images in the Beaufort Sea," in IEEE Journal of Selected Topics in Applied Earth Observations and Remote Sensing, vol. 13, pp. 1540-1550, 2020, doi: 10.1109/JSTARS.2020.2977506.
13. M. Moctezuma-Flores, F. Parmiggiani and L. Guerrieri, "12 Years of Area Variation by the Drygalski Ice Tongue as Measured With COSMO-SkyMed," in IEEE Journal of Selected Topics in Applied Earth Observations and Remote Sensing, vol. 15, pp. 7839-7845, 2022, doi: 10.1109/JSTARS.2022.3205560.

Enhanced NH_3 Adsorption Capacity of $CuCl_2$ by Changing the pore structure

Cheng Shen, LaiHong Shen*

Southeast University, Nanjing 210096, China

Abstract. Ammonia is an important carrier of hydrogen energy, with the characteristics of high hydrogen content density and no carbon dioxide emission. In recent years, many methods of ammonia synthesis using biomass have been widely studied to improve ammonia yield. Therefore, safe and efficient ammonia capture for ammonia synthesis from biomass is an important way to alleviate the energy crisis and solve the energy problem. In this paper, the NH_3 adsorption properties of $CuCl_2$ were measured, and the composite adsorbents were prepared by using silicon and multi-walled carbon nanotubes respectively to support $CuCl_2$, and the NH_3 adsorption properties of the composite adsorbents were studied. The study found that the NH_3 adsorption capacity of the three adsorbents decreased with the increase in temperature, so metal chlorides were more suitable for the low-temperature adsorption of ammonia. Silicon and multi-walled carbon nanotubes have an enhanced effect on the NH_3 adsorption of $CuCl_2$. The reason is that the porous material itself has a physical adsorption effect on ammonia, and silicon can play the role of skeleton support in cupric chloride particles, which enhances the pore structure of the adsorbent, thereby alleviating sintering.

1 Introduction

Air pollution is one of the world's major environmental problems today, [1,2] although ammonia is also a major polluting gas. NH_3 plays an important role in promoting the development of the world economy because it is used as a basic chemical in a variety of applications, including fuel cells, [3] refrigeration, agriculture, energy, etc. NH_3 is an essential raw material that can be used in the manufacture of detergents, chemicals, fibers, plastics, fertilizers, and explosives. NH_3 is also an important carrier of hydrogen energy, [4] with the advantage of higher hydrogen density and near-zero CO_2 emissions during hydrogen production compared to other carriers. Nitrogen fertilizer is the world's largest fertilizer production and use of fertilizer species. Compared with other chemical fertilizer varieties, nitrogen fertilizer has greater advantages for the production of crops. NH_3 is the most important raw material for chemical nitrogen fertilizer production. The contribution rate of synthetic ammonia to food production has reached 50 %.

According to incomplete statistics, there are 180 million tons of synthetic ammonia produced in the world every year. Synthetic ammonia contributes to the production of food that feeds the 7.2 billion people on the planet. As the world's population continues to increase and the demand for synthetic ammonia expands, the production of synthetic ammonia will continue to increase to meet the needs of human survival and development. In industry, the preparation of synthetic ammonia mainly relies on the Haber method of ammonia synthesis. The main principle is to utilize the direct chemical reaction of N_2 and H_2. The working conditions are demanding, with reaction temperatures in the range of 450~550 °C and reaction pressures in the range of 200~400 atm, and iron-based catalysts need to be added for catalysis. However, due to the high temperature and high-pressure reaction environment, a large loss of energy has been caused, and a series of environmental pollution problems have also been caused. The conversion rate of synthetic ammonia is low, and the conversion rate of one reaction is only about 12 %. In recent years, some researchers have proposed a method to synthesize ammonia from biomass, and its conversion rate is up to 56 %. Therefore, the efficient separation and purification of biomass ammonia synthesis are essential. [5]

There are three main technologies for the separation and purification of NH_3, [6-10] including cryogenic separation, membrane separation and pressure swing adsorption. Cryogenic separation technology utilizes the different volatility of gases and uses compression and other techniques to separate gases with different volatility. Membrane separation technology is generally carried out at room temperature. The principle is to use the different concentration differences of the gas components. Under the driving force of this concentration difference, the gas shows different permeability to the membrane so as to realize the separation of the gas components. Pressure swing adsorption separation is a key separation technology in the industry. According to the different bonding forms of

* Corresponding author: lhshen@seu.edu.cn

the adsorption force, it can be divided into physical adsorption, chemical adsorption and selective adsorption. Among them, selective adsorption has great research and application potential due to its advantages of strong adsorption force, low energy consumption and fast regeneration of adsorbents. [11]

Metal chloride is a promising selective adsorbent, and the basic principle of its adsorption is shown in Figure 1. Metal chlorides combine with NH_3 to form metal-ammonia complexes, [12] allowing ammonia to be stored as a stable chemical species. The coordination numbers of the metal-ammonia complexes obtained by the reaction of chlorides of different metals with ammonia are different. The metal-ammine complexes are decomposed under high-temperature conditions, thereby obtaining high concentrations of ammonia and realizing the cyclic regeneration of adsorbents.

Trudel et al. [13] conducted a series of studies on the reaction between metal chlorides and ammonia and the desorption of metal-ammine complexes. It was found that the adsorption of ammonia by the adsorbent is affected by the expansion of the particle volume and changes in the atomic structure during the NH_3 adsorption reaction.

Therefore, this work proposes to prepare composite adsorbents of metal chlorides and porous materials by using silica gel and multi-walled carbon nanotubes to support $CuCl_2$, respectively. The NH_3 adsorption properties of the composite adsorbent at different temperatures and the effect of porous materials on the NH_3 adsorption capacity were studied.

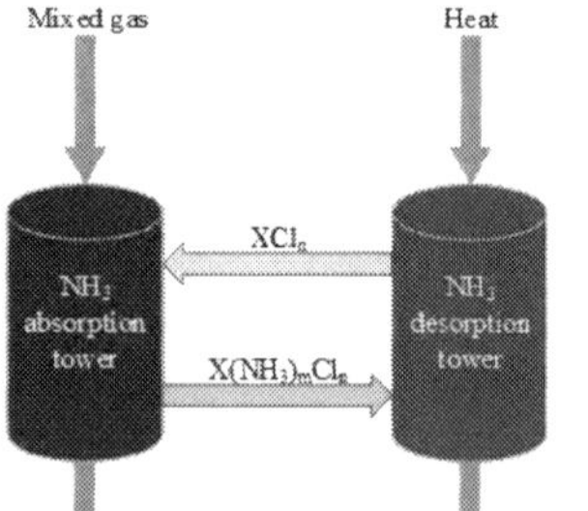

Fig.1. Schematic diagram of the cyclic NH_3 adsorption-desorption of metal chlorides.

2 Experimental section

Commercial copper chloride (anhydrous, powder, purity: 99.9 %) was used in the experiments. The silicon was coarse-pored spherical silicon, purchased from Nanjing Lattice Chemical Technology Co., Ltd., and its specific surface area was about 350-400 m^2/g. The multi-walled carbon nanotubes (MWCNTs) were purchased from Nanjing Wanqing Chemical Glass Instrument Co., Ltd. with a purity of more than 95 % and a specific surface area of about 300-350 m^2/g.

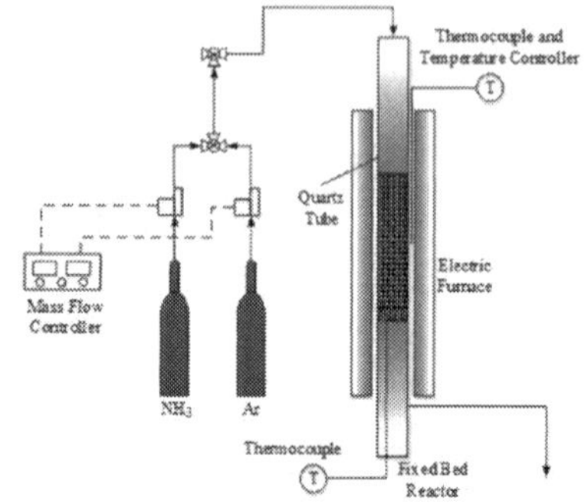

Fig. 2. Fixed bed pedestal device diagram.

Before preparing the composite adsorbent of silicon-supported $CuCl_2$, the silicon was heat-treated using a muffle furnace, and the silicon was pretreated at a high temperature of 300 °C for 2 h to eliminate moisture and other gas impurities contained in the silicon. After the silicon was cooled to room temperature, a certain mass of silicon was weighed and added to the salt solution of $CuCl_2$, and the mass ratio of $CuCl_2$ and silicon was 8:2. The salt solution of $CuCl_2$ was prepared using anhydrous $CuCl_2$ and deionized water. Then, the mixed solution was placed in a magnetic stirrer with a constant temperature water bath and stirred at a rate of 2000 r/min at 80 °C for a certain time. After stirring sufficiently, the mixture was put into an oven to dry for 1 d. Finally, the obtained solid was ground and suitable particles were screened out for experiments. The composite adsorbent prepared by silicon-supported $CuCl_2$ was noted as "$CuCl_2/SiO_2$".

The preparation of the multi-walled carbon nanotube-supported $CuCl_2$ composite adsorbent was similar to the previous method for preparing the silicon-supported $CuCl_2$ adsorbent. The difference was that the pre-treatment temperature of multi-walled carbon nanotubes was different from that of silicon. The multi-walled carbon nanotubes were heat-treated at 200 °C. The purpose was to remove impurities such as moisture in the nanotubes without destroying the structure of the nanotubes. The composite adsorbent prepared by MWCNTs-supported $CuCl_2$ was noted as "$CuCl_2$/MWCNTs".

The experimental setup of NH_3 adsorption properties of different types of composite adsorbents is shown in Figure 2. The height of the fixed bed reactor was 600 mm and the inner diameter was 16 mm. The reactor was heated and temperature controlled by an electric furnace and a temperature controller. The reaction chamber temperature was detected by 2 K-type thermocouples, one placed inside the reactor and the other between the reactor and the heating device. The bottom end of the reactor was 300 mm away from the air distribution plate. Gas flow was controlled by a mass flow meter. In the process of NH_3 adsorption, 100 % of high-purity NH_3 was introduced and continued for 20min to make the adsorbent fully react with NH_3. Then switch to Ar, the reactor was set to heat up to 200 °C with a heating rate of 10 °C/min, and then keep the temperature constant for 20 minutes to carry out the desorption process of NH_3.

A scanning electron microscope (SEM) was used to observe the microstructure of the prepared fresh adsorbents and the adsorbents after the NH_3 adsorption reaction. The phase composition of the prepared fresh adsorbents was tested by X-ray diffractometer (XRD). The specific surface area of the fresh adsorbent was also tested, which was using a Brunauer-Emmett-Teller analyzer (BET) to analyze the effect of silicon and multi-walled carbon nanotubes on particle structure.

3 Results and discussion

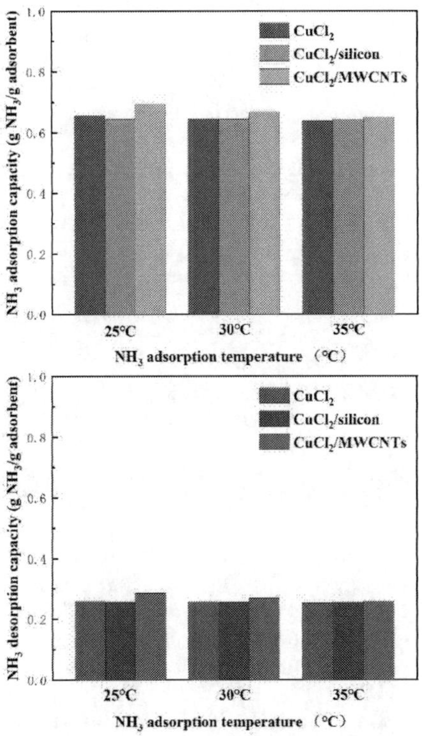

Fig. 3. NH_3 adsorption capacity and desorption capacity of adsorbents.

The variation of NH_3 adsorption capacity of the adsorbents with temperature is shown in Figure 3. Within a certain temperature range, with the increase in temperature, the NH_3 adsorption capacity showed a decreasing trend. The NH_3 adsorption capacity of $CuCl_2$/MWCNTs was the highest at 25 °C, which was 0.69 g NH_3/g adsorbent, which was 4.2 % higher than that of $CuCl_2$. The temperature had the least effect on the NH_3 adsorption capacity of $CuCl_2$/silicon, which was only 0.0009 g NH_3/g adsorbent lower at 35 °C than that at 25 °C.

The addition of porous materials can not only enhance the NH_3 adsorption capacity of the adsorbent but also improve the NH_3 desorption capacity of the metal-ammine complexes. Also, adsorbents with high NH_3 adsorption capacity also have high NH_3 desorption capacity.

The change of the ammonia adsorption temperature had the greatest effect on the NH_3 desorption capacity of copper chloride modified by multi-walled carbon nanotubes, and the NH_3 desorption capacity of 25 °C was 0.2848 g NH_3/g adsorbent, which was 0.0268 g NH_3/g adsorbent higher than that of copper chloride at the same temperature and 0.029 g NH_3/g adsorbent higher than the desorption capacity of $CuCl_2$/silicon at the same temperature. The NH_3 desorption capacity of $CuCl_2$/silicon decreased by 11 % with increasing temperature.

The XRD analysis of the different adsorbents was tested, as shown in Figure 4. The addition of porous materials significantly enhanced the peak of $CuCl_2$. Among them, MWCNTs had the greatest effect on the strength of the peak of $CuCl_2$. It could be seen from Figure 5 that the specific surface area of $CuCl_2$/silicon had the largest specific surface area of 42.3 m^2/g, which was 2.74 times that of $CuCl_2$. The specific surface area of $CuCl_2$/MWCNTs was 168 % higher than that of $CuCl_2$. From Fig. 5, it can be inferred that the high specific surface area is one of the important reasons why the composite adsorbent modified with porous materials possesses high ammonia adsorption energy capacity.

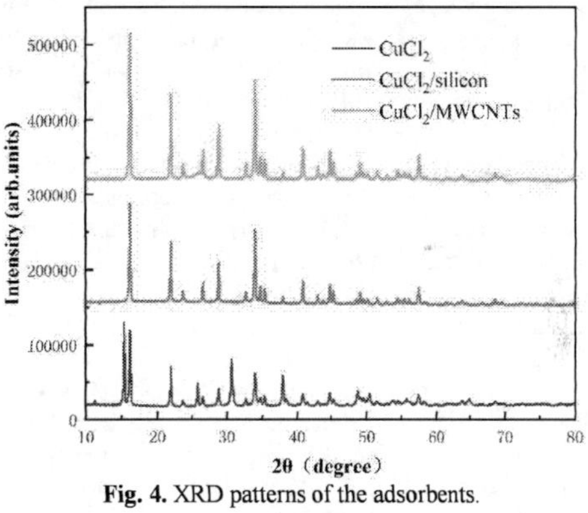

Fig. 4. XRD patterns of the adsorbents.

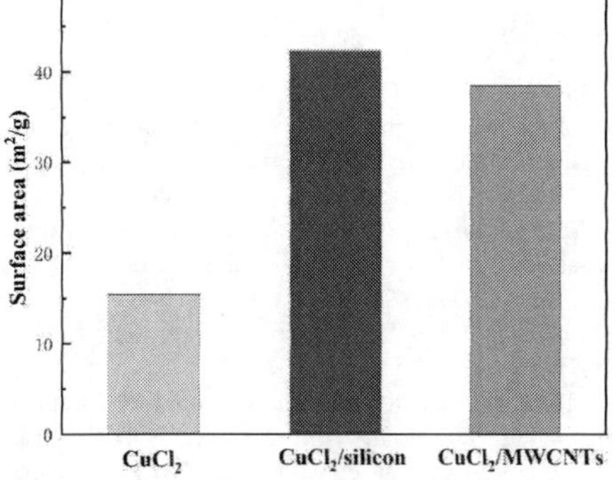

Fig. 5. Specific surface area diagram of the adsorbents.

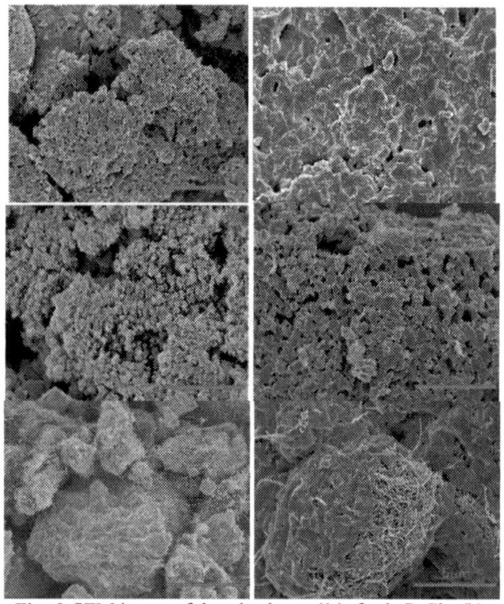

Fig. 6. SEM image of the adsorbents ((a): fresh $CuCl_2$; (b): $CuCl_2$ after NH_3 desorption; (c): fresh $CuCl_2$/silicon; (d): $CuCl_2$/silicon after NH_3 desorption; (e): $CuCl_2$/MWCNTs; (f) $CuCl_2$/MWCNTs after NH_3 desorption).

The apparent morphology of the adsorbents before NH_3 adsorption and after NH_3 desorption was tested and the results are shown in Figure 6. After the NH_3 adsorption and desorption experiments of $CuCl_2$, the particles were severely sintered, and there were almost no pores on the surface. The initial $CuCl_2$/silicon had abundant pores, and some macropores could still be seen on the surface after the NH_3 adsorption and desorption. It can be seen from Figure 6 (e) that $CuCl_2$/MWCNTs were not uniformly dispersed in the adsorbent and were not regularly scattered in the particles. After the NH_3 adsorption and desorption process, the $CuCl_2$ particles were half wrapped around the MWCNTs and the adsorbent surface had some macropores. The structure of MWCNTs was not damaged. The increase in the number of pores also enables the composite adsorbents to have better NH_3 adsorption properties.

4 Conclusion

In conclusion, composite adsorbents of porous material supported with metal chloride were prepared using silicon and multi-walled carbon nanotubes, respectively. The composite adsorbents exhibited higher NH_3 adsorption capacity and more stable NH_3 adsorption properties than $CuCl_2$. The NH_3 adsorption capacity of $CuCl_2$/MWCNTs at 25 °C was as high as 0.69 g NH_3/g adsorbent, which was 4.2 % higher than that of $CuCl_2$. The NH_3 adsorption capacity of $CuCl_2$/silicon at 35 °C was 1.008 times that of $CuCl_2$. Although the doping of silicon did not greatly improve the NH_3 adsorption capacity of $CuCl_2$, silicon played the role of skeleton support in $CuCl_2$ particles, thus alleviating the coalescence of molten salts. The supporting of porous materials increased the specific surface area of the adsorbents and enhanced the pore structure between the particles so that more NH_3 passed through the pores and contacted the surface of the adsorbent particles to achieve more NH_3 adsorption.

Acknowledgments

Financial support from the National Science Foundation of China (52076044) is sincerely acknowledged.

References

1. Shen, C., Luo, C., Luo, T., Xu, J.X., Lu, B.W., Liu, S.L., Zhang, L.Q. ACS omega. 5 (29): 17908-17917 (2020)
2. Shen, C., Shen, L.H., Yan, J.C., Wang, P., Yin, X.L. Energy & Fuels. 35: 13871-13876 (2021)
3. Zamfirescu, C., Dincer, I. J. Power Sources. 185: 459-465 (2008)
4. Aoki, T., Ichikawa, T., Miyaoka, H., Kojima, Y. J. Phys. Chem. C. 118: 18412-18416 (2014)
5. Johannessen, T. US20100062296A1. (2010)
6. Mofidi, S.A.H., Udell, K.S. J. Energy Resour. Technol. 139(3): 032005 (2017)
7. Wang, Z., Yuan, X., Cong, S., Chen, A., Li, Q., Geng, F., Zhao, Z. ACS Appl. Mater. Interfaces. 10(17): 15065-15072 (2018)
8. lexander, G.B., Jan, P., Ondřej, J., Aleksandr, A.S., Lenka, Z. Procedia Engineering. 168: 231-234 (2016)
9. Qajar, A., Peer, M., Andalibi, M.R., Rajagopalan, R., Foley, H.C. Mesoporous Mater. 218: 15-23 (2015)
10. Britt, D., Tranchemontagne, D., Yaghi, O.M. Proc. Natl. Acad. Sci. U.S.A. 105: 11623-11627 (2008)
11. Shen, C., Wang, P., Shen, L.H., Yin, X.L., Miao, Z.W. Ind. Eng. Chem. Res. 61: 8616-8623 (2022)
12. Zhang, T., Miyaoka, H., Miyaoka, H., Ichikawa, T., Kojima, Y. ACS Appl. Energy Mater. 1: 232-242 (2018)
13. Trudel, J., Hosatte, S., Ternan, M. Appl. Therm. Eng. 19: 495–511 (1999)

Comprehensive Benefit Evaluation Analysis And Application Research of Energy Storage

Lei Shi*, Jidong Li, Yuanshen Zhao, Pengyang Li, Pengfei Zhang

State Grid Jiuquan Power Supply Company, No. 8 Dunhuang Road, Suzhou District, Jiuquan City, Gansu Province, 735000, China

ABSTRACT. In recent years, the penetration rate of renewable energy in the power system has increased year by year, and the allocation of energy storage is an important development trend to improve the operation stability of the grid-connected power system containing renewable energy. This paper first analyzes the basic concept and operation principle of energy storage devices, and then explains the costs and benefits of energy storage devices. Finally, the industrial park and energy storage power station are used as practical application scenarios to verify the correctness of the proposed method. The example results show that energy storage should be installed in a place where the system network loss is minimal and the reliability of power supply can be maximized, and the capacity of the energy storage system should match the load capacity of the bus bar of the installation site, so as to achieve the greatest comprehensive benefits.

1 Introduction

With the maturity of production and manufacturing technology of energy storage devices, energy storage has been widely used in power grids, playing an important role in power grid regulation, and has achieved high economic efficiency. In order to better play the value of the energy storage device, it is necessary to optimize the location and capacity of the energy storage device. Literature [1] introduced the capacity optimization allocation method of energy storage power station, but did not analyze the installation position of energy storage power station. Literature [2] conducted a comprehensive evaluation from the technical perspective of energy storage, but did not conduct an evaluation and analysis from the economic perspective. Literature [3] analyzed the comprehensive operating benefits of grid-connected photovoltaic optical storage systems, but did not calculate the comprehensive benefits of energy storage from the perspective of the whole life cycle. Under the above background, this paper first analyzes the cost and benefit of energy storage in the whole life cycle, and then takes industrial parks and energy storage power stations as application scenarios to analyze the energy storage optimization configuration method based on the optimal system comprehensive benefit. The following is a specific analysis and introduction.

2 Overview of Energy Storage Devices

Energy storage devices include many types, such as supercapacitor energy storage, battery energy storage, etc. [4], which have been applied in actual energy storage power stations. Among them, battery energy storage also includes a variety of different types, and the performance parameters of the main battery energy storage technologies are shown in Table 1.

Table 1 Performance comparison of major battery energy storage technologies

Class	Power Upper Limit	Specific Capacity (Wh/Kg)	Specific Power (W/Kg)	Cycle Life (Times)	Charge And Discharge Efficiency (%)	Self-Discharge (%/Month)	Environmental Performance	Technology Maturity
Lead Acid	Tens of MW	35~50	75~300	500~1500	0~80	2~5	Middle	Good
Nimh	Tens of MW	85	160~230	2500	0~65	15~30	Good	Good
Ion	Count MW	150~200	200~315	1000~10000	0~95	0~1	Good	Good
Sodium-Sulfur	A dozen MW	150~300	90~230	4500	0~90	-	Good	Good
Fluid Flow	Count MW	80~130	50~140	13000	0~80	-	Good	Good

*Corresponding author: 769912907@qq.com

For lead-acid batteries, its application experience is relatively rich, and the manufacturing cost is relatively low, but it will have a certain impact on the environment during the application process, so it has been gradually replaced by other battery types. Sodium-sulfur batteries, flow batteries, etc. are important trends in the development and application of energy storage batteries in the future, and their batteries have large capacity, relatively high charge and discharge efficiency, and also have high service life, which are its important application advantages [2-3]. At the same time, for the application scenarios of energy storage, it mainly includes two types: the grid side and the user side, which is also the main way to apply the energy storage regulation system.

3 Comprehensive Benefit Analysis of Energy Storage

For the comprehensive benefits of energy storage, including the cost and benefit of energy storage, the following is a systematic analysis.

3.1 Cost analysis of energy storage

The cost of energy storage mainly includes initial investment costs, later operation and maintenance costs and decommissioning costs. First of all, for the initial investment cost, including energy storage equipment fee, transportation fee, commissioning fee, site fee, feasibility study, initial setup, acceptance and design fee, etc., as shown in (1) below:

$$C_{captial}=C_{s1}+C_{s2}+C_{s3}+C_{s4}+C_{s\ studyable}+C_{s\ initial\ setup}+C_{s\ acceptance}+C_{s\ devise} \quad (1)$$

In the formula: the C_{s1}、C_{s2}、C_{s3}、C_{s4} are energy storage equipment fees, transportation costs, commissioning fees, and site fees.

For O&M costs, the calculation formula is shown in Equation (2) below:

$$C_{OMs,t}=c_{OM,t}P_{smax} \quad (2)$$

Formula: $c_{OM,t}$ the operation and maintenance cost per unit capacity of the energy storage device; Including operation cost C_y and maintenance cost C_w; P_{smax} is the capacity of the energy storage device.

For the cost of decommissioning, when the energy storage device reaches the end of its service life, labor costs and environmental protection costs need to be spent in the disposal process, and the calculation formula is shown in (3) below:

$$C_{res}=C_{labor\ costs}+C_{environmental\ fees}-C_{salvage\ value} \quad (3)$$

wherein, the residual value of the energy storage device is calculated as shown in Equation (4) below:

$$C_{salvage\ value}=\Sigma C_{captial}(1-2/N_i)^{Ni} \quad (4)$$

Formula: N_i is the full life cycle of the energy storage device. At the same time, the average annual cost of energy storage devices is shown in Equation (5) below:

$$C_a=[C_{captial}+\frac{C_{res}}{(1+r)^{N_i}}+\sum_{t=1}^{N_i}\frac{C_{OMs,t}}{(1+r)^t}]\times\frac{r(1+r)^{N_i}}{(1+r)^{N_i}-1} \quad (5)$$

Formula: r is the annual discount rate.

3.2 Benefit analysis of energy storage

The benefits of energy storage mainly include reducing grid expansion, reducing system network loss, low storage and high arbitrage, reliability benefits, etc., and the calculation methods of various benefits are detailed below.

First of all, for reducing grid expansion, when the system is at the peak of load, by discharging the energy storage device, thereby reducing grid expansion, calculated as shown in Equation (6) below:

$$E_{s1}=\lambda_d C_d \eta P_{smax} \quad (6)$$

Formula: λ_d is the depreciation rate of grid assets; C_d is the cost of the unit capacity of the power grid; η is the energy storage efficiency; P_{smax} is the rated power of the energy storage device.

For reducing the efficiency of the system network loss, when the load peaks, the in-situ balance of the load can be achieved through the discharge of the energy storage device, thereby reducing the system network loss, calculated as shown in Equation (7) below:

$$E_{s2}=\sum_{1}^{365}\sum_{i=1}^{24}(\xi_{if}-\xi_{il})e_i \quad (7)$$

In the formula: ξ_{if} and ξ_{il} are the network loss before and after the energy storage device is connected; e_i is the price of electricity.

For low storage and high arbitrage, due to the different on-grid electricity prices in different periods, energy storage can obtain a more objective low storage and high arbitrage [5-6] by charging at low prices and discharging at high prices, calculated as shown in equation (8) below:

$$E_{s3}=n\sum_{i=1}^{24}(P_i^+-P_i^-)e_i \quad (8)$$

Formula: P_i^+ and P_i^- are the discharge and charging power of the energy storage device under the i period, respectively; The number of times the energy storage device is put into operation per year.

For reliability benefits, after connecting to the energy storage device, the power supply reliability of the system can be improved to a certain extent, as shown in Equation (9) below:

$$E_{s4}=(ENSI_{fs}-ENSI_{ls})*IEAR \quad (9)$$

In the formula: $ENSI_{fs}$ and $ENSI_{ls}$ are the lack of power supply in the system before and after the energy storage is added, and IEAR is the evaluation rate of power loss loss.

The annual comprehensive benefits of energy storage devices are:

$$E = E_{s1} + E_{s2} + E_{s3} + E_{s4} \quad (10)$$

3.3 Comprehensive benefit evaluation model and parameter setting of energy storage device

For the comprehensive benefit evaluation model of energy storage device, the maximum comprehensive benefit of energy storage device is the objective function, as shown in equation (11) below:

$$\max\ f = E - C_a \quad (11)$$

The constraints in the comprehensive benefit evaluation model of the energy storage device include the capacity constraints of the energy storage device, the state of charge constraints, the number of charge and discharge constraints, and the system power balance constraints.

Parameter Settings in the comprehensive benefit evaluation model of energy storage device are shown in the following table.

Table 2 Parameter Settings of energy storage device

Figure	Definition	Value	Figure	Definition	Value
C_d	Unit cost of grid	100（Ten thousand yuan /MW）	IEAR	Customer outage loss evaluation rate	6（Ten thousand yuan /MW•h）
λ_d	Depreciation rate of fixed assets of power grid equipment	3%	$P_{s\max}$	Maximum power stored	0.3MW
η	Energy storage efficiency	80%	r	Discount rate	8%
N	Battery life	20 years	e_i	Feed-in tariff	Peak electricity price is 0.088 Ten thousand yuan /MWh, The trough price is 0.0408 Ten thousand yuan /MWh
n	Annual number of energy storage operations	360	Cy	Unit operating cost of energy storage system	429.8（Ten thousand yuan /MW）
T	Duration of charging the energy storage device	6h	Cw	Unit maintenance cost of energy storage system	73.68（Ten thousand yuan /MW）

4 Example Analysis of Energy Storage Optimal Configuration

4.1 Basic information of the study

The example selected this time is the 10kV outlet line of a substation in a certain area, and its wiring mode is single-radiation overhead line. The 10kV distribution network is connected to the industrial park load and is equipped with an energy storage station, and its wiring diagram is shown in Figure 1 below.

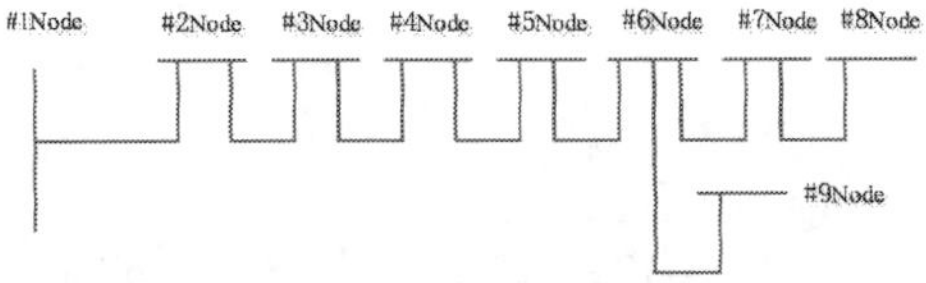

Figure 1 Single-radiation overhead line wiring diagram

In this distribution system, multiple segmented switches are set up, and when a certain section of the distribution network fails, the fault range can be narrowed by the segmented switch, thereby improving the power supply reliability of the distribution network. The transformer connected to the trunk line is 10KV/380V, which can be regarded as a load node. There are a total of 8 load nodes in the figure. The composition of the peak load time of each load node is shown in Table 3.

Table 3 Load composition of the system

Numbering	P(MW)	Q(MW)	Numbering	P(MW)	Q(MW)
1	0.29	0.03	5	0.36	0.12
2	0.38	0.12	6	0.28	0.09
3	0.33	0.11	7	0.36	0.12
4	0.31	0.01	8	0.35	0.12

The following example will be used to analyze the impact of the installation location and installation capacity of the energy storage power station on the comprehensive benefit of energy storage in the distribution network, including the load side, the middle of the line and the power supply side, which are the main application scenarios of energy storage devices.

The peak load time of the distribution network is 13 hours, the load trough is 8 hours [7-8]. When the system is at the peak of the load, the energy storage power station is running in the discharged state as a power source. When the system is in the trough of the load, the energy storage power station is running in the charging state as a load, thereby other functions of regulating the load of the system [9-10]. The total load of the power distribution system is 2.66MW, and for the installed capacity, 0.7MW, 0.9MW and 1.2MW batteries are installed, respectively.

4.2 Analysis of study results

In the above power distribution system, when the energy storage power station is installed at the power side node 2, the middle side node 5, and the load side node 8, the installed capacity is 0.7MW and remains unchanged. Based on the calculation of the above three places, the results show that the reliability benefit of the installation on the middle side of the line is the largest, and the network loss is the smallest. At the same time, the cost-benefit calculation of installation in the above three places is shown in Table 4 below:

Table 4 Results of cost-benefit calculations

Index	Power Side	Middle Side	Load Side
Cost	431 Thousand	431 Thousand	431 Thousand
Benefit	1432.3 Thousand	1620.8 Thousand	1599.8 Thousand
Net Benefits	1001.3 Thousand	1189.8 Thousand	1168.8 Thousand

It can be seen that when installed on the middle side of the line, the net benefit is the largest, because it can reduce the network loss at the same time, the reliability benefit is also the largest, so the installation site of energy storage should be installed in the place where the network loss is the least and the reliability can be maximized. Next, change the installed capacity and invest 0.9MW and 1.2MW respectively to see the net benefit shown in Table 5 below:

Table 5 Cost-benefit calculations

Index	0.7MW	0.9MW	1.2MW
Cost	431 Thousand	526 Thousand	668.5 Thousand
Benefit	1620.8 Thousand	2004.5 Thousand	2580 Thousand
Net Benefits	1189.8 Thousand	1478.5 Thousand	1911.5 Thousand

It can be seen that as long as the energy storage system is reasonably installed, the net benefit is gradually increasing. However, it does not mean that the capacity of the energy storage system can always increase, and the capacity of the energy storage system should match the load capacity of the bus bar at the installation site, so as to achieve the greatest comprehensive benefits. Since energy storage power stations can be invested by both users and users [11]. When the energy storage power station is invested in the power grid, the low storage and high arbitrage benefits obtained by energy storage belong to the power grid. When the energy storage power station invests for the user, the low storage and high arbitrage benefits obtained by energy storage belong to the user. No matter who invests, the construction of energy storage systems can achieve considerable economic benefits.

5 Conclusion

With the increase of the proportion of new energy in the power system, the energy storage system will play an increasingly critical role in the operation control of the power system in the future. In fact, energy storage is a dominant factor in the integration of renewable sources, playing a significant role in maintaining a robust and reliable modern electricity system[12]. In order to apply energy storage more reasonably, this paper constructs a comprehensive benefit evaluation model of energy storage in the whole life cycle, and takes the maximum comprehensive benefit as the objective function to optimize the location and installation capacity of energy storage. The results show that the energy storage should be located in the middle side of the line with the lowest network loss and the highest power supply reliability, and the configured capacity should match the load capacity to achieve the optimal operation of the energy storage.

References

1. Zhang Z, Pi X, Yang D, et al. Optimal decision-making and allocation method for battery energy storage power station replacement substation upgrade[J].Electric Power Construction,2023,44(07):41-49.)
2. Qian X, Zhang S, Chen C, et al. Comprehensive evaluation method of energy storage technology based on cloud model intuitive fuzzy analytic hierarchy method in multiple application scenarios[J].Advanced Technology of Electrical Engineering and Energy,2022,41(09):68-80.)
3. Zhou Z, Yao J, Xiang X, et al. Analysis of comprehensive benefits of energy storage based on multiple photovoltaic penetration rates[J].Northeast Electric Power Technology,2023,44(03):27-31.)
4. Abraham Alem Kebede, Theodoros Kalogiannis, Joeri Van Mierlo, Maitane Berecibar, A comprehensive review of stationary energy storage devices for large scale renewable energy sources grid integration[J]. Renewable and Sustainable Energy Reviews, 2022, 159:112213.
5. Leng Z, Zhao J, Gong C, et al. Research on operation optimization strategy of agricultural grid station area for source-grid-load storage and charging[J].Electromechanical Information,2022(24):11-14.)
6. Zhou T, Zhou R, Huang J, et al. Electric Power Automation Equipment,2023,43(05):171-176.)
7. Yan Q, Shi C, Qin G, et al. Two-layer configuration and operation optimization of electrochemical/hydrogen hybrid energy storage system based on proximal strategy optimization algorithm[J].Electric Power Construction,2022,43(08):22-32.)
8. Pan H, Gao X, Yao Z, et al. Research on benefit distribution mechanism of integrated energy system considering energy storage benefits[J].Smart Power,2022,50(05):25-32.)
9. Zhao L, Gong J, Li Y, et al. Energy storage allocation method of photovoltaic power station considering carbon efficiency and auxiliary peak shaving[J].Geomatics and Engineering Sciences of Wuhan University,2023,56(01):80-88.)
10. Chen Y. Research on operation optimization and benefit evaluation of integrated energy system based

on cooperative game[D].North China Electric Power University (Beijing),2022.)

11. Shi L, Yang F, Liu Y, et al. Optimal allocation of multi-scenario user-side energy storage capacity considering social development[J].Power System Protection and Control,2021,49(22):59-66.)
12. K.C. Divya, Jacob Østergaard, Battery energy storage technology for power systems—An overview[J]. Electric Power Systems Research, 2009, 79(4):511-520)

Research on Key Technologies for Measuring the Recovery Rate and Gas Measurement of SF_6 Gas On-site Maintenance Surface

Minjiang Chen [1,a], Da Chen [1,b], Zhixin Xu [1,c], Yingzhe Cheng [1], Shukai He [2,d], Sanxia Wang [2,e], Xiaozhe Zeng [2,f], Xinghui Wang [2,g*], Tenghui Wang [3,h]

[1] EVH Branch Company of State Grid Fujian Electric Power Co., Ltd., Fuzhou 350000, China
[2] Henan Relations Co., Ltd., Zhengzhou 450001, China
[3] School of Physics, Zhengzhou University, Zhengzhou 450001, China

Abstract. Carrying out full lifecycle management of SF_6 gas recovery, purification, and recycling, developing new digital control technologies, and eliminating SF_6 gas emissions from the source are of great significance for achieving the "dual carbon" goal. Given the low level of digital control of SF_6 gas at present, especially in the situation, the SF_6 gas recovery rate, the effective volume of the gas chamber, and gas weight cannot be accurately measured at an on-site inspection surface, this paper carries out research on SF_6 gas recovery rate measurement at on-site inspection surface and key technologies of gas measurement. An intelligent online monitoring device such as SF_6 gas recovery rate and effective chamber volume of high voltage electrical equipment was developed, which can quickly identify SF_6 gas density, recovery rate, effective chamber volume, weight, and other parameters in high voltage electrical equipment, and evaluate the detection methods such as gas recovery rate and effective chamber volume. The results show that when the recovery rate was about 96.5%, the maximum positive deviation between the recovery rate measured by the detection device and that measured by the weighing method was 0.32% and the maximum negative deviation was -0.24%, both of which are less than 0.50%. When the container to be tested was 1000 L, the maximum indication error was -1.51%, which can meet the requirements of digital control in the SF_6 gas recycling and reuse process of high-voltage electrical equipment.

1. Introduction

SF_6 gas was a kind of strongly electronegative gas, and its molecules have a strong ability to absorb free electrons to form negative ions, so its resistance was very high. In a more uniform electric field, the electric field intensity of SF_6 gas was about 2.5 times that of air. It was the most ideal insulation and arc extinguishing medium at present. However, in the Kyoto Protocol, SF_6 gas was listed as one of the six greenhouse gases with limited emissions, and it was prohibited to discharge [1-3]. The physicochemical properties and application status of SF_6 gas were discussed. On the one hand, new technologies such as SF_6 gas recovery and purification, recharge, and life cycle management are vigorously developed at home and abroad to prevent leakage and emission from the source [4-6]. On the other hand, looking for a new type of insulating gas with excellent environmental protection and electrical properties can fundamentally solve the environmental problems of SF_6 [7]. Although domestic and foreign researchers have done a lot of research on the insulation properties and decomposition characteristics of new environmentally friendly insulating gases 2, 3, 3, 3-4 fluoro-2- (3 fluoromethyl) - propanitrile (4710) and 1, 1, 1, 3, 4, 4-7 fluoro-3- (3 fluoromethyl) -2-butanone (5110) [8-11], the timing of the large-scale application of replacement SF_6 gas was not yet known. From the technical and economic point of view, it was still not as good as SF_6 gas and can not completely replace SF_6 as an insulating medium. SF_6 gas was still the preferred insulating medium for high-voltage electrical equipment [12].

To reduce the consumption and emissions of SF_6 gas, respond to the national call for "carbon peaking and carbon neutrality", and deeply promote the green and low-carbon development of the power grid, the power industry has begun to use SF_6/N_2 mixed gas for promotion and application in electrical equipment. The application and promotion of SF_6/N_2 mixed gas in GIS can effectively reduce the application of SF_6 gas in power equipment. However, the use of mixed gases can not fundamentally solve the greenhouse effect problem caused by SF_6 gas emissions. With the improvement of social awareness of energy conservation and environmental protection and the progress of management and technology, SF_6 gas recovery and purification, recycling, and life cycle management

[a]Chenminjiang@126.com, [b]chenda@126.com, [c]xuzhixin@126.com, [d]heshukai@relations.com.cn,
[e]wangsanxia@relations.com.cn, [f]zengxiaozhe9@126.com, [g*]wangxinghui08@126.com, [h]wangtenghui@126.com

technology are the most effective and feasible methods to control SF_6 gas consumption and emissions.

In 2019, the Ministry of Ecology and Environment required the SF_6 gas recovery rate to reach 96.5% in 2021, an increase of 1.5 percentage points over 2018 in the "Responsibility Letter for the Business Performance of the Head of the Central Enterprise". Based on the change in national policy, SF_6 gas has rapidly changed from independent emission reduction to mandatory emission reduction. It faced more severe challenges in the aspects of potential mining and control of SF_6 gas emission reduction. These include the lack and low level of digital controls. In particular, the SF_6 gas recovery rate, the effective volume of the gas chamber, and the weight of the gas can not be accurately measured on the on-site inspection surface, and there was still a certain gap with the relevant carbon emission regulatory requirements.

To solve the technical problems existing in the recycling process of SF_6 gas circulation for high-voltage electrical equipment, this paper studies the key technologies of SF_6 gas recovery rate measurement and gas metering on-site maintenance surface. An intelligent online monitoring device such as SF_6 gas recovery rate and effective volume of a gas chamber for high voltage electrical equipment has been developed. It can quickly identify SF_6 gas density, recovery rate, the effective volume of the gas chamber, weight, and other parameters in high-voltage electrical equipment. It can evaluate the detection methods such as gas recovery rate and effective volume of gas chamber, and provide important data support for SF_6 gas life cycle management.

2. Measurement principle

2.1. Gas recovery, density, and mass

To achieve a certain insulation capacity, the rated pressure of SF_6 gas in the air chamber of high-voltage electrical equipment was generally not less than 0.5 MPa. SF_6 gas in the gas chamber of high-voltage electrical equipment does not have the conditions of the ideal gas, and its state parameters cannot be determined by Charles' Law. In this paper, based on the Beattie-Bridgman equation of state [13-14], the state parameters before and after SF_6 gas recovery are calculated as shown in Equation (1), and then the inversion model of SF_6 gas recovery rate was derived.

$$P = (RTB - A)d^2 + RTd \tag{1}$$

where P was the absolute pressure (MPa) of SF_6 gas in the high-pressure switching gas chamber; T was the pressure (K) of the gas chamber, and d was the gas density of the SF_6 gas chamber (kg/m^3); A, B, and R are the constants measured by experiment, A=73.882×10^{-5}-5.132×10^{-7}d, B=2.50695×10^{-3}-2.12283×10^{-6}d, R=5.69502×10^{-4}.

The pressure P_1 and P_2, temperature T_1 and T_2 in the high-pressure switching gas chamber before and after recovery are measured by the high-precision pressure sensor and temperature sensor. According to Equation (1), given the state parameters P and T in the high-pressure switching gas chamber, the SF_6 gas densities d_1 and d_2 can be calculated. The volume V of the air chamber remains unchanged before and after recovery. If the volume V_1 of the irregular high-voltage switching gas chamber was known, the mass m_1 and m_2 (kg) of SF_6 gas in the high-voltage electrical equipment before and after recovery are shown in Equations (2) and (3), respectively.

$$m_1 = d_1 \times V_1 \tag{2}$$

$$m_2 = d_2 \times V_1 \tag{3}$$

where d_1 and d_2 are the density of SF_6 gas in the gas chamber before and after recovery, respectively. The recovery rate η (%) of SF_6 gas is shown in Equation (4).

$$\eta = \frac{\mathrm{m}_1 - m_2}{m_2} \times 100\% = \frac{d_1 - d_2}{d_1} \times 100\% = f(P,T) \tag{4}$$

It can be seen that in the SF_6 gas recycling process, the gas mass m and recovery rate in the gas chamber are both state functions of the gas chamber pressure P and temperature T. By measuring the pressure P and temperature T of the gas chamber before and after recovery, the mass and recovery rate of SF_6 gas recycling process can be retrieved.

2.2. Theoretical model of effective volume measurement of high-pressure switching gas chamber

In the process of SF_6 gas recycling, the measurement of SF_6 gas quality on the field inspection surface was of great significance for the realization of gas recycling and digital control. It can be seen from Equations (2) and (3) that the monitoring of SF_6 gas quality was closely related to the effective volume of the gas chamber of high-voltage electrical equipment. Therefore, the measurement of the effective volume of high-voltage electrical equipment becomes a key factor. When the pressure was high, the density of the gas increases, the distance between the molecules decreases, and the interaction of the molecules and their volume cannot be ignored. According to the equation of the state of SF_6 under actual gas behavior, the effective volume of an irregular typical device or model can be derived by introducing the "compression factor" (Z).

$$PV = nRTZ \tag{5}$$

where P was SF_6 gas pressure (Pa), V was the volume of the gas chamber (m^3), n was the amount of matter (mol), and R was the molar gas constant (J/(mol·k)). To obtain the effective volume of an irregular typical device or model, the volume model of the high-pressure switching gas chamber as shown in Figure 1 was established. In Figure 1, V_1 was the volume of the air chamber of the high-pressure electrical equipment and P_1 was the initial pressure of SF_6 gas in the gas chamber to be measured. We use a standard container whose volume was V_2 and vacuum the container, connect the two containers through the pipeline, then open the valve, let the pressure of the two containers balance, and record the balanced pressure P_2. The effective volume of the gas chamber of the high-voltage electrical equipment obtained was shown in Equations (6) and (7).

$$V_1 \times P_1 = (V_1 + V_2) \times P_2 \quad (6)$$

$$V_1 = P_2 V_2 / (P_1 - P_2) \quad (7)$$

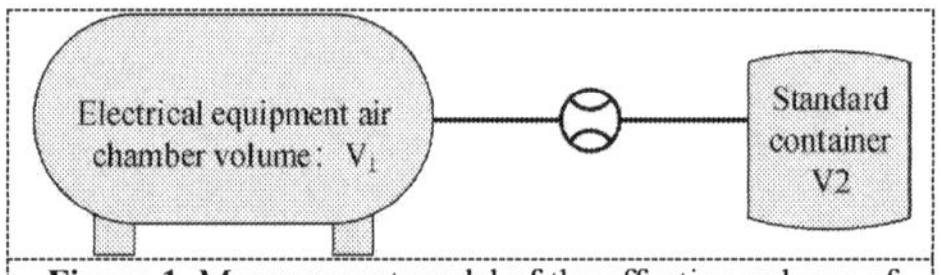

Figure 1. Measurement model of the effective volume of the air chamber

According to the above algorithm, the corresponding standard vessel was designed, and the pressure sensor with a measuring range of -0.1 MPa~0.9 MPa and an error of ±0.1% was selected. The pressure difference between the balanced pressure P_2 and the initial pressure P_1 of the air chamber to be measured was not greater than ΔP. It is ensured that the pressure drop of the air chamber during the measurement process does not influence the working state and the validity of the data. When P_1-P_2=ΔP, Equation (8) can be obtained from Equations (6) and (7).

$$V_2 = V_1 / (P_1 / \Delta P - 1) \quad (8)$$

The volume V_2 of the standard container can be retrieved from Equation (8) as shown in Figure 2. When ΔP=10 kPa, the size of the required standard container V_2 can be calculated by Equation (8). The volume of high pressure switching chamber varies. Relevant statistics show that the volume of gas chambers less than 5 m^3 accounts for about 89.45% of the total, and the design of the above standard container volume meets the application requirements. In the process of application, the multi-valve connection was adopted. According to the volume of different gas chambers to be measured, the corresponding standard container was selected directionally.

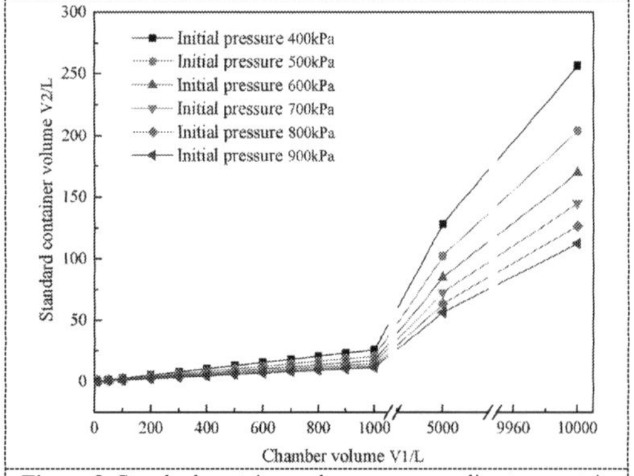

Figure 2. Standard container volume corresponding to a certain SF_6 chamber volume and initial pressure

3. Development of testing device

The design diagram of the detection system was shown in Figure 3. The device was connected to the SF_6 gas chamber of the high voltage electrical equipment through the intake pipe and was connected to the pressure sensing unit, temperature sensing unit, standard container, and SF_6 recovery device respectively after output by the stop valve and pressure regulator valve. The pressure sensing unit and the temperature sensing unit sense the pressure and temperature of SF_6 gas before and after recycling. Combined with the standard vessel, the data processing module based on the Beattie-Bridgman equation of state was used to measure the volume, density, mass, and recovery rate of SF_6 gas in high-voltage electrical equipment under certain temperature and pressure conditions, through the human-computer interaction interface for real-time display, and through Bluetooth, wifi, or LoRa to achieve wireless data transmission.

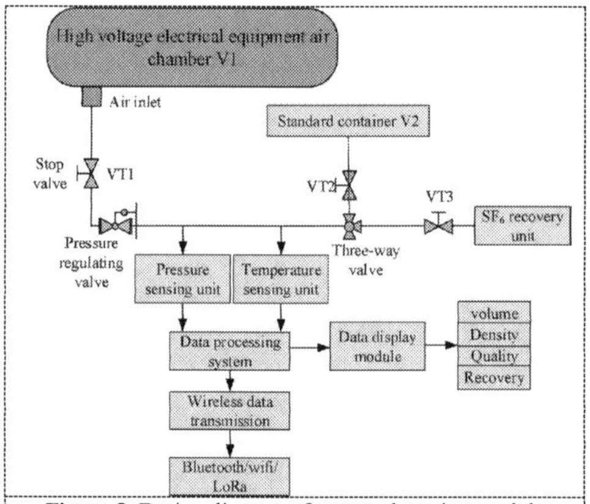

Figure 3. Design diagram of system detection module

4. Performance verification

4.1. Verification of SF6 gas recovery rates

To verify the recovery rate of the SF_6 gas recycling multi-parameter intelligent detection device, an electronic scale with a measuring range of 30 kg and an index value of 0.5 g was selected as a weighing instrument. An 8 L standard cylinder was used as an on-site SF_6-filled gas chamber container for recovery verification as shown in Figure 4. The mass of SF_6 gas filled with 600 kPa in the cylinder was about 300 g, and the accuracy of the electronic scale was about 0.167%. The mass of 1000 kPa gas was about 500 g, and the accuracy of the electronic scale was about 0.1%. All of the above meet the accuracy requirements of the recovery rate verification. The experimental design and procedures are as follows:

(1) After the equipment is ready, the 8 L standard steel cylinder was vacuumed with a two-stage rotary

vane vacuum pump (TRIVACD63L) until the absolute pressure was below 3 Pa, and the quality of the evacuated steel cylinder was accurately recorded by an electronic scale as m_1.

(2) We connect the inflation pipeline, SF_6 gas metering detection module, and SF_6 gas storage tank. Before inflation, the pressure-reducing valve of the standard steel cylinder is closed, and the pipeline system is vacuumed by the vacuum pump until the absolute pressure is below 3 Pa. We fill SF_6 gas to the absolute pressure of 400 kPa, 600 kPa, 800 kPa, and 1000 kPa respectively. Then we remove the gas line and metering module and rest for 3 hours and record the temperature t_1 of the standard cylinder at different pressures and the corresponding SF_6 gas cylinder mass m_2 accurately.

(3) We connect the recovery pipeline, SF_6 gas measurement and detection module, and SF_6 gas recovery device and conduct recovery operations on SF_6 gas separately. During the operation, we check the gas recovery rate displayed in the SF_6 gas measurement and detection module. When the recovery rate reaches around 96.5%, we stop the recovery, end the measurement, and close the recovery pipeline.

(4) Where the temperature t_2, pressure P_2, cylinder mass m_3, and gas recovery rate η_1 in SF_6 gas measurement and detection module corresponding to the standard cylinder after recovery were recorded at different filling pressures.

(5) According to the quality of the standard cylinder before and after recovery, the mass recovery rate η_2 was calculated, as shown in Equation (9).

$$\eta_2 = \frac{m_2 - m_3}{m_2 - m_1} \times 100\% \tag{9}$$

(6) According to the above experimental procedures, repeated sampling was performed, and the experimental data were shown in Table 1.

The experimental results show that when the recovery rate of SF_6 gas was about 96.5%, the maximum positive deviation between the recovery rate measurement results of the detection module and that of the weighing method was 0.32%, and the maximum negative deviation was -0.24%, both of which are less than 0.5%, meeting the application requirements.

(a) Detection module recovery rate measurement (b) Weighing method measurement verification

Figure 4. SF_6 recovery rate validation experiment

Table 1. Recovery rate experimental data

Parameter name	1#	2#	3#	4#	5#	6#	7#	8#
Empty bottle quality (m_1/g)	8036	8036	8036	8036	8037	8037	8037	8037
SF_6 gas pressure before recovery (P_1/kPa)	398.1	594.5	789.3	989.2	398.1	594.5	789.3	989.2
Gas cylinder temperature before recovery (t_1/°C)	28.6	28.4	27.3	25.6	28.6	28.4	27.3	25.6
After filling the cylinder mass (m_2/g)	8236	8341	8462.5	8571	8236	8341	8462.5	8571
SF_6 gas pressure after recovery (P_2/kPa)	15.6	18.8	28.7	38.7	15.6	18.8	28.7	38.7
Gas cylinder temperature after recovery (t_2/°C)	28.1	28.5	27.4	25.8	28.1	28.5	27.4	25.8
Quality of cylinder after recovery (m_3/g)	8044	8046	8050	8055	8044	8046	8050	8055
Detection module recovery (η_1/%)	96.24	97.04	96.69	96.53	96.24	97.04	96.69	96.53
Calculated mass recovery (η_2/%)	96.00	96.72	96.72	96.45	96.48	97.04	96.94	96.63
Measurement deviation (Md/%)	0.24	0.32	-0.03	0.08	-0.24	0.00	-0.25	-0.10

4.2.Validation of the effective volume of SF6 electrical equipment

According to the theoretical model of effective volume measurement of the gas chamber of medium and high voltage electrical equipment, combined with the SF_6 gas metering module, the standard cylinders with 0.73 L, 4 L, 8 L, and 40 L volumes are selected as the standard container ends connected with the three-way valve. The bus bar and stop valve are arranged in the standard container module so that the size of different standard container volumes can be selected according to the volume of the container to be measured. The volume of the simulation vessel of the gas chamber of high-voltage electrical equipment was 40 L, 300 L, 600 L, and 1000 L respectively. SF_6 gas purity was not less than 99.99%

and the output pressure was not less than 0.5 MPa. The process of the verification system was shown in Figure 5. The effective volume verification experiment steps are as follows:

(1) Before the test, the standard vessel and experimental pipeline system were vacuumed with a vacuum pump to an absolute pressure below 10 Pa.

(2) We fill the 300 L, 600 L, and 1000 L containers (V_1) to be tested with about 0.5 MPa SF_6 gas, rest for 3 hours to fully balance it, and accurately record the initial pressure P_1.

(3) We open the valve between the standard sampling bottle and the container to be tested. After the gas pressure was balanced, we record the balance pressure P_2 of the container to be tested and the measured volume V_{1m} of the detection module.

(4) After the test was completed, we open the recharge module for SF_6 gas recharge. The whole measurement process was a closed loop, with no SF_6 gas emission and green environmental protection.

(5) According to the test results, the measurement deviation *Md* and the indication error of the detection module are calculated, as shown in Table 2.

The results show that when the container to be tested was 40 L, the maximum indication error was 2.06%. When the container to be tested was 300 L, the maximum indication error was -1.6%. When the container to be tested was 600 L, the maximum indication error was -2.52%. When the container to be tested was 1000 L, the maximum indication error was -1.51%. The above results can meet the demand for digital control in SF_6 gas recycling and reuse process of high voltage electrical equipment.

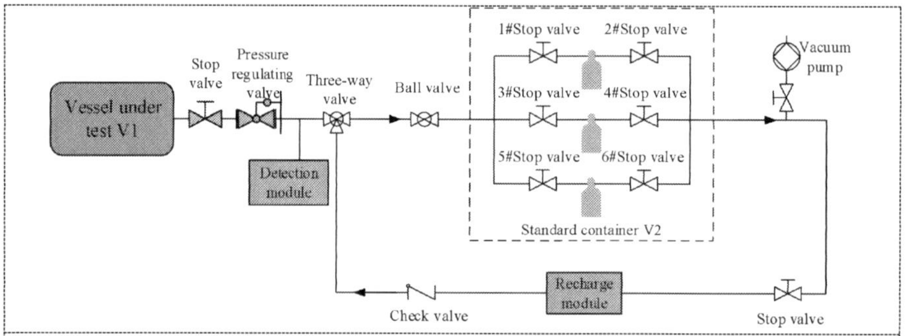

Figure 5. Verification and testing system for irregular chamber volume of high-voltage switch

Table 2. High voltage switch irregular chamber volume verification test data

Initial pressure P_1/MPa	Equilibrium pressure P_2/MPa	Vessel under test V_1/L	Standard sample bottle V_2/L	ΔP/MPa	The measured volume of detection module V_{1m}/L	Measurement deviation *Md*/L	Indication error/%
0.5155	0.5064	40	0.73	0.0091	40.6	0.6	1.56
0.5066	0.4977	40	0.73	0.0089	40.8	0.8	2.06
0.4979	0.4891	40	0.73	0.0088	40.6	0.6	1.43
0.4889	0.476	300	8	0.0129	295.2	-4.8	-1.60
0.4889	0.4762	300	8	0.0127	300.0	0.0	-0.01
0.4889	0.4762	300	8	0.0127	300.0	0.0	-0.01
0.4813	0.4807	600	0.73	0.0006	584.9	-15.1	-2.52
0.4813	0.4801	600	1.49	0.0012	596.1	-3.9	-0.65
0.4813	0.4795	600	2.22	0.0018	591.4	-8.6	-1.44
0.4849	0.4665	1000	40	0.0184	1014.1	14.1	1.41
0.4669	0.4492	1000	40	0.0177	1015.1	15.1	1.51
0.45	0.4328	1000	40	0.0172	1006.5	6.5	0.65

5. Conclusion

This paper analyzes the development and application status of SF_6 gas, the shortcomings of the whole life cycle management, and digital control technology. Based on the Beatty-Bridgman equation of state, the measurement principles of gas density, gas weight, and recovery rate of SF_6 gas in field inspection surface are studied. According to the changes in temperature and pressure before and after recovery, the mathematical models of gas density, gas weight, and recovery were established. The theoretical model of effective volume measurement of high-pressure switching chambers was established, which provides theoretical guidance for the selection of standard vessels under different chamber volumes and pressure. Finally, the SF6 gas recovery measurement and gas metering detection module were developed, and the performance of the detection module was verified. The main conclusions are as follows:

(1) When the SF_6 gas recovery rate was about 96.5%, the maximum positive deviation between the recovery rate measurement results of the gas metering module and that of the weighing method was 0.32%, and the maximum negative deviation was -0.24%, both of which are less than 0.5%.

(2) The effective volume verification results show that different standard containers combined with detection modules can be used to measure different volumes of the gas chamber to be measured. From the 40 L to 1000 L gas chamber to be measured, the indication error of the SF_6 gas measurement and detection module was less than 3%.

(3) The developed SF_6 gas measurement and detection module can simultaneously realize the effective measurement of SF_6 gas key parameters, which can provide important data support for SF_6 gas life cycle management and help realize the dual-carbon goal.

Acknowledgments

The current work is supported by the science and technology project of Fujian Electric Power Co., Ltd., State Grid (No. 52130A220002).

References

1. Li Y, Zhang X, Xiao S, Chen Q, Tang J, Chen D, Wang D. (2018) Decomposition properties of C_4F_7N/N_2 gas mixture: an environmentally friendly gas to replace SF_6. Industrial & Engineering Chemistry Research, 57(14): 5173-5182. https://dio.org/ 10.1021/acs.iecr.8b00010.
2. Chen J. (2014) Research and application of latent defect diagnosis technology for SF_6 electrical equipment based on gas analysis. Wuhan University, 2014. http://ifc.helib.net:80/rwt/CNKI/https/NNYHGLUDN3WXTLUPMW4A/KCMS/detail/detail.aspx?dbname=CDFD1214&filename=1014249823.nh
3. Ji S, Zhong L, Liu K, Li J, Cui Y, Wang Y, Guo J, Ji G. (2015) Research status and development of SF_6 decomposition components analysis under discharge and its application. Proceedings of the CSEE, 35(09): 2318-2332. https://dio.org/ 10.13334/j.0258-8013.pcsee.2015.09.028.
4. Tang F, Liu S, Lü Q, Li X, He S, Zeng X, Xu G, Yue Y. (2019) Real time detection of SF_6 leakage in large area based on Quantum Cascade Laser. Acta Photonica Sinica, 48(10): 193-202. https://dio.org/ 10.3788/gzxb20194810.1030004.
5. Li X, Deng Y, Jiang X, Zhao H, Zhuo R, Wang D, Fu M. (2017) Insulation Performance and Application of Enviroment-friendly Gases Mixtures of C_4F_7N and $C_5F_{10}O$ with CO_2. High Voltage Engineering, 43(3): 708-715. https://dio.org/ 10.13336/j.1003-6520.hve.20170303003.
6. Kim P J , You Y W, Park H , Chang J S, Bae Y S , Lee C H . (2015) Suha J K. Separation of SF_6 from SF_6/N_2 mixture using metal–organic framework MIL-100(Fe) granule. Chemical Engineering Journal, 262:683-690. http://dx.doi.org/10.1016i.cej.2014.09.123.
7. Tang N, He S, Zeng X, Wang H, Sun D, Wu Q, Li J. (2021) Research on Infrared Absorption Characteristics and Detection Technology of New Environmentally Friendly Insulating Gas Trans-1,1,1,4,4,4-Hexafluoro-2-Butene. Spectroscopy and Spectral Analysis, 41(10): 3099-3105. https://dio.org/ 10.13336/j.1003-6520.hve.20170303003.
8. Cui Z, Lin X, Zhong J, Yao Y, Xu J. (2023) Study on the temperature rise characteristics of C_4F_7N/CO_2 mixed gas ultra high voltage bus. Transactions of China Electrotechnical Society, 38(09): 2491-2499. https://dio.org/ 10.19595/j.cnki.1000-6753.tces.220083.
9. Huang Y, Yan X, Gao K, Li Z, Chen T, Wang W. (2023) Experiments on switching performance of grounding switch in C_4F_7N/CO_2 mixture gas-insulated ring main unit. High Voltage Engineering, 49(07): 2858-2867. https://dio.org/ 10.13336/j.1003-6520.hve.20221931.
10. Zeng F, Chen R, Li H, LI Yi1, Dai L, Yao Q, Tang J. (2022) Influence of oxygen on partial over-thermal decomposition of $C_5F_{10}O/N_2$ mixed gas insulating medium. Proceedings of the CSEE, 42(S1): 305-312. https://dio.org/ 10.13334/j.0258-8013.pcsee.213070.
11. Andersen M P S, Kyte M, Andersen S T, Nielsen C J, Nielsen O J . (2017) Atmospheric chemistry of $(CF_3)_2CF$–CN: a replacement compound for the most potent industrial greenhouse gas, SF6, Environmental Science & Technology, 51 (3) : 1321–1329. https://doiorg/101021/acs est.6b03758.
12. Dengming. (2016) Development Prospect of Gas Insulation Based on Environmental Protection. High Voltage Engineering, 42(04): 1035-1046. https://dio.org/ 10.13336/j.1003-6520.hve.20160405020.
13. Zhang C, Cai X, Xiao P, Xiao Z, Wang P. (2022) Study on online measurement method and device development of sulfur hexafluoride gas recovery rate. Hubei Electric Power, 46(01): 37-42. https://dio.org/ 10.19308/j.hep.2022.01.007.
14. Han X, Meng Q, Wang H. (2009) SF_6 gas state parametric equation and its engineering application[J]. Journal of Electrical Engineering, 7: 46-49. https://dio.org/ 10.19308/j.hep.2022.01.007.

Research on Investment Economic Evaluation of Flexible Regulating Resources such as Energy Storage in High Proportion New Energy Environment

Yu Shi[1], Xuefeng Gao[1], Bo Zhao[1], Yiwen Yao[1], Shuai Shao[1], Xueying Yang[1], Xiaotong Zhang[2*]

[1]State Grid Jilin Electric Power Co., Ltd. Economic and Technological Research Institute, No.1427 Pingquan Road, Nanguan District, 130062, China;
[2]North China Electric Power University, Beijing 102206, China

Abstract. With a high percentage of new energy scenarios, it has become a trend for flexible resources such as energy storage systems to participate in long-term planning. In this context, it is important to explore how energy storage systems are configured. Therefore, it is necessary to analyse the economics of energy storage and other resources to ensure that investments in energy storage and other resources are economically valuable under different future models. This paper firstly analyses the current situation of a high percentage of new energy in China, and secondly constructs a model for analysing the economics of flexible regulation resources such as energy storage. Under partial data based on the scenery storage project, the economic analysis under four models is derived, including cost and benefit analysis, investment benefit analysis, and break-even point analysis. Finally, the economic evaluation of investment in energy storage projects under different models is summarized based on the calculation results.It is concluded that different scenario models have different economic benefits, and the operation capacity of the four scenarios is higher than that of the existing models.

1. Introduction

According to the National Energy Administration statistics, from 2006 to 2019, China's installed renewable energy capacity increased from 135 GW to 795 GW, nearly five times, and renewable energy power generation increased from 446.9 TWh to 1998.8 TWh, an increase of 3.5 times [1]. The proportion of installed capacity of renewable energy generation is gradually rising, but new energy generation is affected by changes in wind and light natural resources, and there are problems of difficult prediction, difficult control, and difficult scheduling, and large-scale grid connection has an impact on power quality and transient stability of the grid. At the same time, China's energy structure is gradually transforming, and it needs to deal with the challenges brought by the continuous capacity increase of wind power generation and photovoltaic power generation to the power system and the stable operation of the grid [2].

At present, China 's power system mainly provides automatic generation control (AGC) frequency modulation function by traditional thermal power units. Traditional thermal power units are composed of mechanical devices with rotational inertia, and the conversion of primary energy into electrical energy requires a series of complex processes. Therefore, the response speed of traditional thermal power units to active power regulation is slow. Energy storage technology is applied in the field of AGC frequency modulation. [3]Because of its advantages of fast response speed, accurate control and two-way regulation, it has incomparable advantages over traditional thermal power unit frequency modulation in power grid frequency modulation[4].

Compared with the basic requirements of safety, reliability, and economy of power system operation, flexibility has become an important indispensable indicator for measuring the operating characteristics of the system as the penetration rate of new energy sources has increased significantly. With the rapid development of technology, energy storage has become a source of flexibility for power systems that cannot be ignored . Relying on its flexibility, energy storage technology can help solve a series of challenges brought by the current high percentage of new energy background to the power system. Energy storage technology plays a corresponding role in each link of power system generation, transmission, distribution, and use, which is of great significance to ensure the security of the power grid, improve the quality of power, increase the proportion of renewable energy, and improve the efficiency of energy utilization. In this context, the purpose of this paper is to analyze the investment economy of flexible adjustment resources such as energy storage, so as to ensure that energy storage

*Email: ZXiaoTong99@163.com

 https://doi.org/10.1051/e3sconf/202344101016

investment has a considerable economic value level under current and future conditions.

2. Energy Storage Economic Analysis Model

2.1 Energy storage economic evaluation method

The economic analysis of wind and solar energy storage and storage adopts the technical-economic evaluation method to analyze the initial investment, annual cost, and revenue of energy storage. On this basis, the economic evaluation indexes such as net present value, internal rate of return, and payback period of wind and solar energy storage and storage are measured to draw relevant evaluation conclusions.

2.2 Cost analysis of energy storage

The cost of wind and solar energy storage, including the initial investment of the storage power plant, is apportioned to the annual cost C_1, annual operation, and maintenance fees (including management costs) C_2, annual charging costs C_3, annual financial costs (annual bank loans to be repaid) C_4, annual taxes C_5.

2.3 Analysis of energy storage returns

The annual revenue of wind and solar energy storage includes the revenue of participating in auxiliary services R_1, the annual power generation revenue of the storage power plant R_2, and the equalization to the annual policy subsidy revenue R_3.

2.4 Calculation of economic indicators of energy storage

In this study, the annual cost and annual revenue of the scenic energy storage are analyzed to calculate the annual net revenue; the dynamic payback period is measured to analyze the years required to recover the initial investment of the energy storage project; the economic benefits of the project are measured by calculating the financial net present value and internal rate of return. Considering the time value of money, the three economic evaluation indexes of dynamic payback period, NPV (net present value), and IRR (internal rate of return) are used to evaluate the economics of the energy storage project. The IRR is the discount rate that equates the sum of the present values of a cash flow to zero. If the NPV of a project is zero at a selected discount rate, that rate is the IRR. If the NPV of a project is zero at a selected discount rate, that rate is, by definition, the IRR [5].

3. Economic Analysis of Energy Storage Projects Under Four Modes

3.1 Project basic information and parameters

This paper studies whether the subsequent investment in energy storage has a certain economic value with the development and cost reduction of energy storage technology and the change of business model. It explores the economic situation of energy storage in four modes and provides some reference for the subsequent energy storage investment strategy.[6-8] Therefore, the setting of the basic parameters will be based on the scenic storage project with certain condition adjustments to facilitate the subsequent economic analysis. In addition, with the rapid development of battery energy storage technology, the unit cost level of energy storage has been showing a decreasing trend, so the economic evaluation using the current cost level is more meaningful for future investment strategy. For this reason, this chapter sets 2020 as the construction year and 2021 as the commissioning year and uses the current cost level to calculate the initial investment cost of energy storage, to determine whether the energy storage project has investment value. The specific parameters are set as follows: (1) The capacity of the energy storage project is 19MW/83.5MWh. (2) The first phase capacity of the scenery storage energy storage is 19MW/83.5MWh. The unit cost of about 5500 yuan/kW. The capacity unit cost of about 2000 yuan/kWh measurement. According to the above parameters, the initial investment can be calculated as 271.5 million yuan. The capital of the initial investment is 20%, the loan ratio is 80%, and the loan interest rate is 4.99%. The operation and maintenance overhaul costs are measured at about 2% of the initial investment, so it is measured at 5 million yuan per year. (3) Using the straight-line depreciation method, the depreciable life is the entire operating period, and the benchmark rate of return is i=8%. (4) The discharge tariff is calculated according to the wind turbine feed-in tariff of RMB 0.54/kWh; the charging tariff is calculated according to the average of the PV and wind power feed-in tariffs of RMB 0.456/kW. (5) The loan ratio is 64.94%, the loan interest rate is 5.14%, the repayment period is calculated as 15 years, and the repayment is made in the form of equal principal and interest. The value-added tax rate is calculated at 17%, urban construction, and maintenance tax and education surcharge are 7% and 5%, respectively, the local education surcharge is 3.5%, and corporate income tax is 25%.

3.2 Economic analysis based on the consumption of new energy model

3.2.1 Cost and revenue analysis.

According to the capacity of the scenery storage project is 19MW/83.5MWh, the daily operation of the storage battery is within the range of 10%-90% of its rated capacity, using the abandoned wind energy, the charging electricity price is measured according to 0, the discharging electricity price is measured according to 0.54 yuan/kwh (including tax), one day charging and discharging, thus we can get the cash flow expenditure of the new energy consumption mode as follows: during the construction period, the project investment cost is 271.5 million yuan, the operation and maintenance cost is 5 million yuan per year, and the debt service is 21,472,600 yuan per year. The cash flow inflow revenue of the new energy consumption model comes from the power generation revenue of 11.8497 million yuan per year.

3.2.2 Investment benefit analysis.

Using the technical-economic evaluation method, the internal rate of return of energy storage, net present value, and dynamic payback period indicators can be measured as follows:

Table 1 Energy storage engineering technology economic evaluation indicators

Indicator	Value	Remarks
Net present value	-205,730,000	yuan
Internal rate of return	-10%	
Static payback period		cannot be recovered within the operation cycle
Dynamic payback period		cannot be recovered within the operation cycle

From Table 1, we can see that the NPV of the project is -205.73 million Yuan, which is less than 0; the IRR is -10%, which is less than the benchmark IRR of 8%, which makes the project infeasible and uneconomic from the perspective of economic evaluation.

3.2.3 Break-even point analysis.

First, the energy storage project needs to ensure that the internal rate of return is greater than or equal to 8% and the net present value is greater than or equal to 0 to ensure that the project is economically feasible. After technical and economic calculations, the discharge price of energy storage needs to be increased to at least 2.01 yuan/kwh (including tax) to ensure that the NPV of the project is greater than or equal to 0 and reach the break-even point of the project. Second, since the economic evaluation of energy storage under the current cost does not consider policy subsidies, and at present, China's energy storage power plants can obtain one-time or annual subsidy income according to national, local, or project policies, which can make energy storage projects turn from losses to profits. After technical and economic calculations, it is necessary to obtain a one-time subsidy of 222,192,900 yuan in 2020 to ensure that the NPV of the project is greater than or equal to 0. Third, after technical and economic calculations, the initial investment of energy storage needs to be reduced to at least 43.35 million yuan, which is 84.03% lower than the original initial investment of 271.5 million yuan, in order to ensure that the NPV of the project is greater than or equal to 0. Currently, the power unit cost is 878.34 Yuan/kW, and the capacity unit cost is 319.4 Yuan/kWh[9-10].

3.3 Economic analysis based on peak-shaving and valley-filling model

3.3.1 Cost and revenue analysis.

Referring to the peak and valley split tariff of Jibei Grid, the peak tariff is 0.7033 yuan/kWh, the valley tariff is 0.3151 yuan/kwh, and the difference between the peak and valley tariff is 0.3882 yuan/kWh. The capacity subsidy is calculated according to the reward of 550 yuan/kW in the central and western regions, and the one-time compensation that the scenery storage power station can get is 10.45 million yuan. According to the capacity of the scenery storage project is 19MW/83.5MWh, the daily operation of the storage battery is within the range of 10%-90% of its rated capacity, and it is charged and discharged once a day, so the daily charging volume is 66.8MWh. The charging tariff is 0.3151RMB/kWh, the feed-in tariff is 0.7033RMB/kwh, and the comprehensive efficiency of charging and discharging of the storage battery is 90%. Then the annual power generation revenue of energy storage is 15,433,100 yuan, and the annual charging cost is 7,682,800 yuan. The cash flow expenses of the peak-shaving and valley-filling model can be obtained as follows: the investment cost of the project during the construction period is 271.5 million yuan, the operation and maintenance cost is 5 million yuan per year, the charging cost is 7,682,800 yuan per year, and the debt service is 21,472,600 yuan per year. The cash flow inflow of the peak-shaving and valley-filling model comes from the generated revenue and peak-shaving subsidy. Among them, the generated revenue is 15,433,100 yuan per year and the peak shaving subsidy is 10,450,000 yuan per year of operation.

3.3.2 Investment benefit analysis.

Using the technical-economic evaluation method, the internal rate of return of energy storage, net present value, and dynamic payback period indicators can be measured as follows.

Table 2 Energy storage engineering technology economic evaluation indicators

Indicator	Value	Remarks
Net present value	-224,950,000	yuan
Internal rate of return	-14.49%	
Static payback period		cannot be recovered within the operation cycle
Dynamic payback period		cannot be recovered within the operation cycle

From Table 2, the NPV of the project is -224.95 million yuan, which is less than 0; the IRR is -14.49%, which is less than the benchmark IRR of 8%, making the project infeasible and uneconomic from the perspective of economic evaluation.

3.3.3 Investment benefit analysis.

After technical and economic calculations, first, the peak electricity price of energy storage needs to be increased to at least 2.30 yuan/kwh (including tax) to ensure that the NPV of the project is greater than or equal to 0. Second, a one-time subsidy of $242,428,000 in 2020 is required to ensure that the project NPV is greater than or equal to zero. Third, the initial investment in energy storage needs to be reduced to at least 22.78 million yuan, which is 91.61% lower than the original initial investment of 271.5 million yuan, in order to ensure that the NPV of the project is greater than or equal to 0. Currently, the power unit cost is 461.52 yuan/kW and the capacity unit cost is 167.82 yuan/kW. The cost of a power unit is 461.52 RMB/kW, and the cost of a capacity unit is 167.82 RMB/kWh.

3.4 Economic analysis based on peaking and frequency regulation auxiliary service model

3.4.1 Cost and revenue analysis. According to the current operation of the scenery storage and energy storage, set from the year of commissioning to use the peaking and frequency regulation auxiliary service mode, the FM mileage compensation is determined by the FM mileage, FM performance index, and mileage settlement price, and its model is:

$$R_{f1} = \sum_{t=1}^{Nf} D_t \times K_{AP} \times \lambda_1 \tag{1}$$

where R_{f1} is the FM mileage compensation; D_t is the FM mileage; K_{AP} is the comprehensive FM performance index of the FM unit on the execution date; λ_1 is the number of days of energy storage and FM operation in a year.

Referring to the transaction price of Guangdong FM mileage compensation is taken as 7 Yuan/MW. The enhancement of the original AGC regulation capacity of thermal power plants by electrochemical energy storage is taken as 2.1. thus, the daily revenue of FM service can be estimated as about 6703.2 Yuan, and the annual revenue of FM service is 2.4467 million Yuan.

The peaking compensation parameter is set according to the charging compensation of 0.05 million yuan/MWH of the standard of Southern Power Grid, and the daily operation of the energy storage battery is within the range of 10%-90% of its rated capacity. Assuming that the daily dispatch is the rated power multiplied by 0.8 factor, and the daily power generation is 53.44MWh, then the annual peaking service revenue is 752.8 million yuan. The storage charging tariff is 0.3151 yuan/kWh, the feed-in tariff is 0.7033 yuan/kWh, and the comprehensive efficiency of storage battery charging and discharging is 90%, then the annual power generation revenue is: 12,346,500 yuan and the annual charging cost is 6,146,200 yuan. The cash flow expenditure of the peaking and frequency regulation auxiliary service model can be obtained as follows: the engineering investment cost during the construction period is 271.5 million yuan, the operation and maintenance cost is 5 million yuan per year, the charging cost is 6,146.2 million yuan per year, and the debt service is 21,472.6 million yuan per year. The cash flow inflow of the peaking and frequency regulation auxiliary service model comes from power generation revenue, frequency regulation subsidy, and peaking subsidy. Among them, the generated revenue is 12,346,500 yuan per year, the frequency regulation subsidy is 2,446,700 yuan, and the peak regulation subsidy is 9,752,800 yuan.

3.4.2 Investment benefit analysis. From Table 3, using the techno-economic evaluation method, the internal rate of return of energy storage, net present value, and dynamic payback period indicators can be measured as follows:

Table 3 Energy storage engineering technology economic evaluation indicators

Indicator	Value	Remarks
Net present value	-147,870,000	yuan
Internal rate of return	-3.26%	
Static payback period		cannot be recovered within the operation cycle
Dynamic payback period		cannot be recovered within the operation cycle

From Table 4, the NPV is -147.87 million yuan, less than 0; the IRR is -3.26%, less than the benchmark IRR of 8%, which makes the project infeasible and uneconomic from the perspective of economic evaluation.

3.4.3 Break-even point analysis. After technical and economic calculations, first, the feed-in tariff of energy storage needs to be increased to at least 2.06 yuan/kwh (including tax) to ensure that the NPV of the project is greater than or equal to 0. Second, a one-time subsidy of 159.722 million yuan is needed in 2020 to ensure that the NPV of the project is greater than or equal to 0. Third, the initial investment in energy storage needs to be reduced to at least 102.99 million yuan, which is 62.07% lower than the original initial Third, the initial investment in energy storage needs to be reduced to at least 102.99-million-yuan, 62.07% lower than the original initial investment of 271.5 million yuan, to ensure that the project NPV is greater than or equal to 0. Currently, the power unit cost is 2086.43 yuan/kW, and the capacity unit cost is 758.7 yuan/kWh.

3.5 Economic Analysis Based on Capacity Tariff Model

3.5.1 Cost and revenue analysis. According to the current operation of wind and solar energy storage and storage, the capacity tariff model is set to be used from the year of commissioning to obtain both capacity and power revenue. The capacity of the wind and solar storage project is 19MW/83.5MWh, and the capacity tariff is according to the average value of 865.5 Yuan/kW/year, so the annual capacity cost is 16,444,500 Yuan. The power capacity electricity price is measured according to the average value of 0.312 Yuan/kwh, then the annual power revenue is 6,846,500 Yuan. The daily operation of the energy storage battery is within the range of 10%-90% of its rated capacity, and the FGD benchmark coal-fired tariff in North China is 0.3634 Yuan/kwh, and the charging tariff is measured at 75% of the FGD benchmark coal-fired tariff, then the annual power generation cost is 6.6380 million Yuan. The cash flow expenditure of the capacity tariff model can be obtained as follows: the engineering investment cost during the construction period is 271.5 million yuan, the operation and maintenance cost is 5 million yuan per year, the charging cost is 6.6380 million yuan per year, and the debt service payment is 21.472 million yuan per year. The cash flow inflow of the capacity tariff model comes from capacity revenue and power revenue. The capacity revenue is 16,444,500 yuan per year and the electricity revenue is 6,846,500 yuan per year.

*3.5.2 **Investment benefit analysis.*** Using the technical-economic evaluation method, the internal rate of return of energy storage, net present value, and dynamic payback period indicators can be measured as follows:

Table 4 Energy storage engineering technology economic evaluation indicators

Indicator	Value	Remarks
Net present value	-172,720,000	yuan
Internal rate of return	-5.84%	
Static payback period		cannot be recovered within the operation cycle
Dynamic payback period		cannot be recovered within the operation cycle

If there is no relevant capacity electricity policy subsidy subsequently, the NPV is -17.272 million yuan, which is less than 0; the IRR is -5.84%, which is less than the benchmark IRR of 8%, making the project infeasible and uneconomic from the perspective of economic evaluation.

*3.5.3 **Break-even point analysis.*** After technical and economic calculations, first, the electricity tariff of energy storage needs to be increased to at least 1.56 yuan/kwh (including tax) to ensure that the NPV of the project is greater than or equal to 0. Second, a one-time subsidy of 186,542,100 yuan needs to be obtained in 2020 to ensure that the NPV of the project is greater than or equal to 0. Third, the initial investment in energy storage needs to be reduced to at least 77.38 million yuan, which is 71.5% lower than the original initial investment of Third, the initial investment in energy storage needs to be reduced to at least 77.38-million-yuan, 71.5% lower than the original initial investment of 271.5 million yuan, in order to ensure that the NPV of the project is greater than or equal to 0. Currently, the cost of the power unit is 1567.55 yuan/kW and the cost of the capacity unit is 570.02 yuan/kWh.

4. Economic Evaluation of Energy Storage Investment

Through the evaluation and analysis of the economics of flexible regulation resources such as energy storage, the following conclusions are drawn:

First, the economics of different scenario modes are different, and the operating capacity of all four scenario modes is higher than that of the existing modes. The peak and frequency regulation auxiliary service model has the highest benefit, with an annual increase in revenue of 18.31 million yuan, an increase in net present value of 37.98 million yuan, and an increase in internal rate of return of 2.30% compared with the existing model.

Second, the analysis of the investment income calculation of energy storage shows that although the initial investment is significantly reduced, the NPV and IRR of the projects under the four business models are still negative. Among them, the peak and frequency regulation auxiliary service mode is the least negative, with a net present value of -144.99 million yuan and an internal rate of return of -3%, which cannot be recovered during the operation period of the project, and the project is still uneconomic. Under the existing technical conditions of energy storage and construction costs, the cost of various types of battery energy storage system is still on the high side, investment in energy storage systems do not yet have a significant economic, so it is not yet appropriate to promote battery energy storage technology on a large scale. At this stage of the development of battery energy storage, but also need the support and guidance of the government.

Third, the break-even point analysis shows that if the construction period is considered, such as the Golden Sun and other one-time government subsidies, in order to achieve the break-even of energy storage projects, the least number of subsidies required under the four commercial scenarios is the peak and frequency regulation auxiliary service mode, and the greatest number of subsidies required is the peak-shaving and valley-filling mode. If we do not consider the policy subsidies, from the perspective of improving the price of electricity, in order to achieve the break-even of energy storage projects, the lowest feed-in tariff is the capacity tariff mode and the highest is the peak-shaving and valley-filling mode. From the perspective of reducing the cost of energy storage, in order to achieve the break-even of energy storage projects, the least reduction in the unit cost required under the six commercial scenarios is the peak and frequency regulation auxiliary service mode, and the greatest reduction in the cost required is the peak and valley reduction mode.

5. Conclusions

This article explores four different scenario modes of high-proportion new energy storage and consumption. In addition, it calculates the technical and economic indicators of each scenario mode during the subsequent operation cycle of wind and solar energy storage, while other boundary conditions remain unchanged. Different scenario models have different economic benefits, and the operational capabilities of the four scenario models are higher than those of the existing models. Although the four modes can greatly improve the operational capacity of energy storage, it has not yet reached the project profitability threshold. Government subsidies and the reduction of energy storage costs are still important ways to achieve the critical profitability of the project.

Acknowledgment

This research was funded by State Grid Jilin Electric Power Co., Ltd. 2022 science and technology “Jie Bang Gua Shuai” project, grant number 2022JBGS-05.

References

1. Ssekulima, E. B., Anwar, M. B., Al Hinai, A., & El Moursi, M. S. (2016). Wind speed and solar irradiance forecasting techniques for enhanced renewable energy integration with the grid: A review. IET Renewable Power Generation, 10(7), 885-989.
2. Reboredo, J. C., & Wen, X. (2015). Are China’s new energy stock prices driven by new energy policies? Renewable & Sustainable Energy Reviews, 45, 624-636.
3. ZHAO Shuqiang, Wu Bo, Li Zhiwei et al. Operation Economics Analysis of high-proportion wind power system with wind-energy storage participating in frequency modulation [J]. Southern power grid technology, 2023 (4) : 69-76 + 89. DOI: 10.13648 / j.carol carroll nki issn1674-0629.2023.04.007.
4. Zhang Xiang, Chen Zixiang, Wang Kun and so on. Economic analysis method of electrochemical energy storage on grid side [J]. Sino-foreign Energy, 2023,28 (03) : 84-89.
5. Patrick, M., & French, N. (2016). The internal rate of return (IRR): Projections, benchmarks and pitfalls. Journal of Property Investment & Finance, 34(6), 664-669.
6. Luo Yi, Wang Gang, Wang Longjun. Research on reliability evaluation index of microgrid [J].Power system automation, 2013,37 (2) : 9-14
7. Li Chongyang. Research on reliability evaluation model of distribution network-natural gas interconnection system [D].Chongqing University, 2018.
8. Liu Shihao, Wu Gengwei, Zhao Shuqiang, Fu Zelai, Zhang Xian. Reliability evaluation of distribution network considering external power supply of microgrid [J]. Journal of North China Electric Power University (Natural Science Edition), 2019,46 (01) : 9-16.
9. Chen Ziyuan. Reliability assessment of grid-connected microgrid considering demand response [D]. Chongqing University, 2016.
10. Yu Zhicheng. Research on security evaluation model and method of independent microgrid [D]. Hunan University, 2014.

Progress in the application of Cu-based catalysts in the electroreduction of CO_2 to produce C_{xs} products

Lei Wang[1,*], Mengling Liu[2], Yichen Yu[1], Mingzhen Wei[1], Chunlei Li[2]

[1]Department of Chemical Engineering and Technology, Lanzhou University of Technology, Qilihe District, Lanzhou, Gansu, China
[2]Department of Chemical Engineering, Lanzhou University of Technology, Qilihe District, Lanzhou, Gansu, China

Abstract: Since the industrial revolution, human dependence on fossil fuels has resulted in increasingly severe atmospheric carbon dioxide (CO_2) emissions. Electrocatalytic CO_2 RR has attracted much attention because of its many advantages, such as mild reaction conditions, controllable reaction rate, low cost and scalable electrolyzer.[7] More importantly, electrocatalytic CO_2 RR can control the selective generation of target products by rationally adjusting the reaction potential, electrolyte and catalyst species. However, the CO_2 molecular structure is stable, and obtaining the key CO_2 -based anionic intermediate CO_2^- from activated CO_2 molecules requires overcoming a huge energy barrier. Copper (Cu) catalysts are the only monometallic catalysts to date that can form two-carbon (C_2) products, which establishes their special position in the field. However, the poor selectivity of Cu for a particular product has led to a wide distribution of reduction products, covering the range from carbon monoxide (CO), a reduction product at $2e^-$, to ethanol (C H_{25} OH), a reduction product at $12e^-$. Meanwhile, Cu catalysts are generally less stable, which can seriously affect their commercialization. Currently, in order to obtain highly selective C_{2+} products (mainly ethylene and ethanol), [16]this article reviews the various methods developed by different scholars in China and abroad to modify Cu catalysts. Finally, this article suggests ways to improve the performance of Cu-based catalysts to enhance the Faraday efficiency of the C_2 product from the electroreduction of $C0_2$ and gives an outlook on the future direction of Cu-based catalysts.[19]

1. Introduction

The carbon cycle of nature can no longer accommodate the excessive amount of CO_2 , and the resulting ecological problems such as global warming have had a significant impact on the natural environment and human development. Reducing greenhouse gas emissions and lowering the level of CO_2 gases in the atmosphere has become a broad consensus of the international community. Utilizing solar energy or other forms of renewable energy to create an artificial carbon cycle by converting CO_2 into valuable chemicals is an effective strategy for eliminating excess CO_2 . This not only alleviates the energy and environmental crises caused by excess carbon emissions, but also creates a carbon economy of great economic value. Currently, the utilization of CO_2 reduction reaction (CO_2 RR) based on biological, thermochemical, photochemical and electrochemical methods has been widely studied.

Compared with other methods, it is therefore very difficult to achieve the reduction of chemicals, even selective reduction, which is one of the major reasons why CO_2 RR has not been industrially applied so far. Therefore, there is an urgent need to prepare efficient and stable electrocatalysts to improve the selectivity, conversion, economy and stability of CO_2 RR. , including surface engineering, organic modification and addition of other elements to them.[14]The combination of other metal elements with copper may modulate the product selectivity by modulating the affinity of the product for carbon and oxygen, thus affecting the intermediate adsorption. However, due to the complexity of the mechanistic pathways, how many metal elements to add is still largely a matter of trial-and-error screening, which is very time-consuming. For this reason the development of high performance electrocatalysts remains one of the research priorities in this field.

2. Status of research

This chapter gives the current status of domestic and international efforts for the electroreduction of CO2 over copper-based catalysts.

2.1. Current status of foreign research

(Lin et. al., 2020) study operando time-resolved x-ray absorption spectroscopy reveals the chemical nature enabling highly selective co_2 reduction. A systematic time-resolved study can provide key insights on selective carbon dioxide electro-reduction. (Wu et. al., 2020)

* Corresponding author: 1063456257@qq.com

report that a nanodendrite configuration for Cu catalyst can improve the electrocatalytic performance of Cu catalysts, especially multicarbon product formation, while suppressing HER and methane production. [8](Arán-Ais et. al., 2020) report improved efficiency towards ethanol by tuning the morphology and oxidation state of the copper catalysts through pulsed CO_2 electrolysis. [9](Li et. al., 2021) demonstrate a facile cation exchange method to synthesize Ag/Cu electrocatalysts for CO_2 reduction using Cu sulfides as a growth template. [11](Xiao et. al., 2021) summarize recent progress that has been achieved by tailoring the electrocatalyst architecture for efficient electrochemical CO_2 conversion to the target C_2 products.[12](Peng et. al., 2021) study double sulfur vacancies by lithium tuning enhance co2 electroreduction to n-propanol. The work suggests an attractive approach to create anion vacancy pairs as catalytic centers for multi-carbon-products. (Cao et. al., 2021) offer an effective avenue to design electrocatalysts for the highly selective CO_2RR to produce multicarbon products via a tandem route. Since the initial Cu^{2+} active sites are hard to retain, there have been few studies about Cu^{2+} catalysts for CRR.[13] (Zhou et. al., 2022) propose a solid-solution strategy to stabilize Cu^{2+} ions by incorporating them into a CeO_2 matrix, which works as a self-sacrificing ingredient to protect Cu^{2+} active species. [14]

2.2. Current status of domestic research

(Shengnan Wang et al.,2023)pointed out that the electrocatalytic reduction of CO_2 is expected to allev-iate the problems of depletion of fossil fuels and in-creasing concentration of CO_2 in the atmosphere at the same time.However,there is still a great challeng-e to improve the selectivity of the high-value-added multicarbon products for electrocatalytic CO_2 reductio-n.Therefore,they designed an activation strategy for oxide-derived copper electrodes to construct the Cu_2-O-Cu interface,and synergistically promoted the elec-trocatalytic CO_2 reduction with $Cu^+$and CuO to gen-erate $C_{(2+)}$products with high efficiency and high sele-ctivity.They demonstrated that the synergistic effect of Cu^G and Cu^+at the Cu_2 O-Cu interface could inde-ed enhance the selectivity of specific high-value-added multicarbon products in the electrocatalytic reduct-ion of CO_2.[2](Yongda et al.,20,21,22,23).(Y.D.Zhang et al.,2021)showed that Cu-based nanocatalytic material-s were prepared to facilitate the reduction of CO_2 t-o C_2 products at low overpotentials.The catalysts wer-e prepared,characterized,structurally and compositiona-lly analyzed and electrochemically tested.This stabili-zes the generation of C_2 products as well as improv-es the electrocatalytic activity of Cu-based catalysts.[5](Niu Di et al.,2021)pointed out that the introduction of another metal to form a tandem catalyst on top o-f a Cu-based catalyst is an effective way to improv-e the selectivity of multicarbon products.The tandem catalyst can convert CO_2 to carbon monoxide(CO) over the introduced metal,which is subsequently redu-ced to the multicarbon product by the copper-based catalyst.Based on cuprous oxide(Cu_2O)catalysts,they introduced silver(Ag)and gold(Au)to form tandem cat-alysts with Cu_2O with the aim of achieving highly selective electrocatalytic reduction of CO_2 to C_2 prod-ucts and providing a new idea for the macro prepar-ation of nano-sized Cu_2O tandem catalysts.[6]

3. Commentary

The application of catalysts for the electroreduction of CO_2 to produce the C_{2+} product has been analyzed in detail by our scholars, and it has been found that most of the catalysts suffer from the problems of low selectivity for the C_2 product and weak catalytic activity. On the other hand, Cu is widely used as a CO_2 RR catalyst due to its moderate affinity for CO and can be used for subsequent C-C coupling, which can generate a large amount of C_2 product in aqueous solution with high efficiency and kinetic activity. [18]

Existing studies by our scholars show that the selectivity of Cu-based catalysts for highly reduced products is still low; on the other hand, Cu is the only active component reported so far that can be used for the production of highly reduced products, but it still suffers from the drawbacks of high overpotentials and low Faraday efficiencies (slow kinetics of the conversion of CO_2 to CO on Cu), which leads to the low efficiency of the CO_2 RR for the production of highly reduced products.

In this review, we summarize the recent strategies used by the scientific community (as shown in Table 1 below) including oxide and halide-derived copper nanoparticles. Specifically for the reduction of electrochemical CO_2 to polycarbon hydrocarbons and related oxygenated compounds. Obtaining C_{2+} products is a challenge because copper-based catalysts have different selectivities for different products.The researchers involved are trying to control the properties associated with copper-based catalysts to obtain the products we need.

Table 1 Kinetic activity, and is widely used as a CO_2RR catalyst.

Types	Catalyst	Electrolyte	Synthesis method	Performance parameters	Ref.
Oxide derived	cv treated copper electrode	0.1mol $KHCO_3$	Cu electrodes were prepared by PVD, then treated with Cv and electrodeposited on silicon wafers, and the prepared Cu electrodes were	Main products electrodeposited copper H_2, methane, formic acid finite elements 50%, 30% and 10%	[14]

			cycled and chemically deposited in the potential region of 0.55-0.3v, respectively.		
Halide derived	CuX (X = Cl, Br or I)	0.1 mol $KHCO_3$	Electropolished copper foil synthesized by electrophoresis in different 0.1 M potassium salt solutions	FE C_{2+}: 80 % on cuprous iodide at 0.9 V, about 66-73 % on cuprous bromide and cuprous chloride at 1.0 V; partial geometric current density: about 31.2 mA cm 2 on cuprous iodide electrode	[17]

To address such issues, this article reviews the modification strategies of Cu-based catalysts to improve the performance of the C_{2+} product from the electroreduction of CO_2 .

(1) The multi-step PCET (proton-coupled electron transfer) treatment of sequential intermediates suggests that tuning the adsorption energy of specific reaction intermediates on the catalyst surface to the appropriate level is an effective means of facilitating the selectivity of the C_2 product. In addition, the ease of C-C coupling depends on the surface coverage and adsorption sites of CO. Based on the above theoretical principles, researchers have carried out extensive modification and design of Cu-based catalysts to tune the adsorption capacity of the key intermediates.[1]

(2) Researchers can also construct environments with high CO concentrations, including designing unsaturated coordination atoms, controlling the oxidation state of Cu, preparing Cu nanocubes with different crystalline surfaces, and constructing tandem catalytic systems. The size, morphology and structure of the nanomaterials determine the average number of coordinating atoms in the catalyst.[3]

(3) The coordination number and atomic environment affect the orbital overlap between the surrounding electrons, changing the electronic structure of the active component, which can directly affect the reactivity of the catalyst and substrate. Monoatomic copper has the smallest coordination number.[4]

(4) Since the electronic structure of monoatomic Cu is fundamentally different from that of native Cu, this also leads to a general lack of C-C coupling ability of monoatomic Cu, whose electrolysis products are mainly C_1 products such as CH_3OH, CH_4 , CO. In oxide studies, it has been found that the near-surface oxygen of oxidized derivatives of Cu can reduce the σ-repulsion between CO and surface Cu atoms by extracting the electron density from the surface atoms, thus increasing the bonding strength of CO. CO dimerization is a key step in the formation of the C-C bond and is accompanied by the Cu surface after In addition, oxidized derivatives of Cu can also induce positive interactions between Cu0and Cu^+ through the formation of a mixed-valence state, thereby significantly increasing the C-C coupling capacity of CuO. The positive interaction between CuO and Cu^+ can be induced by the formation of mixed valence states, which significantly improves the selectivity of the C product.[10]

4. Summary and outlook

(1) In this article, we consider that the C-C coupling reaction, which determines the selectivity of single-carbon (C_1) or multicarbon (C_{2+}) products, is a thermocatalytic step that does not involve the transfer of protons and electrons, and thus the catalyst material and surface structure play a decisive role in it. [15]Therefore, our next research direction utilizes nanotechnology principles to design fine and rational electrocatalyst structures that can significantly affect the adsorption energies of some key intermediates and thus the inherent reaction pathways of Cu catalysts;

(2) The combination of other metal elements with Cu may affect the intermediate adsorption by modulating the product selectivity through regulating the affinity of the product for carbon and oxygen. However, due to the complexity of the mechanistic pathway, the amount of metal elements to be added is still largely a matter of trial-and-error screening. Therefore, in order to develop high-performance Cu-based electrocatalysts, our next step could be to explore composite Cu-based catalysts with better performance for the electroreduction of CO_2 product C_{2+} product by comparing the activities of different metal additions into the Cu-based catalysts.[20]

References

1. Zekun Zhang, Xiaosheng Jing, Hao Xu et al. Progress in the design and reaction mechanism of copper-based catalysts for electrocatalytic carbon dioxide reduction[J/OL]. Analytical Chemistry:1-20[2023-03-25].
2. Shengnan Wang, Dan Wang, Benqiang Tian et al. Synergistic promotion of electrocatalytic CO_2 reduction by Cu^+ and CuO at the Cu_2O-Cu interface for efficient and highly selective generation of $C_{(2+)}$
3. products (in English) [J]. Science China (Materials), 2023,66(05):1801-1809.
4. Huanhuan Yang,Shiying Li,Qun Xu. Regulatory strategies for the preparation of $C_{(2+)}$ products by

https://doi.org/10.1051/e3sconf/202344101017

electrocatalytic reduction of CO_2 with high efficienc-y over copper-based catalysts (in English)[J].Chinese Journal of Catalysis,2023,48(05):32-65.

5. Y.D. Zhang. Preparation of nano-copper-based catalysts and their electrocatalytic CO_2RR to C_2 products [D]. China University of Petroleum (Beijing), 2021.DOI:10.27643/d.cnki.gsybu.2021.001320.
6. Niu Di. Study on electrocatalytic reduction of CO_2 to C_2 products over Ag and Au doped Cu_2O catalysts [D]. University of Science and Technology of China, 2021.DOI:10.27517/d.cnki.gzkju.2021.001050
7. Meng Li, Yue Hu, Tianci Wu, Afriyanti Sumboja, Dongsheng Geng,How to enhance the C_2 products selectivity of copper-based catalysts towards electr-ochemical CO_2 reduction?-A review, Materials Today, 2023, ISSN 1369-7021,.
8. Minfang Wu; Chang Zhu; Kang Wang; Guihua Li; Xiao Dong; Yanfang Song; Jiamin Xue; Wei Chen; Wei Wei; Yuhan Sun; *"Promotion of CO2 Electrochemical Reduction Via Cu Nanodendrites"*, ACS APPLIED MATERIALS & INTERFACES, 2020. (IF: 3)
9. Rosa M. Arán-Ais; Fabian Scholten; Sebastian Kunze; Rubén Rizo; Beatriz Roldan Cuenya; *"The Role of in Situ Generated Morphological Motifs and Cu(i) Species in C2+ Product Selectivity During CO2 Pulsed Electroreduction"*, NATURE ENERGY, 2020. (IF: 5)
10. Gracita M Tomboc; Songa Choi; Taehyun Kwon; Yun Jeong Hwang; Kwangyeol Lee; *"Potential Link Between Cu Surface and Selective CO2 Electroreduction: Perspective on Future Electrocatalyst Designs"*, ADVANCED MATERIALS, 2020. (IF: 3)
11. Jinghan Li; Junrui Li; Chaochao Dun; Di Zhang; Jiajun Gu; Jeffrey J. Urban; Joel Ager; *"Copper Sulfide As The Cation Exchange Template for Synthesis of Bimetallic Catalysts for CO2 Electroreduction"*, 2021.
12. Changlong Xiao; Jie Zhang; *"Architectural Design for Enhanced C2 Product Selectivity in Electrochemical CO2 Reduction Using Cu-Based Catalysts: A Review"*, ACS NANO, 2021. (IF: 3)
13. Rong Cao; Dong-Li Meng; Meng-Di Zhang; Duan-Hui Si; Min-Jie Mao; Ying Hou; Yuan-Biao Huang; *"Highly Selective Tandem Electroreduction of CO2 to Ethylene Over Atomically Isolated Nickel-Nitrogen Site/Copper Nanoparticle Catalysts"*, ANGEWANDTE CHEMIE, 2021. (IF: 3)
14. Xianlong Zhou; Jieqiong Shan; Ling Chen; B. Xia; T. Ling; Jingjing Duan; Y. Jiao; Yao Zheng; S. Qiao; *"Stabilizing Cu2+ Ions By Solid Solutions to Promote CO2 Electroreduction to Methane"*, JOURNAL OF THE AMERICAN CHEMICAL SOCIETY, 2022. (IF: 3)
15. T.-C. Chou, C.-C. Chang, H.-L. Yu, W.-Y. Yu, C.-L. Dong, J.-J. Velasco-Vélez, C.-H.Chuang, L.-C. Chen, J.-F. Lee, J.-M. Chen, Controlling the oxidation state of the Cu electrode and reaction intermediates for electrochemical CO2 reduction to ethylene, J. Am. Chem. Soc. 142 (2020) 2857–2867.
16. W. Zhang, C. Huang, Q. Xiao, L. Yu, L. Shuai, P. An, J. Zhang, M. Qiu, Z. Ren, Y.Yu, Atypical oxygen-bearing copper boosts ethylene selectivity
17. toward electrocatalytic CO2 reduction, J. Am. Chem. Soc. 142 (2020)11417–11427.
18. B. Zhang, J. Zhang, M. Hua, Q. Wan, Z. Su, X. Tan, L. Liu, F. Zhang, G. Chen, D.Tan, Highly electrocatalytic ethylene production from CO2 on nanodefective Cu nanosheets, J. Am. Chem. Soc. 142 (2020) 13606–13613.
19. D. Gao, I. Sinev, F. Scholten, R.M. Arán-Ais, N.J. Divins, K. Kvashnina, J.Timoshenko, B. Roldan Cuenya, Selective CO2 electroreduction to ethylene and multicarbon alcohols via electrolyte-driven nanostructuring, Angew.Chem. Int. Ed. 58 (2019) 17047–17053.
20. M.G. Kibria, C.T. Dinh, A. Seifitokaldani, P. De Luna, T. Burdyny, R. Quintero-Bermudez, M.B. Ross, O.S. Bushuyev, F.P. García de Arquer, P. Yang, Asurfacereconstruction route to high productivity and selectivity in CO2 electroreduction toward C2+ hydrocarbons, Adv. Mater. 30 (2018) 1804867.

Effects of late-Holocene fires on the carbon sequestration capacity of peatlands in the Sanjiang Plain

Lu Zhang*

School of Geographic and Environmental Sciences, Tianjin normal University, Beichen, Tianjin, China

Abstract. Peatland fires are frequent, but the effect of fire on soil organic carbon in peatlands is still currently variable. Especially at long time scales, it has not been clarified. In order to reveal the relationship between carbon accumulation patterns in peatlands at long time scales in response to local fire changes, high-resolution peat sediments from peatlands in the Three River Plain were selected and analysed for stratified charcoal debris and organic carbon in this paper. The results showed that the stronger carbon sequestration capacity in the periods of 0.6, 0.75, 1.25, 2.3, 2.75, 2.9, 3.6 and 5 ka BP corresponded to weaker fire intensity or even no fire. Late-Holocene fires in the Sanjiang Plain reduced the rate of carbon sequestration, suggesting that fires have been an important driver of the reduction in carbon sequestration capacity over the last 6000 years.

1 Introduction

In recent decades, frequent fires have been occurring in several countries as a result of global warming and the increasing impact of human activities. Frequent fires can significantly alter the global carbon balance by affecting vegetation community succession, soil structure and nutrient cycling. The long-term effects of fires on ecosystem-atmosphere carbon exchange depend mainly on biogeochemical processes such as accumulation and depletion in vegetation and soils after fires[1]. Of these, carbon emissions from biomass burning account for only a fraction of the overall fire emissions. Soil, as one of the largest carbon reservoirs on the Earth's surface, contributes 2,344 Pg of global carbon[2]. The accumulation of carbon-containing sediments in soils is the long-term result of an imbalance in the production and decomposition of vegetation biomass[3]. Wildfires, by burning vegetation, not only affect soil carbon content during combustion but also change the previous state of carbon accumulation long after combustion[4].

Most of the current studies only compare the changes in soil carbon content before and after a single fire, and there is a lack of long-term fire and soil carbon dynamics records, which can lead to uncertainty in the effects of fire on soil carbon sequestration. In addition, a large number of studies have attempted to reveal the effects of fire regimes on surface organic carbon. Some studies have shown that fire regimes have a significant negative impact on surface carbon sequestration, e.g. a study in the Sydney Basin, Australia, found a negative correlation between forest carbon "sinks", AGC, and frequent and severe wildfires, suggesting temperature-mediated reductions in forest carbon stocks under projections of future climate change[5]. In contrast, another subset of studies yielded little to no effect at millennial scales, e.g., peatlands in northern North American peatlands are likely to be less sensitive to burning, and the lack of a clear correlation between fire regimes and carbon accumulation suggests that fire regimes are not a driver of carbon sequestration at millennial time scales[6]. Even the opposite trend exists, as soil organic carbon (OC) content increased in the first two growing seasons after fire in the northeastern part of the Three Rivers Plain[7]. Similarly, these studies usually attribute the effect of fire on soil carbon accumulation to climate change and to changes in vegetation cover and soil water content in response to climate change. Therefore, it is crucial to select records of prolonged fire and soil organic carbon changes from climate-sensitive areas. The Sanjiang Plain is located in northeastern China and is sensitive to climate change due to its mid-to-high latitude and monsoon edge position[8]. In this study, we quantify the charcoal debris and soil organic carbon content based on peat sediments from the Sanjiang Plain, and then explore the effects of late Holocene fires on long-term changes in carbon accumulation.

2 Study region

The Sanjiang Plain (129°11′-135°05′ E, 43°49′-48°27′ N) is located in the northeastern Heilongjiang Province, China (Fig. 1). It is a huge alluvial plain crossed by three major rivers (Heilong River, Wusuli River and Songhua River), with a total area of 10.9×106 ha, an altitude of <200 m and a slope grade of <1:10,000. The present climate of the plain is temperate humid or sub-humid

*Corresponding author: zl1965774519@163.com

https://doi.org/10.1051/e3sconf/202344101018

continental monsoonal. Until recent decades, fresh-water wetlands covered over 70% of the plain.

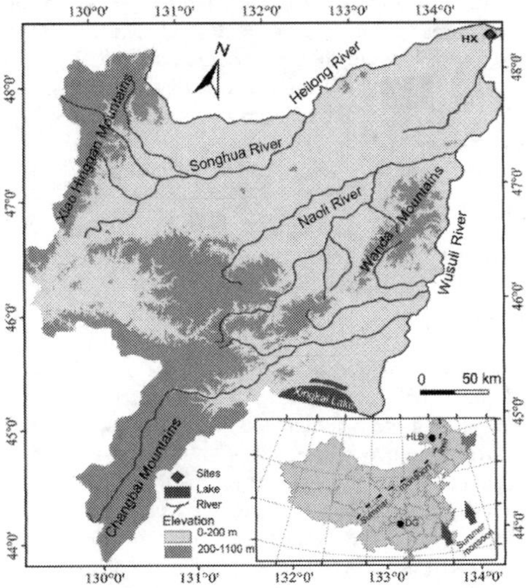

Fig. 1. Digital elevation model of the Sanjiang Plain.The solid diamond in black color indicates the sampling site[9].

3 Materials and research methods

The rock profile studied, HE (47° 35.096'N, 133° 30.006' E, 71 m a.s.l.), has a thickness of 148 cm, and all samples were collected at 1 cm thick intervals, for a total of 148 samples. As the results of sample collection and dating have been published in a separate article, they will not be repeated here[9](fig. 2). In addition, previous articles have conducted experiments related to the samples in this study, which provided charcoal debris and organic carbon data to support this study[9].

One hundred and forty-eight (148) 1 cm interval samples collected from cores were used for loss on ignition (LOI) and charcoal debris analyses. For LOI, samples with a volume of 2 cm^3 were sequentially burned in a muffle furnace at 500 °C and 900 °C to estimate organic matter and carbonate content. For carbon chip analysis samples were processed according to standard acetylation procedures. For each sub-sample with a volume of 2 cm^3 of air-dried sediment, 10 ml of 18% hydrochloric acid was added to remove carbonates. Then 10 ml of 10% NaOH was added and the solution was heated in a water bath at 90 °C for 30 min to remove organic matter. The residue was kept in 5 ml of 45% HF for 12 hours to remove silicates. After this treatment, the charcoal chips were separated from the macroscopic (> 125 μm) and microscopic (< 125 μm) fractions by means of a wet sieve and the charcoal and pollen pellets were concentrated using an ultrasonic cleaner and centrifuge. Slides were then prepared for mounting pollen and microscopic (< 125 μm) charcoal chips. Charcoal chips were identified and counted using an Olympus light microscope, and the total number of microscopic (< 125 μm) charcoal particles exceeded 300. All macroscopic charcoal particles (> 125 μm) from each sample were identified and counted under a stereomicroscope at 80x magnification. To calculate the charcoal concentration, a known number of exotic Staphylococcus spp. spores were added to each sample prior to pretreatment and the charcoal concentration of both fractions was calculated with the aid of Staphylococcus spp. spore counts.

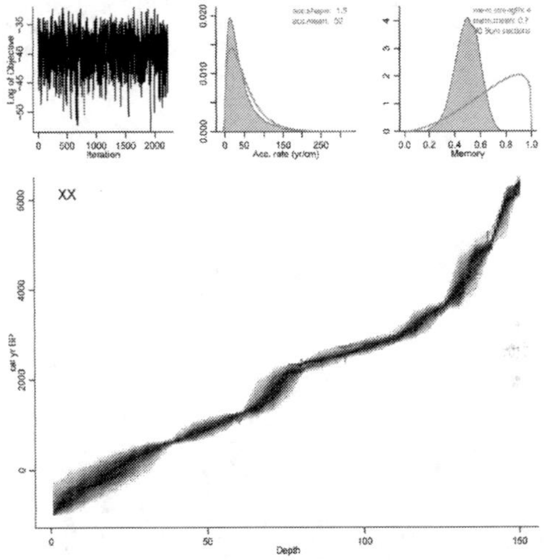

Fig. 2. Age-depth modelling of HE[9]

4 Discussion and analysis of results

The age-depth model shows that the sediment deposition process of HE is approximately 6000 years old. The charcoal debris in the sediments can indicate the scale and extent of fire occurrence, and the evolutionary history of fires during the geological history can be reconstructed by the different grain size contents of the charcoal debris. Typically, after a fire, a portion of the fine charcoal debris travelled downwind as the smoke rose into the air, while the coarse-grained charcoal debris was deposited in situ or in close proximity. In terms of propagation and deposition patterns, coarse-grained charcoal debris travelled shorter distances and was dispersed closer to the source area; in contrast, fine-grained charcoal debris was more easily transported over long distances by air currents, resulting in its dispersal in areas farther away from the source area. Therefore, the cumulative velocity of charcoal debris of different grain sizes in the depositional profile indicates different ranges of fire intensity[10](Fig. 3). Usually, the particle size of charcoal debris is divided by 125 μm, and it is believed that the large charcoal debris >125 μm is mostly deposited in the area of less than 3 km, which can reflect the local fire events, while the small charcoal debris <125 μm is transported by air currents over long distances, which can be indicative of the regional fire events. As shown in Fig. 3, the accumulation rates of large, small and total charcoal debris are generally similar. In addition, according to the trend of the accumulation rate of carbon debris of different grain sizes, the fire intensity can be classified as high or low, with stable changes in the periods of 6-5.7 and 3.7-1.5 ka BP, and lower fire intensity, and larger fluctuations in the periods of 5.7-3.7 and 1.5-0 ka BP, which resulted in higher fire intensity.

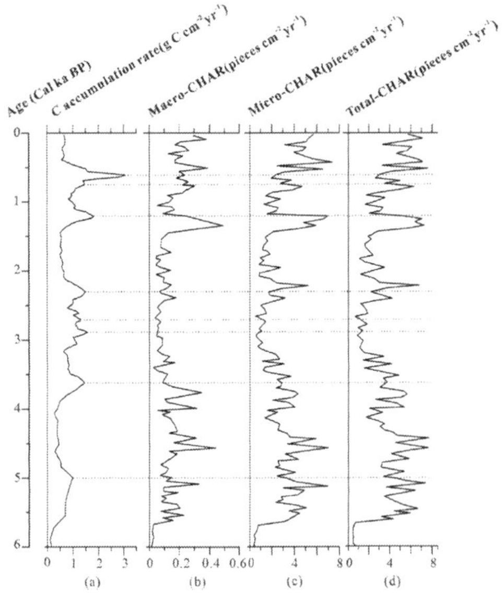

Fig. 3. C accumulation rate(a), charcoal analysis results(b,c and d) from the studied profile HE.

The apparent rate of organic carbon accumulation in the Late Holocene varied from 0.02 g C $cm^{-2}yr^{-1}$ to 3.07 g C $cm^{-2}yr^{-1}$, with an average accumulation rate of 0.95 g C $cm^{-2}yr^{-1}$. According to the rate of organic carbon deposition, it can be used as an indication of the strength of the soil's capacity to sequester organic carbon, and the rate of organic carbon sequestration in the periods of ~6-4, 2.2-1.25 ka BP tended to change steadily, with the capacity of sequestration being relatively weak, while that in the periods of 4-2.2, 1.5-0 ka BP fluctuated more, with the intensity of fire being generally lower. -2.2, 1.5-0 ka BP period changes are more dramatic, the overall carbon sequestration capacity is stronger. In the periods of 0.6, 0.75, 1.25, 2.3, 2.75, 2.9, 3.6, and 5 ka BP, the carbon sequestration capacity was strong, and the intensity of fires was weak or even did not occur. From this, we infer that the Late Holocene fire events in the Sanjiang Plain were unfavourable for organic carbon accumulation. It is widely believed that fires directly burned forest vegetation, causing the reduction of tree productivity and vegetation loss, which directly reduced the carbon sequestration capacity of vegetation in forest ecosystems[11]. At the same time, fire reduces the carbon pool of apoptosis and accelerates the decomposition of apoptosis by directly affecting the amount of apoptosis and indirectly affecting the decomposition rate of apoptosis. In addition, the effect of fire on forest soil carbon pools was manifested in the increase of soil organic matter decomposition, the increase of soil respiratory carbon release, the decrease of carbon input to the soil from above-ground vegetation, and the increase of black carbon as a carbon sink[12]. Therefore, we concluded that the occurrence of Late-Holocene fires in the Sanjiang Plain reduced soil carbon sequestration capacity.

The response of fire to carbon sequestration capacity varies across geography. It is generally accepted that fire depletes carbon sequestered at the surface through direct combustion and net post-fire emissions, but the effects of fire in marshy wetlands may be different. Due to the presence of a certain water table elevation in marshes, it is difficult for wildfires to penetrate the water table and cause damage to carbon stocks in the soil. Thus, the relatively low fire frequency in marshy wetlands, coupled with the dominance of surface fires, allows fire occurrence to play a special role in the long-term carbon accumulation process. In herbaceous-dominated wetland ecosystems, wildfires typically consume most of the biomass above the water table only through intensive, continuous surface fuel movement. Instead, dead plant roots remaining below the surface are decomposed by soil animals and incorporated into the soil. In addition, wildfires increase species diversity and the abundance of marsh soil formers, thereby increasing the amount of plant carbon accumulated on the surface of marsh soils by plant dieback. Thus due to fire-induced increases in surface biomass and subsurface carbon pools (from dead roots), more frequent and severe fires result in increased soil carbon content.

In addition to the direct effects of fire on the process of carbon accumulation in marshy wetland soils, large quantities of fire products (e.g., charcoal debris and charcoal black) can potentially affect carbon accumulation due to their physicochemical properties (e.g., recalcitrance and high surface area). Fire products in wetland soils not only act as the most stable portion of the wetland carbon pool, but medium to high levels of fire products can also promote plant growth and produce more carbohydrates from plant dieback residues in the soil. In addition, the high presence of carbon black in wetland soils makes it easier to retain stabilised carbon (e.g. aromatics) in the soil. And the high level of aromatic content means that the soil has higher stability. Therefore, we suggest that wetland fires not only promote the accumulation of soil carbon, but also the residual fire products increase the stability of soil carbon in marshy wetlands, which is very favourable for soil carbon sequestration on long time scales.

5 Conclusion

This study deepens our understanding of soil carbon sequestration in marshy wetlands as affected by fire. Based on the Late Holocene sedimentary charcoal debris and organic carbon records in the Sanjiang Plain, the studied aspects of fire conditions were linked to the rate of carbon accumulation in soils. The changes were more stable during the periods ~6-5.7, 3.7-1.5 ka BP its fire intensity was overall lower, and fluctuated more during the periods ~5.7-3.7, 1.5-0 ka BP, with an overall higher fire intensity. However, the change in the period of ~6-4, 2.2-1.25 ka BP tends to be stable with weaker carbon sequestration capacity, and the change in the period of 4-2.2, 1.5-0 ka BP is more drastic with overall stronger carbon sequestration capacity. In the periods of 0.6, 0.75, 1.25, 2.3, 2.75, 2.9, 3.6, and 5 ka BP, the carbon sequestration capacity was stronger, and the corresponding fire intensity was weaker or even no fire occurred. Thus, it can be demonstrated that fire in marshy wetlands greatly reduces long-term soil carbon sequestration.

References

1. Kaye, J. P., Romanyà, J., Vallejo, V. R. (2010) Plant and soil carbon accumulation following fire in Mediterranean woodlands in Spain. Oecologia, 164: 533-543. doi: 10.1007/s00442-010-1659-4.
2. Jobbágy, E. G., Jackson R. B. (2000) The vertical distribution of soil organic carbon and its relation to climate and vegetation. Ecological applications, 10(2): 423-436.doi:10.1890/1051-0761(2000)010[0423:TVDOSO]2.0.CO;2.
3. Yu, Z., Loisel J., Brosseau D. P., Beilman, D. W., Hunt S. J. (2010) Global peatland dynamics since the Last Glacial Maximum. Geophysical research letters, 37(13), L13402. doi:10.1029/2010GL043584.
4. Armas, C., Pérez Lambá, F., Badía-Villas, D., Peña Monné, J. L., González-Pérez, J. A., Picazo Millán, J. V., Jiménez, N. T., Sampietro , M. M., Gracia, M. A. (2019) Pyrogenic organic matter from palaeo-fires during the Holocene: A case study in a sequence of buried soils at the Central Ebro Basin (NE Spain). Journal of Environmental Management, 241, 558-566. doi:10.1016/j.jenvman.2018.09.104.
5. Gordon, C. E., Bendall E. R., Stares M. G., Collins L., Bradstock R. A. (2018) Aboveground carbon sequestration in dry temperate forests varies with climate not fire regime. Glob Chang Biol, 24, 4280-4292. doi:10.1111/gcb.14308.
6. van Bellen, S., Garneau M., Ali A. A., Bergeron Y. (2012) Did fires drive Holocene carbon sequestration in boreal ombrotrophic peatlands of eastern Canada? Quaternary Research, 78(1), 50-59. doi:10.1016/j.yqres.2012.03.009.
7. Zhao, H., Tong D. Q., Lin Q., X. Lu X., Wang G. (2012) Effect of fires on soil organic carbon pool and mineralization in a Northeastern China wetland. Geoderma, 189-190, 532-539. doi:10.1016/j.geoderma.2012.05.013.
8. An Z S. (2000) The history and variability of the East Asian paleomonsoon climate. Quaternary Science Reviews, 19(1-5): 171-87. doi:10.1016/s0277-3791(99)00060-8.
9. Zhang Z., Zhong J., Lv X., Tong S., Wang G. (2015) Climate, vegetation, and human influences on late-Holocene fire regimes in the Sanjiang plain, Northeastern China. Palaeogeography Palaeoclimatology Palaeoecology, 438: 1-8. doi:10.1016/j.palaeo.2015.07.028.
10. Peters M E., Higuera P E. (2007) Quantifying the source area of macroscopic charcoal with a particle dispersal model. Quaternary Research, 67(2): 304-310. doi:10.1016/j.yqres.2006.10.004.
11. Beaulne, J., Garneau, M., Magnan, G., Boucher, É. (2021) Peat deposits store more carbon than trees in forested peatlands of the boreal biome. Scientific Reports, 11(1), 2657. doi:10.1038/s41598-021-82004-x.
12. Magnan, G., Garneau, M., Le Stum-Boivin, E., Grondin, P., Bergeron, Y. (2020) Long-Term Carbon Sequestration in Boreal Forested Peatlands in Eastern Canada. Ecosystems, 23, 1481-1493, doi:10.1007/s10021-020-0048.

Production technology of adiponitrile

Xuan Luo*, Jing Dou, An Bai, Lei Wang, Mingzhen Wei

School of Petrochemical Engineering, Lanzhou University of Technology, Qilihe District, Lanzhou, Gansu, China

Abstract: This paper introduces the technical routes and characteristics of four traditional adiponitrile production processes, including acrylonitrile dimerization method, butadiene method, adipic acid catalytic ammonification method and caprolactam method, were introduced and compared in terms of reaction principle, process, production cost and environmental protection. The production process with low cost, low energy consumption and good environmental benefits was found to provide reference for the production and development of adiponitrile in China.

1. Introduction

Adiponitrile (ADN), also known as 1,4-dicyanobutane, has the molecular formula of $NC(CH2)_4CN$. It is a colorless transparent oily liquid with a slight bitter taste at normal temperature and pressure. The substance is flammable and toxic, and it is mostly stored in a cool and ventilated warehouse[1].

Adiponitrile undergoes addition reaction with H_2 under the action of a related catalyst to generate hexamethylene diamine(HMD)[2], and then hexamethylene adipate (nylon 66 salt) can be generated by adipic acid and hexamethylene diamine through strict material ratio, which is the most important industrial use since its development[3].Due to the gradual enhancement of people's awareness of environmental protection, hexamethylene diisocyanate (HDI), a light industrial product with the characteristics of non-yellowing, color preservation, light preservation and sun exposure resistance, is increasingly favored by people. It is found that HDI can be synthesized from hexamethylene diamine by phosgene method, which is another very important use in the downstream products of adiponitrile at present[4]-[5]. In addition, adiponitrile is also widely used in electronic batteries, organic synthesis and other fields, and is also used to make rocket fuel, washing additives, pesticides and so on [6].

At present, the main production processes of adiponitrile include electrolytic dimerization of acrylonitrile (AN), catalytic ammonification of adipic acid (ADA), butadiene (BD) and caprolactam degradation[7]. Adiponitrile production technology is currently in the hands of a few large foreign companies, foreign companies monopolize the supply of raw materials in China, seriously restricting the development of nylon 66 and related industries in China, this paper introduces four traditional adiponitrile production technology routes and puts forward the future research direction of adiponitrile production technology according to the current situation of the domestic adiponitrile market.

* Corresponding author: 3523409775@qq.com

2. Acrylonitrile dimerization

2.1 Electrolytic dimerization of acrylonitrile

The electrolytic dimerization of acrylonitrile is that propylene is oxidized in the presence of ammonia to produce acrylonitrile, and acrylonitrile is polymerized and dimerized at the cathode to produce adiponitrile. Its production process is mainly divided into diaphragm electrolysis method and diaphragm-free electrolysis method, while diaphragm electrolysis is divided into solution method and emulsion method.

As early as 1961, Monsanto Company of the United States for the first time used the solution method in the diaphragm electrolysis method to prepare adiponitrile by electrolysis of the cathode liquid consisting of 20-40% of the mass fraction of acrylonitrile and electrolyte salts. In the later period, Asahi Kasei Company of Japan used emulsion method to electrolyze low-concentration acrylonitrile solution with cathode solution composed of high-conductivity salt (tetraethylamide phosphate)and emulsifier (polyvinyl alcohol) to produce adiponitrile and realize industrialization. Later, the Belgian Chemical Joint Research Company found that the main reaction to generate adiponitrile occurred at the cathode, and the anode reaction did not interfere with each other, so it took the lead in canceling the diaphragm and developing the direct electrolytic dimerization method (diaphragm-free method). Germany's BASF modified the electrolytic bath and adopted a new capillary electrolytic bath with PbO_2-plated graphite as the anode and graphite as the cathode, eliminating the ion exchange membrane [8].Because the diaphragm-free electrolysis method further amplifies the advantages of electrolytic preparation of adiponitrile in terms of cost and three wastes treatment, it has the concept of green chemical industry [9].At this point, the diaphragm-free electrolysis method began to develop[10].Later generations continued to study and devote on this basis, and committed to improving the yield of adiponitrile and the current efficiency of the electrolytic cell. Luo Zhonglin[11] used lead as cathode and carbon steel as anode to electrolytic dimerization of acrylonitrile into adiponitrile in a diaphragm-free electrolytic cell, and discussed the basic conditions such as the optimum acrylonitrile content, tetraethyllammonium hydroxide content, temperature and PH. The process flow of electrolytic dimerization of acrylonitrile is shown in Figure 1.

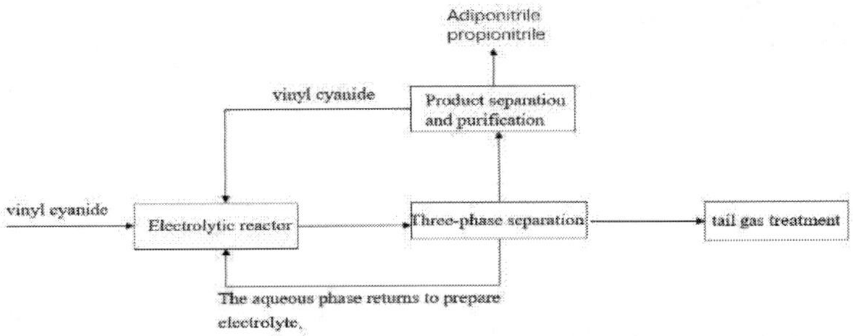

Fig. 1 Process flow of electrolytic dimerization of acrylonitrile

2.2 Acrylonitrile catalytic dimerization

Acrylonitrile catalytic dimerization is a commonly used organic synthesis method, the principle is by adding the catalyst to the acrylonitrile reaction solution, the dimerization between acrylonitrile molecules to form high polymers. This method is widely used in organic synthesis and can be used to prepare various organic compounds, such as polymers, drugs, fuels, etc.

Acrylonitrile catalytic dimerization has become an important part in the field of modern industrial chemical reaction because of its advantages of simple reaction, high yield and controllable molecular weight distribution. By optimizing for different influencing factors, the reaction efficiency and product quality can be improved to the greatest extent, which provides strong support for the development of related industries.

3. Butadiene method

3.1 Cyanidation of butadiene

Butadiene cyanide chlorination is also known as the "two-step method" [12]-[14]. In the first step, 1, 3-butadiene and chlorine gas undergo conjugate addition reaction to produce 1,4-dichloro-2-butene, In the second step, 1,4-dichloro-2-butene and metal cyanide (preferably sodium cyanide) undergo SN2 nucleophilic substitution reaction to produce 1,4-dichloro-2-butene. Finally, with RandyNi as catalyst, 1,4-dichloro-2-butene was added to hydrogen to produce adiponitrile.

3.2 Butadiene direct cyanide method[15]

In the 1970s, DuPont Company of the United States developed the butadiene direct hydrocyanide process based on the butadiene cyaniding process, and established the first production plant in Texas and realized industrialization[16].In this method, 1, 3-butadiene and hydrocyanic acid are reacted in liquid phase at 100 ℃ in the presence of catalyst [17] composed of zero-valent nickel and phosphate-containing ligands to produce an isomer mixture of pentenenitrile. After separation and isomerization of the isomer into linear pentenenitrile, the isomer is then added to hydrocyanic acid to produce adiponitrile. The actual reaction process is divided into three steps: primary hydrocyanidation, isomerization and secondary hydrocyanidation [18]-[20].

The process flow of this method is shown in the figure 2:

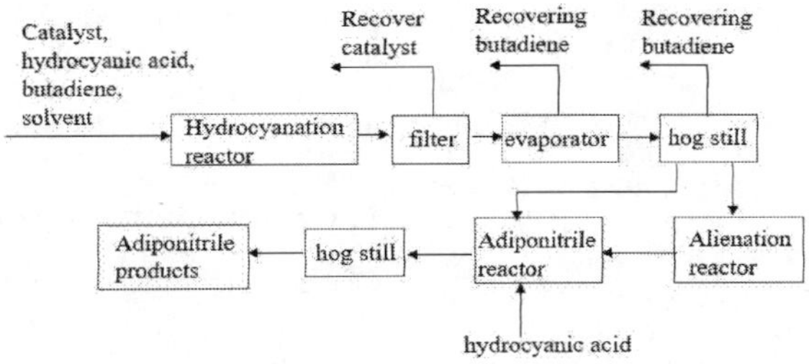

Fig. 2 Process flow of butadiene direct cyanidation

Butadiene, hydrocyanic acid, solvent andcatalyst (complex composed of phosphorus-containing ligand and zero-valent Ni) are added into a reactor with a stirrer for hydrocyanation reaction, the temperature is about 100 ℃, and the reactant is kept in the liquid phase with sufficient pressure, which is about 6.8atm[21]. Pure 3- pentenenitrile (3-PN) obtained by rectification is sent to secondary hydrocyanation, and 2- methyl -3- butenenitrile (2M3BN) distilled from the top of distillation tower is sent to isomerization reaction section.

The isomerization reaction section is also homogeneous reaction, the reaction temperature is 80-120 ℃, the isomerization reactor and the catalyst and intermediate products obtained in the filter and evaporator reaction to obtain 4-PN and 3-PN, and the single-pass conversion rate of 4-PN and 3-PN is 26.4% and the selectivity is 79.8%.

The raw materials for the second-stage hydrocyanation reaction are HCN and 3-PN generated by the first-stage hydrocyanation and isomerization reaction. The 3-PN, 4-PN, hydrocyanic acid and aromatic solvent are put into the adiponitrile reactor for hydrocyanation reaction to generate adiponitrile, and then the final product adiponitrile is obtained through the refining system[22].

3.3 Butadiene carbonylation method

Butadiene first reacts with carbon monoxide and methanol under the action of catalyst to produce 1,6-methyl adipate, and then through ammonolysis and dehydration under the action of catalyst to synthesize adiponitrile, in which the by-product methanol can be recycled as raw material.

4. Adipic acid catalytic ammonification method

The production process of adipic acid (ADA) catalytic ammonification was first developed and successfully by Rhone-Planck in France in the late 1960s, and then industrialized production was formed by BASF, Lattice Chemical Plant and Liaohua in China[23].The process is divided into liquid phase method and gas phase method. The reaction principle of the two methods is basically the same, both are based on phosphoric acid or its salts or cheese as the catalyst, adipic acid and ammonia react to produce diammonium adipate, and then heat and dehydrate to obtain crude adiponitrile, after refining to get the finished product. The reaction formula is as follows:

The production process flow of adipic acid is shown in Figure 3:

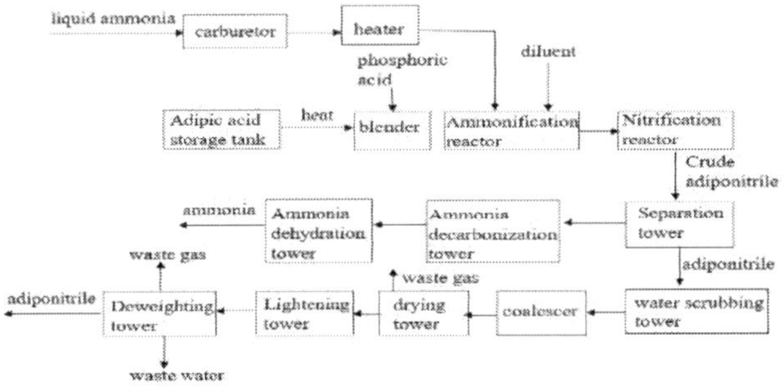

Fig. 3 Process flow of catalytic ammonification of adipic acid

4.1 Adipic acid liquid phase method

The liquid phase method is about 200 ~ 300 ℃ molten adipic acid, ammonia gas, catalyst (phosphoric acid or acid salt) and diluent (semi-nitrile compound) mixed into the reactor for ammonification reaction. After washing, settling and filtering, the organic layer is mainly composed of crude adiponitrile and water. The bottom product of the tower is a semi-nitrile compound containing adipic dinitrile. Part of the liquid phase from the rectification column enters the scraper film evaporator for ammonification and dehydration. All the products at the top of the evaporator are used as diluents, and the residue is discharged from the bottom[24]. The crude adiponitrile enters the refining system to obtain the product adiponitrile. The liquid phase method has a long history but the product quality is poor. The yield of adiponitrile is 84% ~ 93% when diluent is used. The yield of adiponitrile was 65%-73.5% without diluent.

4.2 Adipic acid gas phase method

The reaction temperature of adipic acid gas phase method is 350 ~ 420 ℃, the powder adipic acid is vaporized, the ammonia gasification dehydration reaction is carried out, and the yield of adipic acid is 92% ~ 96%. After further treatment. The recovered ammonia goes into the reactor for recycling, and the final product is obtained by the refining section of adiponitrile. The gas phase method is divided into BASF method and Monsanto method, the main difference is that the reactor type is different.

5. Caprolactam method

Caprolactam method[25] was developed by Toray Company in Japan to produce adiponitrile from waste lactam as raw material by first degrading lactam and then hydrolyzing it. The scale of the law is small, and the products are only used by the downstream products of his

 https://doi.org/10.1051/e3sconf/202344101019

company. The process focuses on the recycling of waste raw materials, which is a good development idea, and the production cost has been reduced, but it cannot be large-scale production due to the lack of raw materials.

6. Comparison of adiponitrile production process

In summary, the four production process pairs of adiponitrile are shown in Table 1 below:

Table 1 Comparison table of adiponitrile production process

Project	Acrylonitrile method		Butadiene method		Adipic acid process	
	diaphragm valve	Diaphragm-free method	Chlorination cyanidation method	Direct cyanidation method	liquid phase method	Gas phase method
source of feed	extensive	extensive	extensive	extensive	extensive	extensive
Raw material cost	tall	tall	tall	low	tall	tall
Technological process	common	common	complex	common	complex	complex
Energy consumption	tall	lower	tall	lower	common	common
production capacity	small scale	small scale	large scale	large scale	moderate scale	moderate scale
product quality	common	tall	common	tall	common	common
yield coefficient	lower	tall	higher	tall	lower	lower
environmental protection	heavy pollution	heavy pollution	serious pollution	Pollution is average	pollution is average	pollution is average
investment	higher	higher	tall	lower	lower	lower

Through the comparative analysis of the production process route of adiponitrile, it can be seen that the acrylonitrile method has a wide range of raw materials, and the diaphragm-free electrolysis method has the characteristics of low energy consumption, high product quality and high yield, which is a very promising process route, but due to the current market price of raw material acrylonitrile and the difficulty of catalyst research and development, there is no fundamental breakthrough, and it is urgent to develop efficient and low-cost catalysts[26]; In the process route of butadiene process, the cyanidation process of cyanidation is complex, the corrosion is serious, the investment is large, and a large amount of chlorine and hydrocyanic acid need to be consumed, which is basically eliminated. Direct cyanidation is the most advanced production technology in the industrial adiponitrile plant, suitable for large-scale industrial production, and is also a good way to treat hydrocyanic acid by-product of acrylonitrile plant, but its technology is monopolized by DuPont; Butadiene carbonylation method has the advantages of simple process, low production cost and clean process, and is a very promising production route for adiponitrile, but it has not yet achieved industrial equipment, and this technology deserves the attention of domestic enterprises[27]; The adipic acid catalytic ammonification method has a long process route, and the market price of adipic acid before 2012 was high, which was not an ideal process route at that time, but with the overcapacity of adipic acid, the price has fall en sharply, and the adipic acid catalytic ammonification method is also an important production process[28].

7. Conclusion

With the development of China's economy, the demand for adiponitrile will increase, although INVISTA may expand production globally, but it will leave about 70% of its own use, and the international market will remain tight in supply for a long time. In order to ensure the healthy and stable development of China's nylon 66 industry and its downstream related fields, it is urgent for China to develop and build its own adiponitrile production equipment[29].

Before 2019, there was no enterprise in China that could produce adiponitrile on a large scale, and all the required adiponitrile relied on imports. In 2020, domestic Huafeng Group took the lead in breaking through the industrial production technology of adiponitrile and put into production of 50,000 tons[30].

According to statistics, the raw material cost accounts for 80%~85% of the production cost of adiponitrile, which has a decisive role in the economy of the process technology route. With the change of supply and demand in the market, the price of adipic acid was high before 2012, resulting in the lack of competitiveness of the adipic acid process[31]. On the whole, the domestic should combine the raw material supply and demand market situation, coordinate the layout, on the basis of accelerating the existing butadiene cyanide route and acrylonitrile electrolysis technology research, under the condition that the domestic adipic acid price remains low, it is feasible to re-enable the adipic acid process route to produce adiponitrile in China, and realize the industrial production and technological autonomy of adiponitrile in China as soon as possible.

References

1. Guangsheng Wang. Petrochemical raw materials and product safety manual[M]. Beijing: China Petrochemical Press,2010.
2. Huasen Zhang. Process and kinetic study of hydrogenation of adiponitrile to the preparation of hexamethylenediamine[D]. Henan: Zhengzhou University,2011.
3. Qinghua Tu. Analysis of adiponitrile market and raw material route[J]. Chemical Industry,2012,30(12):26-30.
4. Mingkai Zhang. Synthesis of hexamethylene diisocyanate (HDI) by non-phosgene method[D]. Hunan: Hunan University,2005.
5. Peng Chang, Jun Yang. New method for preparing HDI and IPTS[J]. Materials Science and Technology,2008,16(6):843-846.
6. Qingen Yu. Analysis of production process of adiponitrile[J]. China Chemical Trade.2011, (9):16-18.
7. Yishan Zhao. Analysis of adiponitrile production technology and market trend[J].Hebei Chemical Industry,2012,35(2):34-36.
8. Jun Xi, Xinsheng Zhang. Simulation and optimization of acrylonitrile recovery process of adiponitrile by electrolysis[J]. Chemical Production and Technology,2006, (4):34-36.
9. Yunhua Chen, An Lin, Fuxing Gan. Research status of chemical modification method of nanoparticles[J].China Surface Engineering,2005,(2):5-11.
10. Lihai O, Genshi Luan, Production of hexanephene by electrolytic dimer method of acrylic eye[J].cultivation. Journal of Fushun Petroleum Institute,1994,33(1):19-24.
11. Zhonglin Luo. Research on the process of electrolytic dipolymerization of acrylonitrile to adiponitrile[D].Hangzhou: Zhejiang University of Technology,2015.
12. Qibo Liu, Guangfei Huo, Jian Tong, et al. Technology development status of synthesis of adiponitrile and hexamethylenediamine from butadiene[J]. Chemical Industry and Engineering Progress,2009,28 (5): 832-835.
13. FREURE B T , DECKER H J . Preparation of nitriles : US3133114[P].1964-05-12.
14. Chong Liu. Basic Organic Raw Materials of Petrochemical Handbook No. 3[M].Beijing: Chemical Industry Press.1987:556-561.
15. Dongkui Deng, Yuemin Jiang, Xicai Zhou. Progress in process technology for the production of adiponitrile by butadiene direct cyanide[J].Chemical Design Communications,2022,48(05):62-64.
16. Fangzhen Sun. Production process of butadiene direct cyanide to hexenediamine[J]. SyntheticFiber Industry,1985,(5):43-46.
17. FREURE B T , DECKER H J . Preparation of nitriles : US3133114[P].1964-05-12.
18. Jianxun Dong, Xiaoyan Feng, Huawen Wu, et al. Research progress on directed addition of conjugated diolefins[J].Organic Chemistry,2011, 31(9) : 1357-1368.
19. Jianxun Dong, Yukun Shi, Xiaojing Hu, et al. Equilibrium reaction of 3-pentenenitrile in zero-valent nickel catalyst catalytic system[J].BLfL#, 2012, (3) : 627-631.
20. BINI L,MULLER C,WILTING J,CHRZANOWSKI L,et al.Highly selective hydrocyanation of butadiene toward 3-pentenenitrile[J]. Journal of the American Chemical Society,2007,129(42):12622- 12633.
21. Wanglin Yu. Adiponitrile production process:101676261A[P].2010-03-24.
22. Zengchao Gao. Research and optimization of preparation of adipsitrile by butadiene hydrocyanide[J].China Petroleum and Chemical Industry Standards and Quality,2023,43(12):183-185.
23. Yuan Ma, Baowei Yu, Haiyan Zhang. Comparison of production process of adiponitrile[J].Henan Chemical Industry,2007,24(8):4-6.
24. Honglin Zhang, Anchao Geng. Environmental Engineering,1995,(4):44-46.
25. Lun Zhang, Zhonglong Ma. Research progress of caprolactam technology by butadiene route[J].Chemical Industry and Engineering,2004,21(1):48-52.
26. Honglin Zhang, Anchao Geng. Environmental Engineering,1995,(4):44-46.
27. Ying Meng, Qiuhua Yuan, Yanhui Han, et al. Industrialization status and development suggestions of adiponitrile in China[J].Synthetic Fiber,2022,51(07):10-12.)
28. Guanglei Shi. Research and optimization of preparation of adionitrile by butadiene hydrocyanide[D].2017..
29. Qinghua Tu. Analysis of adiponitrile market and raw material route[J]. Chemical Industry, 2012, 30(12):26-30.
30. Yingzhi Xiao. Huafeng. Group's 300 kt/a adiponitrile project is under construction in Chongqing[J]. Synthetic Fiber Industry, 2020, 43(6): 19.
31. Guozhong Zhao. Progress and prospect of adiponitrile production technology[J]. Chemical Design,2022,32(04):10-13+50+1.

Short literature review of Li^+ batteries recycling

Mingyuan Wei[a]

University College London, Faculty of Mathematical and Physical Sciences, London, UK

Abstract. Currently, lithium-ion batteries are widely used in advanced electronic devices and electric vehicles due to their low weight, high energy density and long service life. Therefore, the number of waste lithium batteries is also increasing, and it is very important to recover the valuable resources in them. This paper reviews some conventional means of recycling lithium-ion batteries and mentions future directions. The application and recycling of all-solid-state lithium batteries will help to solve the problem of waste lithium batteries.

1. Introduction

For a long time, the consumption of traditional fossil energy has given rise to many social problems, such as air pollution, climate change and environmental damage[1]. At the same time, many countries around the world are facing the challenges posed by the energy crisis due to uncontrollable factors such as wars and natural disasters. In response to the common crisis, the world's major countries have reached a consensus on controlling carbon emissions. The green energy automotive industry is becoming the focus of attention around the world under the goal of "carbon neutrality" . With the development of renewable energy technologies, lithium, as a material with promising applications, will play an important role in reducing carbon emissions[2]. The low mass and high energy density of lithium-ion batteries (LIBs) have made them a favourite in the new mobile electronics industry since Japan's Sony developed the first commercially available high-performance LIB back in 1991[4]. The cathode of a common LIB consists of a lithium-containing metal oxide, such as LCO ($LiCoO_2$), while the anode is dominated by copper or graphite[3]. As related technologies continue to be refined, high-capacity rechargeable power batteries are being used in electric vehicles.

As a core component of new energy vehicles, the production and disposal of lithium-ion power batteries are increasing along with the growing new energy vehicle industry. Global demand for lithium will be around 12,000 tonnes in 2020 and is expected to increase to 25,000 tonnes by 2025. Waste batteries are expected to reach 101 Gwh by 2023, about 1.16 million tonnes[5-8]. The amount of valuable recyclables is positively correlated with the amount of discarded lithium batteries. The global lithium-ion battery market was valued at US$56 billion in 2021; the recycling and dismantling of used lithium-ion batteries is expected to reach a value of US$17,796 million by 2025[9]. But the number of those discarded LIBs that are recycled is very small. Relevant data show that the recycling rate of discarded lithium-ion batteries is insufficient (lower than 1%)[10]. So that recycling these valuable metallic elements, such as cobalt, lithium, manganese and nickel, can reduce the cost of producing new batteries and create economic benefits. On the other hand, the chemicals in LIBs at the end of their useful life must be disposed of in a reasonable manner. Although the recovery efficiency of traditional and reliable lithium battery recycling processes, whatever physical or chemical processes, has been satisfactory, the disposal of highly corrosive waste liquids that accompany the extraction of valuable items can be troublesome. Especially in chemical treatment processes, highly acidic or alkaline liquids can be generated[42]. Moreover, the flammable electrolyte solution poses a potential hazard when conventional Li^+ batteries are being recycled[18]. Besides, pre-processing of LIBs is usually performed manually, which is recognized in many current studies. This illustrates how these methods may still need to be changed when they are used in industry[43]. Therefore, it is important to find excellent ways to recycle discarded lithium batteries, or to design new varieties of LIBs, such as all-solid-state batteries (ASSBs)[11].The purpose of this paper is to synthesize the relevant research results in recent years, call for the recycling of lithium-ion batteries to be taken care of, and look forward to the development of all-solid-state batteries.

2. Working principle of lithium battery

Since the commercialization of LIB in the 1990s, there has been a continuous search for safer, more stable materials with higher conductivity to meet the growing technological demands of lithium-ion batteries.

[a] weimy2023@163.com

Rechargeable batteries all consist of a cathode, anode, conductive electrolyte and separator.

2.1 Composition of lithium-ion batteries

The composition and mass percentage of lithium batteries in common use today are shown in Table 1

Table 1: The chemical composition of LIBs[3,12-14]

Cell components	Chemical composition		%wt.
External casing	Fe-Ni alloy		20-26
	Al		10
Cathode			25-30
Aluminium	Al	Current collector foil	5-8
Binder	Usually PVDF		1-2
Metal oxide	Lithium		1.5-7
	Co	$LiCoO_2$	5-20
	Nickel	$LiNiO_2$	5-10
		$LiNi_{0.5}Co_{0.15}Al_{0.05}O_2$	5
	Manganese	$LiMnO_2$	
	$LiNi_4Co_3Mn_2O_2$		
Separator	Microporous PP or PE		4-10
Electrolyte			10-15
Lithium materials	$LiBF_4$, $LiClO_4$, $LiAsF_6$, $LiPF_6$		
Organic solvents	DMC-EC,PC-DME, BL-THF		
Anode			15-25
Copper	Copper	Current collector foil	8-10
Binder	Mostly PVDF		1-2
Graphite			15-17

Sony used LCO as electrode when they invented the lithium battery, while today's giants of the electric car industry, Tesla, use NCA and NMC[15]. The organic liquid electrolyte commonly used in lithium-ion batteries consists of lithium salt ($LiPF_6$, $LiAsF_6$), organic solvents like DMC (dimethyl carbonate), EC (ethylene carbonate) and additives[16]. In contrast, the electrolyte for solid-state batteries (SSE) is either a conductive solid-state polymer (like polyethylene oxide, PEO), or an oxide/sulfide based chemical (like $LiNbO_3$, Li_3PS_4) [17-19]

2.2 The electrode reaction equation

Take $LiMO_2$ type as an example, where M represents Co, Ni, Mn or the mix.

The cathodic half reaction:

$$LiMO_2 \underset{\text{Charge}}{\overset{\text{Discharge}}{\rightleftharpoons}} Li_{1-x}MO_2 + xLi^+ + xe^-$$

The anodic half reaction:

$$6C + xLi^+ + xe^- \underset{\text{Charge}}{\overset{\text{Discharge}}{\rightleftharpoons}} Li_xC_6$$

Overall reaction:

$$LiMO_2 + 6C \underset{\text{Charge}}{\overset{\text{Discharge}}{\rightleftharpoons}} Li_xC_6 + Li_{1-x}MO_2$$

During the discharge process of a battery, lithium ions produced at the negative electrode move through the electrolyte across the separator toward the cathode. Vice versa, during charging, electrons provided by an external power source combine with lithium ions to form lithium metal, causing it to grow on the anode graphite[20]. The presence of electrolytes will aid in the movement of Li^+ during the chemical reactions described above.

3. Wasted LIB recycling process

Current common methods of lithium-ion battery recycling include physical and chemical processes. In conventional LIBs the liquid electrolyte can be cleaned off by using solvents like N-methyl-2-pyrrolidon after the batteries has been dismantled[21]. But the situation is different for ASSBs. Because of the striking differences in both chemistry and structure of ASSBs compared to conventional LIBs, there are some special points of recycling that need to be considered.

3.1 Physical Processes

Physical processes include disassembly, dissolution and pyrolysis. Of them all, the dissolution process has been the most popular of late, while also maximising the recovery of valuable materials[3]. As selected organic

 https://doi.org/10.1051/e3sconf/202344101020

solvents capable of dissolving adhesives (PVDF, PTFE), N, N-dimethylformamide (DMF), N, N-dimethylacetamide (DMAC), N-methylpyrrolidone (NMP) and dimethylsulfoxide (DMSO) were used to separate the lithium salt from the binder. The solid mixture obtained by separation will be used to carry out the next steps. For instance, Jin et al. used flotation to recycle valuable metals in end-of-life LIBs, firstly, the crushed materials were screened out by air separation, the screened mixture was put into a high-temperature calciner for roasting, and finally, the obtained products were classified for recycling. The recovery rate of lithium and cobalt reached 92%[22].

3.2 Chemical Processes

Chemical processes are primarily hydrometallurgical methods which include processes such as acid or alkali leaching, biological processes and so on. Multiple steps will result in greater costs[3].

3.2.1 Acid leaching

Inorganic acid leaching usually involves the use of strong inorganic acids such as HCl[23-25], H_2SO_4[26-28], HNO_3[29], and H_3PO_4[30,31]. In addition, reducing agents such as H_2O_2 and Na_2SO_3 are used to further enhance leaching efficiency. HCl showed the best leaching results. Wang et al. leached cathode materials in HCl system and showed that the leaching rate of Ni, Co, Mn and Li was 99%[32]. In order to avoid toxic gases (NO_X, Cl_2) produced by the inorganic acid leaching process, the organic acid leaching process is of interest. Organic acids (such as acetic acid, ascorbic acid or malic acid) are biodegradable. Not only do they not pollute the environment, they are also easily recycled[33]. Based on citric acid leaching, an atom-economical process for recycling waste NCM batteries was developed. Under optimal conditions, with the help of the reducing agent D-glucose, 99% of Li, 91% of Ni, 92% of Co and 94% of Mn can be recovered by leaching[34]. Although organic acids have shown many advantages in leaching cathode materials from used lithium-ion batteries, their high expense costs have led to their not being widely used in industry.

3.2.2 Alkaline leaching

The acid leaching process is less selective and is accompanied by high concentrations of H^+, which can easily damage the equipment. Ammonia leaching is based on the complexation of M^{2+} with ammonium ions in alkaline systems to form stable metal-ammonia complexes. Some scholars have used ammonium sulphate ($(NH_4)_2SO_4$) as a leaching agent and sodium sulphite (Na_2SO_3) as a reducing agent to selectively leach valuable metals[35]. Ku et al. used a mixed system of ammonia, ammonium carbonate, and ammonium bisulfate for alkaline leaching of cathode materials from used batteries, which can achieve the effect of selective separation of different metal ions[36].

3.2.3 Biological processes

Biological leaching process has more obvious environmental protection, safety and other advantages, and the economic cost is small. At present, due to the difficulty of microbial cultivation and the long period of strain cultivation, it is still in the stage of laboratory research community, failing to realise the application in industry. But there was good news from the laboratories. For example, acidithiobacillus ferrooxidans uses elemental sulphur and Fe^{2+} to produce metabolites, sulphuric acid and Fe^{2+} in the solution to recover Li and Co from $LiCoO_2$ of LiBs[37].

3.3 Recycle of ASSBs

In addition to conventional lithium-ion battery recycling methods, all-solid-state batteries can be disposed of using direct recycling methods[38].The main step in direct recovery is the lithium regeneration process. In this process, the lithium-depleted cathode material is reacted with a lithium source to restore its original stoichiometry, and then heat-treated to restore the surface structure and morphology[39-41]. Direct recycling therefore has the advantage of low energy costs and environmental friendliness.

4. Discussion

A review of research findings in recent years reveals that hydrometallurgy, as the most commonly used recycling method, has the advantages of less pollution and high recovery. At the same time, its high recycling cost and potential danger are still the reasons that hinder the development of lithium battery recycling industry. Biological leaching is an environmentally friendly and efficient process, but the long culture cycle of the colonies causes it to remain in the lab stage at present. All-solid-state batteries, as a potential wildly used battery, can help to solve the above problems.

5. Conclusion

Recycling of lithium-ion batteries is very important today. Processing methods should be developed that have the lowest environmental impact and the highest recycling efficiency leading to savings in primary raw materials, improved economic efficiency, reduce energy consumption and the cost of safe management of waste and hazardous substances. At the same time, the application and recycling of newer lithium-ion batteries has a bright future. The unique structure and recovery method of ASSBs will be future research directions.

Reference

1. Energy Institute Statistical Review of World Energy 2023
2. Liu, D., et al. (2019). Resources, Conservation and Recycling 145: 311-321.

3. Bankole, et al 2013.. J. Environ. Ecol. 4, 14–28.
4. Mossali, E., et al. (2020). J Environ Manage 264: 110500.
5. Jiefeng, Xiao, Jia, et al. [J]. Environmental science & technology, 2019.
6. Du Z, Lin B, Guan C. [J]. Resources Conservation and Recycling, 2019, 143: 17-26.
7. Xiao J, Li J, Xu Z. [J]. Environ Sci Technol, 54 (2019) 9–25.
8. IEA. Glob. EV Outlook 2020, (2020).
9. Ojanen S, et al. [J]. Waste Management, 2018, 76: 242-249.
10. Shuaibing Ma. [D].Institutes Of Technology Of Jiangxi, 2022. DOI:10.27176/d.cnki.gnfyc.2022.000497.
11. Morscher, A., et al. (2022). Journal of the American Chemical Society 144(48): 22178-22192.
12. Arora, Shashank, Kapoor, Ajay, Shen, Weixiang, 2018. Batteries 4 (30), 1–25.
13. Barik, S.P., Prabaharan, G., Kumar, L., 2017. J. Clean. Prod. 147, 37–43.
14. Beolchini, F., Papini, M.P., Toro, L., Trifoni, M., Veglio, F.,2001. Miner. Eng. 14 (2), 175–184.
15. Blomgren, G.E., 2017. Journal of The Electrochemical Society 164 (1), A5019–A5025.
16. Contestabile, M., Panero, S., & Scrosati, B. (1999). Journal of Power Sources, 83, 75-78.
17. Mageto, T., et al. (2022). Journal of Energy Storage 55.
18. C. Yu, L. van Eijck, S. Ganapathy, M. Wagemaker, Electrochim. Acta 215 (2016) 93–99.
19. R. Chen, Q. Li, X. Yu, L. Chen, H. Li, Chem. Rev. 120 (2020) 6820–6877.
20. Chen, Y., Liu, N., Hub, F., Ye, L., Xi, Y., Yang, S., 2018. Waste Manag. 75, 469–476.
21. Jafari, M., Torabian, M.M., and Bazargan, A.(2020). J. Energy Storage 31, 101564.
22. Yongxun Jin, Komyo Matsuda, Xiaohui Dong, et.al. [J]. Foreign metal ore mineral processing, 2003(7): 32-37.
23. Wen X, Otsuki A, Chagnes A. [J]. RSC Advances, 2019, 2019(9): 38612.
24. Joulié M, Laucournet R, Billy E. [J]. Power Sources, 247 (2014) 551–555.
25. Barik S P, Prabaharan G, Kumar L. [J]. Journal of Cleaner Production, 2017, 147(MAR.20): 37-43.
26. Fan X, Song C, Lu X, et al. [J]. Alloys Compd, 863 (2021) 158775.
27. Li J, Yang X, Fu Y, et al. [J]. Hydrometallurgy, 190.
28. Zhang J, Hu J, Zhang W, et al. [J]. Journal of Cleaner Production, 2018, 204(PT.1-1178): 437-446.
29. Cognet M, Condomines J, Cambedouzou J, et al.[J]. Journal of Hazardous Materials, 2020, 385(Mar.5): 121603.1-121603.8.
30. Jiang Y, Chen X, Yan S, et al. Zhou. [J]. Chem Eng, 426 (2021) 131637.
31. Chen X, Li J, Kang D, et al. [J]. Green Chem, 2019, 21(23): 6342-6352.
32. A R C W, A Y C L, B S H W. [J]. Hydrometallurgy, 2009, 99(3–4): 194-201.
33. Jadhav, U.U., Hocheng, H., 2012. J. Achiev. Mater. Manufact. Eng. 54 (2), 159–167.
34. Chen X, Fan, et al. A[J]. Journal of Cleaner Production, 2016.
35. Zheng, X., et al., 2017. Waste Manag. 60, 680–688
36. Ku H, Jung Y, Jo M, et al.[J]. Journal of Hazardous Materials, 2016, 313(Complete): 138-146
37. Xin, B., Zhang, D., & Zhang, X. (2009). Bioresource Technology, 100(24), 6163-6169.
38. Azhari, L., et al. (2020). Matter 3(6): 1845-1861.
39. .Shi, Y., Chen, G., and Chen, Z. (2018). Green Chem. 20, 851–862.
40. Sloop, S.E., Trevey, J.E., Gaines, L., Lerner, M.M., and Xu, W. (2018).. ECS Trans. 85, 397–403.
41. Kim, K.M., Shin, D.O., and Lee, Y.-G. (2015). . Electrochimica Acta 176, 1364–1373.
42. Zhang, X., Xie, Y., Lin, X., Li, H., Cao, H., 2013. . J. Mater. Cycles Waste Manag. 15 (2), 420–430.
43. Pagnanelli, F., et al., 2014. . J. Ind. Eng. Chem. 20 (5), 3201–3207.

Research on the acceleration effect of horizontal wind in mountainous terrain

He LIN[1], Zhong'an ZHENG[1,*], Xiaojun LI[1], Wenshu YU[1], Xiang XIA[2], Wei JI[3], Jichao YE[3] and Jianyong WU[3]

[1]Lishui Zhengyang Electric Power Design Institute Co., Ltd, Lishui, China
[2]State Grid Zhejiang Electric Power Co., Ltd, Hangzhou, China
[3]State Grid Lishui Power Supply Company, Lishui, China

Abstract. The utilization of mountain wind resources are important issues in the development of wind energy. However, when wind flows over mountainous terrain, its velocity and direction undergo certain changes, resulting in more complex mountain wind fields and induced structural wind effects. To study the distribution characteristics of horizontal wind in mountainous areas, in this study, wind tunnel experiments were conducted on regular mountain models with different slopes to investigate the wind field characteristics at various locations in the mountain range. The research shows that near the mountain surface, the wind speed and velocity ratio follow the order of mountain top > mountain waist > mountain foot, with a wind acceleration effect at the mountain top (wind speed ratio greater than 1) and a wind deceleration effect at the mountain foot and waist (wind speed ratio less than 1). As the height increases, the wind speed ratios gradually approach 1. Moreover, with an increase in slope ratio, the wind speed decreases at the mountain foot and increases at the mountain top. And with the increase of the slope leads to an increasing difference in wind speed ratio between the mountain top and the foot.

1 Introduction

Currently, the efficient use of wind energy and clean and environmentally friendly natural energy is of great concern, which can significantly reduce carbon dioxide and prevent global warming. The terrain in most parts of China is rugged and complicated. In recent years, the construction site of wind turbines has been transferred from coastal areas to mountainous areas. Therefore, wind energy resource assessment, especially in complex terrain areas, is one of the important topics of wind energy development. On the one hand, the abundant wind resources in mountainous areas are conducive to the collection and utilization of wind energy; on the other hand, when the natural wind passes through the complex mountain environment, a more complex wind field will be formed, which brings great difficulties to the construction of wind farms in mountainous areas. Therefore, the study of mountain wind characteristics is of great significance.

As the starting point of flow field modeling over complex terrain, the flow characteristics over simple two-dimensional hills have been extensively studied by numerous researchers. Jackson and Hunt [1] were the first to analyze the wind velocity distribution over two-dimensional smooth and low hills, which established the main parameter for mountain wind field research: wind speed-up ratio. Taylor et al. [2] made corrections to the theoretical calculation formula, simplifying the calculation of wind speed-up ratios at different heights for gentle slopes. More recently, Kamada et al. [3] conducted wind tunnel experiments to investigate the turbulent flow structure around a two-dimensional hill model in a boundary layer flow, revealing wind speed increases along the upstream slope and flow separation over the hilltop. Liu et al. [4] used large eddy simulation to study the influence of wind direction on the flow field over a two-dimensional (2D) hill, finding that the wind behind the hill is not parallel to the incoming wind direction when the wind direction is not perpendicular to the ridge line. However, these studies focused on two-dimensional hill shapes, while actual terrains are three-dimensional. Therefore, considerable research has been conducted internationally to study the wind fields near three-dimensional hills. Sun et al. [5] investigated the influence of mountain height and slope variations on the average wind speed and fluctuating wind speed of a single hill through experiments, revealing that the near-ground average wind speed at the mountain top increases with increasing slope and height. Wei et al. [6] performed numerical simulations to study the impact of different slopes, heights, and topographic features on the acceleration effect of a single hill, proposing a logarithmic calculation model that better describes the near-ground acceleration effect of mountainous terrain and applies to steep slopes. Li et al. [7] used numerical simulations to calculate the acceleration effect of a single hill and found that the maximum acceleration ratio at the

* Corresponding author: 642399553@qq.com

mountaintop increases with the slope and height of the hill while decreasing with the increase of the ground roughness index. Flay et al. [8] conducted numerical simulations and wind tunnel tests to measure the wind speed near Belmont Hill, providing corrections to the local code based on their study.

Based on the above background, this study focuses on three-dimensional mountain slopes with different slopes and conducts wind tunnel experiments to investigate the impact of slope on the wind field distribution over mountain surfaces, providing design recommendations for engineering constructions in mountainous regions.

2 Wind tunnel experiment

The wind tunnel experiments on three-dimensional mountain models were conducted in the low-speed test section of Changsha University of Science and Technology of Wind Engineering and Wind Environment Research Center. The cross-section size of the low-speed test section is 10.0 m × 3.0 m × 21.0 m, and the diameter of the turntable is 5.0 m. The wind speed can be continuously adjusted from 1.0 m/s to 18.0 m/s. According to the GB 50009-2012 standard [9], A-class turbulence in the wind field was generated using a sharp-edged and rough-surface element. The wind speeds at each measurement point and the incoming wind speed were measured using the Cobra probe and dedicated software from the TFI company in Australia. Fig. 1 shows the results of the wind field calibration, which are found to be very close to the required conditions, meeting the experimental requirements.

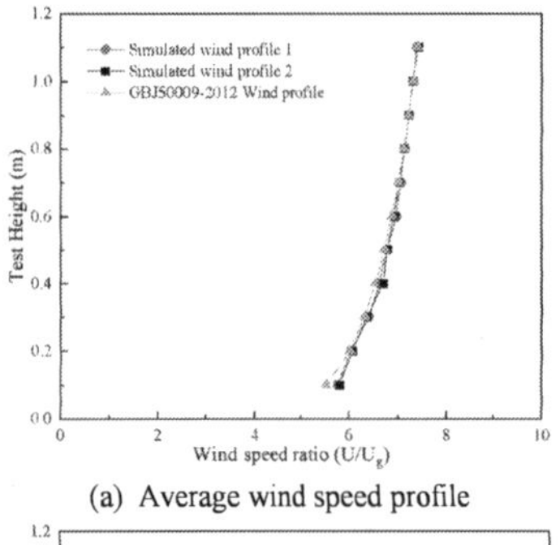

(a) Average wind speed profile

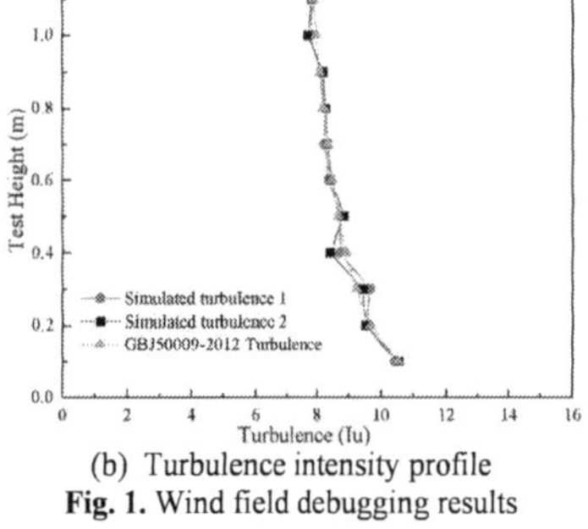

(b) Turbulence intensity profile

Fig. 1. Wind field debugging results

2.1. Model fabrication and measurement point distribution

The experiment used a cosine function to model the cross-section of the mountain and considered the influence factors of the mountain length D. Fig. 2 illustrates the cosine mountain model. The model satisfies the following equations:

$$Z = H\cos^2(\pi r / 2L) \tag{1}$$

Where: r is x, and the distance from the point to the original position o is calculated; H is the height of the mountain, and the distance from the bottom of the mountain to the top of the mountain at the origin o; L is the width of half the mountain, and the horizontal distance from the foot of the mountain to the origin o; z is the coordinate, and the vertical distance from the point to the bottom of the mountain is calculated.

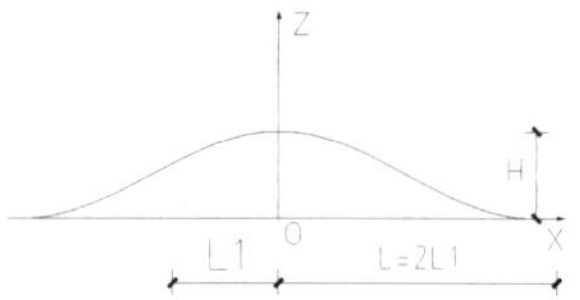

(a) Cross-sectional view of the mountain

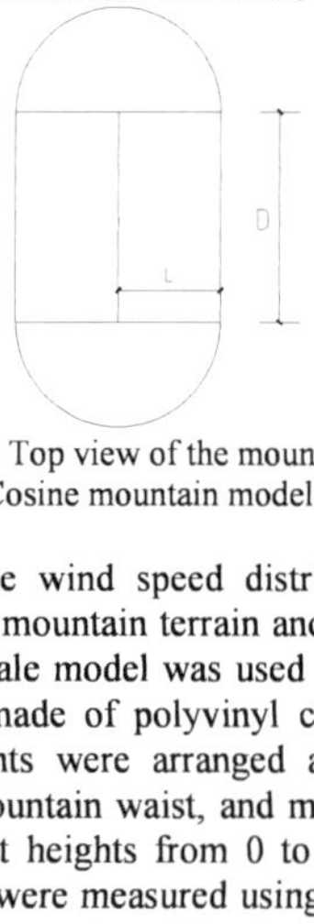

(b) Top view of the mountain

Fig. 2. Cosine mountain model diagram

To analyze the wind speed distribution at various locations near the mountain terrain and its variation with height, a 1:500 scale model was used in the experiment. The model was made of polyvinyl chloride foam, and measurement points were arranged at three locations: mountain foot, mountain waist, and mountain top. Wind speeds at different heights from 0 to 900 m above the mountain surface were measured using the cobra probes. Fig. 3 shows the experimental setup and measurement point distribution.

Fig. 3. Photos of the installation of the regular mountain model in the wind tunnel

 https://doi.org/10.1051/e3sconf/202344101021

2.2. Experimental conditions

In this study, a 1:500 scale regular sine mountain range was constructed, and different slopes were set to investigate the influence of the slope and length on the wind profile. The slope conditions are shown in the following table, and measurement points were set at the mountain top, waist, and foot to obtain wind speed data at distances of 5 m, 10 m, 45 m, 60 m, 90 m, 120 m, 150 m, 180 m, 210 m, 240 m, 270 m, 300 m, 360 m, 420 m, 480 m, 540 m, 600 m, 700 m, 800 m, and 900 m downwind.

Table 1. Different slope mountain model working condition setting (Mountain height 200mm)

Condition	slope	Mountain length	
1	0.1	2000mm	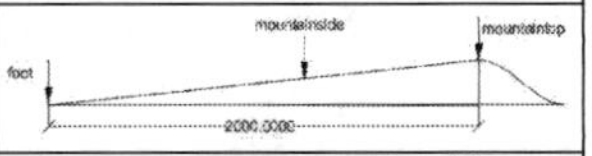
2	0.2	1000mm	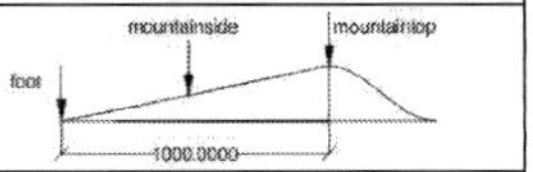
3	0.25	800mm	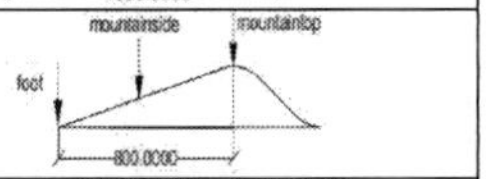
4	0.3	666.66mm	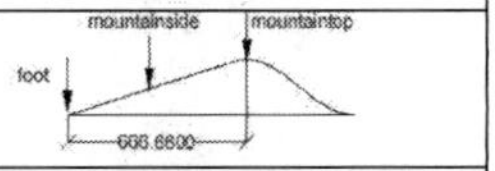
5	0.35	571.43mm	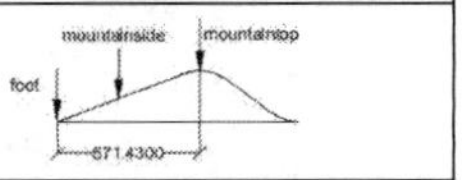
6	0.4	500mm	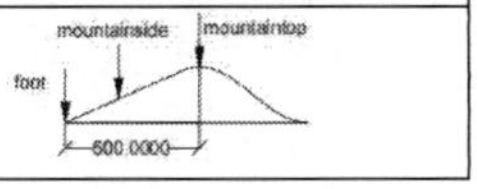

a.Mountain height 200mm

3 Experimental results and analysis

3.1. Definition of wind speed ratio

The acceleration effect refers to the phenomenon where the average wind speed at a certain height in the mountainous terrain is higher than that in the flat terrain at the corresponding height, and it is most pronounced near the ground on the mountaintop [10]. To quantitatively characterize the size of the acceleration effect, the wind speed ratio S is defined as follows:

$$S(Z) = U(Z) / U_0(Z) \quad (2)$$

Where U(Z) is the average wind speed at the height of z from the mountain surface, and U0(Z) is the average wind speed at the corresponding height of the incoming flow. Fig. 4 illustrates the relationship between U(Z), U0(Z), and the acceleration effect.

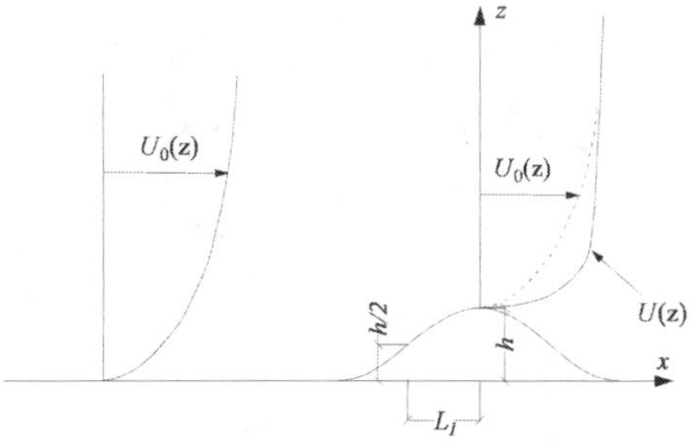

Fig. 4. Horizontal wind speed ratio definition

3.2. Analysis of wind characteristics at different locations near the mountain

To analyze the wind speed distribution of the entire terrain, Fig. 5 compares the changes in wind speed along height at various measurement points in different slopes. It can be seen from Fig. 5 that: near the mountain surface, the wind speed ratio at the top of the mountain is the largest, greater than 1, and it shows a wind acceleration effect here. There is a wind deceleration effect at the foot and side of the mountain, and the wind speed ratio is less than 1. With the increase in height, the wind speed ratio and wind speed ratio gradually get closer. The wind speed ratio is the smallest at the foot of the mountain, which is less than 1, and gradually approaches 1 with the increase in height. Near the top of the mountain, the wind speed ratio is 1~1.15 and gradually decreases to 1 as the height rises. With the increase in slope ratio, the wind speed at the foot of the mountain becomes smaller, and the wind speed at the top of the mountain becomes larger: the wind speed ratio at the foot of the mountain decreases from 0.6 to 0.4; The wind speed ratio at the mountainside is about 0.6; Wind speed increased from 0.9 to 1.15 at the peak. As the slope ratio increases, the difference between the wind speed at the top and the foot of the mountain widens.

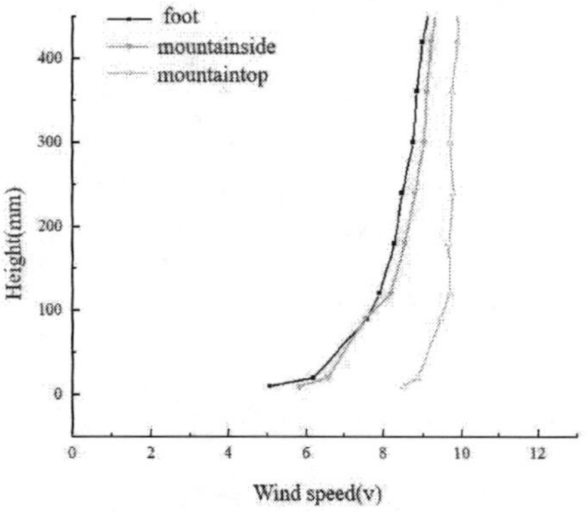

(a) Mountain wind speed (slope 0.1)

https://doi.org/10.1051/e3sconf/202344101021

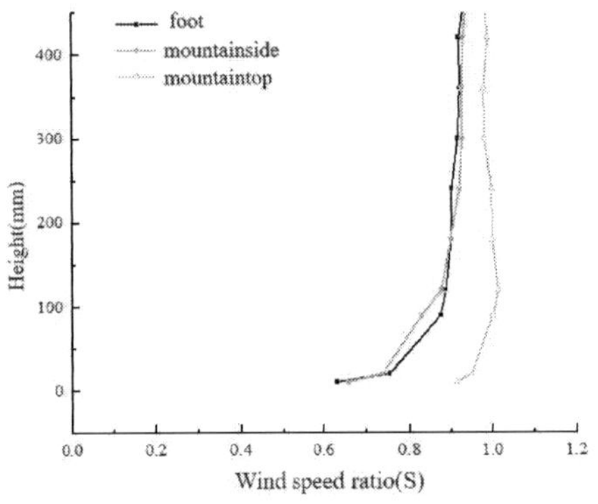

(b) Mountain wind speed ratio (slope 0.1)

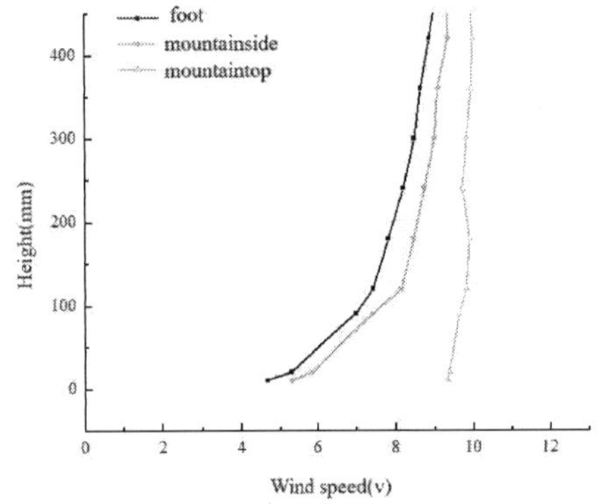

(c) Mountain wind speed (slope 0.2)

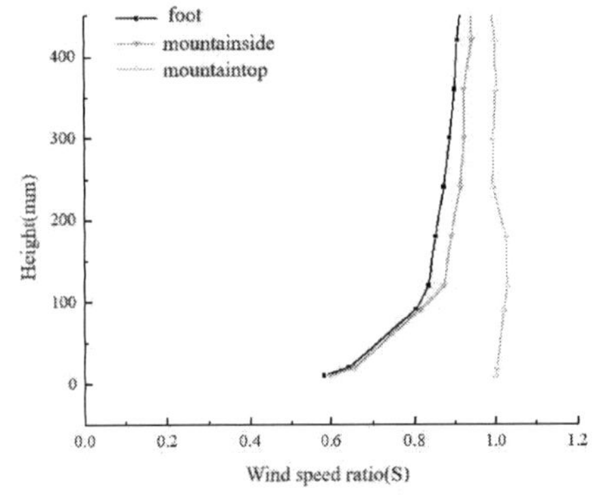

(d) Mountain wind speed ratio (slope 0.2)

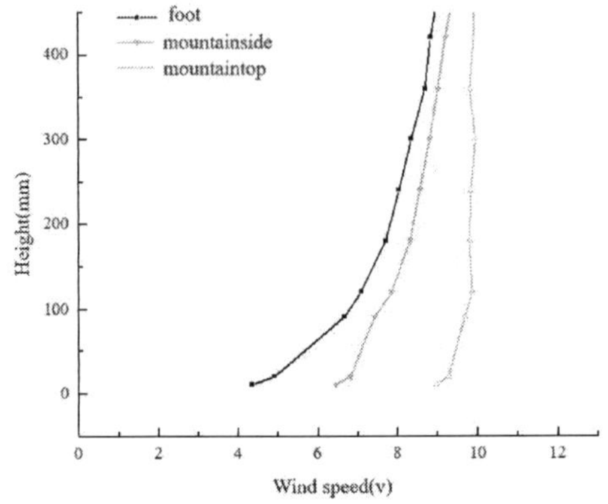

(e) Mountain wind speed (slope 0.25)

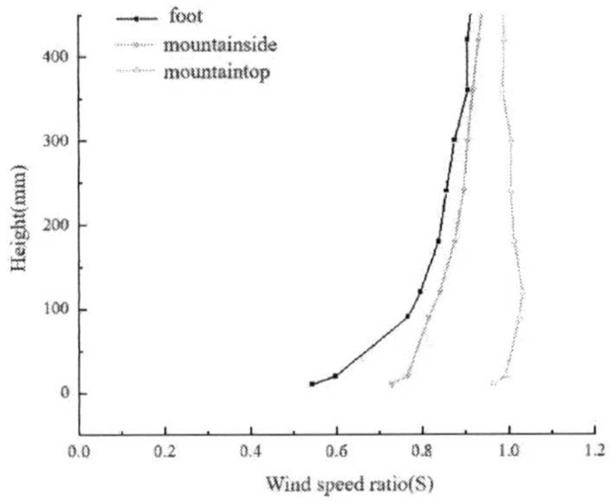

(f) Mountain wind speed ratio (slope 0.25)

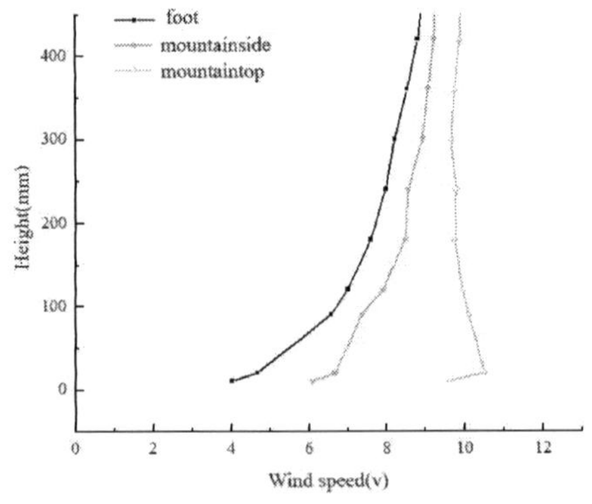

(g) Mountain wind speed (slope 0.3)

https://doi.org/10.1051/e3sconf/202344101021

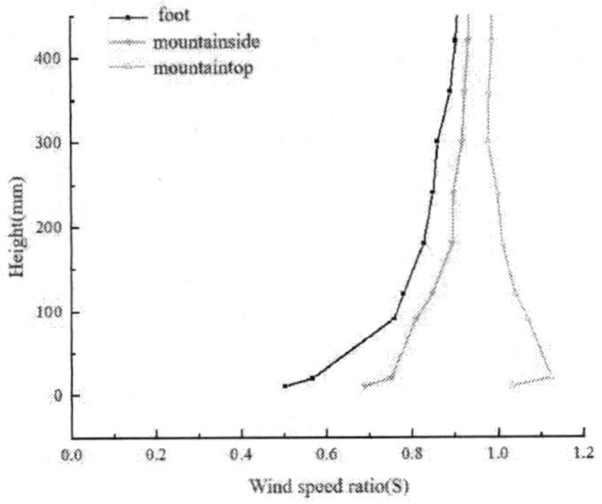

(h) Mountain wind speed ratio (slope 0.3)

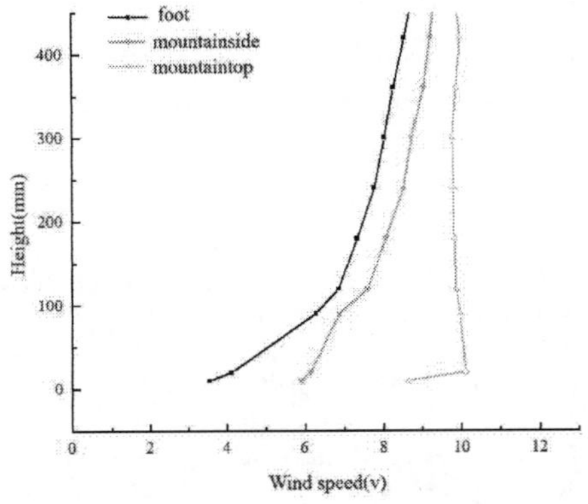

(i) Mountain wind speed (slope 0.35)

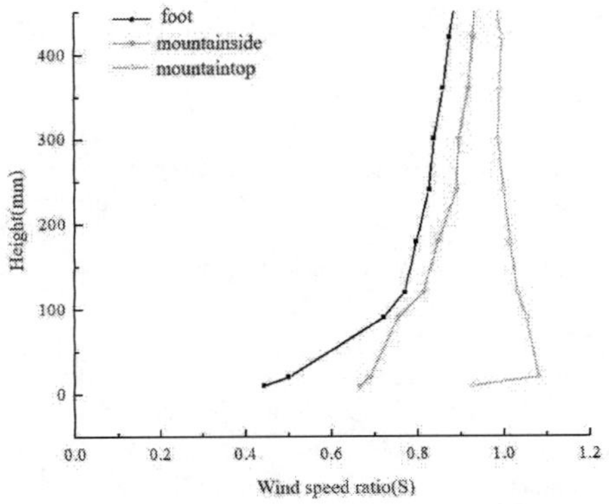

(j) Mountain wind speed ratio (slope 0.35)

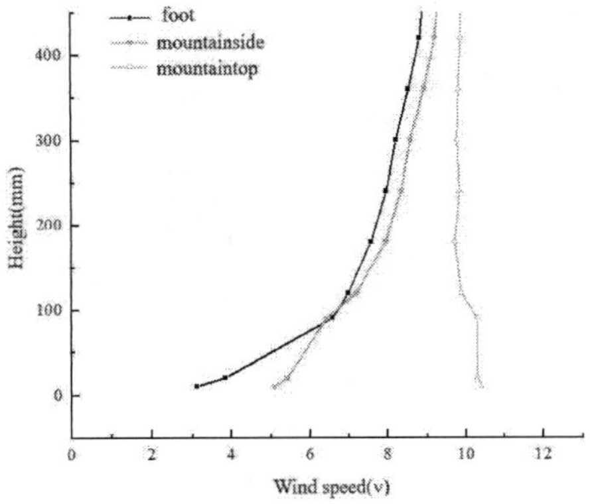

(k) Mountain wind speed (slope 0.4)

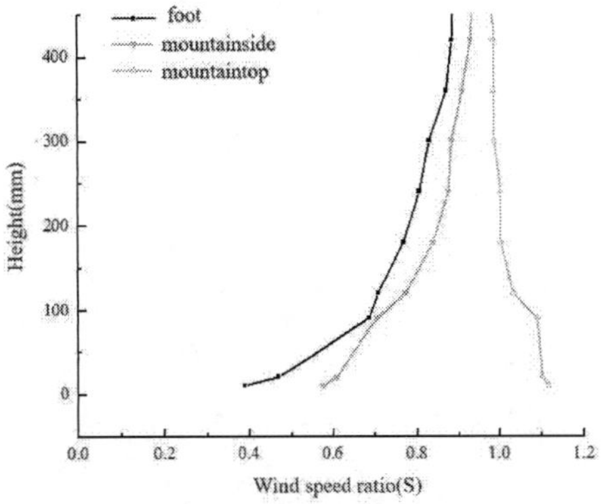

(l) Mountain wind speed ratio (slope 0.4)

Fig. 5. Wind speed and wind speed ratio at the foot, mountainside, and mountaintop of mountains with different slopes

To provide a more detailed analysis of the influence of slope on the wind speed distribution over the mountain surface, Figure 6 presents the wind speeds and velocity ratios at the mountain top for different slope ratios. From the figures, it can be observed that near the mountain surface, the wind speed and velocity ratio are relatively large at the mountain top, and this effect gradually decreases with increasing height. This phenomenon becomes more pronounced with increasing slope ratio. For slopes ranging from 0.1 to 0.25, the wind speed ratio near the ground is approximately 0.9 to 1.0, and the decrease in the velocity ratio with increasing height is not very significant. When the slope reaches 0.3 to 0.35, the wind speed ratio near the ground increases to 1.08 to 1.15, and the decrease in the velocity ratio with increasing height becomes more evident. When the slope reaches 0.4, the wind speed ratio near the ground is 1.15, and the decrease with height becomes more pronounced.

https://doi.org/10.1051/e3sconf/202344101021

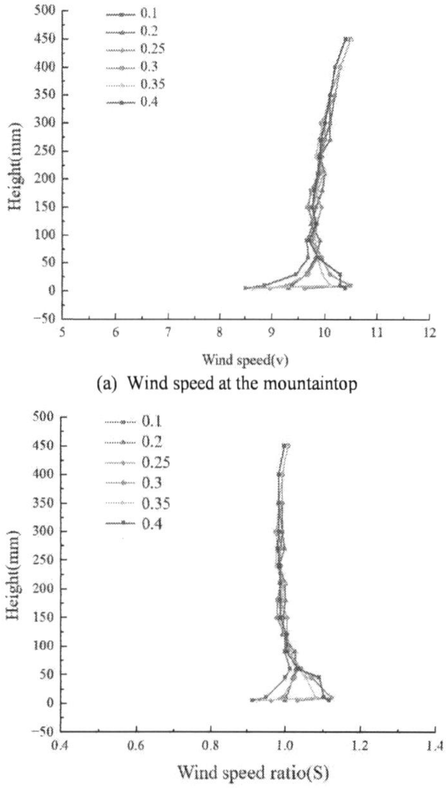

(a) Wind speed at the mountaintop

(b) Wind speed ratio at the mountaintop

Fig. 6. Comparison of wind speed and wind speed ratio at mountaintop with different slopes

4 Conclusions

This study conducted wind tunnel experiments to simulate the wind field distribution over regular continuous mountain ranges in A-class wind fields. By comparing and analyzing the experimental results, the following conclusions can be drawn:

1)Near the mountain surface, the wind speed and velocity ratio follow the order of mountaintop > mountainside > mountain foot. The mountain top exhibits a wind acceleration effect with a velocity ratio greater than 1, the maximum is 1.12, while the mountain foot and waist experience a wind deceleration effect with a velocity ratio less than 1, and the minimum is only 0.4. As the height increases, the wind speed ratios gradually approach 1.

2)With an increase in the slope ratio, the wind speed decreases at the mountain foot and increases at the mountain top: The wind speed ratio at the foot of the mountain decreased from 0.6 to 0.4; The wind speed at the top of the mountain increased from 0.9 to 1.6. The difference in wind speed ratio between the mountain top and foot increases with increasing slope, requiring careful consideration in engineering applications.

3)Near the mountain surface, the wind speed and velocity ratio are relatively large at the mountain top, and this effect gradually decreases with increasing height. This phenomenon becomes more pronounced with increasing slope ratio.

References

1. Jackson P S, Hunt J. Turbulent wind flow over a low hill[J]. Quarterly Journal of the Royal Meteorological Society, 1975, 101(430).
2. Taylor P A, Lee R J. Simple guidelines for estimating wind speed variations due to small scale topographic features[J]. Climatological Bulletin, 1984, 18(2):3-22.
3. Kamada Y, Maeda T, Yamada K. Wind tunnel experimental investigation of flow field around two-dimensional single hill models[J]. Renewable Energy, 2019, 136:1107-1118.
4. Liu Z, Cao J, Yan B, et al. Study of wind-direction effects on flow fields over two-dimensional hills using large eddy simulations[J]. Journal of Wind Engineering and Industrial Aerodynamics,2020, 204, 104285.
5. Sun Y, Li ZL, Huang HJ, et al. Experimental research on mean and fluctuating wind velocity in hilly terrain wind field[J]. Acta Aerodynamica Sinica, 2011, 29(5):593-599 (in Chinese).
6. Wei QK, Li ZL, Sun Y, et al. Calculation Model of Speedup Effect of Wind Velocity in Hilly Terrain[J]. Journal of South China University of Technology (Natural Science Edition), 2010, 38(11):54-58, 73 (in Chinese).
7. Li ZL, Sun Y, Wei QK, et al. Numerical Simulation of Mean Velocity Speed-up Effect Hilly Terrain[J]. Engineering Mechanics, 2010, 27(07):32-37 (in Chinese).
8. Flay R, King A B, Revell M J, Carpenter P, Pirooz AAS. Wind speed measurements and predictions over Belmont Hill, Wellington, New Zealand[J]. Journal of Wind Engineering and Industrial Aerodynamics, 2019, 195,104018.
9. GB 50009-2012, 2012. Load Code for the Design of Building Structures. China Architecture and Building Press, Beijing, China (in Chinese).
10. Ren H, Laima S, Chen WL, et al. Numerical simulation and prediction of spatial wind field under complex terrain[J]. Journal of Wind Engineering and Industrial Aerodynamics, 2018, 180: 49-6.

PtCoNi Alloy Towards Efficient Oxygen Reduction Reaction

Binbin Zhang[1], Jiayun Wu[2], Yuanyuan Cong[2]*

[1] School of Chemistry and Chemical Engineering, Shandong University, Jinan, Shandong, 250199, China
[2] School of Petrochemical Technology, Lanzhou University of Technology, Lanzhou, Gansu, 730050, China

Abstract: A ternary alloy catalyst, CoNi-Pt/C-600, was prepared using a simple polyol method and thermal annealing treatment, and its oxygen reduction reaction (ORR) performance in acidic medium was studied.X-ray diffraction (XRD), X-ray photoelectron spectroscopy (XPS), and transmission electron microscopy (TEM) were used to characterize the structure and morphology of the catalyst particles. These techniques provided information on the structure and morphology of the catalyst. Electrochemical testing was conducted to evaluate the ORR activity and stability of the catalyst. CoNi-Pt/C-600 was compared with JM Pt/C. The results showed that the mass activity (MA) of CoNi-Pt/C-600 was 1.87 times higher than that of JM Pt/C (CoNi-Pt/C-600: 174.02 mA/mg_{Pt}, JM Pt/C: 93.22 mA/mgPt). The specific activity (SA) of CoNi-Pt/C-600 was 3.60 times higher than that of JM Pt/C (CoNi-Pt/C-600: 0.6537 $mA/m^2{}_{Pt}$, JM Pt/C: 0.1814 $mA/m^2{}_{Pt}$). This indicates that CoNi-Pt/C-600 exhibits better ORR activity. Furthermore, durability testing demonstrated the superior electrocatalytic performance of CoNi-Pt/C-600. In conclusion, this study successfully prepared a ternary alloy catalyst, CoNi-Pt/C-600, with excellent ORR activity and stability. This research provides valuable reference for the development of efficient catalysts for the oxygen reduction reaction.

1 Introduction

Hydrogen energy, as an efficient and clean renewable energy source, has attracted significant research attention in recent years[1]. Fuel cells are the primary application of hydrogen energy technology, as they offer high energy conversion efficiency by directly converting the chemical energy of fuel into electricity. Proton exchange membrane fuel cells (PEMFCs) are widely regarded as one of the most efficient fuel cell technologies due to their mature membrane preparation techniques and ability to start quickly at low temperatures[2]. However, the commercialization of PEMFCs is hindered by the use of catalysts. Without catalysts, the oxidation-reduction reactions at the anode and cathode of fuel cells, especially the oxygen reduction reaction (ORR) at the cathode, would proceed very slowly. The most widely used catalyst in PEMFCs is platinum (Pt) catalyst[3,4], which is expensive and scarce, limiting its development. However, since no better catalyst than Pt has been found in the short term, it is necessary to consider modifying Pt catalysts from the perspective of catalyst utilization to improve their efficiency.

To simultaneously reduce the Pt loading and enhance the ORR catalytic performance, researchers have proposed the formation of multi-component Pt-based metal catalysts by alloying Pt with transition metals such as copper, iron, nickel, cobalt, etc[5-8]. Studies have shown that introducing transition metals into the Pt lattice can change the surface electronic structure of the alloy metal particles in the catalyst, further adjusting the bonding energy of reaction intermediates on the surface, thereby improving the catalytic activity of ORR[9]. Lokanathan[10] et al. prepared an ordered Pt_2CoNi ternary alloy nanoparticle catalyst, which modified the electronic structure to optimize the Pt-O bonding energy, exhibiting excellent stability and activity for ORR in acidic medium, with a mass activity (MA) five times higher than that of commercial Pt/C. Additionally, the crystal structure of Pt-based catalysts can be tuned by high-temperature annealing treatment, transforming their atomic disorder structure into an ordered structure with a higher degree of alloying, thereby exposing more active Pt sites and facilitating the smooth progress of the ORR[11, 12]. Wang[13] et al. investigated the influence of the structural changes of PtNiCo catalysts after heat treatment on the ORR activity and stability, and found that the heat-treated PtNiCo catalyst exhibited significantly better electrocatalytic performance than commercial Pt/C, with an MA approximately four times higher.

In this study, we prepared a ternary metal alloy catalyst, CoNi-Pt/C, by doping Co and Ni using a polyol method. The introduction of transition metals into Pt/C catalysts was explored to investigate their effect on the electronic structure of Pt/C catalysts. The electrochemical performance of the Pt-based alloy catalysts before and after high-temperature annealing treatment was compared to examine the role of crystal structure modulation in the ORR.

* Corresponding author: congyuanyuan@lut.edu.cn

https://doi.org/10.1051/e3sconf/202344101022

Based on the above content, we prepared the CoNi-Pt/C ternary metal alloy catalyst using ethylene glycol as a reducing agent and polyvinylpyrrolidone (PVP) as a structure-directing agent. The catalyst was synthesized by reducing a mixture of cobalt hexahydrate nitrate, nickel hexahydrate nitrate, chloroplatinic acid, and carbon carrier, followed by high-temperature annealing of the ground CoNi-Pt/C. The resulting CoNi-Pt/C-600 exhibited an MA of 174.02 mA/mg_{Pt} and an SA of 0.6537 mA/m^2_{Pt}, surpassing the commercial JM Pt/C catalyst (MA: 93.22 mA/mg_{Pt}, SA: 0.1814 mA/m^2_{Pt}).

2 Experimental section

2.1 Materials

Polyvinylpyrrolidone (PVP, $(C_6H_9NO)_n$) was provided by Maclyn Company. $Co(NO_3)_2 \cdot 6H_2O$ was purchased from Shanghai Zhongqin Chemical Reagent Co., Ltd. $Ni(NO_3)_2 \cdot 6H_2O$, H2PtCl6, carbon black, and ethylene glycol were provided by China National Pharmaceutical Group Chemical Reagent Co., Ltd. The equipment used includes an ultrasonic instrument, centrifuge, tubular furnace, and oil bath.

2.2 Preparation of carbon-based Co-Ni alloy Pt electrochemical catalysts

To synthesize CoNi-Pt/C, the following steps were carried out:

Preparation of materials: Add 20 mg of PVP, 20 mg of carbon carrier, 1.62 mg of cobalt hexahydrate nitrate, 1.62 mg of nickel hexahydrate nitrate, and 1.15 mL of chloroplatinic acid (with a theoretical molar ratio of Co:Ni:Pt of 1:1:4, and a Pt loading of 17.4%) to 20 mL of ethylene glycol. Sonicate the mixture for 30 minutes.

Heating and stirring: Transfer the solution to an oil bath and heat it while stirring. Raise the temperature to 130°C and maintain it for 3 hours. Allow the reaction mixture to cool naturally to room temperature.

Centrifugation and washing: Centrifuge the reaction mixture and collect the black precipitate. Wash the precipitate three times with deionized water and ethanol.

Drying: Dry the collected black precipitate at 60°C for 12 hours to obtain CoNi/Pt-C.

For comparison, Pt-C was prepared without the involvement of cobalt hexahydrate nitrate and nickel hexahydrate nitrate. This synthesis method allows for the preparation of CoNi-Pt/C catalyst with the desired composition, and by excluding cobalt and nickel, Pt-C catalyst can be obtained.

We conducted a heat annealing treatment on the obtained CoNi-Pt/C catalyst to promote the surface structural rearrangement of the catalyst, aiming to form a Pt layer surface that can expose more active sites and enhance the electrocatalytic performance. For this purpose, the CoNi-Pt/C catalyst was placed in a tubular furnace and annealed under a N2 atmosphere for thirty minutes. We tried three different annealing temperatures: 600 °C, 700 °C, and 800 °C. The resulting samples were labeled as CoNi-Pt/C-600, CoNi-Pt/C-700, and CoNi-Pt/C-800. The Pt/C catalyst annealed at 600 °C was labeled as Pt/C-600.

2.3 Physical characterization

TEM images were obtained using a JEM-2100 electron microscope at an accelerating voltage of 200 kV. High-angle annular dark-field (HAADF) scanning transmission electron microscopy (STEM) images, energy-dispersive X-ray spectroscopy (EDS), and elemental mapping were acquired using a 300 kV Titan microscope equipped with a probe aberration corrector. X-ray diffraction (XRD) analysis of the lattice structure was performed using Cu Kα radiation (λ= 1.5405 Å) with a scanning rate of 8°/min within the 2θ range of 30° to 90°.

2.4 Electrochemical measurements

The electrochemical performance of the sample was evaluated on an electrochemical workstation using a three-electrode setup with a rotating disk electrode (RDE) to test the performance of the oxygen reduction reaction (ORR). A glassy carbon electrode was used as the working electrode for loading the prepared electrocatalyst. The working electrode was prepared by dispersing 3 mg of electrocatalyst powder in a mixture solution of 2.7 mL ethanol, 0.3 mL water, and 18 μL Nafion through ultrasonic dispersion. Cyclic voltammetry (CV) curves were recorded at a scan rate of 50 mV/s to calculate the electrochemically active surface area (ECSA) of the electrocatalyst. Furthermore, linear sweep voltammetry (LSV) analysis of the ORR activity of the electrocatalyst was conducted at a scan rate of 10 mV/s and 1600 rpm under O_2-saturated 0.1 M $HClO_4$ solution at 25 °C, obtaining LSV curves. The durability of the electrocatalyst was evaluated through chronoamperometry (CA) testing. The test conditions involved a 0.1 M N_2-saturated $HClO_4$ solution, a cyclic potential scan range of 0.6-1.0 V, a scan rate of 500 mV/s, and 5000 cycles.

3 Results and discussion

3.1 Synthesis and Characterization of Electrocatalysts

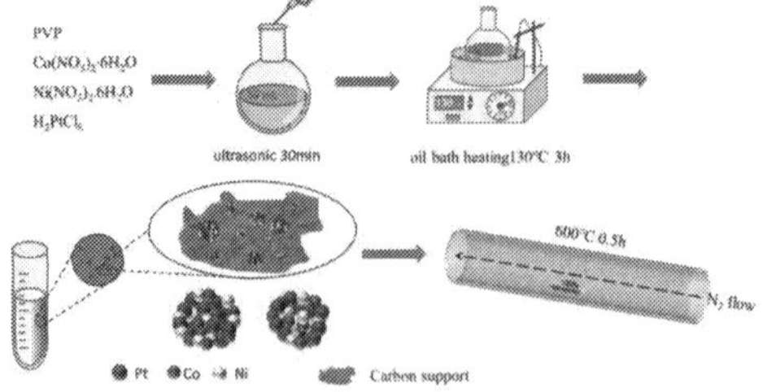

Figure 1 The preparation of CoNi-Pt/C-600

CoNi-Pt/C is synthesized using ethylene glycol as the reducing agent and PVP as the surfactant. Carbon support, nickel hexahydrate nitrate, cobalt hexahydrate nitrate, and chloroplatinic acid are mixed together and sonicated. The mixture is then heated at 130 °C for 3 hours, cooled to room temperature, and centrifuged. After three washes with pure water and ethanol, the CoNi-Pt/C catalyst is dried. To improve the electrochemical performance by enhancing the aggregation of the metal support, the CoNi-Pt/C catalyst is further calcined at 600 °C in a tube furnace for half an hour. The resulting catalyst powder is named CoNi-Pt/C-600. Figure 1 illustrates the specific synthesis steps.

For the as-synthesized CoNi-Pt/C catalyst before heat treatment, its TEM image is shown in Figure 2a with an average size of 2.63 nm. As seen from Figure 2b, the alloy particles are evenly distributed on the carbon support, indicating that the catalyst has a good morphology.

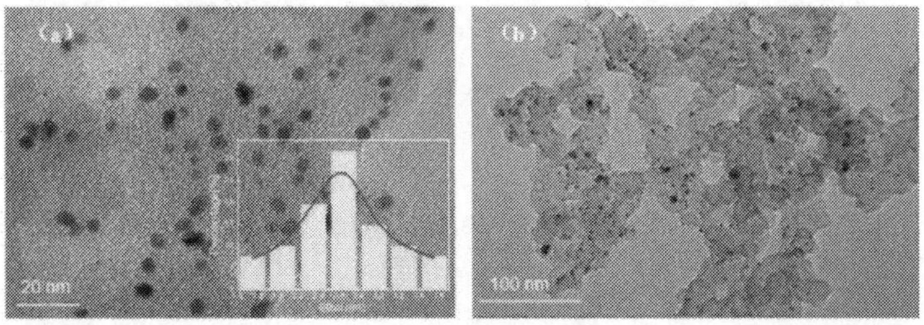

Figure 2 (a) The TEM image and size distribution histogram of the catalyst before heat treatment, (b) The TEM image of CoNi-Pt/C

After further heat treatment, the CoNi-Pt/C catalyst was calcined at 600 °C for half an hour. The resulting catalyst powder was named CoNi-Pt/C-600, and its HAADF-STEM image is shown in Figure 3a. The analysis suggests that the average size of CoNi-Pt/C-600 is 4.80 nm. Compared to the as-synthesized CoNi-Pt/C, the particle size of the catalyst has increased. The HAADF-STEM image and corresponding elemental mapping illustrate that the metal particles of Co, Ni, and Pt are evenly distributed within CoNi-Pt/C (Figure 3b). Furthermore, clear lattice fringes spaced at 0.224 nm can be observed, which corresponds to the Pt [14] crystal plane (Figure 3c).

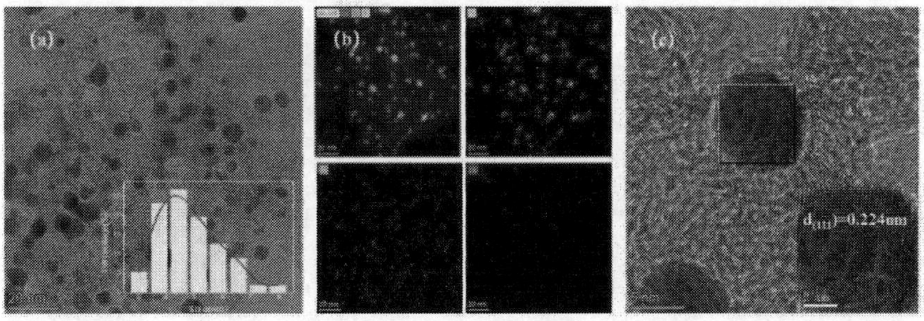

Figure 3 (a) The TEM image obtained after heat treatment with its size distribution histogram, (b) The HAADF-STEM image and the corresponding elemental mapping, (c) The high-resolution HAADF-STEM image with aberration correction is shown, where the red highlighted regions indicate the lattice spacing of CoNi-Pt/C-600 (111).

The crystal structures of CoNi-Pt/C, Pt/C-600, CoNi-Pt/C-600, CoNi-Pt/C-700, and CoNi-Pt/C-800 electrocatalysts were studied using XRD analysis. As shown in the figure, four characteristic peaks corresponding to the (111), (200), (220), and (311) crystal planes of FCC structure were observed. Compared to Pt/C-600, the Pt (111) peak of CoNi-Pt/C-600 shifted towards higher 2θ angles. This could be attributed to the formation of an alloy between Co, Ni, and Pt. The smaller atomic radii of Co and Ni entering the lattice of Pt results in lattice contraction, causing the diffraction peak to shift towards higher 2θ angles. After thermal annealing, the diffraction peaks corresponding to Pt crystal planes became sharper, indicating a higher degree of crystallization and surface restructuring, which favors the exposure of more active Pt sites. With increasing annealing temperature, the diffraction peaks of all crystal planes became sharper and narrower due to particle aggregation and increased atomic radii. The main peak positions were located between Pt (111) and Ni (111) or Pt (111) and Co (111), indicating that thermal annealing enhanced the degree of alloying, which effectively improves the ORR performance of the Pt-based catalyst.

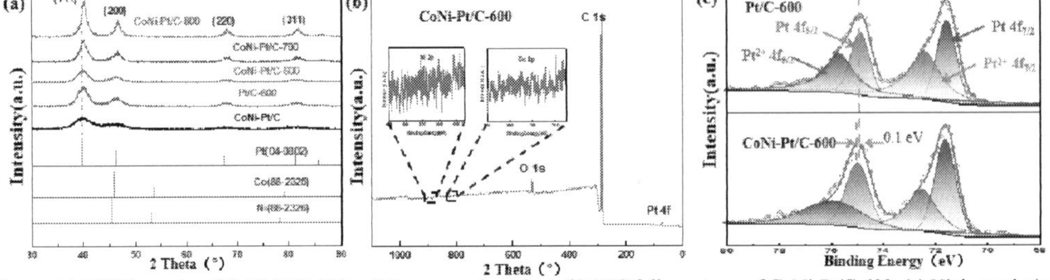

Figure 4 (a) XRD spectra of CoNi-Pt/C-600 and the prepared catalyst, (b) XPS full spectrum of CoNi-Pt/C-600, (c) High-resolution Pt 4f spectra of CoNi-Pt/C-600 and JM Pt/C electrocatalyst.

The electronic structure and surface composition of CoNi-Pt/C-600 were analyzed using XPS. Figure 4 shows the XPS spectrum of the catalyst, revealing the coexistence of Co, Ni, Pt, and C elements. However, due to the lower content of Co and Ni, their peak intensities are weaker. In the high-resolution spectrum, the Pt 4f spectrum of CoNi-Pt/C-600 exhibits two sets of doublet peaks, corresponding to metallic Pt peaks at 75.0 eV and 71.6 eV, and oxidized Pt peaks at 76.0 eV and 72.5 eV. Compared to Pt/C-600, the Pt 4f peak of CoNi-Pt/C-600 shifted by 0.1 eV. This shift could be attributed to the alteration of electronic structure and crystal contraction effects caused by the alloying of Pt with Co and Ni, further confirming the conclusions from the XRD analysis. According to XPS analysis, the surface atomic ratio of Co, Ni, and Pt in the CoNi-Pt/C-600 catalyst is 0.46:0.38:1.

3.2 The effect of the metal components in synthetic catalysts on the catalytic activity of ORR (oxygen reduction reaction)

The electrochemical performance of the catalyst was determined using a three-electrode system in 0.1M HClO4. Figure 5a shows the CV curves of CoNi-Pt/C-600, CoNi-Pt/C-700, CoNi-Pt/C-800, and JM Pt/C in N2-saturated HClO4. The ECSA was calculated from the hydrogen reduction peak, and the ECSA values for CoNi-Pt/C-600, CoNi-Pt/C-700, CoNi-Pt/C-800, and JM Pt/C were 26.62 m^2/g_{Pt}, 26.23 m^2/g_{Pt}, 25.24 m^2/g_{Pt}, and 51.49 m^2/g_{Pt}, respectively. It can be seen that the difference in ESCA values between the catalysts with different heat treatment temperatures is small, indicating that the starting performance of the catalyst is not significantly affected by the heat treatment temperature. The ORR performance of CoNi-Pt/C-600 was found to be better than that of JM Pt/C, as shown by the oxygen reduction peak in the CV curve (Figure 5a). To demonstrate the good ORR performance of CoNi-Pt/C-600, we tested the LSV curves of CoNi-Pt/C-600 and JM Pt/C in O_2-saturated $HClO_4$ (Figure 5b).

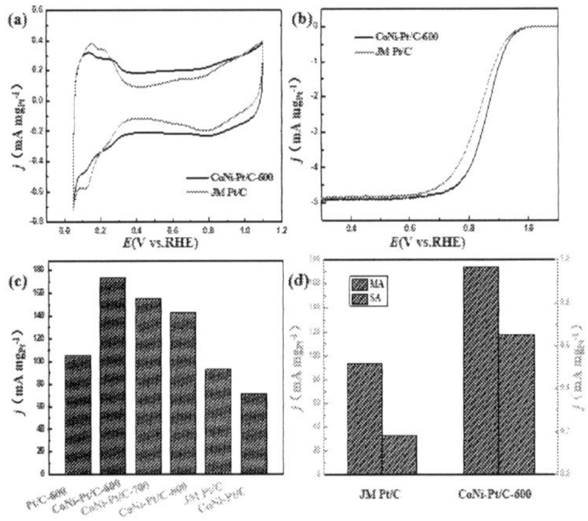

Figure 5 (a) The CV curves of JM Pt/C and CoNi-Pt/C-600 in N_2-saturated 0.1 M $HClO_4$ at a scan rate of 50 mV/s, (b) The LSV curves of JM Pt/C and CoNi-Pt/C-600 in O_2-saturated 0.1 M $HClO_4$ at 1600 rpm with a scan rate of 10 mV/s, (c) A comparison of the mass activity (MA) between CoNi-Pt/C-600 and the prepared catalyst, (d) A comparison of the MA and surface area activity (SA) between JM Pt/C and CoNi-Pt/C-600.

The half-wave potentials of CoNi-Pt/C-600 catalyst were measured to be 0.860 V, significantly higher than the 0.835 V (vs. RHE) of the JM Pt/C catalyst. Despite different annealing temperatures (Figure 5c), CoNi-Pt/C-600 still exhibited the highest half-wave potential, while the half-wave potentials of CoNi-Pt/C-700 and CoNi-Pt/C-800 were similar but still higher than that of JM Pt/C. This indicates that annealing treatment is advantageous for improving the ORR performance of CoNi-Pt/C, with 600°C being the optimal annealing temperature. Although the surface atomic composition of annealed CoNi-Pt/C is slightly inferior to that of JM Pt/C, their ORR electrocatalytic activity is higher than that of JM Pt/C, suggesting that annealed alloy catalysts possess better surface activity. Therefore, based on the Koutecky-Levich (K-L) equation, we calculated the dynamic current at 0.9 V and used this to determine the specific activity (SA) and mass activity (MA) of the catalyst.

The mass activities of CoNi-Pt/C-600, CoNi-Pt/C-700, and CoNi-Pt/C-800 are 174.02 mA/mg_{Pt}, 155.60 mA/mg_{Pt}, and 43.42 mA/mg_{Pt}, respectively, while JM Pt/C is only 93.22 mA/mg_{Pt}. It can be seen that the mass activity (MA) of CoNi-Pt/C-600 is 1.87 times that of JM Pt/C. Furthermore, based on the previously calculated effective surface area (ECSA) values, we determined the corresponding specific activity (SA). The specific activity of CoNi-Pt/C-600 is 0.6537 mA /m^2_{Pt}, which is 3.60 times that of JM Pt/C (0.1814 mA /m^2_{Pt}) (Figure 5c). It can be inferred that the increase in MA and SA is likely due to the optimization of the catalyst's electronic structure and crystal geometry after annealing treatment.

Based on the data obtained from the above tests, it can be concluded that the CoNi-Pt/C-600 catalyst treated at 600°C for 0.5 hours exhibits the highest electrocatalytic activity. Therefore, we conducted further durability tests on CoNi-Pt/C-600 and JM Pt/C, as durability is an important indicator for evaluating the catalytic performance of electrocatalysts.

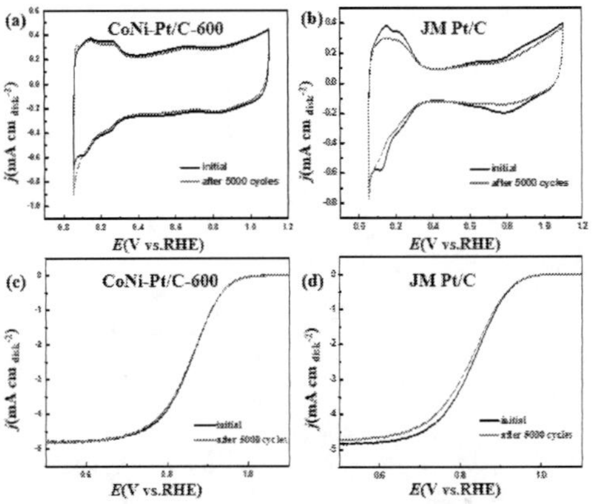

Figure 6 (a) CV curves of CoNi-Pt/C-600 before and after durability testing, (b) CV curves of JM-Pt/C before and after durability testing, (c) LSV curves of CoNi-Pt/C-600 before and after durability testing, (d) LSV curves of JM-Pt/C before and after durability testing.

The CV curves (Figure 6a, b) and LSV curves (Figure 6c, d) of CoNi-Pt/C-600 and JM Pt/C before and after the durability test are shown in the figures. For JM Pt/C, after 5000 cycles of durability testing, the CV curves exhibited a significant contraction, and the ORR polarization curves shifted towards lower potentials. The half-wave potential decreased by 7 mV after 5000 cycles. In contrast, CoNi-Pt/C-600 showed almost no decrease in half-wave potential after 5000 cycles, and the CV curves did not exhibit a noticeable contraction trend. This indicates that the electrocatalytic activity of CoNi-Pt/C-600 remained almost unchanged.

Comparing the ORR curves of CoNi-Pt/C-600 and JM Pt/C before and after durability testing, it was found that the half-wave potential of JM Pt/C was initially 25 mV lower than that of CoNi-Pt/C-600. After durability testing, the difference increased to 28.5 mV. These data indicate that CoNi-Pt/C-600 exhibits better ORR activity and superior durability compared to JM Pt/C.

4 Conclusion

We synthesized a ternary Pt-based alloy catalyst, CoNi-Pt/C-600, which displayed outstanding electrocatalytic performance in acidic solutions following a heat treatment at 600°C. X-ray diffraction (XRD), X-ray photoelectron spectroscopy (XPS), and transmission electron microscopy (TEM) analyses confirmed the presence of Pt, Co, and Ni metals in the catalyst particles. Electrochemical testing revealed that the MA and SA values of the catalyst were 1.87 and 3.60 times higher than those of JM Pt/C, respectively. Additionally, after undergoing durability testing, the electrochemical performance of JM Pt/C significantly deteriorated, while CoNi-Pt/C-600 maintained its relatively stable performance, which was further supported by TEM images. These findings clearly demonstrate that CoNi-Pt/C-600 outperforms JM Pt/C as a superior catalyst. This study highlights the ability to fabricate three-metal alloy catalysts with improved oxygen reduction reaction (ORR) activity and stability through the manipulation of the catalyst particles' electronic and crystal structures. It provides a reliable pathway for the advancement of electrocatalyst development.

References

1. Berger, M. Goldfarb, J. Understanding our energy footprint: Undergraduate chemistry laboratory investigation of environmental impacts of solid fossil fuel wastes. *Journal of Chemical Education* **2017,** *94* (8), 1124-1128.
2. Javed, A. Palafox Gonzalez, P. Thangadurai, V. Critical review of electrolytes for advanced low-and high-temperature polymer electrolyte membrane fuel cells. *ACS Applied Materials & Interfaces* **2023,** *15* (25), 29674-29699.
3. Gerber, I. Serp, P. A theory/experience description of support effects in carbon-supported catalysts. *Chemical Reviews* **2020,** *120* (2), 1250-1349.
4. Shao, M. Chang, Q. Dodelet, J. Chenitz, R. Recent advances in electrocatalysts for oxygen reduction reaction. *Chemical Reviews* **2016,** *116* (6), 3594-3657.
5. Yang, T. Ma, Y. Huang, Q. He, M. Cao, G. Sun, X. Zhang, D. Wang, M. Zhao, H. Tong, Z. High durable ternary nanodendrites as effective catalysts for oxygen reduction reaction. *ACS Applied Materials & Interfaces* **2016,** *8* (36), 23646-23654.
6. Huang, C. Zhang, W. Li, J. Wei, Z. Pt alloy nanocrystal catalysts: Diminishing sizes and improving stability. *The Journal of Physical Chemistry C* **2023,** *127* (27), 12827-12840.
7. Deshpande, P. Prasad, B. Alloying with Mn enhances the activity and durability of the CoPt catalyst toward the methanol oxidation reaction. *ACS Applied Materials & Interfaces* **2023,** *15* (22), 26554-26562.
8. Bhalothia, D. Yan, C. Hiraoka, N. Ishii, H. Liao, Y. Chen, P. Wang, K. Chou, J Dai, S. Chen, T. Pt-mediated interface engineering boosts the oxygen

reduction reaction performance of Ni hydroxide-supported Pd nanoparticles. *ACS Applied Materials & Interfaces* **2023,** *15* (12), 16177-16188.

9. Zhang, P. Dai, X. Zhang, X. Chen, Z. Yang, Y. Sun, H. Wang, X. Wang, H. Wang, M. Su, H. Li, D. Li, X. Qin, Y. One-pot synthesis of ternary Pt–Ni–Cu nanocrystals with high catalytic performance. *Chemistry of Materials* **2015,** *27* (18), 6402-6410.
10. Lokanathan, M. Patil, I. M. Mukherjee, P. Swami, A. Kakade, B. Molten-salt synthesis of Pt_3Co binary alloy nanoplates as excellent and durable electrocatalysts toward oxygen electroreduction. *ACS Sustainable Chemistry & Engineering* **2020,** *8* (2), 986-993.
11. Huang, J. Sementa, L. Liu, Z. Barcaro, G. Feng, M. Liu, E. Jiao, L. Xu, M. Leshchev, D. Lee, S. Li, M. Wan, C. Zhu, E. Liu, Y. Peng, B. Duan, X. Experimental sabatier plot for predictive design of active and stable Pt-alloy oxygen reduction reaction catalysts. *Nature Catalysis* **2022,** *5* (6), 513-523.
12. Zhang, C. Zhang, R. Li, X. Chen, W. PtNi nanocrystals supported on hollow carbon spheres: Enhancing the electrocatalytic performance through high-temperature annealing and electrochemical CO stripping treatments. *ACS Applied Materials & Interfaces* **2017,** *9* (35), 29623-29632.
13. Wanjala, B. Loukrakpam, R. Luo, J. Njoki, P. Mott, D. Zhong, C. Shao, M. Thermal treatment of PtNiCo electrocatalysts: Effects of nanoscale strain and structure on the activity and stability for the oxygen reduction reaction. *The Journal of Physical Chemistry C* **2010,** *114* (41), 17580-17590.
14. Gao, P. Cai, Y. Wang, F. Zhu, H. Pu, M. Pt-based trimetallic nanocrystals with high proportions of M (M=Fe, Ni) metals for catalyzing oxygen reduction reaction. *International Journal of Hydrogen Energy* **2020,** *45* (32), 16039-16048.

 https://doi.org/10.1051/e3sconf/202344101023

Study on regeneration performance of carbon fluoride adsorbent in SF_6 gas

Wei CHENG [a*], Feng ZHU [b], Chen HANG [c], ZhengJie XU [d]

State Grid Anhui Electric Power Co., Ltd. Electric Power Science Research Institute, Hefei, Anhui, China

Abstract: To study the regeneration of CF-100 type microcrystalline material after adsorbing carbon fluorides in SF_6, an experimental device was built that can achieve both vacuum desorption of carbon fluorides and auxiliary heating to promote vacuum desorption of carbon fluorides. Thermogravimetric analyzer and Fourier transform infrared spectrometer were used to analyze the optimal treatment temperature and time of regeneration technology for promoting vacuum desorption of fluorocarbon by auxiliary heating. A comparative study was conducted on the two regeneration techniques mentioned above by analyzing the relationship between adsorption performance and the number of regenerations. The regeneration effects were also validated by using a specific surface area tester to examine the test data. The results demonstrate that desorption of fluorocarbons from CF-100 microcrystalline material is more efficient using a vacuum desorption method aided by auxiliary heating, as opposed to pure vacuum desorption. The CF-100 type of microcrystalline material achieves the best desorption effect when heated to 300°C with an auxiliary heating duration of 2 h, with an activation energy of 293.933 KJ/mol. After activation, CF-100 microcrystalline materials exhibit stable adsorption and desorption performance for decarbonization and possess excellent recyclability.

1 Introduction

Sulfur hexafluoride (SF_6) is extensively employed in high-voltage electrical apparatus owing to its exceptional insulating and arc-extinguishing characteristics, alongside its chemical stability[1, 2]. The operation and maintenance experience shows that electrical equipment utilizing SF_6 gas generates a noteworthy quantity of carbon fluorides (e.g. C_2F_6 and C_3F_8) during extended periods of functioning. This results in a diminished capacity for both insulation and arc extinguishing properties of SF_6 gas. Therefore, in order to ensure the safe and stable operation of electrical equipment, it is necessary to replace the SF_6 gas in the gas chamber when the impurity gas content is high as mentioned above. However, SF_6 gas, as one of the six greenhouse gases explicitly required to be restricted in Kyoto Protocol, is obviously not in line with the target of "double carbon" if it is directly discharged after being replaced. To achieve green development, current grid enterprises will purify the retired SF_6 gas to meet the new gas standards in GB/T 12022 Industrial Hexafluoride, which requires C_2F_6 content in SF_6 gas to be ≤200 ppmw and C_3F_8 content to be ≤50 ppmw, in order to accomplish SF_6 gas recycling[3,4]. Currently, deep cryogenic distillation is mainly used to remove fluorocarbons from SF_6 gas. However, this method has a slow purification rate and high energy consumption, which is not in line with the concept of low carbon[5]. In comparison, the utilization of adsorbent adsorption presents numerous benefits including prompt purification velocity, reduced energy consumption, convenience in operation and regeneration, which largely render it an optimal purification means. Nonetheless, in the present, there exists a limited number of investigations pertaining to adsorbents amenable to adsorbing carbon fluoride in SF_6 gas, and these have yet to be employed in practical engineering applications. The CF-100 microcrystalline material adsorbent was examined and evaluated by the author, and the outcomes of the static adsorption test are presented in Table 1. This particular adsorbent displays remarkable aptitude in averting the co-adsorption occurrence during the separation process of SF_6 gas from fluorocarbon, while exhibiting commendable performance in terms of adsorption and separation[6-8]. The adsorbent will reach its saturation point upon adsorbing a considerable amount of impurity gas, necessitating the replacement or regeneration of the adsorbent to uphold its capacity for adsorption and separation efficacy. Nevertheless, the exorbitant cost of microcrystalline materials proves to be a significant impediment to their direct substitution with new adsorbents. Furthermore, the typical placement of adsorbents within the adsorption tower in the purification equipment necessitates the need for frequent disassembly, which not only entails intricate operations but also exacerbates the risk of gas leakage. The rejuvenation of the adsorbent primarily employs technological methods

[a*]Corresponding author: ali88945@sina.com, [b] feng342524@126.com, [c] whhangchen@126.com, [d] XuZhengJie129@163.com

such as nitrogen replacement, heating, and vacuuming[9]. This process enables the adsorbent to transition from a state of adsorption saturation to a non-adsorption state, or a state of minimal adsorption, without necessitating the replacement of the adsorbent. This method is characterized by its simplicity of operation and its marked efficacy in addressing the problem of adsorbent saturatio.

Table 1 Experimental data of CF-100 microcrystalline material adsorbing fluorocarbon in SF_6.

Adsorbent dosage /kg	Adsorption pressure /Mpa	Adsorption state	C_2F_6 /ppm	C_3F_8 /ppm
2	0.4	Before	550.81	88.60
		After	425.32	55.99

Currently, the predominant application of the purification and recovery device is to achieve the adsorption and segregation of carbon fluoride in SF_6 gas. However, nitrogen purging necessitates an external gas source, consequently augmenting the burden of manual upkeep and failing to meet the genuine engineering prerequisites. Therefore, this paper studies the regeneration technology of fluorocarbon adsorbent in SF_6 gas based on vacuum regeneration and vacuum desorption regeneration of auxiliary heating: An experimental device that can realize vacuum regeneration and high temperature regeneration at the same time was built, and the desorption regeneration of fluorocarbon adsorbent was studied, and the optimal desorption regeneration mode was selected by studying the relationship between adsorption performance and regeneration times of adsorbent. By conducting an analysis of the TG/DTG thermogravimetric curve and Fourier infrared spectrum structure, the optimal desorption temperature and duration were determined. Compare the specific surface area and pore size of the adsorbent before and after regeneration to evaluate the desorption performance, so as to achieve efficient regeneration of microcrystalline material adsorbent and realize the removal of carbon fluoride in SF_6.

2 Experiment

2.1 Sample Treatment and Characterization Experiment

2.1.1 Sample Pretreatment of Microcrystalline Materials

The specimens employed in the investigation were microcrystalline materials denoted as CF-100, which were acquired from Hubei Shentan Environmental Protection New Materials Co., Ltd. The microcrystalline materials without adsorption, after adsorption saturation and after regeneration experiment are respectively marked as three groups: A, B and C. Each group of adsorbents is divided into several parts, each part is 10 g.

The main experimental and analytical instruments used are shown in Table 2:

Table 2 Main experimental and analytical instruments

Instrument name	Model	Manufacturer
FESEM	Hitachi SU8010	Hitachi, Japan
Thermogravimetric analyzer(TG)	TGA/DSC3+	Mettler Toledo, Switzerland
Fourier transform infrared spectrometer(FTIR)	Nicolet IS10	Thermo Fisher Scientific
Full-automatic specific surface area and void analyzer	ASAP 2460	Micromeritics

2.1.2 Experimental steps of adsorbent characterization

(1) SEM experiment

A conventional cold field emission scanning electron microscope (SEM, Hitachi SU8010, Hitachi, Japan) was employed to analyze the micro-morphology of group A microcrystalline materials without adsorption and group B microcrystalline materials with adsorption. The conductive tape was affixed onto the sample stage, and subsequently, a small portion of microcrystalline material from groups A and B was placed onto the surface of the conductive double-sided tape separately. Lastly, the samples were observed following gold spraying using a gold spraying instrument.

(2) TG experiment

Temperature is the most direct factor affecting the desorption effect of adsorbent. Thermogravimetric analyzer (TGA/DSC3+, Mettler Toledo, Switzerland) was used to analyze the pyrolysis characteristics of group B microcrystalline materials at different heating rates (5, 10, 20 and 30 °C/min), and it was found that the desorption effect changed significantly at different temperatures, so as to obtain the best desorption temperature of group B microcrystalline materials. Due to the discernible impact of numerous factors on the thermogravimetric experiment, it is imperative to devise an experimental arrangement that can eliminate extraneous variables and guarantee the validity of the results afore conducting the experiment. Hence, prior to embarking upon the experimental procedures, it is imperative to perform calibration of the balance, desiccate the samples, and guarantee that the sample size is within the range of 1/3 to 1/2 of the crucible volume for the execution of repeated experiments. Take about 10 mg of the sample and put it in a ceramic crucible with holes. The carrier gas flow is high-purity nitrogen (N_2), the flow rate is set to 50 mL/min, and the final temperature of the experiment is set to 900°C. After the thermogravimetric experiment, the thermogravimetric (TG) and differential thermogravimetric (DTG) curves of group B microcrystalline materials are obtained respectively.

(3) FTIR experiment

The optimal desorption time of microcrystalline materials in groups A and B was analyzed using a Fourier transform infrared spectrometer produced by Thermo Fisher Scientific. Samples from group A and group B, with a particle size below 2 μm and mass ranging from 1-2 mg, were co-ground together with 200 mg of potassium bromide (KBr). After uniformly mixing the adsorbent and

KBr, the resulting mixture was then shaped into transparent sheets through the use of an oil press for subsequent determination. Set the wave number range of infrared scanning: 4000~400 cm^{-1}; The temperature of the infrared gas delivery pipe and the gas detection cell is set to 250°C; The scanning frequency is: 8 times/min; Resolution of scanning: 0.4 cm^{-1}.

2.2 Experimental study on regeneration performance of adsorbent

2.2.1 Composition and process of experimental device

(1) Composition of experimental device

This experiment designed and built a Carbfluoride adsorbent desorption experimental device as shown in figure 1, which can simultaneously achieve the functions of vacuum desorption with auxiliary heating and vacuum desorption. It mainly consists of an adsorption column, heating system, vacuum pumping system, analysis system, tail gas treatment system, and data processing system. The adsorption column is an experimental container for microcrystalline materials to adsorb carbon fluoride. The gas outlet end of the adsorption column is equipped with a needle valve and a quick-release connector, which is easy to disassemble; The heating system consists of a heating and thermal insulation steel sleeve wrapped on the surface of the adsorption column, which plays the role of thermal insulation and temperature control; The vacuum pump in that vacuumize system is used for vacuum regeneration to vacuumize the device to remove impurity gas; The gas concentration detector in the analysis system is used to detect the desorbed fluorocarbon gas in the adsorption column pumped by the vacuum pump; The tail gas treatment system is used to treat the desorbed tail gas containing C_2F_6, C_3F_8, etc. In the data processing system, the computer is used to collect and sort out the data of pressure, temperature and gas detection system.

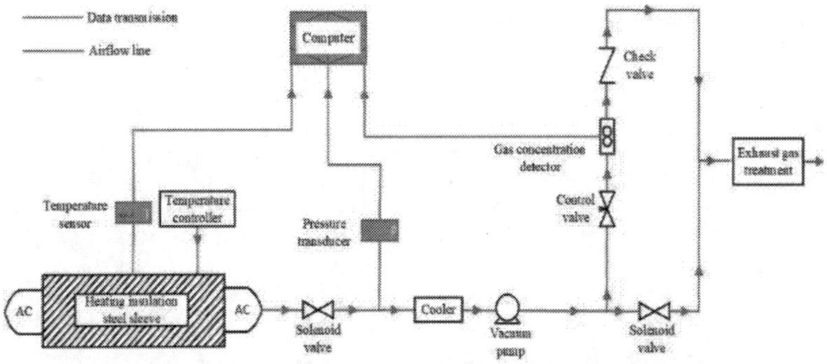

Fig. 1 Experimental device for desorption of fluorocarbon adsorbent

(2) Experimental process

In the vacuum desorption regeneration experiment, the adsorption column is pumped to vacuum by a vacuum pump, and the adsorbent is desorbed and regenerated in a vacuum environment. The specific operation flow is as follows: ① After the adsorption saturation experiment, the adsorption column filled with microcrystalline material is installed on the desorption experimental device; ② Turn on the vacuum pump to vacuum the adsorption column to 133 Pa for 60 min; ③ Close the vacuum pump and the outlet valve at the back end of the adsorption column, and the vacuum regeneration is finished.

In the vacuum desorption regeneration experiment employing auxiliary heating, the microcrystalline material undergoes rejuvenation utilizing the method of vacuum desorption with auxiliary heating. Initially, the heat preservation steel casing is subjected to high temperature in accordance with the predetermined temperature (established through thermogravimetric analysis) to activate the carbon fluorides adsorbent existing within the adsorption column. Following this, a vacuum pump is utilized to conduct evacuation of the adsorption column.

During the above-mentioned desorption experiment, the volume fraction changes of C_2F_6 and C_3F_8 in the desorption gas were monitored in real time to judge the desorption degree of C_2F_6 and C_3F_8 on the adsorbent. When the gas concentration is detected that the volume fraction of C_2F_6 and C_3F_8 becomes 0, the regeneration is terminated, and both the detected tail gas and the desorbed gas pass through the tail gas treatment device, so as to solve the problems of safety hazards and emission pollution caused by carbon-containing fluoride gas.

3 Results and discussion

3.1 Micro-morphological analysis

In order to explore the difference of micro-morphology of CF-100 microcrystalline materials before and after adsorption of fluorocarbon in SF_6 gas, it was characterized by SEM. As shown in Figure 2, scanning images with scanning widths of 1 μm, 10 μm and 50 μm and working voltage of 3.0 kV were selected during the experiment. As can be seen from the figure, the microcrystalline material before adsorption shows a loose and irregular rod-like structure, and the microcrystalline material after adsorption of fluorocarbon has a smooth and compact columnar structure with uniform distribution and smaller

 https://doi.org/10.1051/e3sconf/202344101023

pores, which conforms to the adsorption characteristics of the adsorbent.

Fig. 2 SEM images of microcrystalline materials before and after adsorption of fluorocarbon: (a) before adsorption; (b) After adsorption.

3.2 The optimal processing temperature for thermogravimetric analysis.

In thermal analysis, the reaction rate of microcrystalline materials varies with temperature, which is obtained by observing the change of its mass and the decomposition rate varying with temperature when heating the sample at a fixed heating rate[10]. According to the thermal analysis kinetics theory, the pyrolysis process of microcrystalline material adsorbent is generally expressed as[11]:

$$A(s) \rightarrow B(s) + C(g) \tag{1}$$

In heterogeneous systems and non-isothermal conditions, the reaction kinetics equation is expressed as[12]:

$$\frac{d\alpha}{dT} = \left(\frac{1}{\beta}\right) k(T) f(\alpha) \tag{2}$$

Where: T is the thermodynamic temperature, K; β is the heating rate (generally constant), °C/min; α is the conversion rate, generally, $\alpha = \frac{W_0 - W}{W_0 - W_\infty}$, W_0 is the original mass of the adsorbent, W is the residual mass of the adsorbent at the temperature of T, and W_f is the residual mass of the adsorbent at the end of pyrolysis; $f(\alpha)$ is the expression of reaction mechanism.

The relationship between $k(T)$ and thermodynamic temperature T can be expressed by Arrhenius equation[13]:

$$k = A \exp\left(-\frac{E}{RT}\right) \tag{3}$$

Where: A stands for pre-exponential factor; E represents apparent activation energy; R represents the molar gas constant, 8.314 J/(mol•K).

The equation obtained by simultaneous expression (2) and expression (3) is:

$$\frac{d\alpha}{dT} = \frac{A}{\beta} \exp\left(-\frac{E}{RT}\right) f(\alpha) \tag{4}$$

After performing differentiation and integration on equation (4), the parameters can be determined. The differential method calculates the first derivative of temperature by using the conversion rate. The process of integration necessitates the introduction of a temperature incorporation formula to integrate equation (4) on both sides. Nonetheless, the integral form frequently leads to imprecise outcomes, hence an estimated formula is commonly employed in lieu of the integration formula to facilitate computation[14, 15].

3.2.1 Pyrolysis Characteristics of Microcrystalline Materials

Temperature is one of the important factors affecting the desorption performance of adsorbents. Thermogravimetric analyzer was used to heat the adsorbed microcrystalline material to 900°C at four different heating rates of 5, 10, 20 and 30 °C/min. As shown in Figure 3, the weight loss rate of microcrystalline material adsorbent is about 20% between room temperature and 300°C, which is a rapid pyrolysis stage. After the temperature reaches 300°C, the TG curve is gradually flat, which shows that the gas adsorbed in microcrystalline materials has basically been desorbed in the rapid pyrolysis stage, and the desorption effect is not obvious if the temperature continues to increase. Therefore, it can be preliminarily determined that the optimal treatment temperature of CF-100 microcrystalline material adsorbent is 300°C.

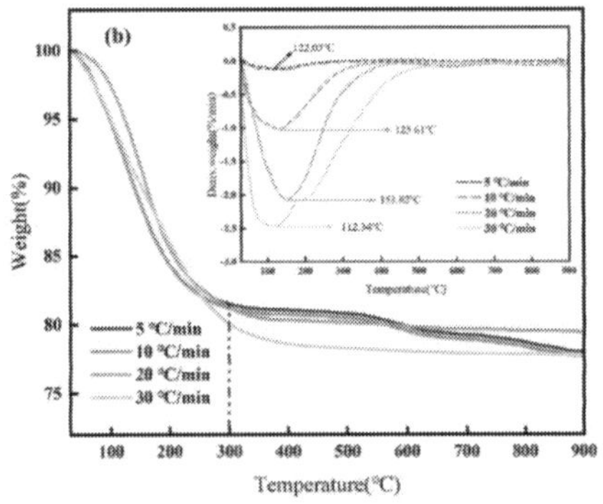

Fig. 3 TG/DTG curves of microcrystalline materials before and after adsorption at different heating rates.

3.2.2 Solution of thermodynamic parameters

Based on the recommendations of the ICTAC kinetics committee, due to the complex nature of the carbon fluorine adsorbent material in this experiment, a non-model method was employed to directly obtain kinetic parameters for different conversion rates. This approach avoids the limitations of conducting thermokinetic research using a single TG curve and reduces errors caused by differences in thermokinetic mechanisms.

In this study, the Flynn-Wall-Ozawa (FWO) integral method was selected for the calculation of activation energy, and the corresponding equation is shown below[16]:

 https://doi.org/10.1051/e3sconf/202344101023

$$\lg\beta = \lg\left(\frac{A_F E_F}{RG(\alpha)}\right) - 2.315 - 0.4567\frac{E_F}{RT} \quad (5)$$

Where E_F is the average activation energy calculated by FWO method, KJ/mol. Through the utilization of Equation (5), with varying conversion rates and four distinct heating rates, and the corresponding thermodynamic temperatures, a determination of activation energy was made by analyzing the slope of the curve produced by the lgβ-1/T relationship. The resulting calculation for E_F yielded a value of 293.933 KJ/mol. The high linearity of the fitting degree at different conversion rates in Figure 4 illustrates the accuracy of the FWO method in analyzing the pyrolysis process results. This method avoids the use of reaction mechanism functions and directly calculates the activation energy. The range of α is selected as 0.05 to 0.7, and linear fitting is performed under different heating rates of 5, 10, 20, and 30 °C/min. The data presented in the diagram reveals that when the value of α exceeds or equals 0.4, there is observed variance in the activation energy with respect to α. This phenomenon is indicative of a shift in the decomposition mechanism of the sample, which raises the possibility of the presence of a multi-stage decomposition mechanism having multiple types of bonds and a multitude of phase conversion characteristics.

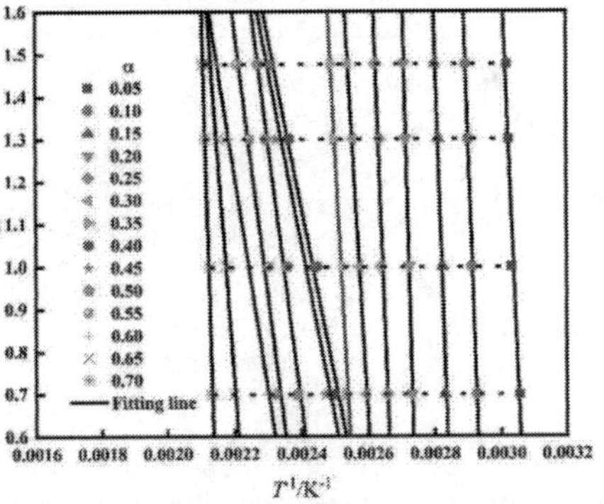

Fig. 4 Relationship between apparent activation energy and different conversion rates

3.3 Optimal processing time for infrared analysis

The adsorption capacity measurements for Group B microcrystalline materials with varying processing durations are illustrated in Figure 5(a). When the treatment duration is below 2 h, the specific surface area and specific volume both elevate as the heat treatment time increases. They culminate at 2 h, implying that the optimal restoration of the adsorption capacity in microcrystalline materials is achieved after 2 h of heat treatment.

Subsequently, Fourier transform infrared spectrometer (FTIR) was used to analyze the chemical composition of group B microcrystalline materials with heat treatment time of 0 h and 2 h, respectively. As shown in Figure 5(b), there were strong absorption peaks at wave numbers of 979.27 and 949.39 cm^{-1}, which were the main characteristic peaks of fluorocarbon, and the wave number peaks of microcrystalline materials with treatment time of 0 h were higher than those of 2 h, indicating that the longer the treatment time, the decrease or even disappearance of C_3F_8 and C_2F_6.

Absorption peaks are observed at wavenumbers around 3475, 1640, and 660 cm^{-1}, which are attributed to the asymmetric stretching and bending vibrations of the O-H bond in water molecules. The microcrystalline material treated for 2 h shows a significant decrease in water content. Absorption peaks between 440 and 754 cm^{-1} are decomposition products of SF_6, such as SOF_4 and SO_2F_2. With the prolongation of processing time, the absorption peaks of characteristic products weaken until they disappear. This indicates that under the conditions of a thermal treatment temperature of 300°C and a time of 2 h, the desorbed gas inside the adsorbent can be significantly reduced.

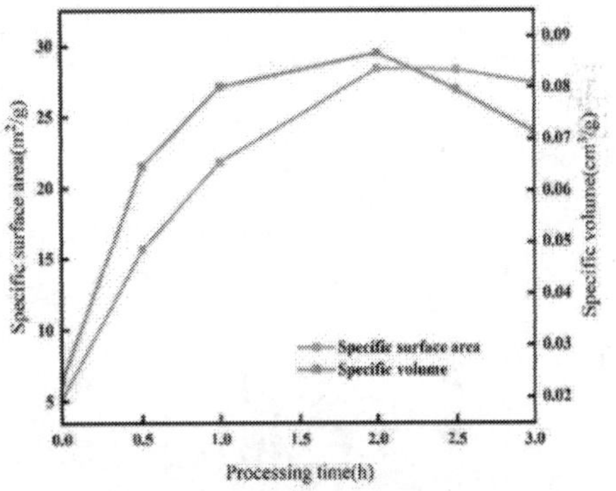

(a)

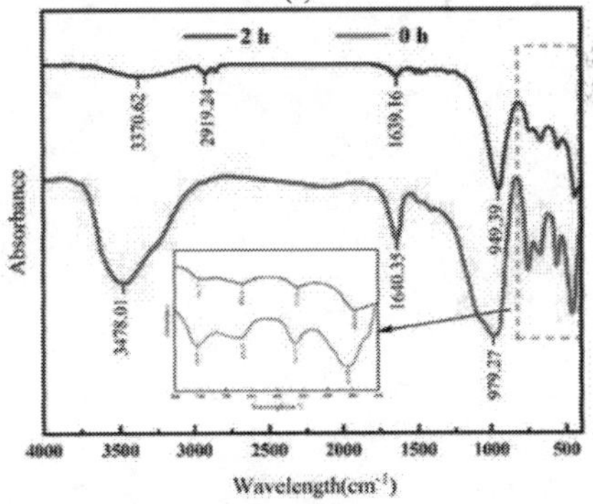

(b)

Fig. 5 Adsorption capacity index and infrared spectrum of group B microcrystalline materials at different treatment times

3.4 Test results of regeneration performance of microcrystalline materials

3.4.1 Cycle stability evaluation of different regeneration methods

The cycling stability of adsorbents is one of the important factors for evaluating the regeneration performance, and it is closely related to the regeneration method of the adsorbent. Therefore, two regeneration methods, vacuum desorption with vacuum pumping and vacuum desorption with auxiliary heating, mentioned in the previous text, were selected to evaluate the cycling stability of the microcrystalline material.

Saturated adsorption capacity refers to the adsorption capacity of fluorocarbon under the equilibrium pressure of 1.0 MPa. The relationship between the regeneration capacity of CF-100 microcrystalline material adsorbent and the regeneration times is shown in Figure 6. From the figure, it can be seen that with the increase of the regeneration times, the adsorption capacity of fluorocarbon changes after two regeneration methods, in which the adsorption capacity of fluorocarbon of microcrystalline material decreases slightly after five times of vacuum pumping and high temperature regeneration, and it is almost ignored, while the adsorption capacity of fluorocarbon decreases greatly after the third vacuum pumping regeneration. The experimental device can realize two kinds of desorption and regeneration experimental functions well. The experiment proves that the desorption and regeneration mode of microcrystalline material adsorbent for fluorocarbon in SF_6 is better than vacuum regeneration. Therefore, the vacuum desorption regeneration performance of auxiliary heating is explored later.

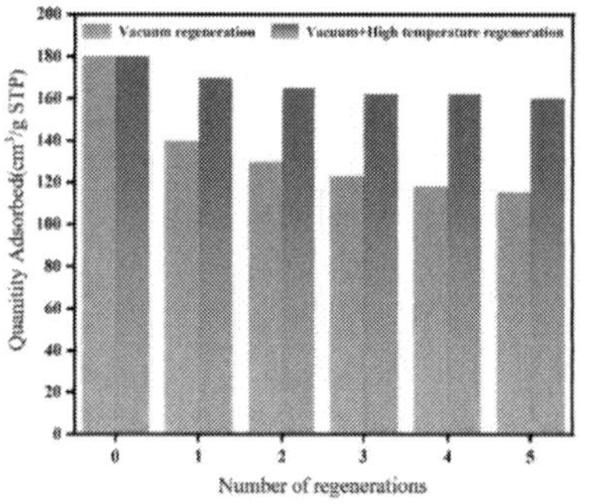

Fig. 6 Relationship between regeneration ability and regeneration times of CF-100 microcrystalline material adsorbent

3.4.2 Study on regeneration performance of high temperature desorption

The adsorption isotherm is a curve depicting the variations in adsorption quantity or adsorption rate of an adsorbent with respect to changes in pressure under constant temperature conditions[17]. It is an essential tool for evaluating the separation process between adsorbates and adsorbents. IUPAC classifies the latest adsorption isotherms into six types, namely, type I Langmuir isotherm, type II S isotherm, type III isotherm (without B inflection point) (the adsorption force between sample and adsorbate is smaller than that between adsorbate), type IV isotherm (with hysteresis loop) (the curve at the beginning of the curve bulges upward, and the isotherm at the end of the curve rises sharply due to capillary condensation), and type V isotherm (with a higher bulge toward the opposite pressure axis), VI type step isotherm[18]. Since the fluorocarbon in SF_6 gas in high-voltage electrical equipment is a mixed gas containing C_2F_6 and C_3F_8, the adsorption isotherm curve equation of the mixed gas, that is, the extended Langmuir equation[19], can be derived by the principle of adsorption potential energy and adsorption heat.

$$q_i = \frac{q_{mi} B_i p_i}{1 + \sum_{j=1}^{n} B_j p_j} \tag{6}$$

In the equation, q represents adsorption amount; q_m represents the saturated adsorption amount; p represents total pressure; B represents Langmuir constant. When correlating data, the Langmuir linear form is commonly used:

$$\frac{1}{q} = \frac{1}{q_m} + \frac{1}{q_m B p} \tag{7}$$

The adsorption parameters B and q_m of single gas can be obtained by using the relationship curve of $\frac{1}{q}$ to $\frac{1}{p}$, and then the adsorption amount of mixed gas can be further predicted by formula (6). At the same time, the specific surface area S_g and specific volume V_m of the adsorbent can be obtained respectively according to the measured adsorption parameters q_m of mixed gas combined with formula (8) and formula (9):

$$Sg = q_m \cdot A \cdot \sigma_m \tag{8}$$

$$Vm = \frac{q_m \times 44}{22400 \rho_m} \tag{9}$$

Where: A represents Avon Gadereau constant, which is 6.02×10^{23}; σ_m is the cross-sectional area of adsorbate molecules; ρ_m is the density of adsorbed phase molecules at 273 K.

The specific surface area and pore size of microcrystalline materials undergo variations with changes in their adsorption capacity. By examining the adsorption data of three groups of microcrystalline materials, namely A, B, and C, with varying adsorption capacities before adsorption, after reaching saturation, and after a 2-hour desorption process, the ability of the microcrystalline material to regenerate after high temperature desorption can be assessed. Figure 7 shows the adsorption-desorption isotherms and pore size distribution charts of three groups of microcrystalline materials A, B, and C. The curves exhibit type IV isotherms and H3 hysteresis loops, indicating that these materials possess mesoporous structures. They undergo monolayer adsorption before proceeding to multilayer adsorption.

The specific surface area and pore structure of groups A, B and C microcrystalline materials are shown in Table 3. From the experimental results in the table, it can be seen that the specific surface area and pore volume of CF-100 microcrystalline materials decreased from 31.669 m^2/g and 0.1074 cm^3/g to 5.1280 m^2/g and 0.0231 cm^3/g, respectively, and the pore size increased from 12.111 nm to 17.525 nm. It shows that the adsorbent has reached the threshold of adsorption capacity, the number of adsorption molecules is low, and the adsorption capacity has reached a great loss. After desorption at 300°C for 2 h, the specific surface area rebounded rapidly, and the specific surface area was slightly lower than that without adsorption, which indicated that the regeneration recovery performance was high.

Table 3 List of specific surface area and pore structure of three groups of microcrystalline materials A, B and C.

Parameters	Samples		
	Non-adsorption	After adsorption	Desorption 2 h
$S_{BET}(m^2/g)$	31.669	5.1280	28.365
$V_{mes}(cm^3/g)$	0.1074	0.0231	0.0867
$D_{ave}(nm)$	12.111	17.525	13.163

S_{BET}: specific surface area measured by the BET method, V_{mes}: mesopores pore volume, D_{ave}: average pore diameter

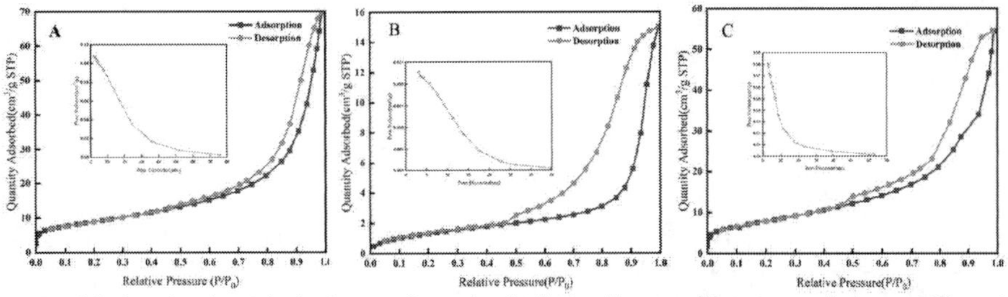

Fig. 7 A, B and C adsorption-desorption isotherms and pore size distribution diagrams of three groups of microcrystalline materials.

4 Conclusion

Based on the experimental device of fluorocarbon adsorbent desorption, which can simultaneously realize Vacuum desorption of auxiliary heating and vacuum desorption, The relationship between adsorption performance and regeneration times, thermogravimetric analysis, infrared spectrum and other characterization methods were used to compare the regeneration performance of CF-100 microcrystalline materials with two regeneration technologies. Get the following conclusions:

(1) Before adsorption, the microcrystalline material had a loose and irregular rod-like structure. After adsorption, the microcrystalline material exhibited a smooth and more compact cylindrical structure that was evenly distributed, with smaller pores, which is in line with the adsorption characteristics of adsorbents.

(2) Analyze the relationship between regeneration ability and regeneration times of CF-100 microcrystalline material adsorbent: The stability of two regeneration methods for desorption is excellent. After five times of vacuumizing and high temperature regeneration, the adsorption capacity of fluorocarbon in microcrystalline materials decreased slightly. However, the adsorption capacity of carbon fluoride decreased greatly after the third vacuum regeneration. It demonstrates that the desorption and regeneration method of carbon fluorides in SF_6 by using microcrystalline material adsorbents shows that the stability of vacuum desorption and regeneration cycle with auxiliary heating is better than that of vacuum regeneration.

(3) According to the data of thermogravimetry and Fourier infrared spectrum, it is concluded that the high temperature desorption reaches the optimal desorption condition when heating treatment is carried out at 300°C for 2 h, Through the relationship between apparent activation energy and different conversion rates, it is known that when $\alpha \geq 0.4$, the activation energy fluctuates with the change of α, and the decomposition mechanism of the sample changes, indicating that there is a multi-stage decomposition mechanism with the diversity of bonds and the multiphase characteristics of their conversion, and the activation energy is 293.933 KJ/mol.

(4) By examining the adsorption data of three groups of microcrystalline materials, with varying adsorption capacities before adsorption, after reaching saturation, and after a 2-hour desorption process, it is concluded that the specific surface area of microcrystalline materials decreases from 31.669 m^2/g to 5.1280 m^2/g after adsorption of fluorocarbon. It shows that the adsorption threshold has been reached at this time, the number of adsorbed molecules is low, and the adsorption capacity has reached a great loss. After desorption at 300°C for 2 h, the specific surface area increased to 28.365 m^2/g, indicating that the regeneration recovery performance was high.

Acknowledgments

This paper is a general project of the National Social Science Fund "Science and Technology Project of State Grid Anhui Electric Power Co., Ltd." (52120522000L)

References

1. Xiao, S., Shi, S.Y., Lin, J.T. (2023) Analysis on the Control Strategy of the Strong Greenhouse Insulating Gas SF6 in High-voltage Electrical Equipment Under the Goal of "Emission Peak and Carbon Neutrality". Proceedings of the CSEE., 43(01):339-358.
2. Luo, Q. (2022) Xi Jinping's Important Statements on Peak Carbon Dioxide Emissions and Carbon Neutrality: Logical Thought, Value Implication and Practice Path. Governance Modernization Studies., 38(04):5-13.
3. GB/T 12022-2004 National Standard for Industrial Sulfur Hexafluoride.
4. Ji, Y.S., Zhang, M., Wang, C.Y., et al. (2021) Study on Decomposition Products of SF6/N2 Gas Mixture Under

the Action of Electric Arc. High Voltage Apparatus., 57(03):145-151+156.

5. Peng, W.W., Shen, X.L., Qian, Y.Q., et al. (2023) Research on Recognition and Diagnosis of Latent Faults of GIS Equipment. NORTHEAST ELECTRIC POWER TECHNOLOGY., 44(08):51-55.
6. Ye, Q.B. (2022) Integrated Technology of Deep Desulfurization and Naphthalene Removal of Coke Oven Gas Based on Microcrystalline Adsorption Process. Shaxi Metallurgy., 45(05):69-70+90.
7. Lu, H., Li, B.X., Sun, B.P., et al. (2023) Crystal Growth and Laser Performance Research Based on Nd: LiLuF4 Microcrystalline Materials. JOURNAL OF SYNTHETIC CRYSTALS., 52(07):1250-1257.
8. Chen, S.H., Liu, X.W., Zou, J.P., et al. (2017) Experimental Study of the Adsorption of CH4/CO2 Mixture by the Regeneration of Molecuar Sieve. The Journal of New industrialization., 7(11):68-72.
9. Chen, P. (2022) INEOS Polypropylene Plant Start-up and Shutdown Process Safety Risks and Solutions. Chemical Enterprise Management., (28):111-114.
10. Chen, Y.X., Zhou, D.S., Hu, W.B. (2021) Progress of Differential Scanning Calorimetry and Its Application in Polymer Characterization. ACTA POLYMERICA SINICA., 52(04):423-444.
11. Vyazovkin, S., Burnham, A.K., Criado, J.M., et al. (2021) ICTAC Kinetics Committee recommendations for performing kinetic computations on thermal analysis data. Thermochimica Acta., 520(1-2):1-19.
12. Huang, Z., Feng, B.B., Zhang, X.R., et al. (2014) Introduction of Thermal Decomposition Kinetic Analysis and Calculation Models. CHINA PRINTING AND PACKAGING STUDY., 6(06):4-17.
13. Minkoo, K., Sunjae, L., Ho, J.K., et al. (2022) Remaining useful life estimation using accelerated degradation test, a gamma process, and the arrhenius model for nuclear power plants. Journal of Mechanical Science and Technology., 36(10):4905-4912.
14. Liu, X.F., Yang, P.B., Wang, J., et al. (2021) Co-Pyrolysis Characterisistics and Kinetic Analysis of Wheat Straw and Lignite. ACTA ENERGIAE SOLARIS SINICA., 42(09):410-415.
15. Zhang, D.R., Lu, L.G. (2021) Pyrolysis and Thermokinetic Characteristics of Calixarene. CHINA PLASTICS., 35(11):104-110.
16. Jiang, G.D., Wei, L.P., Teng, H.P., et al. (2017) A kinetic model based on TGA data for pyrolysis of Zhundong coal. CIESC Journal., 68(04):1415-1422.
17. He, X.D., Zhu, M., Chen, G.Q., et al. (2021) Study on Isotherm Distridution of Low Temperature and Low Pressure Adsorption of Low Temperature Adsorbent. China Special Equipment Safety., 37(04):18-23+74.
18. Cheng, Y.T., Li, B.Y., Li, Z. (2023) MOFs-Based Carbon Materials for Adsorption and Removal of Antibiotics in Water Body. Water Purification Technology., 42(04):31-38.
19. Wei, G., Wang, J.G., Xia, Y.Y., et al. (2014) Regeneration technology of SF6 -molecular-sieve adsorbent using vacuum heat treatment. CHEMICAL ENGINEE R ING(CHINA)., 42(11):35-3.

Effects of vegetation restoration on ecosystem carbon storage and distribution pattern in semi-arid mining area

Yi Wang[1], Xin Zhang[2]*, Qiang Li[1], Xiaodong Ye[1]

[1]China Energy Shendong Coal Group Co., Ltd, Ordos, P.R. China
[2]Institute of Water Resources for pastoral Area Ministry of Water Resources, Hohhot, P.R. China

Abstract: In order to explore the effect of artificial planting to promote vegetation restoration on ecosystem carbon storage in coal mining subsidence area, the distribution pattern of ecosystem carbon storage in mining area during vegetation restoration was studied by space-time substitution method.The results showed that : (1) The biomass of vegetation layer increased gradually with vegetation restoration, and the biomass distribution ratio of each component was shrub layer (92.1%) > litter layer (5.5%) > herb layer (2.4%). The biomass of shrub and herb layer was mainly distributed in the aboveground part, the former accounted for 73.2 %, and the latter accounted for 83.4 %. (2) The carbon content of the vegetation layer was shrub (45.5%) > herb (43.6%) > litter (38.7%). The total carbon storage was in the early initial of vegetation restoration (1657.1 kg hm-2) < mid-term vegetation restoration (4830.6 kg hm-2) < mature stage of vegetation restoration (8446.4 kg hm-2), and shrub was still the main storage part of carbon storage in this layer. In summary, vegetation restoration can significantly increase the carbon storage of the ecosystem, the proportion of carbon storage in the vegetation layer gradually increases.

1 Introduction

Since the industrial revolution, human activities have resulted in the release of large amounts of carbon into the atmosphere and a significant increase in atmospheric CO^2 concentrations[1]. Terrestrial ecosystems are one of the Earth's most important carbon sinks. Enhancing the carbon sink capacity of terrestrial ecosystems can effectively absorb and remove excess CO^2 from the atmosphere and mitigate the greenhouse effect[2] Degradation or restoration of vegetation in terrestrial ecosystems is often accompanied by carbon emissions and carbon sinks[3]. It is estimated that China's terrestrial ecosystems sequestered an average of about 1.1 billion tons of carbon per year between 2010 and 2016, reaching about 45% of carbon emissions in the same period, with ecological projects such as the vegetation restoration of degraded land and the return of farmland to forest and grass contributing the vast majority of ecosystem carbon sinksK. An accurate and valid assessment of carbon stocks as a key measure of the carbon sink function of a carbon pool is of great value to regional carbon cycle and source/sink studies.

The Shendong mining area, located in the southern part of the Ordos Plateau, is a typical ecologically fragile zone. Wind erosion is strong and hydraulic erosion is frequent in the area, especially coal mining activities over the years have caused intense gravitational erosion, with large areas of land subsidence, accompanied by the development of a large number of ground cracks[4]. The multi-dynamic erosion coupled with arid climatic conditions has led to the degradation of the ecosystem in the Shendong mining area, and soil erosion is very serious. A large amount of carbon in the carbon pool was lost as a terrestrial carbon source.

This paper uses the spatial and temporal substitution method to-analyses the changes of carbon stocks in the shrub layer, herb layer, and litter layer in the ecosystem to elucidate the effects of artificial vegetation restoration on the carbon stocks of the ecosystem in the semi-arid coal mining subsidence area.

2 Materials and methods

2.1 Study location

The study area is located in Shangwan Town, Ijinholuo Banner, Ordos City, Inner Mongolia, which belongs to Shenfu-Dongsheng Mining District, and is a typical ecological degradation area formed by the subsidence of the land after coal mining. The southern part of the region is loess hilly area, and the northern part is wind erosion and sanding area of Mao Wu Su. The climate is continental arid and semi-arid, with the characteristics of rain and heat in the same period, rainfall is mostly concentrated in summer and autumn, with an average annual precipitation

*Corresponding author: nmmkszx@163.com

https://doi.org/10.1051/e3sconf/202344101024

of 396 mm and an average annual evaporation of 1790 mm. It is windy and dry in spring and winter, with an average annual wind speed of 3.5 m/s and an average annual temperature of 8.9°C. The soil type in the study area is mainly wind-sandy, with poor soil structure and low organic matter content, and poor water and fertilizer retention. The main plant cover types are shrubs, semi-shrubs and perennial herbs, and common species include Artemisia ordosica, Salix cheilophila, *Hippophae rhamnoides*, Setaria viridis, Stipa bungeana, etc.

2.2 Measurement of indicators

The basic growth parameters (plant height, crown width, basal diameter, etc.) were recorded in each shrub sample plot by plant species, and three standard plants were selected accordingly, and the fresh weight of above-ground and below-ground parts of the standard plants were obtained by the full digging method and sampled proportionally. Herbaceous samples were obtained by mowing method to measure the fresh weight of above and below ground parts and sampled. All litter was harvested, fresh weight was determined, and samples were taken. Plant sample were heat at105°C for 2 hours to sterilizing enzyme,and then dried to constant weight at 85 °C.50 g of the treated samples were taken separately and the plant organic carbon content was determined by the external heating method using potassium dichromate-concentrated sulfuric acid oxidation.

2.3 Data analyses

Calculation of carbon stock in vegetation layer:

$$C = Ws\times CFs + Wh\times CFh + Wl\times CFl$$

where C is the carbon stock of vegetation layer(kg hm-2), Ws is the shrub layer biomass (g m-2), CFs is the shrub layer carbon rate (g kg-1), Wh is the herb layer biomass (g m-2), CFh is the herb layer carbon rate (g kg-1), Wl is the litter biomass (g m-2), and CFl is the litter carbon rate (g kg-1).

Experimental data processing and statistical analysis were performed using Excel and SPSS 20.0 software. Sigmaplot 14.0 software was used for in-text inset mapping. The Least Significant Difference (LSD) method was selected to test the significance of differences (α=0.05) in the spatial and temporal distribution of carbon stocks in the mining ecosystems at different vegetation restoration stages.

3 Results

3.1 Biomass characteristics of different vegetation restoration stages

Table 1 Biomass allocation characteristics of plants layer during vegetation restoration

Sample site Type	Shrubs			Herbaceous		Litter /(g m^{-2})	Total biomass /(g m^{-2})
	Plant species	Aboveground vegetation /(g m^{-2})	Underground vegetation /(g m^{-2})	Aboveground vegetation /(g m-2)	Underground vegetation /(g m-2)		
Initial vegetation restoration	*A. ordosica*	158.3±23.6c	45.8±8.9c	42.5±11.3a	8.3±2.6a	28.4±7.5b	366.2±46.2c
	H. mongolicum	52.6±8.9d	30.3±5.5c				
Mid-term vegetation restoration	*S. psammophila*	566.7±54.1b	278.6±74.1b	19.3±8.2b	4.1±1.3b	60.7±14.9a	1014.3±274.1b
	A. ordosica	58.6±8.2d	26.3±4.2c				
Mature vegetation restoration	*H. rhamnoides*	1233.4±225.4a	369.7±54.9a	2.8±0.7c	0.5±0.2c	77.2±22.1a	1745.5±355.8a
	A. ordosica	40.1±6.7d	21.8±3.7c				

As shown in Table 1, the vegetation biomass differed significantly ($P<0.05$) at different restoration stages and showed a trend of gradual increase with vegetation restoration. The total biomass showed that initial vegetation restoration (366.2 g m-2) < mid-term vegetation restoration (1014.3 g m-2) < mature vegetation restoration (1745.5 g m-2). The shrub layer is the main part of the vegetation layer biomass, with an average proportion of up to 92.1%, and the proportion will gradually increase with the recovery of vegetation. The percentage of biomass in the litter layer was relatively small, averaging 5.5%, and likewise tended to increase with the proportion of vegetation restoration. The herbaceous layer accounted for the smallest proportion, accounting for only 2.4% of the biomass of the entire vegetation layer, and the proportion gradually decreased as the depression of shrubs increased. In terms of the spatial distribution of biomass, the shrub layer biomass was mainly concentrated in the aboveground part, accounting for an average of up to 73.2%, while the Underground part was relatively small, accounting for only 26.8%. The herbaceous layer biomass was likewise mainly concentrated in the aboveground part, with an average of 83.4%, and the Underground part allocated only 16.6% of the biomass.

3.2 Carbon stock characteristics of vegetation layers in different vegetation restoration stages

The overall carbon content of the vegetation layer showed shrubs (45.5%) > herbs (43.6%) > litter (38.7%). The spatial distribution of plant carbon content showed that shrubs were aboveground (48.6%) > underground (42.9%), while herbs were underground (44.9%) > aboveground (42.3%). In terms of variability of carbon content of plant species, S. psammophila (47.8%) > H. rhamnoides (46.4%) > A. ordosica (45.4%) > H. mongolicum (44.2%). From the different stages of vegetation restoration, the average carbon content of the vegetation layer in the middle stage of vegetation restoration was the highest (44.9%), followed by the mature stage of vegetation restoration (44.0%) and the lowest in the early stage of vegetation restoration (43.5%).

As shown in Table 2, there were significant differences ($P<0.05$) in the carbon stocks of the vegetation layer at different restoration stages, and there was a tendency to increase with the restoration of vegetation. The total carbon stock showed that the initial vegetation restoration (1657.1 kg hm-2) < mid-term vegetation restoration (4830.6 kg hm-2) < mature vegetation restoration (8446.4 kg hm-2). The shrub layer is the main storage part of carbon storage in the vegetation layer, with an average proportion of 93.4%, and the proportion of shrub carbon storage is increasing in the process of vegetation restoration. Litter layer is a smaller carbon reservoir, with an average of 4.3% of carbon storage, and the proportion will also increase with vegetation restoration. The herbaceous layer has the weakest carbon storage capacity due to the small stock and insufficient biomass accumulation, accounting for only 2.3% of the carbon stock of the whole vegetation layer. Moreover, as the mature forest of shrubs has encroached on most of the ecological niches, the herbaceous plants have gradually disappeared in the mature stage of vegetation recovery, and the carbon stock is very small. From the spatial distribution of carbon storage, the carbon storage in the shrub layer is dominated by the aboveground part, accounting for up to 76.1%, while the underground part only accounts for 23.9%. The carbon storage in the herbaceous layer is also dominated by the aboveground part, accounting for 82.6% on average, and the underground part accounts for less carbon storage, accounting for only 17.4%.

Table 2 Carbon storage allocation characteristics of plants layer during vegetation restoration

Sample site Type	Shrubs			Herbaceous		Litter /(kg hm-2)	Total carbon stock /(kg hm-2)
	Plant species	Aboveground vegetation /(kg hm-2)	Underground vegetation /(kg hm-2)	Aboveground vegetation /(kg hm-2)	Underground vegetation /(kg hm-2)		
Initial vegetation restoration	A. ordosica	763.0±125.3c	194.7±33.6c	183.2±23.6a	36.7±5.4a	108.5±16.7c	1657.1±223.9c
	H. mongolicum	243.5±36.5d	127.6±21.1cd				
Mid-term vegetation restoration	S. psammophila	2805.2±311.5b	1284.3±253.9b	82.0±16.3b	18.4±2.6b	246.4±37.2b	4830.6±583.4b
	A. ordosica	282.5±47.6d	111.8±14.7d				
Mature vegetation restoration	H. rhamnoides	6327.3±884.1a	1530.6±213.1a	11.6±1.7c	2.3±0.4c	288.7±25.8a	8446.4±672.8a
	A. ordosica	193.3±22.5d	92.7±12.8d				

4 Conclusion

The semi-arid mining area has formed a large area of hollow areas due to coal mining, and the massive appearance of land subsidence and ground cracks, under the combined effect of wind and water erosion, has resulted in serious damage to soil structure, reduced stability of plant communities, and loss of ecosystem carbon pool[5]. Artificially established vegetation restoration measures are an effective means to curb this degradation trend and help to re-establish a good vegetation system in the coal mining subsidence treatment area, reverse land degradation, and protect and increase the ecosystem carbon pool[6]. In this study, the vegetation layer and soil layer are two important components of the carbon pool of the mining ecosystem. In the process of vegetation restoration, the remodeling of microhabitats directly changed the carbon storage capacity of the components, increased the carbon storage of the vegetation-soil system, and formed a distinct spatial and temporal distribution pattern of carbon storage.

In arid and semi-arid regions, where moisture conditions are the main limiting factor for ecosystems, shrubs often become the mainstay of regional ecological restoration and vegetation construction because of their strong adaptability to drought and barrenness[7]. Because of their rapid growth and large root systems, they form an effective plant protection system after a few years of establishment, effectively controlling ground erosion and soil, and constantly changing the soil microcosm[8] The results of the study showed that the shrub layer was the main embodiment of the biomass and carbon storage of the vegetation layer in the mining area, and more than 92% of the vegetation dry matter accumulation and carbon storage were contributed by shrubs. This is consistent with the findings of Wang B et al[9]on carbon stock allocation in desert ecosystems during the reversal of desertification. A study by Sandan et al[10] on the restoration of vegetation in semi-arid grassland mining areas also showed the dominance of shrubs in similar habitats. These results fully demonstrate that the ecological restoration of mining sites with difficult site characteristics should prefer the configuration of native shrubs with excellent traits. It can not only promote positive community succession and maintain stability, but also significantly increase the ecosystem carbon stock and form a good biomass carbon sink. The data analysis showed that 76.1% of carbon storage in the shrub layer was allocated to the aboveground stem and leaf fraction, and the underground root system accounted for a relatively small proportion. This distribution pattern reflects the adaptability of plants to adversity and survival strategies. With low precipitation and shortage of water resources in the Shendong mining area, H. rhamnoides and others can only use a small amount of water infiltrated from the shallow soil after rainfall, which restricts the growth of deep-rooted types, and most of the resources are used for aboveground branch and leaf growth, and root growth is restricted and the carbon storage capacity is relatively weak[11]. It is noteworthy that the cover, biomass and carbon stock of herbaceous plants showed a significant decreasing trend with the shrub establishment process. This may be due to the presence of chemosensory effects of shrubs such as H. rhamnoides and A. ordosica. Compounds such as alkaloids inhibit the seed germination and seedling growth of herbs through root secretion or rain shower, weakening the share of herbaceous layer in the vegetation layer[12-13].

References

1. Zhang Pingping,Li Yan-Hong,Yin Hao-ran, etal.Temporal and spatial changes and dynamic simulation of ecosystem carbon storage in the south-north transition zone of China [J].Journal of Natural Resources,2022,37(05):1183-1197
2. Dubrovina, I. A., Moshkina, E. V., Sidorova, V. A., Tuyunen, A. V., Karpechko, A. Y., Genikova, N. V., Kulakova, L. M. (2021). The Impact of Land Use on Soil Properties and Structure of Ecosystem Carbon Stocks in the Middle Taiga Subzone of Karelia. Eurasian Soil Science, 54(11), 1756-1769. doi:10.1134/s1064229321110053
3. Li miaoyu, Shangguan Zhouping, Denglei. Spatial distribution of carbon storages in the terrestrial ecosystems and its influencing factors on the Loess Plateau. Acta Ecologica Sinica, 2021, 41(17): 6786-6799
4. Zhou Tingting, Su Lijuan, Liu Hui, et al. Variation law and influencing mechanism of surface movement parameters in Shendong Mining Area. Coal Geology & Exploration, 2021, 49(03): 189-198
5. Wang Qi, Ye Yao, Han Yu, et al. The Characteristics of Soil and Vegetation Carbon Cycle and Conversion of Carbon Source and Sink in Mining Subsidence Area of the Semi-arid Region. Journal of Soil and Water Conservation, 2016, 30(04): 166-172+205. doi:10.13870/j.cnki.stbcxb.2016.04.028
6. Zhang Xin, Liu Tiejun, Guo Jianying, et al. Effects of Simulated Erosion Damage on Photosynthetic and Physiological Characteristics of Caragana microphylla in Coal Mining Subsidence Area. Journal of Northwest Forestry University, 2022, 37(05): 24-29+79
7. Wang Bo; Liu Jing; Wang Chenglong; et al. Effects of drawing damage on root growth and soil reinforcement of Hippophae rhamnoides in a coal mining subsidence area[J]. International Journal of Phytoremediation, 2022, 24(04): 409-419
8. Wang Bo, Bao Yuhai, Liu Jing, et al. Effects of erosive agent on root growth and self-healing of Artemisia ordosica in semi-arid mining site. Chinese Journal of Ecology, 2022, 41(02): 263-269. doi:10.13292/j.1000-4890.202202.006
9. Wang Bo; Liu Jing; Zhang Xin; et al. Changes in soil carbon sequestration and emission in different succession stages of biological soil crusts in a sand-binding area[J]. Carbon Balance and Management, 2021, 27(16), DOI: 10.1186/s13021-021-00190-7
10. Shan Dan, Xing Ende, Rong Hao, et al. Ecological restoration of different vegetation collocations of coal mine dump in typical steppe. Chinese Journal of Ecology,2019, 38(02): 336-342. doi:10.13292/j.1000-4890.201902.019
11. Zhan Jin, LiYulin, Han Dan,Yang, et al. Biomass allocation and its ecological significance of three dominant sand-fixing shrubs in the semi-arid desert area. Journal of Desert Research, 2020, 40(05): 149-157
12. Tang Cuiping, Guo Feng, Li Xiaoting, et al. Study on Allelopathy of Hippophae rhamnoides Extract on Thymus serpyllum L.var asiaticus Seeds. Seed, 2014, 33(08): 48-51. doi:10.16590/j.cnki.1001-4705.2014.08.044
13. Wang Fanglin, Xu Xianying, Wang Lide , et al.Research Progress of Allelopathy of Arid Desert Plants in Northwest. Chinese Agricultural Science Bulletin, 2016, 32(05): 165-172

https://doi.org/10.1051/e3sconf/202344102001

The Influence of Plateau Environment on Diesel Engine Starting Process

Fengjun Qi, Changting Li*

Vehicle Application Department, Army Academy of Armored Forces, School of Noncommissioned Officers, Changchun, Jinlin, 130117, China

Abstract: In view of the difficulty of a diesel engine starting in a plateau environment, this paper analyzed three factors that affect diesel engine starting by literature review and theoretical analysis, deeply analyzed the mechanism of the plateau environment on the diesel engine starting process, and introduced auxiliary measures for diesel engine starting in a plateau environment so as to further study how fast and reliable starting has laid a solid theoretical foundation.

1. Introduction

In southwest China, the average sea wave height is 4500 m. Compared with the Central Plains, its environmental characteristics are mainly cold weather, thin air, low oxygen content, a large temperature difference, and low air pressure. Due to its special geographical and climatic characteristics, it is difficult to start the diesel engine of heavy vehicles, and it cannot be started in serious cases. In addition, if the diesel engine is forced to start in a cold state, it will aggravate the wear of the cylinder liner, piston, crankshaft, and other key components of the diesel engine, which will cause diesel engine failure for a long time, affecting the working performance and service life of the diesel engine. Through consulting relevant literature, it is found that when the diesel engine starts at an altitude of 4000 m and an ambient temperature of -25 °C, its starting difficulty is equivalent to the starting condition at -40 °C in plain areas [1-3]. Therefore, it is of great significance to further study the influence mechanism of plateau environment on the starting process of diesel engines and continuously develop the auxiliary device for starting diesel engines at low temperatures in a plateau environment to improve the adaptability of heavy vehicle diesel engines to a plateau environment.

2. Influencing Factors of Diesel Engine Starting Process

The starting process of a diesel engine refers to the whole process from the crankshaft starting to rotate under an external force until the diesel engine starts to automatically rotate. At the initial stage of a diesel engine's starting, the starting motor drives the crankshaft to rotate, forcing the piston to move up and down in the cylinder and compressing the fuel in the cylinder, and the gas expansion works to push the piston to move up and down, thus driving the crankshaft to rotate. When the diesel engine speed reaches the specified starting speed range, the expansion of the combustible mixture in the cylinder works to make the crankshaft rotate. At this stage, the diesel engine's working cycle can be completed automatically, enabling it to work on its own. In the process of a diesel engine starting, whether the engine can start smoothly is mainly determined by the following three factors.

2.1 Impact of Temperature and Pressure in the Cylinder at the End of the Compression

To ensure that the diesel injected into the cylinder can ignite automatically, the temperature of the mixture in the cylinder at the end of compression is required to be about 200 °C higher than the fuel spontaneous ignition point. The auto-ignition point of diesel oil is about 335 ℃ under normal pressure. When the gas pressure at the end of compression in the cylinder reaches 2.0–2.8 MPa, the temperature is 200–235 °C. Therefore, the diesel engine can be started quickly and reliably only when the mixture temperature reaches a high temperature above 400 °C.

In a plateau environment, there are two main characteristics of a diesel engine starting. One is that when starting the diesel engine, the end temperature of air compression in the cylinder often cannot reach the auto-ignition point temperature of diesel; second, the end compression pressure in the cylinder is lower than the normal value.

2.2 Influence of Mixture Uniformity and Concentration Suitability in Cylinder

The precondition for diesel fuel to self-combust in the cylinder is whether the mixture in the cylinder is uniform

*Corresponding author's e-mail: leect1991@163.com

and the concentration is appropriate. If the amount of diesel fuel injected into the cylinder is too small or the quality of the spray is poor, the engine cannot form self-combustion conditions or work continuously. To characterize the concentration of the mixture, it is usually expressed by the excess air coefficient. For a supercharged diesel engine, the excess air coefficient is usually 1.8–2.2 under the maximum power working condition, which is called the ignition limit of the diesel engine. If the mixture is too rich or too thin, it is not conducive to the ignition of the mixture.

2.3 Effect of Minimum Starting Speed of Diesel Engine

Through the analysis of the above two starting conditions, it can be found that the diesel engine starting is related to the starting speed. Under normal conditions of oil and gas supply, to ensure that the diesel engine can start normally, the starting speed should always be greater than or equal to the minimum starting speed. If the starting speed of the diesel engine is less than the minimum starting speed, the fuel injection pressure is insufficient, the atomization quality of diesel fuel is poor, and the compression time becomes longer, which increases air leakage and heat loss and makes it difficult to reach the autoignition temperature of the fuel, so the diesel engine cannot be started.

3. Influence of Plateau Environment on Diesel Engine Starting Process

3.1 Influence of Plateau Environment on Temperature and Pressure in the Cylinder at the End of Compression of Diesel Engine

The plateau environment is characterized by low air pressure, low temperatures, thin air, and low oxygen content. With the increase of altitude, atmospheric pressure Pa and atmospheric density ρ reduce. The corresponding reduction is shown in Equations (1) and (2) [4]:

$$P_a = P_0(1-0.02257\,H)^{5.256} \tag{1}$$

$$\rho = \frac{1.293}{1+0.00376(T_a-273.15)} \times \frac{P_a}{0.101325} \tag{2}$$

where P_0 is the atmospheric pressure at sea level, MPa; H is the altitude, km; ρ Is the atmospheric density, kg/m^3; and T_a is the ambient atmospheric temperature, K.

The temperature and pressure at the end of compression of the diesel engine can be calculated by Equations (3) and (4):

$$T_{co} = T_{ca}\varepsilon_{cc}^{n_1-1} \tag{3}$$

$$P_{co} = P_{ca}\varepsilon_{cc}^{n_1} \tag{4}$$

where T_{ca} is the compression starting point temperature, K; T_{co} is the compression endpoint temperature, K; ε_{cc} is the effective compression ratio, ε cc = (0.8–0.9) ε_c, ε_c is the compression ratio; N_1 is compression polytropic index; P_{ca} is the starting point pressure of compression in the cylinder, Pca = (0.8–0.9) Pa, Pa is the atmospheric pressure, MPa; and P_{co} is the final compression pressure of the cylinder, MPa.

It can be seen from Equations (3) and (4) that the final compression temperature and pressure of a diesel engine are mainly determined by ambient temperature, pressure, and the compression polytropic index. The polytropic index of a diesel engine is mainly affected by the heat exchange between the working fluid and the cylinder wall and the cylinder leakage. When the ambient air temperature is low, the intake air temperature decreases, which increases the oil viscosity, deteriorates the fluidity, and reduces the starting speed, leading to longer compression time, increased mixture leakage, reduced polytropic index of the diesel engine, and reduced temperature and pressure in the cylinder at the end of compression. When the temperature and pressure in the cylinder at the end of compression drop to a certain extent, the mixture will not reach the minimum ignition critical temperature, the mixture will not ignite, and the diesel engine will not start normally.

3.2 Effect of Plateau Environment on Diesel Engine Mixture Characteristics

Three conditions are required for fuel combustion. In addition to temperature and pressure, there is also mixture concentration. These three factors are not independent but interact with each other. See Figure 1 [5]. With the decrease in temperature and pressure, the concentration limit of combustion narrows.

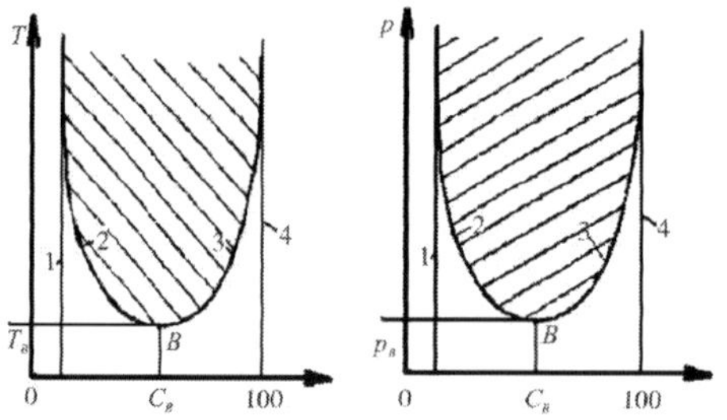

Figure 1. Relationship between ignition temperature, pressure, and ignition limit of mixture

In Figure 1, 1 is the lean oil limit, 2 is the lean oil lower limit, 3 is the rich oil upper limit, 4 is the rich oil limit, TB is the limit temperature, PB is the limit pressure, CB is the concentration value of combustible mixture under the limit temperature and pressure, and the shadow area is the ignition area.

It can be seen from Figure 1 that with the increase in altitude and the decrease in temperature, the temperature and pressure of the diesel engine at the end of compression show a downward trend, and the corresponding ignition mixture concentration range gradually decreases. Therefore, the mixture concentration is likely to exceed the ignition concentration range due to the reduction of the atmospheric density in the plateau, resulting that the mixture in the diesel engine cylinder cannot be ignited.

3.3 Impact of plateau environment on diesel engine starting speed

The starting speed of a diesel engine is mainly affected by the starting power and starting torque. The higher the starting power is, the smaller the resistance torque is, and the higher the starting speed of the diesel engine is, on the contrary, the lower it is.

3.3.1 Effect of low temperature on starting power of diesel engine

The output power of the heavy vehicle starting system directly affects the starting speed of the diesel engine. The greater the output power, the higher the starting speed is, and the diesel engine can start at a lower temperature, which can improve the low-temperature starting performance of the diesel engine. However, with the continuous drop in ambient temperature, the power supply device of the starting system, namely, the battery, will discharge with a large current under low-temperature conditions, resulting in a sharp drop in battery capacity and output power, a decrease in starting speed, and poor starting performance.

In addition, with the temperature decreasing, the viscosity of the electrolyte (diluted sulfuric acid solution) in the battery will increase, which will increase the resistivity of the battery. It can be seen from Table 1 that when the temperature of the electrolyte drops from 30 °C to 40 °C, the resistivity of the electrolyte increases 7.36 times.

Table 1. Effect of Temperature on Resistivity of Sulfuric Acid Solution

Temperature/°C	30	20	10	0	-10	-20	-30	-40
Resistivity/ (Ω·cm)	1.140	1.334	1.602	1.990	2.600	3.570	5.290	8.390

The increase of the resistivity of the battery increases the internal resistance of the battery, which leads to a sharp drop in the starting output power of the battery.

$$U = E_b - I_{st} \cdot R_b = I_{st} \cdot R_L + I_{st} \cdot R_{st} \quad (5)$$

where U is the terminal voltage of the storage battery during starting, V; E_b is the electromotive force of the battery, V; I_{st} is the starting current, A; R_b is the internal resistance of the battery, Ω; R_L is the starting line resistance, Ω; and R_{st} is the starter armature resistance, Ω.

Through Equation (5), it can be found that the line resistance and starter armature resistance change little with the temperature drop, while the internal resistance of the battery increases with the temperature drop, so the terminal voltage of the battery will decrease with the temperature drop when starting. On the other hand, the starting current will increase 3–4 times due to the increase in starting resistance during a low-temperature start.

According to the above analysis, under low-temperature conditions, the terminal voltage of the battery and the starting output power decrease due to the increase in the internal resistance of the battery and the rapid increase in the starting current. The starting speed decreases, and finally, the diesel engine cannot be started [6].

3.3.2 Effect of Low Temperature on Engine Starting Resistance Torque

When the diesel engine starts, the actual speed n that the engine crankshaft can reach depends on the starting torque that the starting device gives to the crankshaft and the starting resistance torque that prevents the crankshaft from rotating. Starting resistance torque refers to the torque that needs to be overcome when the diesel engine is dragged from a static state to starting speed, which can be expressed by Equation (6).

$$M_r = M_t + M_j + M_k \quad (6)$$

where M_t is the friction resistance moment, N·m; M_j is the inertia resistance moment, N·m; and M_k is the compression resistance moment, N·m.

The friction resistance torque of a diesel engine is the main component of starting resistance torque. When the temperature decreases, the friction resistance between various parts increases due to the increase in oil viscosity. Through consulting relevant literature, it is found that when the ambient temperature is 0–5 °C, the friction resistance torque accounts for about 60% of the starting resistance torque, and when the ambient temperature is -20–10 °C, it accounts for about 80%–95% [7].

The compression resistance torque refers to the compression stroke. The piston is formed by the resistance of compressed gas. It mainly depends on the power, displacement, and compression ratio of the engine.

The inertia resistance moment includes the crankshaft, connecting rod, piston, and various auxiliary mechanisms associated with the crankshaft, such as the oil pump, fan, generator, etc. It mainly depends on the inertia force of each component or mechanism in the acceleration process.

Therefore, when the ambient temperature decreases, the oil viscosity will increase rapidly, which will lead to an increase in the friction resistance of the diesel engine and a rapid rise in the starting resistance of the diesel engine, thus affecting the normal starting of the diesel engine.

4. Auxiliary Measures for Diesel Engine Starting at Low Temperature in a Plateau

At present, auxiliary starting measures are used to ensure that the diesel engine can start quickly and reliably when it starts at a low temperature. At present, auxiliary measures for low temperature starting of diesel engines are formulated for plain areas, as shown in Table 2, while those for plateau areas are not ideal.

Table 2. Auxiliary Measures for Low-Temperature and Low-Pressure Starting of Vehicle Diesel Engine

Classification of Starting Aids	Low-temperature Starting Aids	Explanation
Auxiliary Measures for Emergencies Starting	Flame Intake Preheating Device	Low atmospheric pressure and low oxygen content in the air in plateau areas make it unsuitable to use air inlet preheating device.
	Electric heating preheating device	
	Glow plug	
	Cold starting fluid	Forced start, serious engine wear, affecting emissions
	Oil injection in the cylinder	
	Supercapacitor	The ideal starting power supply under low temperature and low-pressure conditions needs to be combined with the heating starting auxiliary measures to reduce engine wear.
	Storage battery heating and insulation device	
	Gelled high-energy battery	
	Low temperature battery	
Auxiliary Measures for Heating and Starting	Coolant heating device	It can not only ensure the smooth starting of the diesel engine but also reduce the starting wear.
	Oil and fuel preheating device	
Other Auxiliary Measures	EGR	Effective in plateau low temperature and low-pressure environment
	Increase cetane number and use additives	High cost, high cost, and limited effect

Given the difficulty in starting diesel engines in a plateau environment, new starting aids currently used mainly include fuel heaters, battery insulation devices, gel high-energy batteries, supercapacitors, etc.

4.1 Fuel Heater

The fuel heater is an auxiliary measure for heating and starting the whole machine and lubricating oil by burning the fuel through the burner and transferring the heat to the coolant and air through the heat exchanger [8]. After the engine is heated, the temperature of the cylinder, piston, piston ring, and bearings increases, and the oil temperature between these friction pairs also increases, reducing the starting resistance and increasing the starting speed, thus increasing the temperature and pressure at the end of the compression and making the engine easy to start. At the same time, the oil viscosity decreases, the lubrication conditions are improved, and the wear on parts is reduced. The fuel heater is a very effective cold-start auxiliary device for diesel engines at low temperatures and is also a commonly used starting aid mode.

4.2 Storage Battery Incubator

The battery insulation box is composed of a box cover, a box, a heating pipe, a sealing strip, fastening bolts, and a battery-fixing device. The basic principle of the battery insulation box is to heat the battery by using the thermal fluid formed by heating the coolant with a fuel heater, then heat the heated battery through the insulation function of the insulation box so that the output power of the battery can meet the needs of the starting system.

4.3 Supercapacitor

The supercapacitor is a new type of energy storage container between traditional capacitors and batteries. The supercapacitor has good discharge ability and reliability in extreme temperature environments, making it an ideal starting power supply for diesel engines in a low-pressure and low-temperature environment. The supercapacitor is characterized by large capacity, small internal resistance, and long life, and can be charged and discharged quickly with a large current. The operating temperature range is 41 to 70 °C. The battery is not suitable for large current discharges, and its low-temperature performance is poor. Due to the limited capacity of the supercapacitor, it cannot provide a long time of large current discharge when used alone, and it is difficult to charge when installed on the vehicle. Therefore, both are not suitable for starting the power supply alone. If the supercapacitor is used in parallel with the battery as the starting power supply, the advantages of the supercapacitor and the battery can be brought into play.

4.4 Colloidal High-Energy Battery

The gelled high-energy battery uses high-energy compounds as an electrolyte, which has the following significant advantages: small self-discharge, good maintenance-free, and easy to be stored for a long time; The excellent charging and discharging characteristics can better meet the operational requirements. Specific energy is high and relative mass is light; No lower limit of discharge voltage is allowed, and deep discharge is allowed. Strong self-recovery ability, which is conducive to emergency use.

The gel high-energy battery has good environmental adaptability. For one thing, the high-energy and environment-friendly battery is different from the lead-acid battery in its mechanism, and there is no gas produced during the charging (discharging) process. For another thing, the battery has very small internal resistance (micro-ohm level), and it will not generate a lot of heat during the charging (discharging) of a large current. The electrolyte usually does not appear gasified, so it is a truly fully sealed battery. Therefore, the high-energy and environment-friendly battery can ensure that the army can use it in any complex environment, such as high temperature, cold, or high altitude, and especially can adapt to the performance requirements of the special vehicle environment. According to the test, it can be used normally in the range of -40 to 70 °C. After being stored at -40 °C for 24 hours, it can be charged (discharged) normally, and its voltage is not lower than 12.5 volts (battery rated at 12 volts).

4.5 Combined Starting Auxiliary Measures

To better improve the starting performance of diesel engines in plateau areas and ensure the service life of diesel engines, two or more starting auxiliary measures can be combined to form a combined starting auxiliary measure. The fuel heater is used as the preheating system of the diesel engine to increase the temperature of the diesel engine body, oil, fuel, or battery by heating the circulating coolant, to realize the starting of the diesel engine under the hot engine state. At the same time, battery insulation devices, gel high energy batteries, or supercapacitor parallel batteries are used as starting power to ensure the starting power of the diesel engine. Combined cold starting auxiliary measures can not only improve the starting ability of diesel engines but also not reduce the reliability and durability of diesel engines, which can completely solve the cold starting problem of diesel engines in plateau-cold regions.

5. Conclusion

(1) The obvious climatic characteristics of the plateau environment are low temperatures and low air pressure, which seriously affect the starting process of diesel engines. The low temperature will reduce the starting power, increase the starting resistance torque, and reduce the engine starting speed, making the diesel engine unable to start normally; low air pressure and low temperature will reduce the pressure and temperature at the end of compression, which cannot reach the autoignition point of the fuel. At the same time, the concentration range of the ignition mixture gradually decreases, which makes the mixture in the cylinder unable to ignite, affecting the starting of the diesel engine.

(2) At present, the starting aids for diesel engines suitable for plateau areas mainly include fuel heaters, battery incubators, supercapacitors, and gel high-energy batteries. The combined cold-start auxiliary measures can not only improve the starting ability of the diesel engine but also increase its reliability and durability.

References

1. Xu Xiang. Research on the Impact and Adaptability of Plateau Environment on Support Equipment [J]. *Equipment Environment Engineering*, 2010,7 (5): 100-103.
2. Benjumea P, Agudelo J, Agudela A. Effect of altitude and palm oil biodiesel fuelling on the performance and combustion characteristics of a HSDI diesel engine[J]. *fuel*, 2009, 88(4): 725-731.
3. Perez P L, Boehman A L. Performance of a single-cylinder diesel engine using oxygen-enriched intake air at simulated high-altitude conditions[J]. *Aerospace Science and Technology*, 2010, 14: 83-94.
4. Jiang Deming. Principles of Internal Combustion Engine [M]. Beijing: China Agricultural Machinery Press, 1988.
5. Liu Ruilin, Dong Surong. Research Report on Evaluation of Cold Start Auxiliary Measures for Vehicles and Equipment in Plateau and Alpine [R]. 2009.
6. Jin Shangjie. Research on the Effect of Cold Start Measures on the Starting Performance of Military Vehicle Diesel Engines [D]. Tianjin: Military Transportation Institute, 2008.
7. Li Dongjiang. Requirements on engine oil for low temperature starting of diesel engines [J] *Diesel Engine*, 1995, (1): 39-42.
8. Liu Ruilin, Jin Shangjie, Sun Wuquan, et al. Cold starting auxiliary measures to improve the low temperature starting performance of diesel engines [J]. *Automotive Technology*, 2007, (6): 5-8.

Analysis of 750 kV Conductors in 750 kV Substations

Xiaowen Zhang*, Yu Wang, Yunhui Jia, Qinyi Jiang

China Energy Engineering Group Gansu Electric Power Design Institute Co., Ltd, Lanzhou, 730050, China

Abstract: Two-bundled conductors are generally used in the design of 750 kV substations. Compared to two-bundled conductors, multi-bundled conductors have higher load capacity and greater start-up voltage. This paper investigates the possibility of using multi-bundled conductors instead of two-bundled conductors in 750 kV substations from both technical and economic aspects.

1. Introduction

During the evaluation of the preliminary design of the 750 kV Yongdeng-Jinchang transmission project and 750 kV Jinchang-Jiuquan-Anxi transmission project, researcher Sun Zhusen of the Infrastructure Department of State Grid Corporation attended the meeting and instructed that a thematic study should be carried out in 750 kV substations and lines to optimize the study of substation fittings and conductors for reducing corona and noise. At present, the 1000 kV substation has been operating through the optimization of the calculation of the fittings and connection conductors. The results are remarkable, and the audible noise is even smaller than that of the 500 kV substation.

Gansu Electric Power Company carried out the 750 kV substation fitting optimization study in the 750 kV Yongbai project, and optimized the equipment fixtures and equal voltage shield rings of the substation except for the connecting wires. Since the connection wire is in the form of two bundles, the effect of reducing the corona and noise of the wire is very limited by optimizing the connection fixture and spacer bar, which can only be improved by increasing the number of conductor bundles. According to the initial calculations of the professor of Western Jiaotong University, if the 750 kV conductors use four bundles, the field strength on the surface of the conductors will be improved by more than 50% compared to that of the two bundles. Therefore, it is necessary to conduct an in-depth study on the conductor bundle of the 750 kV substation and propose a reasonable conductor bundle scheme.

2. Status of 750 kV conductors in 750 kV substations

At present, China has already commissioned several 750 kV substations, and there are more than a dozen 750 kV substations soon to be built. In these substations, the 750 kV busbar and 750 kV intra-string conductors are designed as one type of JLHN58K-1600. JLHN58K-1600 is a new type of conductor specially developed by Xinjiang TBEA Cable Factory for a 750 kV substation according to the research results of the 750 kV substation project. As JLHN58K-1600 is only used in the 750 kV substation and 1000 kV substation, it has less consumption, fewer manufacturers and a higher price [1].

The two-bundled conductors used in 750 kV transmission lines are derived from China's current 330 kV substation and 500 kV substation conductor program. The difference between them is that the conductor cross-section is increased, and the intra-bundle distance is increased from 200 mm to 400 mm. However, the problem of corona noise in the two-bundled conductors has fundamentally unchanged.

In the 750 kV substation fitting optimization project conducted by Gansu Provincial Electric Power Company, all the substation fittings of 750 kV substation were simulated and optimized, and the optimized fittings will significantly reduce the corona and noise of the substation. However, for the two-bundled conductors, the corona and noise generated by the conductors can not be reduced only by optimizing the connection fittings and spacers. In this way, the main source of noise in the whole substation will be the two-bundled conductors, which can be completely solved only by increasing the number of conductor bundles.

Based on the weaknesses above, the two-bundled conductor scheme was abandoned in the three 1000 kV substations in China. All three substations used a four-bundled conductor, which was successful.

Due to the superiority of three-bundled and four-bundled conductors compared with two-bundled conductors, together with the successful use of the four-bundled conductors in 1000 kV substations, this study explores the possibility of using three-bundled or four-bundled conductors for 750 kV conductors in 750 kV substations through calculations, and gives conclusions and recommendations.

*Corresponding author's e-mail: 010699128@163.com

At present, the Baiyin 750 kV substation is under construction, and the following discussion takes Baiyin 750 kV substation as an example.

3. 750 kV conductor design

The 750 kV conductors are mainly controlled by the corona inception voltage and the load capacity of the conductors. We compared various conductors from these two aspects, selected the conductors that can meet the requirements, and calculated the forces of various conductors under the frame.

3.1. Calculation of the minimum diameter of 750 kV conductors with different bundles.

Through the calculation of different conductor's starting corona voltage, the minimum diameters of 750 kV conductors in the case of two bundles, three bundles and four bundles were calculated to ensure that the conductors can not develop a comprehensive corona under a clear sky, and the calculation results are shown in the following table [2].

Table 1. Minimum diameter of 750 kV conductors

Bundle number	Altitude (m)	Interphase distance (m)	Intra-bundle distance (cm)	Minimum diameter of conductor (mm)
2	2000	11	40	60.0
3	2000	11	40	41.6
4	2000	11	40	24.4

According to Table 1 together with the diameter of various conductors, it is known that if the three-bundled conductors are used, the wires need to be above LGJK-630 or NRLH58GJ-1000/125; if the four-bundled conductors are used, the wires need to be above LGJ-630/80 or NRLH58GJ-630/80[3].

3.2 Calculation of the conductors meeting the load capacity requirements

3.2.1 750 kV busbar conductor selection

According to the system design, the maximum power penetration of the 750 kV bus is 5700 MW and the power factor is 0.9, which means the maximum current of the 750 kV bus is 4875 A[4].

Table 2. 750 kV busbar load capacity comparison

	2X(JLHN58K-1600)	3X(LGJK-1000)	3X(LGJK-1250)	3X(NRLH58GJ-1000/125)	4X(LGJ-630/80)	4X(LGJ-800/100)	4X(NRLH58GJ-630/80)
Load capacity	6500 A	4836 A	5499 A	6690 A	4360 A	5176 A	6792 A
Can the load capacity be met	Yes	No	Yes	Yes	No	Yes	Yes
Corona inception voltage	912 kV	——	928 kV	824 kV	——	1115 kV	1031 kV

3.2.2 750 kV intra-string conductor selection

The maximum power penetration in the 750 kV interval string is 3000 MW, which means the maximum current in the 750 kV string is 2567 A[5].

Table 3. Comparison of the load capacity of the 750 kV intra-string conductor

	2X(JLHN58K-1600)	3X(LGJK-630)	3X(NRLH58GJ-1000/125)	4X(LGJ-630/80)	4X(NRLH58GJ-630/80)
Load capacity	6500 A	3741 A	6690 A	4630 A	6792 A
Can it meet the requirements	Yes	Yes	Yes	Yes	Yes
Corona inception voltage	912 kV	880 kV	928 kV	1031 kV	1031 kV

3.3 750 kV conductor schemes

According to Table 2 and Table 3, the 750 kV conductor schemes can be derived.

Table 4. 750 kV conductor schemes

Conductor scheme	750 kV Busbar	750 kV in-string conductor
Two-bundled conductor	2X(JLHN58K-1600)	2X(JLHN58K-1600)
Three-bundled conductor	3X(LGJK-1250)	3X(LGJK-630)
	3X(NRLH58GJ-1000/125)	3X(NRLH58GJ-1000/125)
Four-bundled conductor	4X(LGJ-800/100)	4X(LGJ-630/80)
	4X(NRLH58GJ-630/80)	4X(NRLH58GJ-630/80)

4. Comparison of various schemes of conductor forces

We calculated the forces corresponding to each of the aspects in Table 4 to determine the substation configuration[6].

1) Comparison of busbar force calculation with a busbar span of 42 mm.

Table 5. Calculation of force on busbar frame for various schemeswith a span of 42 meters

		Maximum load	Maximum wind speed	Minimum temperature	Construction and Installation	Three-phase human live maintenance	Single-phase human live maintenance	Lateral wind pressure
2X(JLHN58K-1600)	Horizontal	3502	3140	2736	2718	2969	3155	214
	Vertical	2229	1984	1769	1773	1873	1948	
3X(LGJK-1250)	Horizontal	4112	3648	3246	3227	3483	3671	153
	Vertical	2507	2213	2004	2007	2107	2182	
3X(NRLH58GJ-1000/125)	Horizontal	3594	3171	2828	2810	3061	3247	134
	Vertical	2271	1996	1813	1816	1916	1991	
4X(LGJ-800/100)	Horizontal	3894	3449	2997	2982	3233	3420	115
	Vertical	2413	2126	1891	1895	1995	2070	
4X(NRLH58GJ-630/80)	Horizontal	3539	3128	2709	2694	2942	3127	105
	Vertical	2250	1979	1758	1761	1861	1936	

2) The calculation and comparison of the force within the span show that the span is 98 mm.

Table 6. Calculation of force on busbar frame for various schemeswith a span of 98 meters

		Maximum load	Maximum wind speed	Minimum temperature	Construction and Installation	Three-phase human live maintenance	Single-phase human live maintenance	Lateral wind pressure
2X(JLHN58K-1600)	Horizontal	7016	6240	5451	5107	5821	6215	347
	Vertical	2776	2489	2165	2170	2270	2345	
3X(LGJK-1250)	Horizontal	6331	5691	4712	4518	5109	5479	235
	Vertical	2668	2397	2021	2026	2126	2201	
3X(NRLH58GJ-1000/125)	Horizontal	7114	6216	5577	5252	5945	6331	216
	Vertical	2831	2497	2221	2225	2325	2400	
4X(LGJ-800/100)	Horizontal	6708	5892	5035	4805	5422	5796	171
	Vertical	2773	2450	2112	2116	2216	2291	
4X(NRLH58GJ-630/80)	Horizontal	6808	5965	5134	4869	5511	5887	171
	Vertical	2788	2461	2126	2131	2231	2306	

According to Table 5 and Table 6, if the bus adopts 3X(LGJK-1250), the forces of the frame will increase by about 20%. If the bus adopts 3X(NRLH58GJ-1000/125), the forces of the frame will increase by about 5%. If the bus adopts 4X(LGJ-800/100), the forces of the frame will increase by 10%. If the bus adopts 4X(NRLH58GJ-630/80), the forces of the frame are basically the same.

If busbar 3X(LGJK-630), 3X(NRLH58GJ-1000/125), 4X(LGJ-630/80) and 4X(NRLH58GJ-630/80) are used for the intra-string conductors, the forces of the frame are basically the same.

Through the frame force calibration, if the 750 kV substation adopts a three-bundled conductor or four-bundled conductor, the 750 kV conductor forces will be slightly increased or remain the same, yet within the frame tolerance range, and the 750 kV frame does not need to be replaced[7].

5. Economic and technical comparison of conductors

The costs of the technology schemes are listed below.

Table 7. Comparison of costs for various technical solutions

Conductor type		Length (m)	Cost (million yuan)	Cost savings (million yuan) (relative to the two-bundled conductors)
Busbar	2X(JLHN58K-1600)	2800	102	
	3X(LGJK-1250)	4200	51	51
	3X(NRLH58GJ-1000/125)	4200	68	34
	4X(LGJ-800/100)	5600	49	53
	4X(NRLH58GJ-630/80)	5600	61	41
Intra-string conductors	2X(JLHN58K-1600)	7000	240	
	3X(LGJK-630)	10500	72	168
	3X(NRLH58GJ-1000/125)	10500	153	87
	4X(LGJ-630/80)	14000	78	162
	4X(NRLH58GJ-630/80)	14000	134	106

According to Table 7, if the 750 kV conductors are replaced by other conductors, the maximum saving of the 750 kV Baiyin substation in this period is 2.21 million yuan (64%), and the minimum saving is 1.21 million yuan (35%). Considering the best situation of the corona, the bus bar is selected 4X (LGJ-800/100), and the intra-string conductor is selected 4X (LGJ-630/80), by which savings of 2.15 million yuan (63%) can be achieved. As there is no precedent of using multi-bundled conductors in 750 kV substations, and the four-bundled conductors are much higher than the three-bundled conductors in terms of corona inception voltage, for insurance purposes, it is recommended that 4X (LGJ-800/100) conductors be used for 750 kV busbars[8]. At the same time, in order to reduce the types of fittings, it is recommended that 4X (LGJ-800/100) conductors are also used in 750 kV intra-string conductors. If 4X(LGJ-800/100) conductors are used for both 750 kV busbar and 750 kV intra-string conductors, the whole Baiyin 750 kV substation can save 1,755,000 yuan investment, with a saving ratio of 51%.

6. Conclusion

Through the calculation and analysis above, using three-bundled conductors or four-bundled conductors, the 750 kV substation 750 kV conductor can be completely feasible economically and can meet the requirements technically, therefore it can be applied in the construction.

References

1. Liu Z.Y. Extra high voltage power grid [M] Beijing: China Economic Press, 2005.
2. Wang X., Wu D.R., et al. Calculation regulations for conductor and appliance selection [M] Beijing: China Electric Power Press, 2005.
3. Feng C.Z., Ma X.K. Introduction to Engineering Electromagnetic Fields [M] Beijing: Higher Education Press, 2000.
4. He J , Zhu P , Gan B ,et al.Study on rational choice of intra-bundle conductor spacing for 750 kV transmission lines[J]. 2012.
5. Jesenik M., Trlep M., Hamler A., et al. Electric Field Distribution Under Transmission Lines Dependent on Ground Surface [J].IEEE Transactions on Magnetics, 2009, 45(3).
6. Wu C.L., Li B.J., Liu G.F., et al. Power Engineering Design Manual Electrical Primary Design for Thermal Power Plants [M] Beijing: China Electric Power Press, 2018.
7. Trinh N G , Maruvada P S , Poirier B .A Comparative Study of the Corona Performance of Conductor Bundles for 1200 KV Transmission Lines[J].IEEE Transactions on Power Apparatus and Systems, 1974, PAS-93(3):940-49.DOI:10.1109/TPAS.1974.294013.
8. Min S W , Kim Y J , Sin G Y ,et al.Analysis on Electric Field Distributions and Corona Characteristics of Special Conductor Bundles for 765kV Transmission Line[J]. 2001.

 https://doi.org/10.1051/e3sconf/202344102003

The Supply-Demand Analysis Method for Power System with Considering the Demand Response and Multiple Scenarios

Xin Zhao*, Wanlei Xue and Yan Li

Economic and Technology Research Institute, State Grid Shandong Electric Power Corporation Ltd, Jinan, Shandong, 250000, China

Abstract. This paper proposes a mathematic model of time series production simulation, and presents a case study of the Shandong power grid and investigates the role of demand-side resources in ensuring power supply. Based on the supply-demand analysis method and the design of the demand-side resource regulation plan, the demand response of the Shandong power grid can essentially meet the power balance requirements by 2025 under the baseline scenario. Under the extreme scenario, the power deficit can also be addressed by orderly power consumption. It is suggested to pay more attention to the development and utilization of demand-side resources.

1. Introduction

Given the introduction of the "dual-carbon" goals and the accelerated construction of a novel power system, relying solely on regulation resources on the power generation and grid sides is no longer sufficient to meet the increasingly stringent requirements for system safety and efficiency. As a result, there is a pressing need to tap into the flexible regulation capabilities of demand-side resources, which have garnered growing attention. The Action Plan for Carbon Dioxide Peaking Before 2030, issued by the State Council in October 2021, explicitly calls for the development and utilization of demand-side resources: To actively promote the integration of source, grid, load, and storage, use multiple energy sources to supplement each other, make vigorous efforts to enhance the overall regulation capabilities of the power system, incorporate loads of traditional energy-intensive industries, interruptible industrial and commercial interruptible loads, electric vehicle charging networks, and virtual power plants into the power system regulation, and make grids more secure and reliable. The development and utilization of demand-side resources can yield multiple benefits, such as reducing redundant investment, ensuring grid security, lowering energy costs for users, promoting the development of energy storage industries across the source, grid, load, and storage sides, and reducing carbon dioxide emissions. These benefits will aid China in achieving its "dual-carbon" goals.

Currently, both domestic and foreign experts have conducted research in this area. [1] analyzed the comprehensive value of demand-side resources from the perspectives of the resources themselves, the market and society. [2] proposed a double-level planning model for an active distribution network that considers carbon emissions and demand-side management. This model optimized the planning of the active distribution network through demand-side management to minimize the comprehensive annual cost. [3] analyzed the price elasticity and cost-effective value associated with users' participation in power demand response. [4] analyzed the benefits for power generation companies, grid companies, and society involved in different demand-side management investment projects. [5] calculated the cost-effectiveness of demand-side shared energy storage stations, operators, and users. [6] developed demand response models for both electricity retailers and users and analyzed the long-term benefits of each entity. [7] analyzed the economic benefits of load aggregators and their contribution to the grid based on PJM market data in the United States. [8] calculated the total cost of comprehensive demand response service providers, as well as the benefits of user participation in demand response. In [9], a model of the electricity market that captures the uncertainties on both the operator and user sides is proposed, where an explicit characterization of the optimal user behavior using the Bayesian Nash equilibrium solution concept is derived. In [10], the optimal design of demand response and peak-time rebate programs in the electricity sector is studied. It is showed that compared to a conventional demand response program, the proposed approach significantly improves the targeting of demand response payments. In [11], an overview of AI approaches used for DR applications is presented. Both the Artificial Intelligence and Machine Learning algorithm(s) are employed while discussing commercial efforts (from both new and existing businesses) and large-scale innovation projects that have applied AI technologies for energy DR. In [12], a novel self-reported baseline mechanism (SRBM) where each agent reports its baseline and marginal utility is proposed. The authors show that SRBM is almost optimal in the

*Corresponding author's e-mail: zhaoxin_sd_pg@163.com

metric of average cost of DR provision faced by the aggregator. However, these studies have not yet established a systematic and quantitative method to analyze the value of demand-side resources. This has resulted in an incomplete assessment of the value of demand-side resources in promoting the construction of a novel power system and achieving the "dual-carbon" goals.This paper establishes a time series production simulation model that considers the potential of demand response based on the actual demand response in China. Then we present a case study of the Shandong grid and investigates the role of demand-side resources in ensuring power supply through a supply-demand balance analysis under two scenarios: baseline and extreme scenarios. It proposes to mitigate the power deficit through demand response and orderly power consumption.

2. The Mathematic Model of Time Series Production Simulation

The time series production simulation of the power system aims to minimize the operating cost, meet the constraints of power and electricity balance and equipment operation characteristics at each time period, simulate the operation of units, and calculate the generation cost and electricity profit and loss. Demand response can bring flexibility to the operation of the power system. Firstly, the objective function is:

$$\min_{p_{it},q_t} \sum_{t=1}^{T} \left(\sum_{i=1}^{I} (f_i(p_{it}) + h_i s_{it}) + g(q_t) \right) \quad (1)$$

Where I is the number of units; T is the time periods; p_{it} is the power output of unit i at hour t, which is a continuous variable; f_i is the fuel cost function of unit i, which is a quadratic function generally; h_i is the start-up cost of a thermal power unit (if unit i is not a thermal power unit, this item is 0), s_{it} is a binary variable, if unit i starts up at hour t, it is 1, otherwise it is 0. q_t is power of demand response, which is a positive value for peak shaving demand response, it is a negative value for valley filling demand response. g is the cost function of demand response, the typical function of g is:

$$g(q_t) = c^+ q_t^+ + c^- q_t^- \quad (2)$$

$$q_t^+ \geq 0,\ q_t^+ \geq q_t \quad (3)$$

$$q_t^- \geq 0,\ q_t^- \geq -q_t \quad (4)$$

Where q^+_t and q^-_t are auxiliary variables, c^+ and c^- are prices of peak shaving and valley filling demand response, respectively. When peak shaving demand response is executed at hour t, from (1)、(3) and (4), q^+_t is equal to q_t, q^-_t is equal to 0. Similarly, When valley filling demand response is executed at hour t, q^+_t is equal to 0, q^-_t is equal to q_t. Therefore demand response power q_t is the sum of q^+_t and q^-_t, that is

$$q_t = q_t^+ + q_t^- \quad (5)$$

Then, in terms of constraints, each time period satisfies power balance.

$$\sum_{i=1}^{I} p_{it} + q_t = l_t,\ \forall t \quad (6)$$

Where l_t is the load at hour t. p_{it} is between the maximum and minimum power outputs

$$p_i^{\min} u_{it} \leq p_{it} \leq p_i^{\max} u_{it}, \forall i, t \quad (7)$$

Where p^{max}_i 和 p^{min}_i are maximum and minimum power outputs of unit i at hour t. u_{it} is a binary variable, if unit i operates at hour t, it is 1, otherwise it is 0. s_{it} and u_{it} are correlated by

$$s_{it} \geq u_{it} - u_{it-1} \quad (8)$$

The power of demand response is constrained by its potentiality.

$$-q_t^{\text{vmax}} d_t \leq q_t \leq q_t^{\text{pmax}} d_t \quad (9)$$

Where q_t^{pmax} is the maximum potentiality of peak shaving demand response, q_t^{vmax} is the maximum potentiality of valley filling demand response; d_t is a binary variable, if demand response is executed at hour t, it is 1, otherwise it is 0.

The operation of demand response organizations requires a significant investment in costs, and usually can be implemented at most for a certain period of time throughout the year. Therefore, the implementation duration of demand response has the following constraints

$$\sum_{t=1}^{T} d_t \leq D \quad (10)$$

Where D is total time periods of demand response.

3. Scenario Design

Two scenarios, baseline, and extreme scenarios, are designed, considering factors such as economic development and meteorological conditions. The baseline scenario assumes that power demand will maintain relatively fast growth during the "14th Five-Year Plan" period, considering current stock and approval, ongoing construction, planned power sources, as well as measures such as demand response, reserve sharing, and inter-provincial assistance. The extreme scenario assumes that the probability of extreme weather events will increase due to global climate change, resulting in a significant increase in cooling and heating loads and unexpected growth in maximum load. Coupled with extreme cold and hot weather, wind and solar power may not be able to achieve their generation peak, and therefore may not participate in the power balance.

The application scenario of demand response has shifted from "mitigating the power supply-demand gap" to "improving the level of energy and power fine management and grid operation flexibility, and promoting the consumption of clean energy". This effectively alleviates grid regulation pressure and addresses new issues currently facing the power system. Orderly power consumption has played a significant role in the rational adjustment of China's supply-demand balance and better prioritizing normal power consumption for people's livelihoods, key enterprises, key industries, and key locations.

Orderly power consumption and demand response are both important means of developing and utilizing adjustable resources during the "14th Five-Year Plan" period. The implementation plan should prioritize demand response and use orderly power consumption as a fallback to ensure it is more scientifically precise and effective.

Demand response should be the preferred method, and its large-scale development depends on the maturity of the electricity market mechanism. Priority should be given to adjusting well-performing air conditioning loads and actively developing energy storage resources such as electricity-based thermal energy storage and ice-based thermal energy storage to create the most cost-effective response plan.

Orderly power consumption is an important fallback measure to address power supply-demand tension, which requires government management, grid implementation, and user cooperation. Users should be classified reasonably and treated differently. For example, orderly power consumption should be prioritized for "two-high" users; while for other users, it should be executed in a classified order.

4. Supply-Demand Balance Results

Based on the time series production simulation model of the power system, we investigate the supply-demand balance of the case study system. The supply-demand balance situation of the Shandong grid was analyzed based on the two scenarios mentioned above. It is proposed to alleviate the power deficit through demand response and orderly power consumption.

Baseline Scenario: It is expected that during the "14th Five-Year Plan" period, Shandong will maintain rapid economic growth, achieve significant results in optimizing and adjusting its industrial structure, and experience a steady increase in cooling and heating loads. The maximum load of the Shandong power grid is projected to reach 145 million kilowatts by 2025, as shown in Figure 1, with an average annual growth rate of 4.9% during the "14th Five-Year Plan" period. Considering the current stock and approval, ongoing construction, planned power sources, as well as measures such as demand response, reserve sharing, and inter-provincial assistance, Shandong is expected to achieve a tight balance in 2025.

Extreme Scenario: By 2025, taking into extreme weather factors, the maximum load is expected to increase by 4.50 million kilowatts, accounting for about 3% of the maximum load. New energy sources will not participate in the balance, which will affect the output by 1.40 million kilowatts, accounting for about 5% of the total output. As a result, there will be a maximum power deficit of 7.90 million kilowatts as shown in Table 1 (considering the implementation of demand response, which will reduce the deficit by about 6 million kilowatts).

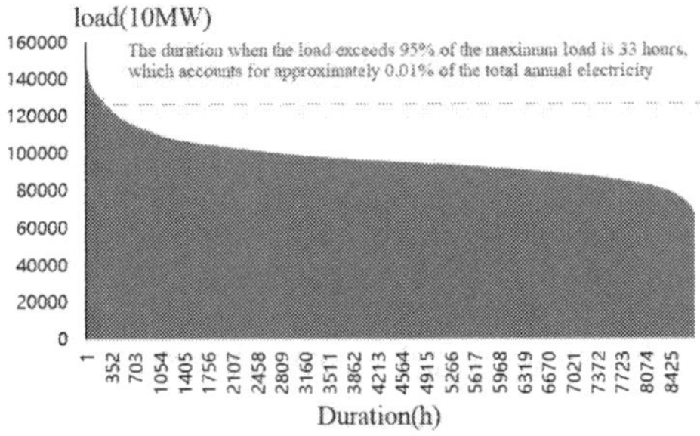

Figure 1 Continuous Load Curve of Shandong in 2025

Table 1 Power Surplus/Deficit in Shandong during the "14th Five-Year Plan" Period (unit: 10,000 kW)

	2023	**2024**	**2025**
Baseline Scenario	0	10	20
Extreme Scenario	-760	-780	-790

5. Demand-Side Resource Regulation Plan

Baseline Scenario: Considering factors such as policies, costs, and technological maturity, the demand response of the Shandong grid can achieve a scale of approximately 6 million kilowatts by 2025, which can meet the requirements of power balance. When formulating demand response plans, priority should be given to aggregated air conditioning loads in commercial buildings and residential areas, followed by industrial and commercial production loads and emerging loads. The order and scale of load dispatch are determined based on the response capacities and response costs of different users, and the most cost-effective load dispatch plan is formulated.

Extreme Scenario: In 2025, there is a power deficit of approximately 7.90 million kilowatts in Shandong's grid. The deficit is addressed through orderly power consumption, and according to the principle of "guaranteed supply with limitations", priority is given to the implementation of orderly power consumption for businesses that are subject to development restrictions. The specific plan is divided into three tiers, which are executed in sequence, as shown in Table 2. The first tier involves "two high" projects and industrial power users. The second tier involves high-energy-consuming enterprises such as the papermaking, rubber, and plastics industries that are not categorized as "two high" projects. The third tier involves users in industries such as construction, wholesale and retail, accommodation and catering, real estate, leasing, and commercial services.

Table 2 Demand-Side Resource Regulation Plan

	Scale (10, 000 kW)	**User Category**
Demand response	290	Air conditioning load regulation

		260	Industrial and commercial production load regulation
		50	Emerging load regulation
Orderly power consumption	**First tier**	709	"Two high" projects and industrial power users
	Second tier	1747	High-energy-consuming enterprises such as the papermaking, rubber, and plastics industries that are not categorized as "two high" projects
	Third tier	1207	Users in industries such as construction, wholesale and retail, accommodation and catering, real estate, leasing, and commercial services

6. Conclusion

Based on the supply-demand analysis and the design of the demand-side resource regulation plan, the demand response of the Shandong grid can essentially meet the power balance requirements by 2025 under the baseline scenario. Under the extreme scenario, the power deficit can also be addressed by orderly power consumption. The development and utilization of demand-side resources are of great importance in ensuring power supply and building a novel power system, as they effectively reduce the load during peak hours and ensure reliable power supply.

It is suggested to pay more attention to the development and utilization of demand-side resources. China should fully recognize the important role and value of developing and utilizing demand-side resources in promoting the smooth energy transition and achieving the goal of "dual-carbon", and garner recognition of the whole society for this concept through multi-channel publicity.

Acknowledgment

This work was supported by the Science and Technology Project of State Grid Shandong Electric Power Company (520625210017): Study on the Energy Transformation Path and Effect Evaluation of Shandong Province under the Target of Carbon Peak and Carbon Neutralization.

References

1. WANG Meibao. Study on Comprehensive Value Evaluation and Utilization Strategy for Power Demand-Side Resources [D]. North China Electric Power University (Beijing), 2020. DOI: 10.27140/d.cnki.ghbbu.2020.000286.
2. WANG Beibei et al. Analysis and Enlightenment of Renewable Energy and Demand Response Participating in UK and US Capacity Markets Considering Capacity Credibility [J]. Power System Technology, 2022, 46(04):1233-1247. DOI: 10.13335/j.1000-3673.pst.2021.1704.
3. Burke, Paul J., and Ashani Abayasekara. "The price elasticity of electricity demand in the United States: A three-dimensional analysis." The Energy Journal 39.2 (2018).
4. XIAO Yong, WANG Yan, HE Hengjing, WU Haowen, JIAN Ganyang, ZHAO Bo, GAO Ciwei. Avoidable Cost Model Study in Power Demand Side Management [J]. Power Demand Side Management, 2020, 22 (04): 56-61.
5. LI Jianlin, CUI Yilin, MA Suliang, WU Yiwen, JIANG Yerong, XIN Dixi. Operation Mode Optimization and Economic Benefit Analysis of Demand-side Shared Energy Storage [J/OL]. Power System Technology: 1-18 [2022-09-29]. DOI: 10.13335/j.1000-3673.pst.2022.1218.
6. FENG Xiaofeng, XIE Tiankuo, GAO Ciwei, LIN Guoying, CHEN Liang, LU Shixiang. A Demand Side Response Strategy Considering Long-term Revenue of Electricity Retailers in the Electricity Spot Market [J]. Power System Technology, 2019, 43 (08): 2761-2769. DOI: 10.13335/j.1000-3673.pst.2019.0651.
7. SUN Weiqin, ZHENG Yuqi, XUE Guiting. Design and Effect Analysis of Demand Response Transaction Mechanism Considering Response Rate Difference [J]. Automation of Electric Power Systems, 2021, 45 (22): 83-94.
8. Qi Bing, ZHENG Shunlin, SUN Yi, LI Bin, TIAN Shiming, SHI Kun. A Model of Incentive-based Integrated Demand Response Considering Dynamic Characteristics and Multi-energy Coupling Effect of Demand Side [J]. Proceedings of the CSEE, 2022, 42 (05): 1783-1799. DOI: 10.13334/j.0258-8013.pcsee.202351.
9. C. Eksin, H. Deliç and A. Ribeiro. Demand Response Management in Smart Grids With Heterogeneous Consumer Preferences [J]. IEEE Transactions on Smart Grid, 2015, 6(06): 3082-3094.
10. I. Schneider, M. Roozbehani and M. Dahleh. An Online Learning Framework for Targeting Demand Response Customers[J]. IEEE Transactions on Smart Grid, 2022, 13(01): 293-301.
11. M. A. Khan, A. M. Saleh, M. Waseem and I. A. Sajjad. Artificial Intelligence Enabled Demand Response: Prospects and Challenges in Smart Grid Environment. IEEE Access, 2023, 11:1477-1505. DOI: 10.1109/ACCESS.2022.3231444.
12. D. Muthirayan, D. Kalathil, K. Poolla and P. Varaiya. Mechanism Design for Demand Response Programs[J]. IEEE Transactions on Smart Grid, 2020, 11(01):61-73. DOI: 10.1109/TSG.2019.2917396.

Life Cycle Environmental Impact of Pumped Hydro Energy Storage

Keshuo Zhang[1,2], Meiyan Zheng[1*], Jiancheng Mo[1]

[1]Guangdong Hydropower Planning & Design Institute Co. Ltd, Guangzhou, China
[2] Guangdong Key Laboratory of Environmental Pollution and Health, School of Environment, Jinan University, Guangzhou 511443, China

Abstract. Pumped hydro energy storage (PHES) is one of the energy storage systems to solve intermittent renewable energy and support stable power generation of the grid. About 95% of installed capacity of the global energy storage system is contributed by PHES. Life cycle assessment (LCA) is used to analyse the environmental impact of PHES construction and operation phase in this study, and 1 MWh of electricity delivered from PHES to the power grid is set as the functional unit. The results show that the electricity power structure and electricity loss caused by the charging-discharging of PHES are the main environmental burden contributors, contributing 80 to 99% of the total environmental emissions. And environmental impacts during the construction phase is mainly due to the use of concrete, steel, and cement. In the future, as the proportion of renewable energy in the grid structure increases, the environmental impacts caused by PHES will decrease accordingly.

1. Introduction

Nowadays, with the limitation of fossil energy resources and the deterioration of the global climate, renewable energy is widely used [1-2]. However, there are intermittency problems with using renewable energy (such as wind energy, solar energy)[3], which might cause erratic output of the electricity grid [4]. Setting up energy storage systems can effectively solve this intermittency problem[5] and ensure the stability of grid power supply [6]. Energy storage systems can be divided into mechanical storage system, electrochemical systems, chemical storage and thermal storage systems[7]. Pumped hydro energy storage (PHES) is the dominating energy storage technique worldwide[8], which is belonged to the mechanical storage systems[9]. As of 2021, the installed capacity of PHES is about 181273 MW, accounting for 95% of the installed capacity of the global energy storage system [10]. PHES pumps water from a lower basin to a higher-level basin to store gravitational potential energy of water transfered from the low-cost electric power (electricity in off-peak time), and the stored water is released through hydro turbines to produce electric power during the periods of high energy demand [11].

Life cycle assessment (LCA) is a extensively employed methodology for evaluating the comprehensive environmental impact of product systems throughout their entire life cycle.[12] Within the field of energy storage, numerous studies have been conducted utilizing the LCA approach. Manal AlShafi et al[13] undertook a thorough analysis employing life cycle assessment to evaluate three energy storage technologies, namely compressed air energy storage, vanadium redox flow battery, and molten salt thermal storage, with the aim of addressing environmental sustainability concerns. Laurent Vandepaer et al[14]conducted a life cycle assessment specifically focused on lithium-ion battery energy storage to ascertain the environmental impact associated with different specifications and types of lithium-ion batteries. Lydia Stougie et al[15] conducted a multidimensional environmental impact assessment on five energy storage systems, including PHES, which was found to cause the least damage to human health, ecosystem diversity, and resource availability. However, many current studies are based on inventories established by predecessors, which may result in certain discrepancies from the actual circumstances.

In this study, we quantify the potential environmental impacts of PHES based on life cycle assessment model and original data to determine the main environmental burden contributors in construction and operation phases, which might further support the construction and operation of the PHES systems.

*zheng.my@gpdiwe.com

2. Methods and Modelling

2.1. Scope and overview

This study analyses the environmental impacts of the construction and operation of Huizhou pumped hydro energy storage in Guangdong Province, China under a life cycle perspective. The goal is to (1) determine the environmental impacts of PHES, (2) analyse the main factors that caused environmental impacts in the construction and operation phases.

2.2. System boundary & functional unit

Figure 1 shows the system boundary selected in this study, and the analysis includes the construction phase and operation phase of PHES. The construction phase includes the construction of lower basin, upper basin, water conveyance system, underground plant, electricity and diesel input. Electricity loss is considered due to charging and discharging efficiency during operation phase. Due to the lack of data in PHES' end of life (EOL) phase, and there are other functional values such as irrigation and tourism after EOL period for PHES[16]. Thus, the EOL phase is not considered. In this study, 1 MWh electricity delivery during the 60-year lifetime of the PHES was used as the functional unit.

2.3. Life cycle inventory and assumptions

Data on raw material input, energy consumption, and charge-discharge efficiency during construction and operation were provided by Guangdong Hydropower Planning & Design Institute Co. Ltd. In addition, according to relevant literature, this study assumes that the PHES has a lifetime of 60 years [17].

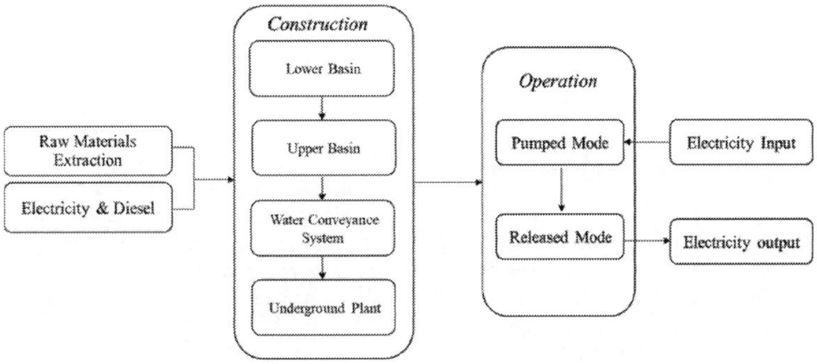

Figure 1. The system boundary of PHES for life cycle assessment

2.4. Life cycle assessment

SimaPro software was used to model the life cycle impacts assessment, and the most widely used ReCiPe 2016 method in the LCA model is applied for the midpoint environmental impacts assessment of the PHES construction and operation periods[18]. The midpoint assessment includes 18 environmental impact categories, global warming (GW), Stratospheric ozone depletion (SOD), ionizing radiation (IR), Ozone formation, human health (OFHH), fine particulate matter formation (FPMP), ozone formation, terrestrial ecosystems (OFTE), terrestrial acidification (TA), freshwater eutrophication (FE), marine eutrophication (ME), terrestrial ecotoxicity (TET), freshwater ecotoxicity (FET), marine ecotoxicity (MET), human carcinogenic toxicity (HCT), human non-carcinogenic toxicity (HNCT), land use (LU), mineral resource scarcity (MRS), fossil resource scarcity (FRS), water consumption (WC).

3. Results and discussion

3.1. Environmental impacts analysis

The LCA results reflect the environmental impacts of PHES (Table 1). In order to better understand the environmental impacts of different phases, Figure 2 shows the proportions of PHES environmental impacts contribution. Due to the characteristics of the raw materials used in the construction phase, the proportion of mineral resource scarcity in the construction phase is higher than other indicators, but it is the lowest contribution in operation phase of all indicators. However, the influence of the operation phase accounted for more than 90% in the remaining 17 indicators. For example, the 1MWh energy delivery causes about 350 $kgCO_2eq$ emissions, while the operation phase accounts for more than 99% of this value. The environmental impacts of the operation phase are mainly related to the charge-discharge efficiency and electricity sources. Compared to other electricity sources, Chinese electricity may cause higher carbon emissions [19].

https://doi.org/10.1051/e3sconf/202344102004

Table 1. Environmental impacts of delivery 1MWh electricity by PHES

Impact category	Unit	Total	Construction ratio (%)	Operation ratio (%)
GW	kg CO2 eq	349.88	0.64	99.36
SOD	kg CFC11 eq	0.00	0.75	99.25
IR	kBq Co-60 eq	5.46	1.04	98.96
OFHH	kg NOx eq	0.97	0.60	99.40
FPMP	kg PM2.5 eq	0.53	0.56	99.44
OFTE	kg NOx eq	0.97	0.62	99.38
TA	kg SO2 eq	1.19	0.53	99.47
FE	kg P eq	0.06	0.94	99.06
ME	kg N eq	0.00	0.91	99.09
TET	kg 1,4-DCB	178.64	4.11	95.89
FET	kg 1,4-DCB	3.23	2.07	97.93
MET	kg 1,4-DCB	4.51	2.12	97.88
HCT	kg 1,4-DCB	9.74	2.54	97.46
HNCT	kg 1,4-DCB	75.64	2.74	97.26
LU	m2a crop eq	4.15	1.04	98.96
MRS	kg Cu eq	0.12	19.19	80.81
FRS	kg oil eq	68.30	0.71	99.29
WC	m3	0.91	6.16	93.84

3.2. Environmental impacts of the construction

Figure 2 shows the environmental impact contributions of construction phase. The construction phase contains four processes, the constructions of water delivery system and the lower basin contribute relatively high environmental impacts during the whole construction phase. The use of concrete and steel is the two most influential factors during the construction of the water delivery system, the lower basin and the underground plant. And a large amount of sand, copper, and concrete are used in the construction of the upper basin. In general, the extensive use of concrete is the main cause of environmental impacts of the construction phase, and the impacts of steel and cement consumption should not be ignored.

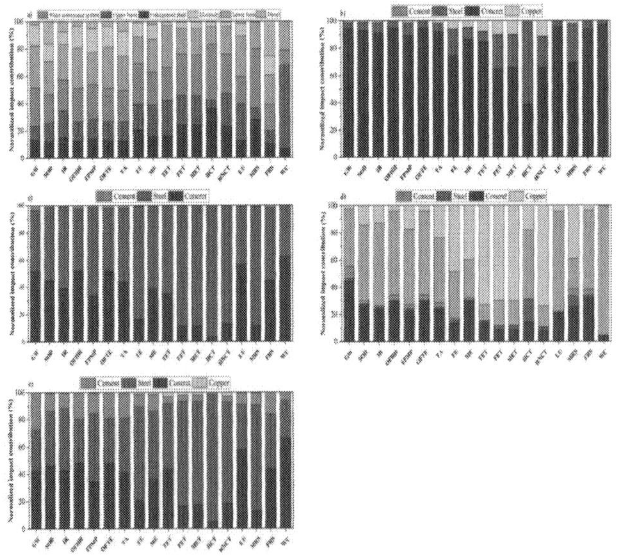

Figure 2. Environmental impact contributions of PHES construction, lower basin, underground plant, upper basin and water conveyance system (a-e).

https://doi.org/10.1051/e3sconf/202344102004

3.3. Discussion

The results show that electricity loss during operation phase is the main factor of environmental impacts for PHES. In 2020, China's electricity power structure is 67.9% thermal power, 17.0% hydropower, 6.0% wind power, 3.5% photovoltaic power, and 5.6% other power generation. It can be seen from the current power structure that China's power grid structure largely relies on thermal power generation which may creates more serious environmental impacts than renewable energy power generation. The research results conducted by Oliveira et al.[20] on the environmental impact of energy storage systems applied in the power grid under different power combinations prove that the use of renewable energy for power generation significantly reduces environmental impact. In the future, thermal power proportion will decrease to 32.0% and 15.0% in 2040 and 2050, respectively, and the proportion of renewable energy used in the grid will increase [21]. So the environmental impacts caused by electricity loss during the operation of PHES in the future will also be reduced accordingly. Besides, the concrete and steel used in a large number of construction phase can be considered to use alternative materials to reduce environmental impacts. For example, Ueda et al [22] consider replacing the concrete factory building structure with a timber structure with the same architectural requirements as the concrete block building in micro-hydropower construction. This material substitution can significantly reduce the potential environmental burdens. In this study, the lifetime of PHES is conservatively assumed to be 60 years, which may underestimate the actual lifetime of PHES and may lead to an overestimation of its environmental impact. For examples, Flury et al. conducted a study on the environmental impact of PHES over a period of 150 years lifetime[23], Immendoerfer et al [24] analysed the environmental impacts of different years lifetime of PHES, 80 and 150 years respectively, and found that the environmental impacts caused by PHES are lower as the lifetime increases.

4. Conclusion

Based on the original data, this paper assesses environmental impacts of the construction and operation for PHES. The results show that concrete, steel, sand and other raw materials cause main environmental impacts during the construction of PHES. The electricity loss caused by the charge-discharge efficiency is directly related to the environmental impacts of the operation phase. For the construction, one of the ways to reduce environmental impacts is to consider the use of alternative materials; this paper uses the Chinese power grid that is highly dependent on thermal power generation to simulate the operation of PHES. In the future, as the proportion of renewable energy in the power grid increases, the resulting environmental impacts will also be reduced. This study quantifies the environmental impact of PHES, identifies the main environmental hotspots, and provides some future improvement recommendations, serving as a reference for the life-cycle management of PHES.

Acknowledgments

This study was supported by the Research on carbon balance and sustainability evaluation method system of pumped hydro energy storage (Contract No. CE682022000386).

References

1. Bouman, Evert A., Martha M. Øberg, and Edgar G. Hertwich. 2016. Environmental impacts of balancing offshore wind power with compressed air energy storage (CAES), *Energy*, **95**: 91-98.
2. Bazdar, E., Sameti, M., Nasiri, F., & Haghighat, F. 2022. Compressed air energy storage in integrated energy systems: A review. Renewable and Sustainable Energy Reviews, 167.
3. Gowrisankaran, G., Reynolds, S. S., & Samano, M. (2016). Intermittency and the value of renewable energy. *Journal of Political Economy*, **124**(4), 1187-1234.
4. Yudhistira, Ryutaka, Dilip Khatiwada, and Fernando Sanchez. 2022. A comparative life cycle assessment of lithium-ion and lead-acid batteries for grid energy storage, *Journal of Cleaner Production*, 358.
5. Mitali, J., Dhinakaran, S., & Mohamad, A. A. 2022. Energy storage systems: a review. *Energy Storage and Saving*, 1(3), 166-216.
6. Tan, Kang Miao, Thanikanti Sudhakar Babu, Vigna K. Ramachandaramurthy, Padmanathan Kasinathan, Sunil G. Solanki, and Shangari K. Raveendran. 2021. Empowering smart grid: A comprehensive review of energy storage technology and application with renewable energy integration, *Journal of Energy Storage*, 39.
7. Olabi, A. G., C. Onumaegbu, Tabbi Wilberforce, Mohamad Ramadan, Mohammad Ali Abdelkareem, and Abdul Hai Al – Alami. 2021. Critical review of energy storage systems, *Energy*, 214.
8. Olympios, A. V., McTigue, J. D., Farres-Antunez, P., Tafone, A., Romagnoli, A., Li, Y., . . . Markides, C. N. (2021). Progress and prospects of thermo-mechanical energy storage—a critical review. *Progress in Energy*, **3**(2).
9. Mahmoud, M., Ramadan, M., Olabi, A.-G., Pullen, K., & Naher, S. 2020. A review of mechanical energy storage systems combined with wind and solar applications. *Energy Conversion and Management*, 210.
10. Sandia National Laboratories and U.S. Department of energy (DOE) global energy storage database. https://www.energystorageexchange.org/projects.

11. Rehman, Shafiqur, Luai M. Al-Hadhrami, and Md Mahbub Alam. 2015. Pumped hydro energy storage system: A technological review, *Renewable and Sustainable Energy Reviews*, **44**: 586-98.
12. Varun, Bhat, I. K., & Prakash, R. (2009). LCA of renewable energy for electricity generation systems—A review. Renewable and Sustainable *Energy Reviews*, **13**(5), 1067-1073.
13. AlShafi, M., & Bicer, Y. 2021. Life cycle assessment of compressed air, vanadium redox flow battery, and molten salt systems for renewable energy storage. *Energy Reports*, **7**, 7090-7105.
14. Vandepaer, L., Cloutier, J., & Amor, B. 2017. Environmental impacts of Lithium Metal Polymer and Lithium-ion stationary batteries. *Renewable and Sustainable Energy Reviews*, **78**, 46-60.
15. Stougie, L., Del Santo, G., Innocenti, G., Goosen, E., Vermaas, D., van der Kooi, H., & Lombardi, L. 2019. Multi-dimensional life cycle assessment of decentralised energy storage systems. *Energy*, **182**, 535-543.
16. de Almeida, A. T., Moura, P. S., Marques, A. S., & de Almeida, J. L. 2005. Multi-impact evaluation of new medium and large hydropower plants in Portugal centre region. *Renewable and Sustainable Energy Reviews*, **9**(2), 149-167.
17. Denholm, Paul, and Gerald L. Kulcinski. 2004. Life cycle energy requirements and greenhouse gas emissions from large scale energy storage systems, *Energy Conversion and Management*, **45**: 2153- 72.
18. Huijbregts, Mark A. J., Zoran J. N. Steinmann, Pieter M. F. Elshout, Gea Stam, Francesca Verones,Marisa Vieira, Michiel Zijp, Anne Hollander, and Rosalie van Zelm. 2016. ReCiPe2016: a harmonised life cycle impact assessment method at midpoint and endpoint level, *The International Journal of Life Cycle Assessment*, **22**: 138-47.
19. Ou, Xunmin, Yan Xiaoyu, and Xiliang Zhang. 2011. Life-cycle energy consumption and greenhouse gas emissions for electricity generation and supply in China, *Applied Energy*, **88**: 289-97.
20. Oliveira, L., Messagie, M., Mertens, J., Laget, H., Coosemans, T., & Van Mierlo, J. 2015. Environmental performance of electricity storage systems for grid applications, a life cycle approach. *Energy Conversion and Management*, **101**, 326-335.
21. CEC, 2021. Annual report on development of China power industry 2021. https://www.cec.org.cn/menu/index.html?688.
22. Ueda, T., E. S. Roberts, A. Norton, D. Styles, A. P. Williams, H. M. Ramos, and J. Gallagher. 2019. A life cycle assessment of the construction phase of eleven micro-hydropower installations in the UK, *Journal of Cleaner Production*, **218**: 1-9.
23. Flury K, Frischknecht R. Life cycle inventories of hydroelectric power generation. Uster, Switzerland: ESU-services Ltd; 2012.
24. Immendoerfer, Andrea, Ingela Tietze, Heidi Hottenroth, and Tobias Viere. 2017. Life-cycle impacts of pumped hydropower storage and battery storage, *International Journal of Energy and Environmental Engineering*, **8**: 231-45.

 https://doi.org/10.1051/e3sconf/202344102005

Design and Selection of Pipelines for Compressed Air Energy Storage Power Plants

Ning Luo*, Bin Xiao, Wei Liu, Yanglong Duan, Dongsheng Niu

Power Construction Corporation of China Northwest Survey, Design and Research Institute Co., Ltd Xi'an, China

Abstract: The principle of Compressed-air energy storage is that the compressed air energy storage system uses compressed air as the energy storage carrier, which is a physical Energy storage that uses mechanical equipment to realize energy storage, transfer and utilization across time and space. At present, Compressed-air energy storage is the second largest technology that is considered suitable for GW level large-scale electric energy storage after pumped storage. Compressed air energy storage has outstanding advantages such as large scale, low cost, long service life, and short construction period. As a key link connecting compressors, expanders, and gas storage devices, the compressed air main pipeline has characteristics such as high operating pressure, low internal fluid temperature, large temperature difference between indoor and outdoor pipelines, and frequent start and stop. The design, calculation, and installation of the compressed air main pipeline will affect the economy and reliability of the entire system, and even bring serious destructive accidents. This article comprehensively introduces the selection method and process of compressed air energy storage pipeline design, and further verifies the feasibility and accuracy of the design method through case studies of specific projects. It provides convenience and calculation methods for the large-scale development of compressed air energy storage projects in the future[1].

1. Introduction

The compressed air energy storage system utilizes the peak valley electricity difference for energy storage and generation, achieving the transfer of electrical energy in time and space. As a key link connecting compressors, expanders, and gas storage devices, the compressed air main pipeline has characteristics such as high operating pressure, low internal fluid temperature, large temperature difference between the inside and outside of outdoor pipelines, and frequent startup and shutdown. The design, calculation, and installation of the compressed air main pipeline will affect the economy and reliability of the entire system, and even bring serious destructive accidents. Therefore, it is required that the selection of compressed air main pipeline should comprehensively consider the characteristics of safety, economy, corrosion resistance, oxidation resistance, aging resistance, cracking resistance, irritation resistance, etc. At the same time, the compressed air pipeline should meet the user's requirements for flow, pressure, temperature, quality, etc., while considering the needs of later expansion and renovation. Therefore, the design and selection of the compressed air main pipeline play a crucial role in the entire project[2].

* Corresponding author: 19991218060@163.com

2. Material selection factors

2.1 Corrosion resistance

First, confirm that you have the correct template for your paper size. This template has been tailored for output on the US-letter paper size. If you are using A4-sized paper, please close this file and download the file for "MSW_A4_format".

Pipeline corrosion is divided into external corrosion and internal corrosion. Pipeline inner wall corrosion refers to the corrosion of the pipeline inner wall caused by factors such as temperature, pressure, flow rate, and alternating stress due to the presence of corrosive impurities in the pipeline medium.

The medium used in compressed air energy storage pipelines is high-pressure and normal temperature air, and the corrosion resistance of pipelines is an important factor and indicator that affects the selection of compressed air pipeline materials. The relative humidity in the atmosphere is generally over 65%. After being compressed and cooled by a compressor, the air becomes wet saturated air. The pressure dew point temperature range of saturated compressed air after cooling and drying treatment is generally -70 °C~10 °C. The more water content in the air, the higher the dew point temperature. When the ambient temperature is below the dew point temperature, liquid water will precipitate from the moisture in the humid air. The liquid water that has not been cooled and separated and is generated due to temperature reduction is carried by compressed air in the pipeline system, becoming the main cause of equipment, pipeline, and valve corrosion[3].

For pipeline materials with good corrosion resistance, the current national standard GB50029-2014 "Code for Design of Compressed Air Stations" has made specific provisions, clarifying that pipeline materials are related to the quality of compressed air transported, as shown in Table 1.

Table 1. Selection regulations for compressed air pipes

Solid particle level or humidity level	Regulations on the selection of pipes
≤5 level	Carbon steel pipes can be used
>5 level, ≤3 level	Hot-dip galvanized steel pipes and stainless steel pipes can be used
>3 level	Stainless steel or copper pipes should be used

This specification specifies the pipeline materials applicable to compressed air of various quality levels, and the selection principle is technical and economic, with the aim of ensuring stable gas supply quality and not being contaminated by the transmission pipeline.

2.2 Price

Pipeline prices are influenced by various factors, including raw material prices, production costs, quality control, transportation costs, maintenance costs, installation costs, etc. The purpose of price analysis is to understand the main components of price, in order to make wise decisions in the actual selection process.

2.3 Construction convenience

The connection of metal pipelines generally adopts welding, threaded connection, or clamp connection, while non-metallic pipelines generally adopt hot melt or electric fusion connection. The construction speed of welded connections is slower than that of threaded, clamp, or hot melt connections, but the sealing stability of welded connections is the most reliable, and the interface is not prone to leakage. The advantages of clamp connection are convenient installation and disassembly, fast construction, and the construction workload is only one-third of that of welding, and it is suitable for situations where on-site hot work is not allowed.

2.4 Design temperature

《The 20G steel and other metal pipes specified in Seamless Steel Tubes for High Pressure Boilers》（CB5310-2017）are mainly used for pipes and components with high pressure and temperature. Due to its good oxidation resistance, plastic toughness, welding performance and other cold and hot processing properties, and considering economic factors, this material is often used at about 400C. However, in the cold regions of northern China, the average monthly minimum temperature over the years has even been lower than the minimum design temperature of -20C. If the device encounters a startup and shutdown process and corresponding process measures or operating rules cannot be taken to avoid the simultaneous occurrence of low temperature and high stress conditions, considering the impact of low temperature conditions in the atmospheric environment on metal temperature, according to national standards, the use of 20G steel and other metal pipelines will be restricted. If the design temperature of the pipeline exceeds 350 ℃ after the device is started, only materials with a wider temperature range such as stainless steel can be selected, which will inevitably increase costs and affect economic efficiency.

2.5 Other factors

The selection of pipes is also influenced by factors such as supply cycle, maintenance cycle, pipeline installation and lifting conditions, pipeline working environment, aesthetics, etc. Among them, pipeline maintenance refers to matters such as strengthening the external anti-corrosion of pipelines and replacing vulnerable parts of pipe joints.

The following table provides a comprehensive comparative analysis of 9 different specifications of pipelines, and their advantages and disadvantages are described in the table 2[4-5].

Table 2. Comparative analysis of pipelines of different specifications

Serial Number	Piping Standards	Pipe Material	Connection method	Pipeline price/(¥/m)	advantage	Disadvantages
1	GB/T3091-2015《Welded steel pipes for low-pressure fluid transportation》	Q235-B steel	Welding threaded connection flange connection	8~220	Low price	Easy to rust and high resistance

2	GB/T8163-2018《 Seamless steel tubes for liquid service 》	20 steel	weld flange connection	9~260	Low price and wide application	Easy to rust and high resistance
3	GB/T8163-2018《 Seamless steel tubes for liquid service 》	20 steel, galvanized	weld flange connection	15~420	Relative corrosion resistance	Zinc coating is prone to damage and corrosion
4	GB/T14976-2012《 Stainless steel seamless steel pipes for fluid transportation 》	06Cr19Ni10	weld flange connection	35~1000	Corrosion resistance and high strength	High price
5	GB/T4437.1-2015 《 Aluminum and aluminum alloy hot extruded tubes - Part 1: Seamless circular tubes 》	aluminum alloy	Clamp connection flange connection	10~200	Fast construction, corrosion resistance, and low resistance	Pipe fittings are expensive
6	CJ/T184-2012 《 Stainless steel lined plastic composite pipes and fittings 》	Stainless steel lined plastic composite pipe	Hot melt socket sleeve connection flange connection	28~530	Anti corrosion, low resistance, beautiful and hygienic	High price
7	CJ/T151-2016 《 Thin walled stainless steel pipe 》	06Cr19Ni10	Compression connection welding	14~460	Corrosion resistant and aesthetically pleasing	low strength
8	CJ/T321-2010《 Aluminum alloy lined plastic composite pipes and fittings 》	Aluminum alloy lined plastic composite pipe	Hot melt socket	35~850	Anti corrosion, low resistance, beautiful and hygienic	High price
9	GB15558.1-2015 《 Buried polyethylene (PE) pipeline system for gas use 》	PE100 polyethylene	Hot melt socket and electric melt socket	5~200	Corrosion resistance, good toughness, and low price	Low intensity, visible light aging

2.6 Comparison of Pipe Materials

Among the 9 common pipes mentioned above, items 1-4 are commonly used. The pressure bearing capacity of items 5-9 of the pipeline is below 1.6MPa, among which aluminum alloy seamless pipeline has been promoted and applied due to its advantages of low corrosion resistance and fast installation speed. Items 6-9 are less commonly used for compressed air pipeline transportation.

Taking into account various factors such as working environment, corrosion resistance, price, construction convenience, and design temperature, the compressed air energy storage power station delivers non clean air supply quality. Under indoor and outdoor overhead or underground laying conditions, 20G is preferred as the compressed air energy storage power station gas pipeline.

20G seamless steel pipe is widely used in high-pressure boiler, heating heat exchanger, petroleum, chemical industry, electric power, shipbuilding, paper making, food and other industries. The advantages of 20G seamless steel pipe mainly include high strength, high temperature resistance, corrosion resistance, and other characteristics, which are suitable for high temperature and high pressure environments.

3. Pipeline Design

The compressed air pipeline of the compressed air energy storage power station should meet the user's requirements for compressed air flow rate, pressure, and quality, and appropriately consider the needs of later expansion and renovation[6].

3.1 Pipeline design pressure and temperature

The design pressure and temperature of compressed air pipelines shall be taken according to the following regulations:

1. The design pressure is taken as the maximum value of the following three items:

1) 105% of the compressor outlet design pressure;

2) 105% of the design pressure at the inlet of the expander;

3) The design pressure at the compressor outlet plus the pressure drop in the pipeline from the compressor outlet to the gas storage tank;

4) Design pressure at the inlet of the expander plus pressure drop in the pipeline from the expander inlet to the gas storage tank;

2. The design temperature should be based on the rated working temperature at the inlet and outlet of the gas storage tank, plus the allowable temperature deviation value during normal operation of the power plant. When the host factory does not provide temperature deviation, the temperature deviation value can be taken as 5 ℃.

3.2 Internal diameter of pipeline

With reference to the requirements of relevant specifications 《Code for Design of Steam and Water Piping in Fossil fuel power station 》 (DL/T5054-2016), the

https://doi.org/10.1051/e3sconf/202344102005

pipe diameter of single-phase fluid such as compressed air can be calculated according to the following formula:

$$Di = 594.7\sqrt{\frac{GV}{v}}$$

In the formula:

Di—Inner diameter of compressed air pipeline, mm;

G— Calculated flow rate of pipeline under normal temperature and pressure, t/h;

V— cut-off specific volume, m3/kg;

v—flow rate of compressed air in the pipeline, refer to relevant regulations in 《Code for Design of Oil and Gas Pipeline in Fossil fuel power station》 (DL-T5204-2017). table 3

Table 3. Recommended Flow Rates for Compressed Air Pipelines

Fluid type	Pipeline type		velocity of flow (m/s)
compressed air	Compressor intake pipe		~20
	Compressor air storage tank		~30
	General pipeline	d≤50	≤18
		d≥50	≤15

3.3 Pipeline components and materials

1. The material selection of compressed air pipelines should comply with the following regulations

1) The steel used for the pipeline should comply with the current national standards or power industry standards;

2) The selection of foreign materials should be based on the latest relevant foreign material standards and confirmed to be suitable for use conditions through analysis before use;

3) The selection of pipes should be based on the properties and parameters of the medium inside the pipes, as well as the safety and economy of operation under various working conditions.

4) Pipeline components should be selected from units that have obtained the qualification for manufacturing pressure pipeline components.

5) For compressed air pipelines without drying and purification requirements, carbon steel pipes can be used;

6) For pipelines with general purification requirements, galvanized steel pipes can be used;

7) For pipelines with a nominal diameter less than or equal to 150mm, welded steel pipes (galvanized) for low-pressure fluid transportation can be selected, and the material is Q235B;

8) For pipelines with a nominal diameter greater than 150mm, seamless steel pipes (galvanized) for fluid transmission can be selected, made of 20G (20 # steel);

9) For conveying pressure with a dew point below -40 °C or dust containing particle size less than or equal to 1 μ The purified compressed air pipeline of m can be made of stainless steel or copper pipes[7].

3.4 Pipeline accessories

3.4.1 Bends and Elbow

1. The main pipeline should use a bend, and the bending radius of the bend should be 3-5 times the outer diameter of the pipe. According to the layout, hot-pressed elbows that meet national or industry standards can also be used.

2. Pipes with nominal pressure below PN10 and nominal size below DN50 can use cold bending bends;

3. Long radius elbows should be used. Short radius elbows are only used for special arrangements;

4. Seamless hot pressed elbows should be used on pipelines with a nominal pressure greater than PN16, and straight pipe sections should be included.

3.4.2 Flange

The applicable pressure and temperature of flanges and accessories should comply with the provisions of the current national standard "Steel Pipe Flanges" GB/T 9124 regarding pressure temperature levels. When connecting flanges of different pressure levels, the usage conditions of flanges and accessories should be based on the lower level flange. [8]

3.4.3 Valves

Valves should be selected based on system parameters, diameter leakage level, opening and closing time, etc., and should meet the requirements of system shutdown, regulation, operation, and layout design. The form and operation mode of the valve should be selected based on the structure, installation, operation and maintenance requirements of the valve.

The commonly used shut-off valves on compressed air pipelines include butterfly valves, gate valves, globe valves, and ball valves.

1) When the nominal diameter is greater than or equal to 100mm, butterfly valves should be selected;

2) When the nominal diameter is less than 100mm, gate or globe valves can be selected;

3) Stop valves are often used in situations where appropriate flow regulation is required;

4) Ball valves are quick to open and close, with high reliability, and are commonly used in situations that require frequent opening and closing.

3.5 Water pressure test

Hydrostatic testing can be used to verify the strength of pipeline components and the tightness of pipeline systems.

1. The tightness test of pipeline system shall meet the relevant requirements of 《Code for Design of Steam and Water Pipeline in Fossil fuel power station 》(DLT5054-2016) and 《Technical Code for Construction of Electric Power Construction Part 5: Pipes and Systems 》(DL/T5190.5).

2. The strength test of pipeline components should comply with the following regulations

1) The strength test pressure (gauge pressure) of pipeline components should be calculated according to the following formula and the maximum value of both should be taken:

$$P_t = 1.25P\frac{[\sigma]^T}{[\sigma]^t} \text{ or } P_t = 1.5P$$

$$P_T = 0.1 + P$$

In the formula:

Pt— test pressure, MPa;

P—Design pressure, MPa;

[σ]T—allowable stress of steel at test temperature, MPa;

[σ]t—allowable stress of steel at design temperature, MPa.

2) Under the hydraulic test, the circumferential stress value inside the specimen should not exceed 90% of the material's yield limit at the test temperature. The circumferential stress should be calculated using the following formula:

$$\sigma_t = \frac{P_t[D_i + (S - C - C_1)]}{2(S - C - C_1)\eta}$$

In the equation:

σ^t— circumferential stress of pipeline components under test pressure (MPa);

P_t—Test pressure, MPa;

D_i— Pipe inner diameter, mm;

S— Pipe wall thickness, mm;

C—additional thickness considering corrosion and wear, mm;

C_1- additional value of negative deviation of wall thickness, mm;

3. After the pipeline installation is completed, the tightness of the pipeline system should be inspected;

1) The pressure of the hydraulic test is the gauge pressure, which should not be less than 1.5 times the design pressure and should not be less than 0.2MPa;

2) The circumferential stress of the pipeline under hydraulic testing, as well as the axial stress caused by internal pressure, live load, and dead load during pressure testing, should not exceed 90% of the material yield limit at the test temperature.

3) The temperature of water used for hydraulic testing should not be lower than 5 °C and should not exceed 70 °C. The test environment temperature should not be lower than 5 °C, otherwise measures should be taken to prevent freezing and cold brittle fracture;

4) The water quality for hydraulic testing should be clean and have minimal corrosion to pipeline system materials;

5) All welds of large-diameter pipes can also be tested for tightness by nondestructive testing instead of hydrostatic testing. The specific requirements of nondestructive testing should comply with the current power industry standard Technical Specification for Welding of Fossil fuel power station DL/T869

6) Tightness testing is not required for exhaust pipes and other pipelines leading to the atmosphere[9].

3.6 Corrosion allowance

According to the properties of soda and related research reports, for 20 G, the corrosion rate is 0.12mm/a. In the calculation process, the corrosion is considered based on a 30 year design life.

3.7 Pipeline Layout

1. Compressed air pipelines should be laid overhead and generally not buried. Indoor pipelines should be installed along the factory building, corridors, walls, and columns, while outdoor pipelines should be laid along cable trays or shared with other pipelines. Generally, single independent supports are not used for installation[10];

2. The pipeline layout should be combined with the equipment layout and building structure of the main plant, and the pipeline direction should be consistent with the axis of the plant. In areas where horizontal pipelines intersect more, it is advisable to delineate the elevation range of vertical and horizontal directions according to the pipeline direction, and arrange the pipelines in layers. The pipeline layout should not cause the main flow of the medium to change direction within the tee;

3. The pipeline system should prevent the occurrence of strain concentration in local areas due to elastic transfer of parts with high stiffness or low stress;

4. The pipeline installation should have a drainage slope of not less than 0.002. The low point of the pipeline should be equipped with a drainage pipe, and a DN15 ball valve should be installed on the drainage pipe. The distance between the drainage pipe mouth and the ground or platform should not be less than 150mm; High points of pipelines with a nominal diameter greater than or equal to 32mm should be equipped with exhaust ports. The exhaust port should be equipped with a pipe cap with a nominal diameter of no less than 15mm, or a DN15 ball valve should be connected to the ground. Thermal compensation should be considered for overhead compressed air pipelines in the factory area. Thermal compensation methods include natural compensation and compensator compensation, and in practical applications, natural compensation is generally used;

5. When pipelines pass through walls and floor panels, galvanized steel sleeves should be installed. Flexible

waterproof materials should be used to fill the gap between the steel sleeve and the pipeline, and cement mortar should be used to fill the gap between the sleeve and the wall. The end face of the sleeve should be flush with the wall or extend 30mm beyond the floor.

3.8 Pipeline Welding

1. Welding rods, wires, and fluxes should be selected based on the chemical composition and mechanical properties of the base material, as well as the crack resistance of the welded joint, carbon diffusion, pre welding preheating, post welding heat treatment, welding process qualification results, and usage conditions;

2 The selection of welding materials and relevant technical requirements shall comply with the relevant provisions of the current industry standard of 《Technical Specification for Welding in Fossil fuel power station》（DL/T 869）and《Technical Specification for Welding of Dissimilar Steel in Fossil fuel power station 》(DL/T 752).

3.9 Pipeline insulation and identification

External insulation treatment is required for pipelines in the northern region. External insulation can be achieved through three methods: rock wool insulation, steam tracing, and electric tracing insulation. Compressed air pipelines are generally insulated with rock wool. The thickness of the rock wool outer insulation layer is 30mm, and the protective layer is made of 05mm thick galvanized iron sheet.

To facilitate the identification of pipelines within the factory area, it is necessary to label the pipelines. For pipelines with a nominal diameter greater than or equal to 50mm, pipeline markings should be applied at approximately 20 intervals near the valve, branch, intersection with other pipelines, workshop entrance, and straight section of the pipeline. For pipelines with a nominal diameter less than 50mm, marking plates should be hung near the valve, branch, intersection with other pipelines, workshop entrance pipelines, and straight section every 30~100m.

3.10 Pipeline Construction

3.10.1 Pipeline construction process

Construction preparation→inspection and cleaning of pipeline sections and accessories→removal of equipment nozzle closures→installation of pipeline supports and hangers→assembly and welding of pipelines→welding of pipeline interface welds→weld inspection and non-destructive testing→tightening and displacement inspection of equipment piping handles→pipeline pressure testing→internal cleaning, blowing, and airtightness testing of pipelines→adjustment of supports and hangers→bolt tightening.

3.10.2 Common problem handling

During pipeline construction, when it needs to pass through walls or floors, steel sleeves should be added according to the on-site situation. The inner diameter of the sleeve should be more than 50mm larger than the outer diameter of the pipeline, and the sleeve should be flush with the wall; When passing through the floor slab, the exposed length of both ends of the casing should not be less than 30mm to prevent water seepage.

When the direction of local pipelines needs to be changed due to construction site reasons, the pipeline direction should be adjusted under the principle of not affecting the passage of personnel, equipment, vehicles, and equipment maintenance, ensuring that the drainage slope of the pipeline is not less than 0.002. An exhaust outlet should be added at the local high point of the pipeline (when the nominal pipe diameter is greater than or equal to 32mm), and a drainage pipe should be added at the local low point.

Since there are many kinds of Industrial gas and liquid pipelines in the plant area, and there are many specialties of pipeline design, the medium pipelines of different specialties are easy to interfere in the details. When interference occurs between pipelines with different media, it is common to choose pipelines with smaller diameters and partially redirect them to avoid pipelines with larger diameters. At present, many enterprises have introduced BIM 3D collaborative design, which involves modeling different media pipelines within the same factory building model. This allows for early detection of any interference between pipelines or between pipelines and buildings, and provides the location of the interference to facilitate timely adjustment of pipeline layout before construction, thereby ensuring construction progress.

4. Case analysis

Case Overview

A 300MW/1800MWh compressed air energy storage project is located in the northwest region of China, with an average annual temperature of 10°C(extreme maximum temperature of 37°C, extreme minimum temperature of -23°C).

The underground gas storage capacity of the project is approximately 370000m3, with a design pressure of 10 MPa, a design temperature of 40°C, and a design flow rate of 2.7 million Nm3/h.

4.1 Design pressure

The selection shall be made according to the system flow chart, referring to the requirements of 3.2.7 in the 《Code for Design of Steam and Water Piping in Fossil fuel power station》 (DL/T5054-2016) and the Chemical Design Manual.

According to relevant regulations, the design pressure of the pipeline should not be lower than the maximum continuous working pressure that may occur in the internal medium during pipeline operation; At the same time, the design pressure should consider the maximum deviation of the equipment, and the design pressure should also include the static pressure of the water column. When it is less than 3% of the rated pressure, it may not be included.

Therefore, for the compressed air main pipeline, the design pressure of the gas storage tank is set at 10MPa. Considering the possible maximum pressure deviation, the design pressure of the compressed air main pipeline is set at 12MPa.

4.2 Design temperature

According to relevant regulations, the design temperature of the pipeline should not be lower than the maximum continuous working temperature of the internal medium during pipeline operation; At the same time, the design temperature should consider the maximum deviation that may occur in the equipment.

Therefore, for the compressed air main pipeline, the design temperature of the gas storage is set at 40°C; Considering the possible maximum temperature deviation, the design temperature of the compressed air main pipeline is set at 50°C.

4.3 Internal diameter of pipeline

The inner diameter of the compressed air pipeline can be obtained from the following equation:

$$D_i = 594.7\sqrt{\frac{GV}{v}}$$

$$D_i = 594.7\sqrt{\frac{3483 * 0.00793}{15}} = 806.99mm$$

4.4 Pipeline wall thickness

The pipe diameter and wall thickness of this project shall be calculated according to the Code for Design of Steam and Water Piping in Fossil fuel power station (DL/T 5054-2016).

(1) Calculation of straight pipe wall thickness

When Do/Di ≤ 1.7, the minimum wall thickness calculation of straight pipes under internal pressure should comply with the following regulations:

Wall thickness determined based on the inner diameter of the pipe:

$$S_m = \frac{PD_i + 2[\sigma]^t\eta C + 2YPC}{2[\sigma]^t\eta - 2P\ (1 - Y)}$$

In the formula,

Sm—minimum wall thickness of the pipe, mm;

Di—inner diameter of the pipe, taking the maximum inner diameter including the positive deviation of the pipe diameter, mm;

Y—Temperature correction coefficient, as shown in Table 4 below;

η—The allowable stress correction coefficient is taken as 1.0 for seamless steel pipes;

C—Additional thickness required for corrosion wear and mechanical strength;

P— Design pressure, MPa;

$[\sigma]^t$—allowable stress of the material, MPa.

(2) The calculated wall thickness of the pipe should be calculated using the following formula:

$$S_c = S_m + C_1$$

In the formula,

C1—additional value of negative deviation of pipe wall thickness, mm.

For pipe specifications based on outer diameter × The steel pipe with wall thickness identification shall be calculated according to the following equation:

$$C_1 = \frac{m}{100 - m} S_m$$

In the formula, m—the allowable negative deviation of wall thickness specified in the technical conditions of the pipe product, taken as a percentage.

Table 4 Temperature Correction Factors

Material	**Temperature (℃)**					
	≤482	***510***	***538***	***566***	***593***	***621***
Ferritic teel	0.4	0.5	0.7			
austenitic	0.4				0.5	0.7

After calculation, the wall thickness of the compressed air main pipeline in this project Sc=42mm.

Therefore, the pipeline specification DN900×42.

4.5 Pipeline components and materials

Refer to 《Typical Design of Parts and Components of Steam and Water Pipelines in Fossil fuel power station》 (GD2016), according to the design pressure of compressed air pipe 12MPa, design temperature 50 °C, pipe specification DN900×42. Seamless pipe fittings are used, with a material of 20G and a nominal wall thickness of no less than 42mm.

5. Conclusion

(1) In this paper, a design and material selection method of compressed air pipeline is proposed by referring to relevant design specifications such as Fossil fuel power station, and various links such as pipe fittings, test, construction, etc. are comprehensively considered;

(2) This article focuses on a 300MW/1800MWh compressed air energy storage project located in the northwest region of China, with an average annual temperature of 10 ℃. The underground gas storage capacity of the project is about 370000 m3, with a design pressure of 10MPa and a design temperature of 40°C. The design pressure of the compressed air main pipeline is set at 12MPa and a design temperature of 50℃. By referring to Typical Design of Parts and Components of Steam and Water Piping in Fossil fuel power station (GD2016), the pipe specification is DN900 × 42. Seamless pipe fittings are used, with a material of 20G and a nominal wall thickness of no less than 42mm.

(3) This article discusses and analyzes the design and selection of compressed air energy storage pipelines in the design of compressed air energy storage power plants, which can provide design and construction reference for the large-scale application of compressed air energy storage power plants in the future.

References

1. Wang Teng, Zhang Bin, Li Junxi, et al. Method for calculating the length of compressed air cooling pipes in pneumatic braking systems [J]. Heavy Duty Truck, 2020 (1): 2. DOI: CNKI: SUN: ZXQC.02201-021
2. Li Ting. Design of Compressed Air Station Equipment and Pipeline System [J]. Engineering Construction, 2020, 52 (8): 6. DOI: 10.13402/j.gcjs.2020.08.009
3. Peng Yansong, Xie Ying, Fang Minghui. Insulation Design of Compressed Air Pipelines in Cement Plants in Cold Regions [J]. Cement Engineering, 2020 (2): 2. DOI: CNKI: SUN: SNGC.0.2020-02-012
4. Li Zaiyou. Design of Compressed Air Systems in Production Workshops [J]. Hydraulic Pneumatic and Sealing, 2020, 40 (3): 3. DOI: CNKI: SUN: YYQD.0.2020-03-014
5. Wan Lipeng, Wang Qiang, Zhou Yang. Discussion on Material Selection for Compressed Air Pipelines [J]. New Industrialization, 2019, 9 (11): 4. DOI: CNKI: SUN: XXHG.20.2019-11-027
6. Duan Jianfeng. Analysis of Compressed Air Pipeline Process Design [J]. Shanxi Chemical Industry, 2018, 38 (3): 3. DOI: CNKI: SUN: SDHW.0.2018-03-040
7. Luo Ning, He Qing, Liu Wenyi. Research Status and Analysis of Gas Storage Devices in Compressed Air Energy Storage Systems [J]. Energy Storage Science and Technology, 2018, 7 (3): 6. DOI: 10.12028/j.issn.2095-4239.2017.0180
8. Economic analysis of gas storage devices in compressed air energy storage systems based on full life cycle cost He Qing; Ronin; Liu Wenyi. Chemical Progress, 2018
9. Ding X , Duan L , Zhou Y ,et al.Thermodynamic analysis and economic assessment of a novel multi-generation liquid air energy storage system coupled with thermochemical energy storage and gas turbine combined cycle[J].Journal of Energy Storage, 2023.
10. Liu Y , Li J , Chen Y .Thermodynamic Analysis of Liquefied Air Energy Storage System Coupled Liquefied Natural Gas and ORC[C]//International Workshop of Advanced Manufacturing and Automation.Springer, Singapore, 2023.DOI:10.1007/978-981-19-9338-1_42.

Study of the high-frequency characteristics of grounding materials

Zhuoran Liu[1,*], Chengwei Xu[2], Song Wei[3], Wei Lan[2], Wuxi Bi[2] and Feng Zhang[2]

[1]State Grid Energy Research Institute, Beijing, China
[2]PipeChina Institute of Science and Technology, Langfang, China
[3]Tsinghua University, Beijing, China

Abstract. Long-term safe and stable grounding system is the fundamental guarantee to maintain the stable operation of equipment and ensure the safety of staff. This paper investigates the high-frequency characteristics of several common grounding materials. Due to the skin effect, the axial impedance modulus, resistance part and reactance part of the conductor are all increase as the frequency. The results could provide reference and guidance for designing long-term reliable and stable grounding systems.

1 Introduction

Long-term safe and stable grounding system is the fundamental guarantee to maintain the stable operation of equipment and ensure the safety of staff [1,2]. The key to long-term stable operation of grounding system lies in the quality and reliability of grounding materials. In order to ensure the equipment and personal safety, the reliability of grounding grid has been put forward for a higher requirement.

For grounding engineering, corrosion is the most critical factor affecting the safety and service life of the grounding system on the premise that the design meets the electrical performance requirements [3]. However, the existing design sometimes does not consider the differences in soil corrosion in different regions, resulting in a short service life of the traditional steel grounding grid, and adversely affecting the safety and reliability of the grounding system. Judging from the excavation and maintenance data of the municipal bureaus, there are a large number of substations re-laid every year in my country due to the serious corrosion of the grounding grid, and that may not be economical and effective.

In this paper, based on the CDEGS simulation and FEM [4,5], the high-frequency characteristics of several common grounding materials is analysed firstly, and the trends of skin effect and impedance with the frequencies of current are obtained. Then the influence of the high-frequency characteristics on the grounding resistance is also discussed. Finally, considering the combined effect of soil resistivity and grounding grid, the trend of impulse grounding resistance and power frequency grounding resistance are obtained when different factors change simultaneously.

2 Analysis of high-frequency characteristics of grounding materials

2.1 Skin effect

Under an AC condition, the current density is greatest near the surface of the conductor, and then decreases exponentially from the surface to the interior. Since the current is mainly in the "skin" of the conductor, this phenomenon is called skin effect. The internal current density and axial impedance of the conductor can be calculated by skin effect, and the distribution of internal current density, impedance-frequency relationship and conductivity of the conductor can be analysed. Skin depth δ is defined as a depth where the current density is only 1/e (about 37%) of the surface value, depending on the frequency of the current and the electrical and magnetic properties of the conductor. The δ of grounding conductor can be calculated by formula (1).

$$\delta = \sqrt{\frac{1}{\pi f \sigma \mu}} \tag{1}$$

σ is the electrical conductivity of the material, μ is the magnetic permeability of the material, and f is the frequency. According to formula (1), the skin depth of different grounding materials under the action of current with different frequency are given, as shown in Table.1.

* Corresponding author: liuzrlaonanren@163.com

Table 1. Skin depth of the different grounding materials.

Materials	Volume resistivity ($10^{-8}\Omega\bullet$m)	relative magnetic permeability	Skin depth (mm)						
			1Hz	50Hz	100Hz	1kHz	10kHz	100kHz	1MHz
Brass	7	1	133.159	18.831	13.316	4.211	1.332	0.421	0.133
Fine copper	1.75	1	66.579	9.416	6.658	2.105	0.666	0.211	0.067
Steel	9.78	636	6.241	0.883	0.624	0.197	0.062	0.020	0.006
Zinc	5.96	1	122.869	17.376	12.287	3.885	1.229	0.389	0.123
Stainless steel	48.9	10	111.295	15.739	11.129	3.519	1.113	0.352	0.111
Carbon	1375	1	1866.257	263.929	186.626	59.016	18.663	5.902	1.866

It can be seen that the skin depth decreases with the increase of frequency. The smaller the skin depth, the more obvious the skin effect, which reduces the effective cross-section of the conductor and increases the effective resistance.

2.2 Experimental study of high-frequency impulse characteristics of grounding materials

2.2.1 Impedance-frequency characteristics of copper

Theoretical and simulated calculations were carried out for a cylindrical copper grounding material with a radius of 10mm and a length of 1 m, and the results of the theoretical and simulated comparison are shown in Fig.1.

The impedance of the cylindrical conductor can be calculated by formula (2):

$$Z^c = \frac{j\omega\mu_c}{2\pi\sqrt{j\omega\sigma_c\mu_c}}\frac{I_0(a\sqrt{j\omega\sigma_c\mu_c})}{I_1(a\sqrt{j\omega\sigma_c\mu_c})} \qquad Z=Z^c l \qquad (2)$$

where, Z^c is impedance per unit length of conductor determined by material, l represents the length, σ_c and μ_c are conductivity and magnetic permeability, respectively. a is the respectively, I_0 and I_1 represent modified zero order and first order Bessel function of the first kind.

As can be seen from Fig.1, the copper impedance increases gradually as the frequency increases. When the simulation software is used to simulate the grounding material, the real and imaginary parts of the impedance are not much different from the theoretical values, so it is considered that this simulation model can be used to simulate the impedance-frequency characteristics of the ground material.

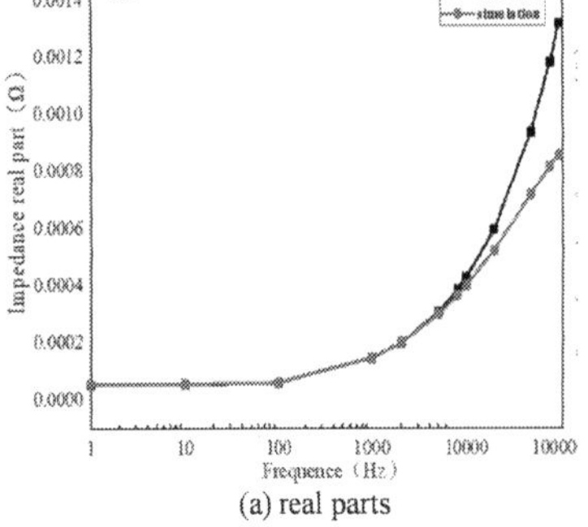

(a) real parts

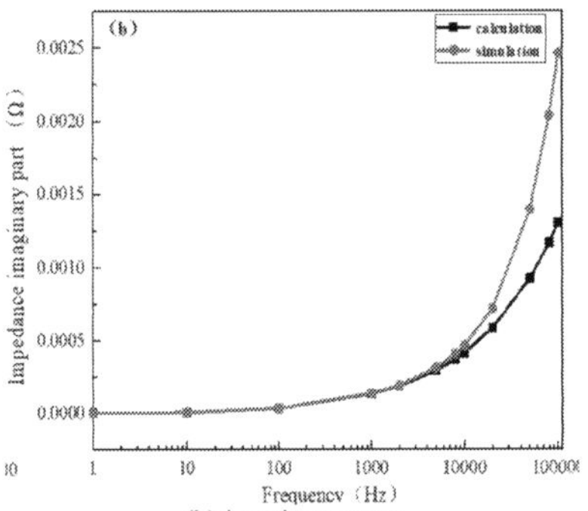

(b) imaginary parts

Fig.1. Theoretical and simulation values of (a)real and(b) imaginary parts of copper impedance

In addition, the current density distribution of copper end face at low frequencies was analyzed by the FEM to obtain the initial effective frequency of the skin effect, as shown in Fig.2. From Fig.2(a) and (b), it can be seen that the skin effect of copper starts to work from a frequency of 10Hz, and the skin effect becomes more obvious until the frequency increases to 100Hz. Meanwhile, it also can be seen from Fig.1, when the frequency increases to 100Hz, both the real and imaginary parts of the impedance start to increase more obviously.

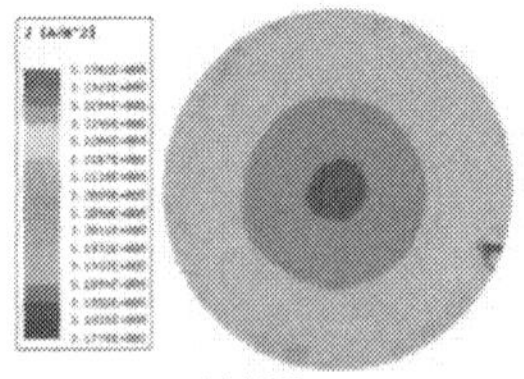

(a)10Hz

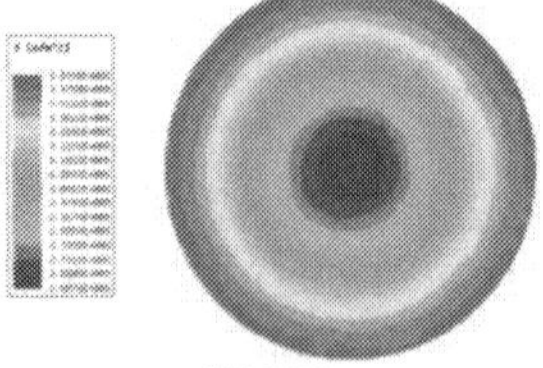

(b)100Hz

Fig.2. Current density distribution of end surface of copper at different low frequencies with (a)10Hz and (b)100Hz.

Then, the current density distribution of copper end face at different high frequencies was analysed to obtain the changing rule of copper impedance and the influence of skin effect, as shown in Fig.3.

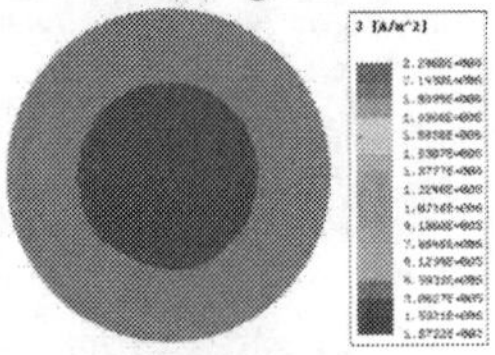

(a)100Hz

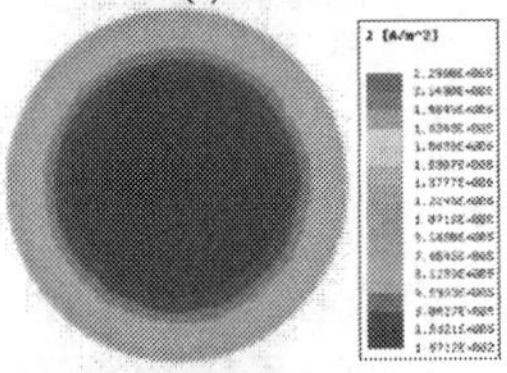

(b)1000Hz

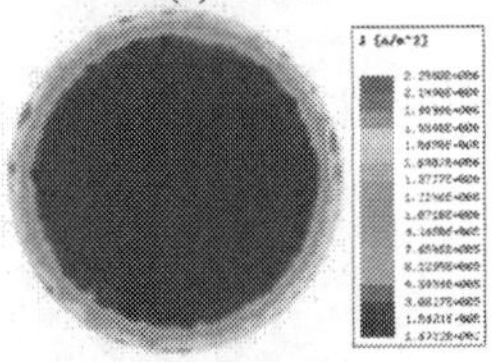

(c)10000Hz

Fig.3. Current density distribution of end surface of copper at different high frequencies with (a)100Hz, (b)1000Hz and (c)10000Hz.

It can be seen that with the increase of frequency, the skin effect is gradually obvious, so that the current tends to flow near the surface. However, since the total injected current is constant, when the current flows along the edge, the current density increases, resulting in more concentrated current, which is equivalent to the equivalent area of the conductor decreases, resulting in a corresponding increase in impedance. To sum up, with the increase of frequency, the impedance of copper will gradually increase. Meanwhile, because the imaginary part of impedance has a great relationship with frequency, the imaginary part of impedance increases relatively fast.

2.2.2 Impedance-frequency characteristics of copper clad steel

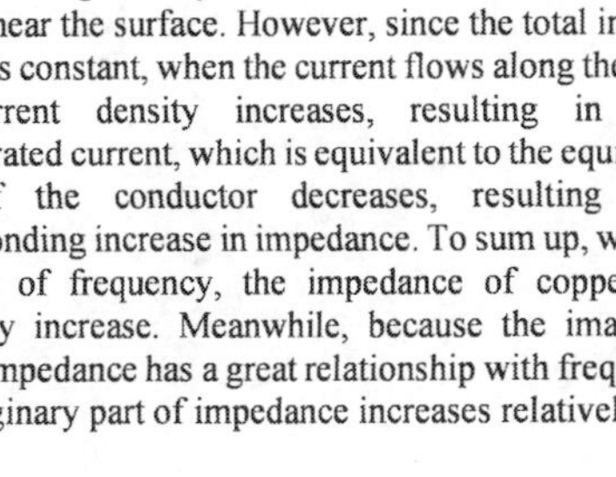

(a) real parts of cylindrical copper clad steel

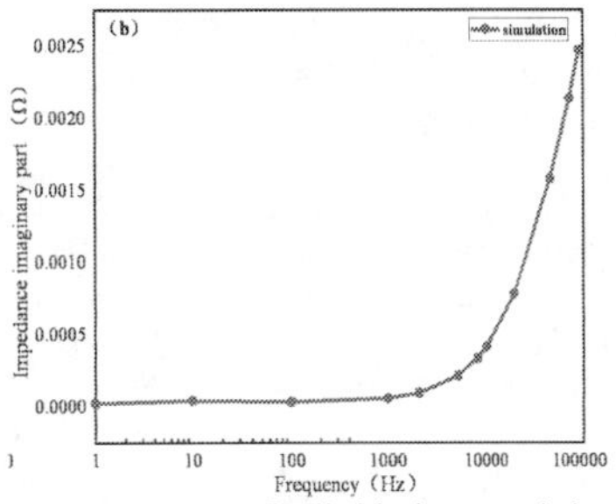

(b) imaginary parts of cylindrical copper clad steel

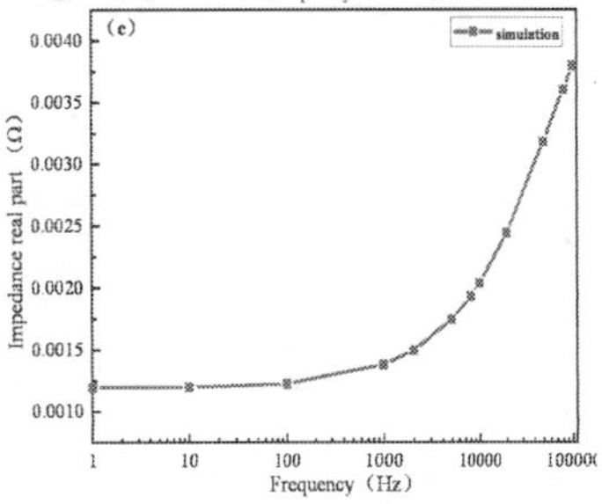

(c) real parts of flat strip copper clad steel.

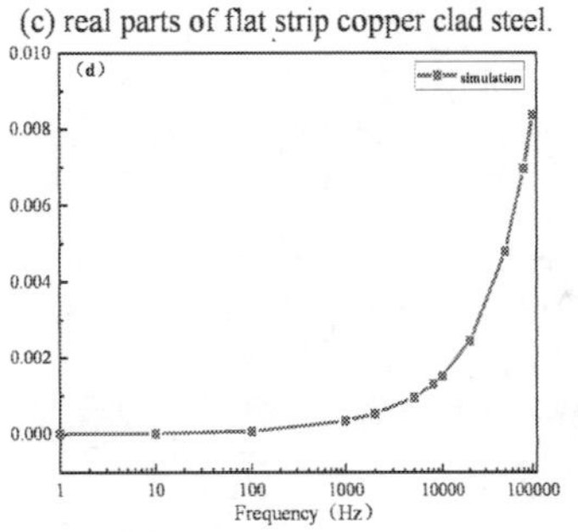

(d) imaginary parts of flat strip copper clad steel.

Fig.4. Simulation value of impedance of copper clad steel. (a) real and (b) imaginary parts of cylindrical copper clad steel. (c) real and (d) imaginary parts of flat strip copper clad steel.

Theoretical and simulated calculations were carried out for cylindrical copper clad steel (the radius of steel is 10mm, the thickness of outer copper coating is 0.7mm, and the length is 1m) and flat strip copper clad steel (steel with a cross-section of 12mm×6mm, outer copper with a thickness of 0.7mm and a length of 1 m), and the simulation results of impedance obtained are shown in Fig.4.

It can be seen that with the increase of frequency, the impedance real part and imaginary part of different shapes of copper clad steel will gradually increase, when the frequency is low, due to the current only has an impact on the impedance real part, the impedance imaginary part is basically 0. With the increase of frequency, the real part of the impedance changes little, until the frequency reaches a certain value (1000Hz), the impedance will have a significant change.

3 Influence of high-frequency characteristics of grounding materials on ground performance

The grounding resistances of three types of grounding grids, namely copper, steel and copper clad steel, were simulated and compared by using CDEGS under different soil resistivity conditions. Taking the grounding grid of a series compensation station as an example, the simulated area of the grounding grid is 278m×175m, the mesh is 15m×15m, and the grounding body is buried in a depth of 0.8m in the uniform soil. The simulation calculation results of grounding impedance are shown in Tab.2.

Table 2. Comparison of grounding resistance of copper, copper clad steel and steel grounding grids under different soil resistivity.

Soil resistivity (Ω•m)	Copper (Ω)	Copper clad steel (Ω)	Steel (Ω)
100	0.193 2	0.193 6	0.212 2
500	0.961 5	0.961 9	0.980 5
1000	1.921 9	1.922 3	1.940 9
2000	3.842 7	3.843 1	3.861 7

Tab.2 illustrate that the grounding resistance of the copper is slightly lower than that of copper clad steel and steel when the soil resistivity is the same, no matter in areas with high soil resistivity or low soil resistivity. For example, when the soil resistivity is 100Ω•m, the ground resistance of the copper is 0.1932Ω, which is 0.2% smaller than copper clad steel and 9.8% smaller than steel. Therefore, whether in areas with high soil resistivity or low soil resistivity, the grounding resistance of the grounding grid is similar when pure copper or copper clad steel are used.

4 Conclusion

In this paper, FEM and CDEGS are used to analyse the high-frequency characteristics of several common grounding materials and the influence of high-frequency characteristics on the ground performance. As the frequency increases, due to the skin effect, axial impedance modulus, resistance section and reactance section of the conductor all increase. The results could provide reference and guidance for designing long-term reliable and stable grounding systems.

Acknowledgments

PipeChina targeted project"Research on hazard assessment and mitigation of pipeline caused by lightning strike and transmission line fault" (AQWH202201).

References

1. G. Parise, L. Parise, L. Martirano, "Intrinsically Safe Grounding Systems and Global Grounding Systems," IEEE Transactions on Industry Applications, vol. 54, no. 1, pp. 25-31, Feb. 2018.
2. M. Camara, F. Atalar, A. E. Ylmaz, "A new grounding cake to improve the safety performance of grounding systems," Journal of Electrostatics, vol.108, pp.10351, Nov. 2020.
3. G. L. Wu, Y. Q. Luo, "Research on soft grounding material with high conductivity in transmission line tower grounding network," in ICMITE 2016, Taiyuan, China, 2016, pp. 346-350.
4. A. Habjanic, M. Trlep, "The simulation of the soil ionization phenomenon around the grounding system by the finite element method," IEEE Transactions on Magnetics, vol. 42, no. 4, pp. 867-870, Apr. 2006.
5. M. Trlep, A. Hamler, B. Hribernik, "The analysis of complex grounding systems by FEM," IEEE Transactions on Magnetics, vol. 34, no. 5, pp.2521-2524, Sep. 1998.

Study on the Preparation and CO_2 Adsorption Performance of Carbon material

ShaSha Wang[1], Jianlin Zhao[2], Bin Ren[1], Zhi Tian[2], Yangguang Zhang[2], Lihui Zhang[1*]

[1]Institute of Energy Resources, Hebei Academy of Sciences, Shijiazhuang, 050081, China;
[2]Hebei Baoli Engineering Equipment Corporation Limited, Hengshui, 053000, China

Abstract. Carbon materials are valuable for adsorbing CO_2. In this work, carbon materials with large specific surface area and pore volume were prepared through the combination of sodium alginate and zinc chloride, were applied to adsorb CO_2 at 273 K and 298 K, showing a large adsorption capacity 2.75 mmol g^{-1} and 2.19 mmol g^{-1}, respectively. Zinc chloride acts as an activator and pore forming agent.

1. INTRODUCTION

Sodium alginate is rich in hydroxyl, which is easy to produce intramolecular hydrogen bond, thus forming gel [1-2]. In recent years, researchers have found that sodium alginate is a good precursor. By adding templates and activators to sodium alginate, carbon materials (CM) with high porosity, large specific surface area (BET) and excellent performance can be prepared. In recent years, researchers have found that sodium alginate is a good precursor. By adding templates and activators to sodium alginate, CM with high porosity, large BET and excellent performance can be prepared[3-4]. This work uses sodium alginate as raw material and zinc chloride as activator to prepare CM with large specific surface area through simple agitation and carbonization, and is used to adsorb CO_2.

2. Methods

Add 10 grams of sodium alginate and 2 grams of zinc chloride into 100 mL of ethanol, stir for 3 hours at room temperature, and then dry in a freeze drying oven for 12 hours. After placing the product in a carbonization furnace and carbonizing at 900 °C for 1 hour, black powder is obtained.

3. Results and discussion

BET is important for properties of substances, and its size is relational to the pore structure. The size of the specific surface area has a significant impact on many other physical and chemical properties of a substance, especially as the particle size decreases, the specific surface area has become a very important parameter for measuring material properties, such as the widely used nanomaterials.The BET physical adsorption instrument is currently mainly used for analyzing pore structure.

As is well known, large BET and appropriate pore size are key factors determining the performance of CM. We conducted N_2 adsorption desorption experiments on samples at different carbonization temperatures, as shown in Figure 1. We calculated the BET (573 cm^2/g), pore volume (0.53 cm^3/g) of each sample using the BET method. The results show that all isotherm are IV type, and the isotherms of the samples rise significantly when p/p0 is 0.5-1, indicating that the samples contain big mesopores. The adsorption and desorption curves of the isotherm curve coincide well, indicating a narrow range of pore size distribution in mesopores[5]. These results can be verified from the aperture distribution map.

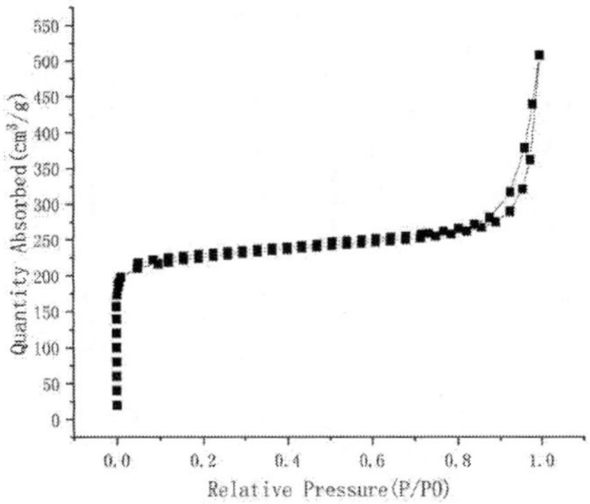

* Corresponding author: 517263436@qq.com

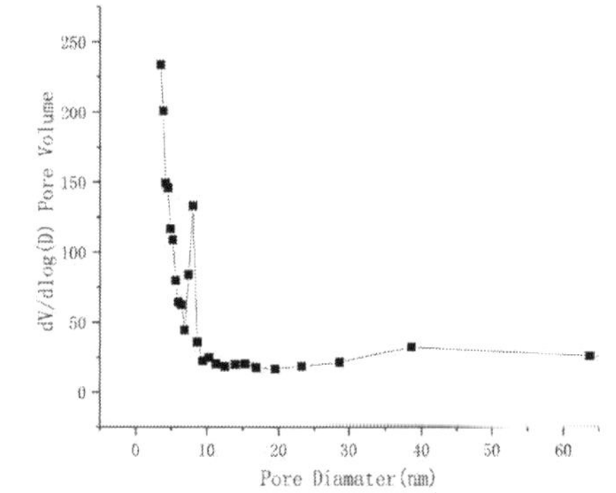

Fig 1. N_2 adsorption-desorption curve

Thermal gravimetric analysis is used to measure the relationship between the mass of a substance and temperature in Figure 2. In order to study the pyrolysis process of organic aerogel at high temperature, thermogravimetric analysis was carried out on the synthetic organic aerogel in this project. By analyzing the thermogravimetric curve, we found that the weight loss of the sample before 100 °C; The weight loss is obvious at 350 °C - 370 °C and 550 °C - 700 °C in Figure 2, where the thermal weight loss rate of organic aerogel reaches the maximum, and the weight loss rate is about 34%. In this range, the cured organic aerogel undergoes decomposition reaction, and the molecular chain gradually splits, producing water vapor, carbon monoxide, carbon dioxide and other substances. After 700 °C, the curve shows little change, and the degree of carbonization deepens, gradually forming a heat-resistant and solid glassy carbon material. During this process, deep carbonization mainly produces small molecule gases such as hydrogen gas[6].

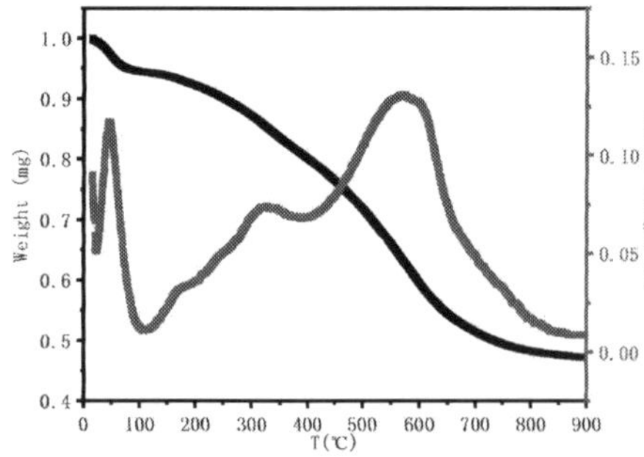

Fig 2. TG images

The project conducted SEM testing on the prepared samples in Figure 3. Figure 3 exhibits a typical three-dimensional nanonetwork structure, with nanoparticles cross-linked and stacked together to form a three-dimensional interconnected network, which is a unique structure of CM. EDS energy spectrum is an instrument for analyzing material elements, elemental distribution of CM were studied in this work. N and O elements were discovered in CM.

Fig 3. SEM images and mapping

FT-IR spectra of sodium alginate and obtained carbon material are displayed in Fig. 4. The peaks of obtained carbon material at 1639, 1579, 1411 and 1060 cm^{-1} were attributed to C=N, C=C, C-N and C-O bonds[7], respectively. The broad bands at 3130–3430 cm^{-1} could be ascribed to O-H and N-H bonds[8]. These findings reveal that N and O elements have been successfully doped into carbon material.

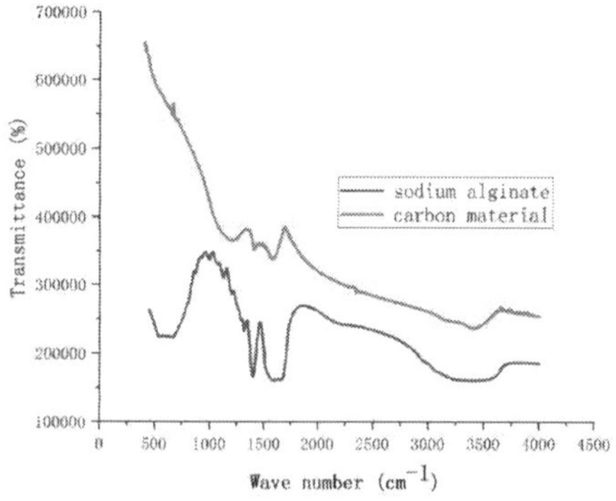

Fig 4. FT-IR spectra

XPS is widely used in the element composition and valence analysis of material surface. XPS combined with imaging function and ion sputtering etching, can be used for surface distribution and depth analysis of element composition and valence state on solid surface[9-10]. Fig. 5 (a), (b), (c) showed magnified C1s, N1s and O1s spectra, respectivityly.

As presented in Fig. 5(a), C1s of CM could be deconvoluted into 4 individual component peaks. Besides, the peaks of N1s could be deconvoluted into 4 peaks[11]. The binding energies for O1s could be divided 4 peaks.

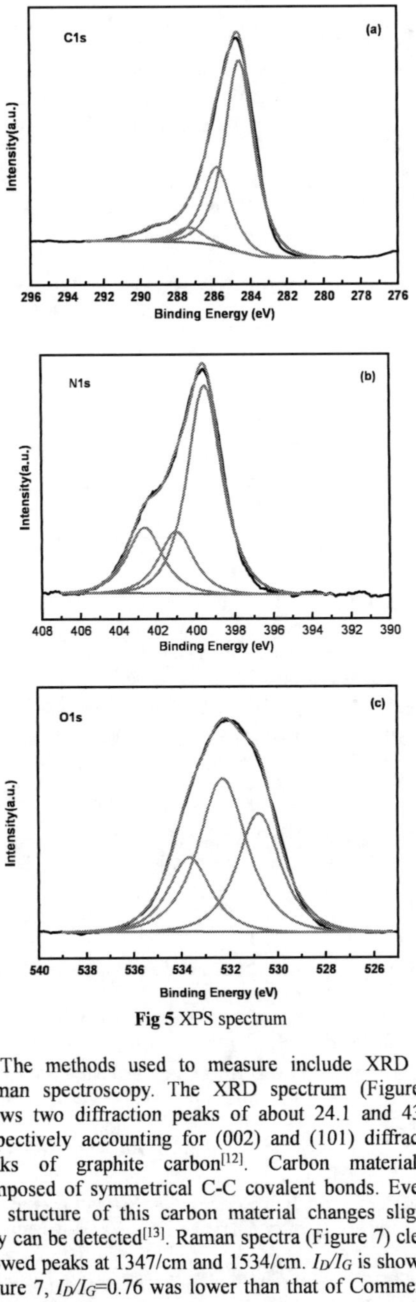

Fig 5 XPS spectrum

The methods used to measure include XRD and Raman spectroscopy. The XRD spectrum (Figure 6) shows two diffraction peaks of about 24.1 and 43.5°, respectively accounting for (002) and (101) diffraction peaks of graphite carbon[12]. Carbon material is composed of symmetrical C-C covalent bonds. Even if the structure of this carbon material changes slightly, they can be detected[13]. Raman spectra (Figure 7) clearly showed peaks at 1347/cm and 1534/cm. I_D/I_G is shown in Figure 7, I_D/I_G=0.76 was lower than that of Commercial activated carbon.

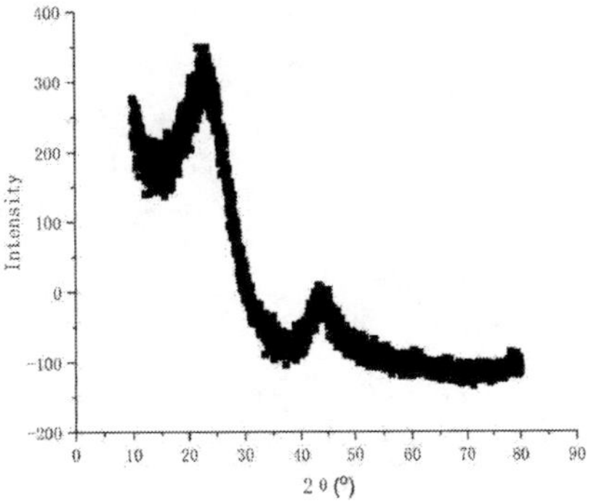

Fig.6 XRD

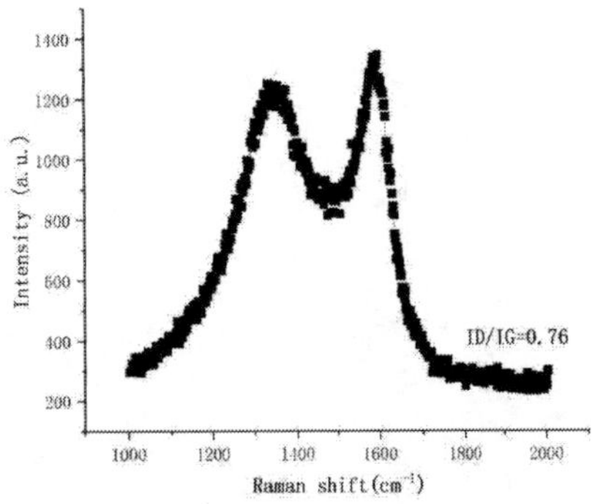

Fig. 7 Raman spectra

This article mainly introduces in detail the method for measuring the CO_2 equivalent adsorption heat of activated carbon materials. In the experiment, the activated carbon sample was tested for the isotherm of CO_2 adsorption by the 3Flex gas adsorption instrument of Mack Instruments. The constant temperature of the sample during the test was controlled by the iso controller low-temperature thermoelectric refrigeration dewar. After the test is completed, calculate the equivalent adsorption heat for all surface coverage ranges (from zero to saturation) using MicroActive software from Mac Instruments. With the increasing attention to climate change and environmental protection, the capture and storage of CO_2 has received significant attention from media, scientific research, and public policies[14-15].

The CO_2 adsorption performance of samples was measured, and the isotherm is shown in Figure 8. All isotherm have no hysteresis, showing the reversibility of adsorption and desorption, indicating that these materials can easily desorb adsorbate during the desorption process. Therefore, the materials obtained can be recovered without need others. The CO_2 adsorption isotherm shows a growth trend, especially at lower pressure, because CO_2 has excellent polarizability and quadrupole moment, and has strong interaction with the prepared materials[16]. From Figures 4 and 5, adsorption capacity of material for

CO_2 at 273 K and 298 K is 2.75 and 2.19 mmol g^{-1}, respectively.

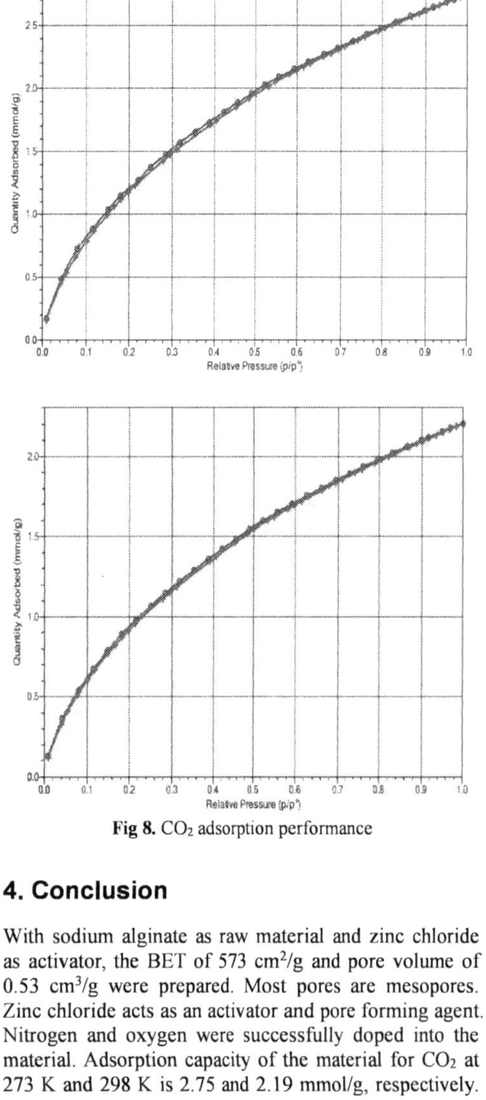

Fig 8. CO_2 adsorption performance

4. Conclusion

With sodium alginate as raw material and zinc chloride as activator, the BET of 573 cm^2/g and pore volume of 0.53 cm^3/g were prepared. Most pores are mesopores. Zinc chloride acts as an activator and pore forming agent. Nitrogen and oxygen were successfully doped into the material. Adsorption capacity of the material for CO_2 at 273 K and 298 K is 2.75 and 2.19 mmol/g, respectively. Therefore, the obtained carbon material is a CO_2 adsorption material with potential application value.

Acknowledgments

This work was financially supported by the Science and Technology Plan Project of Hebei Academy of Sciences (23A09, 22A04).

References

1. Yang, J S. (2013) Benzimidazole grafted polybenzimidazoles. J. Mater. Chem. A. , 4: 4768-4775. https://pubs.rsc.org/en/content/articlelanding/2013/py/c3py00408b
2. Lin H B. (2017) Electrocatalysis of polysulfide conversion. Energy Environ. Sci., 10: 1476-1486. https://pubs.rsc.org/en/content/articlelanding/2017/ee/c7ee01047h#!
3. Zhang, M. (2012) Facile synthesis of water-soluble. J. Mater. Chem., 22:7461-7467. https://pubs.rsc.org/en/content/articlelanding/2012/jm/c2jm16835a
4. Ma, Y Z. (2018) The relationship between the chemical state. Catal. Sci. Technol., 8: 2567-2577. https://pubs.rsc.org/en/content/articlelanding/2018/cy/c8cy00208h
5. Zhu, L. (2017) An carbon aerogels the removal of oils, Micro Meso Mater. 241: 285-292. https://www.sciencedirect.com/science/article/abs/pii/S1387181116305923
6. Zulamita, Z.B. (2013) Cu- and Ag-doped carbon aerogels, Mater. Chem. Phys. 138: 870–876. https://www.sciencedirect.com/science/article/abs/pii/S0254058413000357
7. Lui A.P. (2017) In situ catalytic fast pyrolysis of crude and torrefified, Energy. 128: 701–712. https://www.sciencedirect.com/science/article/abs/pi/S0360544217305893
8. Chmiola, J. (2006) Anomalous increase in carbon capacitance at pore sizes less than 1 nanometer, Science 313: 1760–1763. http://dns2.asia.edu.tw/~ysho/YSHO-English/1000%20WC/PDF/Science313,%201760.pdf
9. Liang, C.D. (2008) Mesoporous CM: synthesis and modifification, Angew. Chem. Int. Ed. 47: 3696–3717. https://onlinelibrary.wiley.com/doi/abs/10.1002/anie.200702046
10. Hu,Y.J. (2016) Electrode and CO2 adsorption. RSC Adv. 6 :15788–15795. https://pubs.rsc.org/en/content/articlelanding/2016/ra/c6ra00822d
11. Paul, A. W. (2014) Adsorption technology for CO2 separation and capture: a perspective. Adsorption. 20: 225–231. https://link.springer.com/article/10.1007/s10450-014-9603-2
12. Bo, G. (2006) Adsorption of Carbon Dioxide on Activated Carbon. Journal of Natural Gas Chemistry. 15 : 223-229. https://www.sciencedirect.com/science/article/abs/pii/S1003995306600303
13. Liu, D, f. (2012) Experimental and molecular simulation studies of CO2 adsorption on zeolitic imidazolate frameworks: ZIF-8 and amine-modified ZIF-8. Adsorption, 19: 25–37. https://link.springer.com/article/10.1007/s10450-012-9407-1

14. Cheng, H. Y. (2012) A Review of CO2 Capture by Absorption and Adsorption. Aerosol and Air Quality Research. 12: 745-769. https://doi.org/10.4209/aaqr.2012.05.0132
15. Rashidi, N. A. (2014) Experimental and modelling studies of carbon dioxide adsorption by porous biomass derived activated carbon. Clean Technologies and Environmental Policy volume 16: 1353-1361. https://link.springer.com/article/10.1007/s10098-014-0788-6
16. Lochan, K. B. (2023) The implications of forest resources depletion, agricultural expansion, and financial development on energy demand and ecological footprint in BRI countries. https://link.springer.com/article/10.1007/s10098-023-02575-3

Research and Selection of Sorbents for Volatile Organic Compounds (VOC) Sampling Tubes

Junyann Hu[a*], Jinna Wu[b], Qian Yao[c], Chong Wang[d], Lixuan Li[e], Chuanjie Li[f]

CATARC Automotive Test Center (Tianjin) Co., Ltd., Tianjin 300300, China

Abstract—Volatile organic compounds (VOCs) are common emitting substances in vehicle interior parts and materials, with a wide range of boiling points (50-250°C) and a polarity range ranging from medium to non-polar, at least 200 species. Currently, VOC sampling mainly uses Tenax TA sorbent for collection, which has good stability and is suitable for weakly polar to non-polar substances between C6-C16. But there are many types of VOC substances, and each substance has its unique characteristics. Tenax TA is not suitable for all substances, and it is not the best sorbent choice for substances with medium polarity or low carbon number. In recent years, the vehicle and indoor environmental standards are open to the selection of sorbent, no longer limiting the type of sorbent sorbent, and more sampling tubes mixed with different sorbent types have emerged to achieve the best sampling results. After summarizing the common emitted substances in the car, this article selects 6 common types of sorbents or mixed sorbents for substance collection, and searches for the optimal sorbent type for each substance.

1. Introduction

The World Health Organization (WHO, 1989) defines Total Volatile organic compound (TVOC) [1][2] as a class of organic compounds whose boiling point ranges from 50°C to 260°C, the saturated Vapor pressure at room temperature exceeds 133.32 Pa, and they exist in the air as vapor at room temperature. According to the chemical structure of volatile organic compounds, they can be further divided into 8 categories: alkanes, aromatics, olefins, halohydrocarbons, esters, aldehydes, ketones, and other compounds.

The main sampling method for volatile organic compounds (VOCs) in China is Tenax TA packing sorbent. However, the latest testing standard GB/T18883-2022 "Standards for indoor air quality" uses Tenax TA or equivalent packing sorbent tubes for VOC sampling. In the standard ISO16000-6:2021, VOC is sampled through active sorbent, and the sorbent packing is no longer specified. Different VOC substances can be sampled using more suitable sorbents, Refer to Table 1 and Table 2.

Table 1 Foreign VOC Main Standards and Scope of Application [3][4][5][6]

NO.	Standard number	Sampling method	Test method	substances
1	ISO16000-6-2021	Solid sorbent/unlimited sorbent	Gas chromatography-mass spectrometry	VOC
2	EN ISO 16017-1-2000	Tenax TA	Gas chromatography/gas chromatography-mass spectrometry	VOC
3	EPA TO-1	Tenax TA	Gas chromatography-mass spectrometry	Non polar organic compounds with boiling points between 80 and 200°C
4	EPA TO-17	Solid sorbent/unlimited sorbent	Gas chromatography-mass spectrometry	VOC

Table 2 Main Domestic VOC Standards and Scope of Application[7][8][9][10][11][12]

NO.	Standard number	Sampling method	Test method	substances
1	HJ583-2010	Tenax TA	Gas chromatography	Benzene derivatives
2	HJ584-2010	Activated carbon	Gas chromatography	Benzene

[a*]corresponding author email: hujunyan@catarc.ac.cn, [b]email: wujinna@catarc.ac.cn, [c]email: yaoqian@catarc.ac.cn, [d]email: wangchong@catarc.ac.cn, [e]email: lilixuan@catarc.ac.cn, [f]email: lichuanjie@catarc.ac.cn

				derivatives
3	HJ644-2013	Stainless steel and glass materials, with an inner diameter of 6mm, equipped with Carbopack B, Carbopack C, and Carboxen 1000, with lengths of 13, 25, and 13 mm respectively	Gas chromatography-mass spectrometry	VOC
4	HJ734-2014	Combination 1 sorbent tube, equipped with Tenax GR and Carbopack B, with lengths of 30 and 25 mm, respectively. Combination 2 sorbent tubes, containing Carbopack B and Carboxen 1000, with lengths of 30 and 10 mm, respectively. Combination 3 sorbent tubes, containing Carbopack C, Carbopack B, and Carboxen 1000, with lengths of 13, 25, and 13 mm, respectively.	Thermal desorption gas chromatography-mass spectrometry	VOC
5	GB/T18883-2022	Tenax GC or Tenax TA or other sorbents	Gas chromatography	TVOC
6	HJ/T400-2007	Tenax TA	Gas chromatography-mass spectrometry	VOC

The testing of volatile organic compounds (VOC) in the vehicle industry involves a wide range of fields, including vehicle VOC (HJ/T400, ISO12219-1), vehicle interior parts and materials VOC (ISO12219-2), materials (VDA278), and evaporative pollutant emissions (hydrocarbon emissions from non fuel systems in (CHINA VI)). There are many standards involved, and the testing methods and equipment are also different. However, the main sampling method is Tenax TA sorbent sorbent, without considering the different types of sorbents applicable to different substances. Currently, research and selection are conducted on the types of sorbents for VOC substances in vehicles.

From domestic and international VOC standards and sampling methods, In some standards, the selection of VOC sorbents has gradually shown an open state. The previously designated sorbents have gradually been transformed into different substances using different sorbents for sampling, and some sampling tubes have been filled with various combinations of sorbents, such as the appearance of combination tubes 1, 2, 3, etc. in standard HJ734-2014, It can select appropriate combination solid sorbents based on specific emitting substances. Specific types of sorbents can be refer to Table 3.

Table 3 Specific Application Scope of Different sorbents[3]

Type	Sorbent	Analyte rang	Conditioning at max.(°C)	Desorption max.(°C)	Hydro-phobic	Notes
Weakly polar graphitized carbon black	Carbopack C	C8-C20	400	360	Yes	Single VOC<1ng; 40/60 Surface area 10m²/gmesh;
	Carbotrap C		400	360	Yes	
	CarboTenax Graph 2TD		400	360	Yes	
Weakly porous polymer	Tenax TA	C6-C22	320	280	Yes	Single VOC1-2ng;Good inertness and analytical efficiency (narrow peak shape);35-60 mesh;Surface area 30m²/g
	Tenax GR		320	280	Yes	
Weak - medium type graphitized carbon black	Carbopack B	C5-C14	400	360	Yes	Surface area 100m²/g;
	Carbotrap		400	360	Yes	
	CarboTenax Graph 1TD		400	360	Yes	
	CarboTenax Graph 4TD		400	360	Yes	
Medium strength porous polymeric sorbent						Single VOC<10ng;Background and batch differences make sorbents rarely used
Medium to strong graphitized carbon black	Carbopack X	1,4-Butadiene Benzene	400	360	Yes	Single VOC<1ng; 40/60 mesh;
	CarboTenax Graph 5TD		400	360	Yes	Surface area 240m²/g;
	CarboTenax Graph 4TD		400	360	Yes	Poor desorption efficiency (wider spectral band)
Carbon molecular sieve	Carboxen 1003	C2/3-C5/6	350	330	No	Single VOC<1ng; 40/60 mesh;
	SulfiCarb		350	330	No	Surface area 400-1000m²/g;
	Carbosieve SIII		350	330	No	Poor desorption efficiency (broadened spectral band);
	Carboxen 1000		350	330	No	When the humidity is greater than 80, it affects the sorbent effect
Other (non sampling end)	Glass wool	C30			Yes	
	Stone wool				Yes	
	Glass bead				Yes	

From the types of common sorbent sorbents that there are many types of sorbents, and different sorbents are suitable for different test ranges and use temperatures. Different substances can choose appropriate sorbent types according to their characteristics. If the composition of VOC is complex and the range of substances (temperature or polarity) is large, each sampling tube can be filled with 2-4 different sorbent types to achieve the best sampling effect.

2. Experimental Part

2.1 Instruments and reagents

2.1.1 Instrument and equipment

Thermal desorption gas chromatography-mass spectrometry (TD-GCMS); Model: Marks TD100 Agilent GC7890B MS5977B.

2.1.2 Main reagent

Dichloromethane (HPLC/UPLC grade).

2.1.3 Standard selection

According to the WTO definition of VOC, the categories include alkanes, aromatics, olefins, halohydrocarbons, esters, and other compounds (acids, alcohols, amines, etc.) (aldehydes and ketones are collected using DNPH sampling tubes and analyzed by liquid chromatography, which is not within the scope of this study). VOC standards are all pure products, and specific substances are shown in Table 5-8.

2.2 Instrument operating conditions [9]

Gas phase conditions: chromatographic column Agilent DB-624 60m× 0.32mm×1.4μm; Column flow rate: 1mL/min Heating program: 40°C(2min) -3°C/min-92°C-5°C/min-160°C-10°C/min-260°C(10min)

Mass spectrometry conditions: voltage: 70ev; Scanning method: SCAN; Scanning range: 33-450amu; Scanning frequency: 3.5 times/second; The test results were matched based on the NIST2017 spectral library, and the qualitative and quantitative ions of each substance were confirmed.

2.3 Preparation of Standard solution

Accurately weigh 0.5000 g of each substance's single standard standard (accurate to 0.0001 g), dissolve it with dichloromethane and transfer it to a 100 mL volumetric flask for constant volume, prepare 5000 mg/L mixed standard stock solution, and store it in a refrigerator at 4°C. When using, dilute the standard stock solution with dichloromethane and prepare 1000 mg/L mixed standard solution for test and analysis.

2.4 Selection of sorbent

The selection of sorbent mainly considers medium polar and non polar substances with boiling points between (50-250)°C. The main characteristic of polar substances is hydrophilicity, and their test results are greatly affected by moisture content, making them unsuitable for sorbent sampling. Therefore, this sorbent selection mainly considers medium to non polar substances, and selects more common types of sorbents that can be purchased on the market.

Select 6 different types of monomers or combinations of sorbents based on the current standard's sorbent types and applicability. Please refer to Table 4 for specific types.

Table 4 Different types of monomers or combinations of sorbents[3][13][14][15][16]

No.	sorbent	Analyte rang	Manufacturer	characteristics
1	Tenax TA	C5-C22	Marks Mix	Good hydrophobicity, suitable for most non-polar substances
2	Tenax GR	C5-C22	Marks Mix	
3	Combination 1 sorbent tube, equipped with Tenax GR and Carbopack B, with lengths of 30 and 25 mm, respectively	C6-C20	CNW	Good hydrophobicity, suitable for most non-polar substances
4	Combination 3 sorbent tubes, containing Carbopack C, Carbopack B, and Carboxen 1000, with lengths of 13, 25, and 13 mm, respectively	C3-C16	CNW/Camsco	For some polar and medium type substances, the sorbent is good, but the hydrophobicity is poor
5	Graphitized carbon	C5-C20	Marks Mix	Good hydrophobicity, suitable for most non-polar substances
6	Graphitized carbon mixing tube	C3-C20	Marks Mix	Combination type, wide application range

3. Results and discussion

The test material selection includes ester, alcohol ether, amine, alkane, olefin, Aromatic hydrocarbon, Cycloalkane and other material categories. Each type of substance shall be tested at least 6 times, and the test results shall be statistically significant. The test results are based on the maximum chromatographic peak area as the reference value, with the deviation being the difference between the peak area of other sorbent sorbents and the maximum value. The statistical result is (100 deviation value), and the closer the value is to 100, the smaller the deviation.

 https://doi.org/10.1051/e3sconf/202344102008

3.1 Acid ester substances

3.1.1 The chromatogram and substance list of ester substances are shown in Figure 1 and Table 5.

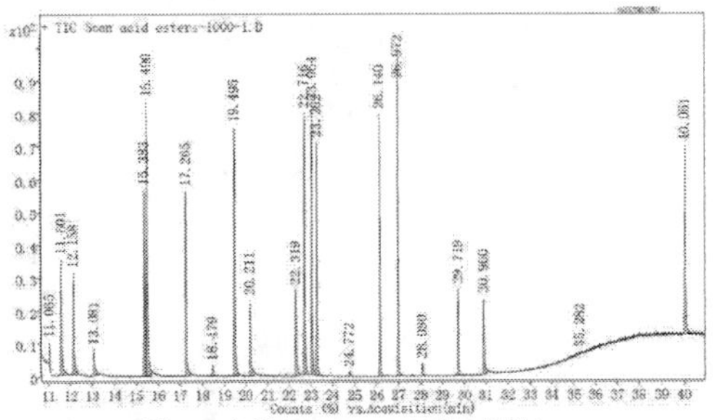

Figure 1 Chromatogram of Acid Esters

Table 5 List of 18 Common Acids Esters in Vehicles[17]

NO.	Substance	CAS NO.	RT	Quantitativeion	Qualitativeion
1	Ethyl acetate	141-78-6	11.601	43	61,70
2	tetrahydrofuran	109-99-9	12.158	71	42,72
3	Acetic acid	64-19-7	13.086	45	60
4	Methyl methacrylate	80-62-6	15.384	69	41,100
5	N-propyl acetate	109-60-4	15.49	61	88,73
6	1, 4-dioxane	123-91-1	15.5	88	43,61
7	Sec-butyl acetate	105-46-4	17.265	87	56,43
8	Butyl acetate	123-86-4	19.495	56.1	43,73
9	Isovalerate butyrate	107-92-6	20.215	60	73
10	Butyl acrylate	503-74-2	22.314	60	87
11	Butyl propionate	141-32-2	22.716	55	73
12	Ethylene glycol ethyl ether acetate	590-01-2	23.064	75	57,87
13	Ethyl 3-ethoxy propionate	111-15-9	23.277	59.1	72,87
14	Ethylene glycol diacetate	763-69-9	26.135	117	71,101
15	Carbonate allyl ester	111-55-7	26.972	86	73,116
16	2-ethylhexanoic acid	108-32-7	29.715	57	87,102
17	2,2, 4-trimethyl-1, 3-pentanediol diacetate	149-57-5	30.9	88	73
18	Ethyl acetate	6846-50-0	40.046	71	43

3.1.2 The test results of ester substances are as follows, and the source of the results is shown in Figure 2.

a) For ester like substances, the best sorbent tube is the Graphitized carbon mixing tube with deviations within 11%, which shows the most stable performance;

b) Tenax GR sorbent has a significant deviation in high boiling point substances, with other deviations within 10%, indicating good performance;

c) The performance of graphitized carbon adsorption tube is relatively extreme, and the adsorption effect of tetrahydrofuran, acetic acid, butyric acid, isovaleric acid and 2,2, 4-trimethyl-1, 3-pentanediol diacetate is poor, but it is the adsorbent with the highest peak area and can be used in a certain range;

d) Tenax TA has poor sorbent performance for low-carbon and highly polar substances, with deviations within 15% for other substances;

e) Combination tubes 1 and 3 have poor sorbent performance for esters, but some substances with deviations within 20% can be selected for use.

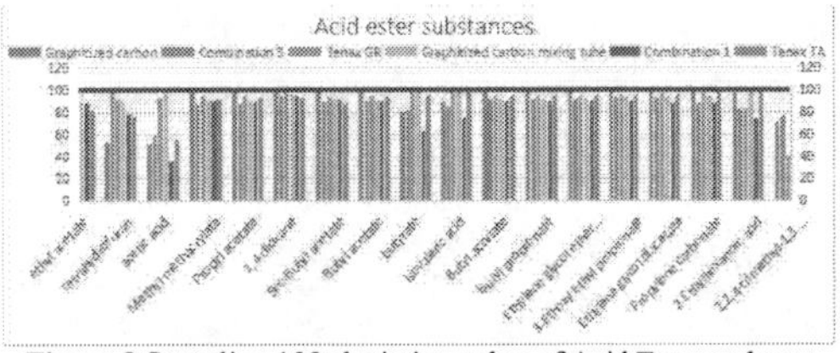

Figure 2 Sampling 100-deviation value of Acid Ester sorbents

3.2 Alcoholic Ethers and Partial Ketones

3.2.1 Chromatograms and substance lists of alcohol ethers and some ketones are shown in Figure 3 and Table 6.

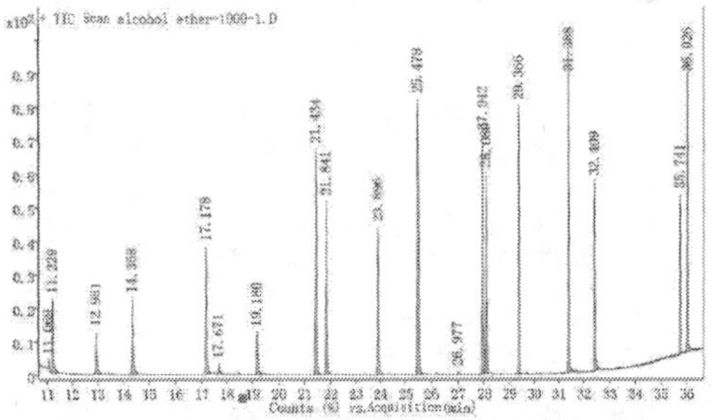

Figure 3 Chromatogram of Alcohol Ethers and Ketones

Table 6 List of 18 Common Alcohol Ether and Ketone Substances in Vehicles[17][18]

NO.	Substance	CAS NO.	RT	Quantitativeion	Qualitativeion
1	2, 3-butanedione	431-03-8	11.234	86	43
2	2-methyl-1-propanol	78-83-1	12.955	43.1	74,33
3	1-butanol	71-36-3	14.358	41.1	56
4	4-methyl-2-penta	108-10-1	17.178	58	43,85

	none				
5	Ethylene glycol	107-21-1	17.676	62	43
6	Propanediol	57-55-6	19.18	45.1	61
7	Dibutyl ether	142-96-1	21.434	57.1	87,41
8	Propylene glycol methyl ether acetate	108-65-6	21.836	72.1	58,87
9	Ethylene glycol butyl ether	111-76-2	23.896	57.1	87,45
10	1, 3-dichloro-2-propanol	96-23-1	25.473	79	45,59
11	Diethylene glycol monomethyl ether	111-77-3	25.482	45.1	79,59
12	1, 2-propylene glycol diacetate	623-84-7	27.939	43	87
13	2-ethyl-1-hexanol	104-76-7	28.08	57.1	70,83
14	Ethylene glycol butyl ether acetate	112-07-2	29.361	87	57,100
15	2-phenylethanol	60-12-8	31.388	91	122
16	Diethylene glycol monobutyl ether	112-34-5	32.408	45	57,75
17	Hexadiol diethyl ester	141-28-6	35.741	157.1	111,128
18	2-butyl-1-octanol	3913-02-8	36.021	57.1	71,85

3.2.2 The test results of Alcohol Ethers and Ketones are as follows, and the source of the results is shown in Figure 4.

There are highly polar substances in alcohol ether substances, and different types of sorbents have different substances used. Overall, it can be seen that:

a) For alcohols and ethers, the best comprehensive sorbents are graphitized carbon and Tenax TA, with deviations within 20% except for a few substances from 1-2;

b) Graphitized carbon mixing tube is the sorbent with the highest maximum value, but its deviation exceeds 30% in high carbon substances;

c) The other four sorbent materials have corresponding usage conditions within different substance ranges.

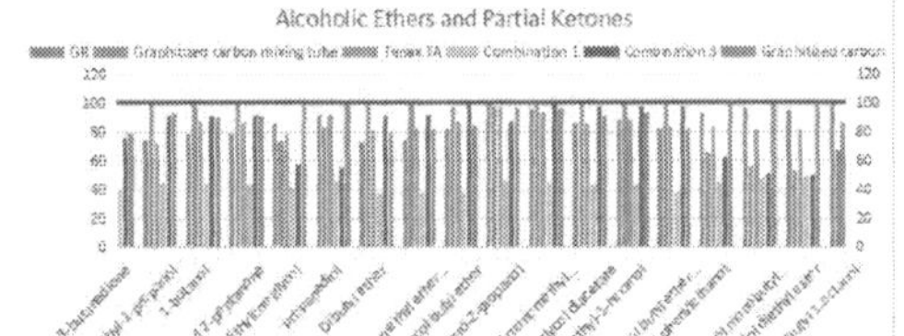

Figure 4 Sampling 100-deviation value of Alcohol Ethers and Ketones sorbents

3.3 Amine substances

3.3.1 Chromatograms and substance lists of Amine are shown in Figure 5 and Table 7.

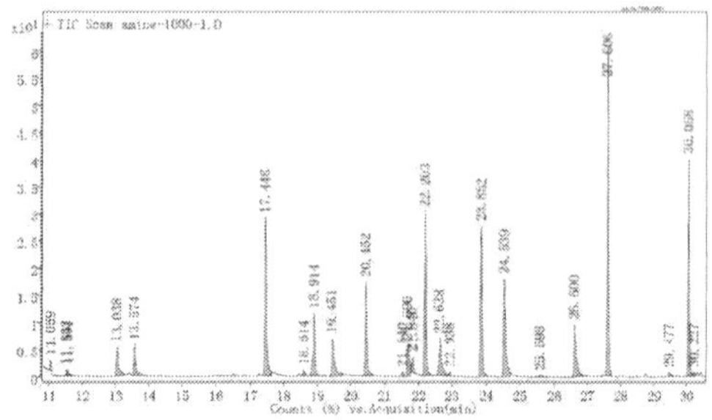

Figure 5 Chromatogram of Amine Substances

Table 7 List of 16 Common Amine Substances in Vehicles[17][18]

NO.	Substance	CAS NO.	RT	Quantitativeion	Qualitativeion
1	Triethylamine	121-44-8	13.56	86.1	101,28
2	Pyridine	110-86-1	17.444	79	52
3	2-butanone oxime	96-29-7	18.914	87.1	42,58
4	Morpholine	110-91-8	19.441	57.1	87
5	N,N-dimethylformamide	68-12-2	20.452	73	44
6	Cyclohexylamine	108-91-8	21.686	56.1	99,43
7	N-ethylmorpholine	100-74-3	22.198	100.1	115
8	Dimethylacetamide	127-19-5	23.852	87.1	44,72
9	N-butylamine	109-73-9	24.534	86.1	44,129
10	1-methylimidazole	616-47-7	26.6	82	54
11	Aniline	62-53-3	27.606	93	66
12	Bis (dimethylaminoethyl) ether	3033-62-3	29.473	86	56
13	Methylpyrrolidone	872-50-4	30.058	99	44
14	Ethyl pyrrolidone	2687-91-4	31.601	98	113,70
15	Benzothiazole	95-16-9	33.855	135	108
16	Caprolactam	105-60-2	35.301	113.1	55,85

3.3.2 The test results of Amine Substances are as follows, and the source of the results is shown in Figure 6.

Amine substances contain highly polar substances, and there is a significant deviation in the sorbent effect between different sorbents. For amine substances:

a) The best comprehensive sorbents are graphitized carbon and Tenax TA, with deviations within 20% except for 1-2 individual substances;

b) The deviation of the Graphitized carbon mixing tube is within 20%, and this sorbent can be used according to different substances;

c) The other three sorbent materials have relatively poor performance, and it is recommended to use them as appropriate.

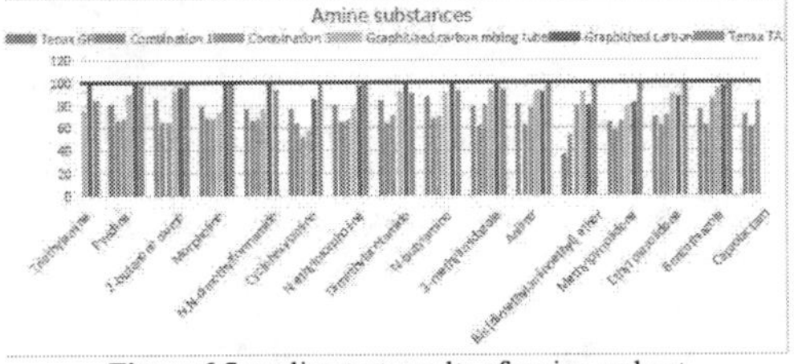

Figure 6 Sampling test results of amine sorbent

3.4 Alkanes, Cycloalkane, olefins, Aromatic hydrocarbon and other substances

3.4.1 Chromatogram and list of alkanes, Cycloalkane, olefins, Aromatic hydrocarbon and other substances are shown in Figure 7 and Table 8

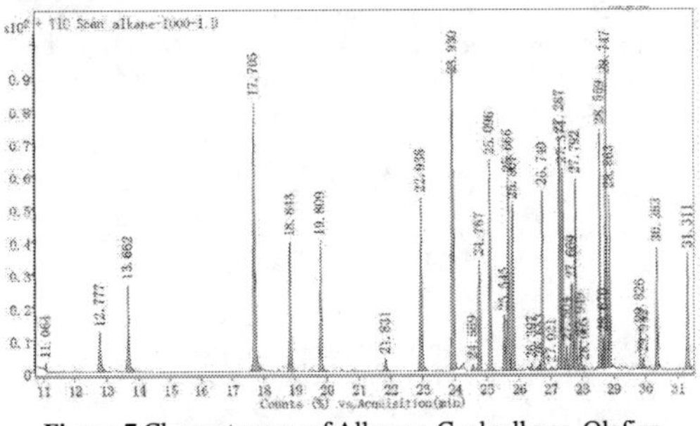

Figure 7 Chromatogram of Alkanes, Cycloalkane, Olefins, Aromatic hydrocarbon and Other Substance

Table 8 List of Common Alkanes, Cycloalkane, Olefins, Aromatic hydrocarbon and Other Substances in Vehicles[16][17][18][19]

NO.	Substance	CAS NO.	RT	Quantitativeion	Qualitativeion
1	Cyclohexane	110-82-7	12.777	84.1	56,69
2	1-heptene	592-76-7	13.667	56.1	56,71
3	1-octene	111-66-0	17.695	55.1	70
4	Toluene	108-88-3	17.71	91	65
5	2,4-Dimethylheptane	2213-23-2	18.847	43.1	85,57
6	2,4-Dimethyl-1-heptene	19549-87-2	19.809	70.1	43,55
7	O-xylene	95-47-6	22.938	91.1	106
8	Alpha pinene	80-56-8	23.92	93.1	105
9	Isopropyl benzene	98-82-8	23.94	105	120
10	Octamethylcyclotetrasiloxane	556-67-2	24.771	281	73,155
11	Propyl benzene	103-65-1	25.096	91	120
12	2,2,4,6,6-pentamethyl heptane	13475-82-6	25.666	57.1	41,71
13	Beta-pinene	127-91-3	25.807	93.1	69
14	Carene	13466-78-9	26.74	93.1	77
15	Butylcyclohexane	1678-93-9	27.287	83.1	55,170
16	Limonene	138-86-3	27.378	68.1	93,79
17	1,2,3-trimethyl benzene	526-73-8	27.79	105.1	105,120
18	Naphthane	91-17-8	28.559	138.1	67,82
19	Phenol	108-95-2	28.67	94	66
20	Undecane	1120-21-4	28.747	57.1	43,71
21	Indene	95-13-6	28.863	115	89
22	Azodiisobutyronitrile	78-67-1	29.826	69.1	41,54
23	1,2,4,5-tetramethylbenzene	95-93-2	30.353	119.1	134
24	Triethyl phosphate	78-40-0	31.286	155	99
25	2,6-dimethylphenol	576-26-1	31.311	107	122
26	2-methylnaphthalene	91-57-6	34.977	141.9	
27	Dimethyl phthalate	131-11-3	38.038	163	77,194
28	2,6-di-tert-butyl-4-methylphenol	128-27-0	38.14	205.1	220
29	2,4-di-tert-butylphenol	96-76-4	38.638	191.1	206,57

3.4.2 The test results of alkanes, Cycloalkane, olefins, Aromatic hydrocarbon and other substances are as follows, and the source of the results is shown in Figure 8, 9.

Alkanes, alkenes and Aromatic hydrocarbon are mainly non-polar substances with boiling points between 50-250°C, and there are many types of sorbents that can be used:

a) For alkanes, olefins and Aromatic hydrocarbon, the best sorbent in comprehensive effect is graphitized carbon, Tenax TA, and mixed sampling tube 1. Except for 1-2 individual substances, other deviations are within 20%;

b) Most of the other three sampling tubes (boiling points between 100 and 200°C) have good sorbent effects, while only some substances have poor sorbent effects. They can be used according to specific substances for reference.

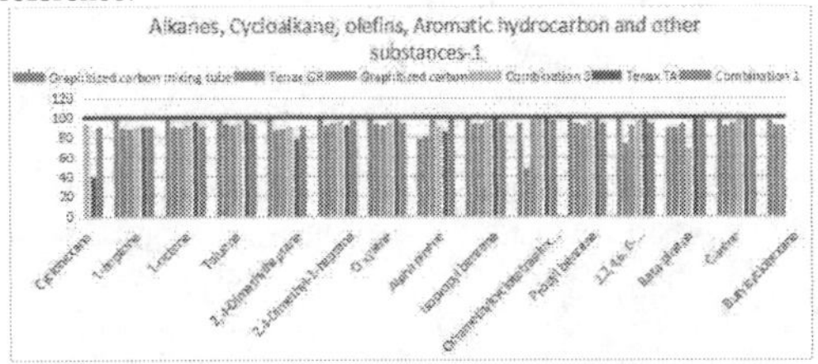

Fig.8 Sampling Test Results of sorbents for Alkanes, Cycloalkane, Olefins, Aromatic hydrocarbon and Other Substances-1

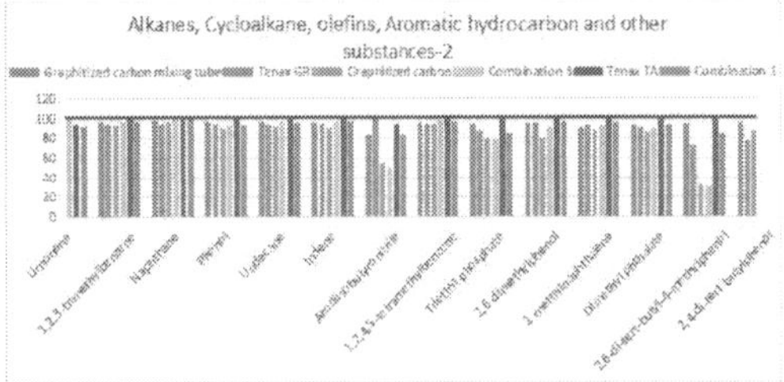

Fig.9 Sampling Test Results of sorbents for Alkanes, Cycloalkane, Olefins, Aromatic hydrocarbon and Other Substances-2

4. Conclusion

From the Conclusion, it can be seen that for substances with strong or medium polarity between boiling points (50-200)°C, Graphitized carbon mixing tube is the most effective sorbent, and its sorbent effect is the best; For substances with boiling points between 100 and 250°C, graphitized carbon and Tenax TA with weaker polarity have better effects; During the specific use process, appropriate sorbent types can be selected based on relevant substances. Achieving optimal testing results.

References

1. Volatile Organic Compounds; Researchers at University of Florence Report New Data on Volatile Organic Compounds (The network of plants volatile organic compounds)[J]. Ecology Environment & Conservation.
2. Environmental Pollution; Recent Findings from University of Trento Has Provided New Information about Environmental Pollution [Characterisation of volatile organic compounds (VOCs) released by the composting of different waste matrices][J]. Ecology Environment & Conservation.
3. ISO16000-6:2021 Indoor air — Part 6: Determination of organic compounds (VVOC, VOC, SVOC) in indoor and test chamber air by active sampling on sorbent tubes, thermal desorption and gas chromatoTenax GRaphy using MS or MS FID
4. EN ISO 16017-1:2000 Indoor, ambiant and workplace air - Sampling and analysis of volatile organic compounds by sorbent tube/thermal desorption/capillary gas chromatoTenax GRaphy - Part 1: Pumped sampling
5. EPA TO-1 Method for the Determination of Volatile Organic Compounds in ambient Air Using Tenax ® Absorption and Gas ChromatoTenax Graphy/Mass Spectrometry (GC/MS)
6. EPA TO-17 Determination of Volatile Organic Compounds in ambient Air Using Active Sampling On to sorbent Tubes Compendium of Methods for the Determination of Toxic Organic Compounds in ambient air
7. HJ583-2010 Ambient air.Determination of benzene and its analogies using sorbent adsorption thermal desorption and gas chromatography
8. HJ584-2010 Ambient air.Determination of benzene and its analogies by activated charcoal adsorption carbon disulfide desorption and gas chromatography
9. HJ644-2013 Ambient air - Determination of volatile organic compounds - Sorbent adsorption and thermal desorption / gas chromatography mass spectrometry method
10. HJ734-2014 Stati onary source emi ssi on -Determi nati on of vol ati l e organic compounds-Sorbent adsorpti on and thermal desorpti on gas chromatography mass spectrometry method
11. GB/T18883-2022 Standards for indoor air quality
12. HJ/T400-2007 Determination of Volatile Organic Compounds and Carbonyl Compounds in Cabin of Vehicles
13. Xing Yuewen. The Tenax tube sampling was used to analyze the TVOC in the indoor air by the secondary thermal analysis of thermal desorption instrument and gas chromatography — mass spectromet [J]. Environmental Protection and Circular Economy, 2021,41 (10): 73-76
14. Li Jian. Research on the Influence of Tenax Ta Adsorption Tube on TVOC Detection [J]. China Science and Technology Information, 2014, No.507 (24): 36-37
15. Kuang Shaoning. Process control and optimization of the analysis method for determining TVOC concentration in indoor air using thermal desorption gas chromatography [J]. Guangdong Science and Technology, 2009, No.212 (10): 80
16. Wang Xiaoxu, Qian Mingyuan, Zhang Xiaobo. Study on the Determination of Semivolatile Organic Compounds in Indoor Air by Ionic Liquid-Thermal Desorption Gas Chromatography/Mass Spectrometry [J]. Environmental Pollution and Prevention, 2021,43 (01): 73-78. DOI: 10.15985/j.cnki.1001-3865.2021-01.014
17. Zhu Xiaoping, Ma Huilian, Zhu Xiuhua, et al. Determination of 67 volatile organic compounds in ambient air by thermal desorption gas chromatography-mass spectrometry [J]. Chromatography, 2019,37 (11): 1228-1234
18. Cui Jin, Liu Chengxin, Chen Shan, et al. Detection of 35 Volatile organic compounds in children's floor mats using air bag sampling/thermal desorption gas chromatography-mass spectrometry [J]. Analytical Laboratory, 2020,39 (06): 700-705. DOI: 10.13595/j.cnki. issn1000-0720.020902

19. Xue Junhai, Qiu Zhaojun, Lv Huanming, et al. Determination of Emission of 7 Terpenes from Furnitures by Thermal Desorption Analysis-gas Chromatography/Mass Spectrometry [J]. Forestry Industry, 2020,57 (07): 35-38. DOI: 10.19531/j.issn1001-5299-202007009

https://doi.org/10.1051/e3sconf/202344102009

Mine-used Flameproof Inverter Heat Dissipation Design and Its Performance Analysis

Yue Wang [1, 2, *], Xiang Rong [1, 2]

[1] Tiandi (Changzhou) Automation Co., Ltd, Changzhou, China.
[2] CCTEG Changzhou Research Institute, Changzhou, China.

Abstract—Flameproof inverter for mine-used have a significant warming effect during operation, resulting in reduced reliability and susceptibility to thermal degradation and thermal failure. However, there is a lack of thermal analysis research on the operation mechanism and process evaluation of mining inverters. This paper analyzes the electrical characteristics of internal power devices and calculates the power loss. Secondly, on the basis of comparing the influence of flow rate, structural parameters and installation location on the thermal performance, the forced water cooling + air cooling + natural cooling method is used to optimize the design of the thermal system. Finally, ANSYS Icepak is used to conduct a comprehensive numerical simulation analysis of the internal temperature field characteristics and convective heat transfer characteristics. The simulation verifies the effectiveness of the heat dissipation-based design, and the results show that: the internal components of the mining explosion-proof inverter are not more than 80 °C, which is much lower than the value specified in the relevant standards, and it has a good heat dissipation performance, which effectively guarantees the safe, stable and reliable operation of the mining explosion-proof inverter.

1. Introduction

Mine-used flameproof inverter plays a vital role in promoting intelligent mining in coal mines, enhancing safe coal mine production, and helping coal mines to develop in a green way [1]. Its compact internal structure, variable working conditions and high power density lead to significant heating effects, reduced reliability and susceptibility to thermal degradation and thermal failure [2-3]. Therefore, the thermal performance of mine-used flameproof inverter is particularly important.

[4]analyzed the temperature and fluid distribution of contactor and busbar under normal working conditions and fan failure conditions. In [5], the temperature field distribution of transformer winding is obtained by finite element thermal-flow coupling calculation. [6] simulated the temperature rise process of the reactor and the temperature field distribution at steady state. In addition, the researchers also used finite element software to design and optimize the key parameters of fin radiator [7], heat pipe radiator [8] and water-cooled radiator [9]. However, the above research is a separate analysis of a certain type of power device or radiator, and there are many power devices in the mine-used flameproof inverter, and the heat exchange between them should be considered, and there are a variety of heat dissipation methods mixed application.

In this paper, the electrical characteristics of internal power devices are analyzed, and the power loss is calculated according to the relevant theory. On the basis of comparing the influence of flow rate, structural parameters and installation position on heat dissipation performance, the heat dissipation system is optimized by forced water cooling + air cooling + natural cooling. Finally, ANSYS Icepak is used to conduct a comprehensive numerical simulation analysis of the internal temperature field characteristics and convective heat transfer characteristics, and the effectiveness of the heat dissipation design is verified.

2. Loss Analysis of Internal Power Devices

In this paper, a 630kW/1140V four-quadrant mine-used flameproof inverter is taken as the research object. Among them, vacuum AC contactor, busbar, IGBT module and output filter are not only large power consumption devices, but also high thermal sensitivity devices, which need to be considered in the design process of heat dissipation system.

2.1. The power loss of IGBT module

The power loss of IGBT module mainly includes IGBT on-state loss, turn-on and turn-off loss and freewheeling diode loss. According to the electrical performance

* Corresponding author: m18635372893@163.com

parameters and output characteristic curve, the theoretical calculation is carried out[10]. The method is as follows:

The on-state loss of IGBT module is :

$$P_{\text{cond,IGBT}} = \frac{1}{T_0}\int_0^{T_0/2}\frac{1}{2}(V_{CE0} + r * i(t)) * I\sin(\omega t) * (1 + m\sin(\omega t + \varphi))dt \quad (1)$$

The turn-on and turn-off losses of IGBT module are :

$$P_{\text{SW,IGBT}} = \frac{1}{\pi} * f_{\text{SW}} * (E_{\text{on,IGBT}}(I_{\text{nom}}, V_{\text{nom}}) + E_{\text{off,IGBT}}(I_{\text{nom}}, V_{\text{nom}})) * \frac{I}{I_{\text{nom}}} * \frac{U_{\text{dc}}}{U_{\text{nom}}} \quad (2)$$

In the formula, U_{CEO} is the threshold voltage, r is the slope resistance; $\tau'(t)$ is the duty cycle of the inverter bridge output; m is the modulation ratio; φ is the phase difference between the output signal and the current; f_{sw} is the switching frequency of IGBT module; Eon, IGBT, Eoff, IGBT are the turn-on and turn-off losses of IGBT under rated conditions, respectively; U_{nom} and I_{nom} are the reference voltage and reference current of IGBT module respectively; I is the output current value of the inverter unit; U_{dc} is the DC bus voltage.

The power loss of the diode is:

$$P_{\text{DIODE}} = \frac{1}{\pi} * f_{\text{SW}} * \left(E_{\text{rec}}(I_{\text{nom}}) * (0.45 * \frac{I}{I_{\text{nom}}} + 0.55)\right) * \frac{U_{\text{dc}}}{U_{\text{nom}}} \quad (3)$$

In the formula, E_{rec} is the turn-off loss of the diode under rated conditions.

According to the formula (1) - (3), combined with the data manual, the power loss of FZ800R33KF2C IGBT module at 125 °C junction temperature is 1700 W.

2.2. Power loss of busbar

According to formula (4), considering the skin effect of the connecting busbar, the power loss is calculated by combining the measured resistance value of the bridge tester.

$$P_{\text{m}} = K_{\text{jf}} I^2 R = \left(\frac{S}{186}\right)^{0.1} I^2 R \quad (4)$$

In the formula: K_{jf} is the skin effect coefficient; I is the rated current of the inverter; R is the resistance value of the busbar; S is the cross-sectional area of the busbar.

According to the calculation, the power loss of the positive phase, negative phase and U / V / W phase of the AC busbar is 5.8 W, 0.7 W and 5.2 W respectively. Similarly, the power loss of the positive and negative phases of the AC-DC connection bus is 19.5 W and 14.7 W; the power loss of the positive and negative phases of the DC connection busbar is 5 W and 5 W; the power loss of the unidirectional output copper bar is 2.5 W.

2.3. Power loss of output filter

The output filter device is composed of reactor and RC absorption circuit. For the absorption resistor, the total power loss is calculated to be 300 W [11].

For the reactor, it can be obtained from the selection manual that the load loss of the reactor is 260 W and the no-load loss is 30 W. At the same time, considering the influence of load fluctuation, the winding loss and core loss of the reactor are 270 W and 30 W respectively, and the power loss is 300 W.

$$P = P_{\text{Cu}} + P_{\text{Fe}} = K_{\text{T}} P_{\text{k}} + P_0 \quad (5)$$

In the formula: $\triangle P$ is the active power loss of the transformer; P_{Cu} is winding loss; P_{Fe} is the core loss; K_T is the load fluctuation loss coefficient, take 1.05; P_k is load loss; P_0 is no-load loss.

2.4. Power loss of vacuum AC contactor

The vacuum AC contactor includes three models : 80A / 1140V, 160A / 1140V and 630A / 1140V. The pull-in time is very short, only 30 ms. Therefore, the thermal effect of the pull-in process is not analyzed in this paper. After suction, the power loss is measured by the holding power. According to the selection manual, the holding power of the aforementioned vacuum AC contactor is 10 W, 18 W, and 15 W, respectively.

3. Design of Heat Dissipation System

The IGBT module and the absorption resistors are cooled by water, the reactor is cooled by air, and the other power devices are cooled by natural cooling.

3.1. Design of Water-cooled Radiator

The thermal performance of different materials, different flow rates, and different channel-substrate distances were comparatively analyzed, and the relationship surface between the above conditions and the maximum temperature of the IGBT module and absorbing resistor was obtained, as shown in Fig. 1.

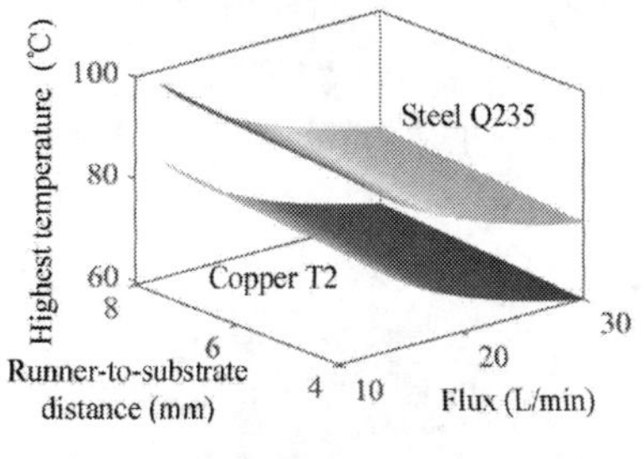

Fig.1 The highest temperature of IGBT module and absorption resistor under different conditions

From fig. 1, it can be seen that under the same conditions, the heat dissipation performance of copper is better than that of steel, but the bearing capacity of steel is better than that of copper, and the material and processing cost of copper is high. In order to meet the same heat dissipation performance, it is necessary to increase the cooling water flow and reduce the distance from the flow channel to the surface. The design requires the IGBT module and the absorption resistor to have a maximum junction temperature of 75 °C.Therefore, the cooling water flow rate is selected to be 30 L/min, and the distance from the channel to the substrate is 4 mm. At this time, the temperature field distribution of water-cooled radiator, IGBT module and absorption resistors are shown in fig.2.

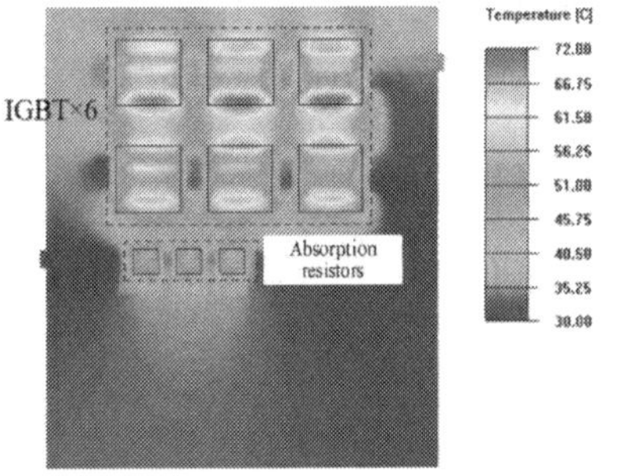

Fig.2 Cloud diagram of temperature distribution of water-cooled radiator, IGBT module and absorption resistor

3.2. Fan Selection and Arrangement

It is assumed that the temperature in the flameproof shell is 40 °C when the reactor is in operation. The design requires that the temperature rise does not exceed 30 K. According to the principle of heat transfer, the flow rate of the fan that meets the heat dissipation requirements is:

$$L = Q/(C_P \cdot \rho \cdot \Delta T) \quad (6)$$

In the formula, Q is the heat transferred from the heat sink to the environment in unit time; C_P is the specific heat capacity of air; ρ is air density.

It can be seen from the Section 1 that the total power loss of the reactor is 300 W. Therefore, when Q=300 W, air density ρ=1.128 kg/m at 40 °C, and air specific heat capacity C_p=0.2814 Wh/(kg K) at 40 °C are substituted into formula 6, the fan flow rate is 30 m^3/h.

Comparing the influence of transverse and longitudinal air inlet on the heat dissipation performance of the reactor, the temperature field distribution of the reactor in different air inlet directions is obtained, as shown in Fig.3.

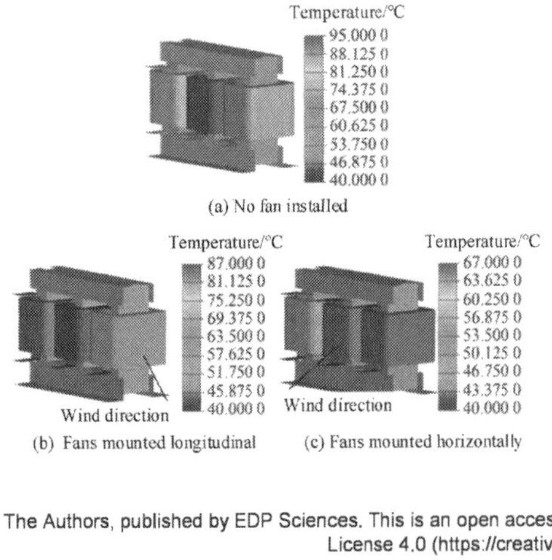

Fig.3 Reactor temperature distribution in different air inlet directions

It can be seen from Fig.3 that the temperature rise of longitudinal air inlet is 7 °C higher than that of transverse air inlet, and the heat dissipation effect of transverse air inlet is better.

4. Temperature Field Simulation Analysis

The integrated model for research is shown in Fig.4. Except that the reactor and the absorption resistors are arranged on the left side, and the vacuum AC contactor is arranged on the right side, the remaining power devices are symmetrically arranged.

ANSYS Icepak is used to conduct a comprehensive numerical simulation analysis of the internal temperature field characteristics and convective heat transfer characteristics. The power loss is substituted into the finite element model and the corresponding material properties are configured. When the ambient temperature is 30 °C, the temperature field distribution of the flameproof shell and internal components is shown in Fig.5 and 6.

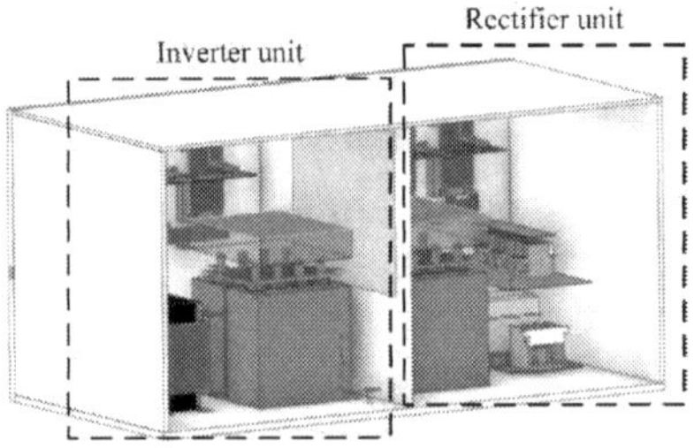

Fig.4 Integrated model of mine-used flameproof inverter

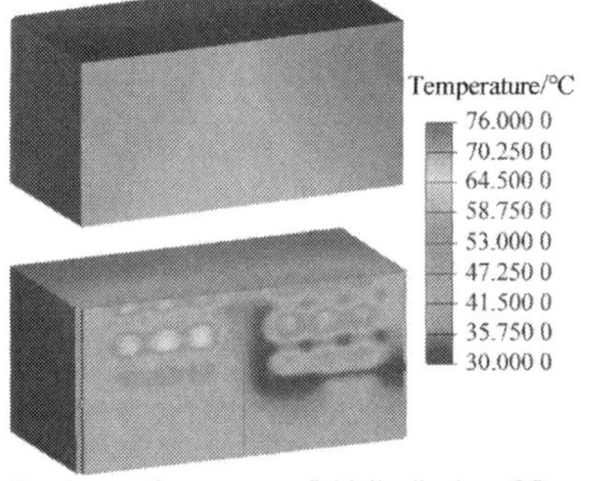

Fig.5 Cloud map of temperature field distribution of flameproof shell

It can be seen from Fig.5 that due to the conduction, convection and radiation heat transfer of the internal power device, the temperature of the flameproof shell is higher than the ambient temperature, the lowest is 36 °C, and the temperature of the rear substrate is higher than other flameproof surfaces, up to 70 °C.

It can be seen from Fig.6 that the temperature of the IGBT module is the highest, followed by the core busbar assembly, and the temperature of the DC filter capacitor assembly is the lowest. For Fig.6 (a), because the two water-cooled radiators are connected in series, the temperature increases with the extension of the cooling water flow path, resulting in the temperature of the IGBT module on the inverter side higher than that on the rectifier side. For Fig.6 (b) and (d), the temperature of the AC-DC busbar is higher than that of the AC busbar, the temperature of the output copper busbar is the lowest, and the temperature of the core busbar assembly on the inverter side is higher than that on the rectifier side. The temperature difference between the two is mainly reflected in the AC-DC busbar. Because the absorption resistors are located on the inverter side, the temperature of the AC-DC busbar adjacent to it will increase under the action of radiation heat transfer. For Fig.6 (c) and (e), the temperature of the DC filter capacitor component on the rectifier side is higher than that on the inverter side. It can be seen that the longitudinal air inlet can also play a role in heat dissipation. For Fig.6 (f), the 80A / 1140V AC contactor has a larger power density and denser arrangement, and the temperature rise is higher than that of the 630A / 1140V AC contactor. Although the 160A / 1140V vacuum AC contactor does not work, there is still a temperature rise phenomenon under the action of conduction and radiation heat transfer. For Fig.6 (g), the temperature of the reactor is higher than the individual analysis value given in Fig.3. This is due to the heat exchange between the internal power devices.

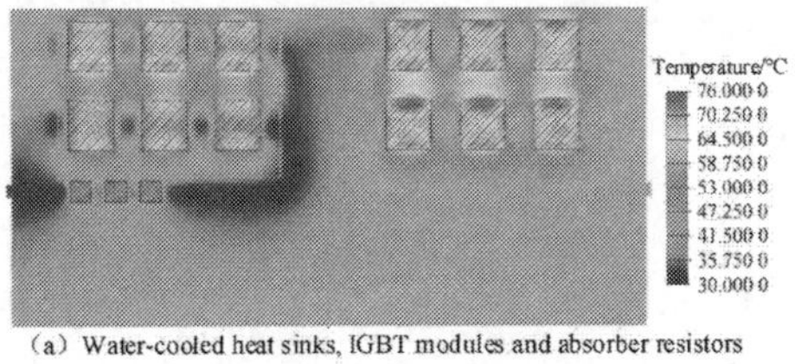

(a) Water-cooled heat sinks, IGBT modules and absorber resistors

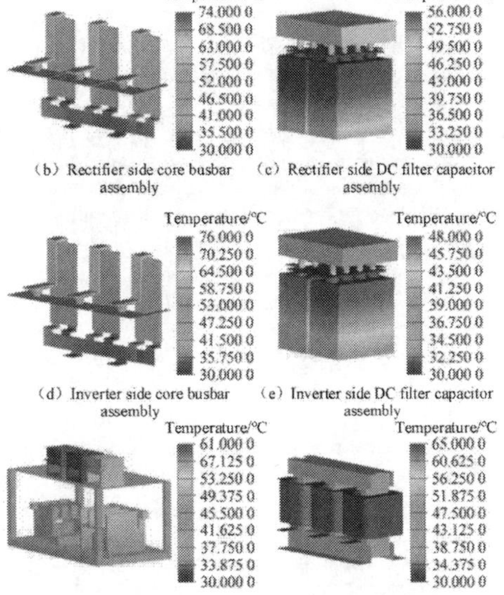

(b) Rectifier side core busbar assembly
(c) Rectifier side DC filter capacitor assembly
(d) Inverter side core busbar assembly
(e) Inverter side DC filter capacitor assembly
(f) Contactor assembly
(g) Reactor

Fig.6 Temperature field distribution cloud diagram of internal components of flameproof shell

On the whole, the internal components of the mine flameproof frequency converter do not exceed 80 °C, which is far lower than the specified value of the relevant standards. It has good heat dissipation performance and can provide guarantee for the safe, stable and reliable operation of the mine-used flameproof inverter

5. Conclusions

A. The electrical characteristics of internal power devices are analyzed and the power loss is calculated. The power loss of IGBT module, absorption resistor and reactor is higher than that of other power devices.

B. The mixed heat dissipation method of water cooling + air cooling + natural cooling is adopted: the selected material is Q235 steel, the cooling water flow is 30L / min, and the distance from the flow channel to the substrate is 4mm ; the temperature rise of longitudinal air inlet is 7 °C higher than that of transverse air inlet.

C. The simulation verifies the effectiveness of the heat dissipation design: the temperature of the internal components does not exceed 80 °C. Among them, the temperature of the IGBT module is the highest, the temperature of the core busbar component is the second, and the temperature of the DC filter capacitor component is the lowest.

Acknowledgment

This research is co-supported by the Special Fund for the Technology Innovation and Entrepreneurship of Tiandi Company Ltd. (Grant nos. 2023-TD-MS005 and 2023-TD-ZD001-006). And the Tiandi (Changzhou) Automation Co., Ltd. R&D project (2023GY1002). The authors would like to thank them.

References

1. ZHU Yongping,XU Xiaojian.Development trend of mine frequency converter[J].Industry and Mine Automation, 2017,43(10):18-23.
2. Ryu, T.; Choi, U.-M.; Vernica, I.; Blaabjerg, F. Effect of thermal loading definitions on the mission profile-based reliability evaluation of power devices in PV inverters. Microelectronics Reliability 2022, 138, 114650.
3. YANG Weilin.Brief discussion on heat dissipation design of explosion-proof frequency inverter[J].Explosion-Proof Electric Machine,2019,54(1):36-39,42.
4. LIU Sijun,HAN Wei,ZHANG Haixing,et al.Research on temperature fluid field simulation and heat dissipation optimization of high voltage switchgear[J].High Voltage Apparatus,2020,56(10):63-69.
5. GU Shengjian,YOU Piaopiao,JIANG Youhua.Thermal performance analysis of dry-type transformer under condition of non-ideal power supply[J].Transformer, 2020, 57(6):19-24.
6. LI Jinzhong,ZHANG Dandan,XU Zhengyu,et al.Heat dissipation performance of ultra-high voltage shunt reactor with sound insulation based on finite element method[J]. High Voltage Engineering,2017,43(3):822-827.
7. XUPengcheng,TAO Hanzhong,ZHANG Hong. Optimization and analysis of IGBT heat pipe heat sink with integral fin[J].Journal of Refrigeration,2014,35(5):101-104,109.
8. DING Jie,ZHANG Ping.Temperature rise test and thermal simulation of heat-pipe radiator of metro vehicel traction inverter[J].China Railway Science,2016,37(3):95-102.
9. YIN Yuxing,ZHU Zhaoxia.Research of water cooling performance of mine flameproof frequency converter[J]. Coal Mine Machinery,2015,36(8):83-85.
10. Das, S. C.; Narayanan, G.; Tiwari, A. Experimental study on the influence of junction temperature on the relationship between IGBT switching energy loss and device current. Microelectronics Reliability 2018, 80, 134–143.
11. RONG Xiang, SHI Han, JIANG Dezhi, et al. A mine-used frequency converter filter device[J].Industry and Mine Automation,2020,46(5):76-81.

Mechanistic Analysis of the Impact of Multidimensional Relationship Networks on Farmers' Behavior of Carbon Emission Reduction

Xiaojuan Luo*, Qingqing Ye

Jiangxi Normal University, Economic Development Research Institute, Nanchang, Jiangxi, China

Abstract. Agricultural carbon emission reduction is indispensable to the achievement of the "double carbon" goal, and the multidimensional relationship network has become the key to the decision-making of farmers' carbon emission reduction behaviour. Based on the perspective of heterogeneous relationship subjects, this paper divides the relationship network into three dimensions: market relationship, government relationship, and social relationship network, and explores the theoretical analysis of the influence of multi-dimensional relationship network on the decision-making of carbon emission reduction behaviours of farmers, and analyses the mechanism of relationship network influencing the carbon emission reduction behaviours of farmers from the three aspects of information flow, demonstration effect, and factor guarantee based on the social capital theory to lay a solid theoretical foundation for the later empirical analyses. It lays a solid theoretical foundation for the later empirical analysis.

1 Introduction

As one of the important sources of greenhouse gas (GHG) emissions, carbon emission reduction in the agricultural production system is attracting much attention, and the Fourth Assessment Report of the United Nations Intergovernmental Panel on Climate Change (IPCC) points out that agricultural GHG emissions account for about 13.5% of the total anthropogenic emissions, and have become the second largest source of GHG emissions, and in China, this proportion has been close to 17%, with an upward trend, Among them, fertiliser is the top carbon source, with a proportion of nearly 60%[1]. In the context of the "double carbon" goal, how to realize the transformation and upgrading from high-carbon agriculture to low-carbon agriculture, and reduce agricultural carbon emissions has become a new direction for the development of modern agriculture. Carbon mitigation technologies are technologies included in the agricultural production process that have a carbon reduction contribution and may have the effect of reducing means of production and increasing agricultural productivity, while the means of achieving emission reductions in agriculture are mainly means of production reductions and technology adoption. However, the adoption rate of carbon emission reduction technologies in rural areas of China is still low. Since farmers are both micro-entrepreneurs of agricultural production and adopters of carbon-reducing technologies, the study of the mechanism of low carbon emission reduction technology adoption from the perspective of farmers is the basis for the establishment of micro incentive mechanism for the promotion of carbon emission reduction technology, which is the key issue to be solved in the agricultural field under the goal of "dual-carbon".

Currently, there are not many studies on agricultural carbon emission reduction at the micro level, but there have been studies related to carbon emission reduction behaviors, such as green technology adoption, ecological production behaviors, etc., which explain the factors that may influence the behaviors of farmers from the perspectives of basic characteristics of farmers and household resources[2]. In addition to the above influencing factors, more scholars have paid attention to the micro influence of relationship networks in the diffusion of new agricultural technologies[3]. Farmer relationship networks is a relatively stable system of relationships gradually formed through the interactions among individual members, and the embeddedness theory suggests that actors are not completely independent in making relevant decisions, and their relationship networks have an impact on their behavioral decisions[4]. Based on the limited knowledge theory, Zhu [5] found that the farmers' relationship network can significantly promote the adoption of resource-saving technologies. Yang [6] found that both clannish and friendship networks have a significant impact on the degree of adoption of green production technologies. Li [7] found that heterogeneous relationships have a more significant effect on technology adoption than homogeneous relationships.

In summary, the existing literature has a certain foundation for the study of relationship networks

* Corresponding author: lxj918@126.com

affecting farmers' behavior, but there are still aspects worth studying: first, in the literature on relationship networks, existing studies have focused on the impact of relational networks on farmers' behavior, but there is not much literature focusing on the heterogeneity of relational subjects. Secondly, the existing literature pays more attention to the mediating role of the actors' own characteristics in the relationship network's influence on farmers' behavior, while ignoring the resourcefulness of the relationship network. Based on the social capital theory, this paper explores the mechanism of the multidimensional relationship network's influence on farmers' behavior from the three perspectives of the flow of information, demonstration effect, and the guarantee of factors.

2 Mechanistic Analysis

2.1 Influence of multidimensional relational networks on farmers' carbon emission reduction behavioral decisions

Research findings have indicated that the relationship network has a positive effect on technology adoption behavior, however, due to the differences in the relationship network, actors have different access to resources, forming differentiated carbon emission reduction behavioral decisions, so it is necessary to clarify the relationship network embedded in the farm household. China's rural areas are a relatively typical relationship-oriented society, the relationship network had the characteristics of high closeness, strong convergence, and low heterogeneity, therefore, the influence of the relationship network on the behavioral decisions of the farmers showed the following characteristics "The one who is close to Zhu is red, and the one who is close to ink is black". Whereas carbon emission reduction technology has significant positive externalities, farmers lack the initiative to adopt it, and the initial promotion of its use requires strong support from the government, At present, in China's important agricultural service system, agricultural suppliers and agricultural technicians, etc. are the main bodies that have the most direct contact with farmers, so their influence on the decision-making of farmers' carbon emission reduction behavior is increasingly significant [8]. Drawing on existing research, this paper classifies the farmers' relationship network into three categories based on the different subjects involved in farmers' behavioral decision-making: (1) social relationship network with farmers linked to relatives, friends, and townspeople, etc. (2) market relationship network with farmers linked to agribusiness merchants, retailers of agricultural production materials, etc. (3) government relationship network with farmers linked to governmental departments, village cadres, and agricultural technology departments.

From the viewpoint of governmental relationship network, the strong connection relationship implies that farmers have access to more comprehensive and scientific technical information, and have greater chances of obtaining technical guidance, financial subsidies and other resources, which positively promotes the decision-making of farmers' carbon emission reduction behavior. The first step in decision-making is to know the technology, which is the biggest obstacle for farmers with limited information, while driven by the pursuit of maximizing the combined economic, social and environmental benefits, the government will take the initiative to improve scientific and comprehensive technical information for farmers, which is conducive to deepening the understanding of farmers on carbon emission reduction technology, and the government will form the farmers' value recognition of carbon reduction technology through the value guidance mechanism, which becomes an internal incentive to guide the adoption of carbon emission reduction technology by farmers. However, the costs and benefits of new technologies are the key factors influencing farmers' willingness to transform into behavioral decisions, and the government's loan subsidies and guarantee fee subsidies can effectively reduce the financial costs of farmers' adoption of carbon emission reduction technologies, especially in the case of "scarcity of resources", and the allocation of resources in the network of relationships can help farmers to obtain these scarce resources in order to support carbon emission reduction behavioral decisions.

In terms of the network of market relationships, strong connectivity implies that farmers have more access to technological information, and that it is easier to allocate market resources such as technology and its accompanying service guidance to promote the adoption of new technologies by farmers. On the one hand, at the initial stage of purchasing new agricultural materials, farmers have less technical information, but in order to promote agricultural products, retailers with resource advantages will provide farmers with detailed knowledge on the use of agricultural products, and voluntarily pay for publicity to organise activities such as visits to experimental plots, technical training, etc., which will help farmers make decisions on carbon emission reduction behaviour. On the other hand, frequent interactions create a cooperative atmosphere of mutual help and friendliness, and agricultural suppliers are more likely to provide credit trial promotion, which to some extent alleviates the financial constraints of farmers in adopting new technologies.

In terms of social relationship networks, dense network connections can shorten the information transmission distance and are characterized by high trust and reciprocal motives, which help farmers adopt new technologies. Firstly, the social relationship network linked by blood and geography is the innate relationship network, so the network members are familiar with each other and have a high degree of trust, which can significantly increase the willingness of farmers to share information and the reliability of information, which is conducive to the transmission of tacit information, and the strong motivation of reciprocity among the network members can help to alleviate the resource constraints encountered in the adoption of technology. Second, the "neighborhood effect" based on social networks can

 https://doi.org/10.1051/e3sconf/202344102010

significantly promote the adoption of technology by farmers.

2.2 Pathways of multidimensional relational networks affecting farmers' carbon emission reduction behavioral decision-making

Social capital theory suggests that the relationship network, as a carrier of social capital, can provide various resources for the actors, thus promoting farmers' technology adoption behaviour. Scholars have already found that information availability, demonstration effect, and factor flow are the key factors that promote the adoption behavior of new technologies by farmers[9-10], and important resources such as capital, technology, and information are embedded in the flow of the relational network, which in turn plays an important supportive role for the diffusion of agricultural technology[11]. In summary, the flow of information, the demonstration of technology, and the guarantee of elements play an important role in the multidimensional relational network influencing farmers' carbon emission reduction behavioral decision-making.

The technological information held by farmers is incomplete, while the relational network can promote information flow, alleviate information asymmetry and revise technology perceptions, which affects carbon emission reduction behavioral decisions. Interactive communication among members of the network can effectively promote the dissemination of information on new agricultural technologies. On the one hand, knowing more carbon emission reduction technologies, it is more likely to select and implement the technologies that match their own endowments and needs, breaking the impact of information barriers on the decision-making of farmers' behaviors, and on the other hand, the flow of information can improve the farmers' technological cognition, which can be used to increase crop yields and yields. On the other hand, the flow of information can improve farmers' technological awareness and more comprehensive understanding of the role of new technologies in increasing crop yields and returns, protecting the ecological environment, etc., to improve the response level of farmers' carbon emission reduction behaviors.

By providing high-quality resources, the relational network increases the learning opportunities of farmers, which in turn creates a "demonstration effect" that influences the decision-making of carbon emission reduction behaviors. Due to the limitations of their own knowledge level and judgment ability, farmers embedded in the relational network will imitate the behaviors of the more successful ones in the relational network, showing obvious demonstration effects and forming the convergence of technology adoption behaviors. On the one hand, many agricultural technological knowledge is acquired through "learning by doing", which has the characteristic of "tacit", and the interaction with the successful ones promotes the transmission of tacit knowledge, and the guidance and accumulation of technological knowledge reduces the technological complexity cognition of farmers and realizes the demonstration effect of technology. On the other hand, the application effect of carbon emission reduction technology is uncertain, and the successful demonstration brings them a clear and visible benefit prospect, which reduces the farmers' perception of technological risk, and then influences the technology preference and adoption behavior decision.

The adoption of new technologies often implies the input of new productive factors, and farmers can obtain capital, labor and other factors through the relationship network, which significantly reduces the cost pressure of farmers' technology transformation and effectively promotes the adoption of carbon emission reduction technologies. Among the many productive factors, capital is a key element in the adoption of new technologies, and the existence of credit constraints weakens the enthusiasm of farmers in adopting technologies, leading to farmers abandoning the selection of new technologies due to the lack of capital, while the relational network can alleviate the problem of credit constraints, which can help farmers to obtain both formal lending and private lending. On the one hand, farmers embedded in strong relational networks are more likely to form informal institutional arrangements of mutual benefit and cooperation and sharing, which effectively reduces the cost of access to finance and alleviates the negative impact of resource constraints on behavioral decisions. on the other hand, the adoption of new technologies has a greater degree of uncertainty, while rural areas lack appropriate financing guarantee institutions, and relational networks as a traditional "invisible guarantee mechanism", which can make it easier for farmers to obtain borrowing and promote carbon emission reduction behavioral decisions.

3 Conclusions

This paper finds that the relationship network can improve the adoption rate of carbon emission reduction technology of farmers, in which the government relationship network is a key factor influencing the decision-making of carbon emission reduction behavior of farmers, and the information, technical guidance, financial subsidies and other resources provided by the government can effectively guide farmers to form the value of carbon emission reduction technology and reduce the cost of adopting the technology, and the market relationship network and the social network can provide the corresponding technical guidance according to the real needs of farmers. The market relationship network and social relationship network can provide corresponding technical guidance according to the real needs of farmers, which reduces the complexity and risk perception of farmers on carbon emission reduction technology. Meanwhile, the high closeness and high trust of the social relationship network promote the reciprocal effect among network members, which in turn promotes the decision-making of farmers' carbon emission reduction behavior to different degrees. In order to further sort out the role of various resources in the multidimensional relationship network in influencing farmers' carbon emission reduction behavioral decision-making, this paper summarizes the important resources flowing in the relationship network based on the social capital theory as three kinds of information flow, demonstration effect, and elemental security, and finds that the relationship network can promote the flow of information, alleviate the asymmetry of information and modify the technical cognition, increase the learning opportunities of farmers by providing high-quality resources, and form the demonstration effect. The demonstration effect contributes to the accumulation of tacit knowledge and the formation of benefit prospects, and at the same time reduces the adoption cost of technology through the provision of capital and other factor guarantees, which in turn promotes the decision-making of carbon emission reduction behavior of farmers. Based on social network theory and social capital theory, this paper provides an in-depth analysis of the mechanism of multi-dimensional relationship network affecting carbon emission reduction decision-making, which helps to further empirically analyse the micro-decision-making mechanism of the diffusion of carbon emission reduction technology, and provides an important reference for the promotion of agricultural carbon emission reduction technology and the realization of agricultural low carbon development and dual carbon goals in China.

Acknowledgment

This paper is one of the phase results of the National Natural Science Foundation of China (NSFC) regional project "Intervention Mechanism and Diffusion Simulation of Dynamic Multi-dimensional Social Network on Farmers' Adoption of Chemical Fertilizer Reduction Technologies and Its Policy Optimization Implications" (No. 72063019)

References

1. Hu, W. L., Zhang, J. X., Wang, H. L. (2020) Characteristics and Influencing Factors of Agricultural Carbon Emission in China. Statistics & Decision, 36: 56-62. DOI: 10.13546/j.cnki.tjyjc.2020.05.012.
2. Zhao, P. P., Zhang, Q. Q., Zhong, Y. W., Liu, T. J. (2022) Impact of social network embeddedness on the adoption ofgreen control technologies by farmers: Mediating effects based on technology perception. Resources Science, 44: 1865-1878. DOI: 10.18402/resci.2022.09.10
3. Bandiera O, Rasul I. (2006) Social networks and technology adoption in northern Mozambique. The economic journal, 116: 869-902. DOI: 10.1111/j.1468-0297.2006.01115.x
4. Granovetter, M. S., （1973） The strength of weak ties. American journal of sociology, 78: 1360-1380. DOI: https://doi.org/10.1086/225469
5. Zhu, Y. J., Yang, Q., Wang, F. (2021) Mechanism of influence of social networks on banana farmers'adoption of resourceconservation technologies: A case study of the drip fertigation system. Resources Science, 43: 1099- 1114. DOI:10.18402/resci.2021.06.04
6. Yang, Z. H., (2018) Ageing, Social Network and theAdoption of Green Production Technology: Evidence from Farm Households in Six Provinces in theYangtze River Basin. China Rural Survey, 142: 44-58. https://kns.cnki.net/kcms/detail/11.3586.f.20180712.1520.008.html
7. Li, Y. B., Lu, Q., Guo, G. (2017) Effects of social network on water-saving irrigation technology adoption: homogeneity or heterogeneity?. Research of Agricultural Modernization, 38: 978-986. DOI: 10.13872/j.1000-0275.2017.0048
8. Li, X. B. (2017) The Study on the Differentiation of Technical Demandof Agricultural Operator. Journal of South China Agricultural University (Social Science Edition), 16: 87-94. DOI: 10.7671/j.issn.1672 -0202. 2017.03.009
9. Qiao, D., Lu, Q., Xu, T., Zhao, M. J. (2017) Informational channel, learning ability and farmers'water-saving irrigationtechnology preference: Based on the investigation of the Minqin Oasi. Journal of Arid Land Resources and Environment,31: 20-24. DOI: 10.13448/j.cnki.jalre.2017.038.
10. Zhang, F. Y.， Yan, T. W.， Zhang, J. B. (2022) Effect of Social Interaction on Farmers' Adoption of Green Technologies: Evidence from 1004 House-hold Survey Data in Hubei Province. Journal of Ecology and Rural Environment, 38: 43-51. DOI: 10.19741/j.issn.1673-4831.2021.0145
11. Kuang, H. Y. (2014) Study on the relation of rural social network and agricultural technology diffusion: taking the technology diffusion of pig raising on the country G as an example. Studies in Science of Scienc, 32: 1518-1524. DOI: 10.16192/j.cnki.1003-2053.2014.10.029.

Measurement and evaluation of the comprehensive utilization efficiency of coal system industrial solid waste resources in Liupanshui City

Jingyang Zhang[a], Yanyuan Yang*, Xiaoyun Xu[b], Yating Li[c]

Beibu Gulf University, Qinzhou, China

Abstract—Liupanshui city is located in the western part of Guizhou province, the region is extremely rich in coal resources, coal industry is more concentrated, coal mining and processing and other aspects of the solid waste resources if the good use of the city of Liupanshui city will get new development opportunities. The article takes the solid waste resources of coal industry in Liupanshui city as the research object, and uses the hierarchical analysis method to measure and evaluate the factors affecting the comprehensive utilisation of the solid waste resources of coal industry in Liupanshui city. The evaluation results show that the comprehensive utilisation of coal-based industrial solid waste resources in Liupanshui city is affected by the investment amount of solid waste per unit, technological maturity, industrial policy, and friendliness to the atmosphere. Countermeasures are proposed for these four constraints in anticipation of improving the comprehensive utilisation of coal-based industrial solid waste resources in Liupanshui city.

1. Introduction

During the "13th Five-Year Plan" period, Liupanshui City has always adhered to the treatment of bulk industrial solid waste, strictly following the principle of harmlessness and resourcefulness of industrial solid waste, reducing the production of industrial solid waste, reducing the pollution of industrial solid waste to the environment, and the comprehensive utilisation of bulk industrial solid waste in the city has been greatly improved. However, solid wastes such as coal gangue, fly ash, coal slurry and coal-fired slag produced in the process of coal mining in Liupanshui City's coal-based industry seriously damage the ecological environment of Liupanshui City, and if the coal-based industrial solid waste resources can be turned into treasures, it can, to a certain extent, promote the ecological sustainable development, and promote the parallel development of environmental governance and economic development.In 2021, Liupanshui City's production of major industrial solid wastes amounted to 29,319,900 tonnes, the comprehensive utilisation volume is 18,442,900 tonnes, and the comprehensive utilisation rate is 62.04%. Based on the existing foundation, we explore the factors affecting the comprehensive utilisation of coal-based industrial solid waste in Liupanshui City, and synergistically develop ecological sustainability.

2. Literature Review

2.1. Current status and recommendations for comprehensive utilisation of coal-based industrial solid waste resources at home and abroad

Du Genjie. [1] by studying the bulk solid waste industry in China, found that the main problems existing in the comprehensive utilization of solid waste in China are the long cycle of industrialization of the comprehensive utilization of solid waste, the imperfect product market regulatory mechanism, the lack of core technology, incomplete policies, and low capital investment.Bhatt A.[2] the utilization rate of fly ash in EU countries is as high as more than 95%, and these fly ashes are mainly utilized in building materials and road construction and other In the field of infrastructure. Tuya and Yang Lei.[3] found that the current situation of China's coal ash utilisation is characterized by unbalanced regional development, large stockpiles, backward technology, imperfect policies, and small scale of waste recycling enterprises. Ren Ruichen et al.[4] concluded that the comprehensive utilisation of tailings is another effective way to deal with the fine coal slurry, and that efficient sorting of the slurry, along with the development and utilisation of kaolin and other resources, is the most effective way to deal with fine coal slurry. The efficient separation of coal slurry and the development and use of kaolin and other resources, turning

[a]840456443@qq.com, * Corresponding author: 1837432910@qq.com, [b]1738863725@qq.com, [c]3220334994@qq.com

waste into treasure, can significantly improve the economic and social benefits of coal enterprises.Wang Ruibin.[5]showed that coal slurry is the primary product of coal enterprises, and local backward areas will make coal slurry into briquettes and activated charcoal, etc., or make coal slurry gangue into construction materials. Although this method can make the solid waste resources to achieve part of the resource, but if the coal slurry gangue is stored in large quantities, the end of the fundamental problem. The development space and potential of coal slurry combustion utilisation is huge, and one of the key ways of coal slurry resource utilisation at this stage is to enhance the added value of coal slurry. Wang Huaguo.[6] through the study of coal combustion slag analysis that it has more pores, large surface area, adsorption, activation performance, the cost is not high, such as a variety of new types of purification materials, solid waste resources coal combustion slag as a raw material for the preparation of purification materials with a high degree of feasibility. Wang Yutao.[7] pointed out through research that at the present stage of China's solid waste resources gangue comprehensive utilisation has a large number of storage and emissions, the production is extremely concentrated, the comprehensive utilisation rate is low, the development of various regions is extremely unbalanced, high value-added use of the current situation of the proportion of not high. Du Jianlei et al.[8] by analysing the reasons for the change of China's bulk industrial solid waste generation and comprehensive utilization in 2021, identified the solid waste resource value of insufficient attention, solid waste energy value is seriously constrained by the policy, cross-industry information asymmetry, the waste-producing enterprises focus on stockpiling and light utilization, the lack of diversified main market-oriented project investment model, the existence of industrial policy is homogeneous and poor landing, and so on. Wang Yingnan et al.[9] proposed that coal-fired slag can be used in the treatment of lead-contaminated wastewater as a kind of chain home and efficient adsorbent.

2.2. Research on the comprehensive utilisation of coal-based industrial solid waste resources at home and abroad

Huang Dingguo et al.[10] summarised and analysed the utilisation areas and utilisation pathways of fly ash as a solid waste resource in Japan by studying the output and utilisation volume of fly ash in Japan, which mainly include cement and concrete mixing, civil engineering, construction materials, agriculture, forestry, aquaculture, and other aspects and so on. Qiao Aiping.[11] pointed out that coal-fired slag can be used for construction, preparation of chemical raw materials, treatment of wastewater purification of water quality, soil improvement and so on. Chen Tingting.[12] found that coal mining enterprises use coal gangue to generate electricity, make building materials with low economic effect, to the Inner Mongolia Jungar use of coal gangue made of a green and environmentally friendly foamed ceramic products as an example, that its future has a broad market demand and obvious economic value. Fang Genliang.[13] in-depth discussion of the generation process of fly ash, chemical properties, composition, the use of silicate cement, ordinary silicate cement blended with fly ash method, concluded that the application of fly ash in the cement industry is the way to prepare fly ash silicate cement, fly ash ultrafine cement, composite silicate cement. Guo Yanqing.[14] found through his research that: fly ash used to make building materials is the low cost and most economical way, made into fertiliser helps to improve the soil, made into adsorbent can be used for sewage treatment. Zhu Hongzheng.[15] used mullite porous ceramics prepared from coal solid waste resources such as coal gangue and coal slurry, manipulated the particle size and addition ratio of coal slurry and wood chips to achieve the pore structure building, and after controlling the sintering temperature and adding glass powder to achieve the optimisation of the pore structure and enhanced mechanical properties. Dong Chenxi et al.[16] developed a new type of curing material using coal-based solid waste gangue, coal-based partial high territory joint arid zone easily accessible solid waste material calcium carbide slag synergistically with ordinary silicate cement.Md. Mohib-Ul-Haque Khan.[18]conducted a sensitivity analysis using Parkland County and its surrounding counties in Alberta, Canada as an example to assess the impact of key techno-economic parameters on the calculation results.Warmadewanthi I.D.A.A.[19] studied the type of biological cover (old and young refuse) and thickness (15, 20, 30 cm) as variables.The results showed that the differences of methane gas and leachate reduction in 15 vs 20 cm bio-cover thickness were significant.

In summary, domestic and foreign research on the current situation of comprehensive utilisation of coal-based industrial solid waste resources and utilisation pathways is very rich, and most scholars conclude from their research that the comprehensive utilisation of coal-based industrial solid waste resources can be applied in the fields of chemical raw materials, construction materials, agriculture, etc. Most scholars conduct research on the gangue and fly ash in power plants or take the regional coal mining industry as the object of research to study the comprehensive utilisation pathways of its solid waste resources, and there are fewer researches on the object of research and evaluation research of the solid waste resources of the coal-based industries, and there is still room for supplementation.

3. Overview of the study area

Liupanshui City is a prefecture-level city in Guizhou Province, located in the Guizhou Department, located in the Yangtze River and Pearl River Basin. At the same time, it is also the junction of four provinces or districts, namely, Sichuan, Yunnan, Guizhou and Guangxi Zhuang Autonomous Region, and has always been called "the crossroads of the four provinces", and has been included in the national "One Belt, One Road" construction and the Yangtze River Economic Belt plan, as it is one of the 66 regional circulation nodes and 196 highway traffic centre cities. It is also included in the national "Belt and Road" construction and the planning of the Yangtze River Economic Belt. Liupanshui city's traffic and location advantages are unique, relying on this good geographical location advantages, Liupanshui city vigorously develop transport, build to the public railway, water and navigation four major transport comprehensive three-dimensional transport network, for the subsequent industrial development of Liupanshui city laid the foundation of transport convenience.

In terms of resource endowment, Liupanshui City, rich in coal resources, in the country's "three lines" construction period, Liupanshui City, as a city of energy and raw materials industry gradually developed. According to statistics, Liupanshui City, coal resources amounted to 24.171 billion tonnes, ranking second in the province, with abundant reserves, coal species, quality and other characteristics, is the most huge coking coal base south of the Yangtze River, has always been "Jiangnan coal capital" title. Since the "13th Five-Year" period, Liupanshui has always been adhering to two concepts, which are "based on coal, do enough coal, not only coal" and "two-handedly, both promote", which is Liupanshui City Energy This is an all-round, deep and historic revolution in the field of energy in Liupanshui City. Various documents and policies introduced during this period have improved the industrial chain structure and promoted industrial transformation and upgrading to a certain extent, actively promoting the development of traditional coal industry to ecological development, the development of speciality coal industry to large-scale development, and the development of emerging coal industry to high-end development, and successfully obtaining the title of the second batch of national industrial transformation and upgrading pilots, which has brought Liupanshui a major strategic task and a great opportunity for development, and made its transformation and development into a brand new period.

During the "13th Five-Year Plan" period and the "14th Five-Year Plan" period, Liupanshui City has continuously increased the investment in the comprehensive utilisation of coal-based industrial solid waste, and has continued to make efforts in terms of policies, measures and platforms to optimise the development environment of the industrial solid waste utilisation industry and guide the comprehensive utilisation of coal-based industrial solid waste industry in the city. Comprehensive industrial solid waste industry to guide the city's comprehensive and healthy development; on the other hand, from the technical aspect to strengthen the development of industrial solid waste synthesis and utilisation industry, one is to strengthen the scientific and technological innovation planning, to set up a platform for the technological innovation of industrial solid waste treatment, and the second is to increase the financial investment on its scientific research, and continuously promote the comprehensive utilisation technology of coal gangue, fly ash and other industrial solid waste resources. These measures have also achieved considerable results, such as Panzhou City from the gangue in the successful extraction of titanium, iron, aluminium and other minerals, the success of the comprehensive utilization rate of the gangue to open to more than 80%, not only expand the scale of treatment of gangue, but also to make use of the technology of the high value of the breakthrough.

Generally speaking, Liupanshui City, coal-based industrial solid waste resources comprehensive utilisation progress has been fruitful, but there are still some of the more prominent problems. These problems are mainly manifested in: first, Liupanshui City, the use of coal-based industrial solid waste resources of the number of enterprises, most of the existing enterprises have a smaller scale of development, development quality and the benefits produced by the enterprise are not high, and most of them are relatively dispersed; second, most of the waste enterprises in the field of building materials industry in the low-end segment, the lack of specialised technical level and technological innovation, which also formed the majority of the Third, the city's waste enterprises have a small product market, waste enterprises have a low level of technology and equipment, the degree of integration of the "two" is not enough, and the technological content of the products is low; fourth, the city's waste enterprises and industrial solid waste emission enterprises have a low degree of coupling coordination and integration; fourth, the city's waste enterprises and industrial solid waste emission enterprises have a low degree of coupling coordination and integration. Fourthly, the degree of coordination and integration between the city's waste enterprises and industrial solid waste emission enterprises is low, the development of industrial clustering is not enough to support the city's industrial chain innovation and entrepreneurship ability is insufficient; Fifthly, the city's waste enterprises are not competitive in the market, the development of the city's waste enterprises are still experiencing operational difficulties, financing difficulties, labour difficulties, electricity costs and logistics costs are not low and other problems, the integration of the support system for innovation urgently needs to be set up, the service system needs to be perfected, the development of the environment needs to be further optimized.

4. Construction of Evaluation Indicator System for Comprehensive Utilisation of Coal-based Industrial Solid Waste Resources in Liupanshui City

4.1. Evaluation index selection basis

The author through the China Knowledge Network, Wipo database and other channels, collects industrial solid waste resources comprehensive utilisation of related literature and carried out research on the evaluation index system of the impact factors, and finally summarised the research results as shown in Table 1.

Table 1. The main research results of some scholars on the comprehensive utilisation of coal-based industrial solid waste resources and evaluation index system of influencing factors

Serial number	ArticleYear	Author	Name of the article	Reference Note
NO.1	2013	Wang qing, Chen anyan, Lan mingzhang , et al	Hierarchical analysis method for comprehensive evaluation of solid waste comprehensive utilisation technology	The article selects technical, economic, social and environmental benefits as the first-level indicators of hierarchical analysis method, and evaluates the comprehensive utilisation of industrial solid waste technology with 12 indicators as the second-level indicators, such as technological maturity, investment per unit of solid waste, operating cost per unit of solid waste, direct benefit per unit of solid waste, utilisation rate of solid waste resources, and friendliness to the weather, water environment, and soil environment.
NO.2	2014	Zhang jiansong, Huang jin, Lin ling	Research on Product Quality and Safety Evaluation Indicator System for Industrial Solid Waste Resource Utilisation	This thesis establishes a quantitative indicator evaluation system for the quality and safety of solid waste resources and products, which mainly contains four first-level indicators for energy, resources, environment and products, and 15 second-level indicators.
NO.3	2022	Wu na, Zhang xiaomin, Wu jia, et al.	Construction of Multi-dimensional Evaluation System for Solid Waste Resource Attributes in Non-ferrous Metal Industry--Tin Tailings as an Example	The paper selected six first-level indicators, including attributes such as material attributes, energy attributes and environmental attributes, and 12 second-level indicators such as solid waste generation, average income from solid waste resourcing technology, and average cost of solid waste resourcing inputs to evaluate solid waste resource utilisation in the metallurgical industry.
NO.4	2022	Qiu juanjuan, Zhai wenchao, Cui chengyue	Hierarchical analysis method for comprehensive evaluation of industrial solid waste comprehensive utilisation technology analysis	The paper uses the hierarchical analysis method to carry out a comprehensive evaluation of the comprehensive utilisation of industrial solid waste technology, which contains four first-level indicators and 12 second-level indicators in various aspects such as technology, investment, cost and environment.

4.2. Establishment of the evaluation indicator system

Referring to the main research results of the scholars in Table 1 above, the article determines that the target layer is the comprehensive utilisation of coal-based industrial solid waste resources in Liupanshui City, and the guideline layer is divided into four layers of economic benefit factors, technological factors, policy factors, and social and environmental benefit factors, with the amount of investment per unit of solid waste there is, the operating cost per unit of solid waste, the direct benefit per unit of solid waste, the added value per unit of solid waste, the maturity of technology, and the technological sophistication, Introduction of technical personnel and other 14 indicator layers. Detailed indicators as shown in Figure 1 Liupanshui City, coal-based industrial solid waste resources comprehensive utilisation impact factor evaluation index system.

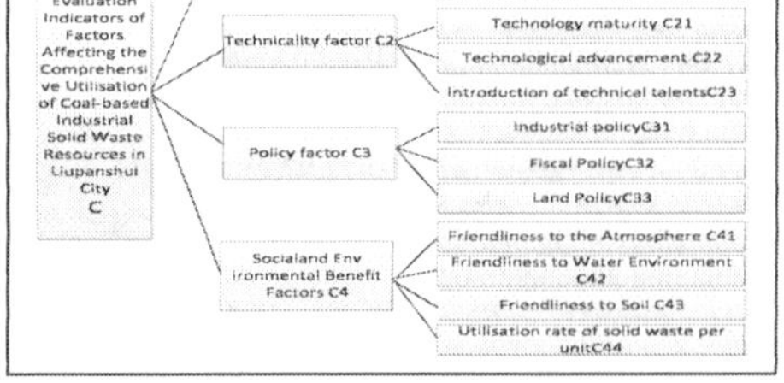

Figure 1. Evaluation index system for influencing factors of comprehensive utilisation of coal-based industrial solid waste resources in Liupanshui city

4.3. Evaluation methodology selection

4.3.1 Introduction to Hierarchical Analysis

Hierarchical analysis was introduced by Satie in 1970. This method gives priority to the organisation of complex problems, sorting out the relevant elements of the problem, and then obtaining an orderly hierarchy according to the connection between the elements, so that the complex problem can be organised. On the basis of the corresponding objective reality to make a judgement, the relative importance of each level to make quantitative expression, relying on mathematical methods to determine the relative importance of the elements of each level, and finally combined with the elements of the ranking of the specific weight of each indicator, after analysis of the results of the ranking, to find out the problem to get a solution strategy[17]. The calculation steps are as follows: the first step is to construct the hierarchical structure; the second step is to construct the judgement matrix; the third step is to calculate the relative weights and make the consistency test; the fourth step is to calculate the weight vector and make the consistency test.

4.3.2 Constructing a judgement matrix

According to the hierarchical analysis method to build the stacked order structure model, to make evaluation of the same level of related factors, the upper level of related factors as a standard, and then with the help of geometric judgement criteria to compare the two factors, so as to determine the importance of the relevant factors, and on the basis of the construction of the relevant judgement matrix. First of all, the first level of factor set $U = \{C_1, C_2, C_3\} =$

{policy factors, technical factors, economic efficiency factors, social and environmental factors}, this paper adopts the "1-9" comparative scale method, which is a systematic analysis method proposed by Prof. Satie in 1970. The "1-9" scale method constructs a two-by-two comparison judgement matrix as shown in Table 2.

Table 2. Criteria for judgement matrices on a scale of 1-9

significance scale	hidden meaning
1	The two elements are equally important
3	Represents two elements where the former is slightly more important than the latter
5	Indicates that the former of two elements is more important than the latter
7	Indicates that the former of the two elements is more strongly important than the latter
9	Indicates that the former of the two elements is more important than the latter
2, 4, 6, 8	denote the compromise between the above two neighbouring judgements
from the bottom (lines on a page)	Assuming that the ratio of the importance of element i to element j is aij, the ratio of the importance between element j and element i is a_{ij} = $1/a_{ij}$

4.3.2.1 Criteria level judgement matrix for target level

In order to make our judgement matrix scientific and reasonable, we invited experts from Liupanshui Industry and Information Bureau, experts from Municipal Environmental Protection Bureau, experts from Municipal Investment Bureau, and experts from the government to make decision-making judgements, and the experts made judgement decisions on them by using their knowledge reserves, experiences and analytical judgement abilities, thus obtaining the judgement matrix in Table 3.

Table 3. Judgement matrix for level 1 indicators

target level A	B_1	B_2	B_3	B_4
B_1 Economic efficiency factors	1	2	3	5
B_2 Technical	1/2	1	2	3
B_3 Policy Factors	1/3	1/2	1	2
B_4 Social and Environmental Factors	1/5	1/3	1/2	1

In accordance with the judgement matrix in table 3, the steps for calculating the weights of the level 1 indicators are as follows:

a.The judgement matrix A of the criterion layer for the target layer:

$$A=\begin{pmatrix} 1 & 2 & 3 & 5 \\ 1/2 & 1 & 2 & 3 \\ 1/3 & 1/2 & 1 & 2 \\ 1/5 & 1/3 & 1/2 & 1 \end{pmatrix}$$

b.Normalise the matrix A:

$$A=\begin{pmatrix} 0.4918 & 0.5217 & 0.4615 & 0.4545 \\ 0.2459 & 0.2609 & 0.3077 & 0.2727 \\ 0.1639 & 0.1304 & 0.1538 & 0.1818 \\ 0.0984 & 0.0870 & 0.0769 & 0.0909 \end{pmatrix}$$

c.Compute the maximum eigenvector W^T:

$$W^T=\begin{pmatrix} 0.4824 \\ 0.2718 \\ 0.1575 \\ 0.0883 \end{pmatrix}$$

d.Calculate the largest eigenvalue λ_{max} :

$$\lambda_{max}=4.0145$$

e.Calculate CI and CR:

RI=0.89, n=4, CI=0.0048, CR=0.0054

Since CR = 0.0054 < 0.1, that is, the judgement matrix passes the consistency test.

4.3.2.2 Judgement matrix from the indicator level to the guideline level

Similarly, the judgement matrix of the indicator level for the criterion level is constructed respectively, and the weights of the first-level indicators are calculated, and the results are shown in Tables 4 to Tables 7.

Table 4. C_{1j} for C_1 judgement matrix (j=1,2,3,4)

C_1-C_{1j}	C_{11}	C_{12}	C_{13}	C_{14}	Weighting (W_{ij})
C_{11}	1	3	2	5	0.4730
C_{12}	1/3	1	1/3	2	0.1425
C_{13}	1/2	3	1	3	0.2978
C_{14}	1/5	1/2	1/3	1	0.0867

λ_{max}=4.0651, CI=0.0217, CR=0.0244<0.1, Passes a one-time test

Table 5. C_{2j} for the judgement matrix of C_2 (j=1,2,3)

C_2-C_{2j}	C_{21}	C_{22}	C_{23}	Weighting (W_{ij})
C_{21}	1	3	7	0.6687
C_{22}	1/3	1	3	0.2431
C_{23}	1/7	1/3	1	0.0882

λ_{max}=3.0037, CI=0.0035, CR=0.0061<0.1, Passes a one-time test

Table 6. C_{3j} for C_3 judgement matrix (j=1,2,3)

C_3-C_{3j}	C_{31}	C_{32}	C_{33}	Weighting (W_{ij})
C_{31}	1	5	2	0.5813
C_{32}	1/5	1	1/3	0.1096
C_{33}	1/2	3	1	0.3092

λ_{max}=3.0037, CI=0.0018, CR=0.0032<0.1, Passes a one-time test

Table 7. C_{4j} for C_4 judgement matrix (j=1,2,3,4)

C_4-C_{4j}	C_{41}	C_{42}	C_{43}	C_{44}	Weighting (W_{ij})
C_{41}	1	3	5	7	0.5579
C_{42}	1/3	1	3	5	0.2633
C_{43}	1/5	1/3	1	3	0.1219
C_{44}	1/7	1/5	1/3	1	0.0569

λ_{max}=4.1185 , CI=0.0237, CR=0.0266<0.1, Passes a one-time test

4.3.3 General ordering and one-off tests

Calculate the evaluation index of the influence factors of each indicator on the comprehensive utilisation of coal-based industrial solid waste resources in Liupanshui city in the target layer, and then carry out the total ranking, and the ranking results are shown in Table 8.

Table 8. Combined weight of each indicator in relation to the evaluation of factors affecting the comprehensive utilisation of coal-based industrial solid waste resources in Liupanshui city.

Primary Indicator	Single sort	Secondary indicators	Single sort	Overall sorting
Economic efficiency factor C_1	0.4842	Unit solid waste investment C_{11}	0.5435	0.3442
		Unit solid waste operating cost C_{12}	0.2675	0.1694
		Unit solid waste direct benefit C_{13}	0.0923	0.0585
		Unit solid waste added value C_{14}	0.0967	0.0613
Technicality factor C_2	0.2718	Technology maturity C_{21}	0.6687	0.1818
		Technological advancement C_{22}	0.2431	0.0661
		Introduction of technical talentsC_{23}	0.0882	0.0240
Policy factor C_3	0.1575	Industrial policyC_{31}	0.5813	0.1514
		Fiscal PolicyC_{32}	0.1096	0.0285
		Land PolicyC_{33}	0.3092	0.0805
Social and Environmental Benefit Factors C_4	0.0883	Friendliness to the Atmosphere C_{41}	0.5579	0.0592
		Friendliness to Water Environment C_{42}	0.2633	0.0280
		Friendliness to Soil C_{43}	0.1219	0.0129
		Utilisation rate of solid waste per unitC_{44}	0.0569	0.0060

4.3.4 Evaluation criteria

4.3.4.1 quantitative standard

In this paper, the coal-based industrial solid waste resources in Liupanshui City as a research object, to explore the degree of influence of the factors affecting the comprehensive utilisation of industrial solid waste resources, we refer to the relevant research results at home and abroad, to get a clear standard, to determine the qualitative and quantitative standards of the selected indicators into the Table 9 shows.

Table 9. Liupanshui City coal-based industrial solid waste resources comprehensive utilisation impact factor evaluation index quantitative treatment standard

Indicators	Quantitative criteria				
	0-2 points	2-4 points	4-6 points	6-8 points	8-10 points
Unit solid waste investment C_{11}	≤50	50-100	150-200	200-250	≥250
Unit solid waste operating cost C_{12}	≤25	25-50	50-100	100-150	≥150
Unit solid waste direct benefit C_{13}	Few	Less	Average	More	A lot
Unit solid waste added value C_{14}	Few	Less	Average	More	A lot
Technology maturity C_{21}	Very weak	Weak	Average	Stronger	Exceptionally strong
Technological advancement C_{22}	Very weak	Weak	Average	Stronger	Exceptionally strong
Introduction of technical talentsC_{23}	Few	Less	Average	More	A lot
Industrial policyC_{31}	Not supportive	Not very supportive	Fairly supportive	More Supportive	Support
Fiscal PolicyC_{32}	Not supportive	Not very supportive	Fairly supportive	More Supportive	Support
Land PolicyC_{33}	Not supportive	Not very supportive	Fairly supportive	More Supportive	Support
Friendliness to the Atmosphere C_{41}	Not good	Not so good	Fair	Better	Very good
Friendliness to Water Environment C_{42}	Not good	Not so good	Fair	Better	Very good
Friendliness to Soil C_{43}	Not good	Not so good	Fair	Better	Very good
Utilisation rate of solid waste per unitC_{44}	Very low	Low	Average	Higher	Very high

In this paper, the comprehensive index will be used to evaluate the comprehensive utilisation of coal-based industrial solid waste resources in Liupanshui city, using the formula $EI = \sum_{i=1}^{n} EI_i \times W_i$

EIi in the formula indicates the score of each index in the evaluation index system of influence factors of comprehensive utilisation of coal-based solid waste resources in Liupanshui city, and Wi represents the weight of the index.

When $0 \leq EI \leq 2$ points, it indicates that the degree of influence of comprehensive utilisation of coal-based industrial solid waste resources is weak; when $2 \leq EI \leq 4$ points, it indicates that the degree of influence of comprehensive utilisation of coal-based industrial solid waste resources is weak; $4 \leq EI \leq 6$ points, it indicates that the degree of influence of comprehensive utilisation of coal-based industrial solid waste resources is general; $6 \leq EI \leq 8$ points, it indicates that the degree of influence of comprehensive utilisation of coal-based industrial solid waste resources is strong; 8 ≤ EI ≤ 10 points, it indicates that the degree of influence of comprehensive utilisation of coal-based industrial solid waste resources is strong. EI ≤ 10 points, indicating that the comprehensive utilisation of coal-based industrial solid waste resources has a strong degree of influence.

4.3.4.2 Evaluation data and quantitative scores

This paper combines the relevant situation and data collection on the factors affecting the comprehensive use of coal-based industrial solid waste resources in Liupanshui City, analyses and calculates the indicators, and quantifies the qualitative and quantitative indicators, as shown in Table 10 below.

Table 10. Quantitative scores of evaluation indicators for impact factors of comprehensive utilisation of coal-based industrial solid waste resources in Liupanshui city

Indicator Layer	gauge	Quantitative criteria	EIi	EIi·Wi
Unit solid waste investment C_{11}	From《 the 2021 Liupanshui City Statistical Yearbook》, we know that the amount of fixed asset investment in Liupanshui in 2020 will be 197.188 billion yuan;《The 14th Five-Year Development Plan of Liupanshui City》	≥250	6	2.0652
Unit solid waste operating cost C_{12}	《Development Plan of Liupanshui City for the 14th Five-Year Plan》	100-150	5	0.847
Unit solid waste direct benefit C_{13}	《Implementation Programme on Accelerating the Comprehensive Utilisation of Bulk Industrial Solid Waste》	usual	4	0.234
Unit solid waste added value C_{14}	《Implementation Programme of the People's Government of Liupanshui City on Accelerating the Comprehensive Utilisation of Bulk Industrial Solid Waste》	usual	5	0.3065
Technology maturity C_{21}	《Implementation Programme on Accelerating the Comprehensive Utilisation of Bulk Industrial Solid Waste》	usual	5	0.909
Technological advancement C_{22}	《Implementation Programme on Accelerating the Comprehensive Utilisation of Bulk Industrial Solid Waste》,《Implementation Programme for the Construction of a Demonstration Base for the Comprehensive Utilisation of Bulk Solid Waste in Liupanshui City (2021-2025)》	usual	4	0.2644
Introduction of technical talentsC_{23}	《Measures for the Introduction of High-level Talents in Liupanshui City》,《Service Guidelines for High-level Talents in Liupanshui City》	usual	5	0.12
Industrial policyC_{31}	《Planning for the Comprehensive Utilisation of Coal-based Solid Waste in Liupanshui City》	General Support	6	0.9084
Fiscal PolicyC_{32}	《Programme for the Implementation of the "Reform of Hundreds of Enterprises Project to Promote Industrial Transformation and Upgrading in Liupanshui City》《the Implementation Programme on Accelerating the Comprehensive Utilisation of Bulk Industrial Solid Wastes》	General Support	5	0.1425
Land PolicyC_{33}	《Programme for the Implementation of the "Reform of Hundreds of Enterprises Project to Promote Industrial Transformation and Upgrading in Liupanshui City》《 the Implementation Programme on Accelerating the Comprehensive Utilisation of Bulk Industrial Solid Wastes》	General Support	4	0.322
Friendliness to the Atmosphere C_{41}	《Implementation Programme on Accelerating the Comprehensive Utilisation of Bulk Industrial Solid Waste》	usual	6	0.3552
Friendliness to Water Environment C_{42}	《Implementation Programme on Accelerating the Comprehensive Utilisation of Bulk Industrial Solid Waste》	usual	5	0.14
Friendliness to Soil C_{43}	《Implementation Programme on Accelerating the Comprehensive Utilisation of Bulk Industrial Solid Waste》	preferably	6	0.0774
Utilisation rate of solid waste per unitC_{44}	《The 14th Five-Year Plan for the Development of Industry in Liupanshui City》	preferably	6	0.036

4.3.4.3 Discussion of evaluation findings

According to the quantitative score of the evaluation index of comprehensive utilization of industrial solid waste resources in Liupanshui coal system in Table 10, the score of each index will be calculated and analysed and the evaluation results will be analysed. The economic efficiency factor index scores 3.4527 points, and the constraints are added value per unit of solid waste and direct benefit per unit of solid waste; the technical factor index scores 1.2934 points, and the constraints are technological sophistication and introduction of technical talents; the policy factor index scores 1.3729 points, and the constraints are financial and tax policies; the social and environmental efficiency factor index scores 0.6086 points, and the constraints are the degree of friendliness to the soil and the unit solid waste utilisation rate; the comprehensive evaluation index scored 6.7276 points. It can be seen that the economic benefit factor is the main factor affecting the comprehensive utilisation of coal-based industrial solid waste resources in Liupanshui City, followed by policy and technical factors, and the social and environmental benefit factor has the least influence. From the evaluation results of the secondary factors in Table 8, it can be seen that from the perspective of economic efficiency factors, the influence of unit solid waste investment amount factors is the largest, and the influence of unit solid waste direct benefit factors is the smallest; from the perspective of technical factors, the influence of technical maturity factors is larger, and the influence of introducing technical talents factors is the smallest; from the perspective of policy factors, the influence of industrial policy factors is the largest, and the influence of fiscal policy factors is the smallest; from the perspective of social and environmental benefits, the influence of social and environmental benefits factors is the largest. From the perspective of social and environmental benefits, the factors of atmospheric friendliness have the greatest impact, and the factors of unit solid waste utilisation have the least impact. Thus, it is known that the investment amount of solid waste per unit, the maturity of technology, industrial policy, and the friendliness to the atmosphere are the key factors affecting the comprehensive utilisation of coal-based industrial solid waste resources in Liupanshui city. From the analysis, it can be seen that the quantitative score of the evaluation of factors affecting the comprehensive utilisation of coal-based industrial solid waste resources in Liupanshui city is 6.7276, which belongs to the category of good development.

5. Conclusions

The study shows that the comprehensive utilisation of coal-based industrial solid waste resources in Liupanshui city belongs to the category of good development, with a better development foundation and greater development prospects. Among the influencing factors of the comprehensive utilisation of coal-based industrial solid waste resources, the economic benefit factor is the most important factor affecting the efficiency of the comprehensive utilisation of coal-based industrial solid waste resources in Liupanshui City, followed by the policy and technical factors, and the social and environmental benefit factors have the least influence. Among the secondary indicators, unit solid waste investment amount, technology maturity, industrial policy, and friendliness to the atmosphere are the key factors affecting the comprehensive utilisation of coal-based industrial solid waste resources in Liupanshui city. There are the following suggestions for this:

5.1. Build a multi-faceted investment and financing system to enhance financial security

The government of Liupanshui City should actively apply for more special funds from the national or provincial level, and increase the financial support for the comprehensive utilisation of solid waste resources in the coal industry, to solve the problem of shortage of funds for enterprises and to ensure the normal operation of these enterprises. Can be created through the development of solid waste resources comprehensive utilisation of development guidance fund, to lead the social capital investment to the comprehensive utilisation of solid waste resources enterprises; improve the financial support for the comprehensive utilisation of solid waste resources enterprise development system, improve the capital turnover system; to promote the green solid waste resources utilisation of enterprise project financing library construction, improve the financial system, increase the supply of financial resources, enhance the security of funds.

5.2. Improve the maturity of technology and promote the comprehensive utilisation of solid waste resources in the coal industry

On the basis of these achievements in the "13th Five-Year Plan" period, the Liupanshui municipal government should unite colleges and universities and scientific research institutes, take the road of cooperation between industry, academia and research institutes, and continue to research and forge ahead to overcome the bottleneck technology limiting the comprehensive utilisation of industrial solid waste of the coal system in Liupanshui city, improve the maturity of the technology, promote the comprehensive utilisation of solid waste resources in Liupanshui, and accelerate the "waste to treasure". "Waste to treasure". The improvement of technical maturity is inseparable from technical personnel, Liupanshui City, according to the needs of the comprehensive utilisation of coal-based industrial solid waste resources enterprises, and actively introduce various types of high-level technical personnel, innovative technology teams, scientific research teams, etc., to grow the original talent team of Liupanshui City, solid waste enterprises, to improve the level of scientific and technological innovation, and provide technical support for the comprehensive utilisation of coal-based industrial solid waste resources technology.

5.3. Strengthen policy propaganda, strengthen the implementation of industrial policy

Meticulous practice of national resources comprehensive utilisation of the relevant superior policies, the state actively incentivise the enterprise itself to carry out the evaluation of the comprehensive utilisation of solid waste resources of the enterprise, encourage eligible enterprises to declare the tax reduction and exemption preferential policies, encourage the research and development of new products of the comprehensive utilisation of solid waste resources, the development and promotion of new technologies and technological improvements, key demonstration project construction. In addition to this, the Liupanshui municipal government can also introduce corresponding tax incentives and reward policies. In the tax aspect of tax reduction and tax cuts, improve the coal industry solid waste resources comprehensive utilisation enterprises live, in the policy support, the industrial revitalisation; in the reward aspect, if the enterprise can deal with the relationship between the solid waste in the environment, for example, piles of solid waste resources occupies an area of small, then the rewards for this type of enterprise, in fact, it is a kind of investment in the comprehensive utilisation of solid waste resources enterprises.

5.4. Promote the green transformation of solid waste utilisation and build a healthy ecological environment

The government should encourage the comprehensive utilisation of solid waste resources enterprises to actively carry out their own green manufacturing production and comprehensive utilisation of solid waste resources and other evaluations, and promote the implementation of standardised management of key solid waste resources utilisation and comprehensive utilisation enterprises, and actively participate in the creation of national as well as provincial-level green factories, green development of products and green industrial parks, and promote the development of the comprehensive utilisation of solid waste resources of the coal system industry in Liupanshui City, and build a healthy ecological environment.

Acknowledgment

This paper is supported by Collaborative Education Project for Industry and Education Cooperation of the Ministry of Education of the People's Republic of China (Project No.: 220601766021213).

References

1. Du, G J. (2019)The main problems of bulk industrial solid waste industry in China. Resource Recycling,208:34-36.
2. Bhatt, A.,Priyadarshini ,S., Mohanakrishnan, A A.,et al.(2019)Physical, chemical,and geotechnical properties of coal fly ash:A global review. Studies in Construction Materials,11:1-11.
3. Tu,Y.,Yang,L.(2020)Development status and suggestions of comprehensive utilisation of fly ash. Chemical Management,555:11-12.
4. Ren,R C.,Hu X M.,Li,C X., et al.(2021) Current status of power coal sludge treatment and utilisation.J. Journal of Guizhou University (Natural Science Edition),38:1-6.
5. Wang, R B. (2021)Current status and prospect of resource utilisation of coal sludge combustion. Clean Coal Technology,27:23-26.
6. Wang,H G. (2021)Physicochemical properties of coal-fired slag and feasibility study of its use in wastewater treatment. Energy and Energy Conservation,190:105-106.
7. Wang,Y T. (2022)Current situation and prospect of harmless disposal and comprehensive utilisation of coal gangue solid waste[J]. Coalfield Geology and Exploration,50:54-66.
8. Du, J L., Du, G J., Gao, J Y., et al. (2023)Comprehensive Utilisation of Bulk Industrial Solid Waste in China. Modern Mining Industry,39:23-26.
9. Wang,Y N., Jiang,L L., Huang J.,et al.(2023) Adsorption of lead in water by coal-fired slag. Inorganic Salt Industry,2023:1-9.
10. Huang, D G.,Wu, Y M.,Wang, W H.(2006) Comprehensive utilisation of fly ash in Japan. Clean Coal Technology,2006(02):92-95.
11. Qiao, A P. (2019)Introduction to coal-fired boiler slag treatment method. Shanxi Science and Technology,34(03):119-121.
12. Chen, T T. (2020)Exploring the economic value of coal mine solid waste resource utilisation. Environment and Development,32(06):255-256.
13. Fang, G L. (2020)Comprehensive utilisation of fly ash in cement industry. Sichuan Cement,288(08):4+8.
14. Guo, Y Q. (2020)Analysis of comprehensive utilisation of fly ash under new situatio. Energy and Energy Conservation,178(07):191-192.
15. Zhu, H Z.(2021)Anhui Technology:New progress in high value utilisation of coal-based solid waste. China Mining News,2021-10-26(004).
16. Dong, C X.,Wang, N Q.,L, J S., et al.(2023) Mechanical characteristics and mechanism of saline soil cured by coal solidification.Coalfield Geology and Exploration,2023:1-9.
17. Xia, S B.,Li, Z M., Zhuo Y.(2016) Analysis of current situation and outlook of comprehensive utilisation of industrial solid waste resources. Low Carbon World,2016:18-19.
18. Md. Mohib-Ul-Haque Khan,Siddharth Jain,Mahdi Vaezi,Amit Kumar. Development of a decision model for the techno-economic assessment of municipal solid waste utilization pathways[J]. Waste Management,2016,48.
19. Warmadewanthi I.D.A.A.,Chrystiadini Gian,Kurniawan Setyo Budi,Abdullah Siti Rozaimah Sheikh. Impact of degraded solid waste utilization as a daily cover for landfill on the formation of methane and leachate[J]. Bioresource Technology Reports,2021,(prepublish).

Mathematical model and dynamic simulink simulation of PEM electrolyzer system

Danxiong Fei[a], Wenwen Fan[b], Zhenlan Dou[c*], Chunyan Zhang[d]

State Grid Shanghai Municipal Electric Power Company of Fengxian, Shanghai, China

Abstract. Hydrogen, being the most abundant element in the universe, holds great promise as an energy carrier for decarbonizing various economic sectors. In particular, green hydrogen production through water electrolysis is essential for achieving this goal, with polymer electrolyte membrane (PEM) water electrolyzers playing a crucial role. PEM water electrolyzers are known for their rapid response, enabling them to effectively adapt to fluctuations in renewable energy sources. Nevertheless, rapid load changes can result in the rapid build-up of heat within the electrolytic cell, leading to a sharp increase in temperature and potentially harming the cell. To address this challenge, we developed an electrolysis water system model using MATLAB and validated its accuracy through experiments. This model allowed us to explore the factors influencing stack temperature and propose a fast and secure dynamic process control strategy. By laying the groundwork for subsequent control studies on PEMEC (Proton Exchange Membrane Electrolysis Cell) stacks and systems, this research facilitates further progress in their control and regulation.

1. Introduction

In the 21st century, humanity faces one of its most significant challenges: mitigating greenhouse gas emissions to curb climate change while meeting the rising energy demands to support economic growth. Renewable energy technology emerges as a crucial solution to address global environmental pollution issues and tackle the energy shortag[1]. Energy sources such as tidal energy, solar energy, and wind energy are location-dependent, intermittent, and hence, less reliable[2]. Hence, efficient energy storage plays a vital role in enabling widespread and dependable utilization of renewable energy technology. Hydrogen serves as an environmentally friendly fuel option[3]. Utilizing surplus renewable power to operate water electrolyzers for water-splitting represents a promising approach for storing excess renewable energy[4].

In recent years, much of the research has been centered around proton exchange membrane (PEM) electrolysis technology due to its numerous advantages over alkaline electrolyzer cells. These advantages include higher current density, enhanced hydrogen purity, non-hazardous liquid electrolyte, and most notably, a rapid transient response, making it highly suitable for applications in renewable energy systems.[5].

There are numerous experimental studies investigating the PEM electrolyzers in different aspects[6]. Indeed, models play a vital role in electrolyzer development as they provide valuable insights into the impact of various parameters on electrolyzer performance, enabling efficient simulation, design, and optimization of electrolyzer systems. Moreover, experimental data collection can be challenging, especially in complex systems or specific conditions. To tackle the dynamic challenges of connecting an intermittent electrical source to an electrolysis system, modeling emerges as a critical and powerful tool for understanding phenomena, conducting control analysis, sizing components, managing energy, and optimizing system performance.[7].

Concerning the state-of-the-art PEM electrolyzer modeling, while there are well-established simulations available in the literature, there remains a limited number of dynamic modeling studies specifically related to PEM electrolysis. Á. Hernández-Gómez[8] constructed an equivalent circuit model for the electrolytic cell and studied its dynamic response process. They accurately simulated the voltage dynamic behavior of the electrolytic cell. Math model also was developed by many works[9]. Models have been perfect simulated the effects of pressure, temperature and current density in stack[10]. Transient operations pose higher demands on system control, and models serve as effective tools for studying dynamic system operations. Therefore, dynamic PEM water electrolysis has sparked considerable interest among researchers. Haluk Görgün explored a model to predict the stack voltage, gas tank pressure and stack efficiency under dynamic operation[7b]. Furthermore, the literature also presents more complex models that consider mass transfer and multi-physics field coupling[11].

[a]feidanxiong@sh.sgcc.com.cn, [b]fanwenwen@sh.sgcc.com.cn, [c*]douzhl@126.com, [d]zhcytongji@126.com

Although many researchers have made extensive efforts in modeling PEM water electrolysis, there has been limited research on the extreme conditions experienced by the electrolytic cell under fluctuating inputs. Investigating extreme conditions is crucial for ensuring the safe operation of the electrolytic cell. To assess the performance of a PEM electrolyzer system, gain insights into the dynamic response characteristics of the system and its individual components, and design superior systems for the future, we developed a dynamic model of a PEM electrolyzer system using MATLAB. This model was utilized to study the system's response patterns under extreme fluctuating conditions. Additionally, the paper proposes corresponding control strategies to effectively address these challenging scenarios.

2. Model description

2.1. Voltage

The electrolyzer voltage (V) using the following expression:

$$V = N_C(E_{Cell} + V_{ACt} + V_{Diff} + V_{Ohm}) \quad (1)$$

where Nc is the number of electrolyzer cells, E_{Cell} is the open circuit voltage V_{ACt} are the anode activation overpotentials and cathode was neglect, V_{Ohm} is the electrolyzer cell resistance (ohmic losses). Similar expressions are used by almost all the authors[12].

The open circuit voltage is typically determined using the Nernst equation, which is a widely used equation found in numerous works reported in the literature, or variations of it are utilized for the same purpose.

$$E_{cell} = E^0_{rev} + \frac{RT}{2F}[\ln(\frac{P_{H_2}P_{H_2}{}^{1/2}}{P_{H_2O}})] \quad (2)$$

where P is the partial pressure of reactants/products, T is the temperature, F is the Faraday constant and E^0_{rev} is the reversible cell potential at standard temperature and pressure. It is calculated by

$$E^0_{rev} = 1.229 - 0.9 \times 10^{-3}(T - 298) \quad (3)$$

Activation losses occur, as mentioned, due to the need to allocate some potential for activating the electrochemical reactions occurring at the anode and cathode sides. One of the more commonly used expressions for calculating activation overpotential is based on the Butler-Volmer equation.

$$V_{Act} = \frac{RT_a}{\alpha_a F}\text{arcsin}(\frac{i}{2i_{0,a}}) + \frac{RT_c}{\alpha_c F}\text{arcsinh}(\frac{i}{2i_{0,c}}) \quad (4)$$

where $i_{0,a}$ and $i_{0,c}$ are the exchange current density at anode and cathode, respectively and α_a and α_c are the charge transfer coefficients at anode and cathode, respectively. i_0 increase with temperature, Therefore, to relate it with the temperature, an expression is presented by applying an Arrhenius expression, as follows:

$$i_0 = i_{0,ref}exp[-\frac{E_{act}}{R}(\frac{1}{T} - \frac{1}{T_{ref}})] \quad (5)$$

where $E_{act}$is the activation energy of the electrode and $i_{0,ref}$ is the exchange current density measured at a reference temperature T_{ref}.

M. Lebbal used an expression based on i_{lim} , The maximum current density that the electrolyzer can handle determines the maximum production rate allowed by the electrolyzer. [13]:

$$V_{Diff} = -\frac{RT}{nF}\ln(1 - \frac{i}{i_{lim}}) \quad (6)$$

The ohmic overpotential is associated with the material's resistance to the flow of protons. The standard Ohm's law is applied in the analysis of the work.

$$V_{Ohm} = RI = \frac{\delta}{\sigma}I \quad (7)$$

where δ is the material thickness and σ the material conductivity. The membrane conductivity, considering Nafion, can be expressed as

$$\sigma = \left(0.005139\lambda - 0.00326\right)\exp\left[1268\left(\frac{1}{303} - \frac{1}{T}\right)\right] \quad (8)$$

where T is the electrolyzer temperature and λ is the membrane water content[14].

2.2. Lumped parameters thermal models

Temperature positively influences the reversible potential value. Additionally, since activation, ionic diffusion, electronic diffusion, and matter transfer are thermally activated processes, temperature also has a beneficial impact on the overall cell voltage. In this study, the stack is treated as a single thermal capacitance (lumped thermal capacitance), and its temperature is assumed to be uniform.

Considering the various input-output enthalpy flows on the electrolysis stack and assuming no water evaporation, the stack's enthalpy balance can be expressed as follows: [Please provide the specific equation for the enthalpy balance if available, as it was cut off in the given text[14]:

$$C_{th}\frac{dT}{dt} = \sum h_i^{in}N_i^{in} - \sum h_i^{p_j}i_i^{put} + Q_{net} \quad (9)$$

where h_i is the specific enthalpy of species i, and N_i is the molar flow rate of the corresponding species in or out of the electrolyzer. The lumped thermal capacitance or overall thermal capacity, C_{th} , is the sum of the component thermal capacities

$$C_{th} = \sum \rho_j V_j C_{p,j} \quad (10)$$

in which ρ_j, V_j, and $C_{p,j}$ are the values for density, volume, and heat capacity of each component, respectively. The net heat generation, Q_{net}, is composed of varied heat sources and sinks in the system.

$$Q_{net} = Q_{gen} - Q_{loss} - Q_{cost} - Q_{misc} \quad (11)$$

the terms described in the net heat generation of the system include Q_{gen}, Q_{loss}, Q_{cost}, Q_{misc} which describe heat generation, loss, cooling, and other miscellaneous phenomena, respectively.

3. Results and discussion

We initially conducted experiments using a 10 kW system to validate the reliability of the model. The verification was done by comparing the I-V (current-voltage) curves obtained from the experiments with the model predictions(Figure 1). This method is commonly used to assess the reliability of PEM water electrolysis models. In the 10 kW system, experiments were carried out at 50°C, and from the comparison shown in the graph, it can be concluded that the model exhibits good accuracy and provides reliable predictions.

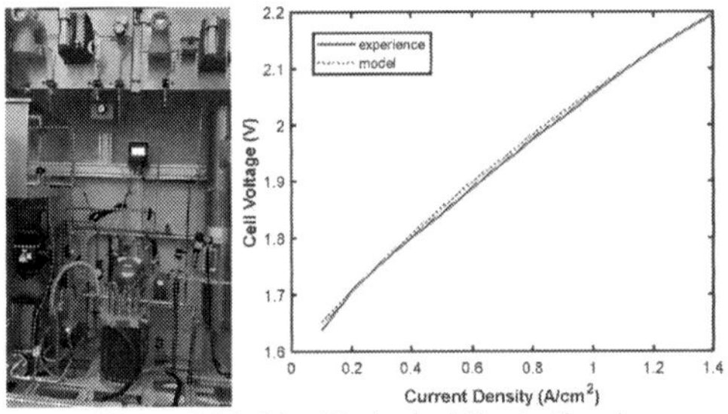

Figure 1. Model validation by 10kw test bench

Figure 2 illustrates the impact of current step changes on the electrolytic cell temperature and the electrolytic cell outlet temperature. From the graph, it can be observed that when the current density jumps from 0A/cm^2 to 1 A/cm^2, the electrolytic cell temperature rapidly increases within 5 seconds, rising from around 323K to 329K, and finally stabilizes. In contrast, the electrolytic cell outlet water temperature shows only a slight increase of approximately 1K. This is because when the current increases, the heat generation in the cell stack also increases. Unlike fuel cells that have dedicated cooling fluids, water serves as both a reactant and a coolant for the electrolytic cell. Due to slow heat transfer and the presence of thermal resistance, the heat generated in the electrolytic cell cannot be fully dissipated, leading to its accumulation inside the cell, especially on the Membrane Electrode Assembly (MEA). This heat accumulation causes the rapid rise in the electrolytic cell temperature. Therefore, when the current input suddenly increases, the outlet water temperature of the electrolytic cell rises slowly and with a smaller magnitude, while the electrolytic cell temperature rapidly increases.

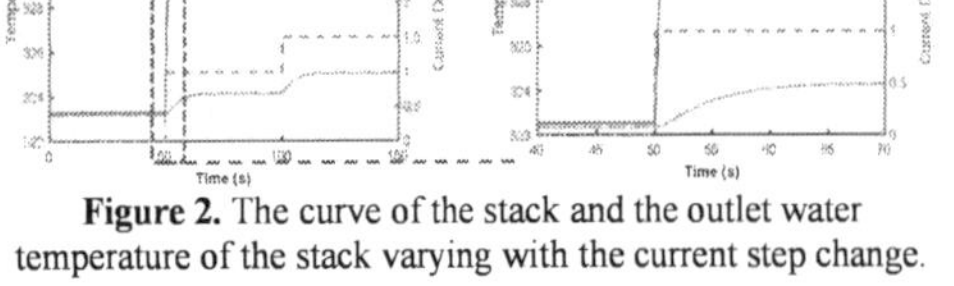
Figure 2. The curve of the stack and the outlet water temperature of the stack varying with the current step change.

Figure 3 illustrates the influence of current step changes on the voltage. The graph shows a voltage difference of 10-20mV between the periods before and after the 1 A/cm^2 step change. It can be observed that when the current undergoes a step change, the voltage instantaneously increases, followed by a slow decrease within a few seconds until it reaches a stable level. This phenomenon occurs because the electrolysis voltage responds instantaneously to the abrupt change in input current. Subsequently, within a few seconds, the electrolytic cell temperature rapidly increases, which leads to a reduction in voltage. Eventually, when the temperature stabilizes, the voltage also stabilizes, and further changes are no longer observed.

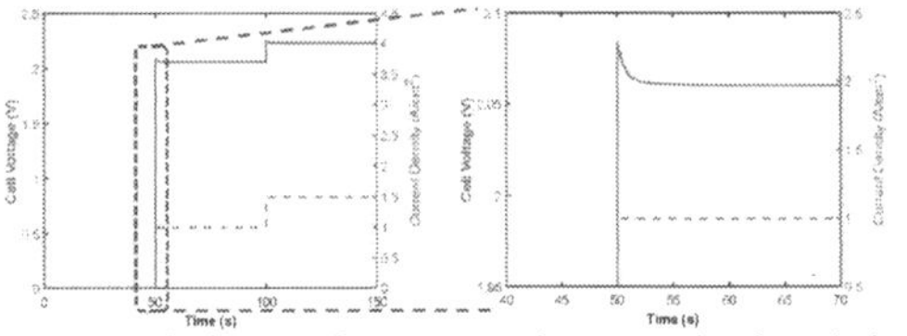
Figure 3. The impact of current step changes on the electrolytic cell voltage.

The electrolytic cell temperature significantly affects the performance and lifespan of the cell. Therefore, studying the temperature variation during fluctuations is of crucial importance. Figure 4 illustrates the trend of electrolytic cell temperature under different current step changes. It can be observed that while the step change in current density exceeds 3.5A/cm^2, the electrolytic cell temperature will rapidly increases and exceed 373K, which greatly exceeds the allowable usage temperature of the Nafion membrane. While temperature spikes are also influenced by factors such as water inflow rate and the material of the electrolytic cell, this indicates that the adaptability of the electrolytic cell to fluctuations has its limits.

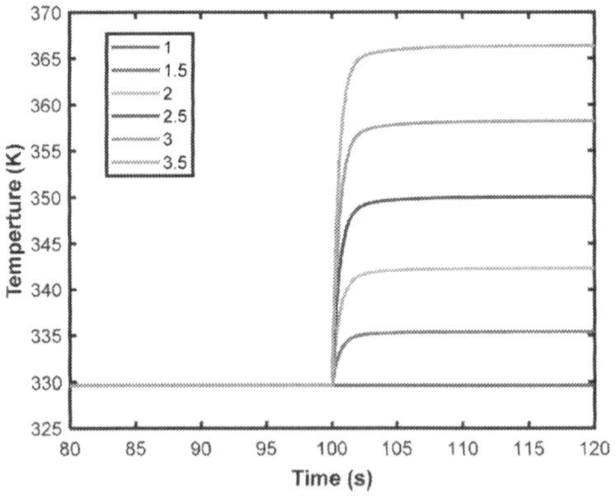

Figure 4. The effect of different step changes in current density(A/cm^2) on the electrolytic cell temperature.

Figure 5 illustrates the impact of different water inflow rates (5 kg/s, 7 kg/s, 9 kg/s) on the electrolytic cell temperature during current step changes. It can be observed that increasing the water inflow rate helps to lower the electrolytic cell temperature. However, when the water inflow rate becomes excessively high, the cooling effect of further increasing the water inflow rate becomes very limited. For instance, when the water inflow rate nearly doubles (from 5 kg/s to 9 kg/s), the change in the electrolytic cell temperature is only 2K.

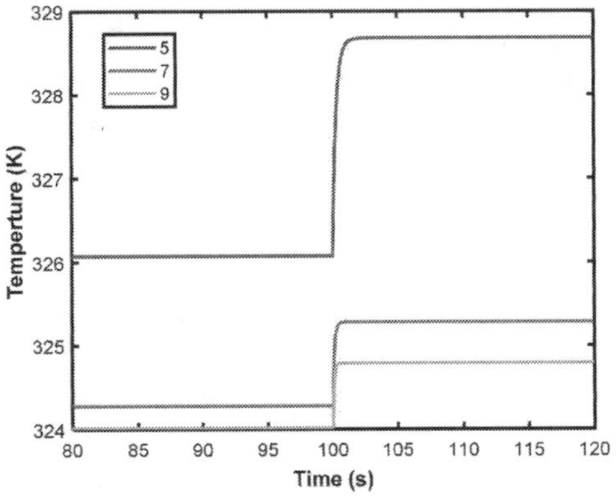

Figure 5. The effect of water inflow rate on the stack temperature(The current jumps from 0A/cm^2 to 1A/cm^2.).

Figure 6 illustrates the impact of different water inflow temperatures (293.15K, 303.15K, 313.15K, 323.15K) on the electrolytic cell temperature during current step changes. It can be observed that lowering the water inflow temperature effectively reduces the electrolytic cell temperature. Compared to adjusting the water flow rate into the electrolytic cell, reducing the water inflow temperature is more effective in mitigating the risk of excessive temperature rise in the electrolytic cell, especially when dealing with larger current fluctuations.

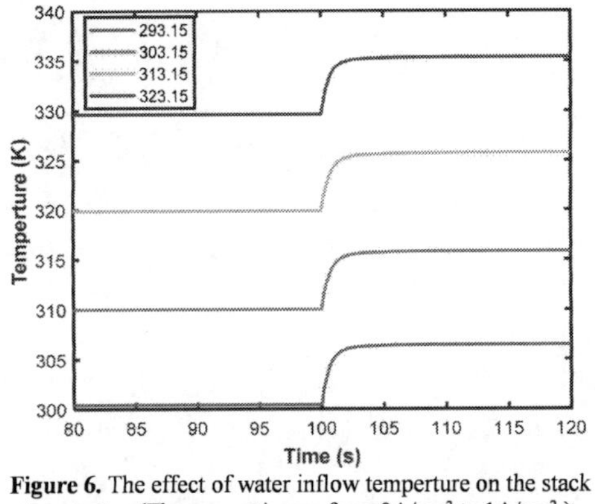

Figure 6. The effect of water inflow temperture on the stack temperature(The current jumps from 0A/cm^2 to 1A/cm^2.).

4. Conclusions

This study developed a system-level simulation model to assess the impact of current input fluctuations on the temperature of PEM water electrolytic cells. The results indicate that fluctuating input currents have a significant effect on the electrolytic cell temperature, especially when the current suddenly increases, leading to a rapid temperature rise in the electrolytic cell. Excessive current fluctuations can cause the electrolytic cell temperature to exceed the allowable temperature of the Nafion membrane, resulting in cell failure and potential safety issues.

To mitigate the rapid temperature rise caused by current step changes, increasing the water inflow can be effective, but it might be costly and less effective for small current step changes. However, for larger current step changes, the effect becomes more pronounced. Therefore, to prevent the electrolytic cell temperature from soaring and damaging the cell due to fluctuating inputs, two approaches can be considered. On one hand, setting an upper limit for the electrolytic cell input can be implemented, and on the other hand, increasing the water inflow appropriately during periods of significant current fluctuations may be necessary.

References

1. C. Intergovernmental Panel on Climate, Climate Change 2013 – The Physical Science Basis: Working Group I Contribution to the Fifth Assessment Report of the Intergovernmental Panel on Climate Change, Cambridge University Press, Cambridge, 2014.
2. aR. Clarke, S. Giddey, F. Ciacchi, S. Badwal, B. Paul, J. Andrews, International journal of hydrogen energy 2009, 34, 2531-2542; bF. Grueger, O. Hoch, J. Hartmann, M. Robinius, D. Stolten, International Journal of Hydrogen Energy 2019, 44, 4387-4397.
3. aS. Griffiths, B. K. Sovacool, J. Kim, M. Bazilian, J. M. Uratani, Energy Research & Social Science 2021, 80, 102208; bA. M. Oliveira, R. R. Beswick, Y. Yan, Current Opinion in Chemical Engineering 2021, 33, 100701.
4. aA. Boretti, B. K. Banik, Advanced Energy and Sustainability Research 2021, 2, 2100097; bM. Oertel, J. Schmitz, W. Weirich, D. Jendryssek-Neumann, R. Schulten, Chemical Engineering & Technology 1987, 10, 248-255.
5. aR. E. Clarke, S. Giddey, F. T. Ciacchi, S. P. S. Badwal, B. Paul, J. Andrews, International Journal of Hydrogen Energy 2009, 34, 2531-2542; bB. Laoun, A. Khellaf, M. W. Naceur, A. M. Kannan, International Journal of Hydrogen Energy 2016, 41, 10120-10135; cA. Mraoui, B. Benyoucef, L. Hassaine, International Journal of Hydrogen Energy 2018, 43, 3441-3450.
6. aP. Millet, in Electrochemical Technologies for Energy Storage and Conversion, 2011, pp. 383-423; bP. Millet, S. Grigoriev, in Renewable Hydrogen Technologies (Eds.: L. M. Gandía, G. Arzamendi, P. M. Diéguez), Elsevier, Amsterdam, 2013, pp. 19-41; cK. Zeng, D. Zhang, Progress in Energy and Combustion Science 2010, 36, 307-326.
7. aZ. Abdin, C. J. Webb, E. M. Gray, International Journal of Hydrogen Energy 2015, 40, 13243-

13257; bA. Awasthi, K. Scott, S. Basu, International Journal of Hydrogen Energy 2011, 36, 14779-14786; cF. d. C. Lopes, E. H. Watanabe, in 2009 Brazilian Power Electronics Conference, 2009, pp. 775-782.

8. aÁ. Hernández-Gómez, V. Ramirez, D. Guilbert, B. Saldivar, Renewable Energy 2021, 163, 1508-1522; bÁ. Hernández-Gómez, V. Ramirez, D. Guilbert, International Journal of Hydrogen Energy 2020, 45, 14625-14639.
9. aP. Colbertaldo, S. L. Gómez Aláez, S. Campanari, Energy Procedia 2017, 142, 1468-1473; bB. Han, S. M. Steen, J. Mo, F.-Y. Zhang, International Journal of Hydrogen Energy 2015, 40, 7006-7016.
10. T. M. Brown, J. Brouwer, G. S. Samuelsen, F. H. Holcomb, J. King, Journal of Power Sources 2008, 182, 240-253.
11. aM. Maier, K. Smith, J. Dodwell, G. Hinds, P. Shearing, D. Brett, International Journal of Hydrogen Energy 2022, 47, 30-56; bP. Olivier, C. Bourasseau, P. B. Bouamama, Renewable and Sustainable Energy Reviews 2017, 78, 280-300.
12. aZ. Abdin, C. Webb, E. M. Gray, International journal of hydrogen energy 2015, 40, 13243-13257; bA. M. Abomazid, N. A. El-Taweel, H. E. Farag, IEEE Transactions on Industrial Informatics 2021, 18, 5870-5881; cE. Afshari, S. Khodabakhsh, N. Jahantigh, S. Toghyani, International Journal of Hydrogen Energy 2021, 46, 11029-11040.
13. M. Lebbal, S. Lecœuche, International journal of hydrogen energy 2009, 34, 5992-5999.
14. G. S. Ogumerem, E. N. Pistikopoulos, Smart and Sustainable Manufacturing Systems 2018, 2, 25-43.

The Effect of Antibiotics on Consortium Structure in Effluent and Sludge from a Municipal Wastewater Treatment Plant

Xiangjin Li[1], Jincheng Hu[2], Chongyang Wang[1], Wangyan Ge[1*]

[1]Miami College, Henan University, Kaifeng, Henan, China.
[2]Peking Univ Shenzhen Grad Sch, Sch Environm & Energy, Shenzhen, China

Abstract. The distribution and diversity of antibiotic-resistance bacteria (ARB) from the excess sludge and effluent of Kaifeng wastewater treatment plant was studied using a culture-dependent method. In comparison with other phylum, microbial community analysis confirmed Proteobacteria was more susceptible to possess the ability to tolerate antibiotic pressure. With the addition of antibiotics from single erythromycin to three antibiotics, the relative abundance of Proteobacteria increased to 92.98%, 94.35%, 95.74% in water sample and 84.33%, 90.99%, 90.84% in sludge samples. *Acinetobacter* in Proteobacteria has stronger antibiotic resistance. This research identified that in the process of sewage treatment, attention should be paid to the removal of species in Proteobacteria to avoid potential resistance gene contamination in the receiving environments.

1 Introduction

Antibiotics have been widely used in veterinary and human medicine, as well as in agricultural disease control, due to their significant role in controlling microbial infections [1]. However, since the broad application of antibiotics, large amounts of these drugs are often overused and enter wastewater treatment plants (WWTPs) through the sewage collection system [2]. Multiple studies have reported high concentrations of antibiotics at the mg/L level within these environments [3,4]. The prolonged exposure to antibiotics creates selective pressure, leading to the widespread distribution of antibiotic resistant bacteria (ARB) and antibiotic resistance genes (ARGs) [5].

ARB are microorganisms that can withstand the effects of antibiotics. If a microorganism can resist three or more types of antibiotics, it is classified as a multidrug-resistant organism (MDRO). These genes are capable of replication and transfer. For example, Jin et al. [6] studied urban wastewater in China and found that sulfonamide-resistant bacteria had the highest proportion, and tetracycline and erythromycin resistant bacteria were also prevalent. Huang et al. [7] also detected a large number of multidrug resistant bacteria in the effluent of a sewage treatment plant.

In recent years, the occurrence of ARB has been extensively studied in WWTPs in developed countries and cities [8]. However, few studies have investigated the occurrence of ARB in underdeveloped areas and small cities [9]. Many small-scale sewage treatment plants have focused only on the removal of biological oxygen demand (BOD) and nutrients (N and P), without incorporating tertiary treatment. Secondary treatment has been found to have a relatively low effectiveness in removing ARB and ARGs. Moreover, reports suggest that WWTPs may even promote the proliferation of ARBs and the horizontal transfer of ARGs [10,11]. Therefore, the effluent and excess sludge of small-scale sewage treatment plants are expected to contain large amounts of ARBs, which pose a significant environmental risk.

This research focuses on a small-scale municipal sewage treatment plant using oxidation ditch process to identify the potential for ARB pollution. Three antibiotics (erythromycin, tetracycline, and norfloxacin) were added to effluent and excess sludge samples to select for resistant communities. The primary objectives of this study were to (1) identify the ARB present in the activated sludge and effluent water, (2) describe the diversity and abundance of bacterial communities at the selected wastewater treatment plant.

2 Materials and Methods

2.1 Sample collection

Effluent water and excess sludge samples are collected from a small municipal wastewater treatment plant located in Kaifeng. The collected water samples were stored in a large plastic bucket, while excess sludge was wrapped in black plastic. Samples are immediately transported to the

* Wangyan Ge: author@e-mail.org

laboratory in an icebox and stored at 4°C until the experiments are completed.

2.2 Reagents

The standards of antibiotics about tetracycline (Te), erythromycin (Er), norfloxacin (No) were purchased from Sinopharm Chemical Reagent co., Ltd (Shanghai, China). The DNAiso Reagent, Taq DNA polymerase, RNAiso Plus, DNase I, cDNA synthesis kit and SYBR® Premix Dimer Eraser™ were purchased from TaKaRa (Dalian, China).

2.3 Cultivation of antibiotic-resistant microbial communities

The antibiotic resistant bacterial consortium were enriched into six flasks (500mL) by SSDM as previously described [12]. Three of these flasks were added with effluent water (5mL) and the other three were added with excess sludge (1g). In order to obtain antibiotic resistant bacteria consortium, erythromycin, tetracycline and norfloxacin were selected and added into these flasks respectively. Apart from these six flasks, two flasks without antibiotics addition were also settled. The settled treatments were shown in Table 1.

Table 1 Treatment settling

Treatment	Name	Sample	Antibiotic
1	ew-D	Effluent water	None
2	ew	Effluent water	None
3	ew-e	Effluent water	Er
4	ew-e-t	Effluent water	Er+Te
5	ew-e-t-n	Effluent water	Er+Te+No
6	es-D	Excess sludge	None
7	es	Excess sludge	None
8	es-e	Excess sludge	Er
9	es-e-t	Excess sludge	Er+Te
10	es-e-t-n	Excess sludge	Er+Te+No

2.4 DNA extraction

The genome DNA were extracted from bacterial consortium ef-D, ef, ef-e, ef-e-t, ef-e-t-n, ex-D, ex, ex-e, ex-e-t, ex-e-t-n (The specific naming method is shown in Table 1). After transferred one time, using FsatDNATMSPIN Kit according to the manufacturer's instructions. The DNA was used for Illumia Miseq sequencing analysis with the V3-V4 region. The V3 and V4 region was amplified using forward primers containing the sequence "CCTACGGRRBGCASCAGKVRVG AAT" and reverse primers containing the sequence "GGACTACNVGGGTWTCTAATCC" (TISINGKE Biological technology). The raw data was controlled and analyzed using QIIME (1.9.1). All reads were classified into operational taxonomic units (OTUs) at 97% similarity.

2.5 consortium structure analysis

The ACE, Chao1, Shannon index, and Simpson index were used to evaluate the diversity of consortiums structure. The ACE index was calculated according to the following formula using R programming language (4.2.0).

$$S_{ACE} = S_{Ab} + \frac{S_{ra}}{C_{ACE}} + \frac{F_1}{C_{ACE}}\gamma^2 \quad (1)$$

Among formula 1:

S_{Ab} stands for the number of OTU above 10 reads

Sra stands for the number of OTU less than or equal to 10 reads

F_1 stands for the number of OTU with 1 reads

$$C_{ACE} = 1 - \frac{F_1}{S_{ra}}$$

$$\gamma^2 = max\left[\frac{N_{ra}\sum_{i=1}^{10} i(i-1)F_i}{C_{ACE}N_{ra}(N_{ra}-1)} - 1{,}0\right]$$

The Chao1 index was calculated using formula 2:

$$S_{chao1} = S_{obs} + \frac{F_1(F_1-1)}{2(F_2+1)} \quad (2)$$

Among formula 2:

S_{obs} stands for the number of OTUs with more than or equal to 1 reads

F_1 stands for the number of OTUs with 1 reads

F_2 stands for the number of OTUs with 2 reads

3 Results and Discussion

3.1 Compositions of microbial communities

In the case of adding of zero to three antibiotics, there was a significant change in the microbial communities in effluent samples and excess sludge samples, as depicted in Fig. 1. The figure shows the relative abundance of the different phyla (Fig. 1a) and the bacterial genus with an abundance of more than 1% (Fig. 1b), while abundances below 1% were classified as "others".

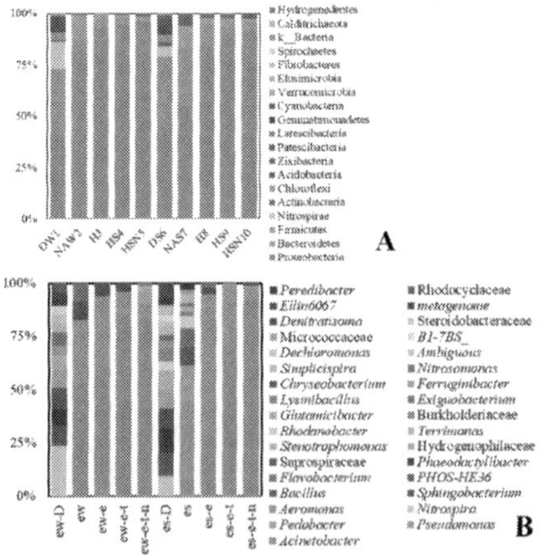

Fig. 1 Composition of microbial community structure at phylum (a) and genus (b) levels

The analysis of effluent water samples (ew-D) revealed a high diversity of genera (Shannon diversity index of 7.536), with Nitrospira (Nitrospirae), Phaeodactylibacter (Phaeophyta), Saprospiraceae (Bacteroidetes), Hydrogenophilaceae (Proteobacteria), and Terrimonas (Bacteroidetes) as dominant genera. In the absence of antibiotics, the predominant phylum was Proteobacteria, accounting for approximately 81.29% of the total in ew, with *Acinetobacter* and *Pseudomonas* being the predominant genera (80.84%). However, with the addition of antibiotics, the relative abundance of Proteobacteria increased to 92.98%, 94.35%, and 95.74% in water samples from single erythromycin to three antibiotics (erythromycin, tetracycline, norfloxacin), respectively. On the other hand, the relative abundance of Bacteroidetes decreased from 5.72% to 0.02%, while the relative abundance of *Acinetobacter* increased. Additionally, *Stenotrophomonas* had the highest abundance in ew-e-t-n (7.14%).

In excess sludge samples (es-D), there was also a high diversity of genera (Shannon diversity index of 7.696), with the dominant genera being similar to those in ew-D. In the absence of antibiotics, the predominant phylum was Proteobacteria, accounting for approximately 54.33% of the total in es, with *Acinetobacter* and Pseudomonas being the predominant genera (36.94%). However, with the addition of antibiotics, the relative abundance of Proteobacteria increased to 84.33%, 90.99%, and 90.84% from single erythromycin to three antibiotics (erythromycin, tetracycline, norfloxacin), respectively. Conversely, the relative abundance of Bacteroidetes decreased, ranging from 28.45% to 4.78%. The relative abundance of *Acinetobacter* increased. Furthermore, *Aeromonas* had the highest abundance in es-e-t Therefore, it is important to pay attention to the elimination of species belonging to the Proteobacteria in the oxidation ditch process of the wastewater treatment plant in order to avoid potential contamination of resistance genes in the receiving environment.

When multiple antibiotics were added to the effluent and sludge samples, the Nitrospirae phylum almost disappeared and the relative abundance of several bacterial phyla, including Firmicutes and Actinobacteria, significantly decreased. This may be attributed to the strong impact of the antibiotics on microbial structure.

3.2 Correlation analysis between microbial communities

Principal components analysis (PCA) is a powerful analytical tool for comparing differences between multiple samples. Based on the results presented in Fig. 2a, it can be observed that there is a significant similarity between the microbial communities of the effluent water and excess sludge samples after being cultured without antibiotics. The communities of the ew and es-e samples are almost identical, indicating that they share a high degree of taxonomic similarity. Moreover, the degree of overlap between the es-e-t and es-e-t-n samples is higher, indicating that the addition of two or more antibiotics has a minimal impact on the microbial community composition of the excess sludge samples at the phylum level. The proximity of the es-e-t-n and ew-e-t-n, as well as the es-e-t and ew-e-t samples at the genus level (Fig. 2b), supports th e conclusion that the addition of multiple antibiotics has no significant effect on the microbial community composition of both effluent water and excess sludge samples at the genus level. Notably, es are located at a significant distance from all other samples at both the phylum and genus levels, indicating a unique microbial community composition.

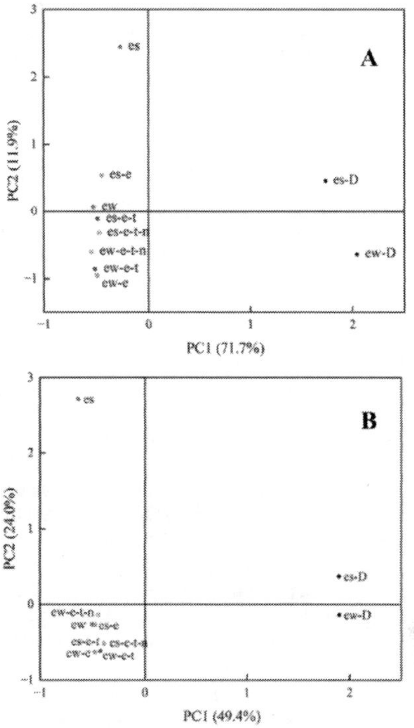

Fig. 2 The result of microbial communities PCA at phylum (a) and genus (b) levels

3.3 Diversity of microbial communities

To characterize the microbial community structures under different antibiotic pressures, Chao1, ACE, Shannon index of the community structures were calculated as shown in Table 2. As the values of these indices increase, so does the estimated total number of species and the diversity of the related consortium [13]. In this study, the addition of antibiotics caused a notable decrease in the ACE and Chao1 indices of both effluent and sludge samples. The bacterial sample of effluent water was more affected by the addition of antibiotics than excess sludge. It is believed that this is due to a higher concentration of antibiotics in the excess sludge, which could cause microbes to become less sensitive. The Shannon and Simpson indices, used to reflect the diversity of species within a community [14,15], showed a much greater decrease for effluent samples than for excess sludge samples when exposed to antibiotics. This suggests that the complexity of bacteria and biomass of dominant bacteria in sludge samples is more moderately affected by the pressure of antibiotics.

Table 2 Diversity estimation of bacterial communities in excess sludge and effluent water after biological treatment

Sample	ACE	Chao1	Shannon	Simpson
ew-D	**797.149**	**793.04**	**7.536**	**0.985**
ew	98.712	100	3.211	0.811
ew-e	50.494	50.333	2.644	0.739
ew-e-t	35.146	33.6	2.289	0.593
ew-e-t-n	43.146	35.25	1.612	0.425
es-D	**794.647**	**786.585**	**7.696**	**0.988**
es	614.33	600.786	5.166	0.917
es-e	588.127	578.714	4.269	0.867
es-e-t	515.434	499	4.533	0.914
es-e-t-n	538.972	535.962	3.403	0.765

Overall, regardless of whether it was water or excess sludge samples, the addition of SSDM medium caused the dominant functional bacteria to become more centralized. The diversity indices demonstrated a clear decline in the diversity of microorganisms after the addition of antibiotics, confirming the expectation that the selected antibiotics would disturb multiple members of the original microbial communities. Similar results were found as a significant decrease in microbial diversity emerged with the addition of antibiotics [16].

3.4 Microbial community structure

Proteobacteria was found to be the most dominant phylum in the studied microbial communities. The addition of multiple antibiotics caused an increase in the relative abundance and proportion of Proteobacteria. Most of the phosphorus-dissolving and nitrogen-fixing bacteria in sludge belong to Proteobacteria, which plays a crucial role in removing organic matter and nutrients from sewage.

Moreover, the metabolic diversity of Bacteroidetes has enabled them to thrive under the atypical conditions prevalent in wastewater treatment [17]. Bacillus exhibits remarkable versatility, including denitrification, organic matter degradation, and the maintenance of activated sludge granules. In this study, it represented the second largest proportion by relative abundance. However, in the presence of multiple antibiotics, its relative abundance sharply declined, consistent with previous findings.

Actinobacteria is associated with sludge swelling, while Firmicutes can utilize hydrolytic enzymes to decompose proteins and sugars in wastewater and sludge. Previous studies have identified four bacterial groups, Proteobacteria, Actinobacteria, Bacteroidetes, and Firmicutes, as major contributors to the microbial community structure of municipal wastewater treatment plant activated sludge [18], effluent [19], and excess sludge [20] samples.

At the genus level, *Acinetobacter*, known for its resistance to antibiotics, exhibited the highest concentration when exposed to antibiotics. Conversely, *Pseudomonas*, a widely recognized clinically resistant bacteria [21], experienced a decline in relative abundance following antibiotic treatment. The inhibitory effect of *Acinetobacter* secretions on *Pseudomonas* has been established [22], supported by our findings that *Acinetobacter* increased while *Pseudomonas* decreased. Conversely, *Pseudomonas* can inhibit *Acinetobacter* [23]. It is pertinent to note that among the various groups, Stenotrophomonas had the highest relative abundance in ef-e-t-n (7.14%), potentially due to the presence of ARGs [24]. meanwhile, using plate coating method and line culture method, several bacteria species were isolated and identified able to tolerate the stress of mixed antibiotics. As shown in Fig. 3, *Escherichia coli* spp., *Citrobacter* spp., *Aeromonas* spp., *Achromobacter* spp., *Brevundimonas* spp., and other species from 16 genus were identified able to survive from the combined antibiotic stress.

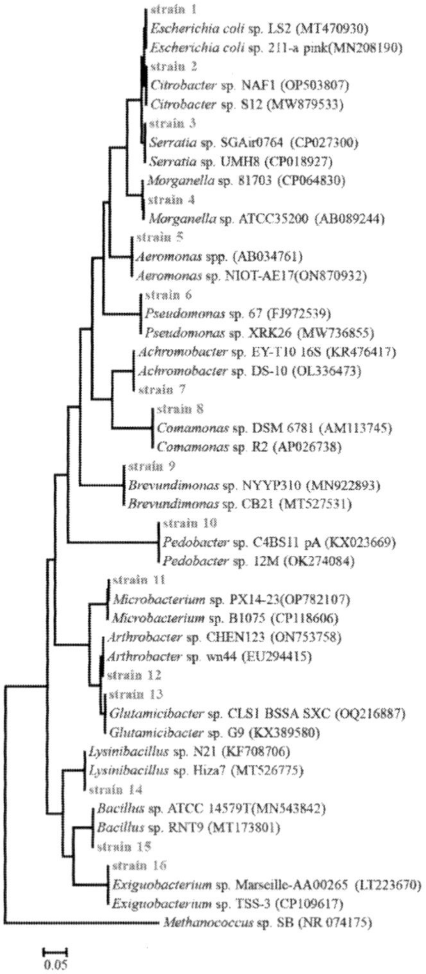

Fig. 3 Phylogenetic tree of isolated ARBs from the effluent and sludge

Additionally, *Stenotrophomonas* may have had a competitive advantage over *Pseudomonas* [25]. Furthermore, *Aeromonas* exhibited the highest relative abundance in ex-e-t. A study has reported that *Aeromonas* is resistant to 12 antibiotics, including erythromycin, tetracycline, neomycin, ampicillin, and ciprofloxacin, and is a multidrug-resistant bacteria [26]. As can be seen, the relationship of these bacteria needs to be further explored based on the above contradictions and speculation.

3.5 Removal of microbial communities

In order to control the microbial risk of WWTP effluent and avoid negative effects of increasing disinfectant dosage on the receiving water body, it is necessary to reduce the biomass of ARB, especially Proteobacteria, to ensure the safety of the effluent water quality. Therefore, for this particular WWTP, the enhanced oxidation ditch process plays a crucial role in the removal of ARB. Oxidation ditch process enhances the contact between wastewater and sludge, leading to better treatment performance, and it is widely applied in urban wastewater treatment. However, the extended retention time of wastewater favors bacterial variation and proliferation, accelerating the horizontal transfer of ARGs, resulting in a significant increase in the variety of ARGs in the effluent. Activated sludge process mainly utilizes adsorption and interception principles to remove ARB from wastewater, through the physical or biological adsorption of the filtering medium and direct interception [27, 28]. Meanwhile, a thorough investigation of the influence and mechanism of hydraulic and water quality parameters on the transfer of resistance genes is needed, and by adjusting the operational parameters, the spread of resistance genes in conventional biological treatment systems can be minimized. The surplus sludge generated from the oxidation ditch process is enriched with a large number of ARB and ARGs, which can enter the soil environment through sludge reuse, posing a threat to human health. Therefore, the subsequent treatment of sludge in the activated sludge process is of great importance, and sludge digestion under high-temperature anaerobic conditions has shown significant removal effects on resistant bacteria and resistance genes [29].

4 Conclusions

The study revealed that antibiotic pressure significantly decreased the microbial diversity, leading to an altered composition and structure of the microbial community. Proteobacteria exhibited a higher ability to tolerate antibiotic pressure in comparison to other phyla. Additionally, the relative abundance of Proteobacteria increased to 92.98%, 94.35%, and 95.74% in water samples and 84.33%, 90.99%, and 90.84% in sludge samples upon the addition of antibiotics from single erythromycin to three (tetracycline, erythromycin, norfloxacin).

Acknowledgments

This work was supported by the key Scientific Research Project in Colleges and Universities of Henan Province of China (22B610001).

References

1. YNAG X, ZHANG Y, LI J, et al. (2010) Distribution of antibiotics, antibiotic resistant bacteria, antibiotic resistance genes in environmental system with reclaimed water irrigation: a review. Asian Journal of Ecotoxicology, 15(6): 43-56.
2. WANG G, FENG J, LUO L, et al. (2019) Distribution, diffuse, and removal of tetracyclines and sulfonamide antibiotic resistance genes in wastewater treatment plant A review. Chinese Journal of Applied Ecology, 30(8): 2875-2882.
3. SIM W J, LEE J W, LEE E S, et al. (2011) Occurrence and distribution of pharmaceuticals in wastewater from households, livestock farms, hospitals and pharmaceutical manufactures. Chemosphere, 82(2): 179-186.
4. ZHANG Y, XIE J, LIU M, et al. (2013) Microbial community functional structure in response to antibiotics in pharmaceutical wastewater treatment systems. Water Research, 47(16): 6298-6308.
5. CHEN H, ZHANG M. (2013) Effects of Advanced Treatment Systems on the Removal of Antibiotic Resistance Genes in Wastewater Treatment Plants from Hangzhou, China. Environmental Science & Technology, 47(15): 8157-8163.
6. JIN M, LIU K, XU Y, et al. (2015) Characteristics of sulfonamide antibiotics, resistant bacteria, resistance genes in sewage treatment plant. Environment Engineering, 33(11): 1-4.
7. HUANG D, JIN L, LIU J, et al. (2023) The Current Status, Energy Implications, and Governance of Urban Wastewater Treatment and Reuse: A System Analysis of the Beijing Case. Water, 15(4): 630.
8. ZHANG X X, ZHANG T. (2011) Occurrence, Abundance, and Diversity of Tetracycline Resistance Genes in 15 Sewage Treatment Plants across China and Other Global Locations. Environmental Science & Technology, 45(7): 2598-2604.
9. NAQUIN A, SHERPA M, SHRESTHA A, et al. (2015) Presence of antibiotic resistance genes in a sewage treatment plant in Thibodaux, Louisiana, USA. Bioresource Technology, 188: 79-83.
10. MAKOWSKA N, KOCZURA R, MOKRACKA J. (2016) Class 1 integrase, sulfonamide and tetracycline resistance genes in wastewater treatment plant and surface water. Chemosphere, 144: 1665-1673.
11. MHONGOLE O J, MDEGELA R H, KUSILUKA L J M, et al. (2017) Characterization of Salmonella spp. from wastewater used for food production in Morogoro, Tanzania. World Journal of Microbiology & Biotechnology, 33(3): 42.
12. GUO G, HE F, TIAN F, et al. (2016) Effect of salt contents on enzymatic activities and halophilic microbial community structure during phenanthrene degradation. International Biodeterioration & Biodegradation, 110: 8-15.

13. SHU D, HE Y, YUE H, et al. (2015) Metagenomic insights into the effects of volatile fatty acids on microbial community structures and functional genes in organotrophic anammox process. Bioresource Technology, 196: 621-633.
14. DIAO X, LI Y, WANG S. (2015) Effects of Outbreak and Extinction of Algal Blooms on the Microbial Community Structure in Sediments of Chaohu Lake. Environmental Science, 36(1): 107-113.
15. WANG Z, ZHANG Z, ZHANG J, et al. (2012) The fauna structure of benthic macro-invertebrates for environmental restoration in a eutrophic lake using water hyacinths. China Environmental Science, 32(1): 142-149.
16. ZHU Y, WANG Y, JIANG X, et al. (2017) Microbial community compositional analysis for membrane bioreactor treating antibiotics containing wastewater. Chemical Engineering Journal, 325: 300-309.
17. KUMAR A, PAL D. (2018) Antibiotic resistance and wastewater: Correlation, impact and critical human health challenges. Journal of Environmental Chemical Engineering, 6(1): 52-58.
18. GAO P, XU W, SONTAG P, et al. (2016) Correlating microbial community compositions with environmental factors in activated sludge from four full-scale municipal wastewater treatment plants in Shanghai, China. Applied Microbiology and Biotechnology, 100(10): 4663-4673.
19. VIEIRA A, GALINHA C F, OEHMEN A, et al. (2019) The link between nitrous oxide emissions, microbial community profile and function from three full-scale WWTPs. Science of the Total Environment, 651: 2460-2472.
20. FANG D, ZHAO G, XU X, et al. (2018) Microbial community structures and functions of wastewater treatment systems in plateau and cold regions. Bioresource Technology, 249: 684-693.
21. FU Y, ZHU G, LI S. (2011) Drug resistance of 264 Pseudomonas aeruginosa strains. Chinese Journal of Nosocomiology, 21(15): 3253-3254.
22. TAN R, SHE P, CHEN L, et al. (2017) Effect of Acinetobacter baumannii secretions on Pseudomonas aeruginosa cell growth and biofilm formation. Chinese Journal of Clinical Laboratory Science, 35(4): 250-253. .
23. QIN J, LI Z, YANG Y, et al. (2012) Inhibitory activity of Pseudomonas aeruginosa in vitro. Chinese Journal of Microecology, 24(11): 1003-1007.
24. GIL-GIL T, MARTINEZ J L, BLANCO P. (2020) Mechanisms of antimicrobial resistance in Stenotrophomonas maltophilia: a review of current knowledge. Expert Review of Anti-Infective Therapy, 18(4): 335-347. .
25. JACKSON L M D, KROUKAMP O, YEUNG W C, et al. (2019) Species Interaction and Selective Carbon Addition During Antibiotic Exposure Enhances Bacterial Survival. Frontiers in Microbiology, 10: 2730.
26. Rodriguez-Castillo, JG,Pino, C, Nino, LF, et al. (2017) Comparative genomic analysis of Mycobacterium tuberculosis Beijing- like strains revealed specific genetic variations associated with virulence and drug resistance. Infection and Evolution, 54: 314-323.
27. CHENG Y Y, ZHANG C M, WANG X C, et al. (2014) Removal efficiency of host-specific markers and enteric pathogenic bacteria by municipal wastewater treatment processes. Chinese Journal of Environmental Engineering, (3): 887-890.
28. BARRIOS-HERNÁNDEZ M L, PRONK M, GARCIA H, et al. (2020) Removal of bacterial and viral indicator organisms in full-scale aerobic granular sludge and conventional activated sludge systems. Water Research X, 6: 100040.
29. DIEHL D L, LAPARA T M. (2010) Effect of temperature on the fate of genes encoding tetracycline resistance and the integrase of class 1 integrons within anaerobic and aerobic digesters treating municipal wastewater solids. Environmental Science & Technology, 44(23): 9128-9133.

https://doi.org/10.1051/e3sconf/202344102014

Research on Carbon Emissions and Carbon Reduction Paths of Power Generation Enterprises under the Dual Carbon Target

Peng Yang[1,a*], Minfang Yao[2,b], Chunyi Chen[3,c], Huangru Zhu[3,d], Yanfeng Tao[1,e]

[1]East China Electric Power Design Institute Co., Ltd. of China Power Engineering Consulting Group, Shanghai, China
[2]Shanghai Liquefied Natural Gas Co., Ltd, Shanghai, China
[3]Shanghai Power Exchange Center Co., Ltd, Shanghai, China

Abstract. Achieving the dual carbon goal is a profound and challenging social transformation. The difficulty is mainly reflected in the fact that economic growth is usually linked to energy consumption. In today's society, no country or industry can completely abandon traditional energy, and the use of traditional energy, namely fossil fuels, would bring carbon emissions. Therefore, reducing carbon emissions while ensuring economic growth and achieving the dual carbon goals is an important issue at present. This article conducted a series of carbon emission analysis and carbon reduction path research for power generation enterprises, which are the major emitters of carbon emissions, based on the background of dual carbon goals.

1. Introduction

As the power lifeline of a region or a country, power generation enterprises or the power sector have sufficient resources, so it is necessary to promote the reform of power generation enterprises. Therefore, this article hopes to reform some technologies and policies in the power sector to reduce carbon emissions to a certain extent, in order to meet the dual carbon goals.

In today's global warming environment, controlling carbon emissions is a protection of the Earth's ecological environment. Engel-Cox J A conducted modeling work on reducing carbon dioxide and other greenhouse gas emissions [1]. Morris J also established an economic forecasting and policy analysis model to study the factors behind carbon capture and storage deployment and their role in reducing carbon emissions [2]. However, Miner K R proposed that the Permafrost in the Arctic stored 17 million tons of frozen and melted carbon, and climate warming may lead to their release, which needed to be prevented [3]. Bruckner B proposed that the distribution of wealth and income among the global population is disproportionate, which can lead to carbon inequality. In order to promote the development of impoverished areas, developed regions need to significantly reduce emissions [4]. This indicates that emission reduction and development need to coexist.

As the main force of carbon emissions, many scholars have put forward suggestions for the reform of the power sector. Dindi A believed that the public power sector should produce a certain proportion of low-carbon electricity by 2030 [5]. Anderson J J proposed that the main challenge in achieving rapid decarbonization in the power sector in the short term is to achieve net zero emissions while maintaining grid reliability and reducing costs [6]. Borowski P F proposed innovative biomass that can utilize bamboo as a productivity energy source for power plants [7]. Bistline J E T proposed a carbon dioxide removal technology that has an impact on deep decarbonization in the power sector [8]. Psomopoulos C S believed that energy can be recovered from waste to help the power sector implement decarbonization work [9]. Gedam V V believed that the power department still needs to do a good job in internal human resource management [10]. This is a proposal for unrelated technological reforms, but it remains important.

2. Background Description of Dual Carbon Targets

The so-called dual carbon target refers to carbon peaking and carbon neutrality. The definition of carbon peak is that at a certain point in time, the emission of carbon dioxide reaches its peak, stops growing, and begins to decline. Carbon neutrality refers to offsetting carbon emissions over a period of time through afforestation, energy conservation, and emission reduction, forming a relatively "zero emission" situation and achieving neutrality. To achieve these two goals, it is inevitable that traditional fossil fuels will transition to new energy. Institutions that use fossil fuels, from power plants and factories to rural individuals who burn coal and firewood, are all targets that require energy reform in the current context of the dual carbon target.

Ren D W proposed that energy activities are the main source of carbon dioxide emissions, with employee

email: [a*]yangpeng@ecepdi.com, [b]yaomf@shlng.com.cn, [c]chenchy@sh.sgcc.com.cn, [d]zhuhuangru@gmail.com, [e]tyf@ecepdi.com

activities accounting for 80% of the total emissions, with the electricity sector accounting for over 40% of the emissions [11]. The testing formula for the spatial spillover effect of carbon emissions is:

$$I = \frac{n\sum_{a=1}^{n}\sum_{b=1}^{n} s_{ab}(x_a - \bar{x})(x_b - \bar{x})}{\sum_{a=1}^{n}\sum_{b=1}^{n} s_{ab} \sum_{a}^{n}(x_a - \bar{x})^2} \tag{1}$$

In the above formula, I is the local spatial indicator of carbon emissions, and n is the number of cities in the country. s_{ab} is the spatial weight. x_a, x_b are the per capita carbon emissions of city a and city b, respectively. $\bar{x}$ is the per capita carbon emissions of cities nationwide. This formula can be used to comprehensively evaluate and detect the spatial correlation characteristics of carbon emissions in target cities.

3. Carbon Emissions of Power Generation Enterprises

As is well known, thermal power generation, as the main force of traditional energy today, emits a large amount of carbon dioxide and other gases during the production process. As mentioned earlier, power generation enterprises account for the vast majority of carbon emissions in all industries. Therefore, in order to achieve the dual carbon target, it is very necessary for power generation enterprises to reduce emissions. Gao Y proposed to use LCA (Life Cycle Assessment) to establish a carbon emission assessment system for the Prefabricated building industry [12].

Table 1. Life Cycle Assessment Objects

Purpose and scope	System function	System boundary	System units
Inventory analysis	Material input	Energy input	Carbon emission factor
Impact evaluation	Resources consumption	Energy consumption	Carbon emissions

The objects of life cycle assessment are shown in Table 1. It can be seen that the evaluation objects of the lifecycle assessment framework are mainly divided into purpose and scope, inventory analysis, and impact evaluation. The purpose and scope mainly include system functions, system boundaries, and system units, while inventory analysis mainly includes material inputs, energy inputs, and carbon emission factors. The impact assessment mainly includes resource consumption, energy consumption, and carbon emissions.

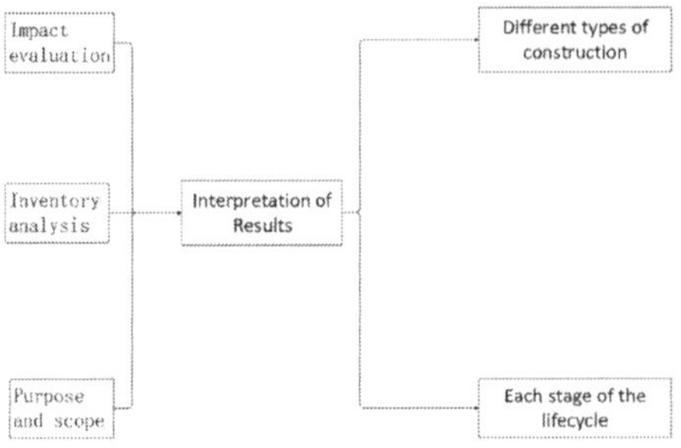

Figure 1. Life Cycle Assessment Framework

The lifecycle assessment framework is shown in Figure 1. The purpose and scope, inventory analysis, and impact assessment of these three evaluation objects explain their respective contents, and finally summarize and analyze the carbon emissions of different stages of the life cycle and different types of construction.

This framework can also be applied in the carbon emission evaluation and analysis of power generation enterprises. Resource consumption, material input, and carbon emission factors are important indicators for analyzing the carbon emissions of target enterprises, and can be calculated using the following formula:

$$\begin{aligned} P &= \lambda \cdot x \\ \lambda &= \eta \cdot y \cdot \mu \cdot \theta \end{aligned} \tag{2}$$

In the above formula, P represents the emission of carbon dioxide. λ is the emission coefficient, and x is the consumption of fossil fuels. η is the carbon emission factor, and y is the low calorific value. μ is the carbon conversion coefficient, and θ is the carbon oxidation rate. After calculating the carbon emissions of the target enterprise, it can be counted and recorded. In the carbon emission analysis system, there is also a need for control and supervision of emission indicators. The government needs to develop emission indicators for each power generation enterprise to limit its excessive emissions. The government's formulation also needs to be based on the actual situation. For example, some areas are underdeveloped and need to obtain more emission indicators. Some areas have serious pollution problems and urgently need to limit emissions, or some areas of a city are densely populated and have more power supply needs. Therefore, emission indicators naturally need to be higher. Ramirez-Meyers K also proposed that the type of power plant needs to be classified and constructed according to different regions [13]. In summary, controlling carbon emissions is urgent in the current context of the dual carbon target, but it is also necessary to achieve a "soft landing" of energy conservation and emission reduction based on actual situations.

4. Carbon Reduction Paths for Power Generation Enterprises

4.1. Substitution of New Energy for Traditional Energy

Generally speaking, reducing carbon emissions leads to a reduction in energy acquisition, resulting in insufficient electricity supply, inadequate heating in some regions, and a slowdown in the development of some heavy industries. Therefore, energy conservation and emission reduction are not limited to simple emission reduction, but should seek various countermeasures to compensate for the energy shortage caused by emission reduction. Clean energy is naturally the preferred solution to this problem. Bhat A attempted to use clean energy in the production of two-dimensional materials [14]. Among many clean energy sources, wind power, solar power, and hydropower are the most representative new energy sources that people first consider. Kudria S pointed out that Ukraine has extremely strong wind energy potential and can carry out wind hydrogen power generation [15]. Alghamdi AS also conducted a series of studies on the efficiency of wind and solar energy [16]. The biggest advantage of these new energy sources is that they are environmentally friendly and renewable, without the need to eliminate greenhouse gases such as carbon dioxide. Moreover, compared to the slow generation rate of fossil fuels, the regeneration rate of new energy is extremely fast. Of course, in addition to these two points, some new energy sources also have great indirect value. For example, Geothermal energy, which is gradually widely used, can be directly used for heating in addition to power generation.

Li K also proposed that using nuclear energy for power generation under the dual carbon target is also worth considering [17]. The biggest problem of nuclear power generation is that at present, mankind only has mastered controllable Nuclear fission technology, not Fusion power, so nuclear power plants supported by Nuclear fission technology are at risk of nuclear leakage. Once a nuclear leakage accident occurs, its pollution and harm level far exceed conventional fossil fuel emissions. In history, such as the Chernobyl incident, it was a major nuclear leakage accident with profound harm. However, although nuclear leakage accidents pose significant risks, the accident rate is not high. Apart from the accident problem, nuclear power generation is considered a clean energy source under the premise of normal operation of nuclear power plants and proper disposal of nuclear waste, as it does not produce any greenhouse gases.

4.2. Improvement of Traditional Energy

Many people believe that renewable energy sources such as wind power and hydropower have considerable limitations. Zhu J H believed that these renewable energy sources did not have resource attributes and have low utilization rates [18]. Due to some technological limitations, the electricity provided by wind power, hydropower and other power generation methods is actually quite limited, and it is also limited by regions, and not all places can introduce these technologies. In his research, he proposed the use of biomass for power generation. In his view, biomass fertilizers such as straw, livestock and poultry discrimination, and household waste can all be used as fuel. China produces more than 5 billion tons of these Organic fertilizer every year. If they are treated by incineration, fermentation, composting or landfill, it would not only waste resources, but also produce more than 1 billion tons of carbon dioxide equivalent of greenhouse gases every year. Pal D B also proposed the use of biomass, and he believed that hydrogen can be produced from biomass. Hydrogen is a renewable energy carrier with great potential [19].

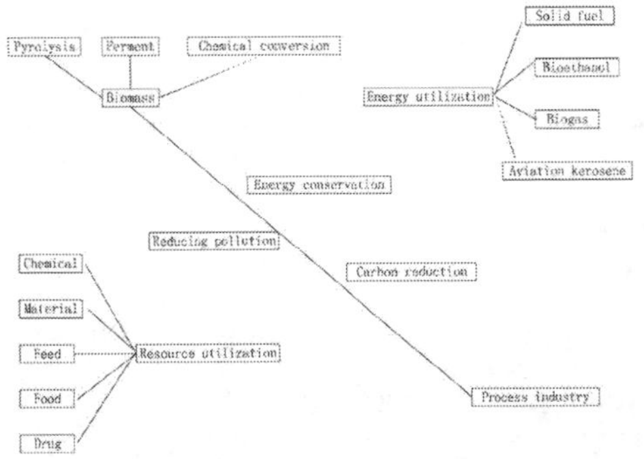

Figure 2. Schematic Diagram of Biomass Utilization

The utilization of biomass is shown in Figure 2. In this scenario, the ultimate goal is to apply biomass to the process industry. Under the overall policy of energy conservation, pollution reduction, and carbon reduction, biomass undergoes this process through high-temperature cracking, biological fermentation, and chemical conversion. In addition, in this process, some additional products can be generated, including chemicals, materials, feed and even food and medicine in terms of resources, and solid fuels, bioethanol, Biogas and aviation kerosene in terms of energy.

In addition, Ma X M also proposed to promote distributed photovoltaic power generation, improve fossil fuels, and appropriately shut down coal-fired power plants to achieve energy conservation and emission reduction in the capital [20]. He also used Shenzhen as an example in his research to analyze the expected emission reduction results of this series of reforms. This article selected a portion to showcase:

Table 2. List of Emission Reduction Technologies for Shenzhen Electric Power Department (Excerpt)

Technical Name	Emission reduction potential (10000 tons)	Emission reduction cost (10000 yuan)
Steam turbine seal modification	0.8	-1.7
Nameplate for unit capacity increase	4.5	-7.2

and modification		
Efficient motor replacement	1.0	1.1
Gas turbine intake cooling	1.6	-9.0
Renovation of Pure Condensing Steam Turbine Units	26.5	-359.5
Newly built gas turbine unit	72.0	775.2
Distributed photovoltaic power generation	415.3	2203.5
Improve the utilization rate of gas turbine units	22.5	-301.3
Other	593.0	1677.4
Total	1137.2	3978.5

The emission reduction technologies are shown in Table 2. In the selected list of technologies, this article lists technologies such as steam turbine seal modification, unit capacity increase and nameplate modification, efficient motor replacement, gas turbine inlet cooling, pure condensing steam turbine unit modification, new gas units, distributed photovoltaic power generation, and improving gas turbine utilization. It can be seen that these technologies either save money or spend money, or introduce new technologies or shut down or improve old technologies, and reform the power sector through various means to achieve emission reduction.

5. Conclusions

This article first explained the goals of each "energy individual" in this context by briefly describing the dual carbon goals, and then proposed that power generation enterprises are the main force of carbon emissions. To control carbon emissions, reforming power generation enterprises is essential. Finally, two methods were proposed, one is to use clean energy, and the other is to reform traditional energy to truly achieve energy conservation and emission reduction. At the end of this article, only a list of reform technologies and emission reduction expectations for the power sector in Shenzhen were listed. However, to truly achieve the dual carbon goal, relying solely on one province, one city, or one district is not enough, and Shenzhen, as a highly developed region, has different development and electricity needs compared to other regions. Therefore, this example cannot fully represent the actual situation in all regions.

Overall, the two major solutions proposed in this article are also the mainstream means of reducing emissions, and implementing emission reduction policies for power generation enterprises is also widely recognized. Therefore, future carbon reduction achievements are worth looking forward to. If all parts of the country can efficiently complete emission reduction work, then the achievement of the dual carbon target is just around the corner.

Acknowledgments

Shanghai Science and Technology Innovation Action Plan Project - Research on Key Technologies and Engineering Solutions for Efficient and Comprehensive Utilization of LNG Cold Energy in Shanghai（Project No.:20dz1205800）

References

1. Engel-Cox J A, Geocaris M. Future Low-Carbon Technology Options for a 100 Percent Decarbonized Power Sector[J]. Climate and Energy, 2023, 39(8): 1-10
2. Morris J, Kheshgi H, Paltsev S, et al. Scenarios for the deployment of carbon capture and storage in the power sector in a portfolio of mitigation options[J]. Climate Change Economics, 2021, 12(01): 1-29
3. Miner K R, Turetsky M R, Malina E, et al. Permafrost carbon emissions in a changing Arctic[J]. Nature Reviews Earth & Environment, 2022, 3(1): 55-67
4. Bruckner B, Hubacek K, Shan Y, et al. Impacts of poverty alleviation on national and global carbon emissions[J]. Nature Sustainability, 2022, 5(4): 311-320
5. Dindi A, Coddington K, Garofalo J F, et al. Policy-driven potential for deploying carbon capture and sequestration in a fossil-rich power sector[J]. Environmental Science & Technology, 2022, 56(14): 9872-9881
6. Anderson J J, Rode D C, Zhai H, et al. Fossil-Fuel Options for Power Sector Net-Zero Emissions with Sequestration Tax Credits[J]. Environmental Science & Technology, 2022, 56(16): 11162-11171
7. Borowski P F. Management of energy enterprises in zero-emission conditions: Bamboo as an innovative biomass for the production of green energy by power plants[J]. Energies, 2022, 15(5): 1-16
8. Bistline J E T, Blanford G J. Impact of carbon dioxide removal technologies on deep decarbonization of the electric power sector[J]. Nature Communications, 2021, 12(1): 1-12
9. Psomopoulos C S, Kiskira K, Kalkanis K, et al. The role of energy recovery from wastes in the decarbonization efforts of the EU power sector[J]. IET Renewable Power Generation, 2022, 16(1): 48-64
10. Gedam V V, Raut R D, Priyadarshinee P, et al. Analysing the adoption barriers for sustainability in the Indian power sector by DEMATEL approach[J]. International Journal of Sustainable Engineering, 2021, 14(3): 471-486

11. Ren D W,Xiao J Y,Hou J M,Du E S,Jin C,Liu Y.Construction and Evolution of China's New Power System Under Dual Carbon Goal[J].Power System Technology,2022,46(10):3831-3839
12. Gao Y,Li Z D,Zhang H,Yu B,Wang J Y.A Carbon Emission Analysis Model for Prefabricated Construction Based on LCA[J].Journal of Engineering Management,2018,32(2):30-34
13. Ramirez-Meyers K, Mann W N, Deetjen T A, et al. How different power plant types contribute to electric grid reliability, resilience, and vulnerability: a comparative analytical framework[J]. Progress in Energy, 2021, 3(3): 1-17
14. Bhat A, Anwer S, Bhat K S, et al. Prospects challenges and stability of 2D MXenes for clean energy conversion and storage applications[J]. npj 2D Materials and Applications, 2021, 5(1): 1-21
15. Kudria S, Ivanchenko I, Tuchynskyi B, et al. Resource potential for wind-hydrogen power in Ukraine[J]. International Journal of Hydrogen Energy, 2021, 46(1): 157-168
16. Alghamdi A S. A Hybrid Firefly–JAYA Algorithm for the Optimal Power Flow Problem Considering Wind and Solar Power Generations[J]. Applied Sciences, 2022, 12(14): 1-24
17. Li K.Research on Optimizing the Schedule of Nuclear Power Generation Equipment Supply Projects[J].Project Management Technology,2020,18(1):88-92
18. Zhu J H,Mu L W,Jiang G C,Liu L,Xiong J J,Lu X H.Biomass integrated industrial processes for system energy conservation, pollution reduction and carbon dioxide mitigation[J].Chemical Industry and Engineering Progress,2022,41(3):1111-1114
19. Pal D B, Singh A, Bhatnagar A. A review on biomass based hydrogen production technologies[J]. International Journal of Hydrogen Energy, 2022, 47(3): 1461-1480
20. Ma X M,Duan Y,Li X,Zhou J P,Ji J P.Research on the Carbon Mitigation Path of Power Sector in Shenzhen[J].Ecological Economy,2018,34(2):24-29

Calculation and Analysis of Carbon Emissions from Road Maintenance Projects ——Taking the Tianjin Avenue Restoration and Maintenance Project as an Example

Liping Guo*, Xiaodong Zhu, Tingting You, Kai Niu, Xingyu Zhang

China Municipal Engineering North China Design and Research Institute Co., Ltd, Tianjin 300074, China

Abstract: In response to the phenomenon that road maintenance construction activities may have a certain impact on passing vehicles and generate additional carbon emissions, based on the Tianjin Avenue Restoration and Maintenance Project, a recommended list of carbon emission factors is proposed. A traffic impact carbon emission calculation model is used to calculate the traffic impact carbon emissions of the supporting project, analyze the characteristics of traffic impact carbon emissions composition, and explore the hot spots of traffic impact carbon emissions of maintenance projects, propose emission reduction strategies for road maintenance engineering traffic organization.

1 Introduction

With the proposal of China's goal of achieving carbon peak and carbon neutrality, various industries are vigorously conducting research on carbon emissions calculation and emission reduction strategies. The transportation industry is an important source of carbon emissions, accounting for 24% of global total carbon emissions. In recent years, road construction has shifted from an incremental market to a stock market, with road maintenance accounting for 99.9% of the total road mileage[1]. The carbon emissions from road maintenance projects cannot be underestimated.In addition to certain carbon emissions caused by construction during road maintenance, the impact on transportation also generates additional carbon emissions. Therefore, this study focuses on calculating the carbon emissions affected by transportation, analyzing the hotspots of carbon emissions, and proposing a recommended plan for optimizing the emission reduction of maintenance engineering traffic organization.

The biggest difference between carbon emissions during road maintenance construction and during the construction phase is the need to consider the additional carbon emissions caused by traffic impacts. In recent years, experts and scholars have conducted relevant research, such as Huang et al. using life cycle assessment methods and VISSIM simulation models to calculate the additional carbon emissions of vehicles caused by road maintenance and construction fences in the UK[2]; However, these studies place more emphasis on using microsimulation software such as VISSIM and AIMSUN to simulate the operation status of vehicles during road maintenance construction[3][4][5]. Through software, the instantaneous operation status of vehicles can be relatively accurately obtained. However, due to the high professionalism of the software, it is difficult for engineering personnel to proficiently operate it in a short period of time, so its promotion is not strong.Chinese scholar Liu Yuanyuan has constructed a carbon emission calculation model for the traffic impact of road maintenance construction[6], which has certain significance for the calculation of carbon emissions. Based on this model, this article proposes a recommended list of carbon emission factors applicable to the Tianjin Avenue Restoration and Maintenance Project. Using the above model, the traffic impact carbon emissions of the supporting project are calculated, the characteristics of the traffic impact carbon emissions are analyzed, the hot spots of carbon emissions in maintenance projects are explored, and a recommended plan for optimizing traffic organization and emission reduction in road maintenance projects is proposed.

2 Relationship between maintenance and construction activities, traffic impact, and carbon emissions

The selection of road maintenance plans is influenced by the location and type of diseases in maintenance and construction activities. The different road maintenance plans determine the location, quantity, and length of closed lanes, while the location of the disease and the degree of

*Liping Guo: 409093646@qq.com

damage determine the construction process. Therefore, it is necessary to analyze the relationship between various environmental factors based on the characteristics of the road and maintenance construction area, as well as the correlation between micro and macro perspectives, And summarize the relationship between maintenance construction engineering, traffic impact, and carbon emissions.

3 Carbon emission factors

Due to the complex road environment, there are differences in the performance of operating vehicles, and the operating states are different. The carbon emission factor list presented here is a recommended carbon emission factor list that combines the characteristics of Tianjin Avenue and the structural characteristics of operating vehicles, and combines the research results of experts and scholars. It is suitable for calculating the carbon emission impact of Tianjin Avenue maintenance and construction traffic. The list is divided into two parts, A list of carbon emission factors for various vehicle types under different speed states is provided in Table 2; The other part is a list of carbon emission factors during the speed change process of different vehicle types, as shown in Table 3; For the convenience of subsequent discussions, first define the code for the vehicle type, as shown in Table 1.

Table 1 Vehicle Type Code Definition Table

TYPE	code
Small and medium-sized passenger cars (fuel)	C1
Small and medium-sized passenger cars (hybrid)	C2
Small and medium-sized passenger cars (electric)	C3
Bus (fuel)	B1
Bus (electric)	B2
Van	T1
Medium truck	T2
Large truck	T3
Extra large truck	T4

Table 2 List of Carbon Emission Factors for Various Vehicle Types under Different Speed States (Unit: g/km/veh)

speed(km/h)	C1	C2	C3	B1	B2	T1	T2	T3	T4
10	261	114	18	424	30	316	472	553	799
20	254	111	17	389	28	308	431	455	630
30	239	105	16	367	27	289	406	444	592
40	233	102	16	354	25	282	394	425	571
50	224	97	15	333	23	276	373	410	546
60	221	92	14	320	21	278	360	399	536
70	217	85	13	354	25	295	396	439	576
80	214	78	12	392	29	316	435	478	622

Table 3 List of carbon emission factors during speed changes of different vehicle types (unit: g/s/veh)

Speed changes (km/h)	C1	C2	C3	B1	B2	T1	T2	T3	T4
80 to 40	1	1	0	35	4	5	38	55	69
40 to 80	9	5	1	225	24	57	275	325	424
40 to 0	1	1	0	35	4	5	38	55	69
0 to 40	12	6	1	253	28	62	264	375	477

4 Calculation model

Using the carbon emission calculation model of road maintenance construction traffic impact constructed by scholar Liu Yuanyuan, based on the characteristics of Tianjin Avenue communication and considering the complex and variable traffic impact factors of the road under construction, the vehicle operation status is divided into five categories, and carbon emissions are calculated and summarized separately. The overall modeling idea is shown in Figure 1.

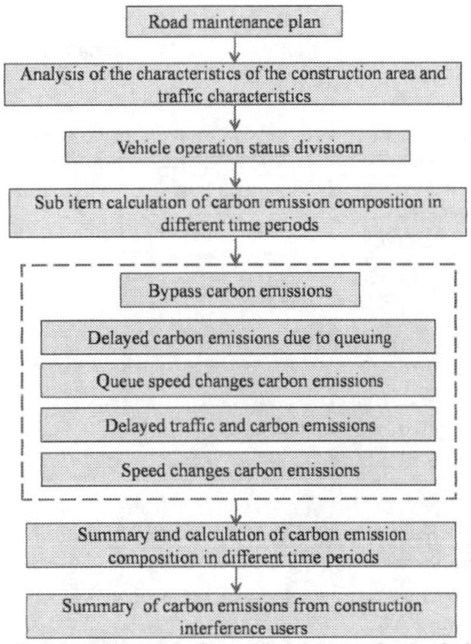

Fig. 1. Calculation ideas for traffic impact carbon emissions during road maintenance construction

5 Calculation and analysis of carbon emissions based on engineering

5.1 Overview of Tianjin Avenue Restoration and Maintenance Project

Tianjin Avenue (S125) starts from the Jingu Interchange on the Outer Ring Road and ends at the Central Avenue in Binhai New Area, with a total length of 36.5 kilometers. The project relies on the maintenance engineering scope of the K28+000 Yingbin Avenue (K34+900) section downstream (from Binhai to the urban area) and the Yingbin Avenue Central Avenue (K36+500) section two-way lanes, with a total length of 8.5 kilometers for the construction section.

The roadbed and pavement engineering is divided into three construction sections. The first section is the K28+000 to Daguhua Railway Qiaoxi Qiaotou (K31+267) section, with a length of 3.267 kilometers; The second section is from the eastern bridgehead of Daguhua

Railway Bridge (K31+785) to Yingbin Avenue, with a length of 3.115 kilometers; The third section is from Yingbin Avenue to Zhongyang Avenue, with a length of 1.6 kilometers. In addition, it also includes drainage and other projects, bridge engineering, traffic and security facilities engineering.

5.2 Roadbed and Pavement Engineering - Calculation of Carbon Emission Composition from Yingbin Avenue to Zhongyang Avenue Section

Due to the relatively complex traffic organization and complete carbon emission composition of the section from Yingbin Avenue to Central Avenue, this section is selected for a detailed discussion on the calculation of traffic impact carbon emissions. The carbon emissions calculation results of its transportation impact are shown in Table 4.

Table 4 Carbon Emission Calculation of Traffic Impact from Yingbin Avenue to Central Avenue Section (Unit: g)

Carbon emission composition	Carbon emission (kg/d)	Carbon emission(t/Duration)	Total carbon emissions(t)
Speed changes carbon emissions	369.44	11.08	26.21
Delayed traffic and carbon emissions	44.06	1.32	
Queue speed changes carbon emissions	35.20	1.06	
Delayed carbon emissions due to queuing	36.50	1.09	
Bypass carbon emissions	388.32	11.65	

The traffic impact carbon emissions of this construction section include: speed change carbon emissions, traffic delay carbon emissions, queue speed change carbon emissions, queue delay carbon emissions, and detour carbon emissions. The second column shows the different carbon emissions caused by construction from 9 to 17 hours a day; The third column shows the different carbon emissions formed during the construction period of 30 days; The last column shows the total carbon emissions from the Yingbin Avenue to Central Avenue section of the roadbed and pavement engineering, which is 26.21 tons. It can be seen that the carbon emissions from detours and speed changes account for the highest proportion, reaching 44.5% and 42.3% respectively. The high carbon emissions from detours are due to the fact that the detour path is generally longer than the original path; The carbon emissions from speed change are mainly composed of carbon emissions from vehicle deceleration and carbon emissions from vehicle acceleration, with vehicle acceleration accounting for a larger proportion of carbon emissions.

5.3 Calculation and analysis of carbon emissions from individual projects

According to the classification of individual projects, the above calculation data can be divided into roadbed and pavement engineering, bridge engineering, traffic and security facility engineering. Due to the fact that only hard shoulders are occupied during the construction process of the drainage project, which has little impact on passing vehicles, the traffic impact and carbon emissions caused are zero, so they are not listed here.

Table 5 Carbon Emission Calculation of Each Single Project (g)

Special engineering	Construction section	Carbon emission (t)	Sum(t)	Proportion (%)
Roadbed and pavement engineering	The first construction section	47.55	75.22	68%
	The second construction section	3.22		
	The third construction section	24.45		
Bridge engineering	Daguhua Railway Bridge	15.21	24.66	22%
	Henan Road and Bridge	9.44		
Traffic and security facilities engineering	——	11.53	11.53	10%
sum	——	111.41		100%

From Table 5, it can be seen that the traffic impact carbon emissions of roadbed and pavement engineering are 75.22t, accounting for 68%, with the highest proportion; The traffic impact of bridge engineering on carbon emissions is 24.65 tons, accounting for 22%; The traffic and security facilities project has a carbon emission impact of 11.53 tons, accounting for 10%.The roadbed and pavement engineering has the characteristics of long construction period and long closed road sections, resulting in the highest carbon emissions.

5.4 Comparative analysis of different carbon emission components

Classify the above calculation data according to the carbon emission composition of the established calculation model, and calculate carbon emissions separately.

Table 6 Calculation of Carbon Emissions from Different Constituents (g)

Carbon emission composition	Carbon emissions (t)	Proportion (%)
Speed changes carbon emissions	67.10	59.21

Delayed traffic and carbon emissions	32.28	28.49
Queue speed changes carbon emissions	1.18	1.04
Delayed carbon emissions due to queuing	1.12	0.99
Bypass carbon emissions	11.65	10.28

From Table 6 , it can be seen that due to the influence of speed restrictions, vehicles generate the most carbon emissions during the speed change process, reaching 67t, accounting for 59.21%; Secondly, due to speed restrictions, the carbon emissions of traffic delays caused by vehicles traveling at low speeds are 32.28 tons, accounting for 28.49%; Next is the traffic congestion in the construction section, resulting in a detour carbon emission of 11.65 tons, accounting for 10.28%; The queuing here is mainly controlled by traffic lights, causing less additional carbon emissions. The carbon emissions due to changes in queuing speed and delayed queuing are both around 1 ton, accounting for about 1%.

5.5 Calculation and Analysis of Carbon Emissions from Different Vehicle Models

For the above calculation data, classify according to vehicle type and separately calculate carbon emissions.

Table 7 Carbon Emission Calculation for Different Vehicle Models (g)

Vehicle model	C1	C2	C3	B1	B2	T1	T2	T3	T4	Sum
Number of vehicles	19937	1692	2537	187	348	7	2	33	40	24782
Vehicle proportion (%)	80.45	6.83	10.24	0.76	1.40	0.03	0.01	0.13	0.16	100
	97.51			2.16		0.33				
Carbon emissions (t)	85.38	5.65	1.36	9.94	1.96	0.08	0.10	2.68	4.24	111.41
subtotal	92.40			11.90		7.11				
Carbon emission proportion (%)	76.64	5.07	1.22	8.93	1.76	0.07	0.09	2.41	3.81	100
	82.94			10.68		6.38				

From Table 7, it can be seen that the proportion of small and medium-sized passenger cars reaches the highest of 98%, while the proportion of trucks is the lowest, less than 1%; The carbon emissions of small buses are 92.40t, accounting for 83%; The carbon emissions of large buses are 11.90t, accounting for 6%; The carbon emissions of trucks are 7.11 tons, accounting for 6%. The number of vehicles is basically positively correlated with carbon emissions, but it is evident that the carbon emissions of a single truck are greater.

From the above table, it can also be seen that the proportion of fuel powered small and medium-sized buses, hybrid small and medium-sized buses, and electric small and medium-sized buses is 82%, 7%, and 11%, respectively. The carbon emission rate of fuel powered small and medium-sized buses reaches 92%, exceeding the proportion of vehicles; The proportion of carbon emissions from hybrid small and medium-sized buses is close to the proportion of vehicles, at 6%; Electric small and medium-sized buses account for the least carbon emissions, only 2%. Therefore, vigorously promoting the use of electric small and medium-sized vehicles is of great significance in reducing vehicle carbon emissions.

5.6 Analysis of Traffic Impact on Maintenance Construction

To further analyze the impact of Tianjin Avenue maintenance construction on traffic, assuming normal traffic, we obtained relevant traffic parameters through simulation software and calculated the carbon emissions without maintenance construction impact to be 394.35 tons. As previously estimated, the carbon emissions from transportation impact were 111.41 tons, and the carbon emissions from transportation impact accounted for 28.25% of the carbon emissions from non transportation impact, indicating that maintenance construction increased traffic carbon emissions by 28.25%.

Calculate the carbon emissions from maintenance and construction operations, including the production, transportation, and on-site construction of raw materials. The total carbon emissions are 1560 tons, and the total carbon emissions from maintenance and transportation are 1671 tons. The carbon emissions from transportation account for 6.67% of the total carbon emissions.

6. Maintenance and construction traffic organization and emission reduction strategies

6.1 HotSpots

6.1.1 Detour

In the roadbed and pavement engineering - Yingbin Avenue to Central Avenue section, the proportion of carbon emissions from detours is the highest, reaching 44.5%. Therefore, detours are one of the hotspots of carbon emissions. However, the overall path of Tianjin Avenue has a certain degree of closure, so there is no reasonable detour plan for other construction sections except for the reasonable detour path available for this construction section. Therefore, the proportion of detour carbon emissions in the total traffic impact of the project is only 10.3%.

6.1.2 Speed limit setting

From the above analysis data, it can be seen that the process of changing vehicle speed due to speed restrictions produces the largest carbon emissions, accounting for 59.21%. Therefore, speed restrictions are the main sensitive point of carbon emissions. However, speed restrictions are influenced by industry standards and other factors, and there are certain restrictions on the flexibility of setting them. The operability of construction organization designers is not high.

6.1.3 Traffic organization mode

Reasonable traffic organization is an important means to reduce traffic delays and carbon emissions. Tianjin Avenue is a two-way eight lane road. Considering that Tianjin Avenue is an important transportation channel connecting the urban area of Tianjin and the Binhai New Area, the interruption of traffic due to construction will inevitably have a huge impact on traffic and negative effects on urban economic development. Therefore, appropriate construction traffic organization methods should be selected to minimize the impact of maintenance construction on road traffic, which has important social and economic significance.

6.1.4 Construction section length

Different lengths of construction sections will have different impacts on traffic. The shorter the construction closed section, the smaller the construction working face, the more scattered the division of construction sections, the slower the construction progress, and the longer the impact of construction on the road may be. The longer the construction closure section, the more traffic delays drivers may experience when driving in the construction area due to the increase in traffic density, the decrease in lane width, and the interference caused by maintenance operations. Drivers will drive cautiously at lower speeds, resulting in increased traffic delays. During this process, the longer the construction section, the more likely it is to have an impact on the traffic of vehicles in the construction area, resulting in queuing and congestion. Therefore, selecting an appropriate length of closed construction section has a positive impact on construction progress and traffic.

6.1.5 Construction period

The impact of road closures on traffic varies at different time periods. The larger the traffic flow, the greater the impact of construction closures, resulting in more additional carbon emissions. Therefore, selecting appropriate construction time periods has certain significance in reducing carbon emissions.

6.1.6 Vehicle composition

Vehicles are the source of carbon emissions, and different power sources generate different carbon emissions. Electric cars generate relatively less carbon emissions. Therefore, promoting the use of electric small and medium-sized vehicles is of great significance in reducing vehicle carbon emissions.

6.2 Emission reduction strategies

6.2.1 Reasonably plan the vehicle guidance path.

In the process of road maintenance, variable information boards and other tools can usually be used to guide the detour path of passing vehicles, reasonably plan the detour path, and determine the appropriate proportion of detour vehicles, which is of great significance for reducing carbon emissions of vehicle detours. Tianjin Avenue is a two-way eight lane road with a certain degree of closure, and in the case of construction occupation, traffic congestion is not obvious. Therefore, there is no need to take traffic guidance measures, allowing passing vehicles to choose whether to detour or not.

6.2.2 Reasonable selection of transportation organization methods.

Tianjin Avenue is a two-way eight lane road with the conditions for construction of a closed part of the lane. To meet the needs of the construction site, it is recommended to choose a traffic organization plan that closes two lanes and passes through two lanes.

6.2.3 Reasonably divide the length of construction sections.

The damage situation of different sections of Tianjin Avenue varies, and the maintenance construction plan is also different. Multiple construction sections are divided according to the pavement disposal plan. For road maintenance and major repair operations, considering the continuity of the construction process, the length of the enclosed area should be $\geqslant$ 2km. Due to the fact that the length of the above construction sections does not meet the standard for dividing them into multiple construction sections, no further division will be carried out according to the above construction plan.

6.2.4 Reasonably select the construction period.

The vehicles passing through the Tianjin Avenue section are mostly passenger and freight flows between the urban area and the Binhai New Area, with obvious tidal characteristics in vehicle travel. From 8 to 9 o'clock in the morning, there are more vehicles from the urban area to the Binhai New Area, and from 17 to 19 o'clock in the evening, there are more vehicles from the Binhai New Area to the urban area. Closed roads should be avoided for maintenance and construction during peak hours in the morning and evening. If there are fewer surrounding residents and the construction content allows, nighttime construction can also be considered.

6.2.5 Optimize vehicle composition.

Measures such as further optimizing policies to support the purchase and use of new energy vehicles, encouraging enterprises to enrich the supply of new energy vehicles, and increasing the configuration of charging facilities can be taken to promote the application of new energy vehicles.

7. Conclusions

(1) Propose a carbon emission list applicable to the traffic impact of Tianjin Avenue maintenance and construction, including a list of carbon emission factors for each vehicle type under different speed states and a list of carbon emission factors for different vehicle types during speed changes.

(2) Using a traffic impact carbon emission calculation model, calculate the traffic impact carbon emissions of supporting projects, and analyze the calculation data from the perspectives of individual projects, different carbon emission components, and different vehicle models.

(3) Explore the hot topics of carbon emissions caused by maintenance engineering traffic impact, and propose targeted emission reduction strategies for road maintenance engineering traffic organization based on the identified hotspots.

(4) The carbon emission factor list proposed in this article is based on the transportation characteristics of Tianjin Avenue, and has a certain regional and engineering significance. In subsequent research, further calculations should be made on the carbon emission factors affected by transportation, and a universal list of carbon emission factors should be proposed.

References

1. Statistical Bulletin on the Development of the Transportation Industry in 2022 [J]. China Water Transport, 2023 (07): 29-33 DOI: 10.13646/j.cnki.42-1395/u.2023.07.009
2. Yue H, Bird R, Bell M. A comparative study of the emissions by road maintenance works and the disrupted traffic using life cycle assessment and micro-simulation[J]. Transportation Research Part D Transport & Environment, 2009, 14(3):197-204.
3. Qian Zhang. Research on Environmental Assessment Method for Asphalt Pavement Maintenance Based on Life Cycle Theory [D] Beijing: Beijing University of Technology, 2015.
4. Galatioto F., Huang Y., Parry T., et al. Traffic modelling in system boundary expansion of road pavement life cycle assessment[J]. Transportation Research Part D, 2015, 36(2015):65-75.
5. Frey H.C. Trends in on road transportation energy and emissions[J]. Journal of the Air Waste Management Association, 2018,68(6):514-563
6. Yuanyuan Liu. Research on the Theory and Method of Carbon Dioxide Measurement for Highway Life Cycle Based on ALCA [D]. Chang'an University, 2019.

Comprehensive Assessment and Intervention: Analyzing and Mitigating Light Pollution Level in Different Regions Using the Entropy Weight Method, TOPSIS, and Fuzzy Comprehensive Evaluation Method

Xiaofan Lu*, Huanle Yin and Lingjun Shao

School of Mechanical Science and Engineering, Huazhong University of Science and Technology, Hongshan District, Wuhan, Hubei, China

Abstract. **Light pollution** is wreaking havoc on a global scale, which affects the environment, the rhythms of living things, and even our health and safety. In order to calculate the light pollution level in different regions, **the entropy weight method** was used to calculate the weight of light pollution evaluation indicators of eight cities of four types. Then, the light pollution risk level of eight cities was calculated using **TOPSIS(the Technique for Order Preference by Similarity to an Ideal Solution)**, which was defined as LPRI. Finally, the degree and order of light pollution in four types of regions are obtained. According to each evaluation indicator of light pollution, we put forward corresponding measures to reduce light pollution, and obtained the most effective measures for each region by **fuzzy comprehensive evaluation method**. The entropy weight method and TOPSIS can make full use of the original information of each indicator data in eight cities to obtain an objective and ac-curate light pollution risk level. When it is impossible to obtain the specific data of the impact of different measures on the LPRI index values of each region, the fuzzy comprehensive evaluation method using fuzzy mathematical concepts is used to objectively obtain the most effective intervention measures for different regions.

1 Introduction

Due to the excessive use of artificial light and other reasons, light pollution has become a new environmental pollution source after waste gas, waste water, waste residue and noise pollution. In order to reduce the impact of light pollution on humans and nature and create a more comfortable living environment, it is necessary to evaluate the level of light pollution in various regions of the world and give corresponding intervention measures. To achieve these goals, it is necessary to establish a universal evaluation model of light pollution level and propose effective intervention measures based on this model. Faichi et al.[1] produced a detailed map of global light pollution based on satellite data and precise measurements of sky brightness, revealing that many parts of the world are affected by light pollution. Kyba et al. (2017)[2] studied the growth of light pollution on a global scale and found that it is on the rise in many regions. Neither study delved into how to quantify and assess the risk of light pollution in specific areas. Gaston et al. (2013)[3] explored the impact of light pollution on biodiversity, highlighting the negative effects of light at night on many biological populations. Holker et al. (2010)[4] described in detail in their study how light pollution affects human communities and their quality of life. While their study discussed in detail the ecological and human impacts of light pollution, there is less discussion of how to quantify this impact and suggest specific measures to reduce it. Taking into account the shortcomings of previous studies, the main contents of this study are as follows: In order to calculate the light pollution level in different regions, we use entropy weight method to obtain the weight of the light pollution evaluation index combined with the data of eight cities of four types. Then we use TOPSIS(the Technique for Order Preference by Similarity to an Ideal Solution)to calculate the light pollution risk level of eight cities, which was called LPRI. Considering each light pollution evaluation indicator, we propose corresponding interventions to reduce the degree of light pollution, and use fuzzy comprehensive evaluation method to obtain the most effective measures for each region.

The following structure is: the third chapter introduces the main algorithm, the fourth chapter introduces the experimental results of the algorithm, and the fifth chapter is the summary.

* Corresponding author: 572216092@qq.com

2 Method

2.1. Selection of cities to be evaluated

To make the assessment universal, we divided four different types of regions[5]: Protected land, rural community, suburban community and urban community. After abstracting the characteristics of these four types of regions, we use eight cities with corresponding characteristics around the world to represent different regions to obtain the data of the light pollution evaluation index. The corresponding cities and regions are as follows: The urban community corresponds to Shenzhen in China and New York in the United States; the suburban community corresponds to Shijiazhuang in China and Rio de Janeiro in Brazil; the rural community corresponds to Lhasa in China and Kigali in Rwanda; the protected land corresponds to Sansha in China and Dili in East Timor.

2.2 Selection of indicators

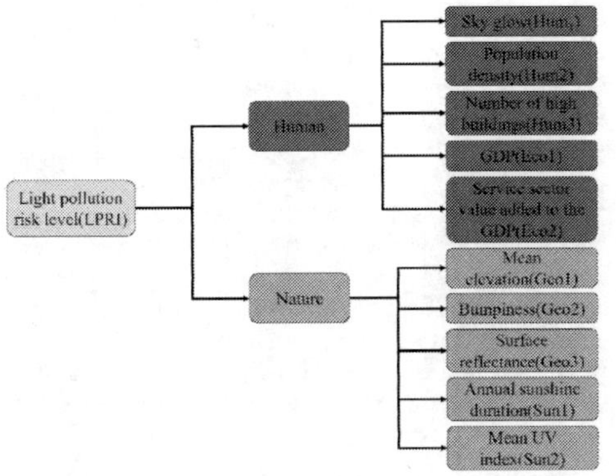

Figure 1 Evaluation indicators of light pollution level

Start with the possible causes of light pollution, the influencing factors of light pollution are divided into human factors[6] and natural factors[7], and then list the specific indicators. The selected indicators are shown in Figure 1.

2.3 Calculation of indicator weights

In general TOPSIS, the weight of all indicators is equal, which is obviously not in line with objective facts. Entropy weight method is an objective weighting method, which believes that the smaller the change degree of data value, the less information reflected, and the corresponding weight value is lower. The specific calculation steps are as follows[8]:

(1) Positivize and standardize. Since the indicators considered in this paper are all of the larger and better indicators, there is no need to positivize(that is, the larger the index value, the higher the degree of light pollution). There are 8 evaluation objects and 10 evaluation indicators, and the original data matrix X obtained is:

$$X = \begin{pmatrix} x_{11} & x_{12} & \cdots & x_{110} \\ x_{21} & x_{22} & \cdots & x_{210} \\ \vdots & \vdots & \ddots & \vdots \\ x_{81} & x_{82} & \cdots & x_{810} \end{pmatrix} \tag{1}$$

Since the index data are all positive, each element in matrix X is standardized using the following method:

$$z_{ij} = \frac{x_{ij}}{\sqrt{\sum_{i=1}^{n} x_{ij}^2}} \tag{2}$$

(2) Calculate the proportion of the i sample in the j index and regard it as the probability used in the calculation of relative entropy. Each term in the probability matrix is calculated as follows:

$$p_{ij} = \frac{z_{ij}}{\sum_{i=1}^{n} z_{ij}} \tag{3}$$

(3) Calculate the information entropy of each indicator. The calculation formula is as follows:

$$e_j = -\frac{1}{\ln n}\sum_{i=1}^{n} p_{ij} \ln p_{ij} \tag{4}$$

(4) Calculate the information utility value and normalize it to get entropy weight. The greater the information entropy of an indicator, the less information it reflects, so the information utility value is defined as:

$$d_j = 1 - e_j \tag{5}$$

The entropy weight of each index is obtained by normalization:

$$w_j = \frac{d_j}{\sum_{i=1}^{m} d_j} \tag{6}$$

2.4 Use TOPSIS to calculate LPRI of eight cities

When there is no data of the evaluation object, the analytic hierarchy process (AHP) is often used. However, when various index data of evaluation objects have been obtained, TOPSIS which can make full use of the original data is often used, and the result can accurately reflect the gap between evaluation objects[9]. The main steps for calculating LPRI using TOPSIS are as follows:

(1) Positivize and standardize. This step is done in the entropy weight calculation. **(2) Calculate the score and normalize.** The data matrix obtained after step 1 is as follows:

$$Z = \begin{pmatrix} z_{11} & z_{12} & \cdots & z_{110} \\ z_{21} & z_{22} & \cdots & z_{210} \\ \vdots & \vdots & \ddots & \vdots \\ z_{81} & z_{82} & \cdots & z_{810} \end{pmatrix} \tag{7}$$

Define the maximum and minimum values Z^+ and Z^- as:

$$Z^+ = (Z_1^+, Z_2^+, \cdots, Z_{10}^+) \tag{8}$$

where Z_j^+ equals $\max\{z_{1j}, z_{2j}, \cdots z_{8j}\}$.

$$Z^- = (Z_1^-, Z_2^-, \cdots, Z_{10}^-) \tag{9}$$

where Z_j^- equals $\min\{z_{1j}, z_{2j}, \cdots z_{8j}\}$.

Calculate the distance between the evaluation object and the maximum and minimum values:

$$D_i^+ = \sqrt{\sum_{j=1}^{10} \omega_j (Z_j^+ - Z_{ij})^2} \tag{10}$$

$$D_i^- = \sqrt{\sum_{j=1}^{10} \omega_j (Z_j^- - Z_{ij})^2} \tag{11}$$

ω_j is the weight of each index determined by entropy weight method. The unnormalized index of each city is calculated as follows:

$$S_i = \frac{D_i^-}{D_i^+ + D_i^-} \tag{12}$$

The LPRI value of each city is obtained by normalizing the score:

$$LPRI = \frac{S_i}{\sum_{i=1}^{n} S_i} \tag{13}$$

2.5 Use fuzzy comprehensive evaluation to get the optimal measure

Considering the indicators of light pollution mentioned above, we put forward ten possible measures to reduce light pollution: (1) Sky glow(Hum_1): Promote the use of low illumination and directional lighting. (2) Population density(Hum_2): Implement strict lighting regulations and standards in densely populated places. (3) Number of high buildings(Hum_3): Specify the materials and design of new buildings to reduce reflection and direct lighting. (4) GDP(Eco_1): Raise public awareness of light pollution through education and publicity to encourage more environmentally friendly ways of living and working. (5) Service sector value added to the GDP(Eco_2): Set up special lighting regulations for commercial areas. (6) Mean elevation(Geo_1): The altitude of a place is immutable, but we can give special management can to higher elevations. (7) Bumpiness(Geo_2): Designing and planning terrain can help reduce light pollution. (8) Surface reflectance(Geo_3): Choosing materials with low reflectivity for roads, buildings and other artificial surfaces can reduce the reflection of light and thus reduce light pollution. (9) Annual sunshine duration(Sun_1): Develop effective lighting strategies based on the number of hours of sunlight in the region. (10) Annual average UV index(Sun_2): Raise public awareness of UV pollution and encourage the use of sunshade devices.

(1) Determine the factor set. Here the factor set consists of each light pollution evaluation index:

$$\mathbf{U} = \{\mathrm{Hum}_1, \mathrm{Hum}_2, \cdots, \mathrm{Sun}_2\} \tag{14}$$

(2) Determine the evaluation set. We set the evaluation set as possible effect of measures on indicator value:

$$\begin{aligned}\mathbf{V} = \{&\text{'substantial increase'},\\ &\text{'slight increase'},\\ &\text{'little effect'},\\ &\text{'slight reduction'},\\ &\text{'substantial reduction'}\}\end{aligned} \tag{15}$$

(3) Determine the weight of each factor. In the previous article, We have used the entropy weight method to get the weights of each index:

$$\mathbf{A} = \{0.0110, 0.1937, \cdots, 0.0300\} \tag{16}$$

(4) Determine the fuzzy comprehensive evaluation matrix. The degree of membership of factors for each evaluation is defined as follows: "bound to occur" as 1, "very likely to occur" as 0.7, "likely to occur" as 0.3, and "almost unlikely" as 0. Taking the impact of measure (1) "Promote the use of low illumination and directional lighting" on urban community as an example, according to the above definition of evaluation set and membership degree, we can obtain the fuzzy comprehensive evaluation matrix. (5) Make comprehensive evaluation. From the weight **A** and fuzzy comprehensive evaluation matrix **R**, the fuzzy vector **B** about the evaluation set **V** can be obtained:

$$\mathbf{B} = \mathbf{A} \cdot \mathbf{R} = [0, 0, 0.8231, 0.1659, 0.0110] \tag{17}$$

Bi represents the degree of membership of the urban community to evaluation i when implementing measure (1). We assign each element in the evaluation set a score of {10, 4, -1, -4, -10}, with the positive number representing the increase and the negative number representing the reduction. The greater the negative value, the better the effect of the measure. Therefore, we can calculate the comprehensive effect value of measure (1) on the urban community:

$$I = [10, 4, -1, -4, -10] \cdot \mathbf{B} = -1.5968 \tag{18}$$

Repeat step(4) and step(5) to calculate the comprehensive effect value I when different measures are applied to the four regions(Urban community, suburban community, rural community and protected land), and we can obtain the comprehensive effect value of each measure in the four regions(The horizontal axis shows ten measures and the vertical axis shows four categories of areas).

3 Results and analysis

3.1 Indicator weight size analysis

The weights of indicators calculated by the steps of 2.2 are as follows in Table 1:

Table 1 Weights of indicators

Indicator	Weight
Hum_1	0.0110
Hum_2	0.01937
Hum_3	0.3581
Eco_1	0.3152
Eco_2	0.0442
Geo_1	0.0258
Geo_2	0.0171
Geo_3	0.0006
Sun_1	0.0043
Sun_2	0.0300

We can find that the weights of the number of high buildings(Hum_3) and GDP(Eco_1) are 0.3581 and 0.3152 respectively, which is large, while the weights of surface reflectance and annual shine duration are 0.0006 and 0.0043, which is very small. The direct reason is that the data of the first two fluctuates greatly while the data of the latter two fluctuates little. We try to analyze the underlying reasons as follows: For the "number of high buildings (Hum_3)", its greater weight may be due to the widespread use of artificial light sources in tall buildings, especially in commercial areas. For the "GDP (Eco_1)", its greater weight may be because GDP tends to correlate with the degree of industrialization and population density of an area, both of which can increase light pollution. For "surface reflectance (Geo_3)", it may have less weight because this indicator mainly reflects the light reflection of natural sur-faces (such as soil, vegetation, etc.). While this factor has some effect on light pollution, it may appear relatively small compared to the effect of artificial light. The lower weight of "annual sunshine duration (Sun_1)" may be because although natural light has some effect on light pollution, the effect of artificial light may be greater. On sunny and rainy days, the effects of natural light can vary greatly, but the effects of artificial light are relatively stable.

3.2 Analysis of LPRI in four regions

The calculated LPRI for each city from highest to lowest are shown in Table 2:

Table 2 LPRI of each city

Name of city	**Type of area**	**LPRI**
New York, USA	Urb	0.3849
Shenzhen, China	Urb	0.2767
Rio de Janeiro, Brazil	Sub	0.0898
Lhasa, China	Rur	0.067
Kigali, Rwanda	Rur	0.0646
Shijiazhuang, China	Sub	0.0563
Dili, East Timor	Pro	0.0508
Sansha, China	Pro	0.01

It can be seen that the light pollution degree is ranked from high to low as follows: Urban community, suburban community, rural community and protected land. We try to analyze the differences in light pollution levels in various regions as follows:

Urban Community: In terms of human factors, urban areas usually have higher population density, higher level of economic development, prosperous service industry, greater demand for artificial light, which will inevitably produce more pollution. **Suburban Community:** Human activities in suburban community have a certain demand for artificial light, and fewer buildings and more open terrain make natural light pollution reach a certain degree **Rural Community:** The low population density and low level of economy in rural communities limit the use of artificial light, but there is some natural light pollution due to the open terrain and few buildings. **Protected land:** The protected land is basically unaffected by human activities, so man-made light pollution can be ignored. However, due to the preservation of the original ecological environment, the roughness of the ground and the average altitude are often higher, so there is some light pollution caused by sunlight.

According to the above analysis, although light pollution can be divided into natural light pollution and artificial light pollution, the proportion of influence of human factors on light pollution is significantly greater than that of natural factors, indicating that human activities are the main cause of light pollution. Therefore, to reduce light pollution, the key point is how to adjust and limit the human behavior that is easy to cause light pollution.

3.3 The most effective measures to reduce light pollution in four regions

As shown in Table 3, the calculated comprehensive effect value of each measure in the four regions are as follows:

Table 3 The comprehensive effect value of each measure in the four regions

	Urb	Sur	Rur	Pro
(1)	-1.5968	-1.5968	-1.0231	-1.0099
(2)	-3.3312	-2.5828	-2.1317	-1.1974
(3)	-2.9724	-2.9724	-1.0231	-1.0099
(4)	.1.4679	-1.3751	-1.0099	-1.0000
(5)	-2.8082	-2.0083	-1.5739	-1.0000
(6)	-1.3322	-1.3322	-1.0000	-1.0000
(7)	-1.1330	-1.2067	-1.1539	-1.1539
(8)	-1.0153	-1.0043	-1.0029	-1.0013
(9)	-1.3535	-1.3235	-1.0000	-1.0000
(10)	-1.0000	-1.0000	-1.0000	-1.0000

According to the results in Table 3, the most effective measure for urban com-munity, rural community and protected land is measure (2) – "implement strict lighting regulations and standards in densely populated places". The most effective measure for suburban community is measure (3) – "specify the materials and design of new buildings to reduce and direct lighting". We analyze the reasons as follows:

Urban Community: Strict lighting time management system is required to reduce unnecessary lighting caused by high light demand. **Suburban Community:** By taking effective measures in architectural design, such as using building materials with lower reflectivity and arranging windows strategically to guide natural light, light pollution can be reduced at the source. **Rural Community:** Implementing strict lighting regulations and standards can help control and manage increasing demand for light. **Protected Land:** By implementing strict lighting regulations and standards, the impact of the activities of tourists and researchers on the natural environment of protected lands can be minimized as much as possible.

4 Conclusion and discussion

4.1 Conclusion

In this paper, the entropy weight method is used to calculate the weight of ten indicators to evaluate light pollution, and then TOPSIS is used to calculate the light pollution risk level of eight cities, and the results are corresponding to four types of regions. The study found that human indicators such as the number of high buildings(Hum_3) and GDP(Eco_1) have a high weight, while natural indicators such as sur-face reflectance(Geo_3) and annual shine duration(Sun_1) have a low weight. The light pollution levels in four regions from high to low are: Urban community, suburban community, rural community, protected land. After proposing ten measures to reduce light pollution corresponding to ten indicators, we use fuzzy comprehensive evaluation method to get the most effective measures corresponding to four regions.

4.2 Discussion

Human activities and natural factors are different in different regions, so after obtaining the quantified impact level of each measure on each region, efficient light pollution prevention and control can be targeted. In future studies, the selection of light pollution evaluation indicators can be optimized through multi-source data, expert consultation and preliminary research, and the membership degree in fuzzy comprehensive evaluation can be improved by adjusting model parameters or introducing new data sources.

References

1. Falchi, F., Cinzano, P., Duriscoe, D., Kyba, C. C., Elvidge, C. D., Baugh, K., ... & Furgoni, R. (2016) The new world atlas of artificial night sky brightness. Science advances, 2(6), e1600377.
2. Kyba, C. C., Kuester, T., Sánchez de Miguel, A., Baugh, K., Jechow, A., Hölker, F., ... & Guanter, L. (2017) Artificially lit surface of Earth at night increasing in radiance and extent. Science advances, 3(11), e1701528.
3. Gaston, K. J., Visser, M. E., & Hölker, F. (2015) The biological impacts of artificial light at night: the research challenge. Philosophical Transactions of the Royal Society B: Biological Sciences, 370(1667), 20140133.
4. Hölker, F., Moss, T., Griefahn, B., Kloas, W., Voigt, C. C., Henckel, D., ... & Tockner, K. (2010) The dark side of light: a transdisciplinary research agenda for light pollution policy. Ecology and Society, 15(4). [5] Pothukuchi, K. (2021) City light or star bright: a review of urban light pollution, impacts, and planning implications. Journal of Planning Literature, 36: 155-169.
5. Cinzano, P., & Falchi, F. (2014) Quantifying light pollution. Journal of Quantitative Spectroscopy and Radiative Transfer, 139: 13-20.
6. Rajkhowa, R. (2014) Light pollution and impact of light pollution. international Journal of Science and Research (IJSR), 3: 861-867.
7. Zhu, Y., Tian, D., & Yan, F. (2020) Effectiveness of entropy weight method in decision-making. Mathematical Problems in Engineering, 2020: 1-5.
8. Papathanasiou, J., Ploskas, N., Papathanasiou, J., & Ploskas, N. (2018) Topsis. Multiple Criteria Decision Aid: Methods, Examples and Python Implementations, 2018: 1-30.
9. Feng, S., & Xu, L. D. (1999) Decision support for fuzzy comprehensive evaluation of urban development. Fuzzy Sets and Systems, 105: 1-12.

Sustainability assessment of a Low-carbon Park building Based on Life cycle assessment(LCA)-Carbon emission method

Junxue Zhang[1,*], Jiankang Han[2], Xiaoting Zhai[2]

[1]Zhejiang Engineering Research Center of Building's Digital Carbon Neutral Technology, Zhejiang University City College, Hangzhou, 310015, China
[2]School of Civil engineering and Architecture, Jiangsu University of Science and Technology, Zhenjiang 212100, China

Abstract: The data shows that the greatest electricity consumption occurs during the first three months of the year, accounting for approximately 31.4% of the total electricity consumption. This is attributed to the high energy consumption associated with heating air conditioners during cold weather conditions. Similarly, the corresponding carbon dioxide emissions follow a similar pattern, with a peak observed in the first quarter of every year. Therefore, it is important to focus on reducing carbon emissions during this period. These findings highlight the need for energy-efficient heating and cooling systems that can help reduce the negative impact on the environment while promoting sustainability. By adopting renewable energy sources and improving energy efficiency in building systems, we can work towards creating more sustainable and environmentally friendly built environments.

1. Introduction

1.1 Background

In the face of global temperature changes, sustainable urban systems need to focus on. As an integral part of urban systems, the ecological and sustainable design of building systems is becoming increasingly important. As a complex system, building systems involve various factors that can be broadly classified into three categories: material flow, information flow, and energy flow. Additionally, low-carbon design of building systems is currently a popular research topic[1-4]. The Chinese government has implemented a series of policies and measures to encourage the development of low-carbon industries. These measures include the formulation of policy documents advocating energy conservation, emission reduction, and low-carbon development, establishing low-carbon demonstration zones and industrial parks, and strengthening research, development, and application of clean energy and environmental protection technologies. Chinese low-carbon industrial parks are typically regions that gather low-carbon industry enterprises and resources, aiming to promote the research, application, and dissemination of low-carbon technologies. These parks provide infrastructure and support services, attracting many low-carbon businesses to settle in. Within these parks, one can find enterprises and projects related to renewable energy, energy conservation, environmental protection, and clean technologies.

Overall, the development of low-carbon industrial parks in China benefits from government policy support and market demand. With increasing global attention on the low-carbon economy, it is expected that China's low-carbon industrial parks will continue to grow and play an important role in promoting sustainable development and addressing climate change.

1.2 Literature review

Different researchers have conducted multidimensional analyses on the sustainability of building systems. Some researchers have applied the life cycle theory to the field of architecture, aiming to analyze the sustainability of building systems. Some scholars have combined building information systems with the life cycle theory to support ecological building design. The ecological exergy approach is also a valuable perspective that can effectively validate the sustainability performance of building systems. In pursuit of carbon neutrality targets, researchers from around the world have conducted various analyses of building systems from a low-carbon perspective, involving areas such as low-carbon building design, zero-carbon buildings, low-carbon building materials, low-carbon building operation modes, and carbon sink analysis. Ecological low-carbon industrial park buildings are a special type of building complex that can integrate various low-carbon technologies and perspectives for analysis to determine their carbon emission status[5-18].

* Corresponding author: zjx2021@just.edu.cn

2. Case and Method

2.1 Building case introduction

The building is situated at No. 26, Hufu Road, Gusu District, Suzhou City, Jiangsu Province, adjacent to the Tiger Hill Scenic Spot. The project aims to design a low-carbon industrial park comprising of three categories: Huqiu Academy (8600 m2), catering hotel (1518 m2), and residential land (1000 m2). The focus on developing a low-carbon industrial park highlights the importance of reducing negative environmental impacts associated with building design and operation. By utilizing sustainable practices and materials and promoting renewable energy sources, we can work towards creating more environmentally friendly and sustainable built environments that minimize our negative impact on the environment while promoting long-term sustainability(Table 1).

Table 1 Basic situation

No.	Picture	Data	Power consumption
1		8600 m^2	Start of use: January Operation period: Open for 365 days a year Air conditioning operation period: The air conditioning runs for 10 months in a year. Summer: Cooling typically starts in late April and ends in late October, lasting for 6 months. Winter: Heating typically starts in early November and ends in late March, lasting for 4 months. Air conditioning control methods: Central air conditioning and split air conditioning. Front desk centralized control: Areas such as the lobby, restaurant, and entertainment rooms. Guest self-control: Guest room air conditioning, which is equipped with panel-style wired controllers. Space lighting: Independent switches are used to manage each area. Public areas: Emergency lighting is on 24 hours a day, and other lights are controlled by timed switches based on the work schedule. Indoor guest rooms: Each guest can turn on/off the lights in their own room. Outdoor lighting: Landscape lighting is automatically controlled by the control system and the operating time is adjusted automatically according to the season and light intensity. The main operating period is roughly from 18:00 to 23:00.
2		1518 m^2	
3		1000 m^2	

2.2 Methodology

The calculation formula of the carbon emission calculation model for the whole life cycle of buildings is as follows:

$$E_\lambda = E_\sigma + E_t + E_c + E_o + E_d$$

Where E_λ =the total carbon emission; E_σ =the carbon emission of material production stage; E_t = the carbon emission of transport stage; E_c = the carbon emission of construction; E_o = the carbon emission of operationalphase; E_d = the carbon emission of demolition stage.

3. Results and Discussion

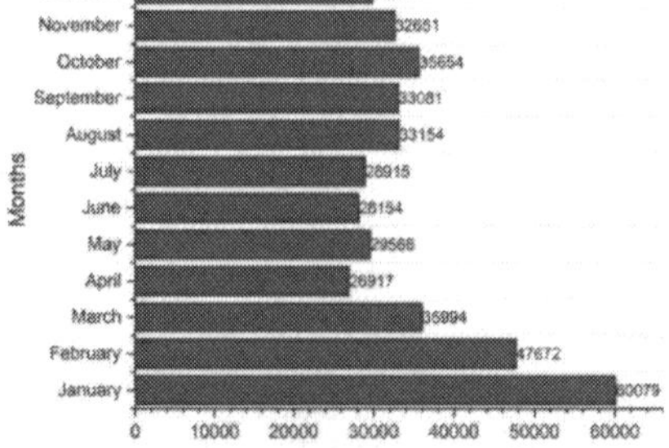

Figure 1 Comparison of building electricity consumption in industrial park

 https://doi.org/10.1051/e3sconf/202344102017

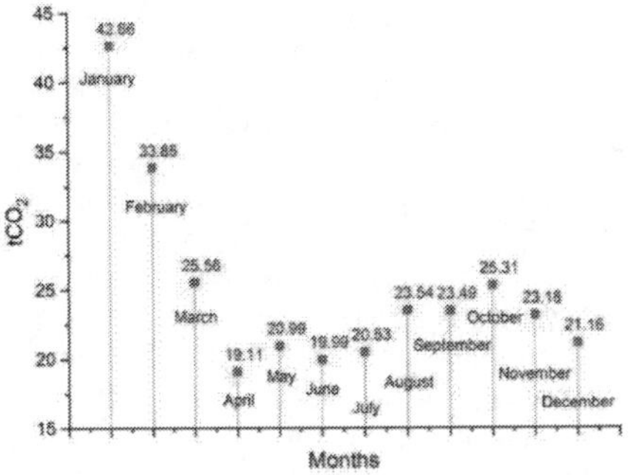

Figure 2 Comparison of building carbon emission in industrial park

In Figure 1 and Figure 2, based on the data presented, it is evident that the greatest power consumption occurs in the first three months, accounting for approximately 31.4% of the total electricity consumption. This is due to the high energy consumption of heating air conditioners during cold weather conditions. Similarly, the corresponding carbon dioxide emissions also follow a similar pattern. The highest carbon dioxide emissions were observed in January (42.66 tCO_2), followed by February (33.85 tCO_2) and March (25.56 tCO_2). These findings highlight the need for energy-efficient heating and cooling systems that can help reduce the negative impact on the environment while promoting sustainability. By adopting renewable energy sources and improving energy efficiency in building systems, we can work towards creating more sustainable and environmentally friendly built environments. Overall, the analysis of electricity consumption and carbon emissions provides valuable insights into the environmental impact of building systems and can guide the development of strategies to promote long-term sustainability and reduce our negative impact on the environment.

4. Improved Measures

The Chinese government is currently promoting the development of clean energy utilization as a means of promoting sustainability. By adopting clean energy sources such as solar, wind, and hydroelectric power energy, the building system can significantly reduce its negative impact on the environment and promote long-term sustainability. Among the three main types of clean energy in China, solar energy is the most popular way of harnessing renewable energy. By utilizing solar panels to generate electricity, building systems can significantly reduce their dependence on non-renewable energy sources and promote environmental sustainability. Overall, by adopting clean energy sources and improving energy efficiency in building design and operation, we can work towards creating more sustainable built environments that minimize our negative impact on the environment while promoting long-term sustainability. Continued research and innovation are needed to identify new solutions to improve the sustainability of building systems and promote the adoption of clean energy sources.

To enhance the sustainability of low-carbon industrial park buildings, the following paths can be taken:

1. Energy efficiency improvement: Reduce energy consumption by using efficient equipment and technologies such as LED lighting, energy-efficient HVAC systems, and smart energy management systems. Enhance building insulation and sealing to minimize energy losses.
2. Utilization of renewable energy: Meet a portion or all of the electricity demand through renewable energy sources like solar power and wind energy. Install facilities such as solar photovoltaic panels and wind turbines, integrating the buildings with renewable energy systems.
3. Water resource management: Adopt water-saving devices and technologies such as low-flow faucets and water-efficient toilets to reduce water usage. Collect and utilize rainwater for non-potable purposes like irrigation and flushing.
4. Building material selection: Choose environmentally friendly and renewable materials, giving priority to those with lower carbon emissions. Minimize resource extraction and pollutant emissions.
5. Waste management: Implement effective waste sorting and disposal systems including recycling, reusing, and proper waste treatment to minimize environmental impact.
6. Green transportation planning: Encourage low-carbon commuting methods such as walking, cycling, and public transportation. Provide convenient infrastructure like bicycle parking facilities and electric vehicle charging stations.
7. Building design and layout: Optimize building design by incorporating standards like Passive House to reduce energy demands. Plan building layouts strategically to minimize energy transmission losses and carbon dioxide emissions.
8. Environmental monitoring and assessment: Establish comprehensive environmental monitoring systems to track real-time energy consumption, water resource utilization, and waste management. Regularly evaluate the effectiveness of sustainability measures and make improvements accordingly.

By comprehensively employing these pathways, the sustainability of low-carbon industrial park buildings can be significantly improved, reducing their negative environmental impact while conserving resources and lowering operational costs.

5. Conclusions

The use of LCA-Carbon emission methodology provides valuable insights into the environmental impact of building systems, allowing architects and designers to make informed decisions to promote sustainability. By adopting a comprehensive approach that integrates sustainable practices and materials into building design and operation, we can work towards creating more environmentally friendly and sustainable built environments. The use of LCA-Carbon emission methodology can guide the development of strategies to reduce negative environmental impacts and promote long-term sustainability. Overall, continued research and innovation are needed to identify new solutions to improve the sustainability of building systems and reduce carbon emissions. By adopting sustainable practices and materials and promoting the adoption of renewable energy sources, we can work towards creating more sustainable built environments that minimize our negative impact on the environment while promoting long-term sustainability.

Acknowledgments

The work described in this paper was supported by the XJTLU Urban and Environmental Studies University Research Centre (UES)(UES-RSF-23030601); the Zhejiang Engineering Research Center of Building's Digital Carbon Neutral Technology (No. ZJ2022YB-04)

References

1. Genevieve L. Noyce, Alexander J. Smith, Matthew L. Kirwan, Roy L. Rich & J. Patrick Megonigal. Oxygen priming induced by elevated CO2 reduces carbon accumulation and methane emissions in coastal wetlands. nature geoscience. 2023, 16, 754-760.
2. Peng Jiang, Christian Sonne, and Siming You. Dynamic Carbon-Neutrality Assessment Needed to Tackle the Impacts of Global Crises. Environmental Science Technology. 2022, 56, 9851–9853.
3. Ruijun Chen, Yaw-Shyan Tsay, Ting Zhang. A multi-objective optimization strategy for building carbon emission from the whole life cycle perspective. Energy, 2023, 262, 125-129.
4. Qiang Du, Yalei Wang, Qiaoyu Pang, Tingting Hao, Yuqing Zhou. The dynamic analysis on low-carbon building adoption under emission trading scheme. Energy, 2023, 263, 248-251.
5. Yubing Zhang, Xiaoyan Jiang, Caiyun Cui, Martin Skitmore. BIM-based approach for the integrated assessment of life cycle carbon emission intensity and life cycle costs. Building and Environment, 2022, 226, 109-111.
6. ZhiWu Zhou, Julián Alcalá and Víctor Yepes. Bridge Carbon Emissions and Driving Factors Based on a Life-Cycle Assessment Case Study: Cable-Stayed Bridge over Hun He River in Liaoning, China. International Journal of Environmental Research and Public Health, 2020, 17, 53-59.
7. Yiting Kang, Wei Xu, Jianlin Wu, Han Li, Ruijie Liu, Shilei Lu, Xian Rong, Xiaolong Xu, Feng Pang. Study on comprehensive whole life carbon emission reduction potential and economic feasibility impact based on progressive energy-saving targets: A typical renovated ultra-low energy office.Journal of Building Engineering, 2022, 58, 105-129.
8. Martin Röck, Marcella Ruschi Mendes Saade, Maria Balouktsi, Freja Nygaard Rasmussen, Harpa Birgisdottir, Rolf Frischknecht, Guillaume Habert, Thomas Lützkendorf, Alexander Passer. Embodied GHG emissions of buildings – The hidden challenge for effffective climate change mitigation. Applied Energy, 2020, 258, 114-117.
9. Petrovi´c; Zhang, X.; Eriksson, O.; Wallhagen, M. Life Cycle Cost Analysis of a Single-Family House in Sweden. Buildings, 2021, 11, 215-220.
10. Xiao-juan Li, Wan-jun Xie, Le Xu, Lu-lu Li, C.Y. Jim, Tai-bing Wei. Holistic life-cycle accounting of carbon emissions of prefabricated buildings using LCA and BIM. Energy & Buildings, 2022, 266, 132-136
11. Hu Li, Zhixing Luo, Xudong Xu, Yujie Cang, Liu Yang. Assessing the embodied carbon reduction potential of straw bale rural houses by hybrid life cycle assessment: A four-case study. Journal of Cleaner Production, 2021, 303, 127-131.
12. Li, Y., Zeng, S., Liu, Y., Dong, L., & Wang, J. Evaluation and optimization of low carbon industrial park development in China: A case study of Suzhou Industrial Park. Journal of Cleaner Production, 2019, 231, 375-386.
13. Wang, J., & Shen, G. Q. Low-carbon industrial parks: A review. Journal of cleaner production, 2017, 166, 1046-1063.
14. Song, Y., Qian, Q. K., & Chen, Q. P. Design strategies for low carbon industrial building in China: A preliminary study on energy performance and thermal comfort. Energy Procedia, 2018, 153, 32-37.
15. Zhang, X., Yuan, X., Luo, H., Yan, C., & Ji, R. Life cycle evaluation framework for low carbon industrial park buildings with renewable energy utilization. Applied Energy, 2020, 260, 114253.
16. Xu, W., Xia, B., & Yang, H. Evaluation model for low-carbon industrial park buildings based on analytic hierarchy process and fuzzy comprehensive evaluation. Sustainability, 11(2), 460.
17. Guo, L., & Ying, Z. Research on the sustainable development of low-carbon industrial parks based on the ecological civilization perspective. Sustainability, 2018, 10(4), 1268.
18. Zhao, M., Wang, J., & Wu, J. Promoting low-carbon development in industrial parks through policy mix: Evidence from China. Journal of Cleaner Production, 2019, 209, 1326-1337.

https://doi.org/10.1051/e3sconf/202344102018

Design and Test Analysis of 1GYF-240 Sugarcane Straw Crushing and Returning Machine

Shuo Wang[1,2], Weihua Huang[1,2,*], Chang Ge[1,2,3], Zhenhui Zheng[1], Lijiao Wei[1,2,4], Ming Li[1,2], Zhaojun Niu[1,2], Dongjie Du[1,2,3] and Changdong Sun[5]

[1]Agricultural Machinery Institute, Chinese Academy of Tropical Agricultural Sciences. Zhanjing, 524001, China
[2]Zhanjiang Conservation Tillage Equipment Engineering Technology Research Center. Zhanjing, 524001, China
[3]Zhanjiang Kaixiang Technology Co., Ltd. Zhanjing, 524001, China
[4]GuangDong Engineering Technology Research Center of Precision Emission Control for Agricultural Particulates. Zhanjing, 524001, China
[5]Chinese Academy of Tropical Agricultural Sciences. Haikou, 570100, China

Abstract. Sugarcane is one of the largest economic crops in China. However, the current domestic sugarcane straw crushing and returning machines have limitations such as short service life, poor crushing effect, and low operating efficiency. In order to solve these problems, a 1GYF-240 sugarcane straw crushing and returning machine was developed in this study. First of all, the power is provided by the tractor, and at the same time, it drives the blade in the crushing mechanism to rotate at a high speed. Next, the profiling leaf collector collects the sugarcane stalks that are close to the ground and feeds them into the crushing chamber. Finally, the swinging knife and the fixed knife are used for crushing, and the crushed straw is evenly thrown to the field along the rear baffle under the action of centrifugal force, thereby completing the operation of crushing and returning the sugarcane straw to the field. The results of field experiments show that the qualified rate of sugarcane straw crushing is 93.8%, and the work efficiency is 0.78hm2·h-1, which shows good operation effect. The results of comparative experiments show that this study not only solves the environmental pollution problem of traditional incineration well, but also has the function of heat preservation and water retention. In addition, compared with similar domestic models, the experimental results show that the crushing and returning machine has obvious advantages in picking up rate and crushing rate of sugarcane straw, and has higher reliability in actual operation. This research can provide the foundation and support for the design and popularization of the sugarcane leaf crushing and returning machine.

1 Introduction

Sugarcane is the main sugar crop in China, and more than 90% of the sugar comes from sucrose. The healthy development of the sugarcane industry is an important guarantee for the effective supply of sugar and the national sugar security[1,2,3]. In the sugarcane industry, a large amount of waste (such as cane leaves, straw, etc.) is often piled up on a large scale or burned on site[4,5]. This phenomenon has caused the environment to suffer serious pollution and has caused non-negligible damage to the quality of the environment. Therefore, how to turn these materials, which are regarded as wastes, into treasures has become an issue that the current government departments attach great importance to and arouses widespread concern in the society.

Crushing and returning sugarcane stalks to the field is the most direct, realistic and easy-to-promote technical measure[6,7]. At present, sugarcane straw crushing and returning machines widely used in China, such as 1GYF-150, 4F-1.8, 3SY-140, 1JH-150, etc., all use wheeled tractors as power supporting equipment to complete the operation[8]. However, the above-mentioned implements have disadvantages such as short service life, poor crushing effect, and low operating efficiency in the actual operation process. To solve the above problems, this study designed a 1GYF-240 type sugarcane straw crushing and returning machine based on the characteristics of sugarcane planting in southern hot areas. Possibility of wind and water erosion, while improving soil fertility and drought resistance. The whole machine has the advantages of light weight and stable operation performance.

* Weihua Huang: 271253449@qq.com

2 Machine structure and working principle

2.1. Machine structure

As shown in Figure 1 and Figure 2, the 1GYF-240 sugarcane straw crushing and returning machine is mainly composed of a transmission system, a frame, a front baffle, a throwing knife, a fixed knife, a knife roller, a rear baffle, a limit wheel and a profiler. Composition of leaf collectors, etc. The machine is hung on a wheeled tractor with a power of more than 88 kW through three points. Among them, the profiling leaf collector is mainly composed of inclined rods and connecting plates. The slant bar consists of a bent rod and a culet etc. bolted together, which allows quick and easy replacement of the culet after wear or impact damage.

Fig. 1. 1GYF-240 sugarcane straw crushing and returning machine physical picture

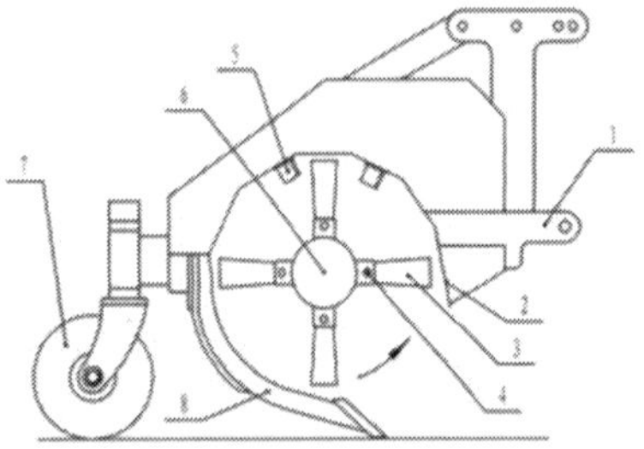

Fig. 2. 1GYF-240 Structural Diagram of Sugarcane Straw Crushing and Returning Machine. 1. Three-point suspension 2. Frame 3. Moving knife 4. Pin shaft 5. Fixed knife 6. Knife shaft 7. Limiting wheel 8. Leaf collector

The 1GYF-240 sugarcane straw crushing and returning to the field has the following characteristics: ① The limit wheel is used as the profiling wheel, and it is rigidly connected with the inclined rod. This makes the oblique bar and the tractor wheel on the same end face, and is arranged in front of the lower flail. ② The design of the machine makes the bottom tip of the inclined rod close to the ground, which can be adjusted adaptively according to the ups and downs of the ground, so as to achieve a roughly profiling effect. ③ Due to the function of the profiling leaf collector, the sugarcane stalks close to the ground can be collected and lifted to a certain height, which facilitates subsequent crushing operations. The main technical parameters of the 1GYF-240 double-roller sugarcane straw crushing and returning machine are shown in Table 1.

2.2 Working principle

In actual operation, the 1GYF-240 sugarcane straw crushing and returning machine is first connected to the tractor through a three-point suspension. Then, the tractor provides power, and the power is transmitted to the gear box through the universal joint shaft, and then to the V-belt pulley through the transmission shaft, and the V-belt variable speed transmission drives the blade in the crushing mechanism to rotate at a high speed. Then, when the tractor is moving forward, the profiling leaf collector collects the sugarcane stalks that are close to the ground and feeds them into the crushing chamber, where the throwing knife and fixed knife are crushed. Finally, under the action of centrifugal force, the crushed straw is evenly thrown into the field along the back baffle, thus completing the operation of crushing and returning sugarcane straw to the field[9,10,11]. Adjusting the height of the limit ground wheel and the frame from the ground can control the working height of the movable knife from the ground and improve the crushing rate of sugarcane straw.

Table 1. Main technical parameters of 1GYF series sugarcane straw crushing and returning machine.

Main parameter unit	Value
Length×width×height/mm	2650×1650×1250
Prototype total mass/kg	750
Supporting power/kW	88.2～117.6
Machine working range/cm	240
Type and number of flails	48 L improved knives
Knife roller speed/r·min-1	1830
Productivity/hm2·h-1	≥0.74
Crushing rate/%	≥93.6
Pick up rate/%	≥93.2

3 Structural Design of Key Components

3.1. Tool selection and design

In general, the blades of the straw crushing and returning machine can be divided into four categories according to the shape, namely, the hammer knife, the straight knife, the T knife, the L-shaped knife and its improved type[12,13]. Among them, the hammer melon crushing effect is good, but the energy consumption is large. The straight knife has a good cutting effect, but it is easy to clog, and the blade needs to be cleaned frequently, which affects the working efficiency. The manufacturing cost and maintenance difficulty of the T-knife are relatively high, and at the same time, it has relatively large vibration and noise during operation, which may affect the stability of the machine and the operating environment. The L-improved knife is mainly used for chopping and cutting, and has the characteristics of good picking effect and not easy to be entangled and blocked by sugarcane leaves. Therefore, this study chooses the L improved knife as the knife of the sugarcane straw crushing and returning machine. At the same time, the throwing knife and the fixed knife overlap a certain amount in the radial direction, so that the whole machine has a higher picking rate and crushing rate, and the power consumption is small.

Fig. 3. 1GYF-240 Sugarcane Straw Crushing and Returning Machine

Through the combination, optimization and arrangement of moving and fixed knives, the moving knives adopt two L improved knife structures (Figure 3), the knives are arranged in a double helix arrangement, the number of blades is increased on both sides, and the crushing chamber is enlarged. The cutter effectively overcomes the problem of straw blockage and winding, the crushing rate is increased by 5%, the efficiency is increased by 3%, and it can be adapted to the crushing operation of different crop straws.

3.2. Design of steerable ground wheels

Sugarcane is a ridge planted crop. In actual operation, the sugarcane straw crushing and returning machine usually needs to be equipped with a ground wheel device. The device is generally composed of wheels, wheel shafts, pin shafts, etc., and adjusts the distance between the tool tool and the ground by controlling the up and down movement of the pin shaft. However, due to the high-speed rotation of the tool rolls, the rotational speed of most of the tool rolls is above 1800 revolutions per minute, which makes the entire machine tool prone to relatively large vibrations. Especially when making headland turns or lifting, this vibration can become more severe. In severe cases, this vibration may cause damage to important components and parts such as cutter roller bearings, and affect the working reliability and service life of the machine tool.

Fig. 4. Limit Wheel Connection

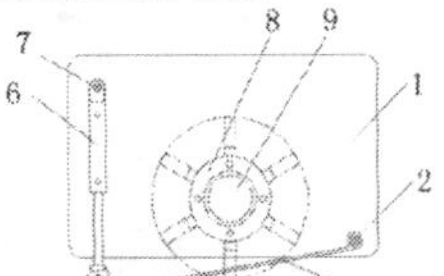

Fig. 5. Automatic adjustment limit device

To solve the above problems, this study designed a steerable profiled ground wheel instead of a fixed ground wheel, and designed a high-standard tire vibration damping configuration. The device effectively solves the problem of vibration caused by jumping or displacement when the machine tool is lifted, and improves the working efficiency and reliability of the machine tool, with a reliability of 92.7%. In addition, the limit wheel connection device (Figure 4) and the automatic adjustment limit device (Figure 5) of the sugarcane straw crushing and returning machine were designed to improve the crushing efficiency of the machine and ensure the reliability of the machine.

3.3. Rack design

In this study, according to the growth characteristics of sugarcane planting, the frame and rear cover of the crushing and returning machine are optimized, and the transmission gearbox is reversed and reversed to adapt to the working environment with many stones and uneven surfaces.

In this study, an adjustable offset sugarcane straw crushing and returning machine connection mechanism is designed. Through the lateral movement of the front beam, the sugarcane leaf crushing and returning machine can still continue to work normally during the horizontal movement of the sugarcane straw crushing and returning machine , which improves the quality and work efficiency of the sugarcane straw crushing and returning machine.

4 Field experiments

In order to verify the effectiveness of the proposed 1GYF-240 sugarcane straw crushing and returning machine, this study designed sugarcane straw crushing and returning experiments and comparative experiments.

4.1. Evaluation index

During the test, the tractor speed was set to the slow second gear of 1 m/s, and 10 1×1 (m) measurement areas were randomly selected in the test area where the sugarcane straw crushing and returning machine passed, and all the crushed particles in the measurement area were picked up. For sugarcane straw, use a balance to weigh out the mass m1, then pick out the sugarcane straw with a length > 20 cm and weigh out its mass m2, and calculate the pass rate of crushing sugarcane straw in each measurement area according to the formula (1) [14], namely:

$$P = \left(\frac{m_1 - m_2}{m_1} \right) \times 100\% \quad (1)$$

In the formula, P is the pass rate of sugarcane straw crushing in each measurement area; m1 is the total mass of sugarcane straw (g) in each measurement area; m2 is the total mass (g) of sugarcane straw with a length > 20 cm in each measurement area.

The working efficiency of the sugarcane straw crushing and returning machine refers to the working area of the machine per unit time. Use a stopwatch to measure and record the time it takes for the machine to complete a sampling test unit in the field, and its work efficiency is calculated according to formula (2). The average value of the recorded 10 sets of values is the working efficiency of the sugarcane straw crushing and returning machine[15].

$$\omega = \frac{3.6LB}{10000q} \qquad (2)$$

In the formula, ω is the working efficiency of the machine (hm2/h); L is the length of the sampling unit (m); B is the working width (mm); q is the time spent on one sampling test unit (s).

4.2. The experiment of crushing and returning to the field

The test of sugarcane straw crushing and returning to the field was carried out at the sugarcane straw crushing and returning demonstration base of the Institute of Agricultural Machinery, Chinese Academy of Tropical Agricultural Sciences, Jinxing Farm, Zhanjiang Nongken, Guangdong, with an area of 2 hm2. Before the test, 5 small experimental areas of 10×10 (m) were randomly selected in the test area, and the sugarcane leaf coverage rate and sugarcane leaf coverage thickness were measured. The field operation scene diagram of the 1GYF-240 sugarcane straw crushing and returning machine is shown in Figure 6.

Fig. 6. 1GYF-240 sugarcane straw crushing and returning machine physical picture

Table 2. Coverage rate of sugarcane straw in experimental plot.

Serial number	1	2	3	4	5	Average
Coverage/ %	95.45	97.24	92.68	99.36	98.82	96.71

Table 3. Covering thickness of sugarcane straw in the experimental plot.

Serial number	1	2	3	4	5	Average
Covering thickness/cm	10.2	9.7	11.6	13.8	11.5	11.36

Table 2 and Table 3 are the calculation results of sugarcane leaf coverage and sugarcane leaf coverage thickness in randomly selected experimental areas, from which it can be seen that the average coverage rate and average straw coverage thickness of sugarcane straw in the experimental area are 96.71% and 11.36cm, respectively. The calculation results show that the coverage rate of sugarcane straw in the field is high, and a crushing and returning machine is needed for efficient crushing and returning to the field. At the same time, the sugarcane straw covering thickness is within the operating range of the 1GYF-240 sugarcane straw crushing and returning machine.

Table 4. Coverage rate of sugarcane straw in experimental plot.

Crushing pass rate/%	Working efficiency/$hm^2 \cdot h^{-1}$
93.8	0.78

It can be seen from Table 4 that in the actual operation process, the pass rate of sugarcane straw crushing is 93.8%, and the working efficiency is 0.78 $hm^2 \cdot h^{-1}$. The field test results are shown in Figure 6. The 1GYF-240 sugarcane straw crushing and returning machine works well, and the crushing pass rate and machine performance are relatively stable.

4.3. Contrast test of crushing and returning to field and burning

In order to further prove the impact of the 1GYF-240 sugarcane straw crushing and returning machine on the field after operation, this section designs a group of comparison experiments between the sugarcane straw crushing and returning method and the traditional incineration returning method. The test was carried out at the sugarcane straw crushing and returning demonstration base of the Institute of Agricultural Machinery, Chinese Academy of Tropical Agricultural Sciences, located in Zhanjiang Nongken Jinxing Farm, Guangdong Province (20.65° north latitude, 110.15° east longitude). The trial will run from April 2022 to December 2022. The soil is red loam, and the sugarcane variety is 'Yuetang 7929', which will be newly planted in 2021.

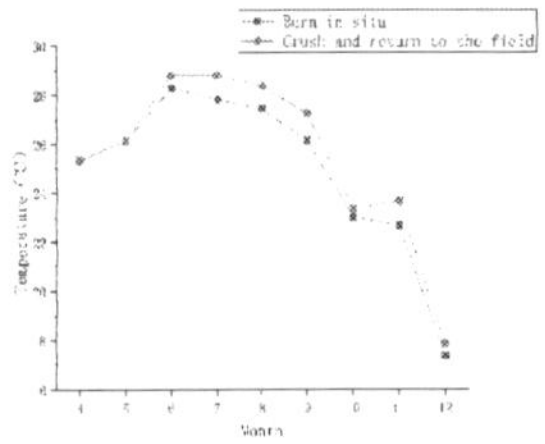

(a) Temperature

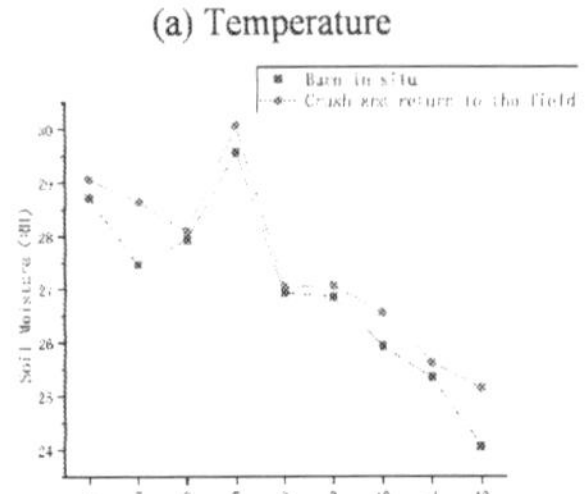

(b) Humidity

Fig. 7. The temperature and humidity change curves of the field under the method of returning to the field by crushing and returning to the field by burning

The comparative experiment included two treatment methods of crushing sugarcane straw and returning it to the field by burning it. Among them, each treatment group includes 5 experimental plots with an area of about 0.26 hm2, and each experimental plot has 5 random temperature and humidity measurement nodes.

Fig. 7 is a graph showing the change of temperature and humidity in the field with time after the sugarcane stalks were crushed and returned to the field and incinerated and returned to the field. As for the soil temperature, it can be seen from Figure 7(a) that the temperature of the test plot using the method of crushing and returning to the field is higher than that of the test plot using the method of burning and returning to the field from April to December. In terms of soil moisture, it can be seen from Figure 7(b) that the humidity of the test plots using the method of crushing and returning to the field was higher than that of the test plots using the method of burning and returning to the field from April to December.

The test results show that the use of crushing and field returning machines not only solves the problems of air pollution caused by traditional incineration, endangering the environment and human health, but also has the function of heat preservation and water retention, which helps to protect the health of the soil ecosystem and provides Good growth environment, thereby increasing sugarcane yield.

Based on the analysis of this conclusion, the main reasons are as follows:

Insulation phenomenon: ①Straw mulch can form a physical barrier to protect the soil surface from direct sunlight and high temperature. This helps to slow down the rate of evaporation of the soil, thereby reducing the loss of soil moisture. ②Straw mulching can slow down the change of soil temperature, especially at night, and help maintain a relatively stable soil temperature.

Water retention phenomenon: ①Straw mulching can reduce the evaporation rate of the soil surface, reduce the loss of water evaporation from the soil to the air, and help to maintain soil moisture. ②The decomposition of straw will release organic matter, which can improve the water retention capacity of the soil. They absorb water and form stable water stores in the soil, making the soil more water-retaining. ③Straw mulching can slow down the erosion and runoff speed of rainwater, help to retain water in the soil and reduce water loss.

4.4. Compared with the domestic sugarcane straw crushing and returning machine

In order to further reflect the superiority of the 1GYF-240 sugarcane straw crushing and returning machine proposed in this study, this section designs a group of comparative experiments between this model and the existing domestic sugarcane straw crushing and returning machine.

Table 5. Comparison of 1GYF-240 sugarcane straw crushing and returning machine with similar domestic models.

Type	1GYF-240	1GYF-150	3SY-140	FZ-100	4F-1.8
Productivity (hm2/h-1)	≥0.74	≥0.15	≥0.133	≥0.2	≥0.8
Workpiece width (cm)	240	150	140	100	180
Pick up rate (%)	≥93.2	≥91	≥90	≥90	≥90
Crushing rate (%)	≥93.6	≥91	≥80	≥80	≥80
Crushing length (cm)	≤20	≤20	≤30	≤25	≤10
Supporting power (kW)	88.2～117.6	36～58	36～58	36～58	58～66
Working reliability (%)	92.7	91.3	89.5	91.4	90.8

It can be seen from Table 5 that, compared with similar crushing models at home and abroad, the 1GYF-240 sugarcane straw crushing and returning machine has higher productivity and operating width, and the sugarcane straw picking rate and crushing rate have obvious advantages. In addition, in this study, the performance of the crushing and returning machine body is more stable, the service life of the V-belt is at least doubled, and the sugarcane head is not damaged, and the reliability is higher in actual operation.

5 Field experiments

1) A 1GYF-240 sugarcane straw crushing and returning machine was developed, which saves energy and has good crushing rate and working efficiency. First of all, the power is provided by the tractor, and at the same time, it drives the blade in the crushing mechanism to rotate at a high speed. Next, the profiling leaf collector collects the sugarcane stalks that are close to the ground and feeds them into the crushing chamber. Finally, the swinging knife and the fixed knife are used for crushing, and the crushed straw is evenly thrown to the field along the rear baffle under the action of centrifugal force, thereby completing the operation of crushing and returning the sugarcane straw to the field.

2) The results of field experiments show that the pass rate of 1GYF-240 sugarcane straw crushing and returning machine is 93.8%, and the working efficiency is 0.78hm2 • h-1. The crushing pass rate of the whole machine and the performance of the equipment are relatively stable, which meets the agronomic requirements of sugarcane leaf crushing.

3) The results of comparative experiments show that this study not only solves the environmental pollution problem of traditional incineration well, but also has the function of heat preservation and water retention.

4) Finally, compared with similar domestic models, the experimental results show that the crushing and

returning machine has obvious advantages in picking up rate and crushing rate of sugarcane straw, and has higher reliability in actual operation.

Acknowledgments

This research was funded by the Hainan Provincial Natural Science Foundation of China (No.320QN328, 122QN386). Central Public-interest Scientific Institution Basal Research Fund (No.1630132023002, 1630132022001, 1630012023009). Zhanjiang Science and Technology Plan Project (No.2020A04011, 2020A05004).

References

1. Chen C., Hu Z. (2021) Considerations on High Quality Development of Sugar in China. Sugarcane and Canesugar, 50(1):1-7.
2. Liu X., Cao F., Li K., et al. (2021) Comparison of Global Sucrose Industry Competitiveness and China's Improvement Path Discussion--A comparative analysis based on Brazil Australia, Thailand, and India. Price:Theory & Practice, (12) : 12-17, 138.
3. Wei L., Huang X., Huang W., etc. (2022) The Analysis on the Supporting Equipment and Technical Mode of Sugarcane Leaf Crushing and Returning to the Field in Southern Sugarcane Region. Modern Agricultural Equipment, 43(05):41-44.
4. Dong X., Zhang J., Li M., et al. (2013) Structural Design of Key Components of 1GYF-240 Sugarcane Leaf Crushing and Returning Machine. China Tropical Agriculture, (6): 39-41.
5. PABLO A. SILVA ORTIZ, DANIEL FLOREZ-ORREGO, ADRIANO PINTO MARIANO, et al. (2022) Exergy and economic assessment of renewable electricity generation from sugarcane straw for improved efficiency of sugarcane biorefineries[J]. International journal of exergy, 38(2):218-237.
6. ITALLO DIRCEU COSTA SILVA, ZIGOMAR MENEZES DE SOUZA, ANA PAULA GUIMAR?ES SANTOS, et al. (2022) Removal of Different Quantities of Straw on the Soil Surface: Effects on the Physical Attributes of the Soil and the Productivity of Sugarcane Yield in Southeast Brazil[J]. Sugar Tech: An international journal of sugar crops and related industries, 24(5):1404-1419.
7. Leônidas C. A. Melo, Puga A P , Coscione A R ,et al. (2016) Sorption and desorption of cadmium and zinc in two tropical soils amended with sugarcane-straw-derived biochar[J].Journal of Soils & Sediments, 16(1):226-234.
8. Niu Z., Li M., Wu L., etc. (2015) Structural Design of Key Components of 1GYF-250 Typed Sugarcane Trash Shredder. China Tropical Agriculture, (06):28-30.
9. Lu J., Li M., Wei L., etc. (2011) Optimized Design 0f 1GYF-200 Sugarcane Leaf Shattering and Returning Machine's Cutter Roller. Applied Mechanics and Materials, (120):172-177.
10. Li M., Wang J., Deng Y., et al. (2008) Structural design and experiments on sugarcane leaf shattering and returning machine. Transactions of the Chinese Society of Agricultural Engineering, 24 (2): 121-125.
11. Li M., Lu J., Wei L., Zhang J., Fu X., Wang J., and Song D. (2011) Optimization Design and Experiments on Sugarcane Leaf Shattering and Returning Machine. ADVANCED SCIENCE LETTERS,(4): 1-7.
12. Sun X., Wang P. (2011) The research status and thinking of straw chopping and returning machine. Xinjiang Agricultural Mechanization,(2): 41-43.
13. Wei S., Li Y., Zhang X., et al. (2019) Design and Experiment of Double-roller Returning Machine with the Type of Cutting and Feeding for Banana Straw. Journal of Agricultural Mechanization Research, (10): 80-84.
14. Huang W., Yan X., Dong X., et al. (2018) Design and Experiment on the 1GYFH Type Returning Machine of Grinding and Mixed-buried Sugarcane Leaves. Journal of Agricultural Mechanization Research, (1): 140-144.
15. Zhang X., Gan S., Zheng K., et al. (2015) Design and experiment of rolling cutting feed type banana pseudostem crushing and returning machine with horizontal axis swing knife. Transactions of the Chinese Society of Agricultural Engineering, 31 (4): 33-41.

Comprehensive Review on Properties and Generation of Biodiesel

Ravindra Mohan[1,2,a], P.N.Ahirwar[1,b], Geetesh Goga[3,c], Santosh Kumar[4,d]

[1]Department of Mechanical Engineering, Madhyanchal Professional University, Bhopal MP, India
[2]Department of Mechanical Engineering, IES College of Technology, Bhopal MP, India
[3]Department of Mechanical Engineering, Bharat Group of College, Sardulgarh, Punjab, India.
[4]Department of Mechanical Engineering, Chandigarh Group of Colleges, Landran, Mohali, Punjab, India

Abstract. Recently, biodiesel has become more attractive owing to its attractive properties such as positive environmental effects, non-toxic, biodegradable, derived from renewable resources. Hence, this paper aims to provide an overview on biodiesel production and generations of biodiesel produced from distinct feedstocks. Thereafter the comparison of fuel properties, oil yield for different edible and non-edible oil and feedstock cost in distinct countries is given. Finally, the properties of biodiesel developed from distinct feedstock are summarized. The outcome demonstrates that the various sources offered varying yields owing to the process factors. However, the yield of biodiesel varies w.r.t. feedstocks owing to the physical and chemical characteristics of the sources and the process factors. Among waste cooking oil, commercial diesel fuel, and biodiesel from waste cooking oil, waste cooking oil offered higher pour point (284K), density (0.924 kg/L), water content (0.42%), kinematic viscosity (36.4 mm^2/s) than commercial diesel fuel and biodiesel from waste cooking oil.

1. Introduction

The world's escalating energy needs are mostly met by non-renewable fossil fuels due to the fast-growing population and accelerating industrialization. Fossil fuel reserves are running out more quickly, and their overuse has caused significant environmental problems that make moving to renewable energy sources necessary. However, owing to its biodegradability, renewability, and ecologically friendly characteristics, biodiesel has attracted significant interest as a feasible and flexible fuel replacement to the currently used non-renewable fuels. Transesterification is the manufacturing method that is most frequently used to produce biodiesel from a variety of sources [1]. Around 9 billion people are projected to live on the earth by 2050 [2]. Natural energy sources are being depleted as a result of rising worldwide prices and increased energy consumption [3-5]. The combustion of fossil fuels has a number of negative effects on the environment, including a rise in the emissions of greenhouse gases (GHG), mainly CO_2 [6,7]. Global primary energy consumption has sharply expanded in recent decades as a result of rapid industrialization and rising living standards [3,8]. To maintain a reasonable quality of life, developing nations like South Africa, the South Asian area, and Brazil need 12–24 gigajoules (GJ) of energy per capita each year [9]. At the moment oil, coal, and natural gas make up over 80% of the world's energy supply, and nearly 98% of that energy is produced by carbon emissions from these fuels [9]. Through the 21st century, it is anticipated that droughts would last longer and be more intense, resulting in a five-fold reduction in water reserves [2].

Hence, this review paper discusses about the various aspects for biodiesel production from distinct feedstocks. Further, this paper includes the comparison of fuel properties, oil yield for different edible and non-edible oil and feedstock cost in distinct countries. Finally, the properties of biodiesel produced from distinct feedstock are summarized. Hence, this review paper will be useful for the futuristic scientists to get the properties of biodiesel produced from distinct feedstock.

2. Biofuels Types, Generation and Production

According to Figure 1, there are four generations of biofuels

[a]ravindra.mohan09@gmail.com, [b]pnarayan220@mail.com, [c]geeteshgoga@gmail.com, [d]santoshdgc@gmail.com

 https://doi.org/10.1051/e3sconf/202344102019

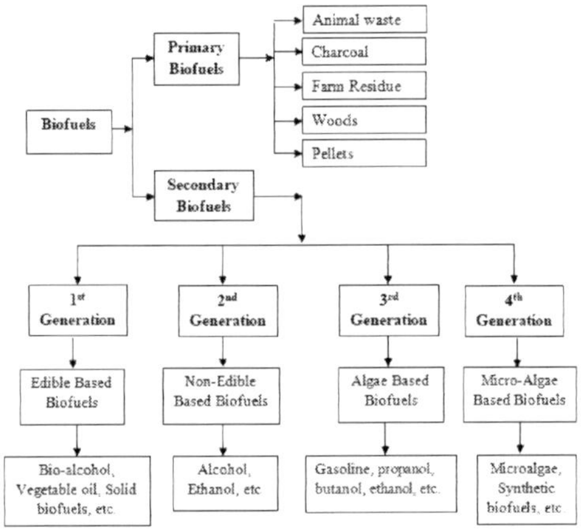

Figure 1. Types and generations of Biofuels [10]

2.1. 1st Generation

1st generation biofuels are traditional biofuels that are primarily produced from two categories of edible feedstock: starch-based feedstocks (such as potatoes, maize, wheat and wheat) and sugar-based feedstocks (such as sugar beet and sugarcane) [11, 12]. Crop availability and relatively easy conversion procedures are the primary benefits of 1st generation raw materials. However, there is a decreased food supply as a result of utilizing edible food crops for the manufacture of biodiesel [13–15]. The use of agricultural land for fuel generation is another issue. Environmental damage could result from using excessive amounts of insecticides and fertilizer for agricultural production [16].

2.2. 2nd Generation

Advanced biofuels known as second-generation biofuels are made from a variety of trees, grass, shrubs, and agricultural waste [10]. The second-generation biofuels include the Cellulosic Ethanol, Algae-Based Biofuels, Algae-Based Biofuels, Alcohol, Biosynthetic Natural Gas (Bio-SNG) etc. [17-20].

2.3. 3rd Generation

Algal biomass and used cooking oil are used to create 3rd generation biofuels. These biofuels have the benefits of greater growth and productivity, using less agricultural land, having more oil, and having less of an influence on the food supply. The main sources of 3rd generation biodiesel feedstocks include microalgae, animal fat, fish oil and used cooking oil [21].

2.4. 4th Generation

The biofuel produced by such a method is referred to be a fourth-generation biofuel since it uses Molecular biology, transdisciplinary physicochemical methods, and genetic engineering, including the utilization of "CRISPR/Cas9" with guided RNA for genetic change in algae [22]. In order to increase biofuel yield, genetically altered algae that accumulate high lipid and carbohydrate content are used in the manufacturing of 4thgeneration biofuels [23-25]. The Edible & non-edible vegetable oils developed by distinct countries is given in **Table 1.**

Table 1. Vegetable oils, both edible and not, by nation of production [26].

Countries: Edible oil	Countries: Non-Edible oil
China, USA: Corn	USA: Jojoba
Thailand, Malaysia, Indonesia: Palm	South Asia: Neem, Karanja
Canada: Canola	China, Canada, Russia: Linseed
France, Finland, Germany: Rapeseed	Asia: Silk cotton
Thailand, Philippines: Coconut	India, China, Brazil: Tobacco, Cotton seed
Sweden, Russia, Italy, UK: Soybean	India, Australia, Sri Lanka: Sandalwood
India, China: Groundnut	China: Cotton seed
Argentina, China, Brazil: Sunflower	India: Mahua, Polanga
Canada, China, Europe: Mustard	India, China, Thailand, Philippines, Indonesia: Jatropha
Russia, USA: Sunflower	Malaysia: Rubber seed

The feedstock comes in three generations. Vegetable oils that can be consumed, like sunflower oils, and soybean are first generation feedstocks [27]. **Table 2** lists the second-generation feedstocks, which are derived from non-edible sources including jojoba oil, used cooking oil, animal fats and jatropha etc. [28].

Table 2. Second-generation biodiesel production feedstock types [29].

2nd generation biodiesel production from distinct feedstock		
Non-edible oil seed	**Waste materials**	**Animal fats**
Jatropha	Acid oils	Pork lard
Kusum	pomace oil	Beef tallow
Karanja	Vegetable oil soap stocks	chicken fat
Botryococcusbraunii	dried distiller's grains (DDG),	yellow grease
Mahua	tall oil	By-products from fihi oil
Ethiopian mustard		waste salmon
Linseed		sorghum bug
Tomato seed		melon bug
Cotton seed, Tobacco, Neem, etc.		

High costs, ineffectiveness, and sustainability are constraints on the development of these feedstocks [30–34]. Number of distinct seed oils were evaluated as a possible fuel source for the diesel engine, with the third-generation feedstock, microalgae, receiving the majority of attention. **Figure 2** compares the characteristics of several fuels.

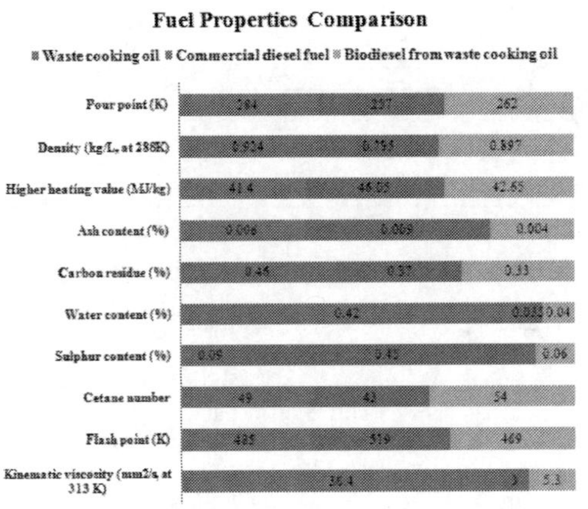

Figure 2. Distinct fuel properties comparison [35].

Further, the comparison of oil yield (%) in case of edible and non-edible oil source is shown in **Figure 3.**

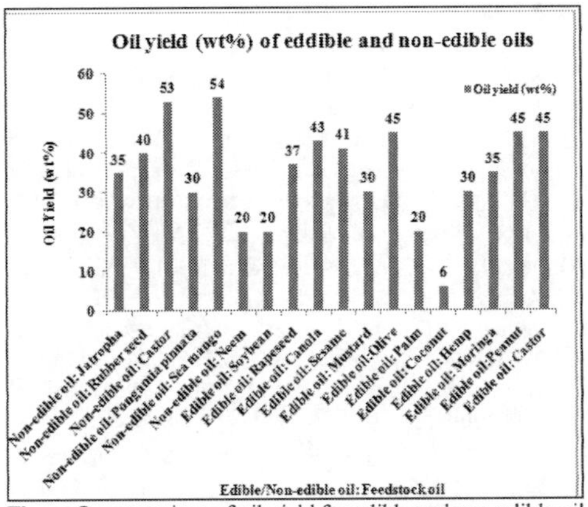

Figure 3. comparison of oil yield for edible and non-edible oil [35-37]

Further, the cost of distinct biodiesel feedstock in distinct countries is represented in **Figure 4.**

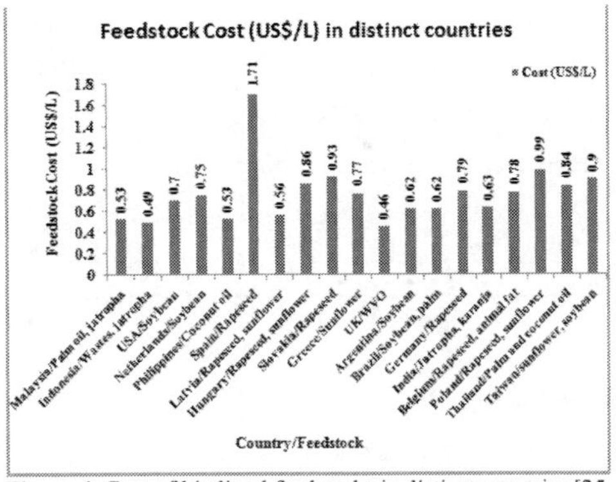

Figure 4. Cost of biodiesel feedstocks in distinct countries [35, 38]

Researchers and scientists have created many processes for producing biodiesel from various biofuels. The majority of academics and experts claimed that using a catalyst increased the production of biodiesel. To create biodiesel, Ahn et al. [39] used a 2-step reaction procedure. By employing, potassium hydroxide, sodium hydroxide as catalysts, canola methyl ester (CME), rapeseed methyl ester (RME), beef tallow ester (BTE) and sunflower methyl ester (SME) were all produced. Cold-pressed rapeseed oil and methanol are transesterified in two stages at low temperatures to produce biodiesel [40-45].

3. Properties of Biodiesel

Fatty acid methyl ester (FAME), sometimes referred to as biodiesel, is created by combining methanol with any triacylglycerol-carrying substance, including vegetable oil, animal fat, or both. The value of FAME's properties is greatly altered by variations in feedstocks [46, 47]. The cetane number, viscosity, calorific value, pour temp., iodine number, cloud temp, flashpoint, specific gravity, and cold filter plugging point are only a few of the attributes that are closely linked to the composition of FAME. However, the FAME characteristics of bio-diesel fuel are impacted by its handling and production procedures in addition to its intrinsic characteristics [48–54]. The water content, acid number, cold soak filtration, ash content, metals, sediment, and methanol concentration are some of these characteristics [47, 55]. Table 3 provides a summary of the main physic-chemical characteristics of biodiesel

Table 3: Distinct characteristics of biodiesel developed from distinct feedstock [55-64]

Feedstocks	**Density at 15°C (kg/m³)**	**Heating Value (MJ/kg)**	**Cloud Point (°C)**	**Flash temp. (°C)**	**Pour Temp. (°C)**	**Viscosity at 40 °C (mm²/s)**	**Cetane Number**	**Iodine Number**	**Sulfur Content (wt.%)**	**Ad Value mg/g**
Babassu	872	31.8	4	117	-	4.2	63.25	-	-	0.425
Bitter almond	884	-	4.5	169	-6	4.60	45.18	117.2	-	0.27
Camelus dromedaries	871	39.52	12.7	158	15.5	3.39	58.7	65.3	0.031%	-
Canola	878	-	-3.25	172.36	-8	4.42	54	113.6	2 ppm	0.49
Chicken fat	883	40.17	-7	172	-	4.98	48	-	23.45	0.22
Variabilis	867	38.7	-	157	-	4.8	58.6	-	0	-

Cotton seed	887	39.7	1.7	210	-12.5	4.19	48.1	120	-	0.5
Groundnut	920	39.8	8	132	3	4.4	59.85	71.8	1.31ppm	-
*J. curcas*L.	865	40.7	5.6	175.5	6	4.52	55.43	95.75	0.008	0.24
Karanja	889	36.56	13.3	157.4	6.4	4.79	56.55	89	0.003%	-
Lard	877	36.50	-	143.5	7	4.84	-	77	-	0.12
Mahua	895	36.9L, 39.4H	4.33	129.5	4.33	4.77	55	74.2	-	0.41
MicheliaChampaca	870	39.5	-	158	-	5.11	50.28	104	-	0.44
Neem	929	-	19	214	10	38.8	41	75.2	-	-

4. Conclusion

According to the literature, biodiesel from sources including sunflower, rapeseed, palm, and jatropha has enormous potential. It has been established that base catalysts outperform acid catalysts and enzymes, and that biodiesel combustion properties are comparable to those of diesel. The usage of vegetable oil/diesel blend is also implied to have resulted in worse engine performance since the high viscosity oil contaminated the lubricating oil and caused coking of the injectors. The literature study led to the conclusion that there is a significant area for research in the manufacturing of biodiesels, as well as in performance improvement and emission reduction. Additionally, there are significant opportunities for study in the field of increasing biodiesel output.

References

[1] Fatima Akram, IkramulHaq, Salehalbadat Raja, AzkaShahzad Mir, SumbalSajidQureshi, AmnaAqeel, Fatima Iftikhar Shah, Current trends in biodiesel production technologies and future progressions: A possible displacement of the petro-diesel, Journal of Cleaner Production, Volume 370, 2022. https://doi.org/10.1016/j.jclepro.2022.133479

[2] Neupane, D.; Adhikari, P.; Bhattarai, D.; Rana, B.; Ahmed, Z.; Sharma, U.; Adhikari, D. Does Climate Change Affect the Yield of the Top Three Cereals and Food Security in the World? Earth, 2022, 3, 45–71. https://doi.org/10.3390/earth3010004

[3] Mehmood, M.A.; Ibrahim, M.; Rashid, U.; Nawaz, M.; Ali, S.; Hussain, A.; Gull, M. Biomass production for bioenergy using marginal lands. Sustainable Production and Consumption, 2017, 9, 3–21. https://doi.org/10.1016/j.spc.2016.08.003

[4] Maheshwari, P.; BelalHaider, M.; Yusuf, M.; Klemeš, J.J.; Bokhari, A.; Beg, M.; Al-Othman, A.; Kumar, R.; Jaiswal, A.K. A review on latest trends in cleaner biodiesel production: Role of feedstock, production methods, and catalysts. Journal of Cleaner Production, 2022, 335, 131588. https://doi.org/10.1016/j.jclepro.2022.131588

[5] Singh, A.R.; Singh, S.K.; Jain, S. A review on bioenergy and biofuel production. Materials Today: Proceedings, 2022, 49, 510–516. https://doi.org/10.1016/j.matpr.2021.03.212

[6] Kumar, A., Singh, V.P., Nirala, A., Singh, R.C., Chaudhary, R., Mourad, A.H.I., Sahoo, B.K. and Kumar, D., Influence of tool rotational speed on mechanical and corrosion behaviour of friction stir processed AZ31/Al2O3 nanocomposite. Journal of Magnesium and Alloys, 2023 7(3), 487-500. https://doi.org/10.1016/j.jma.2023.06.012

[7] Venugopal, V., Singh, V.P. and Kuriachen, B., Underwater friction stir welding of marine grade aluminium alloys: A review Materials Today: Proceedings, 2023 https://doi.org/10.1016/j.matpr.2023.07.182

[8] Singh, V.P., Kumar, A., Kumar, R., Modi, A., Kumar, D., Mahesh, V. and Kuriachen, B., Effect of Rotational Speed on Mechanical, Microstructure, and Residual Stress Behaviour of AA6061-T6 Alloy Joints through Friction Stir Welding. Journal of Materials Engineering and Performance, 2023 pp.1-16. https://doi.org/10.1007/s11665-023-08527-8

[9] Yusuf, M.; Beg, M.; Ubaidullah, M.; Shaikh, S.F.; Keong, L.K.; Hellgardt, K.; Abdullah, B. Kinetic studies for DRM over high-performance Ni–W/Al2O3–MgO catalyst. International Journal of Hydrogen Energy, 2021, 47, 42150–42159. https://doi.org/10.1016/j.ijhydene.2021.08.021

[10] Deora, P.S.; Verma, Y.; Muhal, R.A.; Goswami, C.; Singh, T. Biofuels: An alternative to conventional fuel and energy source. Materials Today: Proceedings. 2022, 48, 1178–1184. https://doi.org/10.1016/j.matpr.2021.08.227

[11] Mat Aron, N.S.; Khoo, K.S.; Chew, K.W.; Show, P.L.; Chen, W.H.; Nguyen, T.H.P. Sustainability of the four generations of biofuels–a review. International Journal of Energy Research, 2020, 44, 9266–9282. https://doi.org/10.1002/er.5557

[12] Aransiola, E.F.; Ojumu, T.V.; Oyekola, O.; Madzimbamuto, T.; Ikhu-Omoregbe, D. A review of current technology for biodiesel production: State of the art. Biomass Bioenergy, 2014, 61, 276–297. https://doi.org/10.1016/j.biombioe.2013.11.014

[13] Mahapatra, S., Kumar, D., Singh, B. and Sachan, P.K., Biofuels and their sources of production: A review on cleaner sustainable alternative against conventional fuel, in the framework of the food and energy nexus. Energy Nexus, 2021, 4, p.100036. https://doi.org/10.1016/j.nexus.2021.100036

[14] Yadav, P.K., Dixit, G., Dixit, S., Singh, V.P., Patel, S.K., Purohit, R. and Kuriachen, B., Effect of eutectic silicon and silicon carbide particles on high

stress scratching wear of aluminium composite for various testing parameters. Wear, 2021, 482, p.203921. https://doi.org/10.1016/j.wear.2021.203921

[15] T. Mizik, G. Gyarmati, Economic and sustainability of biodiesel production—a systematic literature review, Clean Technologies. 2021, 3 19–36. https://doi.org/10.3390/cleantechnol3010002

[16] Obergruber, M.; Hönig, V.; Procházka, P.; Ku˘cerová, V.; Kotek, M.; Bou˘cek, J.; Ma˘rík, J. Physicochemical properties of biobutanol as an advanced biofuel. Materials, 2021, 14, 914. https://doi.org/10.3390/ma14040914

[17] Lynd, L.R.; Sow, M.; Chimphango, A.F.; Cortez, L.A.; Brito Cruz, C.H.; Elmissiry, M.; Laser, M.; Mayaki, I.A.; Moraes, M.A.;Nogueira, L.A. Bioenergy and African transformation. Biotechnology for Biofuels, 2015, 8, 1–18. https://doi.org/10.1186/s13068-014-0188-5

[18] Posten, C.; Schaub, G. Microalgae and terrestrial biomass as source for fuels—A process view. Journal of Biotechnology, 2009, 142, 64–69. https://doi.org/10.1016/j.jbiotec.2009.03.015

[19] Xu, H.; Wang, C. A Comprehensive Review of 2,5-Dimethylfuran as a Biofuel Candidate. Biofuels from Lignocellulosic Biomass: Innovations Beyond Bioethanol, 2016, 105–129. https://doi.org/10.1002/9783527685318.ch5

[20] Zhang, W.; He, J.; Engstrand, P.; Björkqvist, O. Economic evaluation on bio-synthetic natural gas production integrated in a thermomechanical pulp mill. Energies, 2015, 8, 12795–12809. https://doi.org/10.3390/en81112343

[21] Alaswad, A.; Dassisti, M.; Prescott, T.; Olabi, A.G. Technologies and developments of third generation biofuel production. Renewable and Sustainable Energy Reviews, 2015, 51, 1446–1460. https://doi.org/10.1016/j.rser.2015.07.058

[22] Godbole, V.; Pal, M.K.; Gautam, P. A critical perspective on the scope of interdisciplinary approaches used in fourth-generation biofuel production. Algal Research, 2021, 58, 102436. https://doi.org/10.1016/j.algal.2021.102436

[23] Abdullah, B.; Muhammad, S.A.F.a.S.; Shokravi, Z.; Ismail, S.; Kassim, K.A.; Mahmood, A.N.; Aziz, M.M.A. Fourth generation biofuel: A review on risks and mitigation strategies. Renewable and Sustainable Energy Reviews, 2019, 107, 37–50. https://doi.org/10.1016/j.rser.2019.02.018

[24] Lan, E.I.; Liao, J.C. Metabolic engineering of cyanobacteria for 1-butanol production from carbon dioxide. Metabolic Engineering, 2011, 13, 353–363. https://doi.org/10.1016/j.ymben.2011.04.004

[25] Almeida JR, F_avaro LC, Quirino BF. Biodiesel biorefinery:opportunities and challenges for microbial production of fuels and chemicals from glycerol waste. Biotechnology for Biofuels, 2012;5:48. https://doi.org/10.1186/1754-6834-5-48

[26] Ghazali, Wan Nor Maawa Wan, Rizalman Mamat, Haji Hassan Masjuki, and Gholamhassan Najafi. "Effects of biodiesel from different feedstocks on engine performance and emissions: A review. Renewable and Sustainable Energy Reviews, 51 (2015): 585-602. https://doi.org/10.1016/j.rser.2015.06.031

[27] Atadashi, I. M., Mohamed Kheireddine Aroua, AR Abdul Aziz, and N. M. N. Sulaiman. "Production of biodiesel using high free fatty acid feedstocks. Renewable and Sustainable Energy Reviews, 16, 5 (2012): 3275-3285. https://doi.org/10.1016/j.rser.2012.02.063

[28] Pinzi, Sara, David Leiva, Isabel López-García, M. Dolores Redel-Macías, and M. Pilar Dorado. "Latest trends in feedstocks for biodiesel production." Biofuels, Bioproducts and Biorefining, 2014, 8, 1, 126-143. https://doi.org/10.1002/bbb.1435

[29] No, Soo-Young. Inedible vegetable oils and their derivatives for alternative diesel fuels in CI engines: A review. Renewable and Sustainable Energy Reviews, 2011, 15, 1, 131-149. https://doi.org/10.1016/j.rser.2010.08.012

[30] Singh, V.P., Kumar, D. and Kuriachen, B., Effect of Low Welding and Rotational Speed on Microstructure and Mechanical Behaviour of Friction Stir Welded AZ31-AA6061-T6. Transactions of the Indian Institute of Metals, 2023, pp.1-9. https://doi.org/10.1007/s12666-023-02971-9

[31] Kumar, A., Nirala, A., Singh, V.P., Sahoo, B.K., Singh, R.C., Chaudhary, R., Dewangan, A.K., Gaurav, G.K., Klemeš, J.J. and Liu, X., 2023. The utilisation of coconut shell ash in production of hybrid composite: Microstructural characterisation and performance analysis. Journal of Cleaner Production, 398, p.136494. https://doi.org/10.1016/j.jclepro.2023.136494

[32] Singh, V.P., Kumar, R., Kumar, A. and Dewangan, A.K., Automotive light weight multi-materials sheets joining through friction stir welding technique: an overview. Materials Today: Proceedings, 2023. https://doi.org/10.1016/j.matpr.2023.02.171

[33] Kumar, R., Dwivedi, R.K., Singh, V.P., Kuriachen, B. and Krishnan, N.A., Influence of toughness and retained austenite on wear behaviour of carbide-free bainite in high silicon steel. Transactions of the Indian Institute of Metals, 2022, 76, 2425–2434. https://doi.org/10.1007/s12666-022-02818-9

[34] Singh, V.P., Kumar, D., Mahto, R.P. and Kuriachen, B., Microstructural and Mechanical Behavior of Friction-Stir-Welded AA6061-T6 and AZ31 Alloys with Improved Electrochemical Corrosion. Journal of Materials Engineering and Performance, 2023, 32, 4185–4204. https://doi.org/10.1007/s11665-022-07380-5

[35] Selvaraj, R., R. Praveenkumar, and I. Ganesh Moorthy. "A comprehensive review of biodiesel production methods from various feedstocks.

Biofuels 10, no. 3 (2019): 325-333. https://doi.org/10.1080/17597269.2016.1204584

[36] Wan Omar WNN, Saidina Amin NA. Optimization of heterogeneous biodiesel production from waste cooking palmoil via response surface methodology. Biomass and Bioenergy, 2011;35:1329-1338. https://doi.org/10.1016/j.biombioe.2010.12.049

[37] Gui, Meei Mei, Keat Teong Lee, and Subhash Bhatia. Feasibility of edible oil vs. non-edible oil vs. waste edible oil as biodiesel feedstock. Energy, 2008, 33.11 1646-1653. https://doi.org/10.1016/j.energy.2008.06.002

[38] Kumar, Niraj, and Sant Ram Chauhan. Performance and emission characteristics of biodiesel from different origins: A review. Renewable and Sustainable Energy Reviews, 2013, 21, 633-658. https://doi.org/10.1016/j.rser.2013.01.006

[39] Ahn E, Koncar M, Mittelbach M, Man R. A low-waste process for the production of biodiesel. Separation Science and Technology 1995;30:2021–33. https://doi.org/10.1080/01496399508010391

[40] Cvengro J, Povaz AF. Production and treatment of rapeseed oilmethyl esters as alternative fuels for diesel engines. Bioresource Technology 1996;55:145–50. https://doi.org/10.1016/0960-8524(95)00190-5

[41] Kaieda, Masaru, Taichi Samukawa, Takeshi Matsumoto, Kazuhiro Ban, Akihiko Kondo, Yuji Shimada, Hideo Noda et al. Biodiesel fuel production from plant oil catalyzed by Rhizopus oryzae lipase in a water-containing system without an organic solvent. Journal of bioscience and bioengineering, 1999, 88(6) 627-631. https://doi.org/10.1016/S1389-1723(00)87091-7

[42] Uosukainen E, La¨msa¨ M, Linko Y, Linko P, Leisola M. Optimization of enzymatic transesterification of rapeseed oil ester using response surface and principal component methodology. Enzyme and Microbial Technology, 1999, 25, 236–243. https://doi.org/10.1016/S0141-0229(99)00034-4

[43] Ma Fangrui, Hanna Milford A. Biodiesel production: a review. Bioresource Technology, 1999; 70:1–15. https://doi.org/10.1016/S0960-8524(99)00025-5

[44] SamukawaTaichi, Kaieda Masaru, Matsumoto Takeshi, Ban Kazuhiro, Kondo Akihiko, Shimada Yuji, et al. Pretreatment of immobilized Candida Antarctica lipase for biodiesel fuel production from plant oil. Journal of Bioscience and Bioengineering, 2000;90:180–183. https://doi.org/10.1016/S1389-1723(00)80107-3

[45] Ikwuagwu OE, Ononogbu IC, Njoku OU. Production of biodiesel using rubber [Heveabrasiliensis (KunthMuell.)] seed oil. Industrial Crops and Products 2000;12:57–62. https://doi.org/10.1016/S0926-6690(99)00068-0

[46] Crabbe Edward, Nolasco-HipolitoCirilo, Kobayashi Genta, Sonomoto Kenji, IshizakiAyaaki. Biodiesel production from crude palm oil and evaluation of butanol extraction and fuel properties. Process Biochemistry 2001; 37:65–71. https://doi.org/10.1016/S0032-9592(01)00178-9

[47] Sharma, V., Das, L., Pradhan, R. C., Naik, S. N., Bhatnagar, N., & Kureel, R. S. Physical properties of tung seed: An industrial oil yielding crop. Industrial Crops and Products, 2011 33(2), 440-444. https://doi.org/10.1016/j.indcrop.2010.10.031

[48] Singh, Virendra Pratap, and Basil Kuriachen. Experimental investigations into the mechanical and metallurgical characteristics of friction stir welded AZ31 magnesium alloy. Journal of Materials Engineering and Performance 31, no. 12 (2022): 9812-9828. https://doi.org/10.1007/s11665-022-07055-1

[49] Ranjole, C., Singh, V.P., Kuriachen, B. and Vineesh, K.P., Numerical prediction and experimental investigation of temperature, residual stress and mechanical properties of dissimilar friction-stir welded AA5083 and AZ31 alloys. Arabian Journal for Science and Engineering, 2022, 47(12), pp.16103-16115. https://doi.org/10.1007/s13369-022-06808-3

[50] Singh, Virendra Pratap, Surendra Kumar Patel, and Basil Kuriachen. Mechanical and microstructural properties evolutions of various alloys welded through cooling assisted friction-stir welding: A review. Intermetallics 133 (2021): 107122. https://doi.org/10.1016/j.intermet.2021.107122

[51] Singh, V.P., Patel, S.K., Ranjan, A. and Kuriachen, B., Recent research progress in solid state friction-stir welding of aluminium–magnesium alloys: a critical review. Journal of Materials Research and Technology, 2020, 9(3), pp.6217-6256. https://doi.org/10.1016/j.jmrt.2020.01.008

[52] Kumar, A., Singh, R.C., Chaudhary, R. and Singh, V.P., April. Tribological studies and microstructural characterisation of SiC and fly ash particles based aluminium 2024 alloy composites prepared through stir casting route. In IOP Conference Series: Materials Science and Engineering, 2020, 804(1) p. 012025). IOP Publishing. https://doi.org/10.1088/1757-899X/804/1/012025

[53] Singh, V.P., Patel, S.K., Kuriachen, B. and Suman, S., 2019. Investigation of general welding defects found during friction-stir welding (FSW) of aluminium and its alloys. In Advances in Additive Manufacturing and Joining: Proceedings of AIMTDR 2018 (pp. 587-595). Singapore: Springer Singapore. https://doi.org/10.1007/978-981-32-9433-2_51

[54] Singh, V.P., Patel, S.K., Kumar, N. and Kuriachen, B., Parametric effect on dissimilar friction stir welded steel-magnesium alloys joints: a review. Science and Technology of Welding and Joining, 2019, 24(8), pp.653-684. https://doi.org/10.1080/13621718.2019.1567031

[55] Singh, D.; Sharma, D.; Soni, S.; Sharma, S.; Kumari, D. Chemical compositions, properties, and standards

 https://doi.org/10.1051/e3sconf/202344102019

for different generation biodiesels: A review. Fuel, 2019, 253, 60–71. https://doi.org/10.1016/j.fuel.2019.04.174

[56] Neupane, D.; Bhattarai, D.; Ahmed, Z.; Das, B.; Pandey, S.; Solomon, J.K.Q.; Qin, R.; Adhikari, P. Growing Jatropha (Jatrophacurcas L.) as a Potential Second-Generation Biodiesel Feedstock. Inventions 2021, 6, 60. https://doi.org/10.3390/inventions6040060

[57] Patel, S.K., Singh, V.P., Roy, B.S. and Kuriachen, B., 2020. Recent research progresses in Al-7075 based in-situ surface composite fabrication through friction stir processing: A review. Materials Science and Engineering: B, 262, p.114708. https://doi.org/10.1016/j.mseb.2020.114708

[58] Patel, S.K., Singh, V.P. and Kuriachen, B., 2019. Friction stir processing of alloys with secondary phase particles: an overview. Materials and Manufacturing Processes, 34(13), pp.1429-1457. https://doi.org/10.1080/10426914.2019.1662037

[59] Patel, S.K., Singh, V.P., Kumar, N., Kuriachen, B. and Nateriya, R., 2020. Wear behaviour of Al-silicon (LM13) alloy composite reinforcement with TiC and ZrSiO4 particles. Silicon, 12(1), pp.211-221. https://doi.org/10.1007/s12633-019-00114-8

[60] Patel, S.K., Singh, V.P. and Kuriachen, B., 2019. Microstructural, Tribological and mechanical properties evolution of ZrSiO4/A4047 surface composite fabricated through friction stir processing. Transactions of the Indian Institute of Metals, 72(7), pp.1765-1774. https://doi.org/10.1007/s12666-019-01647-7

[61] Patel, S.K., Singh, V.P., Kumar, D., Roy, B.S. and Kuriachen, B., 2022. Microstructural, mechanical and wear behavior of A7075 surface composite reinforced with WC nanoparticle through friction stir processing. Materials Science and Engineering: B, 276, p.115476. https://doi.org/10.1016/j.mseb.2021.115476

[62] Singh, A.K., Mishra, A.K., Gupta, K.K., Bhatnagar, P. and Kim, T., 2020. An integrated converter with reduced components for electric vehicles utilizing solar and grid power sources. IEEE Transactions on Transportation Electrification, 6(2), pp.439-452. https://doi.org/10.1109/TTE.2020.2998799

[63] Singh, N., Gupta, K.K., Jain, S.K., Dewangan, N.K. and Bhatnagar, P., 2020. A flying squirrel search optimization for MPPT under partial shaded photovoltaic system. IEEE Journal of Emerging and Selected Topics in Power Electronics, 9(4), pp.4963-4978. https://doi.org/10.1109/JESTPE.2020.3024719

[64] Mahla, S.K., Goyal, T., Goyal, D., Sharma, H., Dhir, A. and Goga, G., 2022. Optimization of engine operating variables on performance and emissions characteristics of biogas fuelled CI engine by the design of experiments: Taguchi approach. Environmental Progress & Sustainable Energy, 41(2), p.e13736. https://doi.org/10.1002/ep.13736

 https://doi.org/10.1051/e3sconf/202344102020

Research on fault monitoring system for collector ring and carbon brush of generator rotor

Jianlong Zhao*, Xi Zhou and Wei Hou

NR Electric Co., Ltd., Nanjing, Jiangsu, 211102, China

Abstract. The operating status of the generator rotor's collector ring and carbon brush determines the safe and stable operation of the generator. It is critical to promptly detect and handle any abnormalities or faults in the collector ring and carbon brush to prevent accidents from expanding. At present, the status monitoring of the collector ring and carbon brush mainly relies on manual inspection, which has low efficiency and cannot discover faults in real-time. This article studies the establishment of a fault monitoring system for collector ring and carbon brush of the generator rotor. Through real-time collection and analysis of arc light, temperature, and electrical data, abnormalities and faults can be detected and alerted in a timely manner. This system can improve the accuracy and effectiveness of monitoring, and ensures the safe operation of the generator.

1. Introduction

The collector ring and carbon brush of the generator are important components for normal power generation and conduction of current. The good working of the collector ring and carbon brush directly determines the safe and stable operation of the generator. Discharge, vibration, uneven wear, and surface oil stains between the collector ring and carbon brush are prone to occur during the operation of the generator, which will cause local heating of the carbon brush and collector ring[1~5]. If the temperature exceeds a certain value, the performance of the insulation material will deteriorate, accelerating the aging of the insulation and leading to arc faults. If the anomalies can not be detected and manually intervened in a time, it may bring about the shutdown of generator or even primary equipment damage accident. At present, the monitoring of the collector ring and carbon brush is mostly based on a single state, such as magnetic field [6], temperature [7], carbon brush length [1, 2], etc. However, the method based on a single state variable is difficult to comprehensively monitor various faults of the collector ring and carbon brush, and when the collection sensor of a single state is abnormal, all functions of the monitoring system will fail, resulting in low system reliability.

Therefore, it is necessary to establish a real-time fault monitoring system for the generator rotor collector ring and carbon brush based on multiple sensors to improve the level of abnormal monitoring and avoid catastrophic faults.

*Corresponding author's e-mail: zhaojl@nrec.com

2. Overall system scheme

As shown in Figure 1, the real-time fault monitoring system for the collector ring and carbon brush of generator rotor adopts a layered and distributed structure design, consisting of station layer, bay layer and process layer.

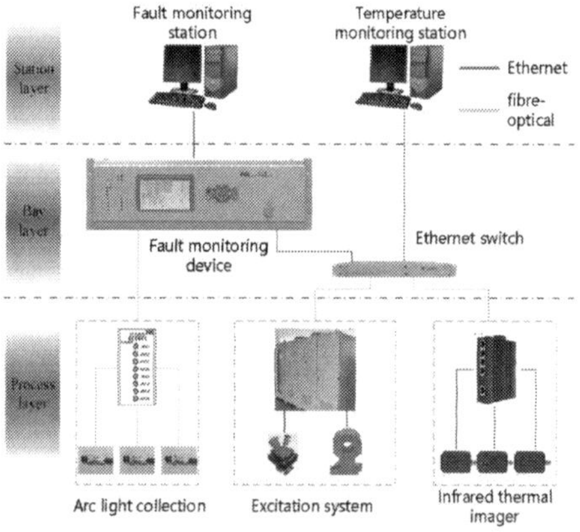

Figure 1. Network structure diagram of collector ring carbon brush fault monitoring system.

2.1. Station layer

The main function of station layer equipment is to achieve device monitoring, alarm, and information exchange, in order to complete data collection, monitoring, management, and storage. The station layer equipment includes the fault monitoring station and the temperature monitoring station. The fault monitoring station displays the data and discrimination results submitted by the fault monitoring device, and storing the data in a history database. The temperature monitoring station analyse and works with the data from the infrared thermal imager, displays temperature images, stores historical data and images, transmitting key data to the fault monitoring device through ethernet communication.

2.2. Bay layer

The bay layer equipment is used for gathering real-time data information from process layer and completing data analysis and logical operation discrimination. The equipment mainly includes fault monitoring device and ethernet switch. Fault monitoring device receives sampling data such as arc light, electrical quantities of the excitation system and temperature. Through analysing and calculating the data, the working status of collector ring and carbon brush is distinguished, which will be sent to fault monitoring station.

2.3. Process layer

The process layer equipment which includes arc light acquisition unit, excitation system and infrared temperature measurement unit is applied to collect real-time operating electrical quantities, monitor equipment operation status and so on.

a) The arc light collection unit consists of arc light sensors and the distributed arc light collection units. The arc light sensors collect the arc light signal of collector ring and carbon brush in case of a fault, and then transmit the signal to arc light collection units through fibre-optical; The distributed arc light collection units are responsible for collecting the sampling signals of all arc light sensors on site and sending them to fault monitoring devices.
b) The electrical datas of the generator, such as excitation voltage and current, active and reactive power, are connected to the excitation system controller and transmitted to the fault monitoring device.
c) The infrared temperature measurement unit is composed of infrared temperature measurement probes and dedicated ethernet switch. The probes collect real-time temperature information of collector ring and carbon brush in the excitation room and the dedicated switch gathers the temperature information and transmits it to the temperature monitoring station.

3. Fault monitoring discrimination methods

The real-time fault monitoring device for collector ring and carbon brush adopts an embedded hardware architecture, which supports arc signal acquisition, infrared temperature measurement data access and electrical quantity access. The arc energy accumulation model and the collector ring and carbon brush temperature rising model are established by arc, temperature, and electrical quantity data. The working condition of collector ring and carbon brush is identified based on the analysis results of arc energy, temperature rise rate, and excitation resistance changes. Independent fault monitoring criteria are designed and optimized according to various sampling signals.

3.1. Fault monitoring criteria based on arc light

During the operation of the generator, the imbalanced current often leads to the over current through some of the carbon brushes. As the operation time increases, the temperature of the carbon brushes, rising up with the strengthening of current, damages the surface structure of the carbon brushes, and causes arcing or fire between the carbon brushes and the collector ring. In severe cases, it can also form a ring fire, posing a direct threat to the safety and stability of the generator.

The fault monitoring system detects the arc light generated during the arc of collector ring and carbon brush via the reasonable arrangement of arc sensors, and extracts the characteristic quantities of equipment status through data analysis. When the characteristic quantities of the arc signal exceeds the threshold, an alarm signal is given. The logic is shown in Figure 2.

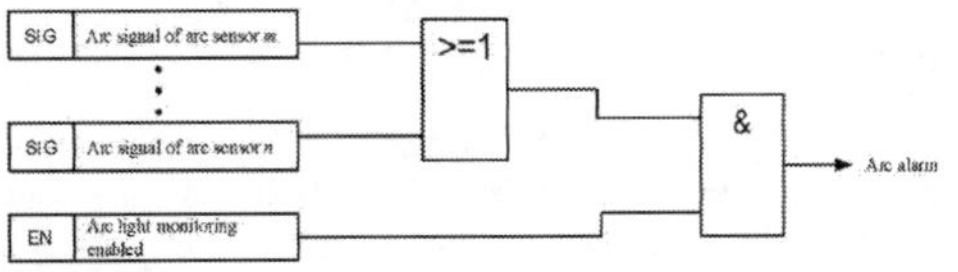

Figure 2. Arc light monitoring logic diagram.

3.2. Fault monitoring criteria based on temperature

When the current distribution of carbon brushes is uneven or even ring fire occurs, the temperature of some carbon brushes will be higher than the temperature during normal operation. The fault monitoring area covers all carbon brushes through the reasonable arrangement of temperature measuring instruments, and monitors the temperature of carbon brushes in real time. By analysing the temperature of carbon brushes within each interval, combined with characteristics such as maximum temperature and temperature rise rate, the operating conditions of the collector ring and carbon brushes are judged using the following criteria.

a) Overtemperature alarm: By recording the temperature characteristics of the collector ring and carbon brush during the entire operation process of the generator, and analysing the historical temperature data, the operating temperature range of the collector ring and carbon brush during normal motor operation is determined, and the overtemperature alarm threshold is rationally set. When the temperature of the carbon brush of the collector ring is detected to be higher than the overtemperature alarm threshold, an overtemperature alarm signal is issued.

b) Abnormal temperature rising alarm: Analysing historical temperature data to determine the safe rising rate and duration of collector ring and carbon brush temperature during normal operation of the generator. When it is detected that the temperature rising rate of the carbon brush is significantly higher than the safety threshold, or when the combined temperature rising rate and duration are significantly higher than normal, an abnormal temperature rising alarm signal is issued.

The logic is shown in Figure 3.

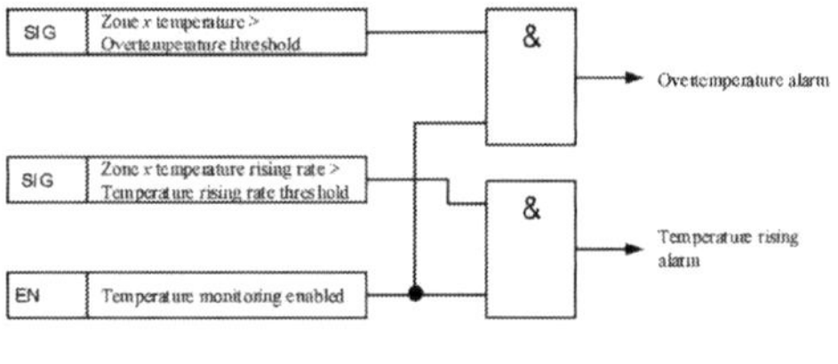

Figure 3. Temperature monitoring logic diagram.

3.3. Fault monitoring criteria based on electrical quantity

The excitation current of the generator is transmitted to the excitation winding through carbon brushes and collector rings. The equivalent circuit of the excitation system is shown in Figure 4.

Where, R_{P1}, ... , and R_{PN} respectively represent the positive electrode of the contact resistance between each carbon brush and the collector ring; R_{Q1}, ... , and R_{QN} is the negative electrode of the contact resistance individually; The resistance of the carbon brush and collector ring, which can be ignored, is much smaller than the contact resistance; R is the excitation winding resistance; U is the excitation voltage, and I is the excitation current. The total resistance of the excitation circuit is:

$$R_{\Sigma}=R_{P1}||...||R_{PN}+R+R_{Q1}||...||R_{QN}=\frac{U}{I} \quad (1)$$

If all carbon brushes are in normal condition and have a good contact with the collector ring, the excitation current can be evenly distributed by each carbon brush current:

$$I_{P1}=...=I_{PN}=I_{Q1}=...=I_{QN}=\frac{I}{N} \quad (2)$$

When a fault occurs in the collector ring and carbon brush, such as improper compression force and looseness of carbon brush, insufficient or uneven pressure of constant pressure spring, mixed use of different types of carbon brushes, poor contact between carbon brush and slip ring, weak connect between brush braid and carbon brush and so on, which expand the contact resistance between the faulty carbon brush and collector ring, leading to an increase in the total resistance of the excitation circuit. In order to maintain the stability of the excitation current, the excitation system will adjust and raise the excitation voltage in real time.

Considering the analysis above, it is apparent that when some carbon brushes experience errors, an increase in the total resistance of the excitation circuit will come together. Via excitation voltage and current data, the total resistance can be calculated in real-time. When it exceeds the safety threshold, an alarm signal of abnormal excitation resistance will be issued. The logic is shown in Figure 5.

In addition, intelligent learning is conducted on the correlation between the collected active and reactive power of the generator, excitation voltage, and excitation current, so a history database is established. After sufficient data samples are available, whether there is a significant difference in the electrical correlation between the abnormal state and normal state of collector ring and carbon brush can be determined, which contributes to the establishment of relevant fault criteria.

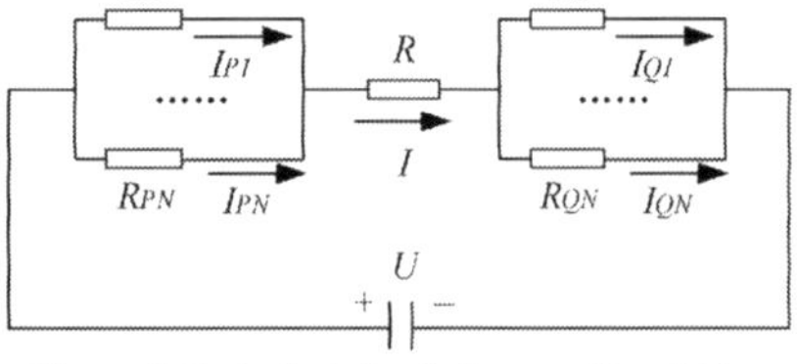

Figure 4. Equivalent circuit diagram of excitation system.

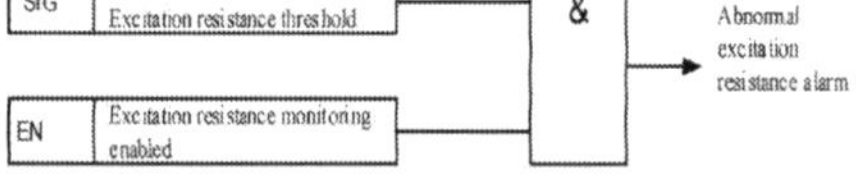

Figure 5. Excitation electrical quantity monitoring logic diagram

3.4. Optimizing fault monitoring criteria

Based on the above analysis, when the carbon brush malfunctions, it will exhibit fault characteristics in arc light, temperature, electrical quantity and other information. By estimating these characteristics in conjunction with the operating conditions of the generator, a single criterion can be optimized to improve the reliability of fault monitoring criteria.

During the initial stage of applying excitation when the generator starts, the temperature of the carbon brush will gradually increase from room temperature to operating temperature, which will have an impact on the monitoring criteria based on temperature rise rate described in section *3.2*. By monitoring the excitation current, this process can be distinguished, and one can choose to temporarily exit the temperature rise anomaly detection function during this process, or establish a

https://doi.org/10.1051/e3sconf/202344102020

temperature rise feature model for the generator excitation stage based on the temperature data of this process, which can be distinguished from the fault temperature rise feature and improve the sensitivity of fault monitoring. The logic is shown in Figure 6.

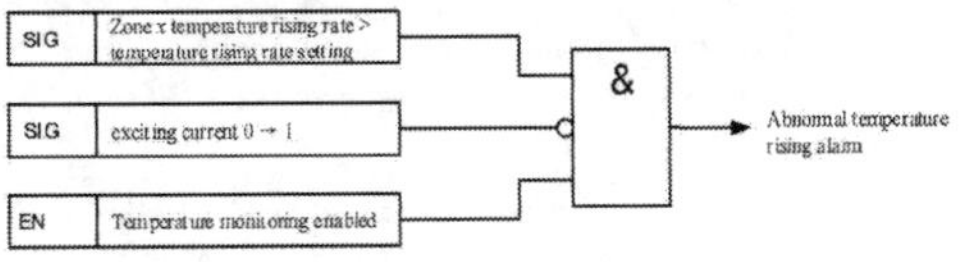

Figure 6. Temperature rise anomaly monitoring optimization logic diagram.

4. Signal acquisition schemes

4.1. Arc light signal acquisition

The earliest arc light sensor used in China was a visible light sensor with a wide spectrum of light response. Although visible light sensors can effectively monitor arc signals, daily light such as lighting, sunlight, and strong electric light during inspections can trigger arc alarm signals.

The energy of arc light is mainly concentrated in the ultraviolet range of 300 - 380nm and the visible range of 380 - 700nm, with over 70% of it being ultraviolet light. Based on this characteristic, arc light sensors based on ultraviolet light sensing technology have been developed in recent years, which can effectively detect ultraviolet light in the wavelength range of 300 - 380nm.

In order to ensure reliable detection of arc light signals and effectively prevent the interference of natural light and strong flashlights during inspection from affecting the normal operation of the device, ultraviolet light sensors are used in this design scheme.

According to the arrangement of the collector ring and carbon brush in the excitation compartment of a thermal power plant, the installation plan of the arc light sensor is shown in Figure 7.

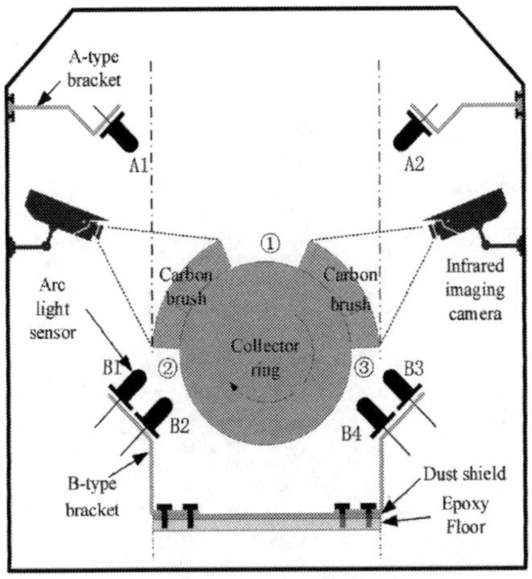

Figure 7. Schematic diagram of side view of equipment installation.

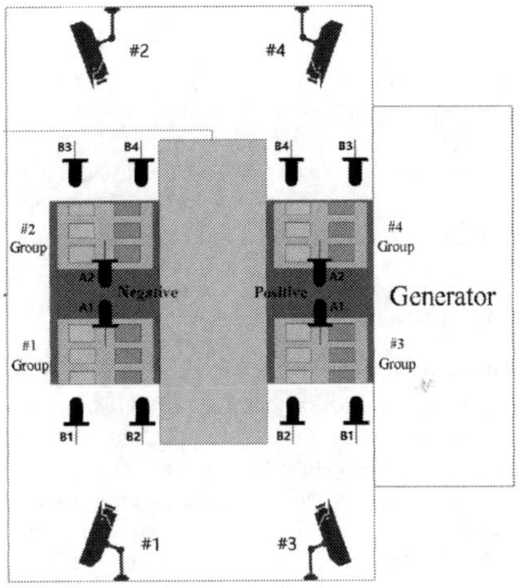

Figure 8. Schematic diagram of look down of equipment installation.

a) Install an A-type bracket symmetrically on the inner wall of the small chamber above the carbon brush side of the collector ring, and install one arc light sensor on each bracket, totally two, numbered A1 and A2, to detect the arc light signal at the contact surface ① of the collector ring and carbon brush.
b) Install a B-type bracket symmetrically on the brush holder base below the carbon brush side of the collector ring, and install 2 arc light sensors on each bracket, totally 4, numbered B1, B2, B3, and B4, to detect the arc light signals at the contact surfaces ② and ③ of the collector ring and carbon brush.

In summary, 6 arc sensors need to be installed for the arc detection of the collector ring and carbon brush of each excitation pole, and a total of 12 arc sensors need to be installed for the positive and negative electrodes of each excitation chamber.

4.2. Temperature signal acquisition

The temperature signal of the collector ring and carbon brush is collected using infrared imaging cameras. For example, a thermal power plant symmetrically installs two infrared imaging cameras above the carbon brush, and the monitoring area is the entire carbon brush group, as shown in Figure 7. The number and location of cameras installed for positive and negative electrodes are consistent, with a total of 4 infrared imaging cameras installed.

The top view diagram of the installed equipment is displayed in Figure 8. The temperature measurement system divides all carbon brushes monitored by each infrared imaging camera into 3 - 6 temperature measurement areas. With each temperature measurement area is independently monitored and analyzed, the system extracts information such as the highest, the lowest, and average temperatures of each area, generates temperature images, and sets various alarm thresholds.

The fault monitoring device obtains the main temperature signals and alarm information from the temperature monitoring workstation through the communication method, such as analog information of the maximum temperature, minimum temperature, average temperature, predicted temperature, etc. in each temperature measurement area, as well as switch information like alarm signals and warning signals.

4.3. Electrical quantity signal acquisition

The collector ring and carbon brush fault monitoring system needs to collect electrical information such as excitation voltage and current of the excitation system, as well as active and reactive power of the generator, to extract fault features and achieve the electrical quantity based collector ring and carbon brush fault detection.

The electrical quantity data of general generators has been connected to the excitation system. In this scheme, the fault monitoring device obtains electrical quantity information from the excitation system in the way of communication.

5. Conclusions

By installing arc sensors and infrared imaging cameras at specific locations to collect arc and temperature information, and using the communication method to obtain electrical information from the excitation system, the fault monitoring system conductes comprehensive logical analysis to detect faults of the collector ring and carbon brush of generator rotor. Compared to the traditional manual inspection method, the fault monitoring system for collector ring and carbon brush of generator rotor has the advance of real-time and reliability, providing strong support for the safe operation of generators.

References

1. Yao, Y., Zhang, C.H.. (2021) Analysis of Common Accidents in the Slip Ring and Carbon Brush of Steam Turbine Generators and Preventive Measures. Guangxi Electric Power., 44(3): 70-74.
2. Yu. N. Samorodov(2011). The physical and chemical causes of damage to generator contact rings and brushes. Power Technology and Engineering.,45(3):227-233.
3. Wang, C.W.. (2012) Analysis and Treatment for Burning Damage on A Generator Rotor Collector Ring. North China Electric Power., 5: 52-56.
4. J.K. SkjlbergHÅvar F. OhmaMagne Runde. (2010) Wear Rates and Current Distribution of Carbon Brushes on Steel Slip Rings. IEEE Transactions on Energy Conversion., 24(4): 835 – 840.
5. Ma, M.Y., Liu, T., Wang, H. (2017) Analysis of collector-ring's fire of turbine generator and its solution. Thermal Power Generation., 46(5): 105-107.
6. Wu, Y.C., Li, Y.G., Feng, W.Z. (2014) A Method for Assessing Turbine Generator Carbon Brush and Collecting Ring States. Large Electric Machine and Hydraulic Turbine., 5: 50-54.
7. Wang, J., Wen, Z.H., Chen, G.Q., et al. (2020) On Line Monitoring System of Collecting Ring Carbon Brush Infrared Thermal Imaging. Electric Power System Equipment., 1: 140-141.

Application of Dynamic Reactive-power Compensator Based on SVG in Coal Mines

Ran LI[1], Peng ZHANG[2], Dian-qing ZHANG[3,*]

[1]XinWen Mining Group Design and Research Institute Co., Ltd, Tai'an, China
[2]Shandong S HANKUANG Machinery Co., LTD, Jining, China
[3]China EPRI Science & Technology Co., Ltd, NARI Group Corporation Beijing, China

Abstract—With the improvement of automation degree of coal mine mechanical and electrical equipments, such as conveyors, crushers, mining motors, inverters, and a large number of non-linear loads, cause large reactive power consumption, power factor of coal mine power grid, power supply quality and stability, and bring greater influence to coal mine safety production and economic benefits.As a new generation of high voltage dynamic reactive power compensator,SVG (Static Var Generator) has fast response and compound functions,such as compensating reactive power, suppressing harmonic and imbalance with FC(Filter Capacitor), which plays a good role in improving the power quality of coal mine.This paper designs the system parameters of SVG in coal mine and introduces its application.

1. Introduction

With the improvement of automation degree of coal mine mechanical and electrical equipments, such as conveyors, crushers, mining motors and large fully mechanized mining equipments, inverters and a large number of non-linear loads, cause large reactive power consumption and high harmonics, equipment starting and working current increase, a large number of unnecessary power loss, lead to coal mine power grid power factor, power supply quality and stability decline, bring great influence to coal mine safety production. When the power factor of coal mine power grid is low, it may face high fines of power supply companies, which affects the economic benefits of coal mine enterprises[1][2].

To improve the power grid power factor, conventional reactive power compensation generally takes parallel capacitor centralized compensation.Its advantage is simple and reliable,and equipment cost is relatively low, but due to the amount of capacitor compensation is fixed, it cannot dynamicly and continuously compensates reactive power according to the grid real-time reactive load.To be worse,it is prone to under compensation or over compensation, and parallel resonance may occur[3]-[5]. As a new generation of high voltage dynamic reactive power compensation device , compared to static reactive power compensator SVC, SVG uses full-controlled power electronic devices IGBT (Insulated Gate Bipolar Transistor), and features fast response(below 10ms).SVG uses the cascade module multi-level wave shift phase technology,and the output harmonic is low.With coordination of FC compensator, SVG has compound functions,such as harmonic supression, which can effectively improve the power factor and power quality[6]-[9].

This paper briefly illustrates the working principle of SVG dynamic reactive power compensation device, and introduces the parameter design and application of SVG system in a coal mine.

2. Working principle of the SVG reactive power compensation device

2.1. Working principle of SVG

As shown in figure 1,the SVG converter valve is connected to the power grid by connecting the reactor. By adjusting the phase and amplitude of the output voltage on the AC side of the converter valve, the SVG device can absorb or emit reactive current, and achieve the purpose of dynamic adjustment of reactive power or voltage.

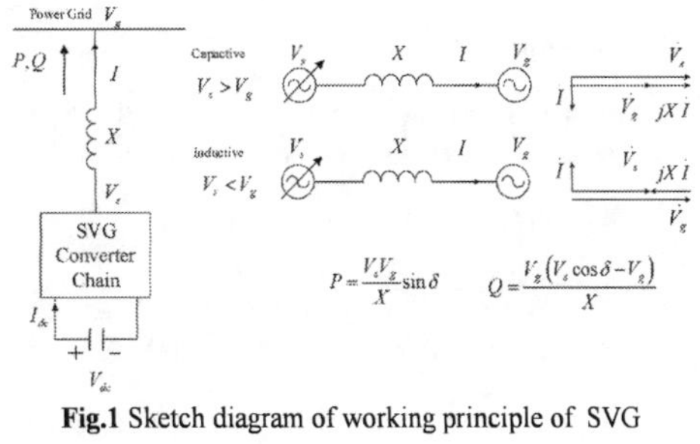

Fig.1 Sketch diagram of working principle of SVG

* 267315049@qq.com

The exchange of reactive power conducted between the SVG and the power grid is achieved by controlling the amplitude Vs of the output voltage of the SVG converter chain. If the SVG output is reactive, the control amplitude of Vs is greater than the amplitude of Vg; if the SVG output is reactive, the control amplitude is of Vs less than the amplitude of Vg.

The reactive power emitted by the SVG to the grid system is indicated by the following equation:

$$Q = \frac{VsCos\delta - Vg}{X} Vg$$

In this equation, Vs indicates the compensation voltage of the SVG output voltage and Vg is the power grid voltage of the power grid system.δis the leading angle of Vs according to Vg,and X indicates the inductance of the connection reactance between the SVG and the power grid system.

2.2. Overall Structure of SVG

The overall structure and main wiring of the SVG device are shown in Figure 2.

The main components of SVG device are: bus switch and breaker, MOV, cascade H-bridge IGBT converter chain, SVG soft starting switch(SST), soft starting resistor and inductor. High-speed optical fiber communication is adopted between SVG converter chain and SVG control and protection device to realize primary and secondary electrical isolation.

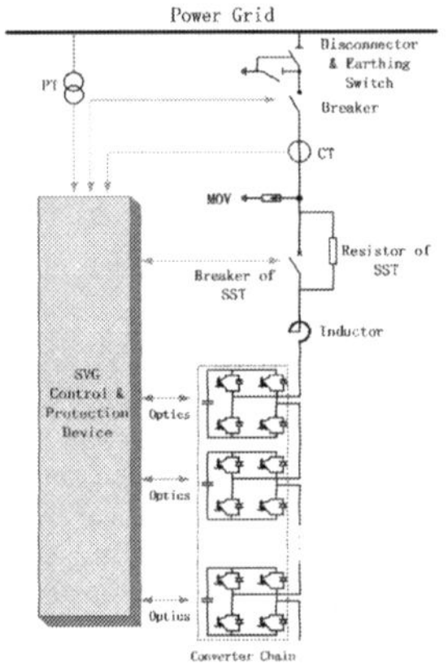

Fig.2 Sketch diagram of main electrical scheme of SVG

3. SVG Design and Implementation

This paper introduces the system design of the SVG dynamic reactive power compensation device in a coal mine.

3.1. Substation Power Data

The substation system of this coal mine adopts 6kV single busbar segment topology, with double power supply. The bus switch is designed between section I and section II bus bars, and each section is equipped with ventilation fans, air compressors, main hoists and power supply transformer in living area.

The active power of section I busbar is 1870kW, and the power factor before compensation is 0.79.

The active power of section II busbar is 1520kW, and the power factor before compensation is 0.81.

When the main equipments of the mine are running, the power grid system mainly has 3rd and 5th harmonics.

3.2. System Parameter Design of SVG

In order to ensure that the buses of segments I and II meet the compensation power factor requirements, section I and II are equipped with an SVG dynamic reactive power compensation device and filter capacitor FC. With FC device, in addition to filtering the grid system harmonic, it can also effectively reduce the SVG device capacity and equipment investment.

SVG directly connects 6kV high voltage bus without step-down transformer. The electrical wiring diagram of the system is shown in Figure 3.

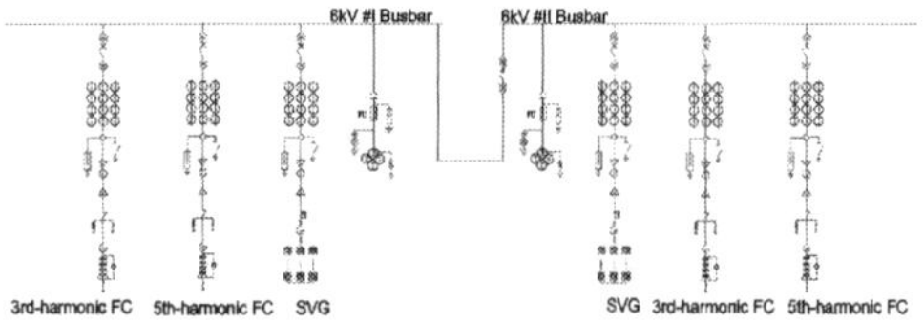

Fig.3 Sketch diagram of electrical scheme of SVG

To meet the requirement of power factor lower than 0.95, the compensation capacity of SVG unit is calculated as follows:

tanφ=tan(arccos(0.95))=0.328

tanφ1=tan(arccos(0.79))=0.775

tanφ2=tan(arccos(0.81))=0.724

Compensation capacity: 1870kW * (tan φ 1-tan φ) = 835 kvar

SVG compensation capacity of Section II bus: 1520kW * (tan φ 2-tan φ) = 601 kvar

Considering the subsequent expansion, the buses of section I and II are equipped with a set of dynamic reactive power compensation device with voltage level 6kV and rated capacity of 2Mvar SVG, and a total capacity of 500kvar FC device (including filtering for 3rd and 5th harmonics), which can realize the continuous adjustment of 1.5 Mvar inductive ~ 2.5 Mvar capacitive power output.

3.3. Design of SVG Modules

The converter chain is the core part of the SVG device, realizing the power exchange between the SVG device and the power grid system.

SVG adopts the cascade H bridge voltage source converter structure, and uses the carrier shift phase SPWM pulse width modulation technology,which can effectively reduce the loss of SVG link unit,obtain a better output voltage waveform and reduce the harmonic output of SVG device. Figure 4 and 5 show the schematic diagram and physical objects of SVG chain unit elements.

The 6kV SVG unit of this project adopts the 8- levels IGBT converter chain unit cascade.

IGBT is the core element of SVG chain units, and each chain unit consists of four single-tube IGBT forming a single-phase full-bridge circuit. Considering the different rated current of SVG with different capacity levels and the design requirements of voltage margin, the IGBT modules with different parameters and specifications are selected.

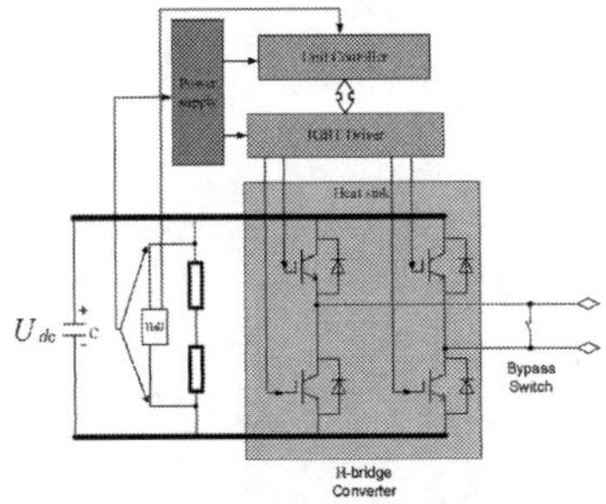

Fig.4 Sketch diagram of SVG chain-unit

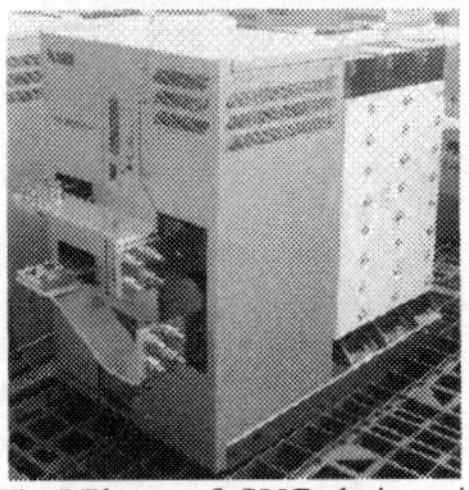

Fig.5 Photo of SVG chain-unit

The rated capacity of SVG device in this project is 2 Mvar and the rated current is 192A. Considering the margin, the rated voltage of 1700V and the rated current of 400A IGBT is selected.

3.4. Design of Control and Protection System of SVG

The control and protection system of SVG device includes SVG control and protection device, valve control device, HMI, etc. Figure 6 shows the photo of control and protection system cabinet of SVG.

Fig.6 Photo of control & protection system cabinet of SVG

The SVG control and protection device receives the analog sampling information and the switching quantity state, performs the SVG control and adjustment strategy, and transfer the adjustment and control instruction to the valve control device through the high-speed optic fiber,as is shown in Figure 7. The CPU board of the control and protection device adopts the high-speed digital signal processor DSP with large-capacity programmable gate array FPGA architecture, which can realize the calculation of dozens of us-level periodic sampling control, and meet the requirement of SVG dynamic response time of less than 10ms.

The valve control unit is responsible for receiving the control regulation command issued by the control protection device, integrating the chain DC voltage detected by the unit controller, completing the underlying control adjustment algorithm, generating the PWM waveform and related control instructions to the IGBT unit controller,and reporting the real-time information of the converter chain.

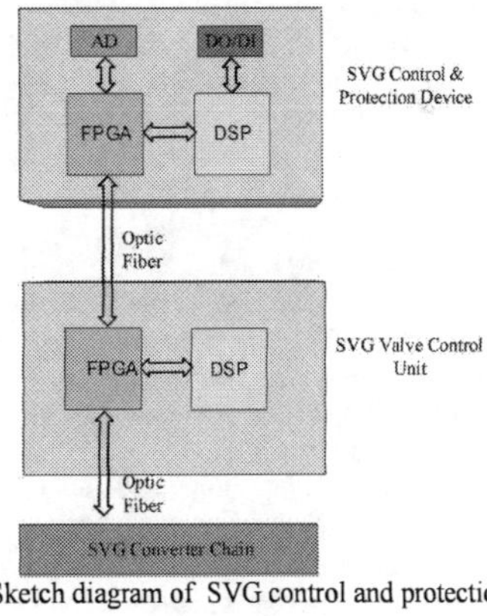

Fig.7 Sketch diagram of SVG control and protection device

3.5. Protection Function of the SVG

In order to quickly protect the SVG device in case of the power grid system failure and the SVG device itself failure, the SVG control and protection device is configured with the following protection functions:

System current: fast over-current protection, over-current protection, overload protection.

System voltage: over-voltage protection, under-voltage protection.

SVG converter chain: the converter chain sub-module level protection function, such as module DC over-voltage protection, over-current protection, over-heat protection, etc.

SVG control and protection device: communication fault protection, etc.

In addition, the SVG device is equipped with a micro-computer protection device at the incoming line switch cabinet side, which is independent of the SVG body protection. The microcomputer protection of the incoming line switch cabinet is mainly equipped with voltage protection, frequency protection, three-stage current, zero-sequence voltage / current and other protection functions.

3.6. Control Mode of SVG

(1) Constant voltage mode

To stabilize the voltage of the power grid system, the voltage of the power grid system is adjusted, and the output reactive power of the SVG device is automatically adjusted.

(2) Constant reactive power mode

The output is constant reactive power, the reactive output target value can be set between rated capacitive reactive power and rated inductive reactive power, automatically adjust the output reactive power of the device.

(3) constant power factor mode

The power factor of the grid system is feedback, the target value of the constant power factor can be set from +1 to-1 (positive stands for capacitive power), and the output reactive power of the SVG device can be automatically adjusted.

(4) Integrated control mode of reactive power voltage

SVG is under controlling according to the interval range of the grid system voltage. The voltage of the grid system voltage is within the set range and operates in the constant reactive power mode. If the voltage exceeds the set value, SVG switches to the constant voltage mode.

(5) Transient compensation mode

The transient compensation mode is that when the power grid fails or the system is disturbed, the SVG device quickly outputs reactive power to support the system voltage. When the fault is removed and the system voltage returns to the normal range, the SVG device exits the transient compensation mode and switches to the original working mode.

The above control mode and control target value can be flexibly set, and the coal mine SVG reactive power compen-sation device uses the constant power factor control mode.

3.7. Cooling system of the SVG unit

The SVG device includes water cooling and forced air cooling.

When SVG capacity is large (generally 35kV and above, capacity greater than 30 Mvar), water cooling and special cooling water machine system shall be configured.While SVG capacity is small (generally 10kV and below, capacity less than 10 Mvar), forced air cooling is adopted to reduce the cost and floor area of SVG device.

The rated capacity of the SVG project is 6kV-2Mvar , forced air cooling is adopted.

4. Application

The SVG device is under operation in a coal mine.In the process of the coal mine load changes, SVG device dynamicly adjusts output reactive current, and section I, II bus power factor meet the requirement of greater than 0.95, and the highest up to 0.98.SVG effectively reduces the line loss and production electricity comsumption of the coal mine, and the power grid system harmonics are below 3%, which meets the national 6kV power grid system requirements standard of no more than 4% harmonic total content.

Figure 8 shows the transition waveform of the inductive power output to capacitive power output of SVG.

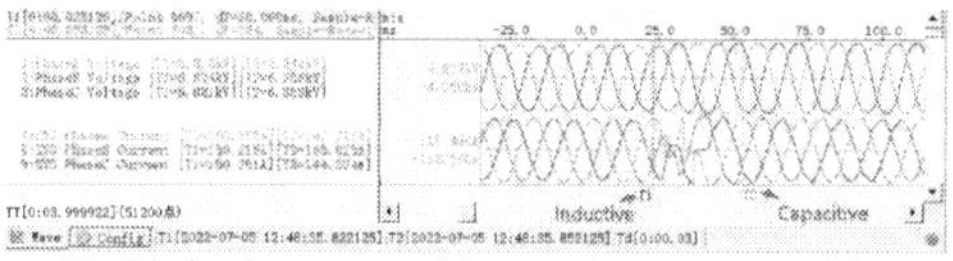

Fig.8 inductive to capacitive reactive power transition waveform of SVG

5. Conclusion

SVG dynamic reactive power compensation device is put into practical application with FC device,which can effectively improve the power factor and power quality of coal mine power grid, reduce the line loss and harmonic content, and improve the economic benefit of coal mine enterprises. This scheme can be a reference for the reactive power compensation and power quality management for coal mining enterprises.

References

1. Fan Yongjie, Huang Jing..Research and application of the mine power factor improvement scheme [J].Technology Information, 2012,25:398-399.
2. D. Sabin, M. Norwalk, K. Kittredge and S. Johnston, "IEEE Power Quality Standards," 2022 20th International Conference on Harmonics & Quality of Power (ICHQP), Naples, Italy, 2022:1-6.Guo Zhiqiang.Research on the underground application of mining stationary reactive power generator (SVG) in coal mine [J].Tongmei Technology, 2019 (01): 31-33.
3. Zheng Haisheng, Chen Huizhen.Application and exploration of SVG reactive power compensation technology in the 6kV power grid of coal mine [J].Jiangxi Coal Science and Technology, 2019 (02): 56-58.

4. Zhu Jianjun, Lu Zhigang.Reactive power compensation of medium and low voltage distribution network [J].Power capacitors and reactive power compensation, 2011,32 (04): 19-23.
5. H. F. Bilgin,M. Ermis,et al., "Reactive power compensation of coal mining excavators by using a new generation STATCOM," Fourtieth IAS Annual Meeting. Conference Record of the 2005 Industry Applications Conference, 2005.Vol. 1: 185-197.
6. Zan De Cheng.Design and optimization strategy of the SVG + FC hybrid reactive power compensation device[J].Electromechanical Information, 2020,14:63-64
7. Yuan Xiaodong.The advantages and disadvantages of SVC and SVG products [J].Electrical applications, 2013,9:16
8. Zheng Junhui, Wang Changchun.Research on the application of active dynamic reactive power and harmonic compensation device of hybrid power in coal mine power grid [J].Coal mine Mechanical and Electrical Engineering, 2012 (06): 5-8
9. GB / T 14549-1993, Quality of electric energy supply Harmonics in public supply network [S].

https://doi.org/10.1051/e3sconf/202344102022

Simulation and optimization of drying and salt formation process for radioactive high-salt waste liquid of medical isotope test reactor

Chunxin Fan*, Feng Luo, Huiqing Yang, Jiheng Fan, Maomao Liu

The First Research Institute, Nuclear Power Institute of China, Chengdu, 610000, China

Abstract. Medical isotope test reactor is the first reactor in the world. In order to implement the principle of minimizing radioactive waste, the radioactive waste liquid adopts advanced drying and salt formation technology. According to the current dry salt formation technology a number of problems arise due to the complexity of the feedstock waste stream. In this study, a thermodynamic electrolyte simulation model was developed and its predictive accuracy was verified for complex electrolyte systems. Then, sensitivity analysis and discussion of the operating parameters such as temperature, pH and salt adjustment of the dry salting technology are presented. The new approach of the thermodynamic electrolyte simulation model has been used to improve and optimise the safety aspects of the technology and the associated process configuration. It provides a certain general idea and follow-up theoretical guidance for drying salt formation technology.

1 Introduction

Environmental issues in the nuclear industry have always been a major concern[1]. Compared to cement curing technology, the dry salt technology has the advantages of good capacity reduction, low overall costs and high automation[2]. At present, many research institutions at home and abroad are carrying out independent research and development. However, researchers are well aware that there are obvious limitations in the current research on this technology. Although several companies have achieved good results in modelling a variety of radioactive wastes and sludges [3,4]. Subsequently, several research institutes and enterprises in China have carried out research work on drying into salt technology[5-10]. With the rapid development of the nuclear industry, the sources of high-salt waste fluids have increased. The more representative ones include those from isotope extraction, nuclear fuel reprocessing and decommissioning of nuclear facilities. In particular, the medical isotope test reactor is the first reactor in the world. In order to better realize the principle of minimizing radioactive waste, the radioactive waste liquid is dried and salted. As the composition of wastewater is more complex, the existing experimental verification for simple system is no longer applicable.

The main reasons why it is difficult to apply dry salt formation techniques to complex salt systems are as follows: complex salts mean that there is a wide variety of anions and cations in the system. These involve complete ionisation of strong electrolytes, partial ionisation of weak electrolytes, ionic reactions, complex ion formation, precipitation and dissolution of salts. This results in uncertainty in the number of reactions in the system, uncertainty in the extent to which the reactions proceed and uncertainty in the physicochemical properties. Ultimately, the generation, diffusion, escape or accumulation of flammable, explosive, toxic and hazardous substances in complex salt systems during the drying and salt formation process cannot be determined and controlled. The amplification effect of reaction absorption and heat release in industrial application and the technical characteristics of high temperature of drying salt formation will lead to great insecurity and difficulty in control of the technology. Generally speaking, the new drying and salt formation process of complex wastewater system needs to be enlarged from small test to pilot test. However, if the research horizon is extended to isotope extraction, decommissioning of nuclear facilities and changes in production demand, etc., All the above conditions will lead to the continuous complex changes of wastewater components. If the study of new technology parameters is always used and clumsy and exhaustive experiments are used to recustomize the new complex system one to one, it will only cause a lot of waste of manpower, material resources and financial resources. Therefore, it can be seen that the bottleneck of drying salt formation technology to deal with wastewater with complex components is the lack of universality to solve the complexity of chemical reaction.

* Corresponding author: fcx1403010805@163.com

2 Research methods

In this study, the innovation is to use Aspen Plus software to simulate and optimize drying salt formation. The complex electrolyte thermodynamic model was constructed according to the actual waste liquid and waste gas conditions of YD(medical isotope test reactor) project. Among them, ELENRTL physical property is selected as physical property.

2.1 Principle of material balance

According to the Law of Conservation, the general expression of material balance in the system is: Accumulation in the system = Input(F_{in})-Output(F_{out}) + Generation(G_R)-Consumption(C_R). Assuming the system is leak-free, there is the following formula:

$$\frac{dF}{dt} = F_{in} - F_{out} + G_R - C_R \quad (1)$$

In the above formula, the term generated or consumed is the amount generated or consumed due to chemical reaction. The accumulation amount can be positive or negative. When the cumulative quantity in the system is not zero, it is called an unstable state. When the accumulation is zero, it is called stable state The steady state can be simplified as:

$$F_{in} = F_{out} - G_R + C_R \quad (2)$$

2.2 Principle of energy balance

The material balance of the system is based on the law of conservation of energy, and the balance formula of energy is established. The energy balance expression of the system is: Input system energy = output system energy + accumulated system energy. The equipment in this device mainly includes reactor tower equipment, heat exchanger and fluid conveying machinery. The energy input into the whole production system mainly includes electric energy, fuel, chemical energy and enthalpy of materials(H_{in}). The output energy mainly includes the energy of generating steam and the enthalpy of the output material(H_{out}). According to the energy relationship in the system, the energy balance equation is listed as follows:

$$Q + W = H_{out} + H_{in} \quad (3)$$

Where Q is the heat load of the equipment and W is the mechanical energy input to the system.

2.3 Process and chemical reaction system

The model includes the waste stream from seven production lines, the waste gas treatment system alkaline wash stream, the pH adjustment line and the salt solution adjustment line. The treatment unit includes recirculating spray alkaline scrubbing of NOx gas, mixing, drying to salt and condensation cooling. The model was constructed and the reactions considered are shown in Figure 1 and Table 1.

Table 1. Simulation model considers reactions and details

Reactions	Control limitations
$2NO_2 + 2OH^- \leftrightarrow NO_2^- + NO_3^- + H_2O$	Dynamic
$2NO_2^- + O_2 \leftrightarrow 2NO_3^-$	
$HNO_3 + H_2O \leftrightarrow H_3O^+ + NO_3^-$	
$HCL + H_2O \leftrightarrow CL^- + H_3O^+$	
$HNO_2 + H_2O \leftrightarrow H_3O^+ + NO_2^-$	
$NH_3 + H_2O \leftrightarrow OH^- + NH_4^+$	
$2H_2O \leftrightarrow OH^- + H_3O^+$	
$NH_4NO_3(s) \leftrightarrow NH_4^+ + NO_3^-$	Thermodynamic
$NaOH(s) \leftrightarrow Na^+ + OH^-$	
$NH_4CL(s) \leftrightarrow NH_4^+ + CL^-$	
$NaCL(s) \leftrightarrow Na^+ + CL^-$	
$Na^+ + OH^- + H_2O \leftrightarrow NaOH \bullet H_2O(s)$	
$NaNO_3(s) \leftrightarrow Na^+ + NO_3^-$	
$NaNO_2(s) \leftrightarrow Na^+ + NO_2^-$	
$NaNO_3 \rightarrow Na^+ + NO_3^-$	
$NaOH \rightarrow Na^+ + OH^-$	
$NH_4NO_3 \rightarrow NH_4^+ + NO_3^-$	
$NaNO_2 \rightarrow Na^+ + NO_2^-$	

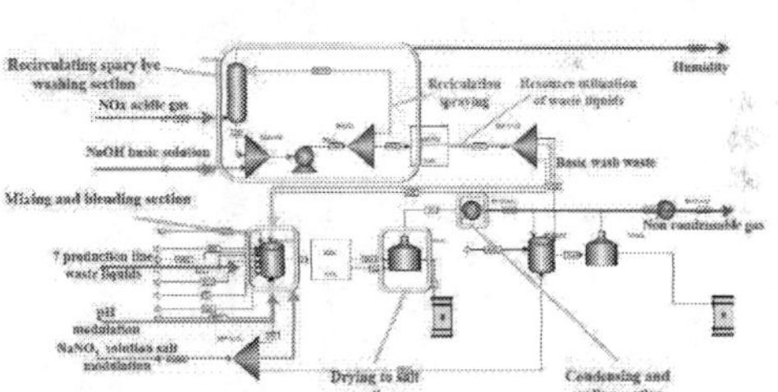

Fig. 1. Thermodynamic electrolyte model process flow.

3 Results and discussion

3.1 Effect of electrolyte system and model validation

According to the experimental data of Chao et al.[5], the electrolyte simulation model is verified by taking sodium nitrate, the main component of crystal, as the system. The validation results are shown in Figure 2.

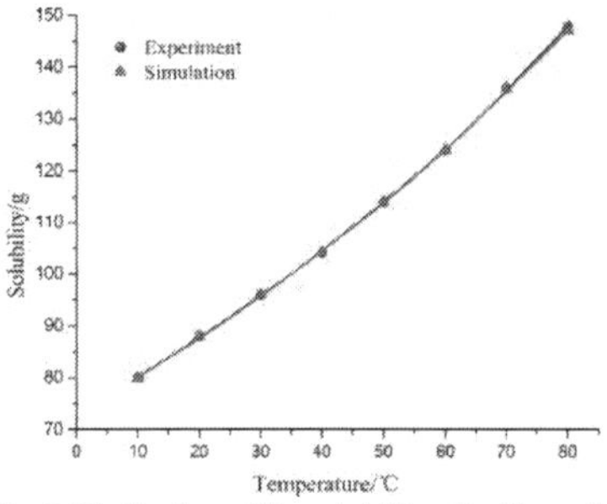

Fig. 2. Verification of the solubility of sodium nitrate.

The graph shows a high degree of agreement between the simulation results and the experimental results. In addition, the removal efficiency of NOx by circulating alkali washing and spraying in acid gas is more than 99.9%. Therefore, it shows that the electrolyte simulation model built in this study has some prediction accuracy. Next, this section simulates a boric acid system[5] containing five electrolytes. The results are shown in Figure 3.

The graphs express the simulated trends in approximately the same way as the experimental results. In particular, the start and end phases of microwave heating are highly consistent. However, there is still a gap between the two processes in the later stages of heating. In the actual heating process, there is a problem of uneven heating and heat loss in the later stages of microwave heated crystallisation. This is due to the growth and accumulation of crystals resulting in moisture in the gaps of the crystal particles is limited by heat transfer.

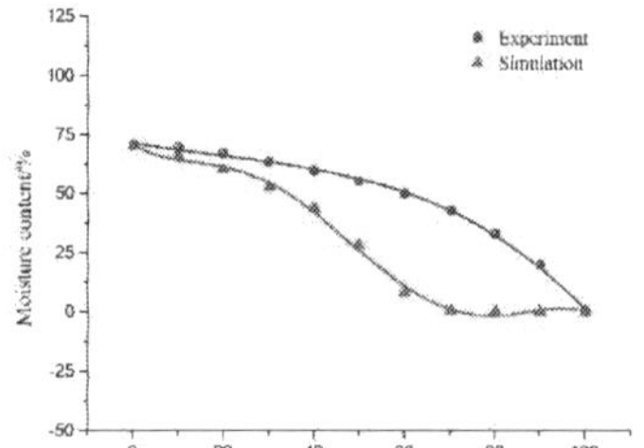

Fig. 3. Crystallisation verification of complex boron-containing electrolyte systems.

In summary, if the microwave heating crystallisation process needs to reduce heat loss and improve energy efficiency. As far as possible, a reasonable trade-off should be made between phased feedstock and the heating process. Controlling the water content to reach the standard temperature within about 20 minutes can effectively reduce the energy loss of microwave heating.

3.2 Effect of temperature on the production of dangerous crystals

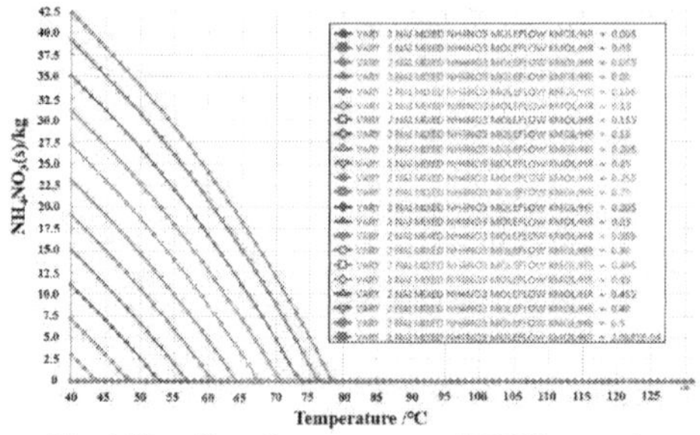

Fig. 4. The effect of temperature on NH_4NO_3 crystals.

NH_4NO_3 crystals are extremely unstable and decompose easily at high temperatures. A sensitivity analysis regarding the production of NH_4NO_3 crystals is shown in Figure 4. The picture shows that NH_4NO_3 crystals are mainly produced in the temperature range below 80°C. Meanwhile, the amount of NH_4NO_3 crystals produced decreases with increasing temperature until they disappear completely. Due to the pyrolytic nature of NH_4NO_3 crystals, it decomposes at high temperatures to produce NH_3 gas. If NH_4NO_3 crystals are produced during the drying to salt process, this will result in the decomposition of NH_4NO_3 crystals at high temperatures and during the crystal build-up. Further, the salt blocks produced by the crystal build-up will have a large void ratio, difficulty in achieving the water content and a low density of the salt blocks. However, due to the inevitable process of rising from low to high temperatures in the dry salt formation technique. Therefore, the solution proposed according to the electrolyte simulation model is to avoid NH_4NO_3 crystal production by controlling the NH_4^+ content, NO_3^- content and pH of the feedstock waste stream.

3.3 Effect of pH on NH3 concentration

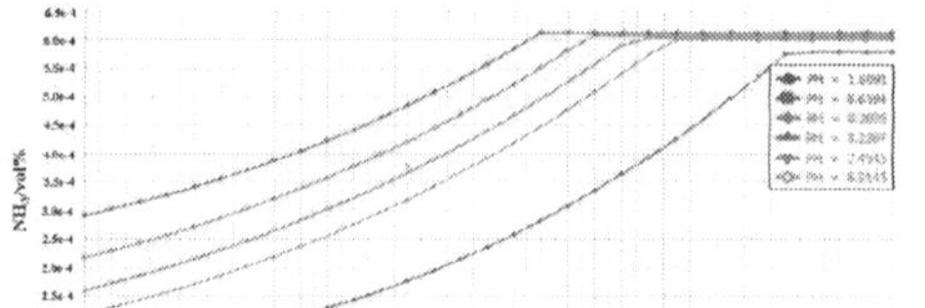

Fig. 5. The effect of pH on the concentration of NH_3 in secondary steam.

The effect of ph on the concentration of NH_3 in the vapour produced by the drying to salt process is shown in Figure 5. The images show that the concentration of NH_3 increases and then levels off with increasing temperature. In addition, the concentration of NH3 also increases with increasing pH and tends to level off at a lower temperature point. NH_3 explosion limit of 16-25 vol%. The concentration of NH_3 produced by the present electrolyte system is less than 1%, therefore it can be concluded that there is no explosion risk in the present study system. However, in order to avoid unpredictable escape and accumulation of NH_3, this process system is still equipped with NH_3 concentration monitoring and a nitrogen dilution interlock.

For the electrolyte system in this study, in addition to ensuring that the water content of the stacked crystal blocks is less than 1%. It is also necessary to avoid NH_4NO_3 crystal production. Furthermore, according to practical engineering experience alkaline waste streams are prone to foaming when heated. This in turn will lead to the purification effect of the waste stream being affected. In summary, a heating temperature of 152°C and a pH of 6.9 is suitable, taking into account energy consumption, safety and purification factors.

4 Conclusion

The current dry salt formation technology has good development momentum and broad application prospects. In this study, the thermodynamic electrolyte simulation model established in this study is verified to be accurate in prediction. On this basis, the optimum operating parameters of drying salt formation technology are determined. These include the modulation of temperature, pH and salt composition. Furthermore, through a sensitivity analysis of the technology, the study has improved the safety and optimisation of the technology and the associated process configuration. This study provides theoretical guidance for the research ideas and in-depth research of drying salt formation technology.

References

1. Feng L, Ming-dong R, Li W, Chen-long L, Lan-ju Y, Li L. Overview and Evaluation of the Source and Treatment Techniques of Radioactive Wastewater. SICHUAN ENVIRONMENT. **38**,108-114(2019)
2. Yi L, Wen-yu L, Zhen-chen L, Li C, Feng L, Ji-heng F. Mathematical Model and Dynamic Simulation of In-Drum Drying System for Radioactive Evaporation Concentrate. SICHUAN ENVIRONMENT. **40**,118-122(2021)
3. R S, G W, D F. Overview of Hybrid Microwave Technology. Journal of the South Carolina Academy of Science. 9,25(2011)
4. C G. Microwave In-Drum Drying. Radwaste Solutions. **14**,21-24(2007)
5. Chao G, Dahai X, Meilan J, Hongxiang A. Pilot Study on Microwave Drum Concentrate-drying Technology. Radiation Protection Bulletin. **33**,42-45(2013)
6. Xiaojun Y, Zhaofeng L, Honghui L, Dong L, Chao G, Meilan J. 2014. PrimaryExperimental on Concentrate Drying in Electrically Heated Drum. Radiation Protection Bulletin. 34,36-38;
7. Meilan J, Chao G, Dong L, Hongxiang A, Zhaofeng L. 2015. Primary study on microwave in-drum drying for simulated borate concentrate. Radiation Protection. 35,36-42;
8. Chao G, Hongxiang A, Dong L, Meilan J, Weibing Y, Xiaojun Y, Zhaofeng L. 2012. The Pilot Study on Microwave Drying Technology for Waste Resin from Nuclear Power Plant. ENVIRONMENTAL SCIENCE AND MANAGEMENT. 37,116-118;
9. Dong L, Zhaofeng L, Chao G, Xiaojun Y, Hongxiang A, Honghui L, Wei L, Shuaiwei Z, Weibing Y, Wendong F, Jiasheng R, Chengming J. 2016. Study on In-drum Drying Test of Simulated Waste Concentrates. ENVIRONMENTAL SCIENCE AND MANAGEMENT. 41,108-112;
10. Dong L, Dunyi Y, Zhaofeng L, Xiaojun Y, Chao G, Hongxiang A, Honghui L. 2018. IMPROVEMENT AND ITS VALIDATION TO IN-DRUM DRYING EQUIPMENT FOR WASTE CONCENTRATES FROM NPPS. Environment Engineering. 36,999-1002;

The Impact of Agricultural Industry Agglomeration in Sichuan Province on Agricultural Carbon Emissions

Yating Yang, Zhengyong Cao*, Maoling Huang, Anqi Jing

College of Economics, Sichuan Agricultural University, Wenjiang, Chengdu, Sichuan, China.

Abstract. This article utilizes panel data from 2005 to 2020, covering 21 cities in Sichuan Province, to empirically examine the relationship between agricultural industry concentration and carbon emissions. The findings reveal a clear inverted U-shaped relationship between agricultural industry agglomeration and carbon emissions. This relationship also exhibits temporal lag and regional disparities. In Sichuan Province, the link between agricultural industry agglomeration and carbon emissions follows this inverted U-shaped pattern, emphasizing the need for a comprehensive understanding of agglomeration's role in shaping emissions.Carbon emissions in agriculture display strong temporal path dependence, underscoring the importance of timely policies for carbon reduction. Local governments should adapt their strategies to regional peculiarities, promoting the growth of local agricultural industries through increased scale and agglomeration. A well-planned distribution of agricultural industries across regions is essential for sustainable development.

1 Introduction

In September of the year 2020, China unveiled its ambitious dual carbon objectives: attaining a "carbon peak" by the year 2030 and subsequently achieving "carbon neutrality" by 2060. This strategic vision is poised to mitigate carbon emissions through the means of technological advancement, innovation, optimization of industrial structures, and the promotion of a lifestyle rooted in low-carbon principles, green ethics, and environmental sustainability.

In conclusion, the trajectory of agricultural industry agglomeration is marked by distinct phases, each yielding its own set of impacts on agricultural carbon emissions. It is a matter of profound debate whether agricultural industry agglomeration, with its multifaceted implications, ultimately exerts a beneficial or detrimental influence on the trajectory of agricultural carbon emissions.

2 Literature review

To sift through the existing literature, we embark on our scholarly voyage by commencing with an exploration of agricultural carbon emissions themselves. The focal points of this inquiry encompass the realm of agricultural carbon emission accounting and its distinctive attributes.

Initially, agricultural carbon emission accounting revolved around six pivotal facets: fertilization, pesticides, agricultural film, diesel consumption, irrigation, and tillage [1,2]. This approach predominantly concentrated on the agricultural land's utilization within the farming milieu. However, as research endeavors evolved, the accounting paradigm broadened its scope. A burgeoning corpus of research now encompasses a comprehensive array of agricultural carbon sources, encompassing facets like agricultural land usage, the planting industry, livestock and poultry husbandry, among others [3,4].

Turning our gaze to the characterization of agricultural carbon emissions, the research terrain shifts to encompass an exploration of their spatiotemporal evolution characteristics [5-8], spatial spillover effects [9-11], threshold attributes [12, 13], structural traits [14-15], and convergence phenomena [17,18].

2.1 The study hypothesis proposes

Based on the combing of the existing literature, the following hypothesis is proposed in order to make a better conclusion.

Hypothesis 1: Agricultural industrial agglomeration has an impact on agricultural carbon emissions by producing agglomeration effects such as scale effect, technology effect, structure effect, social effect and cumulative effect.That is, the relationship between agricultural industry agglomeration and agricultural carbon emission is "inverted u type".

Hypothesis 2: Agricultural carbon emissions may have a certain time path dependence, that is, the agricultural carbon emissions in the early stage may have an impact on the later stage.

*Corresponding Author email: 1036790656@qq.com

3 Data

3.1 Variable selection

This feat is achieved by the multiplication of the carbon emissions of each individual source by their corresponding carbon emission coefficients, which are, in turn, judiciously summed. Hence, the most current methodology embraced within this discourse predicates itself upon the ensuing blueprint:

$$C = \sum g_i \varepsilon_i \quad (1)$$

In this paradigm, the symbol "C" assumes the mantle of representation for the overarching edifice of total agricultural carbon emissions. "G," on the other hand, stands as the venerable envoy, bearing the imprints of emissions emanating from each discrete carbon source. Meanwhile, the Greek letter "ε," with its intrinsic symbolism, serves as the custodian of the carbon emission coefficients, deftly charting the course for each unique carbon source enmeshed in this intricate narrative of environmental reckoning.

3.2 Independent Variables (lq)

The independent variable denoted as 'lq' signifies the agricultural industrial agglomeration index, a parameter of significant relevance.

$$lq_{ij} = (k_{ij} / k_i)/(k_j / k) \quad (2)$$

The symbols "k_{ij}" and "k_i" denote the provincial agricultural output value and the regional output value for the i-th province, respectively. Correspondingly, "k_j" and "k" represent the national agricultural output value and the national GDP, respectively. It is noteworthy that the magnitude of "lq_{ij}" directly mirrors the degree of agglomeration within the agricultural industry.

3.3 Control variables

Agricultural carbon emission intensity, rural per capita net income (*inc*), agricultural industrial structure (*struc*), agricultural mechanization level (*ma*), planting industry structure (*pis*), animal husbandry industrial structure (*ais*), and industrialization level (*il*) may engage in intricate, mutually causal relationships, giving rise to endogenous dynamics. Consequently, these variables are incorporated as control factors, thereby enhancing the precision and robustness of our analytical model.

Notably, the level of urbanization (*ur*) is ascertained by quantifying the proportion of the regional non-agricultural population to the total population. Additionally, we gauge regional economic development by evaluating the unchanged gross domestic product (GDP) per capita at year-end relative to the total population. The agricultural mechanization level (*mach*) is characterized by the cumulative horsepower of agricultural machinery employed, while the structural composition of industry (*struc*) is elucidated by the ratio of total agricultural output value to the aggregate output value encompassing agriculture, forestry, animal husbandry, and fisheries.

3.4 The empirical model

$$\ln c_{it} = a + a_1 \ln c_{it-1} + \beta_1 lq_{it} + \beta_2 lq_{it}^{2} + control + \varepsilon_{it} \quad (3)$$

Esteemed scholars of yore delved into the intricate realm of the "inverted U"-shaped association between the aggregation of the agricultural industry and the emanation of agricultural carbon emissions. They did so by invoking the venerable Environmental Kuznets Curve framework, adorning it with the primary (*lq*) and secondary (*lq*^{2}) terms of the agricultural industry.

3.5 Data sources

The wellspring of our original dataset finds its origins in an array of venerable tomes, including the provincial statistical yearbooks, "China Statistical Yearbook," "China Rural Statistical Yearbook," and the locally cherished "Sichuan Provincial Statistical Yearbook." Further insights were gleaned from the venerable annals of various city and state-level statistical compendia.

4 Methodology

4.1 Descriptive statistics of the variables

Table 1. Descriptive statistics of the variables

N	mean	sd	min	max
336	256.8	188.5	10.38	857.1
336	1.194	0.603	0.0383	3.317
336	0.522	0.0872	0.317	0.740
336	0.512	0.152	0.233	0.896
336	7,550	5,559	1,374	33,195
336	0.556	0.276	0.132	1.416
336	0.361	0.0851	0.112	0.574
336	0.308	0.0895	0.105	0.581
336	0.172	0.172	0.0161	2.344

Upon an attentive perusal of Table 1, it becomes evident that a tapestry of variables, including the degree of agricultural industrial agglomeration, the echelons of agricultural mechanization, the tier of regional economic development, and the dispersion gradient of per capita net income in rural areas, manifests with an air of magnitude.

4.2 Unit root inspection

To circumvent the specter of pseudoregression, we diligently subjected the dataset to rigorous scrutiny. Employing the discerning LLC test and the venerable Fisher-ADF test, we embarked on a quest to ascertain the stability of the sequence.

Table 2. Root in unit and co-integration tests for variables

variable	LLC checkout	The Fisher-ADF test	conclusion
lncit	0.0000	0.5122	non-stationary
Δlncit	0.0480	0.0000	steady
lq	0.0135	0.5084	non-stationary
Δlq	0.0000	0.0000	steady
lq2	O.0045	0.9919	non-stationary
Δlq2	0.0000	0.0000	steady
lninc	0.6222	0.9785	non-stationary

Δlninc	0.0000	0.0000	steady
il	0.9956	0.6427	non-stationary
Δil	0.0000	0.0000	steady
ais	0.3003	0.6037	non-stationary
Δais	0.0000	0.0000	steady
pis	0.9858	0.8303	non-stationary
Δpis	0.0000	0.0000	steady
ur	0.0013	0.07959	non-stationary
Δur	0.0000	0.0000	steady
struc	0.0000	0.0162	steady
Δstruc	0.0000	0.0000	steady
KAO-MDF	0.0000		Correlation exists
KAO-DF	0.0000		Correlation exists
KAO-ADF	0.0000		Correlation exists

Intriguingly, As shown in Table 2, our inquiry into co-integration, as gauged by the outcomes of the KAO-MDF, KAO-DF, and KAO-ADF tests, has yielded noteworthy results. It is with a degree of statistical significance, set at the conventional 5% threshold, that we decisively reject the null hypothesis, thus affirming the presence of a co-integration nexus interlinking these variables.

4.3 Panel estimation results

Table 3. panel estimation results

	fixed effect (fe)	stochastic effect (re)	System gmm two-step
	lnc	lnc	lnc
l.lnc			0.9991***
			(49.5319)
lq	0.4757***	0.7020***	0.1732**
	(4.8164)	(6.7845)	(2.4693)
lq2	-0.0557**	-0.1105***	-0.0461**
	(-2.1534)	(-4.0333)	(-2.3743)
lninc	0.8334***	0.3311***	-0.0768
	(7.9792)	(5.8090)	(-0.9415)
lnma	0.1620***	0.2021***	0.0281
	(4.7575)	(5.7755)	(1.5167)
il	1.1810***	1.6651***	-0.0895
	(7.0438)	(10.1173)	(-0.5184)
ais	1.0037***	0.7712***	-0.4340**
	(3.8908)	(2.9600)	(-2.2491)
pis	-0.2892***	-0.4307***	-0.0261
	(-5.0865)	(-7.2249)	(-0.5670)
ur	0.3013	0.1159	0.5364*
	(1.3606)	(0.5517)	(1.8865)
struc	0.1173	0.1516	0.0724
	(0.4857)	(0.6451)	(0.3006)
_cons	-2.6952***	1.0252**	0.4792
	(-3.2953)	(1.9771)	(0.5990)
N	600	600	570
R2	0.755	0.734	
AR(1)[P]			-3.53[0.000]
AR(2)[P]			-1.45[0.147]
Hasen[p]			18.25[0.148]
symmetry axis			2.1730

In accordance with the findings in Table 3, with the inclusion of a lagged term in the dependent variable within our empirical model, a crucial transformation unfolds. It imparts a nuanced complexion to our modeling, rendering the residuals no longer strictly exogenous and homoscedastic in nature. In such circumstances, the instrumental variable method emerges as the more fitting and judicious approach to estimation. Simultaneously, recognizing the potential interplay between agricultural carbon emission intensity, our core explanatory variables, and the regional level of economic development, we have opted to employ the Systematic Generalized Method of Moments (GMM) to skillfully navigate the endogeneity lurking within our equation.

4.4 Five-point sample estimation results

This division neatly segregates them into two distinctive echelons: the echelon of heightened Agricultural Industrial Concentration ($aq \geq 1.194$) and the echelon marked by diminished Agricultural Industrial Concentration ($aq < 1.194$). Within the lofty stratum of agricultural industrial concentration, a cadre of nine cities emerges, collectively constituting a compendium of 176 meticulously observed instances.

Table 4. Results of the sample systematic GMM test

Variables	High cluster group	Low cluster group
	(1)	(2)
	lnc	lnc
L.lnc	0.8550***	0.9705***
	(8.8504)	(33.3249)
lq	0.0433	0.2768**
	(0.4061)	(2.9125)
lq2	-0.0035	-0.0809***
	(-0.1799)	(-2.9806)
struc	0.2385	-0.2329
	(1.1494)	(-1.2629)
ur	-0.4132	0.5104
	(-0.9053)	(1.6499)
lninc	0.2042	-0.0819
	(1.1848)	(-0.9436)
lnma	0.0542	0.0152
	(0.8984)	(0.8590)
il	0.3721	0.0483
	(0.9227)	(0.2776)
ais	0.4134	-0.5536***
	(1.3165)	(-3.3001)
pis	0.1951	-0.0716***
	(1.2615)	(-2.9679)
N	247	304
AR(1)[P]	-2.45[0.014]	-2.63[0.009]
AR(2)[P]	-0.65[0.513]	-1.53[0.125]
Hansen[P]	27.37[0.159]	20.22[0.164]

Table 4 displays the statistical distribution of survey responses.Through the application of systematic generalized method of moments (SYS-GMM) testing on our samples, we discern a dichotomy within the realm of agricultural industry agglomeration. In one echelon, we observe a salient positivity, while in the secondary echelon, a discernible negativism manifests. However, it is crucial to underscore that this relationship acquires its significance solely within the sphere of low concentration. This observation vividly illuminates the nuanced interplay between agricultural industry agglomeration and agricultural carbon emissions, characterized by a heterogeneous, inverted U-shaped curve.

5 Discussion and Implication

From a Sichuan Province vantage point, a conspicuous "inverted U-shaped relationship" manifests between agricultural industrial agglomeration and agricultural carbon emissions, thus validating Hypothesis 1. This empirical validation underscores the imperative of a nuanced grasp of agricultural industrial agglomeration's role in shaping agricultural carbon emissions. It further accentuates the necessity for region-specific, timely policy formulations.

Temporal entwinements exert a profound influence on agricultural carbon emission intensity. The echoes of heightened agricultural carbon emission intensity in the antecedent period resonate through time, casting a shadow upon subsequent periods, thereby corroborating Hypothesis 2. This underscores the exigency for temporal considerations in addressing carbon fixation and reduction within Sichuan Province's agricultural sector.

Author Information

Yating Yang (1998-), female, from Leshan, Sichuan, master's degree student, research direction: Innovation research on Sichuan agricultural industry development.

Corresponding Author: Zhengyong Cao (1971-), male, from Ya'an, Sichuan, professor and master's supervisor, research direction: Innovation research on Sichuan agricultural industry development.

References

1. Miguel A. López-Navarro,Vicent Tortosa-Edo,Vanesa Castán-Broto.Firm-local community relationships in polluting industrial agglomerations: How firms' commitment determines residents' perceptions.Journal of Cleaner Production,**186** (2018)
2. Zwane Talent Thebe,Udimal Thomas Bilaliib,Pakmoni Lariba.Examining the drivers of agricultural carbon emissions in Africa: an application of FMOLS and DOLS approaches..Environmental science and pollution research international **30**, 19(2023).
3. Tian Yun, Yin Min Hao. Remeasurement of agricultural carbon emissions in China: basic status quo, dynamic evolution and space spillover effect . China's rural Economy,**03** (2022).
4. Ran Jincheng, Su Yang, Hu Jinfeng, Tang Hongsong, Wang Jingjing, Cui Pan. Research on the spatial and temporal characteristics, peak prediction and influencing factors of agricultural carbon emission in Xinjiang. Agricultural Resources and regionalization in China **08**, 16-24(2017),.
5. Liu Huajun, Bao Zhen, and Yang Qian. The distribution dynamics and evolution trend of carbon dioxide emissions in China. Resource Science,**10** (2013).
6. Wu Guoyong, Sun Xiaojun, Yu Fubo, Yang Lisa. Analysis of spatial correlation pattern ofcarbon productivity in Chinese planting industry. Population, resources and environment in China,**30** (2020).
7. Yin Xiyang, Jia Xiaojuan, Li Dongmei. The influence of agricultural industrial agglomeration on agricultural total factor productivity —— Based on the perspective of spatial spillover effect. Agricultural resources and regionalization in China,**11** (2022).
8. Johnson J M F, Franzluebbers A J, Weyers SL. Agricultural opportunities to mitigate greenhouse gas emissions. Environmental Pollution,**150** (2007).
9. Xu Xiaoyu, Dong Huizhong, Pang Min. Analysis of the spatial and temporal evolution characteristics and driving factors of agricultural carbon emission efficiency in the three northeastern provinces.Environmental Management in China,**02** (2023)
10. Gan Tianqi, Liu Mingming, and Zhou Zongyuan. Spatial correlation characteristics of agricultural carbon emissions in China.Journal of Sichuan Agricultural University,**01** (2023)
11. Li Yuanling, Wang Jinlong, Yang Ling. Spatiotemporal characteristics of agricultural carbon emission in Hunan Province based on county scale. Agricultural Resources and regionalization in China, **04**,75-84(2022)
12. Zheng Yangyang, Luo Jianli. Carbon emission effect of agricultural production efficiency: space overflow and threshold characteristics. Journal of Beijing University of Aeronautics (Social Science edition) **01**,96-105(2021).
13. Wang Hui, Bian Yijie. The dynamic evolution and threshold characteristics of agricultural production efficiency and agricultural carbon emission. Agricultural technology and economy **06**,36-47 (2015) .
14. Li Chen, Li Haoyu, Kong Haizheng, Feng Wei. Characteristics of implied carbon emission structure and decomposition of driving factors in China's fishery production system. Resource Science,**06** (2021)
15. Wang Baoyi. Study on the structural characteristics of agricultural carbon emission in China. Research Worl **09**,3-10 (2016)
16. Liu Naidong. Carbon emission structure characteristics and influencing factors of rice production in Jiangsu Province [D]. Tutor: Hu Hao. Nanjing Agricultural University, 2014
17. Guo Si dai, Qian Yupeng, Zhao Rui. Analysis of agricultural carbon emission efficiency and convergence in western China —— based on SBM-Undesirable model. Rural Economy,**11** (2018).
18. Wu Haoyue, He Yanqiu, Chen Rou. Chinese agricultural carbon emission performance evaluation and random convergence study —— based on SBM-Undesirable model and panel unit root test. Chinese Journal of Eco-Agriculture,**09** (2017)

The development of nuclear power and related environmental evaluation due to nuclear accidents

Xuanbo Zhao*

Beijing Etown Academy, Daxing, Beijing, China

Abstract. Nuclear power has played a pivotal role in meeting the growing global energy demand. However, the development of nuclear energy has been punctuated by a series of high-profile accidents, such as Chernobyl and Fukushima, which have raised significant concerns regarding its environmental impact. This paper aims to evaluate nuclear energy through data by using additional methods and to provide an overview of historical development of nuclear power technology and nuclear accidents. In order to evaluate thoroughly, the method of ratio estimation had been applied to find the trend for nuclear energy in previous decades and the potential development in the future. As a result, it emphasizes the effectiveness of nuclear energy changes structure of energy outputs in some countries. France, for instance, has already placed nuclear power as the main source for clear energy output. The paper proofs the development of nuclear power has been marked by significant improvements in safety and environmental consciousness. However, the specter of nuclear accidents looms large, necessitating ongoing vigilance and robust environmental evaluation processes. Despite that, the paper still contains additional assumption needed to be solved or explained.

1 Introduction

The utilization of nuclear power has emerged as a pivotal aspect of modern energy production, promising substantial benefits in terms of efficiency, reliability, and reduced greenhouse gas emissions. However, this technological advancement also brings with it complex challenges, particularly concerning the potential environmental ramifications stemming from nuclear accidents. These accidents have underlined the critical need for thorough and accurate environmental evaluations, not only to gauge the immediate impacts but also to comprehend the long-term consequences on ecosystems, human health, and the broader environment. Three famous nuclear accidents showcase the consequences of using nuclear energy: On March 11, 2011, the accident at the Fukushima Daiichi Nuclear Power Plant (FDNPP) in Japan shocked the world. Approximately one hour after a powerful earthquake of approximately 9.0 magnitude struck the Pacific Ocean near Japan's eastern coastline, two enormous tsunamis were generated, and these subsequently hit Japan. The occurrence of the tsunamis was a direct result of the earthquake [1]. The earthquake caused significant damage to the main electric power grid, rendering it inoperative.

Additionally, the backup power supply was affected by seawater intrusion, resulting in a loss of power. Consequently, the cooling systems of the four nuclear reactors were disrupted. This disruption led to heightened pressure levels and the generation of hydrogen gas due to the extreme heating of the cooling water. Over the subsequent days, the release of large amounts of radioactivity into the atmosphere was triggered by hydrogen explosions. In order to prevent further escalation of damage and more substantial releases of radioactive materials, plant managers made the decision to use seawater as a cooling medium [2]. Besides, Large quantities of radioactive substances were released into the Pacific Ocean, causing irreversible effects on the marine ecosystem. Over 80% of the radionuclides released into the atmosphere during the Fukushima disaster are believed to have dispersed offshore, with subsequent deposition occurring in the Pacific Ocean [3]. Radionuclides originating from Fukushima were detected in seawater and marine organisms across the Pacific Ocean, as evidenced by various studies[4]. The Monticello nuclear power plant on the banks of the Mississippi River suffered a radioactive water leak on Nov. 21 last year, causing 1.5 million liters of tritium-containing effluent to escape; The Chernobyl nuclear accident occurred on April 26, 1986, during a technical test conducted in Unit 4 of the Chernobyl Nuclear Power Plant (NPP). The accident was triggered by inappropriate reactor operation at a low power level, which resulted in the "xenon-poisoning" of the reactor. Unfortunately, the reactor staff did not properly recognize this issue, which led to improper operation of the reactor's control rods [5]. The operating error resulted in a rapid and uncontrolled increase in power within the RBMK-1000 reactor, leading to its thermal destruction. This, in turn, caused at least one

* Corresponding author: xuanbo.zhao@hotmail.com

(steam) explosion and ignited the graphite moderators [6]. Initially, Chernobyl's "exclusion zone" covered a 30 km radius (2800 km2) around the Nuclear Power Plant (NPP). In the months that followed the accident, around 116,000 people residing within the "exclusion zone" were evacuated to areas with lower contamination levels. Unfortunately, the evacuation process started 3 to 11 days after the accident, which, for some of the affected population, was considered delayed and resulted in increased exposure to radiation [7].

Nuclear power provides a reliable baseload source of electricity that is not dependent on weather conditions like solar and wind power [8]. Nuclear plants have very high capacity factors, typically around 90%, meaning they produce close to their maximum output consistently. Moreover, nuclear power emits virtually no greenhouse gases or air pollutants during operation, helping countries meet climate change goals and air quality standards [9]. The IPCC has identified nuclear power as a key technology for reducing CO2 emissions. More importantly, nuclear power diversifies a country's energy mix, reducing dependence on any single source and improving energy security [10]. Many studies have argued for an "all of the above" strategy, which includes nuclear power to stabilize the electricity supply. All these factors demonstrate that nuclear energy can effectively meet the world's energy needs to a large extent. However, the use of nuclear energy does have some environmental impacts and accidents, such as radiation risks, waste disposal, water pollution and other problems. Accidents like Fukushima and Chernobyl released huge amounts of radiation, leaving the region a wasteland; The disposal of nuclear waste has been an unsolved problem, and nuclear waste remains very dangerous for thousands of years, requiring safe long-term storage and environmental isolation. Nuclear power plants require a large amount of cooling water, and untreated direct discharge into the ocean can have a significant negative impact on Marine life.

There are numerous papers to discuss the topic of nuclear energy focused on development and evaluation. *Research status of nuclear power: A review*, its systematic approach to organizing the various aspects of nuclear power research. The categorization of topics enables readers to navigate through the complexities of the subject matter more effectively. By using a theory of Cross-country comparisons to identify the degree of development of nuclear power research in a country, which is impressive; *Comparison of the Chernobyl and Fukushima nuclear accidents: A review of the environmental impacts* establishes that the aftermath of the Chernobyl incident significantly surpassed the consequences of the Fukushima accident across various aspects. Both accidents released a substantial number of volatile radionuclides such as noble gases, iodine, cesium, and tellurium. Besides, the author summarized the release of refractory elements, including actinides, during the Chernobyl incident was notably higher by approximately four orders of magnitude compared to Fukushima.

A partial of research, however, still has a space to improve. For instance, the data shown nuclear generation by countries in a paper named *nuclear power as foundation of a clean energy future: A review* (Published in 2019) is already outdated. Although the author did a comprehensively analysis, that is based on 2013 data analysis, which needs to update a new one. Also, the innovate nuclear reactors are developed some new predictable technologies in recent years, and its functions and advantages are already change; *Research status of nuclear power: A review* is not well understood or investigated, requiring further investigation.

The primary objective of this research is to analyze the annual energy generation data of different nuclear energy developed countries and to evaluate the importance of nuclear energy in the country through different analytical methods. Also, the development of nuclear reactors and different nuclear leakage incidents in the world will be discussed and evaluated.

2 Materials and Methods

2.1 Data

All of the data in this paper is obtained from a data archive at www.ourworldindata.org, where articles and charts were referenced in nearly 50,000 media articles last year; and over 20,000 of these references were in large media outlets with international reach. On this website, we have collected data on renewable energy (Hydro, Wind, solar), fossil fuel consumption (oil, coal, natural gas), and nuclear power generation for seven countries from 1990 to 2021, including China, U.S, France, Canada, Japan, German, and Russia. These seven countries were chosen as the main data analyzers because they are relatively advanced in nuclear energy production and their economies are in a relatively stable and favorable condition.

The annual nuclear power generation from 1990 to 2021 for the seven selected countries is shown in Fig .1. Annual generation of different types of new energy and fossil fuels in China, U.S, France, Canada, Japan, German, and Russia, is shown in Fig 2-8, respectively.

Year	China	U.S.	France	Canada	Japan	Germany	Russia
1990	0.0	607.2	314.1	72.5	194.6	152.5	118.3
1991	0.0	644.8	331.3	84.3	208.7	147.2	120.0
1992	0.0	651.3	338.5	80.0	217.0	158.8	119.6
1993	1.6	642.4	368.2	93.3	247.7	153.3	119.2
1994	14.8	674.1	360.0	107.1	258.2	150.7	97.8
1995	12.8	708.8	377.2	97.2	286.9	153.1	99.5
1996	14.3	710.2	397.3	92.1	296.5	160.0	109.0
1997	14.4	661.7	395.5	82.0	321.2	170.3	108.5
1998	14.1	709.2	388.0	71.0	326.0	161.6	103.7
1999	14.9	766.6	394.2	73.0	317.2	170.0	121.9
2000	16.7	753.9	415.2	69.2	306.0	169.6	122.5
2001	17.5	768.8	421.1	72.9	303.9	171.3	125.4
2002	25.1	780.1	436.8	71.8	280.3	164.8	134.1
2003	43.3	763.7	441.1	71.2	228.0	165.1	141.2
2004	50.5	788.5	448.2	85.9	268.3	167.1	137.5
2005	53.1	782.0	451.5	86.8	280.5	163.1	137.6
2006	54.8	787.2	450.2	92.4	291.5	167.3	144.7
2007	62.1	806.4	439.7	88.2	267.3	140.5	148.0
2008	68.4	806.2	439.5	88.3	241.3	148.5	152.1
2009	70.1	798.9	409.7	85.1	263.1	134.9	152.8
2010	74.7	807.0	428.5	85.5	278.4	140.6	159.4
2011	87.2	790.2	442.4	88.3	153.4	108.0	162.0
2012	98.3	769.3	425.4	89.5	15.1	99.5	166.3
2013	111.5	789.0	423.7	97.6	10.4	97.3	161.4
2014	133.2	797.2	436.5	101.2	0.0	97.1	169.1
2015	171.4	797.2	437.4	96.1	3.2	91.8	182.8
2016	213.2	805.7	403.2	95.7	14.9	84.6	184.1
2017	248.1	805.0	398.4	95.6	27.8	76.3	190.1
2018	295.0	807.1	412.9	95.0	47.8	76.0	181.8
2019	348.7	809.4	399.0	95.5	63.9	75.1	195.5
2020	366.2	789.9	353.8	92.7	41.9	64.4	215.7
2021	407.5	778.2	379.4	87.4	61.2	69.1	222.4

Fig. 1 Nuclear power generation (TWh) between 1990-2021

Year	Geo Biomass Others	Solar	Wind	Hydro	Nuclear	Fossil
1990	0.06	0.00	0.00	126.74	0.00	7563.27
1991	0.06	0.00	0.01	124.69	0.00	7968.69
1992	0.11	0.00	0.13	130.69	0.00	8386.55
1993	0.12	0.00	0.21	151.85	1.60	9003.10
1994	0.46	0.01	0.38	167.43	14.76	9509.28
1995	3.01	0.01	0.62	190.58	12.83	9741.28
1996	1.53	0.01	0.09	187.97	14.34	10305.78
1997	2.72	0.01	0.20	195.98	14.42	10351.44
1998	2.48	0.01	0.36	198.89	14.10	10363.47
1999	2.52	0.02	0.47	196.58	14.95	10752.91
2000	2.54	0.02	0.59	222.41	16.74	11085.26
2001	2.55	0.03	0.72	277.43	17.47	11595.46
2002	2.55	0.05	0.84	287.97	25.13	12657.19
2003	2.54	0.06	1.00	283.68	43.34	14859.57
2004	2.53	0.08	1.28	353.54	50.47	17332.89
2005	5.32	0.08	1.95	397.02	53.09	19713.34
2006	7.13	0.10	3.71	435.79	54.84	21612.99
2007	9.86	0.11	5.48	485.26	62.13	23448.26
2008	14.87	0.15	13.10	636.96	68.39	23943.34
2009	20.86	0.28	27.61	615.64	70.05	25087.56
2010	24.90	0.70	49.40	711.38	74.74	26642.34
2011	27.63	2.61	74.10	688.05	87.20	28883.00
2012	30.13	3.59	103.05	862.79	98.32	29581.56
2013	37.13	8.37	138.26	909.61	111.50	30523.66
2014	46.27	23.51	159.76	1059.69	133.22	30937.32
2015	54.07	39.48	185.59	1114.52	171.38	31034.00
2016	62.13	66.53	240.86	1153.27	213.18	31191.84
2017	79.60	117.80	304.60	1165.07	248.10	31973.58
2018	93.73	176.90	365.80	1198.89	295.00	32888.71
2019	112.73	224.00	405.30	1272.54	348.70	33692.99
2020	135.63	261.10	466.50	1321.71	366.20	34232.88
2021	169.93	327.00	655.60	1300.00	407.50	36222.59

Fig. 2 New energy and fossil fuel generation (TWh) in China (1990-2021)

Year	Geo Biomass Others	Solar	Wind	Hydro	Nuclear	Fossil
1990	57.46	0.37	2.82	292.28	607.22	344.95
1991	60.55	0.48	2.98	287.33	644.81	420.84
1992	64.82	0.41	2.92	251.43	651.34	406.70
1993	66.92	0.48	3.04	279.25	642.41	423.47
1994	66.35	0.50	3.48	259.34	674.15	464.67
1995	63.27	0.51	3.20	311.22	708.84	484.87
1996	64.83	0.54	3.27	347.55	710.24	499.69
1997	65.82	0.53	3.32	355.97	661.73	521.46
1998	65.62	0.52	3.06	322.09	709.16	534.11
1999	66.48	0.52	4.53	316.61	766.58	538.37
2000	66.58	0.52	5.65	272.76	753.89	535.78
2001	66.83	0.57	6.81	210.24	768.83	541.57
2002	71.79	0.60	10.46	258.17	780.06	591.69
2003	71.33	0.61	11.30	269.97	763.73	631.16
2004	71.95	0.70	14.29	262.55	788.53	675.07
2005	72.60	0.75	17.99	266.43	781.99	704.60
2006	73.08	0.82	26.86	285.54	787.22	741.29
2007	73.87	1.10	34.80	243.04	806.42	818.36
2008	73.55	1.63	55.92	251.05	806.21	938.40
2009	73.16	2.08	74.63	271.53	798.85	925.56
2010	75.06	3.01	95.61	257.27	806.97	972.44
2011	75.78	4.74	121.39	316.10	790.20	1020.87
2012	77.04	9.04	142.24	274.03	769.33	1074.03
2013	80.67	16.04	169.54	266.55	789.02	1133.01
2014	84.07	29.22	183.49	255.75	797.17	1120.85
2015	83.74	39.43	192.65	246.45	797.18	1230.11
2016	82.72	55.42	229.29	263.76	805.69	1283.44
2017	82.80	78.06	256.87	296.81	804.95	1286.79
2018	81.89	94.31	275.42	289.51	807.08	1268.25
2019	76.82	107.97	298.87	285.47	809.41	1253.32
2020	74.31	132.04	341.35	282.78	789.88	1167.04
2021	75.49	165.36	383.60	257.69	778.19	1217.94

Fig. 3 New energy and fossil fuel generation (TWh) in the U.S. (1990-2021)

Year	Geo Biomass Others	Solar	Wind	Hydro	Nuclear	Fossil
1990	1.91	0.00	0.00	53.87	314.08	1601.26
1991	2.10	0.00	0.00	57.60	331.34	1707.94
1992	2.13	0.00	0.00	68.96	338.45	1677.70
1993	1.97	0.00	0.00	64.90	368.19	1606.06
1994	2.16	0.00	0.00	78.79	359.98	1551.73
1995	2.33	0.00	0.00	73.12	377.23	1585.17
1996	2.42	0.00	0.01	66.04	397.34	1654.88
1997	2.68	0.00	0.01	63.76	395.48	1624.68
1998	2.65	0.00	0.02	62.09	387.99	1718.93
1999	2.85	0.00	0.04	72.51	394.24	1722.83
2000	2.99	0.01	0.05	66.36	415.16	1719.71
2001	3.33	0.01	0.13	74.27	421.08	1725.44
2002	3.54	0.01	0.27	60.40	436.76	1697.53
2003	3.72	0.01	0.39	58.94	441.07	1730.17
2004	3.78	0.01	0.60	59.56	448.24	1742.17
2005	3.87	0.01	0.96	51.48	451.53	1742.59
2006	3.84	0.01	2.18	56.30	450.19	1711.23
2007	4.22	0.02	4.07	57.60	439.73	1675.24
2008	4.44	0.04	5.69	63.65	439.45	1661.72
2009	4.58	0.17	7.91	56.99	409.74	1587.68
2010	4.92	0.62	9.95	62.71	428.52	1609.97
2011	5.48	2.08	12.05	44.79	442.39	1503.19
2012	6.35	4.02	15.11	58.78	425.41	1500.31
2013	7.81	4.74	16.05	70.84	423.68	1501.05
2014	8.32	5.91	17.25	62.83	436.48	1364.99
2015	8.84	7.26	21.35	54.56	437.43	1392.82
2016	9.64	8.16	21.31	59.92	403.20	1422.17
2017	10.33	9.10	24.54	48.96	398.36	1437.18
2018	10.55	10.41	28.53	63.92	412.94	1404.10
2019	10.74	11.75	34.62	56.03	399.01	1384.47
2020	10.67	12.93	39.75	61.17	353.83	1204.73
2021	11.19	14.61	36.97	57.97	379.36	1303.04

Fig. 4 New energy and fossil fuel generation (TWh) in France (1990-2021)

Year	Geo Biomass Others	Solar	Wind	Hydro	Nuclear	Fossil
1990	3.95	0.00	0.00	295.68	72.46	1916.02
1991	3.98	0.00	0.00	307.28	84.34	1860.18
1992	4.47	0.00	0.06	315.17	80.02	1928.26
1993	4.82	0.00	0.06	322.21	93.28	1946.97
1994	5.73	0.00	0.06	327.86	107.08	2009.36
1995	7.14	0.00	0.06	334.06	97.16	2095.40
1996	7.53	0.01	0.06	354.59	92.12	2139.88
1997	8.17	0.01	0.06	348.68	81.95	2201.38
1998	8.99	0.01	0.06	330.87	70.99	2231.21
1999	8.94	0.01	0.16	345.00	72.98	2293.16
2000	8.92	0.02	0.26	356.76	69.16	2356.58
2001	9.67	0.02	0.33	331.52	72.86	2350.01
2002	10.05	0.02	0.41	349.27	71.75	2428.99
2003	9.42	0.02	0.69	336.14	71.15	2495.39
2004	9.80	0.01	0.95	338.40	85.87	2508.97
2005	9.16	0.02	1.57	361.96	86.83	2469.93
2006	8.94	0.02	2.47	352.89	92.44	2464.81
2007	8.99	0.03	3.01	367.62	88.19	2579.58
2008	7.98	0.04	3.79	377.49	88.30	2506.02
2009	8.94	0.11	6.64	368.69	85.13	2392.14
2010	10.30	0.26	8.72	351.38	85.53	2471.71
2011	10.21	0.57	10.19	375.72	88.29	2535.40
2012	10.91	0.88	11.31	380.27	89.49	2528.76
2013	11.04	1.50	11.15	391.79	97.58	2576.82
2014	9.80	2.12	12.82	382.50	101.21	2607.89
2015	9.99	2.89	26.97	382.19	96.05	2617.33
2016	11.36	4.03	30.93	385.43	95.69	2544.80
2017	10.85	3.57	31.51	394.59	95.57	2591.32
2018	10.54	3.80	33.14	385.89	95.03	2653.27
2019	10.31	4.08	32.88	381.77	95.47	2648.05
2020	9.39	4.28	35.64	386.54	92.65	2421.13
2021	9.70	5.16	35.12	380.85	87.36	2483.22

Fig. 5 New energy and fossil fuel generation (TWh) in Canada (1990-2021)

https://doi.org/10.1051/e3sconf/202344102024

Year	Geo Biomass Others	Solar	Wind	Hydro	Nuclear	Fossil
1990	11.34	0.00	0.00	86.90	194.57	4374.01
1991	11.65	0.00	0.00	94.42	208.69	4428.32
1992	11.85	0.01	0.00	80.05	217.04	4502.18
1993	11.56	0.02	0.00	92.58	247.70	4460.20
1994	13.47	0.03	0.00	64.46	258.25	4712.57
1995	14.30	0.04	0.00	78.79	286.89	4876.48
1996	15.25	0.05	0.00	77.17	296.50	4966.40
1997	16.44	0.10	0.00	86.44	321.16	4982.26
1998	16.17	0.14	0.01	89.08	325.97	4837.62
1999	16.49	0.22	0.04	83.79	317.23	4976.21
2000	16.11	0.34	0.11	84.47	305.95	5035.61
2001	16.03	0.50	0.25	81.54	303.86	4988.12
2002	16.94	0.69	0.41	80.60	280.34	5021.03
2003	17.88	0.95	0.83	92.47	228.01	5200.92
2004	18.05	1.27	1.44	91.67	268.32	5122.87
2005	21.69	1.63	1.91	77.56	280.50	5157.83
2006	21.53	2.00	2.14	88.81	291.54	5086.07
2007	22.20	2.31	2.74	74.51	267.34	5130.28
2008	21.29	2.59	2.95	75.16	241.25	5100.91
2009	20.33	3.05	3.43	70.47	263.05	4527.24
2010	21.83	3.98	3.93	88.50	278.36	4790.78
2011	21.11	5.44	4.46	82.49	153.38	4841.04
2012	22.13	7.37	4.73	77.10	15.12	5189.08
2013	23.18	12.91	5.13	79.33	10.43	5133.33
2014	23.63	23.55	5.01	81.71	0.00	4996.62
2015	28.49	34.54	5.22	85.77	3.24	4853.51
2016	23.67	43.33	5.34	79.43	14.87	4761.28
2017	27.39	54.24	5.83	79.29	27.75	4755.59
2018	30.13	62.11	6.44	81.11	47.82	4643.38
2019	32.16	67.75	6.75	73.64	63.88	4477.79
2020	34.80	75.14	7.82	77.41	41.86	4113.94
2021	35.76	86.27	8.25	77.64	61.22	4205.64

Fig. 6 New energy and fossil fuel generation (TWh) in Japan (1990-2021)

Year	Geo Biomass Others	Solar	Wind	Hydro	Nuclear	Fossil
1990	1.44	0.00	0.07	17.34	152.47	3703.87
1991	1.51	0.00	0.10	15.85	147.23	3620.40
1992	1.56	0.00	0.28	18.64	158.80	3497.78
1993	1.66	0.00	0.60	18.96	153.28	3492.76
1994	1.91	0.00	0.91	20.20	150.70	3456.35
1995	2.05	0.01	1.50	21.56	153.09	3462.44
1996	2.14	0.01	2.03	18.82	160.02	3578.19
1997	2.30	0.01	2.97	18.95	170.33	3496.78
1998	2.72	0.02	4.49	19.00	161.64	3481.88
1999	2.94	0.02	5.53	20.69	170.00	3392.88
2000	3.40	0.00	9.50	24.90	169.61	3391.78
2001	5.20	0.10	10.50	23.20	171.30	3460.38
2002	6.40	0.20	15.80	23.70	164.84	3408.06
2003	8.95	0.31	19.09	18.32	165.06	3405.06
2004	10.64	0.56	26.02	20.75	167.07	3366.81
2005	14.71	1.31	27.77	19.64	163.05	3308.26
2006	18.93	2.27	31.32	20.03	167.27	3360.62
2007	24.62	3.14	40.51	21.17	140.53	3214.11
2008	28.03	4.51	41.39	20.44	148.49	3225.88
2009	30.90	6.72	39.42	19.03	134.93	3018.43
2010	33.95	11.96	38.55	20.95	140.56	3134.53
2011	36.91	19.99	49.86	17.67	107.97	3032.67
2012	43.23	26.74	51.68	21.76	99.46	3052.56
2013	45.59	30.62	52.74	23.00	97.29	3147.40
2014	48.39	34.56	58.50	19.59	97.13	2963.16
2015	50.46	37.17	80.62	18.98	91.79	2981.13
2016	51.10	36.67	79.92	20.55	84.63	3062.10
2017	51.08	37.89	105.69	20.15	76.32	3065.36
2018	50.97	43.46	109.95	17.69	76.00	2951.88
2019	50.32	44.38	125.89	19.73	75.07	2811.78
2020	51.09	48.64	132.10	18.32	64.38	2545.10
2021	50.90	49.00	117.70	19.10	69.13	2655.22

Fig. 7 New energy and fossil fuel generation (TWh) in Germany (1990-2021)

Year	Geo Biomass Others	Solar	Wind	Hydro	Nuclear	Fossil
1990	0.07	0.00	0.00	166.85	118.33	9246.80
1991	0.07	0.00	0.00	168.09	119.98	9100.94
1992	0.06	0.00	0.00	172.59	119.63	8716.63
1993	0.06	0.00	0.00	174.29	119.19	8065.39
1994	0.06	0.00	0.00	175.93	97.82	7340.46
1995	0.06	0.00	0.00	176.26	99.53	6898.25
1996	0.06	0.00	0.00	154.31	109.03	6647.38
1997	0.06	0.00	0.00	157.43	108.50	6237.44
1998	0.06	0.00	0.00	158.50	103.72	6234.79
1999	0.06	0.00	0.00	160.49	121.87	6265.97
2000	0.08	0.00	0.00	164.08	122.46	6357.78
2001	0.11	0.00	0.00	173.90	125.36	6431.02
2002	0.17	0.00	0.01	162.17	134.14	6427.90
2003	0.37	0.00	0.01	155.67	141.17	6577.55
2004	0.45	0.00	0.01	175.68	137.47	6606.42
2005	0.45	0.00	0.01	172.61	137.64	6597.73
2006	0.51	0.00	0.01	173.29	144.65	6915.32
2007	0.48	0.00	0.01	177.05	147.99	6942.08
2008	0.49	0.00	0.01	164.78	152.06	7010.48
2009	0.50	0.00	0.00	174.20	152.78	6595.96
2010	0.54	0.00	0.00	166.48	159.41	6897.14
2011	0.56	0.00	0.01	163.06	162.02	7171.50
2012	0.50	0.00	0.01	163.45	166.29	7182.12
2013	0.48	0.01	0.00	181.15	161.38	7050.27
2014	0.55	0.16	0.10	173.39	169.07	7073.82
2015	0.48	0.34	0.15	167.99	182.81	6920.01
2016	0.45	0.47	0.15	184.61	184.05	7051.23
2017	0.52	0.54	0.14	185.16	190.12	7086.87
2018	0.50	0.62	0.24	190.64	181.76	7372.34
2019	0.52	0.99	0.31	194.38	195.54	7292.44
2020	0.51	1.86	1.14	212.44	215.74	6909.20
2021	0.52	2.31	2.58	214.53	222.44	7556.90

Fig. 8 New energy and fossil fuel generation (TWh) in Russia (1990-2021)

2.2 Ratio Estimation

In order to be able to better evaluate the impact of nuclear energy, the data had been presented by using three ratio estimations: The first one is the ratio of nuclear energy to fossil fuel, which was determined by analyzing the multi-year trend of nuclear energy in seven countries, compared with conventional fossil fuels; The second one is the ratio of nuclear energy to clean energy (Clean energy is equal to the combination of nuclear energies and renewable energies), also based on time development, it focuses on the development status of clean energy and nuclear energy in each country, and determines the overall ratio through data integration analysis; The third one is the ratio of nuclear energy to the whole energy use. By summarizing energy sources other than nuclear energy, the total share of nuclear energy in the seven countries is obtained.

3 Results and Discussion

As described above, the data show the change in nuclear energy per year from 1990 to 2021 for seven different countries. As can be seen from Fig. 9 the trend of growth of nuclear energy in some countries is clear. For example, China accelerates its nuclear energy energy output from about 2020, growing from 16.74 KWh to a staggering 407.5 KWh, with the most pronounced growth from 2015 to 2020, boosting nearly 240 KWh in just a few years; however, there are some countries where the trend in nuclear energy production has a clear plunge. Japan, for example, had relatively stable nuclear energy production until 2011, basically staying in the 250KWh to 330Kwh range, but suddenly dropped to 15.2KWh in 2012 and even stopped nuclear energy production during 2014.

We believe the likely reason for the decline is due to the Fukushima Accident. The Government temporarily shut down the production of this energy source because of a huge nuclear-energy-induced disaster in the country.

On top of that, analyzing the data as well as the images, we were surprised to find that Russia's nuclear leak: the Chernobyl incident, did not drastically affect the country's overall nuclear energy production. The exact reasons for this are unknown to us, and we presume that it was influenced by internal government policy, coupled with the fact that Russia's base of nuclear power plants was relatively large at the time, so that the loss of production from one site would not greatly affect the country's overall production.

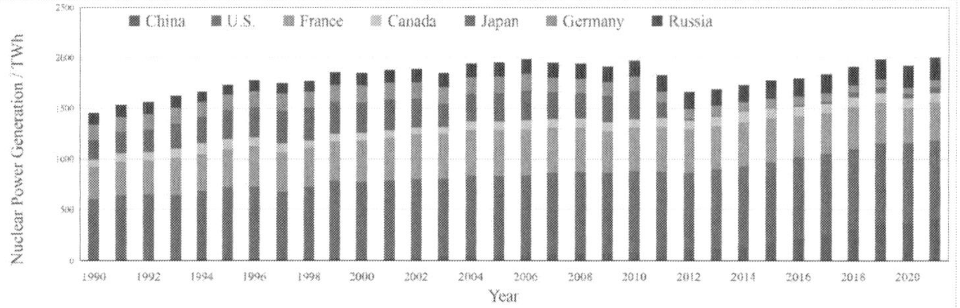

Fig. 9 the trend of growth of nuclear energy in Seven countries from 1990s to 2021s

3.1 Changes of Nuclear Energy

Fig. 10 shows the distribution of the average value of each non-polluting energy source over a ten-year period in seven countries, Fig. 10 for 1990 to 1999 and Fig. 10 for 2012 to 2021. Distributed energy sources include Geo Biomass other energy, Solar Generation, Wind Generation, Hydro Generation, and Nuclear Generation. It can be seen that from 1990 to 2021, the U.S. and France will mostly utilize nuclear energy to generate energy. In the U.S., nuclear energy accounts for 50% of the total, and in France it accounts for more than 75%, while other sources account for relatively little. This shows the value and benefits of nuclear energy to help a country's economy. About 75% of France's electricity comes from nuclear power, which is far ahead of other countries. At the same time, France, through the use of nuclear power, has gradually moved away from carbon emissions in electricity generation, so that France's per capita emissions are less than half that of Germany and Britain. It can also be seen from **Fig. 4** that nuclear energy has become the most important energy structure in France. Meanwhile, the two charts for China also show that although China is relatively new to Hydro Generation, nuclear energy output has been much higher in the last decade, at nearly 30%. Additionally, there is a subtle change in the clean energy supply structure of two of these countries. Japan's share of Hydro Generation is much higher in 2012-2021 than in 1990-1999, accounting for nearly 40% of the total; meanwhile, nuclear energy output declines due to nuclear accidents. In Germany, nuclear energy production has also changed. Whereas nuclear generation dominated overall clean energy in 1990-1999, it has declined to about 30% in the last decade.

In summary, the distribution of clean energy in the seven countries confirms the effectiveness of nuclear energy. With increasing technological maturity, nuclear energy will become the dominant clean energy source, and may even replace fossil fuel generation in the future. However, the decline in the overall share of nuclear energy in Germany is not well explained, and requires further in-depth discussion of the historical context of technological development in each country at the time.

3.2 Fossil Fuels and New Energy

In order to explore the importance of nuclear energy in each country, this section will analyze the situation from three different perspectives: **Fig. 11a** is the ratio of Nuclear Generation to fossil fuel for seven countries from 1990 to 2021; **Fig. 11b** is the ratio of Nuclear Generation to Clean Energy (including Geo-Biomass Other, Solar Generation, Wind Generation, Hydro Generation, and Nuclear Generation); **Fig. 11c** is the ratio of Nuclear Generation to a country's total energy output (including Geo-Biomass Other, Solar Generation, Wind Generation, Hydro Generation, and Nuclear Generation).

Fig. 11a can be concluded that the nuclear energy production of each country from 1990 to 2021 has been on a general upward trend, but the ratio to oil production has been in a declining state, which is due to the fact that nuclear energy production per year is much lower than that of oil production, which side by side reflects the fact that the demand for oil has increased considerably in recent decades. For example, the ratio of nuclear energy production to oil production in the United States has fallen from 1.76 in 1990 to about 0.64 in 2021, and U.S. oil production has risen from 344 TWh to 1,217 TWh. In addition to this, this chart shows that there are many countries that are already trending toward nuclear energy as a primary source of energy, even though oil is currently the more dominant source of energy. For example, in France, although the ratio of nuclear energy production to

https://doi.org/10.1051/e3sconf/202344102024

oil production is on a downward trend, the decrease in the ratio has slowed down since 2008, proving that nuclear energy production is gradually equalizing with oil production.

In addition, **Fig. 11a**'s ratio began to increase upwards between 1998 and 2000 for a number of reasons, notably the acceptance of the benefits of nuclear energy, the construction of nuclear power plants, and the promulgation of relevant policies.

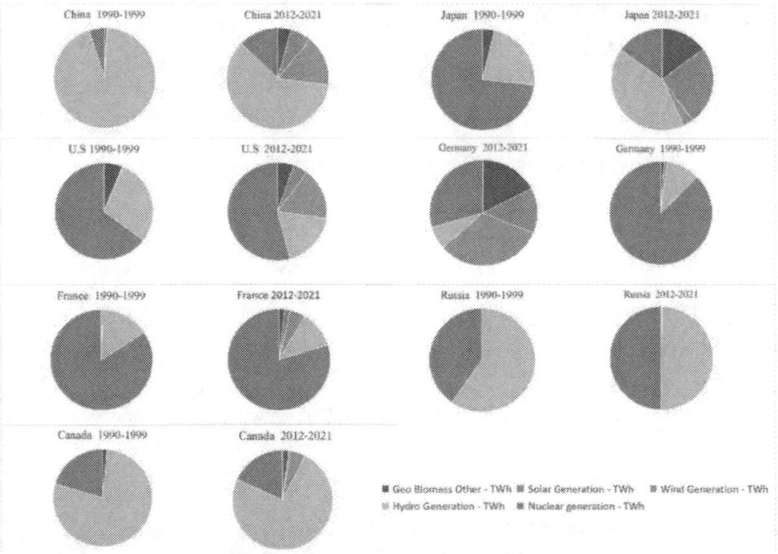

Fig. 10 the distribution of the average value of each non-polluting energy source over a ten-year period in seven countries

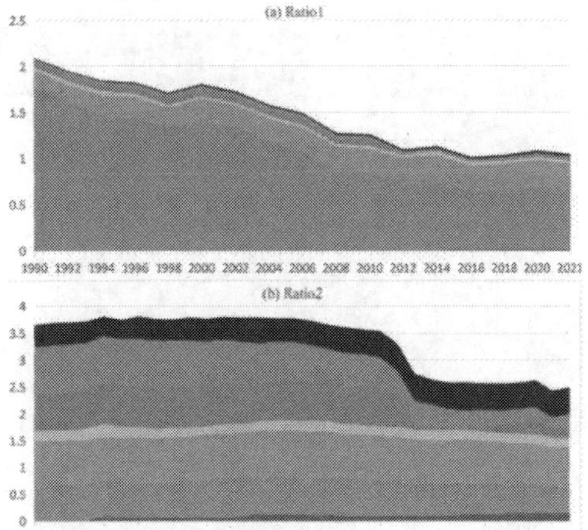

Fig. 11 (a)the ratio of Nuclear Generation to fossil fuel for seven countries from 1990 to 2021 (b) the ratio of Nuclear Generation to Clean Energy (including Geo-Biomass Other, Solar Generation, Wind Generation, Hydro Generation, and Nuclear Generation); (c)the ratio of Nuclear Generation to a country's total energy output (including Geo-Biomass Other, Solar Generation, Wind Generation, Hydro Generation, and Nuclear Generation).

Fig. 11b visually shows that most countries are producing less nuclear energy compared to clean energies year after year. For example, the ratio for the United States has gone from about 1.76 in 1990 to about 0.63 in 2021, indicating that the yearly increase in clean energy production has made the denominator of the ratio larger, and the nuclear energy production in a single year is not as much as that of clean energies, which leads to a decrease in the ratio year by year. However, there are some exceptions, such as China, where the ratio has basically been on an upward trend since 1993. Prior to 1993, the ratio was zero, indicating that China had not yet begun to produce nuclear energy. However, from a formulaic point of view, clean energies remain one of the most important energy sources for China compared to nuclear energy.

The rise in China's nuclear energy production is due to a number of factors, and we hypothesize, based on the historical background, that China was in the midst of its **"reform and opening up"** wave, which brought in many outstanding experts and equipment in science and technology to develop China's nuclear energy business, leading to a general upward trend in nuclear energy production from 1993 to 2021.

Fig. 11c shows that the ratio of nuclear energy to energy use is generally decreasing in the seven countries, but not as much as **Fig. 11a** and **Fig. 11b**. The reason for this is that the annual production of nuclear energy in each country is not as large as the total national production. For example, in the United States, the ratio decreases from 0.46 in 1990 to 0.27 in 2021; however, in France, the ratio tends to increase, although it remains roughly unchanged as shown in the Fig. 11a and Fig. 11b. It increases from 0.15 in 1990 to 0.21 in 2021, which shows that France is gradually shifting the focus of its production to nuclear energy and moving towards the production of clean energy.

3.3 History of nuclear plant construction

The evolution of nuclear plant construction stands as a testament to humanity's unceasing pursuit of innovative and sustainable energy solutions. From its initial steps in the mid-20th century to the present day, the development of nuclear power facilities has borne witness to both technological triumphs and formidable challenges. Since 1950, scientists had been conducted advanced technology of nuclear plants.

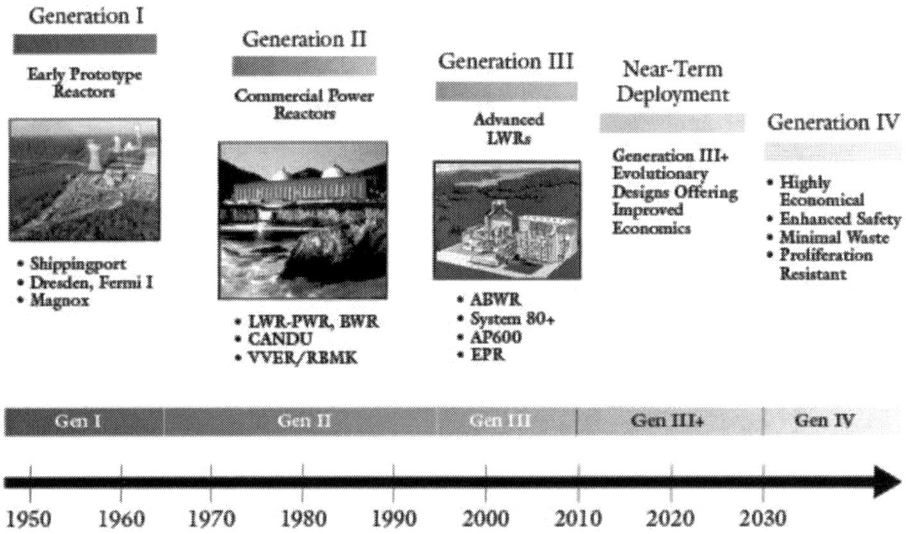

Fig. 12 Reprinted from U.S. Department of Energy, Office of NuclearEnergy, "Generation IV Nuclear Energy Systems:Program Overview" (Department of Energy, n.d.), http://nuclear.energy.gov/genIV/neGenIV1.html.

Generation I

Generation I reactors encompass the pioneering prototype and power reactors that laid the foundation for civilian nuclear power. This generation encompasses early-stage prototype reactors hailing from the 1950s and 1960s, notable examples being Shippingport (1957–1982) in Pennsylvania, Dresden-1 (1960–1978) in Illinois, and Calder Hall-1 (1956–2003) in the United Kingdom. These reactors primarily operated at power levels aimed at showcasing feasibility. In the United States, the Nuclear Regulatory Commission (NRC) oversees Generation I reactors, subject to regulations outlined in Title 10, Code of Federal Regulations, Part 50 (10 CFR Part 50). Wylfa Nuclear Power Station in Wales stands as the sole remaining commercial Generation I facility. While its closure was initially slated for 2010, the UK Nuclear Decommissioning Authority shifted plans in October 2010, extending Wylfa's operation until December 2012.

Generation II

Generation II systems originated in the late 1960s and have since become the predominant type among the world's 400+ commercial pressurized water reactors (PWRs) and boiling water reactors (BWRs). These reactors, often referred to as light water reactors (LWRs),

https://doi.org/10.1051/e3sconf/202344102024

incorporate conventional active safety mechanisms, some of which are automated and can also be initiated by nuclear reactor operators. Notably, certain engineered systems operate passively, functioning independently of operator intervention or auxiliary power availability. Generation II reactors represent a specific category of commercial nuclear reactors meticulously designed to balance economic viability and operational reliability. These reactors typically have an operational lifespan of around 40 years. Notable examples of Generation II reactors include pressurized water reactors (PWR).

Generation III

Generation III nuclear reactors represent a significant advancement over their predecessors, blending cutting-edge innovations with fundamental design elements. These improvements encompass various aspects such as fuel technology, thermal efficiency, modular construction, and safety systems, with an emphasis on passive safety features. The primary goal of Generation III reactors is to achieve extended operational lifespans, often spanning 60 years or more before requiring major overhauls and reactor vessel replacements. However, further research is needed to confirm the feasibility of operating nuclear plants beyond 60 years. Unlike earlier generations, Generation III reactors adhere to Nuclear Regulatory Commission (NRC) standards outlined in 10 CFR Part 52. Prominent Generation III designs include the Westinghouse AP-600 and GE Nuclear Energy's Advanced Boiling Water Reactor (ABWR), with Japan pioneering the introduction of ABWR units in 1996. Other concepts include the Enhanced CANDU 6 by Atomic Energy of Canada Limited (AECL) and System 80+ by Combustion Engineering. Currently, only four Generation III reactors are operational, all falling under the ABWR category, and none are active in the United States.

Generation III+ and Generation IV+

Generation III+ reactor designs represent a significant leap in nuclear safety compared to Generation III counterparts, with the development initiated in the 1990s. These designs incorporate passive safety measures, eliminating the need for human intervention during abnormal events and relying on natural processes like gravity or convection for safety. They also adhere to defined Western safety standards, setting global safety benchmarks. However, challenges include increased nuclear waste production, the need for extensive power grids, and public acceptance issues. This evolution underscores nuclear technology's journey towards enhanced safety while addressing complex challenges. Notably, the last remaining commercial Gen I plant, the Wylfa Nuclear Power Station in Wales, continued operations until December 2012 despite being scheduled for closure in 2010.

3.4 Historical Nuclear Accidents

Throughout the history of nuclear technology, several notable accidents have served as cautionary tales, shaping the global discourse on nuclear safety. From the early days of nuclear experimentation to more recent incidents, each event has contributed to our understanding of the complex challenges associated with harnessing nuclear power. In particular, several nuclear accidents need to be introduced based on **Fig. 13.**

Fukushima (2011):

The Fukushima Daiichi nuclear disaster in 2011 was a catastrophic event triggered by a massive earthquake and tsunami that struck Japan. The Fukushima Daiichi Nuclear Power Plant, located on the northeastern coast of Japan, experienced a loss of power and cooling capabilities, leading to partial core meltdowns in three of its operational reactors. This resulted in the release of radioactive materials into the environment and necessitated the evacuation of tens of thousands of residents from the affected area.

The disaster had profound consequences, not only for Japan but also for the global nuclear industry and energy policy. It raised serious questions about the safety of nuclear power plants, emergency preparedness, and the long-term management of nuclear waste. In their study, Tokonami et al. (2012) examined a group of 62 individuals, including both residents and evacuees from regions severely affected by contamination around the Fukushima Nuclear Power Plant. Their primary aim was to ascertain the thyroid doses resulting from inhalation exposure. The findings revealed that the median committed equivalent dose stood at 4.2 mSv for children and 3.5 mSv for adults. Notably, the highest recorded thyroid doses reached 23 mSv for children and 33 mSv for adults in this cohort[**11**]. In response to the disaster, many countries reevaluated their nuclear energy policies, with some opting to reduce their reliance on nuclear power or phase it out entirely. Fukushima served as a stark reminder of the potential risks associated with nuclear energy and the importance of stringent safety measures and international cooperation in managing these risks. The ongoing decommissioning efforts at Fukushima Daiichi underscore the enduring challenges and complexities involved in dealing with the aftermath of a nuclear disaster.

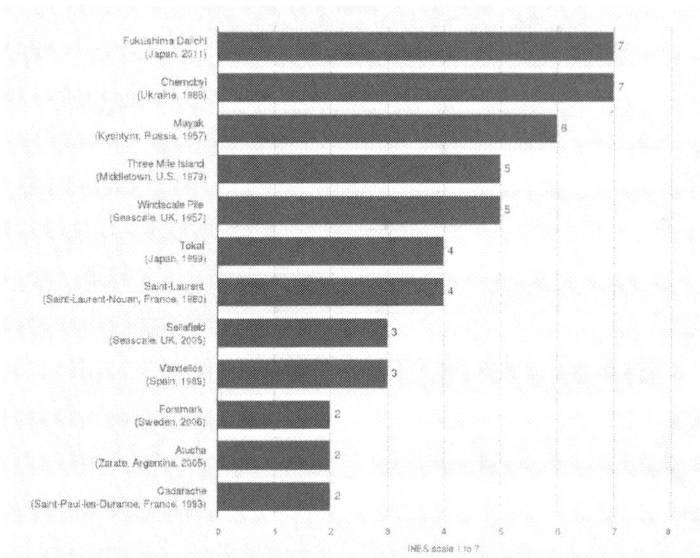

Fig. 13 Nuclear accidents globally between 1957 and 2011, rated by INES(International Nuclear and Radiological Event Scale) scale. This figure is obtained from IAEA(International Atomic Energy Agency).

Chernobyl (1986):

The Chernobyl disaster is widely regarded as the most catastrophic nuclear accident in history. It occurred at the Chernobyl Nuclear Power Plant in the Soviet Union (now Ukraine) during a late-night safety test. Due to design flaws and a series of human errors, an uncontrolled power surge caused a steam explosion, followed by a graphite fire in the reactor core. This led to the release of a massive amount of radioactive material into the atmosphere over several days. The accident resulted in the immediate deaths of two plant workers and the subsequent deaths of many cleanup personnel due to acute radiation sickness. The release of radioactive contaminants had long-term health and environmental impacts, including an increased incidence of cancers and the displacement of nearby populations.

Mayak(1957):

The Kyshtym disaster serves as a stark testament to the perils arising from insufficient safety precautions in the pursuit of nuclear technology. This calamity transpired on the early morning of September 29, 1957, at the Mayak Production Association facility located within the Soviet Union, yielding extensive repercussions for both ecological integrity and human well-being. Stemming from a combination of technical breakdowns, secrecy, and a disregard for the potential risks tied to nuclear substances, this incident was provoked by a catastrophic malfunction within the cooling system of a storage tank that housed high-level liquid radioactive waste. The primary purpose of this cooling mechanism was to avert the overheating of the waste and the subsequent discharge of perilous radioactive particles into the atmosphere. However, due to inadequate maintenance and technical oversight, this cooling system faltered. Consequently, the temperature within the tank surged, ultimately culminating in a chemical explosion that liberated an immense volume of radioactive materials into the surroundings.

4 Conclusion

This paper uses the latest data and formulas to analyze and evaluate nuclear energy from different perspectives, visualizing the status of nuclear energy in seven different countries. At the same time, the paper also provides a detailed description and evaluation of nuclear plantations. At the end of the paper, a review of nuclear accidents and a rating of nuclear accidents are presented.

In this study, it is found (Section 3.1, 3.2, and 3.3) that the use of nuclear energy is gradually increasing in many countries. The efficacy of nuclear energy is being strongly recognized, both in terms of environmental protection and cost. However, there is still instability in nuclear power plants. As can be seen from the nuclear accidents in Section 3.4, the impact of a nuclear energy accident on humans and the environment is enormous. These accidents, nevertheless, are relatively small probability events. With the development of science and technology, the security facilities of nuclear power plants and nuclear reactors have been updated and upgraded, which means that the probability of nuclear accidents will become smaller and smaller in the future, while the production and efficiency will become even better and become the main source of clean energy. Although the paper already evaluated and analyzed the nuclear energy thoroughly, there is still some room for improvement in this paper, such as the assumptions mentioned in 3.2 need to be proved and the data visualization needs to be more diversified.

References

1. Akahane, K., Yonai, S., Fukuda, S., Miyahara, N., Yasuda, H., Iwaoka, K., Matsumoto, M., Fukumura, A., Akashi, M., 2012. The Fukushima Nuclear Power Plant accident and exposures in the environment. Environmentalist 32, 136–143. http://dx.doi.org/10.1007/s10669-011-9381-2.
2. IRSN, 2012. Fukushima, one year later. Initial Analyses of the Accident and Its Consequences. Institut de Radioprotection et de Surete Nucleaire.
3. Kawamura, H., Kobayashi, T., Furuno, A., In, T., Ishikawa, Y., Nakayama, T., ... & Awaji, T. (2011). Preliminary numerical experiments on oceanic dispersion of 131I and 137Cs discharged into the ocean because of the Fukushima Daiichi nuclear power plant disaster. Journal of Nuclear Science and Technology, 48(11), 1349-1356.
4. Lu, Y., Yuan, J., Du, D., Sun, B., & Yi, X. (2021). Monitoring long-term ecological impacts from release of Fukushima radiation water into ocean. Geography and Sustainability, 2(2), 95-98
5. Grishanin, E. I. (2010). The role of chemical reactions in the Chernobyl accident. Physics of Atomic Nuclei, 73, 2296-2300.
6. Michel, R., Handl, J., Ernst, T., Botsch, W., Szidat, S., Schmidt, A., ... & López-Gutiérrez, J. M. (2005). Iodine-129 in soils from Northern Ukraine and the retrospective dosimetry of the iodine-131 exposure after the Chernobyl accident. Science of the total environment, 340(1-3), 35-55.
7. Pröhl, G., Mück, K., Likhtarev, I., Kovgan, L., & Golikov, V. (2002). Reconstruction of the ingestion doses received by the population evacuated from the settlements in the 30-km zone around the Chernobyl reactor. Health Physics, 82(2), 173-181.
8. El-Katiri, L. (2014). A roadmap for renewable energy in the Middle East and North Africa. Oxford institute for energy studies.
9. Change, I. C. (2014). Mitigation of climate change. Contribution of working group III to the fifth assessment report of the intergovernmental panel on climate change, 1454, 147.
10. Lovins, A. B. (2011). Renewable energy's 'footprint'myth. The Electricity Journal, 24(6), 40-47.
11. Tokonami, S., Hosoda, M., Akiba, S., Sorimachi, A., Kashiwakura, I., & Balonov, M. (2012). Thyroid doses for evacuees from the Fukushima nuclear accident. Scientific Reports, 2(1), 507.

Comparative Performance Analysis of Exhaust Air Heat Recovery Devices Based on Microchannel heat pipe

Qianfan Dong[a], Zhiyong Li* and Yuqing Zhao

North China University of Technology, Civil Engineering Department, Shijingshan, Beijing, China

Abstract. In response to the problem of low recovery efficiency of sensible heat recovery devices, this paper designs a microchannel heat pipe heat recovery device (MHPHRD) and compares its heat transfer performance during summer use through experiments and simulation calculations. The experimental results show that the microchannel heat pipe (MHP) has strong temperature uniformity, and the temperatures of the evaporation and condensation sections are basically the same. When the inlet temperature of fresh air changes between 32°C and 40°C, the recovery efficiency of the device varies between 67%-80%, and the linear relationship between the two is obvious. The established calculation model has high accuracy, with a maximum error of 0.7% between measured and simulated values. Simulation calculations show that the cooling capacity recovery efficiency of the MHPHRD is 4% -16% higher than that of the cylindrical heat pipe heat recovery device (CHPHRD), and 24%-30% higher than that of the plate heat recovery device (PHRD). The research in this article can provide reference for the practical application of the device.

1 Introduction

The fresh air energy consumption of air conditioning system accounts for about 25%-50% of the total energy consumption of air treatment [1], and studies have shown that the use of reasonable exhaust air heat recovery technology can reduce air conditioning energy consumption by 40%-50% [2-3]. Exhaust air heat recovery refers to the use of some technical means to transfer the energy in the exhaust air to the new air to pre-treat the new air. However, the traditional method of sensible heat recovery has problems such as additional consumption of electrical energy by auxiliary equipment, not close fit with the heat exchange surface, inflexible arrangement, easy to block, and occupying a large space [4-8], which leads to low heat exchange efficiency.

In order to solve the problem of low efficiency of the above mentioned sensible heat recovery method, this study proposes an exhaust air heat recovery device based on microchannel heat pipe heat recovery device (MHPHRD). The microchannel heat pipe (MHP) is shaped as a plate, with multiple parallel and independent heat transfer microchannels inside. MHP not only has advantages such as large heat transfer coefficient of traditional heat pipe, but also solves the problems of small heat transfer contact area, complex installation and slow response speed of cylindrical heat pipe(CHP), which has attracted more and more attention from scholars.Diao Y H [9] used nanofluid as a work material to test the heat transfer performance of MHP. Later, MHPHRD was designed and experimentally investigated [10], and the experimental results showed that the recovery efficiency of the device was 51%-78% in summer,and 56%-77% in winter, which is a significant recovery effect. Shen Chao[11] also designed a comprehensive performance test device for MHPHRD, and the experimental results showed that the MHP had a fast response speed and the thermal resistance was only 0.06 K/W. Yang Jingang[12] proposed a new type of low-temperature soot heat recovery collector based on the MHPHRD, and carried out experimental research on it, and the results showed that the heat recovery efficiency of the MHPHRD was stabilized at 77-97%, which also corroborated the recovery effect of the device in the summer of 51-78%, and in the winter of 56-77%. The excellent performance of the MHPHRD applied in heat recovery was also corroborated.

Through the above analysis, it can be found that the current research on the MHPHRD is mostly to explore its heat transfer performance, while the comparative research on the MHPHRD and other sensible heat recovery devices is less. Therefore, in order to accurately analyze the performance of this device, this paper builds a performance test device, and establishes mathematical models of different sensible heat recovery methods, and analyzes the advantages of this device relative to other sensible heat recovery devices when used in summer through experiments and simulation comparisons, and the research in this paper can provide references to the practical application of this device.

[a] qianfan19990306@163.com, * lizhiyongnbe@163.com

2 Experimental setup of MHPHRD

2.1 Construction of experimental device

The device consists of inclined flow duct fan, air duct, finned tube heater, regulator, microchannle heat pipe (MHP) and data acquisition system. The heating section of the duct on the fresh air side is made of 3cm high-temperature-resistant aluminum silicate heat-insulating board, the duct on the exhaust side and the remaining part on the fresh air side are made of 5cm polystyrene foam board, polyurethane foam caulking is used in the joints and fan connections to ensure the duct's sealing, and the gaps in the adiabatic section of the heat pipe are filled with aluminum-foil insulating cotton.

When the equipment is turned on, the fresh air is heated by the fin heater and flows to the evaporation section of the MHP, where the work material is heated and boils, and the steam produced by the work material carries heat into the condensation section of the MHP, where the steam passes through the fins and heats the cold air, and then returns to the evaporation section to cool down the hot air under the effect of gravity after the vapor is condensed, and so on and so forth. The data acquisition system is connected to the LU-924U04Y four-channel measurement and control module, and a T-type thermocouple with an accuracy of ±0.2°C and an AR866 hot-wire anemometer are used to collect the temperature and wind speed. The schematic diagram of the experimental setup and the distribution of temperature and air velocity measurement points are shown in Figure 1, where t_1 is the fresh air inlet temperature, t_2 is the fresh air outlet temperature, t_3 is the exhaust air inlet temperature, and t_4 is the exhaust air outlet temperature.

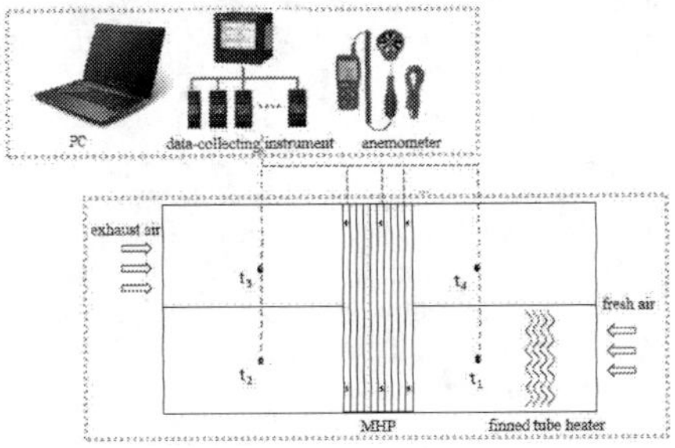

Fig. 1. Experimental procedure and distribution of measurement points

2.2 Experimental conditions

The actual test time of this article is from May 25th to June 20th, and the test location is a typical natural ventilation room, which size of 5.2m×4.5m×3m. The experimental device is placed vertically, through the hot wire anemometer shows the value of real-time control of the air duct wind speed at about 1m / s, the same amount of fresh air and the exhaust volume, the temperature of the room is 26°C. Control the fin heater temperature of the fresh air heating, fresh air inlet temperature were set to 30°C, 32°C, 34°C, 36°C, 38°C, 40°C, and test 6 sets of data, to be stable air temperature every 2 min to record the temperature of the measurement point once. In order to ensure the accuracy of the data, each set of data testing 3 times to take the average value as the point of the working conditions of the test data.

2.3 Performance evaluation indicators

MHPHRD takes MHP as the core heat transfer element, so the evaluation of device performance first evaluates the performance of MHP, and after confirming that the MHP is in good operating condition, then the recovery efficiency is used as the evaluation index to measure the performance of the device, and the definitions of the evaluation parameters and formulas are as follows.

The MHP can be evaluated by the average temperature difference between the evaporation section and the condensation section to evaluate the performance of the heat pipe. The performance of MHPHRD can be evaluated by the cold recovery efficiency of the device, defined as the ratio of the actual heat transfer to the theoretical maximum heat transfer, and the cold recovery efficiency E_C is calculated as follows:

$$E_c = \frac{L_x(t_1 - t_2)}{L_{min}(t_1 - t_3)} \times 100\% \quad (1)$$

In the formula, L_x is the mass flow rate of fresh air, kg/h; L_{min} is the minimum value of mass flow rate of fresh and exhaust air, kg/h.

3 Mathematical modeling

3.1 Analysis of heat transfer process

Sensible heat recovery units are surface heat exchangers where hot and cold fluids do not contact each other. If the inlet temperature t_1 and flow rate m_1 on the fresh air side, the inlet temperature t_2 and flow rate m_2 on the exhaust air side, and the thermophysical parameters of the air are known, we can apply the method of heat exchanger efficiency to establish a mathematical model of the heat transfer process and combine it with the actual heat transfer of the device. Neglecting the phase change heat transfer process inside the heat pipe, we can get the general model of the heat pipe and plate heat recovery device as follows:

The formula for calculating the heat exchange capacity of the unit is as follows:

$$Q = E_c C_{min}(t_1 - t_3) \quad (2)$$

In the formula, E_C' is the heat transfer effectiveness of the device, defined as the ratio of the actual heat transfer to the theoretical maximum heat transfer; C_{min} is the minimum value of the heat capacity on the fresh air side and the exhaust air side, kJ/h·K.

In the above formula minimum heat capacity and heat transfer effectiveness are calculated as follows:

$$C_{min} = \min(m_1 c_{p1}, m_3 c_{p3}) \quad (3)$$

In the formula, C_{p1} is specific heat of air in the fresh air stream, kJ/(kg·K); m_1 is the mass flow rate of fresh air inlet air, kg/h; C_{p3} is specific heat of air in the exhaust stream, kJ/(kg·K); m_3 is the mass flow rate of exhaust air inlet air, kg/h.

$$Ec' = \frac{1-\exp[-NTU(1-C_{min}/C_{max})]}{1-(C_{min}/C_{max})\exp[-NTU(1-C_{min}/C_{max})]} \quad (4)$$

In the formula, NTU is the number of heat transfer units; C_{max} is the maximum value of the heat capacity of the fresh air side and the exhaust air side, kJ/h·K.

The formula for the number of heat transfer units NTU of the heat exchanger in the above equation is as follows:

$$NTU = \frac{KA}{C_{min}} \quad (5)$$

In the formula, K is the heat transfer coefficient between air and heat pipe wall, W/(m^2·K); A is the effective heat transfer area, m^2.

Subsequently, the outlet temperatures of the fresh and exhaust air are calculated from the calculated heat exchanger Q, using the following formula:

$$t_2 = t_1 - \frac{Q}{m_1 C_{p1}} \quad (6)$$

$$t_4 = t_3 + \frac{Q}{m_3 C_{p3}} \quad (7)$$

In the formula, the calculation process first assumes that the exit temperature value of the fresh air side is t_2' (t_2'=t1+(4~6) ℃), and compares the Ec calculated according to Equation (1) with the Ec' calculated according to Equation (4), and if the error between the two is less than 0.1, it is considered that the exit temperature value of the fresh air side is the actual exit temperature value, and if it is not, then reset the t_2' for calculation. It is worth noting that in order for the model to achieve higher accuracy, the thermophysical parameters of the air in the modeling process are not fixed values, but change with the change of the average temperature in the duct, so with the change of the assumed value of t_2', the subsequent values of the parameters change until the error meets the requirements.The heat transfer efficiency algorithm of the three heat exchangers is basically the same,the difference lies in the formula (5) in the heat transfer coefficient K and the calculation of the effective heat transfer area A, the calculation of K can be referred to the "Principles and Design of Heat Exchangers" [13] in the relevant formulas, the calculation of A is relatively simple, and will not be repeated here. The model is simulated by MATLAB software, and the specific calculation steps are shown in Fig. 2.

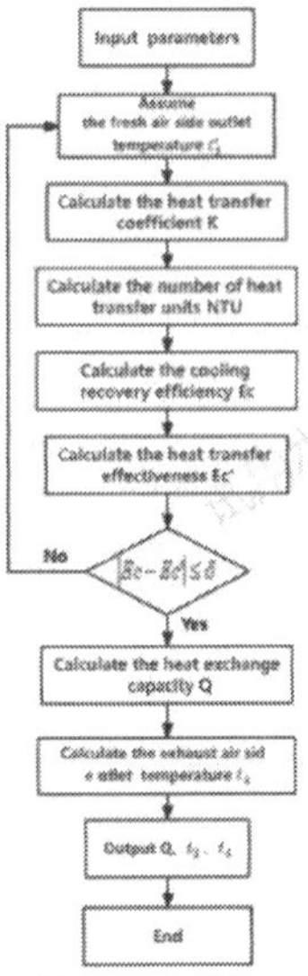

Fig. 2. Flowchart of the model of heat pipe heat recovery device

3.2 Calculation of working conditions

In order to compare the performance of the return air heat recovery device scientifically, according to the principle of heat exchanger volume proximity, cylindrical heat pipe (CHP) and plate heat exchanger (PHE) are selected for comparative analysis.CHP and MHP have the same piping and material, and the length of the fins along the direction of the airflow is 50mm, and the dimensions are different as shown in Fig. 3 and Table 1 below.CHPHRD is arranged in forked rows, with dimensions of 170mm×170mm×660mm and MHPHRD is arranged in parallel rows, with dimensions of 500mm×100mm×190mm after the arrangement. PHE model is B3-60-66, which manufactured by a certain manufacturer, with dimensions of 476mm×140mm×174mm, and an effective heat exchange area of 3.52 m^2.The plate material of PHE is stainless steel, with a thermal conductivity of 14.4 W/(m·K). , there are 66 plates, the distance between the plates is 1.5mm, the width of the channel is 140mm, and the thickness of the plate is 1.2mm.

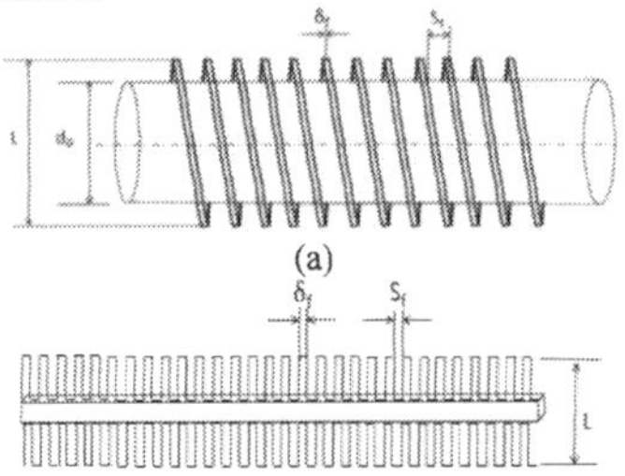

Fig. 3. (a)Calculated dimensions for CHP (b)Calculated dimensions for MHP

Table 1. CHP and MHP size

	CHP	MHP
Dimension	The length of the tube is 660mm and the diameter d_0 is 25mm	500mm×50mm ×3mm
Effective heat transfer area A/m²	2.89	3.12
thickness of fin δ_f /mm	0.2	0.5
Pitch of fin S_f/mm	5	1.6
distance between tubes S_1/mm	60	16
quantities	8	20

3.3 Model accuracy analysis

The accuracy of the model calculation was analyzed through the experimental study of the MHPHRD. As can be seen from Fig. 4,the average relative error between measured and simulated values is 0.38%, and the maximum relative error is 0.7%, which proves that the established model can be used for the simulation and analysis of heat exchange of the device.

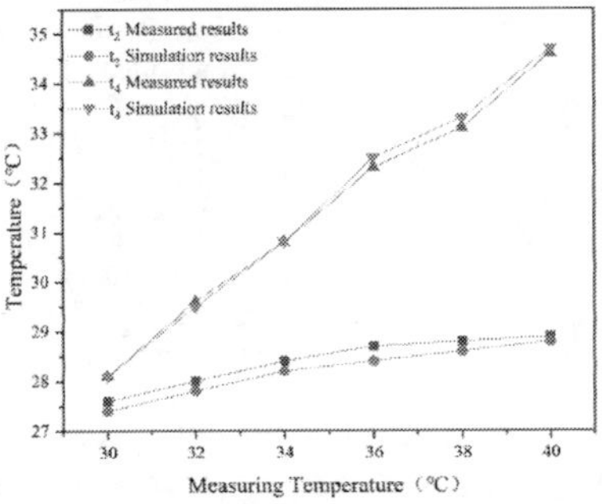

Fig. 4. Comparison results between simulated and measured results of t_2, t_4

4 Analysis of findings

4.1 MHP Thermal homogeneity

The temperature distribution of the evaporation section and condensation section measurement points when the MHP is working is shown in Fig. 5, from which it can be seen that the highest temperature of the evaporation section is 29.6°C and the lowest temperature of the condensation section is 28.6°C under condition 1, with a temperature difference of 1°C.The temperature difference between the evaporation section and condensation section under other conditions is about 0.7°C, which indicates that the MHP has good temperature uniformity and high heat transfer efficiency.

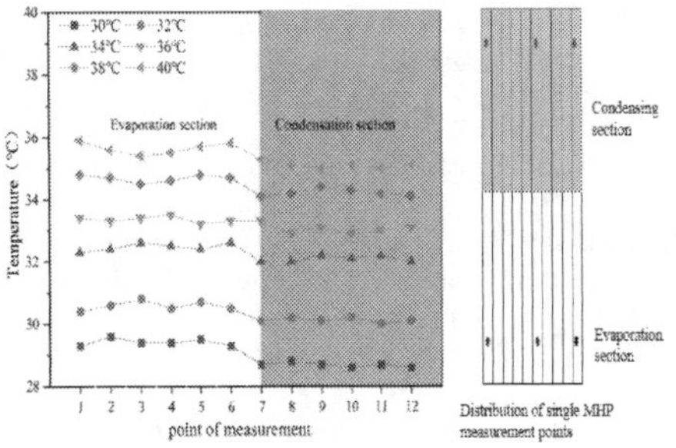

Fig. 5. MHP evaporation section, condensation section measurement point distribution and temperature

4.2 Cooling Recovery Efficiency

Fig. 6 shows the cold recovery efficiencies at different temperatures.

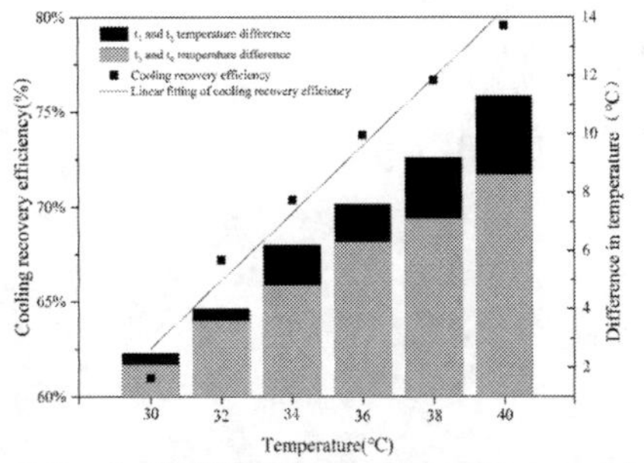

Fig. 6. Results of MHPHRD temperature efficiency variation with fresh air temperature

As shown in Fig. 6, the cold recovery efficiency of the device as a function of outdoor temperature can be derived as:

$$E_c = 0.0176t_1 + 0.091 \quad (8)$$

From the formula (8) can be calculated under any outdoor temperature corresponding to the device cooling recovery efficiency. Combined with Table 2 and Figure 6, it can be seen that in the fresh air inlet temperature of 30°C, the fresh air import and export temperature difference is only 2.5°C, with the increase in fresh air temperature of the import and export temperature difference increases, in the fresh air inlet temperature of 40°C can be up to 11.3°C, the exhaust air inlet and outlet temperature difference has a similar phenomenon. Exhaust air inlet and outlet temperature difference are smaller than the fresh air inlet and outlet temperature difference, which is due to the device in the operation process has a certain heat loss caused by. As the temperature of the fresh air increases, the cooling recovery efficiency also increases, and the efficiency changes linearly with the fresh air inlet temperature.

4.3 Comparative analysis of sensible heat recovery for CHPHRD, MHPHRD, and PHRD

According to the established model, the efficiency of three heat recovery devices was calculated, and the results are shown in Figure 7. The efficiency fitting curves of CHPHRD, MHPHRD, and PHRD efficiencies are from the experimental studies in the literature [6], [10], and [8], respectively. From the figure, it can be seen that the trend of the calculated values of the cooling recovery efficiency of the three devices are all with all increase with increasing temperature. The recovery efficiencies from high to low are: MHPHRD, CHPHRD, PHRD. Among them, the cold recovery efficiencies of MHPHRD and CHPHRD are more obvious with the change of fresh air inlet temperature. the cold recovery efficiencies of MHPHRD are 4%-16% higher than those of CHPHRD, and 24%-30% higher than those of PHRD.

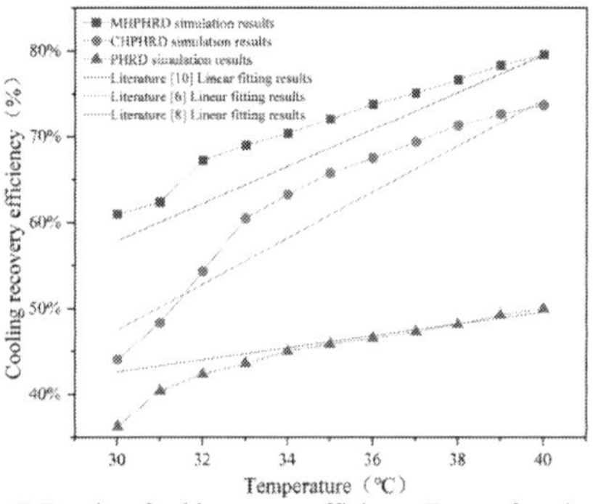

Fig. 7. Results of cold recovery efficiency E_C as a function of outdoor temperature for CHPHRD, MHPHRD,PHRD and literature fitting results

5 Conclution

Through experiments and comparative simulation calculations, it can be found that the temperature of the evaporation section and condensation section of MHP is basically the same, and the average temperature difference is basically 0.7 °C, which proves that MHP has small thermal resistance and high heat transfer efficiency in the process of heat transfer.The heat recovery efficiency of MHPHRD is mainly concentrated in the range of 67%-80%, which is greater than that stipulated in GB/T 21087-2020 Heat Recovery Fresh Air Units in most of the operation time. 65% heat recovery efficiency in summer. The simulation results show that the cooling recovery efficiency of MHPHRD is 4%-16% higher than that of CHPHRD and 24%-30% higher than that of PHRD, indicating that MHPHRD has a more obvious advantage over other device in terms of recovery efficiency.

Reference

1. Fangyuan C, Yang C. Analysis of carbon emission reduction of fresh air heat recovery device used in summer for office buildings in three types of climate areas in China [J]. Building Science, 2013,**29**(12):103-107.DOI:10.13614/j.cnki.11-1962/tu. 2013.12. 020.
2. Yi J. Building energy consumption status and effective energy-saving ways in China. HVAC [J], 2005,35 (5): 30-40.
3. Fehrm M,Reiners W,Ungemach M. Exhaust air heat recovery in buildings[J]. International Journal of Refrigeration,2002,**25**(4).
4. Ewa Zender.A Review of Heat Recovery in Ventilation[J].Energies, 2021, 14(6):1759.DOI: 10.3390/en14061759.
5. Jie Z,Lihong Z, Jianning G, etc. Application of heat pipe in heat recovery of air conditioning [J]. Environmental Engineering, 2009,**27**(03): 72-74 + 107.
6. Xia Y, Wen W, Ruzhu W. Application of heat pipe in air conditioning [J]. HVAC, 2004(05): 26-29 + 46.
7. M.Ahmadzadehtalatapeh，Y.H.Yau.The application of heat pipe heat ex changers to improve the air quality and reduce the energy consumption of the air conditioning system in a hospital ward-A full year model simulation [J].Energy and Buildings,2011,**43**(9)：2344-2355.
8. Zhaoming Q, Zhimei H. Performance test of plate heat recyclator [C] / / Warm and Air Conditioning Academic Committee of Architectural Society of China, fifth Professional Committee of Chinese Refrigeration Society. National HVAC refrigeration 1994 academic annual conference data set. [Publisher unknown], 1994:4.
9. Zhang J , Diao Y.H , Zhao Y.H ,et al.Experimental study on the heat recovery characteristics of a new-type flat micro-heat pipe array heat exchanger using nanofluid[J].Energy Conversion and Management, 2013, 75:609-616.DOI:10.1016/j.enconman.2013.08.003.
10. Diao Y.H , Liang L , Kang Y.M ,et al.Experimental study on the heat recovery characteristic of a heat exchanger based on a flat micro-heat pipe array for the ventilation of residential buildings[J].Energy and Buildings, 2017, **152**(oct.):448-457.DOI:10.1016/j.enbuild.2017.07.045.
11. Chao S, Shaolun Y, Dongwei Z, etc. Experimental study on heat transfer performance of parallel heat flow heat exchanger [J]. Journal of Engineering Thermophysics, 2018,**39**(06): 1339-1343.
12. Jingang Y,Weiyu S, Zhenhua Q, etc. Study on the performance of lampblack heat exchanger based on micro-heat pipe array [J]. Building Science, 2019,**35**(12):49-54.DOI:10.13614/j.cnki.11-1962/tu. 2019.12.0
13. Jianzu Y. (2006) Design and principle of heat exchanger. Beihang University Press, Beijing.

Construction of an evaluation index system for power market considering economic-environmental multi-objectives

Xiaochun Cheng, Qin Wang*

Capital Power Exchange Center Co.LTD, Beijing, 100031, China

Abstract: At present, China's electricity market has been initially built, direct trading of electricity continues to increase, and the mode of "multiple purchase and multiple sale" has emerged in the market, which enables the electricity market to compete effectively and make the optimal allocation of resources. Therefore, it is necessary to evaluate the operation of the electricity market. The article firstly studied the indicators of power market operation from three aspects: economic value, green value and security value; secondly, established a three-level evaluation index system of power system; on this basis, proposed a comprehensive evaluation model of power market based on fuzzy comprehensive evaluation. It provides a reference basis for the evaluation of China's power grid operation and is of great significance to promote the healthy development of China's power market.

1. Introduction

In 2015, the release of the Chinese development [2015] No. 9, the new electricity reform and other related documents showed that the process of China's electricity market reform is accelerating and gradually establishing a fully competitive, open and orderly, healthy development of the electricity market system in the country. The safe, stable, reliable and efficient operation of the market determines the construction of the market and the operation of the market. Therefore, it is imperative to construct a set of electric power market evaluation index system suitable for China's national conditions. In the market scheme design stage, the relatively better market design scheme is evaluated and selected, and in the market operation stage, the market operation and market performance are monitored and evaluated to find out the problems in the market construction and operation. There is not much research work on evaluation system at home and abroad, and the methods and focus of establishing it differ from country to country, and no accepted evaluation system has been formed yet.

Most of the relevant domestic research results focus on the necessity and feasibility analysis of establishing electricity market indicators, improving the market regulation model, and the implementation principles and steps.Indicators such as competitiveness were studied for the European Electricity Market Evaluation System (EEMES) by Taysi et al[1]. Fei Chen et al. started from the international evaluation standards, took into account the construction of the electricity market, the purpose and constraints of government control, and tried to build a set of evaluation index system that meets the actual situation in China[2].Tao Li and Shengyu Wang evaluate the trading in China's electricity market from three perspectives: the fundamentals of the electricity industry, the electricity market setup and the real trading in the electricity market[3].Shuhong Shi et al. established a set of evaluation index system in direct power trading, and a comprehensive evaluation system based on topo cloud theory[4].Bangcan Wang et al. established an evaluation index system for the electricity market trading mechanism from five aspects: market liquidity, stability, economy, development coordination, and openness[5].Guided by the SCP model, Zhixiong Hu constructed indicators for market structure, market operation and market performance to evaluate the new market operation in terms of[6].

Most of the previous researches are still focused on the theoretical support level of index system construction, more inclined to the evaluation model before the operation of the electricity market, but not more applied to the practical operation and effectiveness of the judgment support and feedback regulation. Therefore, in view of the current situation of this kind of research, this paper will establish a set of practical indicators to evaluate the current electricity market mechanism from the perspective of the whole value system of electricity energy, based on the economic value, green value and security value, and make specific evaluation of the indicators based on the fuzzy comprehensive evaluation method, and each province can make market evaluation according to this indicator system and the steps of the evaluation method. In order to provide a basis for the specific investigation of the current electricity market mechanism.

*Corresponding author's e-mail: 1975988911@qq.com

https://doi.org/10.1051/e3sconf/202344103001

2. Electricity market evaluation index system considering economic-green-security

A healthy electricity market mechanism should make it possible to: optimize the allocation of market resources under the action of value law, expand the interests of all market subjects under the premise of meeting social needs; promote the formation of a market system and long-term mechanism for green energy production and consumption, and promote the construction of a new power system with new energy as the main body; improve the efficiency of the power system and the level of safe and reliable supply. This is also the core of this paper from this value perspective, from the three major value systems of economic value, green value and security value of the electric power market, and to reflect the effect of decision-making influence, to build an electric power market assessment index system and form a power market assessment model to assist market decision-making.

2.1 Market economic value category indicators

In order to judge whether the current market mechanism design is conducive to the recovery of economic value, we should first consider the legitimacy of the market force manipulation in the electricity market on the general principle of fairness and equity, i.e., the indicators of market health. Only on this basis can we further consider the interests of market members under this market mechanism and measure the economic value of its electricity market by the indicators of market efficiency.

2.1.1 Market Health

(1) Lerner Index

The Lerner Index (LI), which measures the monopoly position of a market player by the degree of deviation between the electricity market clearing tariff and the marginal cost. The following are specific definitions:

$$LI = \frac{P-MC}{P} \tag{1}$$

In the formula: P is the current market price; MC is the marginal cost of the unit.

(2) Residual Supply Index

The Residual Supply Index (RSI) of a power producer is the sum of the market shares of all units except this one in a given period. It is expressed by the following formula:

$$RSI_i = \frac{\sum_{j=1}^{k} q_j}{D} - \frac{q_i}{D} = \frac{\sum_{j=1}^{k} q_j - q_i}{D} = \frac{\sum_{j=1, j\neq i}^{k} q_j}{D} \tag{2}$$

In the formula: k is the number of all units in the market; q_j is the capacity of the first j unit; q_i is the capacity of the first i unit; D is the market all load demand.

(3) HHI indicators

The HHI metric is measured by the sum of the squares of the market shares held by each supplier in the market. It is expressed by the following formula:

$$HHI = \sum_{i=1}^{l} (100 \times t_i)^2 \tag{3}$$

In the formula, t_i indicates the market share of the first i supplier in the market; l indicates the number of suppliers in the market.

2.1.2 Market Benefits

(1) Producer surplus

In terms of the market, the producer surplus is the additional profit that producers receive for the difference between their true costs and the prevailing tariff, i.e., the difference between the actual revenue that generators earn in the market and the minimum profit they can afford to make.

(2) Consumer surplus

For an individual market, consumer surplus is the difference between the maximum price that a power purchaser (including large customers and power sellers) is willing to pay for a particular electricity and the price that they actually pay. Consumer surplus is measured by the additional economic benefits obtained by purchasing the network.

(3) Market Benefit Distribution

Combining the producer surplus with the consumer surplus yields the overall market surplus. In this paper, from the current electricity market reform in China, the concept of market benefit distribution can be defined as the ratio of supply-side benefits to user benefits.

2.2 Market Green Value Category Indicators

In order to judge whether the current market mechanism design is conducive to the recovery of green value, the first should be considered from the perspective of the nature of energy fuels, coal and other thermal energy in power supply compared to renewable clean energy will cause a greater degree of ecological loss; and from the perspective of the results of green value, environmental sustainability indicators are more relevant.

2.2.1 Energy mix

(1) Fuel Diversity Indicators

By simulating the power grid planning for a specific area, it is possible to count different types of power generation in a certain period of time and calculate the different types of power generation with the calculation formula:

$$P = \frac{\sum_i E_i}{E} \times 100\% \tag{4}$$

In the equation, P is the fuel diversity, E_i represents the amount of electricity generated in a region by other forms of generation besides coal; E represents the total amount of electricity generated in this region during this period.

2.2.2 Environmental Sustainability

(1) CO_2 emission index per unit of power generation

The CO_2 emissions per unit of electricity generated indicator counts the CO_2 emissions per unit of electricity generated and is calculated as follows:

$$Q = \frac{m}{E} \quad (5)$$

In the formula, Q is the CO_2 emission per unit of power generation, m indicates the amount of CO_2 emitted by power generation in a certain region during a certain period of time; E indicates the total amount of power generation in the region within that period of time.

(2) Renewable Energy Efficiency Indicators

Improving the capacity of renewable energy consumption is conducive to reducing environmental pollution and reducing resource consumption. The formula for calculating this indicator is:

$$W = E(Q) \times H \quad (6)$$

In the formula, W indicates the renewable energy benefit, $E(Q)$ indicates the extra electricity generated from renewable energy, and H indicates the environmental benefit per unit of renewable energy electricity.

2.3 Market Safety Value Category Indicators

Real-time balancing of power supply and load of the power system is a prerequisite for ensuring the security and stability of the grid. Therefore, it is important to evaluate whether the current market mechanism design is conducive to the recovery of security value in terms of both capacity value and regulation value. The construction of new power grid with new energy as the core must not be at the cost of ensuring energy security.

2.3.1 Capacity Market

(1) Maximum emergency backup power

Find the sum of the contingency reserve capacity (or contingency reserve capacity set aside for units) in the entire market to find the maximum value of contingency reserves in that market.

(2) Forced out-of-service rate

The forced outage rate is an important indicator of the efficiency of the electricity market operation. Under extreme conditions, the lower the forced outage rate, the lower the price of electricity.

2.3.2 Ancillary Services Market

(1) Active regulation of service costs

The active up or down service cost for the electricity market is the total cost of spinning reserve for the entire market, which includes the operating costs of all spinning reserve units in this area and the opportunity costs borne by units that retain some spinning reserve capacity.

(2) Cost of reactive regulation services

In the electricity market, the cost of reactive services is mainly composed of reactive energy supply and reactive accidental backup.

(3) Up or down service price and FM mileage

Using the U.S. PJM market as a benchmark, the maximum, average and minimum values of daily (monthly) up and down prices and FM miles were selected as indicators for analysis.

3. Comprehensive Electricity Market Evaluation Methodology

Power market evaluation has the characteristic of fuzzy, based on this characteristic, this paper adopts the fuzzy comprehensive evaluation method as the method of power market operation evaluation.

3.1 Overview of fuzzy integrated evaluation method

Fuzzy comprehensive evaluation is a combination of fuzzy mathematics and fuzzy statistics, taking into account various influencing factors, so as to make scientific judgments.

The basic idea of fuzzy comprehensive evaluation is to incorporate various influencing factors into the assessment of the assessment object based on the principle of maximum subordination in fuzzy mathematics. The method combines qualitative and quantitative aspects, and provides objective qualitative analysis of fuzzy attributes that are difficult to analyze quantitatively, thus closely combining qualitative description with quantitative analysis. Thus, it is a more appropriate method for comprehensive evaluation of the electricity market[7].

3.2 Mathematical model of fuzzy comprehensive evaluation method

(1) Determine the set of indicators and evaluation set

First, determine the set of evaluation indicators, that is, which indicators are selected for evaluation. If there are m indicators selected for evaluation, then these m evaluation indicators constitute a finite set of evaluation indicators X:

$$X = [x_1, x_2, \dots, x_m] \quad (7)$$

Then finally, the set of assessments for the goal is determined, that is, the various possible assessments for the goal are assessed. If the scoring criteria are divided into n levels according to the actual needs, it represents a finite set of scoring criteria Y:

$$Y = [y_1, y_2, \dots, y_n] \quad (8)$$

(2) Determine the fuzzy matrix R

Determine the relational affiliation r_{ij} of each ordered pair $(x_i, y_j) \in X \times Y$.

r_{ij} is the affiliation degree of element x_i in X for rank y_j in Y (i=1, 2, ..., m; j=1, 2, ..., n), from which the fuzzy relation matrix $R = [r_{ij}]$ is obtained.

(3) Determine the weight of each indicator

The importance of each indicator x_i in the evaluation index set $X = [x_1, x_2, \ldots, x_m]$ to the evaluation objective varies, and each indicator can be assigned a weighting coefficient $a_i(i = 1, 2, \ldots, m)$ and satisfy $0 \le a_i \le 1$, $\sum_{i=1}^{m} a_i = 1$.

Evaluation index weights:

$$A = [a_1, a_2, \ldots, a_m] \tag{9}$$

In fuzzy comprehensive evaluation, establishing the fuzzy relationship matrix of each factor and determining the weights of each indicator are two important tasks, which can usually be determined by expert evaluation and statistical analysis methods.

(4) Determine the fuzzy integrated evaluation vector B

The fuzzy comprehensive evaluation vector B is based on the weights of the evaluation indexes, and the fuzzy relationship matrix R is synthesized to obtain the comprehensive affiliation degree of each evaluation level.

The fuzzy relationship matrix R is synthesized and calculated on the basis of the evaluation indexes to obtain the comprehensive affiliation degree of the evaluation object on each evaluation level.

$$B = A \cdot R = [a_1, a_2, \ldots, a_m] \cdot \begin{bmatrix} r_{11} & \cdots & r_{1n} \\ \vdots & \ddots & \vdots \\ r_{m1} & \cdots & r_{mn} \end{bmatrix} = [b_1, b, \ldots, b_n] \tag{10}$$

(5) Determine the final comprehensive evaluation grade of the evaluation object

In the comprehensive evaluation vector B, the level of the assessment target can be determined based on the principle of maximum affiliation. And to finalize the level of the assessment target, it is necessary to quantify the individual rubrics in $Y = [y_1, y_2, \ldots, y_n]$.

4. Conclusion

With the implementation of the new round of electricity reform, the speed of market-oriented reform of China's power grid has also been accelerated. Electricity is an important part of the national economy, and the reform of the power market inevitably requires safe, stable and reliable operation of the power system, thus improving the operational efficiency of each link of power generation, transmission and sale. Therefore, it is particularly necessary to construct a complete evaluation system for the electric power market. Based on the research on the evaluation methods of electric power market, this paper constructs the evaluation indexes of electric power market from three aspects: economic value, green value and safety value. On this basis, an evaluation method based on fuzzy comprehensive evaluation is proposed to reflect the overall situation of the electric power market. The market evaluation based on the index system and evaluation method in this paper can effectively judge whether the operation of the current electricity market is economically beneficial to maximize, environmentally sustainable, and safe, and provide a judgment basis for the healthy development of the electricity market. It provides new ideas for the future exploration of electric power market mechanism in each province of China, and also lays a solid foundation for the future construction and development of electric power market.

References

1. Z. C. Taysi, G. Biricik. Evaluation of Price forecast systems for Turkish Electric Market [C].2015.
2. Chen, F., Liu, J., Zhang,Y. (2017) Objectives 、 constraints and evaluation criteria of power market construction.J.Price Theory and Practice.,12:38-43.
3. Li, T.,Wang, S. (2018) Research on the evaluation system of China's power market transactions based on gray correlation degree and fuzzy comprehensive evaluation method.J.Industrial Technology and Economics.,09:130-137.
4. Shi, S.,Wang, L.,Zhang, Q.,Liu, D.,Yang, T.,Liu,Z. (2018) A Comprehensive Assessment of Electricity Direct Trading Market Efficiency - An Analysis Based on Topo Cloud Theory.J.Price Theory and Practice.,06:146-150.
5. Wang, B.,Zhang, M.,Xu, J.,Xu, Y.,Ying, L.,Cui, X.,Huang, S. (2019) Research on comprehensive evaluation model of electricity market trading mechanism.J.Yunnan Power Technology.,01:80-88.
6. Hu, Z. (2020) Research on Electricity Market Operation Evaluation Indexes and Evaluation Methods under the New Electricity Reform.D.North China Electric Power University (Beijing).
7. P.Kumar,S.K.Singh. A comprehensive evaluation of aspect-oriented software quality (AOSQ) model using analytic hierarchy process (AHP) technique[C]. 2016

E3S Web of Conferences **441**, 03002 (2023) https://doi.org/10.1051/e3sconf/202344103002
CELCT 2023

Enhancing the work extraction of the entanglement engine via quantum coherence

Hai Li*, Xiaosong Liu and Yaming Hao

College of Information and Electronic Engineering, Shandong Technology and Business University, Yantai, Shandong 264005, China

Abstract. Studying the relationship between quantum coherence and energy is an intriguing and vital subject in quantum thermodynamics. Quantum engine provides an excellent platform to emphasize it. By constructing a bipartite entanglement engine (BEE) model consisting of a single qubit as the target qubit and a two-qubit as the ancilla, we have investigated the influence of quantum coherence of ancilla on work extraction of BEE at length, and analytically calculated the extractable work of BEE for the ancilla with quantum coherence (QC). We found that the extractable work of BEE is not only dependent on coherence magnitude but also related to the coherence phase of ancilla. Moreover, under certain coherent parameters, the QC can boost the work extraction from BEE. The study reveals the role of QC acting as "fuel" of engine well which deepens the understanding of the energetic effect of QC.

1. Introduction

Quantum effects are unique phenomena in microscopic quantum systems. In recent years, exploring the connection between quantum effects and energy and understanding the physical mechanisms of quantum effects in quantum systems have received extensive attention. Researchers have extensively explored the thermodynamic characteristics of quantum effects based on different quantum system models, such as quantum battery [1-4], quantum engine [5-8], and quantum thermal manager [9-10]. However, we still lack a comprehensive and in-depth understanding of the relationship between quantum effects and energy, as well as the conversion mechanisms in quantum devices, which are worth further investigation.

Recently, Bresque et al. [11] constructed an entanglement engine model based on two detune qubits (a target qubit (TQ) and an auxiliary qubit). The auxiliary qubit is initially prepared in the excited state, while the TQ is in the ground state. They showed that entanglement can serve as the "fuel" for extracting work in the engine. Currently, a series of studies have shown that quantum effects in non-equilibrium environments have an impact on the thermodynamic properties of quantum systems [12-14]. Therefore, a natural question is what kind of impact quantum effects will have on the performance of the entanglement engine. For example, when are quantum effects involved [11]? What are the characteristics of the conversion between quantum effects and energy?

Based on the above questions, a bipartite entanglement engine (BEE) model consisting of an ancilla with a coherent two-qubit and a single TQ has been constructed. The characteristics of extracted work from BEE have been analyzed. It shows that the QC, under certain parameters, can enhance the work extraction of BEE, that is, the QC can serve as the "fuel" of BEE.

2. Model

Here, we model the BEE shown in Figure 1 which includes four steps: entanglement preparation, measurement, feedback control, and initial state resetting. The specifical implement in each step is given below.

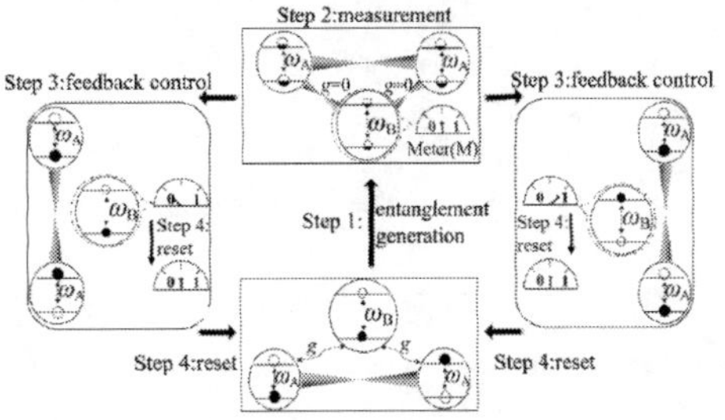

Figure 1. Diagram of bipartite entanglement engine (BEE) cycle.

Step 1: Entanglement preparation. We denote A as the ancilla composed of two identical qubits and B for the TQ, and ω_A and ω_B are the transition frequencies of A and B, respectively. We assume that the detune

*E-mail: lihai@sdtbu.edu.cn

https://doi.org/10.1051/e3sconf/202344103002

$\delta = \omega_B - \omega_A > 0$. The free Hamiltonian of the composite system formed by TQ and ancilla can be expressed as follows:

$$\boldsymbol{H}_0 = \hbar\omega_B\boldsymbol{\sigma}_B^\dagger\boldsymbol{\sigma}_B + \hbar\omega_A\sum_{j=1}^{2}\boldsymbol{\sigma}_j^\dagger\boldsymbol{\sigma}_j \tag{1}$$

where $\sigma_k^\dagger = |1_k\rangle\langle 0_k|$ ($\sigma_k = |0_k\rangle\langle 1_k|$), with $k = \{1, 2, \mathrm{B}\}$, represents the raising (lowering) operators of the qubit k, and $\hbar$ is the Planck constant. The initial state of the composite system is:

$$|\psi_0\rangle = |\psi_A\rangle \otimes |0_B\rangle \tag{2}$$

where $|\psi_A\rangle = \exp(-i\varphi)\sin(\alpha)|10\rangle + \cos(\alpha)|01\rangle$ with α and φ being the probability amplitude parameter and coherent phase, respectively, and $\alpha \in [0, \pi/2]$ and $\varphi \in [0, 2\pi]$. $|\psi_A\rangle$ and $|0_B\rangle$ represents a general pure state of ancilla with coherence and the ground state of TQ. The density matrix of the initial state $|\psi_A\rangle$ can be expressed as:

$$\rho_A = |\psi_A\rangle\langle\psi_A| = P_1^{\mathrm{dia}}|01\rangle\langle 01| + P_2^{\mathrm{dia}}|10\rangle\langle 10| + P_1^{\mathrm{nd}}|01\rangle\langle 10| + P_2^{\mathrm{nd}}|10\rangle\langle 01|, \tag{3}$$

where $P_1^{\mathrm{dia}} = \cos^2(\alpha)$, $P_2^{\mathrm{dia}} = \sin^2(\alpha)$, $P_1^{\mathrm{nd}} = \exp(i\varphi)\sin(2\alpha)/2$, and $P_2^{\mathrm{nd}} = \exp(-i\varphi)\sin(2\alpha)/2$. According to the l_1-norm measure of quantum coherence [15], the coherence of an arbitrary quantum state ρ is equal to the sum of absolute values of all off-diagonal elements in its density matrix, i.e., $C(\rho) = \sum_{i\neq j}|P_{ij}|$, where P_{ij} represents the element of the density matrix ρ . The coherence of ancilla with the state ρ_A is:

$$C(\rho_A) = |P_1^{\mathrm{nd}}| + |P_2^{\mathrm{nd}}| = |\sin(2\alpha)|. \tag{4}$$

We denote $C_A = C(\rho_A)$. According to Equation (4), the coherence of ρ_A satisfies $0 \le C_A \le 1$, where $C_A = 0$ and $C_A = 1$ correspond to ρ_A being an incoherent state and a maximal coherent state, respectively. Here, we consider the entanglement generated by the energy-conserving coupling between the ancilla and TQ. The interaction is given as:

$$\boldsymbol{V} = \hbar g\sum_{j=1}^{2}\left(\boldsymbol{\sigma}_j^\dagger\boldsymbol{\sigma}_B + \boldsymbol{\sigma}_j\boldsymbol{\sigma}_B^\dagger\right) \tag{5}$$

where g represents the coupling strength between the TQ and each qubit of ancilla. The composite system undergoes unitary evolution with the evolution operator:

$$\boldsymbol{U}(t) = \exp\left[-\mathrm{i}\left(\boldsymbol{H}_0 + \boldsymbol{V}\right)t\right] \tag{6}$$

After a time t , the composite system reaches the state:

$$|\psi_t\rangle = \boldsymbol{U}(t)|\psi_0\rangle = |\psi_t^A\rangle|0_B\rangle + \gamma|001_B\rangle \tag{7}$$

where $|\psi_t^A\rangle = (\xi|10\rangle + \kappa|01\rangle)/2$ and $\gamma = -g[\exp(it\Omega) - 1][\exp(i\varphi)\cos(\alpha) + \sin(\alpha)]/\Omega$, $\xi = \cos(\alpha)[X + Y + \exp(it\delta/2)] + \exp(-i\varphi)\sin(\alpha)[X - Y + \exp(it\delta/2)]$, $\kappa = \cos(\alpha)[X + Y + \exp(it\delta/2)] + \exp(-i\varphi)\sin(\alpha)[X + Y - \exp(it\delta/2)]$, $X = [4\exp(it\Omega/2)g^2]/(\Omega^2 - \delta\Omega)$, $Y = [4\exp(-it\Omega/2)g^2]/(\Omega^2 + \delta\Omega)$, and $\Omega = \sqrt{8g^2 + \delta^2}$ is the Rabi frequency. Equation (7) shows that if the coefficient γ of the state vector $|001\rangle$ in $|\psi_t\rangle$ is not equal to 0 or 1, then the entanglement between TQ and ancilla occurs.

Step 2: Measurement. When the evolution time reaches $t_0 = \pi/\Omega$, the entanglement of the composite system becomes the maximum, and the corresponding state is:

$$|\psi_{t_0}\rangle = |\psi_{t_0}^A\rangle|0_B\rangle + \gamma(t_0)|001_B\rangle \tag{8}$$

with $\gamma(t_0) = 2g[\cos(\alpha) + \exp(-i\varphi)\sin(\alpha)]/\Omega$ and $|\psi_{t_0}^A\rangle = [\xi(t_0)|10\rangle + \kappa(t_0)|01\rangle]/2$, $\xi(t_0) = \exp[i\pi\delta/(2\Omega)][-\exp(i\varphi)\cos(\alpha) + \sin(\alpha)] + i[\exp(i\varphi)\delta\cos(\alpha)/\Omega + \sin(\alpha)]$, $\kappa(t_0) = \exp[i\pi\delta/(2\Omega)][\exp(i\varphi)\cos(\alpha) - \sin(\alpha)] + i\delta[\exp(i\varphi)\cos(\alpha) + \sin(\alpha)]/\Omega$. The corresponding density matrix becomes:

$$\rho_{AB}(t_0) = |\gamma(t_0)|^2|001_B\rangle\langle 001_B| + |\psi_{t_0}^A 0_B\rangle\langle\psi_{t_0}^A 0_B| + \gamma^*(t_0)|\psi_{t_0}^A 0_B\rangle\langle 001_B| + \gamma(t_0)|001_B\rangle\langle\psi_{t_0}^A 0_B| \tag{9}$$

At the moment, we first turn off the coupling channel ($g = 0$), and then implement a positive operator valued measure (POVM) on TQ via a measurement device M. After the measurement, the coherence in TQ, i.e., the supposition between the basis $|0_B\rangle$ and $|1_B\rangle$, can be destroyed or disappears. And, the state of the entire system (ABM) can be expressed by:

$$\rho_{ABM} = P_B^1|00\rangle\langle 00| \otimes |1_B 1_m\rangle\langle 1_B 1_m| + |\psi_{t_0}^A\rangle\langle\psi_{t_0}^A| \otimes |0_B 0_m\rangle\langle 0_B 0_m| \tag{10}$$

where we have assumed that when the TQ collapses to the excited (ground) state $|1\rangle$ ($|0\rangle$), the pointer of the device M points to 1 (0) corresponding to $|1_M\rangle$ ($|0_M\rangle$), and:

$$P_B^1 = \frac{4g^2}{\Omega^2}\left[1+\cos(\varphi)\sin(2\alpha)\right] \quad (11)$$

Step 3: Feedback Control. Based on the information obtained from measurement device M, different control protocols are implemented on the system. When the device M remains in $|1_M\rangle$, implying that the TQ is in excited state $|1_B\rangle$, one can perform a π pulse on TQ to flap TQ's state to the ground state $|0_B\rangle$, and the extracted work is:

$$\Delta E = \hbar(\omega_B - \omega_A) = \hbar\delta . \quad (12)$$

When the device M remains in $|0_M\rangle$, no operation is carried out. And, the working medium returns to Step 1 to implement the entanglement evolution. Therefore, the average work in a single cycle can be given as:

$$W = P_B^1\Delta E = \frac{4g^2}{\Omega^2}\left[1+\cos(\varphi)\sin(2\alpha)\right]\hbar\delta . \quad (13)$$

Step 4: Initial state resetting. After the feedback control operations, TQ has returned to its initial state. One needs to reset the ancilla and the device to their initial states where the cost of resetting operation is not the interest of this paper and cannot be considered further. After the resetting process, the BEE repeats the above four cyclic steps to enter the next cycle.

3. The impact of quantum coherence on work extraction.

Here, we mainly focus on the influence of QC on the output work of BEE. Combing with Equations (4) and (13), the average output work of BEE can be further expressed as:

$$W = \frac{4g^2}{\Omega^2}\left[1 \pm \cos(\varphi)C_A\right]\hbar\delta = \frac{1}{2+\Delta^2/4}\left[1 \pm \cos(\varphi)C_A\right]\hbar\delta \quad (14)$$

where $\Delta = \delta/g$. The signs + and − in Equation (14) correspond to $\sin(2\alpha) > 0$ and $\sin(2\alpha) < 0$, respectively. Equation (14) is the main result in this paper. When the coupling strength g and the detune δ are fixed, work extracted from BEE are determined by the coherence C_A and the coherent phase φ. Meanwhile, the extractable work of BEE is bounded by:

$$\frac{1}{2+\Delta^2/4}\left[1-|\cos(\varphi)|C_A\right]\hbar\delta \le W \le \frac{1}{2+\Delta^2/4}\left[1+|\cos(\varphi)|C_A\right]\hbar\delta . \quad (15)$$

Equation (15) shows that QC can not only boost the extractable work but also suppress the work extraction. Since $0 \le |\cos(\varphi)|C_A \le 1$ the maximum and minimum values of the extractable work W are given as:

$$W_{max} = 2W^{dia} = \frac{\hbar\delta}{1+\Delta^2/8} \quad \text{and} \quad W_{min} = 0 \quad (16)$$

where the W_{max} and W_{min} correspond to $\cos(\varphi)\sin(2\alpha) = 1$ and $\cos(\varphi)\sin(2\alpha) = -1$. The ancilla's state for W_{max} can be obtained easily, such as $|\psi_A\rangle = (|10\rangle + |01\rangle)/\sqrt{2}$, which can be regarded as the optimal QC state.

In the similar entanglement engine model in [11], where the ancilla with a single excitation and no coherence is considered, the extractable work is obtained as $W^{single} = \hbar\delta/(1+\Delta^2)$. Therefore, comparing both models of ours with coherence and the one without coherence in [11], the ratio of the maximum extractable work can be expressed as:

$$R = \frac{W_{max}}{W^{single}} = \frac{1+\Delta^2}{2+\Delta^2/4}\left[1+|\cos(\varphi)|C_A\right] \quad (17)$$

which satisfies:

$$\frac{1+\Delta^2}{2+\Delta^2/4} \le R \le \frac{1+\Delta^2}{1+\Delta^2/8} . \quad (18)$$

From Equation (18), we note that the upper bound of ratio R (denoted as R_{up}) is always twice its lower bound (denoted as R_{low}), i.e., $R_{up} = 2R_{low} = (1+\Delta^2)/(8+\Delta^2)$. Moreover, R_{up} is a monotonically increasing function of the parameter Δ, as shown in Figure 2. It also means that the more obvious the advantage of the coherent BEE is the weaker the coupling strength ($g \propto 1/\Delta$) of TQ-ancilla is. Meanwhile, $R_{up} > 1$ always holds for $\Delta > 0$, which suggests that the maximum extractable work W_{max}, under the QC ancilla with optimal coherence parameters ($C_A = 1$ and $\varphi = 0$) is always greater than that under the incoherent single-qubit ancilla with equal energy. In the limit $\Delta \to \infty$ (or weak-coupling limit $g \ll \delta$), R_{up} has the maximum value:

$$R_{up}^{max} = \lim_{\Delta\to\infty} \frac{1+\Delta^2}{1+\Delta^2/8} = 8 \quad (19)$$

which demonstrates that the maximal extractable work of coherent BEE can reach eight times that of incoherent engine in [11].

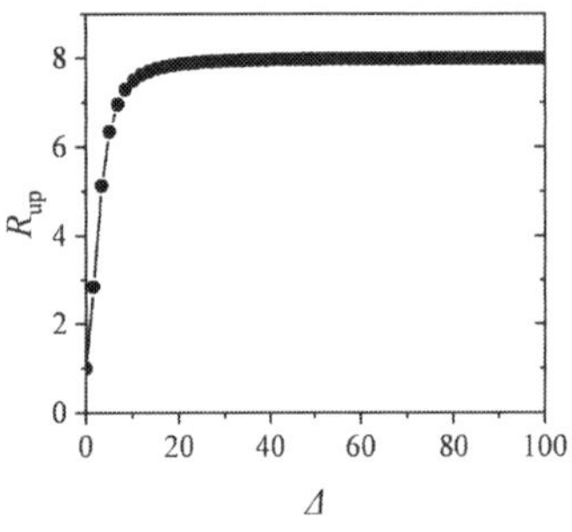

Figure 2. Curve of R_{up} varying with parameter Δ ($\Delta = \delta/g$).

4. Conclusion

In this paper, we have constructed a BEE model with a two-qubit ancilla and investigated the influence of QC of the ancilla on extractable work from the engine. We have analytically derived the extractable work of the BEE with the single-excitation general QC state of ancilla, analyzed the range of the extractable work, and obtained the upper and lower bounds of extractable work. Compared with the incoherent entanglement engine, QC can act as the work resource of the BEE to effectively increase the extractable work of the engine. This study enriches our understanding of the relationship between quantum coherence and energy, providing useful insights for improving the performance of related quantum devices.

Acknowledgments

This work is financially supported by the Natural Science Foundation of Shandong Province of China, Grant Number ZR2011FL009, and the Yantai Foundation for Development of Science and Technology, Grant Number 2022JCYJ044.

References

1. Liu, J.X., Shi, H.L., Shi, Y.H., et al. (2021) Entanglement and work extraction in the central-spin quantum battery. J. Physical Review B., 104(24): 245418.
2. Seah, S., Perarnau-llobet, M., Haack, G., et al. (2021) Quantum speed-up in collisional battery charging. J. Physical Review Letters., 127(10): 100601.
3. Andolina, G.M., Keck, M., Mari, A., et al. (2019) Extractable work, the role of correlations, and asymptotic freedom in quantum batteries. J. Physical Review Letters., 122(4): 047702.
4. Dou, F.Q., Lu, Y.Q., Wang, Y.J., et al. (2022) Extended dicke quantum battery with interatomic interactions and driving field. J. Physical Review B., 105(11): 115405.
5. Camati, P.A., Santos, J.F.G., Serra, R.M. (2019) Coherence effects in the performance of the quantum otto heat engine. J. Physical Review A., 99(6): 062103.
6. Hammam, K., Hassouni, Y., Fazio, R., et al. (2021) Optimizing autonomous thermal machines powered by energetic coherence. J. New Journal of Physics., 23(4): 043024.
7. Latune, C.L., Sinayskiy, I., Petruccione, F. (2021) Roles of quantum coherences in thermal machines. J. The European Physical Journal Special Topics., 230(4): 841-850.
8. Ji, W., Chai, Z., Wang, M., et al. (2022) Spin quantum heat engine quantified by quantum steering. J. Physical Review Letters., 128(9): 090602.
9. Yu, W.L., Li, T., Li, H., et al. (2021) Heat modulation on target thermal bath via coherent auxiliary bath. J. Entropy., 23(9): 1183.
10. Micadei, K., Peterson, J.P.S., Souza, A.M., et al. (2019) Reversing the direction of heat flow using quantum correlations. J. Nature Communications., 10(1): 1-6.
11. Bresque, L., Camati, P.A., Rogers, S., et al. (2021) Two-qubit engine fueled by entanglement and local measurements. J. Physical Review Letters., 126(12): 120605.
12. Shahidani, S. (2022) Thermodynamic forces and flows between a thermal bath and a squeezed thermal bath: Application to optomechanical systems. J. Physical Review A., 105(6): 063516.
13. Latune, C.L., Sinayskiy, I., Petruccione, F. (2019) Heat flow reversals without reversing the arrow of time: The role of internal quantum coherences and correlations. J. Physical Review Research., 1(3): 033097.
14. Man, Z.X., Xia, Y.J., An, N.B. (2020) Heat fluxes in a two-qubit cascaded system due to coherences of a non-thermal bath. J. Journal of Physics B: Atomic, Molecular and Optical Physics., 53(20): 205505.
15. Baumgratz, T., Cramer, M., Plenio, M.B. (2014) Quantifying coherence. J. Physical Review Letters., 113(14): 140401.

https://doi.org/10.1051/e3sconf/202344103003

Exploring the analysis of key factors of new energy consumption and measures to solve them

Yufei Gong [1], Xingyun Li [1], Yuzhi Wang *[2]

[1]CHINA LONGYUAN POWER GROUP CORPORATION.LIMITED, Beijing, 100034, China
[2]School of Control and Computer Engineering, North China Electric Power University, Beijing, 102206, China

Abstract. In order to achieve sustainable and healthy development, power companies must gradually abandon traditional energy sources and actively develop new energy sources, which could achieve green energy-saving and environmental protection effects. This requires an analysis of the key factors of new energy consumption, to clarify the causes of this problem and to take effective countermeasures to solve it. The application of new energy generation technology in power grids is a reasonable solution to the problem of new energy consumption, ensuring a continuous and sufficient supply of electricity resources to meet demand. This paper focuses on the key factors for the consumption of new energy and the measures to address them.

1 Introduction

Now that the country's favourable economic environment has created opportunities for all sectors to develop rapidly, electricity resources have become even more important. The massive use of electrical equipment in various fields has led to higher requirements for electrical energy. For power companies, facing the continuous increase in demand for electricity, they have to take effective measures to cope with it. At this time, if the traditional technology is still used for power generation, the transmission mode of electrical energy is not updated in time, it will not be able to meet the demand. This not only consumes a lot of energy, but also releases many toxic substances that seriously pollute the environment, and it is very detrimental to the future development of the power industry. At present, industrial enterprises are showing new development trends, the innovation process is also accelerating, people's quality of life is gradually improving, a large amount of electrical energy consumption, electricity consumption is also significantly increased. If power companies do not update their equipment and technology in the process of supplying electricity to customers, and use traditional models and traditional power generation technologies, the power companies will not be able to fully meet the current increasing demand for electricity. This can lead to a situation where the supply of energy is not in harmony with the environment and contradictions are becoming increasingly apparent, which can lead to a large amount of energy not being fully utilized and serious environmental pollution. With the emergence of new energy technologies, the rationalization of new energy generation technologies, and the efficient application of clean energy, it is possible to make full use of energy and its advantages, thus improving energy utilization and keeping energy consumption to a minimum, considering the green development of enterprises [1]. No pollutants are emitted during the application of new energy sources and environmental pollution can be effectively avoided. However, However, from the current application of new energy technologies, there are still some problems due to various factors. Especially in northern China, northeast China and northwest China, where new energy is commonly used due to the specificity of the geographical environment, there will be problems with the consumption of new energy. To effectively address the above issues, all aspects must be taken into account when analyzing the key factors in order to respond accordingly and improve the efficiency of the solution.

2 New energy generation technologies

Among the new energy generation technologies, the more common applications are solar power technology and energy generation technology, as follows:

2.1 Solar power technology

The most widely used new energy source in all areas of aggregation is solar energy, which is the ability to collect and convert solar radiation into electrical energy and can achieve good results in power generation technology. Compared to other energy sources, solar energy can be accepted because of its excellent stability in the process of application and, most importantly, its easy availability, inexhaustible trend and inexhaustible use. In the process of solar power technology, sunlight is shone on the connected application panels and converted into

* Corresponding author: 120212227319@ncepu.edu.cn

electricity, which can then be transmitted to the user side for use [2]. In order to obtain the maximum output power, concentrated solar power generation technology has gradually become a research hotspot [13]. However, no matter which industry is involved, there are drawbacks in the use of solar energy, which means that there are limitations when it comes to the use of energy. In order to effectively remedy this, it is necessary to actively research and develop this technology from a practical application point of view, in order that solar energy has a high energy utilization rate.

For example, when designing an intelligent control system for solar street lights, the flow chart is mainly for the charging part and the discharging part. Among them, the charging part makes clear its charging situation by detecting the light intensity and voltage level, and the discharging situation by monitoring the light intensity to control the switch of the street light part. The core of the controller is the microcontrollers, for which the system is initialized and then the internal resource module registers and the relevant flag bits are set. During the day, the solar cells can be collected periodically and the terminal voltage of the batteries can be obtained at the same time; the control process will vary depending on the state of the voltage combination. When it comes to night time, collecting the terminal voltage of the battery can be done periodically while obtaining the power supply current on the LED lamp side to reasonably control the different power of the street light and to protect the battery discharge situation.

2.2 Wind power technology

Among the new energy sources, in addition to solar energy, the use of wind power technology is also very common. The application of this technology has certain environmental requirements, namely the abundance of wind resources and the ability to convert wind energy into mechanical energy and later into electrical energy in specific applications. From the point of view of the utilization of wind energy in various regions of the world, the technical development varies. Western European countries are relatively mature in wind energy technology and mainly use the energy for agricultural and industrial production, making full use of local resources while also reducing traditional energy consumption. Windmills are essential for the collection and use of wind energy, which is transformed into electricity to meet people's production and living needs. In the process of applying wind power technology, wind power generation equipment is installed in the building structure, which transmits the collected wind energy into it and converts it into electrical energy to supply the various electrical equipment installed on the building, thus improving the functional efficiency of the building and making full use of natural wind energy resources. The use of wind energy in the operation of power systems has resulted in a continuous increase in the amount of electricity supplied [3].

China has very rich wind resources, but in the process of applying wind power generation there is a problem that different regions have different geographical environments and different resource conditions, and the distribution of wind energy lacks uniformity. In China's northwest and offshore, windy weather is frequent and windy, and wind power resources are sufficient, while other areas are clearly lacking in this respect, especially in the inland plains, where there is hardly any strong wind and wind power resources are seriously lacking. If power generation company uses wind power, it needs to invest a lot of resources but can't recycle effectively, which results in a waste of resources. Therefore, regional environmental conditions should be taken into account in the application of wind power systems. In the case of a lack of resources, there are certain restrictions on wind power systems. Without good conditions and sufficient resources, even if wind power technology is used, it cannot be implemented effectively, and even if wind power generation can be carried out, good results cannot be obtained, the stability of wind energy use cannot be achieved and its proper value cannot be brought into play [4]. Therefore, there are limitations to wind energy compared to solar energy, and wind power consumption has become one of the urgent problem of new energy sources [14]. In order to effectively manage the wind power generation system and give full play to the value of wind power generation technology, it is also necessary to make full use of modern technology, such as the reasonable adoption of information technology and the combined use of intelligent technology, including the application of information technology to the wind power generation system, the implementation of information management and intelligent operation of various functions, which can achieve unified supervision and control. In view of the instability in the application of wind power generation technology, it is necessary to invest costs for maintenance management, which has become a key research topic in the current new energy. Many regions are now using distributed power sources in their wind energy discoveries and have established connections with the power generation system, which inevitably makes scheduling more difficult, in order that the system cannot be effectively controlled. When the grid is in operation, its security is not guaranteed and it is not possible to maintain a safe and reliable operation, which is a shortcoming of wind energy generation technology [5]. Currently, the application process of wind power discovery technology uses distributed power generation. Due to the complex geographical environment in some areas, if wind power technology is used, higher requirements are placed on the communication data acquisition system, such as smooth information transmission. And the need to maintain the stability of the power to ensure better performance can lead to higher operation and maintenance costs. For the operation and maintenance of distributed generation equipment, the above-mentioned problems can be a great challenge, which has higher technical requirements for the operation of the control system, to improve the technical level, but also to achieve high quality and efficient management.

3 New energy consumption in China

China's abandoned wind and light are mainly in North China, Northeast China and Northwest China, but most of the regions do not have the problem of abandoned wind and light. The total installed capacity of new energy in Jiangsu and Shandong provinces is very large, which can reach 8.34 MW and 8.54 MW, and the new energy is basically for full consumption [6]. In terms of time distribution, when entering the heating period, wind power abandonment is very high, with over 66% nationwide and over 90% in the northeast alone, and when in the low load phase, the proportion of wind power abandoned in the total amount of wind abandoned is very high, with over 80%. Compared to other countries, China's wind and light abandonment is more geographically concentrated, and there is a certain concentration in terms of seasons and also very concentrated periods.

When discussing the problem of new energy consumption in China, an analysis of the abandonment of wind and light in China reveals that the main reason for this phenomenon is the distribution of new energy installations, which has a certain correlation with the load [7]. The distribution of new energy installations in China is not homogeneous due to the geographical environment, which is inverse to the load distribution. The proportion of abandoned wind and light loads in North China, Northeast China and Northwest China is not very high, at 36% of the total load, while these three regions already account for more than 75% of the country's installed new energy capacity. Among them, Gansu, the Inner Mongolia East Region, Xinjiang and Ningxia have a very high penetration of new energy and reach 100%, which is already far more than economically developed countries. As the local load size is not very large, there are high requirements for new energy consumption, requiring power supply regulation performance to match the prescribed requirements and also grid interconnection [8].

In terms of electricity regulation performance, the proportion of electricity flexible regulation in China is not very high. In North China, Northeast and Northwest China, the proportion of electric power flexible regulation in new energy installations is 17.9%. Among them, the Northeast region accounts for 7% of the proportion of electric power flexible regulation in new energy installations and the Northwest region accounts for 2.4% of the proportion of electric power flexible regulation in new energy installations. Economically developed countries, such as Spain, Germany and the United States, account for 30%, 18% and 47% respectively of new energy installations with flexible electricity supply regulation. Among them, the US can reach 8.5 times of the new energy installations for electricity supply regulation and Spain can reach 1.5 times of the flexible electricity regulation. In terms of the level of interconnection of the Chinese grid, the demand for new energy exports from northern, north-eastern and north-western China cannot be met.

Overall, factors related to new energy consumption in China include load size, grid interconnection and electricity regulation. Due to the lack of sufficient conditions for new energy consumption in Northern, Northeastern and Northwestern China, there is room for consumption in the central and eastern regions. But the utilization of grid interconnections is not fully utilized, leading to the problem of abandoned wind and sun. In addition, China has yet to establish a peak regulation compensation mechanism and price response mechanism in the construction of a national electricity market. Local governments have set up generation plans and each region has no willingness to accept new energy sources. Barriers exist between provinces, making it difficult to consume new energy [9].

4 Effective measures to solve the problem of new energy consumption

For the effective solution of the new energy consumption problem and the reasonable use of solar and wind energy, the wind-light-storage-hydrogen-ammonia integration is used to build a power supply of 600,000 kilowatts of new energy, including 400,000 kilowatts of wind power, 200,000 kilowatts of photovoltaic, 90 megawatts/360 megawatt-hours of electrochemical energy storage, in line with the efficient coupling of the side supporting 300,000 kilowatts of point solution water hydrogen production system and 120,000 standard cubic meters of hydrogen storage system. Specific efforts include improving power conditioning capabilities, improving grid interconnection and strengthening demand-side management, as follows:

4.1 Improved power conditioning

For some regions of China, due to the backwardness of the economy, power equipment is not updated in a timely manner and the phenomenon of significant aging occurs, which not only affects the quality and efficiency of power generation, but also leads to high input costs and the emission of various pollutants. For example, in the northeast and northwest regions, when renovating the power grid thermal power, without actively introducing advanced technology, it will lead to the renovation can not achieve the expected results [10]. This work is carried out in accordance with the design requirements, with the minimum technical output of the thermal units operating at a low level of 55% and the minimum technical output of the condensing units operating at the same time, already at 30%. At the same time, the use of new energy technologies in the North-east has led to an accelerated pace of construction of pumped storage power stations, which have reached 4.1 MW in terms of current installed capacity, and a much higher power regulation capacity [5].

4.2 Improving grid interconnection

The use of new energy technologies for electricity has led to a corresponding increase in the level of grid interconnection. Cross-regional and cross-provincial transmission channels have now been established. In order to expand capacity, construction is also being increased, which makes the AC grid at the sending and receiving ends operate efficiently. The generation capacity is significantly increased, which makes the DC power in full power state and ensures the electric energy public [11]. Taking the new energy grid in the Northeast and Northwest regions as an example, from the situation of the new cross-regional DC outgoing channel, it is clear that the application of electric power new energy technology is very good, not only covering a wide area, but also the good effect of electric energy transmission. According to the relevant statistics, the capacity of the electricity from Zalut to Qingzhou is 8 MW, which can last for 6,000 hours and transmit up to 48 (TW-h); the capacity of the electricity from Jiuquan to Hunan is 8 MW, which can last for 6,000 hours and transmit up to 48 (TW-h); the capacity of the electricity from the east Junggar basin to the south of Anhui province is 12 MW, which can be utilized for 6,000 hours and deliver 72 (TW-h); the capacity of the electricity from Shanghaimiao to Shandong is 10 MW, which can be utilized for 6,000 hours and deliver 60 (TW-h); the capacity of the electricity from Ningdong to Zhejiang is 8 MW, which can be utilized for 6,000 hours and deliver; the capacity from Ningdong to Zhejiang is 8 MW, which can be utilized for 6,000 hours and deliver up to 48 (TW-h).North-east has led to an accelerated pace of construction of pumped storage power stations, which have reached 4.1 MW in terms of current installed capacity, and a much higher power regulation capacity [5].

4.3 Enhanced demand-side management

To improve the efficiency of the use of new energy must vigorously promote the replacement of traditional electricity by new energy, which requires strengthening demand-side management according to the actual electricity needs. During the operation of the power grid in the Northeast and Northwest regions, where traditional energy is the main source, if new energy generation is used to replace traditional energy, the new increase in electricity consumption will be significantly increased, even up to 80TW. Good results are obtained for the grid operation in the Northeast and Northwest regions until 2020. The number of electric vehicles is estimated by investigating the use of new energy sources. In comparison with conventional energy sources, new energy sources can supply energy for more than 600,000 electric vehicles. In carrying out this work, demand-side response mechanisms can be introduced into the mix to implement new energy technology management from a demand perspective, where 10% of the total load can be involved in the response [12].

5 Conclusion

It is clear from the above study that traditional energy sources consume a lot of energy and pollute the environment compared to new energy sources. New energy, as a renewable energy source, replaces traditional energy sources and can effectively alleviate the energy crisis. Moreover, new energy can ensure the continuity of electricity supply and generate sufficient electricity to meet the growing demand for electricity, which includes maintaining people's normal production and life, while the cost of energy use is reduced accordingly. This can create higher economic benefits, while obtaining good social and environmental benefits. However, there are disadvantages in the application of new energy. In the face of the existence of the problem of consumption, it is necessary to increase the financial investment in new energy generation technology, from a practical point of view to effectively solve this aspect of the problem, in order to promote the sustainable and healthy development of the industry.

References

1. Mao Chenyi.(2019) Introduction to the key factors of new energy consumption in the power system and measures to solve them[J]. 000(4):296-297. https://www.g3mv.com/ journal/2114
2. Wang Yunzhu.(2020).Analysis of restrictive factors on renewable energy consumption in China and counter-measures. Coal Economic Research(02),4-11. doi:10.13202/j.cnki. cer.2020.02.002.
3. Wang Jingran, Wang Dongsheng, Shi Guirong, Sun Rongfu, Cui Zhengpai, Sun Yamin.(2020).Data Application Architecture and Key Technologies of Intelligent Dispatching for Improving Renewable Energy Accommodation. Proceedings of the CSU-EPSA(11),88-94. doi:10.19635/j.cnki.csu-epsa.000515.
4. Fan Xiaochao, Wang Weiqing, Cheng Zhijiang, Xu Xuetao, Pei Jian.(2021-03-15). Research on capacity enhancement measures and control strategies for large-scale new energy consumption.2023-05-17,https://kns.cnki.net/KCMS/detail/detail.aspx?dbname=SNAD&filename=SNAD000001867873.
5. Zhang Xingyou, Han Deshun, Ma Jie, Yuan Shuai, Wang Xiaobo, Sun Shumin, Wang Nan. (2019).Analysis of Influence Factors and Countermeasures for Renewable Energy Usage in Terminal Power Grid. Journal of University of Jinan (Science and Technology)(01),68-72. doi:10.13349/j.cnki.jdxbn.2019.01.012.
6. Sheng Siqing, Qiu Hao, Zhang Wenzhao, Zhang Zhenyu, Wang Jili, Ke Xianbo. (2019). Comprehensive assessment method of new energy consumption considering steady and dynamic active power equilibrium constraints. Power System Protection and Control(10),51-56. doi:10.19783/j.cnki.pspc.180705.
7. Qiu Weiqiang, Wang Maochun, Lin Zhenzhhi, Yang Li, Wang Lianfang, Sun Jianying. (2021).Comprehensive evaluation of shared energy storage towards new energy accommodation scenario under targets of carbon emission peak and carbon neutrality. Electric Power Automation Equipment(10),244-255. doi:10.16081/j.epae.202110017.

https://doi.org/10.1051/e3sconf/202344103003

8. Wang Ding, Shen Yangwu, Shao Zhu, Xu Ming, Cao Wei, Zuo Jian.(2021).Quantitative Analysis on Key Influencing Factors of New Energy Consumption in Hunan Power Grid. HUNAN ELECICPOWER(02),6569.https://kns.cnki.net/kcms2/article/abstract?v=3uoqIhG8C44YLTIOAiTRKibYlV5Vjs7iy_Rpms2pqwbFRRUtoUImHSPtX8nzccjT_gGkMxzD9HWCSnzWh4Ag7cdFlccWVL&uniplatform=NZKPT
9. Wang Jili, Xue Fei, Huang Yuxiong, Li Hongqiang, Li Gengfeng.(2022).Data-driven Intelligent Identification Method of Hindering Factors for Renewable Energy Consumption in Regional Power Grids.SmartPower(10),95101http://zhdl.paperopen.com/oa/darticle.aspx?type=view&id=2022 10013
10. Su Changsheng, Wang Sen, Sun Yiqian, Li Fengting.(2022). Refined evaluation of consumption resistance of system with high proportion of renewable energy for network nodes. Electric Power Engineering Technology(05),67-75.doi: 10.12158/j.2096-3203. 2022.05.008
11. Cong Yu, Liu Yu, Wang Liqiang, Yuan Shuai.(2022).Analysis on Influence Factors and Promotion Methods of Power Control Capability in New Energy Stations and Its Impoving Method. INNER MONGOLIA ELECTRIC POWER(02),1-6. doi:10.19929/j.cnki. nmgdljs. 2022.0021.
12. Chen Wenjing.(2022).Analysis and Research of Deep Peak Shaving of Thermal Power Generating Units in the Context of Renewable Energy Accommodation.METALLURGICALPOWER(05), 5-8.doi:10.13589/j.cnki.yjdl.2022.05.004.
13. Ashikuzzaman, A.K.M.; Hassan, Md. Shakil; Ahmed, Md. Muhtasim. Introduction of Fractal Hybrid Layout in Concentrated Solar Power Technology. SSRN. 2022, 1556-5068.
14. Y. Zhang, W. Ningbo, K. Ding, Q. Zhou, P. Gao and Z. Zhang, "The Key Technology of the Coordinated Control System of Wind Power and Energy-intensive Load," 2019 4th International Conference on Intelligent Green Building and Smart Grid (IGBSG), Hubei, China, 2019, pp. 349-353, doi: 10.1109/IGBSG.2019.8886294.

Research on Design and Planning Method of Energy Internet Scenarios Under New-type Urbanization

Chen Zhang[1,a], Guangda Wang[1,b*], Zhongyao Jiang[2,c], Zhouyang Wu[1,d], Yufeng Chai[1,e], Wei Dai[2,f]

[1]State Grid Energy Research Institute, Beijing, China
[2]College of Electrical Engineering, Guangxi University, Nanning, China

Abstract—In the general context of urbanization and energy transformation, energy development under new-type urbanization is faced with typical common issues such as a low share of clean energy, high energy intensity and high carbon intensity. To address these issues, this paper, focused on the design and planning method of scenarios in the energy interconnection systems under new-type urbanization, constructs the allocation of physical elements in the four aspects of energy supply, conversion, transmission and storage, establishes the architecture of energy interconnection systems and proposes a 1+N scenario construction method suitable for the diverse scenarios under new-type urbanization. In line with the energy development trends of interconnection, efficiency, low-carbon, digitization and diversification, this paper proposes a planning method of energy interconnection systems that adapts to diverse scenarios under new-type urbanization and verifies the proposed method with a practical computing example in a location. It provides valuable references for the construction of differentiated energy interconnection systems under new-type urbanization.

1. Introduction

Cities and towns are the core of energy supply and consumption. Since the reform and opening up, China has witnessed rapid urbanization. With increasing energy demand, energy security problems caused by heavy dependence on the import of oil, natural gas and other energy sources have brought a series of new problems and challenges to the urbanization development in China [1]. New-type urbanization, as its concept suggests, refers to the coordinated development and mutual promotion of small, medium, and large cities, small towns, and new rural communities [2]. Therefore, the construction of energy systems under new-type urbanization should be based on the basic guideline of green, low-carbon and sustainable development to improve the efficiency of energy utilization and fully leverage the natural advantages of renewable energy.

Under the macro constraints of energy transformation and the new urbanization strategy, urban energy systems will be stimulated to under great changes. With the deep integration of power energy and information technology and with the gradual maturity of market mechanisms, the future energy development under new-type urbanization will navigate in five directions: interconnection, efficiency, low-carbon, digitization and diversification [3]. Due to the lack of energy interconnection pathways, traditional urban energy systems struggle with problems such as low comprehensive energy efficiency, a low share of clean energy, and insufficient interaction between supply and demand [4]. At present, the energy Internet, which attracts the attention of experts and scholars at home and abroad, marks an important direction for the future energy development in China. With the aim to create a comprehensive service system for smart energy, an energy Internet ecosystem should be built to deeply integrate the Internet with the production, transmission, storage and consumption of energy and with the energy market, so that the resulting new development pattern of the energy industry is characterized by multi-energy coordination and open trading. This is an important technical solution to deal with the energy problems arising in the urbanization process of China [5-7]. At the same time, the resulting energy Internet technology under new-type urbanization is also an important part of the energy Internet ecosystem to promote the integration and unification of energy, information and business flows and to enhance the safe, economic and efficient use of energy in urban areas [8].

Planning is the basis for transforming energy interconnection systems from theory to practice, and it plays an important role in promoting the interconnection and close

[a] zhangchen1@sgeri.sgcc.com.cn, [b*] wgd_chongqing@163.com, [c] vit41502@163.com, [d] wuzhouyang@sgeri.sgcc.com.cn,
[e] chaiyufeng@sgeri.sgcc.com.cn, [f] weidai@gxu.edu.cn

 https://doi.org/10.1051/e3sconf/202344103004

interaction of various energy sources [9]. At present, the energy interconnection systems under new-type urbanization are characterized by complex integrated modeling, multifaceted uncertainties, and frequent interactions between supply and demand sides. Traditional planning methods fail to provide comprehensive and coordinated planning of energy systems under new-type urbanization. Meanwhile, in order to ensure the efficient, reliable and stable operation of the energy interconnection systems, the characteristics mentioned above also pose challenges on the research and development of different energy coupling devices and of information and communication technology devices [10].

Therefore, this paper proposes the design and planning method of energy Internet scenarios under new-type urbanization, and verifies the proposed method with a practical example in a location. The rest of the paper is organized as follows: Section II describes the implication and basic architecture of the energy Internet under new-type urbanization and provides a 1+N scenario construction model for it; Section III introduces a modeling method of energy Internet planning that adapts to multiple scenarios under new-type urbanization; Section IV verifies the proposed method with a practical example in a location; Section V ends with the main conclusions.

2. Overview and Basic Architecture of Energy Internet under New-type Urbanization

2.1 Implication of the Energy Internet under New-type Urbanization

Urban energy Internet is a complex urban energy system with multiple network flows rooted in electric systems, based on the Internet and other cutting-edge information technologies, with distributed clean energy as the main primary energy sources, and closely coupled with other systems such as the natural gas network, the transportation network, and the heat/cooling supply network [11]. The supply side of the urban energy Internet includes not only primary energy sources but also distributed energy sources that have accessed, such as solar and wind energy; the transmission side must integrate a variety of industrial control and network information technologies to support efficient energy distribution; conversion is the essence of the energy Internet, and conversion between energy sources has gradually become a reality thanks to the development of technology, which is more favorable to the optimal distribution of energy. On the load application side, the access of active loads such as energy storage and electric vehicles results in load randomness and uncertainty - both temporally and spatially [12].

2.2 Architecture of Energy Internet under New-type Urbanization

The overall architecture of the urban energy Internet can be divided into a physical network layer of multi-energy interconnection, a support layer of energy information fusion, and a service layer of energy operation and management. As shown in Fig. 1, the physical network layer of multi-energy interconnection is a multi-energy network centered on power grids, with various forms of energy deeply integrated. It is strong in structure, flexible in operation, sharing and interactive, laying the material basis for the energy Internet under new-type urbanization. The support layer of energy information fusion is an energy communication network covering the urban energy systems in all the regions and fields, with high bandwidth, high speed, high reliability and high compatibility, providing intelligent support for the energy Internet under new-type urbanization. The service layer of energy operation and management is a comprehensive energy service management platform covering various energy sources in cities to realize the horizontal interaction between different energy systems and the vertical interaction between energy service providers and users, acting as a service management hub for energy Internet under new-type urbanization.

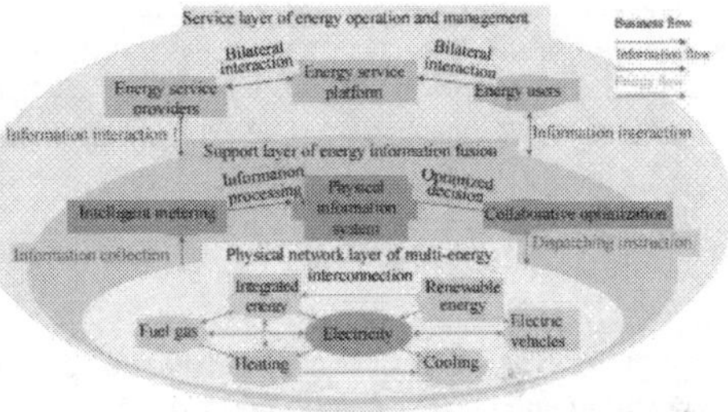

Figure 1. Architecture of Energy Internet under New-type Urbanization

2.3 1+N Scenario Construction Model in Energy Internet under New-type Urbanization

"1" represents the energy supply-demand relationship under new-type urbanization that is unique. The urban energy supply-demand relationships include energy input, energy output, and energy self-balance. For specific new-type cities and towns, the unique supply-demand relationship depends on the energy resource endowment, different power sources and energy consumption.

"N" represents the energy demand characteristics of each functional cluster in new-type cities and towns. For example, a city or a town may be divided by industrial orientation into residential, commercial, industrial, logistics and many other functional zones each of which has different energy consumption characteristics. Specifically, southern and northern residential areas as well as commercial areas may be differed in energy consumption. The energy consumption characteristics in the industrial zone can be classified by industrial types. An urban energy interconnection system can be planned as a cluster of individual functional zones or as a multi-functional cluster or a random cluster.

For example, for the purpose of planning the energy Internet system in a town, the local energy supply-demand relationship is assessed and analyzed and it is concluded that this is an energy input town that meets residential, commercial and industrial energy demand in the north. According to the "1+N" scenario construction model, the town falls into the scenario of "residential, commercial and industrial clusters in the north based on energy input".

3. Modeling of Energy Internet Planning under New-type Urbanization

In order to build a clean, low-carbon, safe and efficient modern energy system, the objective function of the model is set as a comprehensive optimum of economic, environmental, energy efficiency and electrification rate objectives, and the weight factors of the objectives are balanced against the practical development needs of the planned region, taking into account the localized development focus while seeking comprehensive development. The economic objective is to minimize the total investment and the total operating cost; the environmental objective is to minimize the total carbon dioxide emissions; the energy efficiency objective is to maximize the total energy efficiency of the system; and the electrification rate objective is to maximize the proportion of electric energy in the planned region.

3.1 Objective Function

3.1.1 Economic Objective: To Minimize the Total Investment and the Total Operating Cost of the Energy Interconnection System

$$\min F = C_{INV} + C_{OPE} \tag{1}$$

In which: C_{INV} is the total investment cost, and C_{OPE} is the total operating cost.

3.1.1.1 Investment Cost

The investment cost includes the average annual construction cost of alternative lines for the electricity, gas, and heating subsystems, the average annual construction cost of alternative equipment in the energy conversion process, and the average annual construction cost of distributed power and energy storage equipment.

$$\begin{aligned} C_{INV} = & \sum_{(i,j)\in\Omega_{EL}} x_{ij}^{EL} C_{ij}^{EL} + \sum_{(i,j)\in\Omega_{GL}} x_{ij}^{GL} C_{ij}^{GL} + \sum_{(i,j)\in\Omega_{HL}} x_{ij}^{HL} C_{ij}^{HL} \\ & + \sum_{k\in\Omega_{EnH}} x_k^{EnH} C_k^{EnH} + \sum_{l\in\Omega_{DG}} x_l^{DG} C_l^{DG} + \sum_{m\in\Omega_{ES}} x_m^{ES} C_m^{ES} \end{aligned} \tag{2}$$

In which: Ω_{EL}, Ω_{GL}, Ω_{HL} and Ω_{EnH} are the sets of alternative lines for the power distribution network, the natural gas subsystem and the heating subsystem and the set of alternative equipment in the energy hub, respectively; x_{ij}^{EL}, x_{ij}^{GL} and x_{ij}^{HL} are the token 0-1 variables of alternative lines between nodes in the power distribution network, the natural gas subsystem and the heating subsystem, respectively, 0 for no alternative line while 1 for the availability of alternative lines; C_{ij}^{EL}, C_{ij}^{GL} and C_{ij}^{HL} are the average annual construction costs of the alternative lines between Node i and Node j in the power distribution network, the natural gas subsystem and the heating subsystem, respectively; x_k^{EnH} is the token 0-1 variable of certain alternative energy conversion equipment in the energy hub, 0 for no alternative equipment while 1 for the availability of alternative equipment; C_k^{EnH} is the average annual construction costs of each alternative equipment for energy conversion, including: gas generator (GG), combined heat and power (CHP), combined cooling, heat and power (CCHP), gas boiler (GB), electric heater (EH), heat pump (HP), electric cooler (EC), absorption chiller (AC) and power-to-gas (P2G). Ω_{DG} and Ω_{ES} are the sets of alternative distributed power sources and the set of alternative energy storage, respectively; x_l^{DG} and x_m^{ES} are the token 0-1 variables of alternative distributed power sources and alternative energy storage, respectively, 0 for no alternative distributed power source and alternative energy storage while 1 for the availability of alternative distributed power source and alternative energy storage; C_l^{DG} and C_m^{ES} are the average annual construction costs of the corresponding distributed power source and energy storage equipment, respectively.

3.1.1 2 Operating Cost

The operating cost includes the cost of purchased electricity for the power distribution network, the cost of purchased gas for the gas distribution network, the cost of purchased heat for the heat distribution network, and the incentive cost of demand response.

$$C_{OPE} = \sum_s \sum_t \left(\begin{aligned} & \lambda_e \sum_{i\in\Omega_{esta}} SP_i^{E,ts} + \lambda_g \sum_{i\in\Omega_{gsta}} SP_i^{G,ts} + \lambda_h \sum_{i\in\Omega_{hsta}} SP_i^{H,ts} \\ & + \mu_{drp} \sum_{i\in\Omega_{drp}} P_i^{drp,ts} + \mu_{drh} \sum_{i\in\Omega_{drh}} P_i^{drh,ts} + \mu_{drc} \sum_{i\in\Omega_{drc}} P_i^{drc,ts} \end{aligned} \right) \tag{3}$$

In which: Ω_{esta}, Ω_{gsta} and Ω_{hsta} are the sets of substations, gas distribution stations and heat distribution stations, respectively; $SP_i^{E,ts}$, $SP_i^{G,ts}$ and $SP_i^{H,ts}$ are the energy purchased from the substations, gas distribution stations and heat distribution stations at t (hour) in s (season), respectively; λ_e, λ_g and λ_h are the unit prices of the purchased electricity, gas and heat, respectively. Ω_{drp}, Ω_{drh} and Ω_{drc} are the sets of electricity nodes, heat energy nodes and cooling energy nodes with demand response potential, respectively; $P_i^{drp,ts}$, $P_i^{drh,ts}$ and $P_i^{drc,ts}$ are the electric power, the heat power and the cool power at which the demand response is reduced at t (hour) in s (season), respectively; μ_{drp}, μ_{drh} and μ_{drc} are the unit incentive coefficients of electric, heating and cooling demand response, respectively.

3.1.2 Environmental Objective: To Minimize the Total Carbon Dioxide Emissions

$$\min F_{EN} = \sum_t (\rho_{CO2}^{GB} PG_i^{GB,t} + \rho_{CO2}^{GG} PG_i^{GG,t} + \rho_{CO2}^{CHP} PG_i^{CHP,t} + \rho_{CO2}^{CCHP} PG_i^{CCHP,t}) \tag{4}$$

In which: $PG_i^{GB,t}$, $PG_i^{GG,t}$, $PG_i^{CHP,t}$ and $PG_i^{CCHP,t}$ are the power of GB, GG, CHP and CCHP at t, respectively; ρ_{CO2}^{GB}, ρ_{CO2}^{GG}, ρ_{CO2}^{CHP} and ρ_{CO2}^{CCHP} are the proportionality coefficients of the power of GB, GG, CHP and CCHP to carbon dioxide emissions, respectively.

3.1.3 Energy Efficiency Objective: To Maximize the Total Energy Efficiency of the System

$$\min F_{EF} = \sum_{i \in \Omega_{esta}/\Omega_{gsta}/\Omega_{hsta}} \frac{LP_i^{E,ts} + LP_i^{G,ts} + LP_i^{H,ts} + LP_i^{C,ts}}{SP_i^{E,ts} + SP_i^{G,ts} + SP_i^{H,ts}} \tag{5}$$

In which: $SP_i^{E,ts}$, $SP_i^{G,ts}$ and $SP_i^{H,ts}$ are the energy purchased from the substations, gas distribution stations and heat distribution stations at t (hour) in s (season), respectively, and $LP_i^{E,ts}$, $LP_i^{G,ts}$, $LP_i^{H,ts}$ and $LP_i^{C,ts}$ are the electric, gas, heat, and cool energy consumed by users at t (hour) in s (season), respectively.

3.1.4 Electrification Rate Objective: To Meet the Required Proportion of Electric Energy

$$\min F_{EF} = -\frac{SP_i^{E,ts}}{SP_i^{E,ts} + SP_i^{G,ts} + SP_i^{H,ts}} \tag{6}$$

In which: $SP_i^{E,ts}$, $SP_i^{G,ts}$ and $SP_i^{H,ts}$ are the energy purchased from the substations, gas distribution stations and heat distribution stations at t (hour) in s (season), respectively.

3.2 Constraints

The constraints for the planning model of the energy interconnection system under new-type urbanization can be divided into three categories based on the functional modules: energy conversion constraint, energy storage constraint, and energy transmission constraint. Each module is an integral part of the energy interconnection system under new-type urbanization, covering the operating constraints in the whole chain of power supply, grid, load and energy storage and various types of equipment. The energy conversion constraints mainly include the constraint of cooling, heat and electric power balances in the energy hub, the capacity constraint of equipment in the energy hub and the distributed power constraint. The energy storage constraint is mainly the power balance constraint constructed on the basis of the functional characteristics of storage devices. The energy transmission constraints mainly include the power balance constraints, transmission capacity constraints, and demand response characteristics of the distribution network, the gas distribution network and the heat distribution network.

3.2.1 Energy Conversion Devices Constraints

3.2.1.1 Constraint of Cooling, Heat and Electric Power Balances in Energy Hub

$$\begin{cases} LP_i^{C,ts} = \eta_i^{EC} P_i^{E,EC,ts} x_{i,EC}^{EnH} + \eta_i^{AC} P_i^{H,AC,ts} x_{i,AC}^{EnH} \\ \quad + \eta_i^{CCHP} P_i^{G,CCHP,ts} x_{i,CCHP}^{EnH} \\ LP_i^{H,ts} = \eta_i^{GB} P_i^{G,GB,ts} x_{i,GB}^{EnH} + \eta_i^{CHP,H} P_i^{G,CHP,ts} x_{i,CHP}^{EnH} \\ \quad + \eta_i^{CCHP,H} P_i^{G,CCHP,ts} x_{i,CCHP}^{EnH} + \eta_i^{EH} P_i^{E,EH,ts} x_{i,EH}^{EnH} \\ \quad + \eta_i^{HP} P_i^{E,HP,ts} x_{i,HP}^{EnH} \\ LP_i^{E,ts} = \eta_i^{GG} P_i^{G,GG,ts} x_{i,GG}^{EnH} + \eta_i^{CHP,E} P_i^{G,CHP,ts} x_{i,CHP}^{EnH} \\ \quad + \eta_i^{CCHP,E} P_i^{G,CCHP,ts} x_{i,CCHP}^{EnH} \\ LP_i^{G,ts} = \eta_i^{P2G} P_i^{G,P2G,ts} x_{i,P2G}^{EnH} \end{cases} \tag{7}$$

In which: $LP_i^{C,ts}$, $LP_i^{H,ts}$, $LP_i^{E,ts}$ and $LP_i^{G,ts}$ are the cooling load, heat load, electric load and gas load of the energy hub i at t (hour) in s (season), respectively; $P_i^{E,EC,ts}$ and η_i^{EC} are the input electric power of the electric cooler in the energy hub i at t (hour) in s (season) and the efficiency of the electric cooler to convert electric energy into cooling energy, respectively; $x_{i,EC}^{EnH}$ is the token 0-1 variable of the electric cooler, 0 for no electric cooler while 1 for availability of the electric cooler. The variables for the other equipment are similar and not repeated herein.

3.2.1.2 Capacity Constraint of Equipment in Energy Hub

$$\begin{cases} 0 \le \eta_i^{GG} P_i^{G,GG,ts} \le P_i^{GG\max} x_{i,GG}^{EnH} \\ 0 \le (\eta_i^{CHP,H} + \eta_i^{CHP,E}) P_i^{G,CHP,ts} \le P_i^{CHP\max} x_{i,CHP}^{EnH} \\ 0 \le (\eta_i^{CCHP,H} + \eta_i^{CCHP,E} + \eta_i^{CCHP,C}) P_i^{G,CCHP,ts} \le P_i^{CCHP\max} x_{i,CCHP}^{EnH} \\ 0 \le \eta_i^{GB} P_i^{G,GB,ts} \le P_i^{GB\max} x_{i,GB}^{EnH} \\ 0 \le \eta_i^{EH} P_i^{E,EH,ts} \le P_i^{EH\max} x_{i,EH}^{EnH} \\ 0 \le \eta_i^{HP} P_i^{E,HP,ts} \le P_i^{HP\max} x_{i,HP}^{EnH} \\ 0 \le \eta_i^{EC} P_i^{E,EC,ts} \le P_i^{EC\max} x_{i,EC}^{EnH} \\ 0 \le \eta_i^{AC} P_i^{H,AC,ts} \le P_i^{AC\max} x_{i,AC}^{EnH} \\ 0 \le \eta_i^{P2G} P_i^{E,P2G,ts} \le P_i^{P2G\max} x_{i,P2G}^{EnH} \end{cases} \tag{8}$$

In which: $P_i^{GG\max}$, $P_i^{CHP\max}$, $P_i^{CCHP\max}$, $P_i^{GB\max}$, $P_i^{EH\max}$, $P_i^{HP\max}$, $P_i^{EC\max}$, $P_i^{AC\max}$ and $P_i^{P2G\max}$ are the maximum output energy of GG, CHP, CCHP, gas turbine, EH, HP, EC, AC and P2G in the energy hub i, respectively.

3.2.1.3 Distributed Power Constraint

$$0 \le P_i^{DG,ts} \le \sigma_i^{DG,ts} P_i^{DG\max} x_i^{DG} \tag{9}$$

In which: $P_i^{DG\max}$ is the maximum active power output by the distributed power source at Node i, and $\sigma_i^{DG,ts}$ is the output coefficient of the distributed power source at t (hour) in s (season).

3.2.2 Energy Storage Device Constraint

$$P_i^{ES,n} = P_i^{ESout,n} - P_i^{ESin,n} \tag{10}$$

$$SOC_i^{n+1} = SOC_i^n + \varepsilon_i^{loss} P_i^{ESin,n} - P_i^{ESout,n} \tag{11}$$

$$SOC_i^0 = SOC_i^N \tag{12}$$

$$-P_i^{ES\max} x_i^{ES} \le P_i^{ES,n} \le P_i^{ES\max} x_i^{ES} \tag{13}$$

$$0 \le P_i^{ESin,n} \le P_i^{ES\max} x_i^{ES} \tag{14}$$

$$0 \le P_i^{ESout,n} \le P_i^{ES\max} x_i^{ES} \tag{15}$$

$$SOC_i^{\min} \le SOC_i^n \le SOC_i^{\max} \tag{16}$$

In which: n is a time period, and N is the total number of time periods in a dispatching cycle which is 24h in a day. $P_i^{ES,n}$ is the net energy discharge power; $P_i^{ESin,n}$ is the energy charge power; $P_i^{ESout,n}$ is the energy discharge power; SOC_i^n is the stored energy; ε_i^{loss} is the energy loss rate during charging.

3.2.3 Energy Transmission Grid Constraints

3.2.3.1 Power Distribution Network Constraints

① Node Balance Constraint

$$\sum_{(i,j)\in\Omega_{GL}} P_{ji}^{G,ts} + SP_i^{G,ts} = \sum_{(i,k)\in\Omega_{GL}} P_{ik}^{G,ts} + LP_i^{G,ts} + P_i^{G,ts} \tag{17}$$

In which: $P_{ik}^{G,ts}$ is the energy distributed by the gas distribution network ik at t (hour) in s (season); $SP_i^{G,ts}$ is the energy injected from the gas distribution station to Node i at t (hour) in s (season); $LP_i^{G,ts}$ is the gas load at Node i of the gas distribution network at t (hour) in s (season); $P_i^{G,ts}$ is the energy input into the energy hub from the gas distribution network at t (hour) in s (season).

② Gas Distribution Pipe Capacity Constraint

$$0 \le P_{ij}^{G,ts} \le P_{ij}^{G\max} x_{ij}^{GL} \tag{18}$$

In which: $P_{ij}^{G\max}$ is the maximum energy transmitted by the gas distribution pipe ij.

③ Gas Distribution Station Output Constraint

$$0 \le SP_i^{G,ts} \le SP_i^{G\max} \tag{19}$$

In which: $SP_i^{G\max}$ is the maximum energy input to Node i of the gas distribution network.

3.2.3.2 Heat Distribution Network Constraints

① Node Heat Balance Constraint

$$\sum_{(i,j)\in\Omega_{HL}} P_{ji}^{H,ts} + SP_i^{H,ts} = \sum_{(i,k)\in\Omega_{HL}} P_{ik}^{H,ts} + LP_i^{H,ts} + P_i^{H,ts} - P_i^{DR,h,ts} \tag{20}$$

In which: $P_{ik}^{H,ts}$ is the heat transmitted by the heat distribution network pipe ik at t (hour) in s (season); $SP_i^{H,ts}$ is the heat injected into Node i from the heat distribution station at t (hour) in s (season)s; $LP_i^{H,ts}$ is the heat load at Node i in the heat distribution network at t (hour) in s (season); $P_i^{H,ts}$ is the heat input into the energy hub from the heat distribution network at t (hour) in s (season); $P_i^{DR,h,ts}$ is the heat power for demand response reduction at t (hour) in s (season).

② Heat Distribution Pipe Capacity Constraint

$$0 \le P_{ij}^{H,ts} \le P_{ij}^{H\max} x_{ij}^{HL} \tag{21}$$

In which: $P_{ij}^{H\max}$ is the maximum energy transmitted by the heat distribution pipe ij.

③ Heat Distribution Station Output Constraint

$$0 \le SP_i^{H,ts} \le SP_i^{H\max} \tag{22}$$

In which: $SP_i^{H\max}$ is the maximum heat output by the heat distribution station at Node *i*.

④ Demand Response Constraint

$$0 \le P_i^{DR,h,ts} \le (P_i^{DR,h,ts})_{\max} \tag{23}$$

In which: $(P_i^{DR,h,ts})_{\max}$ is the maximum heat power for demand response at Node *i*.

3.3 Model Solution Process

The solution process of the multi-objective optimal programming model built for the energy interconnection system under new-type urbanization in this section is shown in Fig. 2. The model is built with the YALMIP language, transformed into a mixed integer linear programming model suitable for solution through linearization of the nonlinear model, and solved on MATLAB 2017b with the commercial Cplex solver.

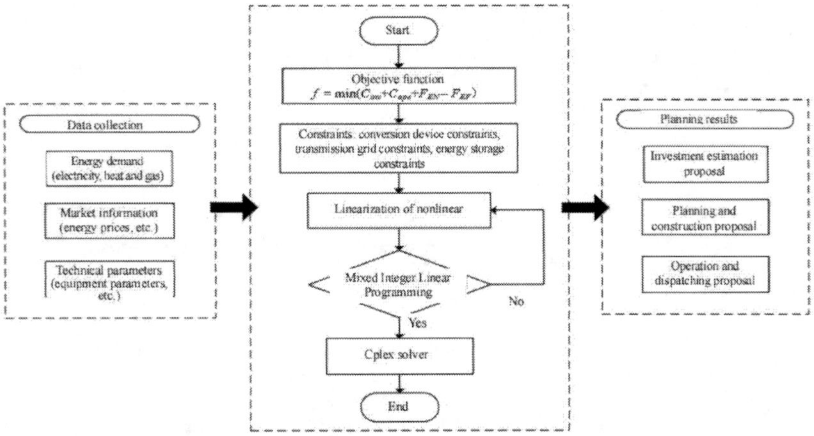

Figure 2. Solution Process of the Overall Planning Model

4. Example Test

4.1 Setting of Background Parameters in the Example

According to the definition of the "1+N" scenarios, "N" in the example is set as a functional cluster of "residential, commercial and industrial areas" where the energy consumption characteristics in the residential area are consistent with the basic characteristics in southern cities. Different scenarios are set according to the abundance of external gas and power supplies. When the external gas supply is insufficient but the local gas supply is abundant and the local power supply is sufficient, the condition for exporting energy is met. In this case, the scenario is defined as energy output. When the external gas supply is limited and the external power supply is insufficient but the local power supply is sufficient, power energy is considered as the main local energy and sufficient. In this case, the scenario is defined as energy self-balance, as shown in Table 1.

Table 1 External Power and Gas Supply Capacity by Scenario

Scenario Setting	External Power Supply Capacity	External Gas Supply Capacity	Type of Energy Supply Scenario	Type of Energy Consumption Scenario
Scenario 1	Abundant	Abundant	Energy input	Residential, commercial and industrial areas
Scenario 2	Abundant	Short external supply, but local abundant supply	Energy output	Residential, commercial and industrial areas
Scenario 3	Short external supply, but local abundant supply	Moderate	Energy self-balance	Residential, commercial and industrial areas

As shown in Fig. 3, the example is composed of an IEEE 30-node power distribution network system, a 7-node gas distribution network system, and a 6-node heat distribution network system. In the example, the fixed electricity price is RMB 0.6/kW·h, the natural gas price is RMB 3/m3, the purchased heat price is RMB 25/kW·h, and the construction cost of the alternative lines for the fixed power distribution network system is RMB 100,000/line.

https://doi.org/10.1051/e3sconf/202344103004

The loads in the IEEE 30-node power distribution network system are divided into loads in residential, commercial and industrial areas, while taking into account the access of EV charging piles. The typical daily curves of electricity loads are shown in Fig. 4, Fig. 5, Fig. 6 and Fig.7. The setting is that the typical days account for 360 dispatching cycles (one day for one dispatching cycle) in a year, and the typical days in each of the spring, summer, autumn and winter account for 90 dispatching cycles. The gas and heat load curves for each typical day are shown in Fig. 8 and Fig. 9, in which the gas and heat load curves for typical days in spring and autumn are integrated as one curve for the convenience of calculation because they are relatively similar. The unit investment cost and conversion efficiency of the equipment, as well as the unit investment and construction cost of the energy storage devices in the energy interconnection system are shown in Table 2, Table 3 and Table 4, respectively.

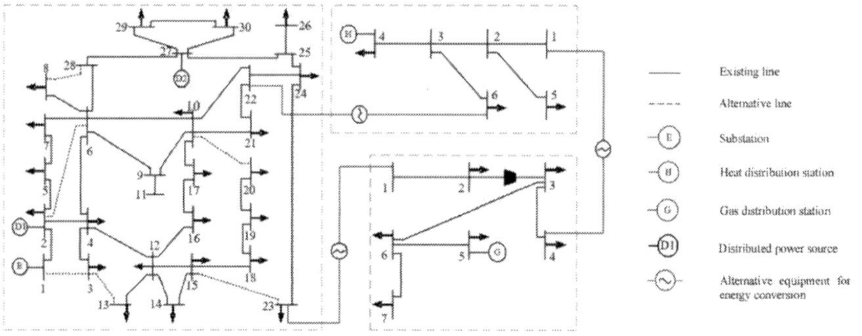

Figure 3. IEEE 30-Node Power Distribution Network - 7-Node Gas Distribution Network - 6-Node Heat Distribution Network Systems

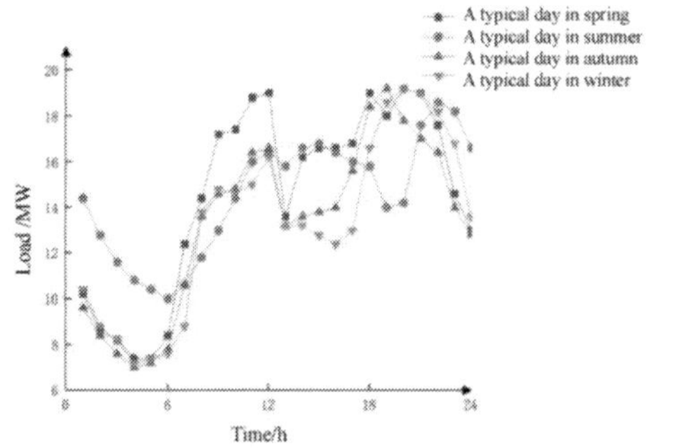

Figure 4. Electricity Load Curves of Typical Days in Residential Area

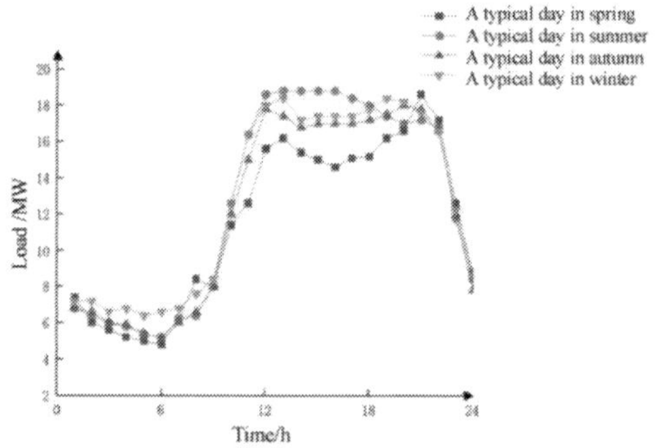

Figure 5. Electricity Load Curves of Typical Days in Commercial Area

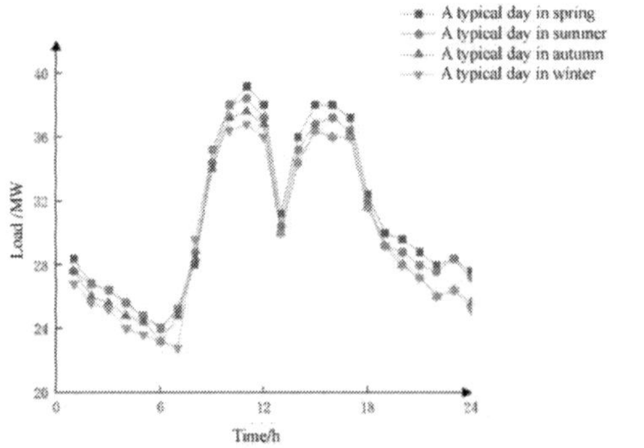

Figure 6. Electricity Load Curves of Typical Days in Industrial Area

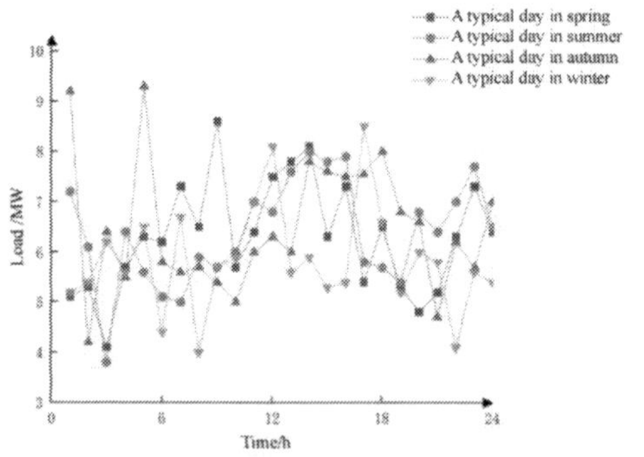

Figure 7. Electricity Load Curves of Typical Days with EV Charging Piles

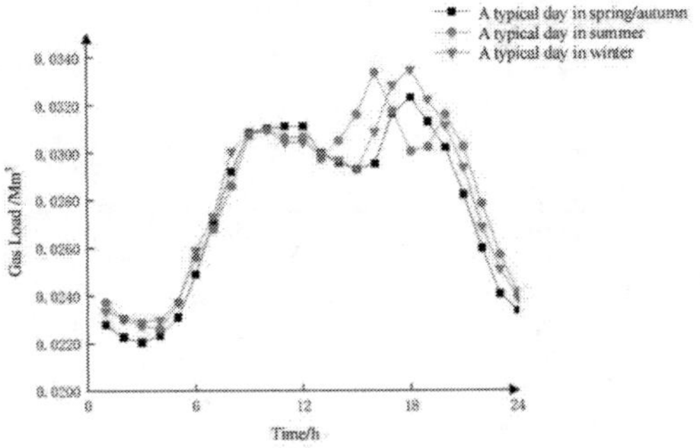

Figure 8. Gas Load Curves of Typical Days

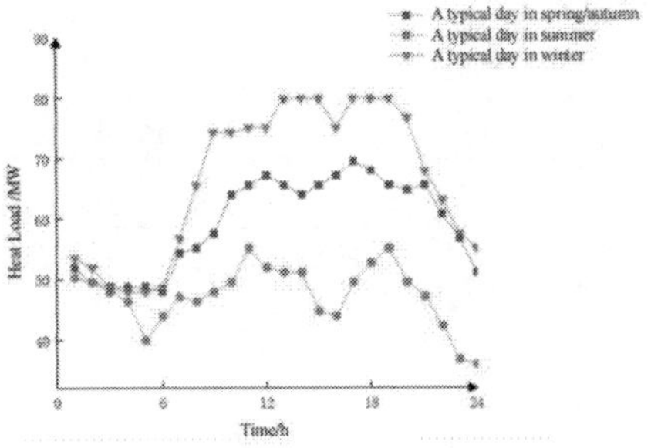

Figure 9. Heat Load Curves of Typical Days

Table 2 Unit Investment Cost of Conversion Equipment (Unit: RMB/kW)

Type	GB	GG	CHP	CCHP	EH	HP	P2G	PV Power Generation	Wind Power Generation
Investment Cost	1700	6500	6000	9800	3000	1000	3500	6500	7500

Table 3 Parameter Settings of Conversion Equipment

Type	GB	GG	CHP		CCHP		EH	HP	P2G
			η_{CHP}^{e}	η_{CHP}^{h}	η_{CCHP}^{e}	η_{CCHP}^{h}			
Conversion efficiency	0.85	0.3	0.4	0.45	0.4	0.4	3.2	3	0.7

Table 4 Unit Investment Cost of Energy Storage Equipment (Unit: RMB/kW)

Type	Power Storage	Gas Storage	Heat Storage
Investment cost	200	3.33	100

4.2 Analysis on Example Results

The overall planning model of the energy interconnection system is simulated and computed in the three scenarios shown in Table 1, and the construction of distribution lines and grids, the construction of energy conversion equipment and distributed power source capacity, the construction of different energy storage equipment and their average annual costs are shown in Table 5, Table 6, Table 7 and Table 8. As shown in the construction of power distribution network grids in Table 5, when the external power and gas supply capacity is abundant, the demand of the power distribution system for energy grid construction is minimal; when the energy supply is insufficient, the security and stability of the system network is reduced. The results in Table 6 show that power storage is the common choice in the three scenarios and operates in coordination with distributed power generation to help the system better achieve the supply-demand balance and effectively reduce the total cost of energy users. As shown in Table 7, the construction scales of HP and GG vary with the external power and gas supply capacity, mainly because they are the key energy supply elements of the electricity-centered energy Internet and the gas-centered energy Internet, respectively. In the case of short gas supply, GG construction will decrease, and P2G construction will increase correspondingly. As shown in Table 8, the total annual average cost is minimal in the three scenarios when the external power and gas supplies are sufficient. Fig. 10, Fig. 11 and Fig. 12 are diagrams of power, gas and heat systems in a typical day in winter in Scenario 1, respectively.

Table 5 New Power Distribution Lines in Three Scenarios

Power Distribution Line No.	Node i	Node j	Scenario 1	Scenario 2	Scenario 3
1	1	3	√	√	√
2	2	6		√	
3	3	13			
4	8	28	√	√	√
5	10	20	√	√	√
6	15	23			√

Table 6 Construction of Different Energy Storage Equipment (Unit: kW)

Energy Storage Type	Scenario 1	Scenario 2	Scenario 3
Power storage	30000	30000	30000
Gas storage	-	3000	-
Heat storage	-	-	-

Table 7 Construction of Energy Conversion Equipment and Distributed Power Sources (Unit: kW)

Energy Conversion Equipment	Scenario 1	Scenario 2	Scenario 3
GB	10000	10000	10000
GG	5000	2000	5000
CHP	-	-	-
CCHP	-	-	-
EH	10000	10000	10000
HP	20000	20000	10000
P2G	-	20000	-
PV power generation	20000	20000	30000
Wind power generation	20000	20000	20000

Table 8 Average Annual Costs in Three Scenarios (Unit: RMB 10,000)

Scenario Setting	Total Cost	Investment Cost	Operating Cost
Scenario 1	164040	35950	128090
Scenario 2	173780	34250	139530
Scenario 3	183910	45450	138460

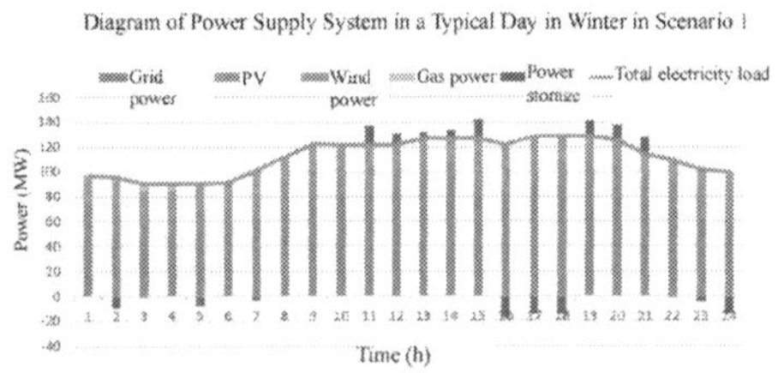

Figure 10. Diagram of Power Supply System in a Typical Day in Winter

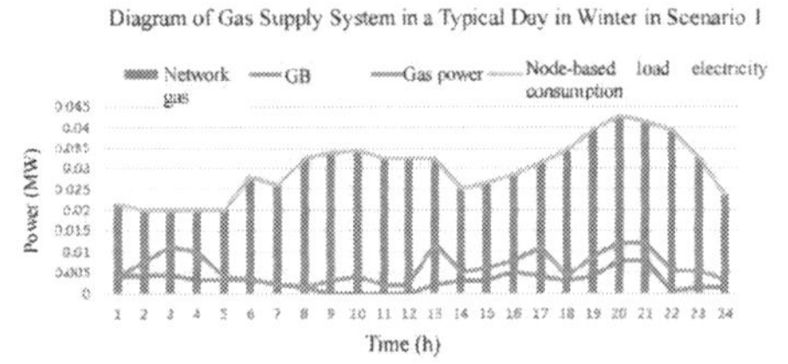

Figure 11. Diagram of Gas Supply System in a Typical Day in Winter

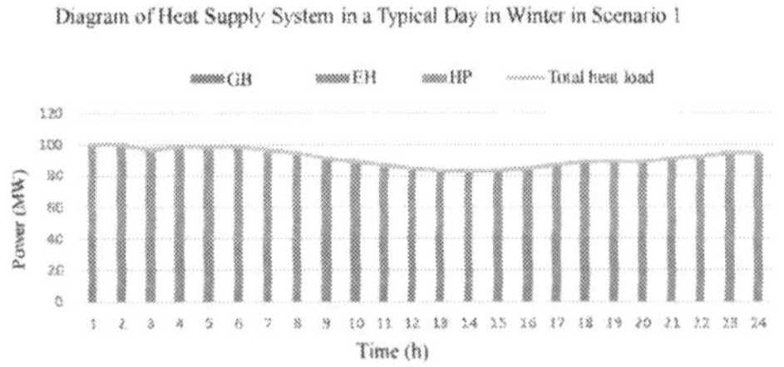

Figure 12. Diagram of Heat Supply System in a Typical Day in Winter

5. Conclusion

This paper first reviews the planning elements involved in the energy interconnection system under new-type urbanization and goes on to build a planning model system applicable to diverse energy use scenarios with the elements and scenarios as the core, with the comprehensive optimum of economic cost, environmental cost, energy efficiency and electrification rate as the aim, with the physical operational limitations of the energy interconnection system as the constraints, with diverse energy supply and consumption scenarios as the boundary conditions, and with the elements of the energy interconnection system as decision variables. It also provides a basic mathematical model, proposes the corresponding solution idea, and builds a comprehensive planning model that covers all the elements and different scenarios. Finally, the proposed method is tested in typical scenarios to prove that it can flexibly adapt to different energy supply and consumption scenarios and planning requirements and will provide general and practical value in the application of energy interconnection system planning under new-type urbanization and play an important role in supporting the high-quality and rapid advancement of new-type urbanization.

References

1. H. Gu, Y. Li, J. Yu and T. Zhang. (2020) Bi-level Combinatorial Optimal Dispatch for New-town with Multi-Energy Industrial Parks under Carbon Emission Constraints. In: 2020 4th International Conference on

https://doi.org/10.1051/e3sconf/202344103004

Smart Grid and Smart Cities (ICSGSC). Osaka. Japan. pp. 154-159.

2. Wei Li and Ming Song. (2018) Report on the Development of New Urbanization (2016): research on featured towns. Beijing. Science press.
3. Haifei Gu, Jie Yu, Yang Li, Hanlin Zhang and Yang Zheng. (2019) Bi-level Joint Optimization Economic Dispatch of New-type Town with Multi-energy Parks Under Environmental Constraints. In: Proceedings of the CSEE.
4. Jianxiao Wang, Haiwang Zhong, Ziming Ma, Qing Xia and Chongqing Kang. (2017) Review and prospect of integrated demand response in the multi-energy system. In: Applied Energy. vol. 202. pp. 772-782.
5. R. J. Flores and J. Brouwer. (2018) Optimal design of a distributed energy resource system that economically reduces carbon emissions. In: Applied Energy. vol. 232. pp. 119-138.
6. Haifei Gu, Yang Li, Jie Yu, Chen Wu, Tianli Song and Jinzhou Xu. (2020) Bi-level optimal low-carbon economic dispatch for an industrial park with consideration of multi-energy price incentives. In: Applied Energy. vol. 262.
7. Yang J, Sun W, Harrison G, et al. (2019) A novel planning method for multi-scale integrated energy system. In: 2019 IEEE Milan Power Tech. pp. 1-6.
8. Y. Wang et al., (2019) Optimal Scheduling of the Regional Integrated Energy System Considering Economy and Environment. In: IEEE Transactions on Sustainable Energy. vol. 10. no. 4. pp. 1939-1949
9. T. Ding, W. Jia, M. Shahidehpour, O. Han, Y. Sun and Z. Zhang (2022) Review of Optimization Methods for Energy Hub Planning, Operation, Trading, and Control. In: IEEE Transactions on Sustainable Energy. vol. 13. no. 3. pp. 1802-1818.
10. Amin S, Ali R. (2016) An integrated steady-state operation assessment of electrical, natural gas and district heating networks. In: IEEE Transactions on Power Systems. pp. 3636-3646.
11. Q. Wang, N. Li, X. Gu, N. Zhang, Y. Chen and G. HE. (2020) Review on the Two-layer Coordinated Architecture for Optimal Regulation and Control of Energy Interconnection System in ew-type Towns. In: 2020 12th IEEE PES Asia-Pacific Power and Energy Engineering Conference (APPEEC). Nanjing, China. pp. 1-6.
12. M. Muratori and G. Rizzoni. (2016) Residential Demand Response: Dynamic Energy Management and Time-Varying Electricity Pricing. In: IEEE Transactions on Power Systems, vol. 31, no. 2, pp. 1108-1117.

Energy-Saving Bias in Green Technology Innovation A Spatial and Temporal Analysis of manufacturing in China

Yang Liu[1,a], Feng Li[2,b], Jingjing Dai[1,c], Tixin Li[1,*]

[1]Kunming University, School of economics and management, Kunming, China
[2]Yunnan Transportation Development Project Management Co., Department of Project Management, Kunming, China

Abstract—This study presents an in-depth analysis of the energy-saving bias in green technology innovation across 30 provinces in manufacturing from 2011 to 2021, utilizing a novel Malmquist-Luenberger multidimensional decomposition index based on the directional distance function. The research reveals that green innovation, characterized predominantly by energy conservation, plays a pivotal role in driving China's green total factor productivity. The impetus for innovation in energy saving is found to surpass that of emission reduction in manufacturing enterprises. Energy-saving biased green technology innovation, originating in economically advanced provinces, has gradually expanded to the northern region, and it encompassed the majority of provinces in China. This type of innovation serves as the primary driver of regional green innovation. The study also identifies a conspicuous spatial aggregation effect of energy-saving biased green technology innovation, linked intrinsically to the degree of industrial aggregation and the spatial correlation effect of innovation.

1. Introduction

As one of the world's largest manufacturing hubs, China is at the forefront of energy-efficiency innovation challenges. The manufacturing industry, while being a significant contributor to China's GDP and employment, is also a major consumer of energy, accounting for nearly 70% of the country's total energy use [1]. Innovation-driven green industrial manufacturing is proliferating [2]. However, the choice of direction for green technology innovation is not clear, resulting in the inability to assess energy-efficient green innovation on a regional macro level. To address this issue, this paper proposes a nonparametric model to measure the energy efficiency bias of industrial manufacturing and its spatial evolution characteristics. The parametric method, which enhances the production function by deriving a comprehensive total factor productivity function [3], allows researchers to unravel the driving forces behind changes in total factor productivity. Non-parametric methods, on the other hand, focus on analyzing the factor bias of technological progress by examining the non-isometric movement of the production frontier across different time periods. This movement induces changes in the marginal output ratio of various factors, providing insights into how technological innovation tend to favor specific input factors such as capital, labor, or energy [4]–[6]. In contrast, the parametric approach has more stringent modeling assumptions and some difficulties in dealing with non-expected outputs [7]. In the context of green technology innovation, it is crucial to account for unexpected outputs and incorporate negative environmental outcomes into the analytical framework [8]. This approach enables a comprehensive analysis of the biased allocation of factors at the input stage, which is of significant importance for exploring energy-saving biases in green innovation. Due to the certain spatial association characteristics of green development [9], [10], a clear understanding of spatial evolution characteristics of green biased innovation in manufacturing will help policy-making departments to suit the remedy to the case.

This study draws on manufacturing data from 30 provinces in China, excluding Hong Kong, Macao, Taiwan, and Tibet over the period from 2010 to 2021. By calculating the degree of green technology innovation bias, decomposing the substitution effects of input factors, and obtaining the results of energy-saving bias for green innovation of each province and year, this study aims to analyze the spatial evolution characteristics of these biases.

2. Method and Model

We draw on Färe and chung's study [11], [12]by imposing a weak disposability constraint on non-desired outputs and combine it with the Malmquist-Luenberger productivity index methodology in order to compute the green technology innovation bias as follows:

$$ML_0^t(x^t,y^t,x^{t+1},y^{t+1}) = D_0^t(x^{t+1},y^{t+1})/D_0^t(x^t,y^t), t = 1,...,T-1$$
$$= \Delta T(x^{t+1},y^{t+1}) \cdot \Delta TE(x^t,y^t,x^{t+1},y^{t+1})$$

[a]2275100825@qq.com, [b]5522505@qq.com, [c]1175487921@qq.com, [*] Corresponding author: 710220459@qq.com

$$= \left[\frac{D_0^t(x^{t+1},y^{t+1})}{D_0^{t+1}(x^{t+1},y^{t+1})}\right] \cdot \left[\frac{D_0^{t+1}(x^{t+1},y^{t+1})}{D_0^t(x^t,y^t)}\right]. \quad (1)$$

$$\text{GII}\left(x^t, y^t, x^{t+1}\right) = \left[\frac{D_i^{t+1}(y^t,x^{t+1})}{D_i^t(y^t,x^{t+1})} / \frac{D_i^{t+1}(y^t,x^t)}{D_i^t(y^t,x^t)}\right] \quad (2)$$

$$\text{GIO}\left(y^t, x^{t+1}, y^{t+1}\right) = \left[\frac{D_0^t(x^{t+1},y^{t+1})}{D_0^{t+1}(x^{t+1},y^{t+1})} / \frac{D_0^t(x^{t+1},y^t)}{D_0^{t+1}(x^{t+1},y^t)}\right] \quad (3)$$

Where x^t denotes the set of input factors, y^t denotes the set of expected output factors, and the set of unexpected output factors as b^t. The ML index can be decomposed into efficiency change (ΔT) and technological change (ΔTE), as in equation (1). Further, the ΔTE index is decomposed to obtain the input bias of green technological innovation (GII) and the output bias of green technological innovation (GIO), as in equation (2) and (3).

In this paper, we further decompose GII so as to construct the substitution relationship between input factors in order to calculate the substitution index of green technology for energy, i.e., the energy-saving biased green technology innovation index (EBGI), as in equation (4).

$$EBGI_{I,E} = \left(\frac{I^{t+1}}{E^{t+1}} / \frac{I^t}{E^t} - 1\right) \times (GII - 1) \quad (4)$$

where I gauges the green technology innovation, represented by number of green patents; E refers to energy input, represented by comprehensive energy consumption per unit of industrial added value; $\frac{I^{t+1}}{E^{t+1}} / \frac{I^t}{E^t}$ is the ratio of the marginal substitution rates of factors I and E from stage t to $t+1$. $EBGI_{I,E} > 0$, it indicates that I realizes the bias of green technology innovation by substituting E, which is the energy-saving green technology innovation of factor E.

3. Data and Varialbes

The data are mainly from China Statistical Yearbook (2011-2022), China Industrial Statistical Yearbook (2011-2022), and China Environmental Statistical Yearbook (2011-2022). The variables are designed in table 1 as follows:

Table 1. Variable description

Types	Variable		
	variable name	***indicators***	***Sign***
Input	Energy	Comprehensive energy consumption per unit of industrial added value	+
	Technology	Number of invention patents of industrial enterprises above designated size	+
	Labor	Manufacturing employment	+
Output	Non-expected outputs	COD emissions per unit of industrial added value	-
		Ammonia nitrogen emissions per unit of industrial added value	-
		SO2 emission per unit of industrial added value	-
		Nitrogen oxides per unit of industrial added value	-
		Solid Waste Generation per unit of industrial added value	-
		Wastewater emissions per unit of industrial added value	-
	Expected outputs	Industrial value added as % of GDP	+

4. Results

4.1. GII and GIO index

As depicted in Fig 1, it is evident that the influence of input bias on green technology innovation was more pronounced prior to 2014. Post-2014, however, the scenario changed significantly, with the output bias of green technology innovation emerging as a key driver of green total factor productivity. Furthermore, an examination of the factor substitution bias at the input end reveals that the Green Technology Innovation Index (GII) has been greater than 1 in the majority of years, suggesting that green technology in the manufacturing sector has had a positive impact on the substitution of intergroup bias of energy and labor input factor groups.

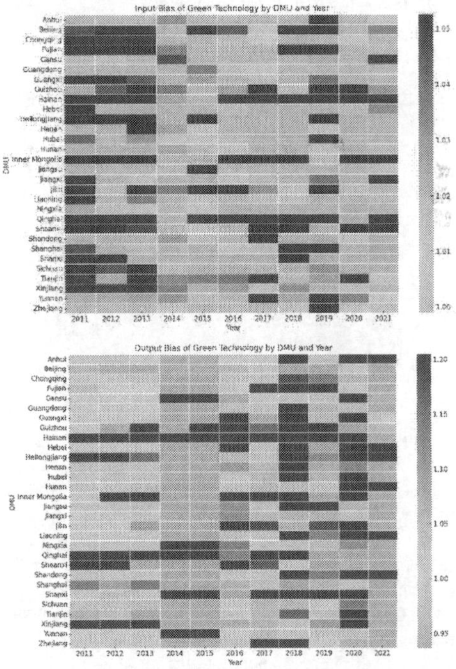

Figure 1. Heatmaps of GII and GIO

On a regional scale, there are marked differences in green technology innovation across various provinces in China. Specifically, between the years 2016 and 2020, the

Green Innovation Index Outputs (GIOs) for most regions were found to be higher than 1.15. This trend aligns with the contemporary concept of low-carbon development. In 2021, the GIOs of provinces such as Anhui, Hebei, Heilongjiang, Hunan, Liaoning, and Shanghai exceeded 1.2. This suggests that innovations aimed at reducing emissions, propelled by environmental regulations, have been driving the adoption of green technologies in these regions. Concurrently, the 2021 data also shows that in provinces like Annex, Hainan, Inner Mongolia, Jiangxi, Qinghai, and Shaanxi, the GII was higher than 1.05. This indicates that under the constraints of resources and environment, input bias has been a significant factor driving green technology innovation in these regions.

4.2. EBGI index

Figure 2 illustrates the intra-group bias of the input factors associated with green technology innovation within the manufacturing sector across 30 provinces in China, spanning from 2011 to 2021. The data reveals that both prior to 2014 and post-2018, green technology innovation displays a clear inclination towards energy conservation. More specifically, when examining the bias in technology innovation between the two factors of "technology" and "energy", where the Energy Bias Green Innovation (EBGI) is greater than zero, it becomes evident that the overall generation of green innovation is characterized by energy efficiency. This suggests that, for manufacturing enterprises, the impetus for innovation in energy conservation surpasses that of emission reduction.

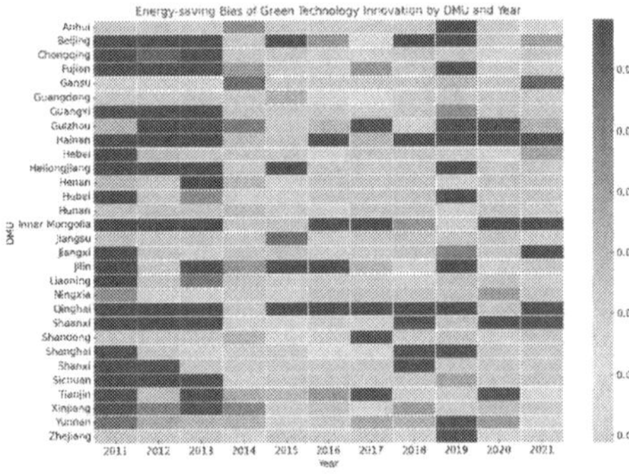

Figure 2. Heatmaps of EBGI

Upon further analysis of the spatial distribution of the energy-saving bias in green innovation, data from the most recent three years (2019, 2020, and 2021) reveal that regions such as Guizhou, Hainan, Inner Mongolia, and Shaanxi exhibit pronounced energy-saving bias characteristics. This is a departure from the traits observed in areas with higher levels of economic development in previous years. These regions have transitioned from an energy-driven technological innovation model to a model where technology drives energy-saving green technological innovation. This shift not only indicates that the technology gap is being bridged but also suggests that green technology innovation is beginning to demonstrate regional spillover effects.

4.3. Characteristics of spatial evolution

Figure 3 provides a visual representation of the spatial patterns of the Green Innovation Index (GII) at the provincial level, illustrating the bias levels for the years 2015, 2018, and 2021.

As depicted in Figure 3, the geographical distribution of Green Innovation (GI) across China is characterized by a steady expansion. The pattern of GII underwent a smooth transition throughout the study period, marked by a continuous distribution. Notably, the bias in green technology innovation in the southern regions is gradually shifting towards the input side, indicating a growing emphasis on resource allocation. Concurrently, the central region is witnessing a gradual intensification of input bias, underscoring a similar trend.

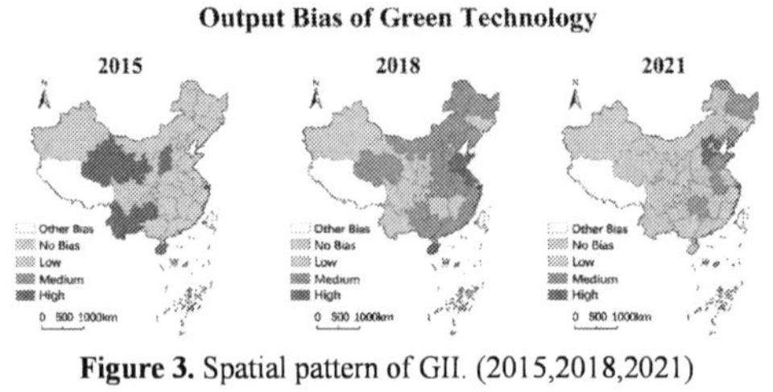

Figure 3. Spatial pattern of GII. (2015,2018,2021)

Figure 4 presents the spatial patterns of the Green Innovation Outputs (GIO) at provincial scales, visualizing the bias level for the years 2015, 2018, and 2021.

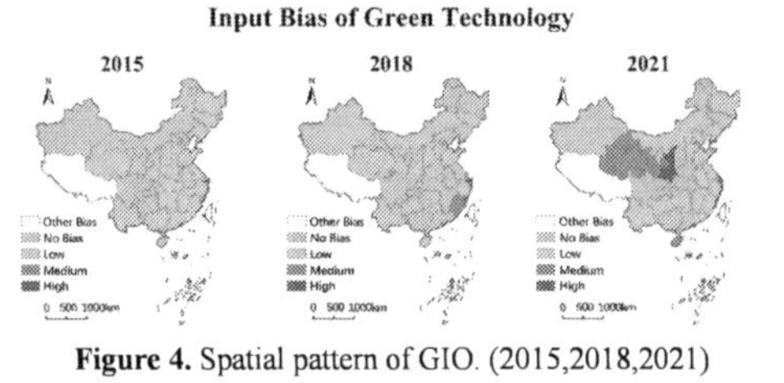

Figure 4. Spatial pattern of GIO. (2015,2018,2021)

As depicted in Figure 4, in contrast to the characteristics of input bias, the output bias of green innovation within the Chinese manufacturing industry is showing a trend of gradual attenuation. Concurrently, an examination of the bias characteristics reveals distinct regional traits of green technology innovation bias within the manufacturing industry. The eastern region predominantly exhibits environment-friendly technological innovation, while other regions are characterized by a focus on resource-saving technological innovation.

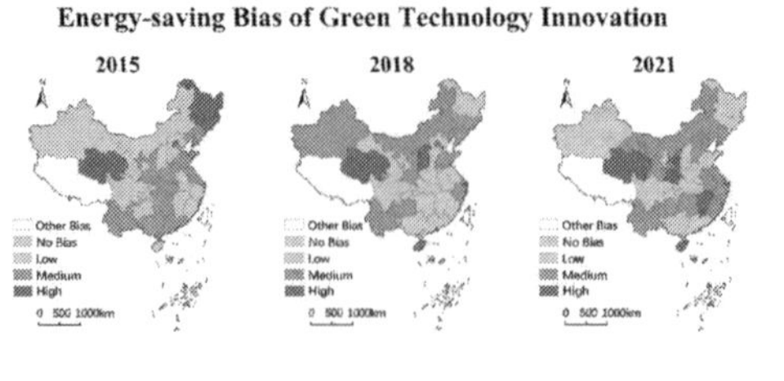

Figure 5. Spatial pattern of EBGI. (2015,2018,2021)

Figure 5 primarily investigates the regional spatial evolution characteristics of the bias towards energy-saving green technology innovation.

As depicted in Figure 5, energy-saving biased green technology innovation emerges as the principal driver of regional green innovation. The spatial characteristics of energy-saving biased green technology innovation are as follows:

(1) In 2015, prior to the introduction of the "green manufacturing" concept, energy-saving biased technology innovation was predominantly concentrated in economically developed provinces, including the eastern coastal region, Beijing, Tianjin, and other areas. This focus gradually shifted towards the northern region, and by 2021, it encompassed the majority of provinces in China.

(2) The spatial aggregation effect of energy-saving biased green technology innovation is pronounced, which correlates with the degree of industrial aggregation and the spatial correlation effect of innovation.

(3) In comparison to labor substitutability, advancements in technology and the reduced costs of green technologies have made them more accessible and feasible for adoption within the manufacturing industries.

5. Conclusions

Leveraging the directional distance function, we have constructed a novel Malmquist-Luenberger multidimensional decomposition index. This tool has been used to scrutinize the energy-saving bias of the input factors of green technological innovation across 30 provinces in China, with a particular focus on its spatial and temporal evolution characteristics. Our comprehensive evaluation explores the pivotal role of green technology innovation in propelling green total factor productivity of manufacturing in China from 2011 to 2021. The findings underscore that green innovation is predominantly characterized by energy conservation, suggesting that for manufacturing enterprises, the impetus for innovation in energy saving surpasses that of emission reduction.

In conclusion, our research illuminates the critical role of energy-saving bias in green technology innovation. The spatial evolution characteristics of this bias indicate a steady expansion of green technology innovation, with an increasingly pronounced input bias in the central region over time. We have also observed that energy-saving biased green technology innovation serves as the primary driver of regional green innovation. This type of innovation, which originated in economically advanced provinces, has gradually expanded to the northern region, and by 2021, it encompassed the majority of provinces in China.

The conspicuous spatial aggregation effect of energy-saving biased green technology innovation is intrinsically linked to the degree of industrial aggregation and the spatial correlation effect of innovation. The regional disparities identified in our study underscore the need for a deeper understanding of the factors contributing to these differences. This understanding could further inform the design and implementation of regional green technology policies.

Acknowledgment

This study was financially funded by the project of Joint Special Funds for Basic Research in Local Undergraduate Universities in Yunnan Province in 2022 (202101BA070001-170): Green-biased technological innovation and sustainable green development in Yunnan: mechanisms of action and policy conditions.

References

1. N. Ansari and A. Seifi, “A system dynamics model for analyzing energy consumption and CO2 emission in Iranian cement industry under various production and export scenarios,” Energy Policy, vol. 58, pp. 75–89, Jul. 2013, doi: 10.1016/j.enpol.2013.02.042.
2. V. De Marchi, “Environmental innovation and R&D cooperation: Empirical evidence from Spanish manufacturing firms,” Res. Policy, vol. 41, no. 3, pp. 614–623, Apr. 2012, doi: 10.1016/j.respol.2011.10.002.
3. C. Antonelli and F. Quatraro, “The effects of biased technological change on total factor productivity: empirical evidence from a sample of OECD countries,” J. Technol. Transf., vol. 35, no. 4, pp. 361–383, Aug. 2010, doi: 10.1007/s10961-009-9134-2.
4. W. L. Weber and B. R. Domazlicky, “Total factor productivity growth in manufacturing: a regional approach using linear programming,” Reg. Sci. Urban Econ., vol. 29, no. 1, pp. 105–122, Jan. 1999, doi: 10.1016/S0166-0462(98)00013-1.
5. P.-C. Chen and M.-M. Yu, “Total factor productivity growth and directions of technical change bias: evidence from 99 OECD and non-OECD countries,” Ann. Oper. Res., vol. 214, no. 1, pp. 143–165, Mar. 2014, doi: 10.1007/s10479-012-1087-4.
6. J. Peng, J. Xiao, L. Wen, and L. Zhang, “Energy industry investment influences total factor productivity of energy exploitation: A biased technical change analysis,” J. Clean. Prod., vol. 237, p. 117847, Nov. 2019, doi: 10.1016/j.jclepro.2019.117847.
7. Z. Shen and X. Zhao, “Evaluation of Resource Utilization Efficiency in the Machining Process Based on the SBM-DEA Model with Non-Expected Output,” Processes, vol. 11, no. 3, Art. no. 3, Mar. 2023, doi: 10.3390/pr11030916.
8. F. Fan, H. Lian, X. Liu, and X. Wang, “Can environmental regulation promote urban green innovation Efficiency? An empirical study based on Chinese cities,” J. Clean. Prod., vol. 287, p. 125060, Mar. 2021, doi: 10.1016/j.jclepro.2020.125060.
9. Y. Wang, W. Chen, Y. Kang, W. Li, and F. Guo, “Spatial correlation of factors affecting CO2

https://doi.org/10.1051/e3sconf/202344103005

emission at provincial level in China: A geographically weighted regression approach," J. Clean. Prod., vol. 184, pp. 929–937, May 2018, doi: 10.1016/j.jclepro.2018.03.002.

10. J. Song, Q. Feng, X. Wang, H. Fu, W. Jiang, and B. Chen, "Spatial Association and Effect Evaluation of CO 2 Emission in the Chengdu-Chongqing Urban Agglomeration: Quantitative Evidence from Social Network Analysis," Sustainability, vol. 11, no. 1, pp. 1–19, 2018, doi: 10.3390/su11010001.
11. R. Färe, E. Grifell-Tatjé, S. Grosskopf, and C. A. Knox Lovell, "Biased Technical Change and the Malmquist Productivity Index," Scand. J. Econ., vol. 99, no. 1, pp. 119–127, Mar. 1997, doi: 10.1111/1467-9442.00051.
12. Y. H. Chung, R. Färe, and S. Grosskopf, "Productivity and Undesirable Outputs: A Directional Distance Function Approach," J. Environ. Manage., vol. 51, no. 3, pp. 229–240, Nov. 1997, doi: 10.1006/jema.1997.0146.

 https://doi.org/10.1051/e3sconf/202344103006

Arduino-based Automatic Light Switching System

Moqi Li*, Ziqian Zhang, Haocheng Yin
Instructor: Chang-huan Liu, Feng-min Jiang

The High School Affiliated to Renmin University of China, Beijing, China

Abstract: This article focuses on the problem of high power consumption caused by long working hours of lights, and designs an automatic light switch system based on the number of people to achieve energy-saving effects. After experiencing the process of selecting hardware, designing circuits, writing programs, structural design, 3D modeling, 3D printing processing, assembly, etc., an automatic light on/off system has been designed, which is low-cost and can achieve automatic light on/off based on Arduino microcontroller technology, playing a role in saving electricity to a certain extent.

1. Introduction

Electric energy is the main energy and power widely used in modern industrial and agricultural production, scientific and technological research, daily life and other fields, and plays a pivotal role in all aspects of human society. But many people do not have a clear concept of saving energy or have difficulty to do so in some circumstances, leading to a lot of wasting electricity in daily life.

According to the formula 'power consumption = power × time "shows that a 40 watts fluorescent tube ignoring the power consumption of the ballast, working for 10 hours a day will also consume 0.4 kilowatt hours of electricity[1], indicating the power consumption of the lamp. Many people have not developed the good habit of turning off the lights before going out in their daily lives, especially in public spcaces, which can cause a certain amount of waste of electrical energy.

The traditional solution to this problem is to develop new materials and produce energy-saving lamps to reduce the energy consumption of electric lamps. But this is one way to solve the problem. The use of energy-saving lamps cannot reduce the working time of the lamp, and if the working time of the lamp cannot be reduced, it cannot effectively reduce the energy consumption of the lamp, and thus cannot fundamentally solve this problem. Some people also want to reduce energy consumption by reducing the lighting time, but arbitrarily controlling the lighting time may cause some inconvenience to use.

At present, research on energy-saving intelligent switches mainly focuses on sound control with delay and pyroelectric infrared detection of the human body. The application scenarios of sound control and delay technology are limited, and the delay time is usually very short and may be triggered by unrelated sounds. The pyroelectric infrared detection of human bodies requires the installation of pyroelectric infrared detection modules in the workplace. The accuracy of switches is related to the number of detection modules, and the cost of a large number of detection modules is not economical.

Few people think of using microcontroller control technology to solve this problem. But using Arduino microcontroller technology, we can combine number detection with automatic light on/off. When the number of people in the room is zero, mechanical devices are used to turn off the lights. When there are people in the room, the same mechanical device is used to turn on the lights, reducing the working time of the lights and reducing energy consumption.

This project is based on Arduino control technology to design an automatic light on/off system according to the number of people in certain space. The system is straightforward and efficient, which does not require a large number of detection modules, and does not need to produce unnecessary sound. It is particularly suitable for public places, such as meeting rooms, canteens, activity rooms. This system can not only be applied to lighting, but also to air conditioning, background music control, etc. It can greatly save unnecessary energy consumption and equipment wear and tear.

2. Control System Design

The control system is based on Arduino, using 3 ultrasonic sensors for input and 2 servos for output. (See Figure 1)

*limomax@126.com

https://doi.org/10.1051/e3sconf/202344103006

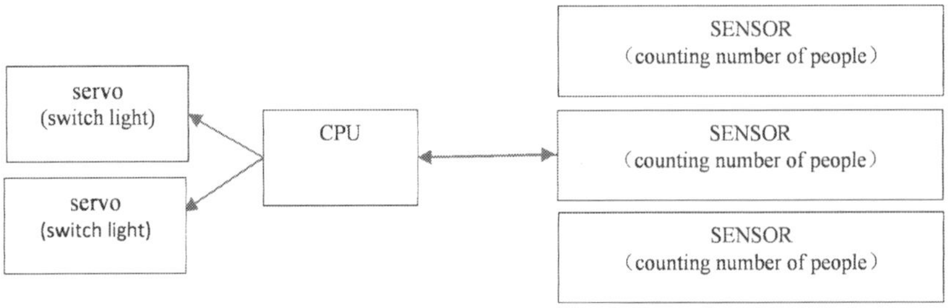

Figure 1. Control System Design

3. Selecting Hardware

This project is based on the Arduino platform and the Arduino Nano development board[2] (See Figure 2-left). This development board has a small size, low energy consumption, and the ability to drive and process other components, making it a very suitable development board for beginners. Its small size makes it very suitable for compact projects, and it is relatively cheap and cost-effective.

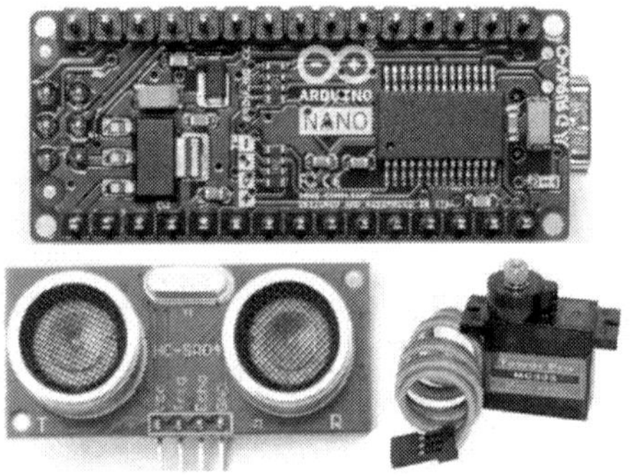

Figure 2. Selected Hardware

This project uses the ultrasonic sensor HC-SR04 to measure distance [3] (See Figure 2-middle). Ultrasonic sensors have the advantages of fast response speed, anti-interference, and long distance. Compared to the applicable distance of infrared sensors within 30cm, the applicable distance of ultrasonic sensors is 2-300cm, which can better meet the requirements of distance measurement at the entrance.

This project uses the MG90S servo (See Figure 2-right) to press the switch of the light to achieve the effect of automatic light on and off [4]. The rotation angle of the MG90S steering gear is within the range of 90 ° to 180 °, which can achieve the operation of turning on and off the lights. The working voltage is 4.8V to 6V, which is suitable for the Arduino Nano board and can well meet the requirements of this project.

4. Circuit Design

Connect input and output terminals of the 3 ultrasonic sensors to digital pin 3,4,5,6,7,8 of Arduino Nano development board respectively [5]. Connect the control wire of the servo to pin 9,10. Connect the positive and negative terminals of the ultrasonic sensor and the servo to the corresponding pins on the development board. After testing on a breadboard, all parts are welded on to a PCB board. (See Figure 3)

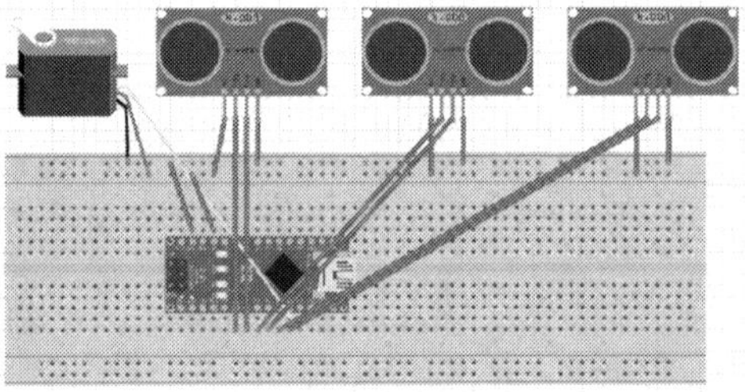

Figure 3. Circuit connection diagram based on Breadboard

5. Programming

This project uses Arduino software to write the program [6].

This project uses three ultrasonic sensors to achieve the function of number detection, and uses a while and for loop structure to write the program.

The program flowchart is as follows(See Figure 4):

Define the number of people as the variable c=0. When the ultrasonic sensor located in the middle detects a distance of less than 30 centimeters, the ultrasonic sensors on both sides begin to detect. When the ultrasonic sensor located on the inner side of the door detects a distance of less than 30 centimeters, it causes the number of people to be c1. At this time, if c=1, the steering gear will rotate and the light will be turned on; When the ultrasonic sensor located on the outside of the door detects a distance of less than 30 centimeters, the number of people is c-1. At this time, if c=0, the steering gear rotates and the lights will be turned off.

 https://doi.org/10.1051/e3sconf/202344103006

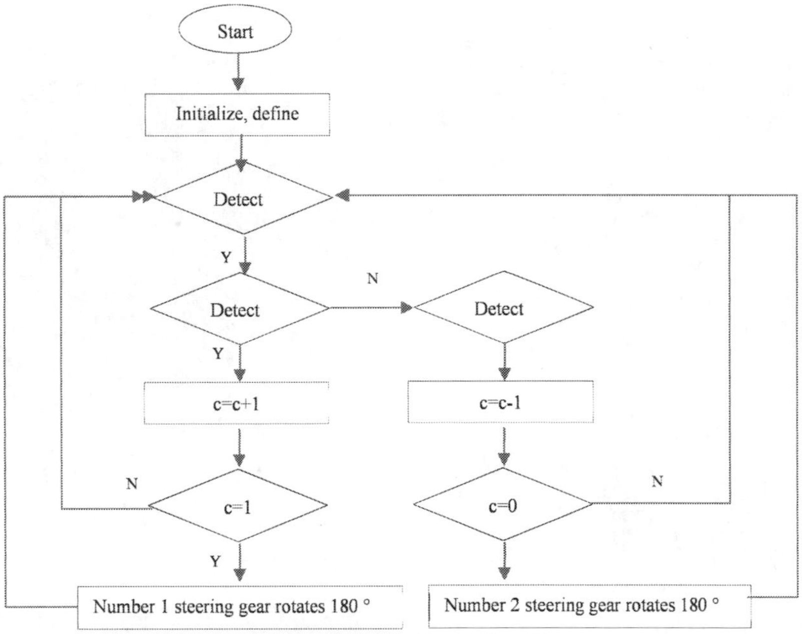

Figure 4. Programming Process Flow Chart

6. Structure Design

Considering that most lights and their switches have already been built upon completion of the building, the control method has been determined to be manual switching, and embedding intelligent control components requires renovation, which is too costly and time-consuming. We decided to use existing manual switches and use external mechanical attachment to press the switches to achieve intelligent control.

To meet the needs of existing switches and provide a platform for intelligent control hardware, we have come up with a solution as shown in the following figure and designed the following device.

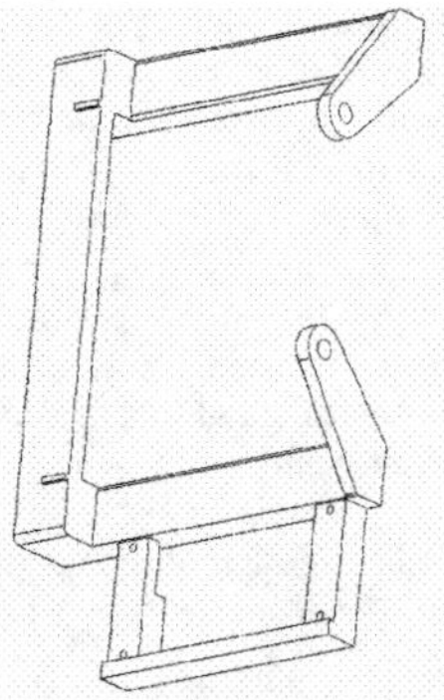

Figure 5. Mechanical part design drawing

This device is a mechanical attachment installed on an existing manual switch, consisting of a frame that plays a fixed role around the manual switch, a vertical rod that plays a fixation role in the steering gear, and two horizontal rods that are responsible for pressing the manual switch to achieve automatic light switch function. (See Figure 5)

This device can be fixed around the switch by relying on tight adhesive, with minimal damage to the wall. Fix the two servo motors on the left vertical pole and connect them to the horizontal rod. When the program runs, the servo motor drives the horizontal rod to rotate, causing the horizontal rod to press the manual switch (86 type switch), achieving the effect of intelligent automatic light switch. Under its support structure, there is a bearing frame for the Arduino Nano microcontroller, which can accommodate the Arduino Nano microcontroller as a control part and its wiring with the sensing part. This device will cover most of the original switch, but still leave enough space for people to manually operate the switch directly.

The scheme for the induction part is as follows:

This device is a sensor holding device, placed at the entrance and connected to the control part through a flexible wire. This component can hold three HC-SR04 ultrasonic modules to meet the requirements of the testing program. (See Figure 6)

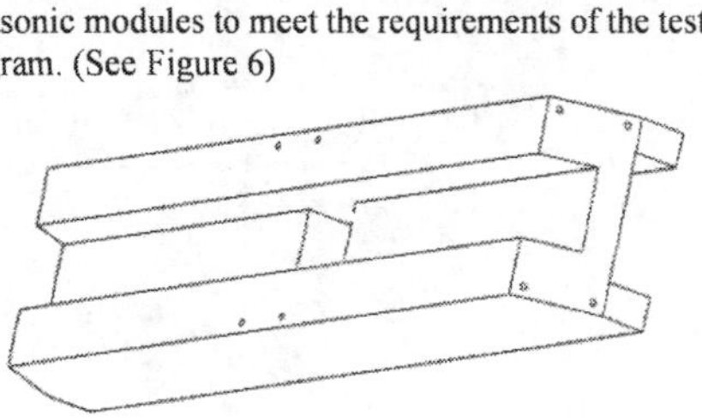

Figure 6. Design diagram of sensor clamping device

https://doi.org/10.1051/e3sconf/202344103006

7. 3d modeling

This project uses SOLIDWORKS software for modeling, which has the characteristics of rich functionality, relatively simple operation, and easy 3D printing. (See Figure 7)

Figure 7. 3D modeling

8. 3D printing, processing and assembly

The mechanical attachment installed on the existing manual switch in this project is made using 3D printing method, while the sensor clamping device has a relatively simple structure and is processed from wooden boards. The finished product diagram is as follows(See Figure 8):

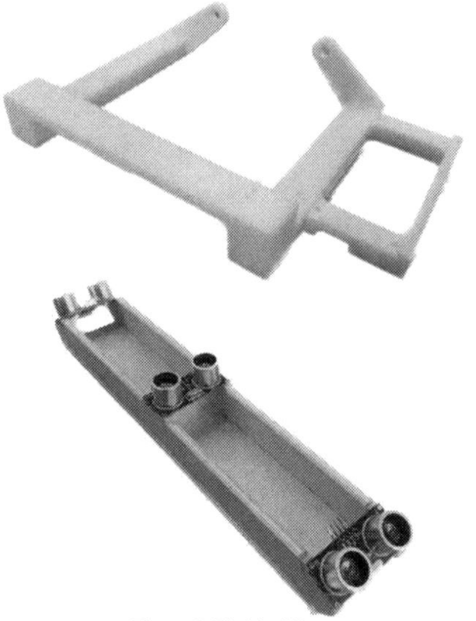

Figure 8. Finished Parts

Arduino Nano finished wiring part is as follows (See Figure 9)

Figure 9. Finished Wiring Parts

Assemble the mechanical attachment, sensor clamping device, and Arduino Nano board, and the final product is as follows (See Figure 10):

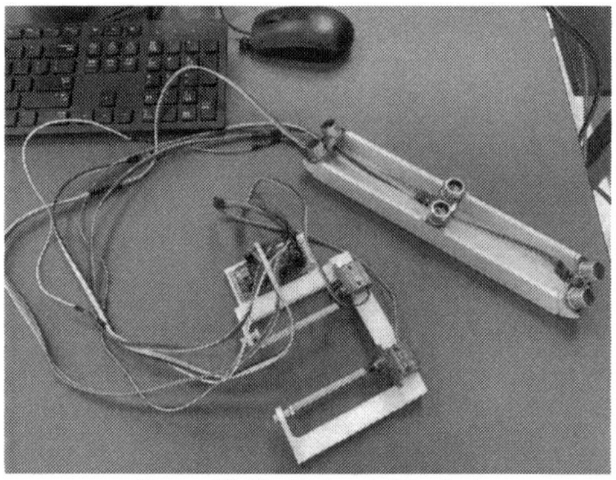

Figure 10. Final Integrated Product

9. Summary and Outlook

This project utilizes Arduino control technology to successfully achieve automatic control of on/off lights based on the number of people in the room, in order to achieve energy-saving effects. However, the measurement conditions of this system are relatively harsh, and the measurement accuracy needs to be improved. In future work, we expect to improve the hardware design of the sensing section to improve performance in crowdy situations to avoid turning light on or off incorrectly. We also strive to avoid situations where the switch is not pressed correctly by enhancing the motor power, as well as improving modular design of the control part to adapt to different types of switches and enhance the versatility of the product.

Acknowledgments

We would like to take this opportunity to express our sincere gratitude to our teachers Liu Changhuan and Jiang Fengmin for their guidance and provision of some hardware. We are also grateful to our parents for their support. Last but not least, we express our special gratitude to our high school-- High School Affiliated to Renmin University of China for providing us with a valuable experience and giving us the opportunity to complete such an interesting project.

References

1. Prashant Kumar Soori; Moheet Vishwas. Lighting control strategy for energy efficient office lighting system design, Energy and Buildings. 2013 ;66:329-337.doi:10.1016/j.enbuild.2013.07.039
2. N. Hey, Nano revolution, IEE Review. 2003 ;49(4): 46-48.doi:10.1049/ir:20030405
3. PingYongXiaoCaiJi, CSDN. Use of Arduino ultrasonic sensors. blog.csdn.net/bolanlanlan/article/details/114003094
4. AnthonyYMH, CSDN.MG90S actuator driving principle and program code. blog.csdn.net/weixin_38075894/article/details/80027600
5. Malcom_110, CSDN. Arduino Nano pin allocation diagram and detailed definition explanation, blog.csdn.net/malcolm_110/article/details/95320094
6. XiangMuShenBaoXiaoKuangRen, CSDN. Arduino control servo detailed explanation (including code) blog.csdn.net/m0_58857684/article/details/125740665

The analysis of the potential through the Mode Change of transport from Road to Waterway

Meizhi Jiang[1,2,*], Benmei Wang[1], Yingjun Hao[1], Shijun Chen[1], Yuanqiao Wen[2]

[1] Port and Navigation Research Institute, Zhejiang Scientific Research Institute of Transport, Hangzhou, China
[2] Intelligent Transportation System Research Center, Wuhan University of Technology, Wuhan, China

Abstract: In 2015, the Chinese government committed to reducing all greenhouse gas emissions, to reach the peak of carbon dioxide before 2030 and carbon neutrality by 2060. Transport accounts for a large share of global greenhouse gas emissions, so greener and low-carbon transport is an important part of achieving the two-carbon goal. Although many researches have examined carbon emissions from the transport sector, few have analyzed the potential for carbon reduction from the perspective of optimizing the transport structure. Based on the analysis and prediction of the logistics system of goods transportation in Zhejiang Province, the potential of changing the mode of transport from road to waterway for CO2 emission reduction is further analyzed. The results show that the implementation of a "mode change of transport from road to waterway" can better reduce the carbon emissions in the field of transportation, to provide a theoretical basis for the energy conservation and emission reduction targets in the field of transportation in Zhejiang Province.

1. Introduction

The Party Central Committee and the State Council attach great importance to addressing climate change. In December 2021, the Central Committee of the Communist Party of China and the State Council issued the "Opinions on Deepening the Battle of Pollution Prevention and Control", which focused on energy, industry, urban-rural construction, transportation and other fields, as well as steel, non-ferrous metals, building materials, petrochemical and chemical industries, to carry out carbon peak activities in depth.

The transportation industry is a leading and fundamental industry for the development of the national economy and has always been a key area in addressing climate change. Research shows that CO2 emissions in the transportation sector account for about 11% of the total social emissions, and the main driver of transportation carbon emissions is road transportation. According to the "China Mobile Source Environmental Management Annual Report (2020)", greenhouse gas emissions from roads, waterways, civil aviation, and railways are about 6.9, 0.7, 0.5, and 0.1 billion tons of carbon dioxide equivalent, accounting for 84.1%, 8.5%, 6.1%, and 1.2%. Promoting transportation structure adjustment is an important way to achieve carbon peak in the transportation field. In December 2021, the General Office of the State Council issued the Work Plan for Promoting the Development of Multimodal Transport, Optimizing and Adjusting the Transportation Structure (2021-2025), which proposed to promote the "transit of Commodity from the public to the railway and from the public to the water", and formed a development pattern of railway and waterway based long-distance transportation of bulk goods and containers [4]. In June 2022, the Office of the Leading Group of Zhejiang Province, a strong transportation province, issued the "Implementation Plan for Carbon Peak in the Transportation Field of Zhejiang Province", proposing to carry out a concentrated campaign of "public transportation to water" for bulk goods and medium to long-distance goods.

Therefore, based on the development status of waterway transportation in Zhejiang Province, this paper comprehensively analyzes the potential of " mode change of transport from road to waterway (MCRW)" in cargo transportation, calculates the emission reduction benefits of MCRW in cargo transportation in the province, and formulates specific implementation paths. It provides technical solutions for the adjustment of the transportation structure in Zhejiang Province and provides strong support for the green and low-carbon transformation of transportation and the construction of a high-level transportation province.

This paper aims to analysis the potential of the mode change from road to waterway transport. The rest of this manuscript is structured as follows. Section 2 reviews the literature on CO2 emission in the transport sector including road and waterway. The analysis of the development status of waterway transportation is proposed in Section3. Section 4 analysis the potential of MCRW in the perspective of four types of transport cargo. The conclusions are summarized in Section 5.

*Corresponding author Email: jiangmeizhi_zrit@163.com

2. Literature review

In the past few decades, with the increasing impact of greenhouse gases, academia and industry have conducted many studies on carbon dioxide emissions reduction in transportation. For example, [6] proposed a method for calculating transportation related carbon dioxide emissions in European tourist cities. [8] proposed an innovative model to predict the potential benefits of carbon dioxide emissions trading in transportation in China. [2] introduced a marginal emission reduction cost curve to identify the connections between all transportation modes, providing a decision-making basis for estimating the cost and potential of reducing carbon dioxide emissions in China's transportation sector.

The research on carbon reduction in the field of highways is also an important part. [1] used the LMDI method to analyze the impact of various influencing factors on carbon emissions reduction, providing feasibility for China to achieve its strategic goals of carbon peaking and carbon neutrality. [3] analyzed the current carbon reduction potential of China's road transportation and found that short-term carbon reduction should focus more on the quantity and structure of transportation, while long-term carbon reduction should focus more on the application of new technologies in the field of highways.

Waterway transportation is also one of the key research directions in the field of transportation. [5] developed a new model to study the relationship between ship speed and carbon emissions from the perspective of the offshore supply chain, and applied the research results to simulate the flow direction of container traffic. [9] analyzed the efforts made by the shipping industry in carbon reduction, providing a scientific basis for achieving global greenhouse gas emissions reduction,

In addition to the research specifically focused on carbon emissions in the fields of waterways and highways mentioned above, there are also some experts and scholars who pay more attention to the carbon emission relationships between different modes of transportation. [7] conducted a comparative analysis of carbon emissions from road, transportation, and waterway transportation. The analysis results showed that the process of converting goods from road transportation to waterway transportation significantly reduces carbon dioxide emissions in the transportation sector. Moreover, through the implementation of a series of policies, this model transformation has great potential. [10] first analyzed the influencing factors of carbon emissions using the LMDI method. The analysis results showed that the optimization of transportation structure has huge implementation space and can reduce carbon emissions to a certain extent. The results confirm that effective changes in traffic structure can help reduce CO2 emissions

3. Analysis of the development status of waterway transportation

In 2008, the per capita GDP of Zhejiang Province exceeded 6000 US dollars, entering the late stage of industrialization development. Throughout developed countries such as the United States, Germany, and France, the demand for transportation of goods during this period showed typical characteristics of increasing scale, slowing down the growth rate, prioritizing highways, and decreasing water transportation. The scale of cargo transportation in our province has increased from 84.6 million tons in 1978 to 3.27 billion tons in 2021, with a growth rate of 10.4% before 2000 to 6.1% after 2010. The proportion of highways has increased from 31.8% to 65.3%, and the proportion of waterways has decreased from 51.5% to 33.4% (including sea transportation).

Since 2018, Zhejiang Province has begun a three-year transportation structure adjustment action. In 2018 and 2019, the growth rate of waterway freight volume and cargo turnover was higher than that of roads, especially in 2018, the freight volume and cargo turnover of waterways were respectively 4% and 8% higher than that of roads, and the results were remarkable. But on the whole, the transport structure has not changed fundamentally. Details can be seen in Fig. 1 to Fig.4.

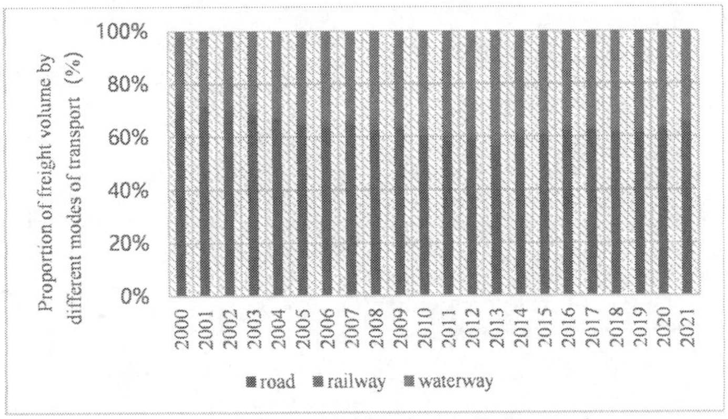

Fig.1 Structural change of freight volume in Zhejiang Province

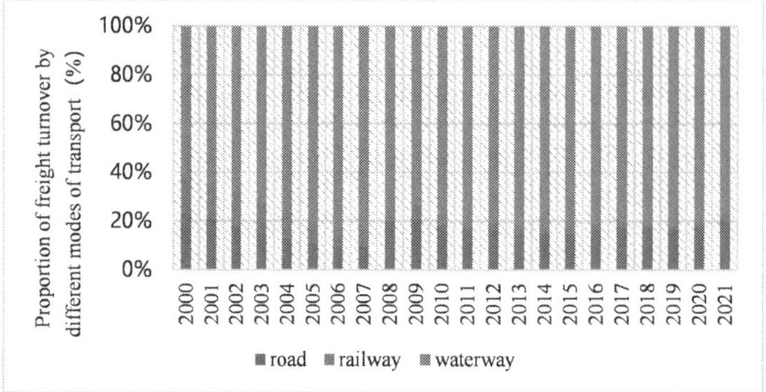

Fig.2 Structural change of cargo turnover in Zhejiang Province

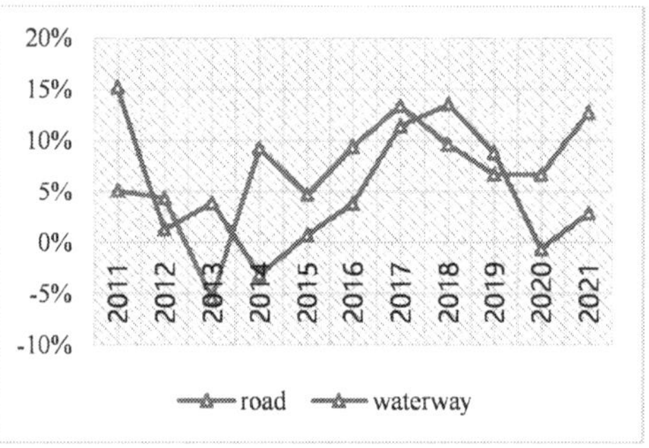

Fig.3 Annual growth rate of freight volume by transport mode

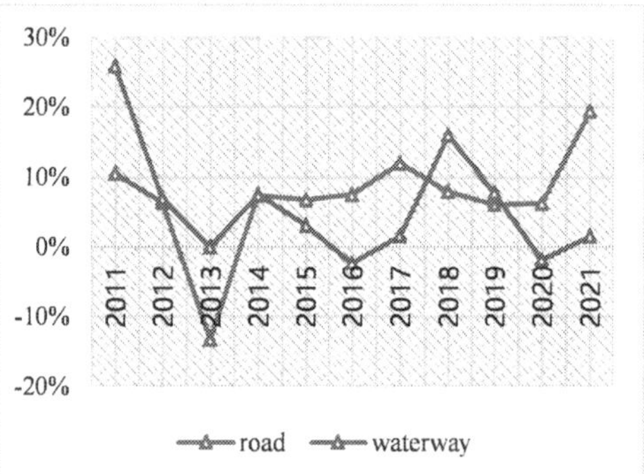

Fig.4 Annual growth rate of freight turnover by transport mode

Overall, in recent years, the demand for water transportation in Zhejiang province has stabilized, and the scale of water transportation has maintained stable growth. Specifically, with the advancement of industrialization, the growth rate of inland waterway transport has gradually slowed down. In 2021, the province completed the inland cargo transport of 250 million tons and a cargo turnover of 37.54 billion tonne-kilometres, and the cargo scale is expected to reach the plateau in 2030. Coastal freight has maintained a rapid growth trend, in recent years, affected by the epidemic and the international situation, the growth rate is also slowing down. In 2021, the province will complete 820 million tons of coastal cargo transport and 879.56 billion tonne-kilometres of cargo turnover, which is expected to reach about 1 billion tons in 2025. The demand for ocean freight is greatly affected by global economic conditions and other uncertainties. In recent years, ocean freight in Zhejiang province has shown a downward trend after a downward trend, but it has also shown a downward trend under the influence of the entire shipping market.

China's industrialization will be fully completed around 2030, and then enter the post-industrialization stage, and economic development will shift from the high-speed growth stage to the high-quality development stage. It is expected that before 2030, the overall trend of waterway transportation in the province will continue to maintain growth will not change, and the growth rate will continue to slow down, and then it will maintain overall stability at a certain scale, with certain annual fluctuations. Among them, ocean freight is synchronized with global economic development, and the growth of inland water transportation mainly depends on the adjustment of transportation structure.

4. Analysis of the potential of MCRW

Coal, building materials (cement, etc.), containers, oil products, natural gas, iron ore, grain and agricultural products, express delivery, less-than carloads and hazardous chemicals and other ten types of cargo constitute the main part of cargo transport in Zhejiang Province, accounting for about 60% of the total social freight volume in our province. In 2019, through the analysis of the current situation and trend of the transport logistics of the ten major cargo types, it was clear that the cargo types with the potential of MCRW in Zhejiang province were concentrated in the four categories of coal, cement, grain and container, and the potential of MCRW research and analysis was carried out for the above cargo types.

4.1 Analysis of coal transport logistics

4.1.1 Analysis of the potential of MCRW of coal transport

The coal in Zhejiang province mainly flows into the province from outside the province and from east and north of Zhejiang. Since 2013, Zhejiang province has officially withdrawn from the field of coal production, and all coal demand has been transferred from outside the province and overseas, and coastal and inland ports have become the main nodes of coal into Zhejiang. From the regional distribution of coal throughput in our province, the coal throughput in eastern and northern Zhejiang regions will account for nearly 85% of the province in 2021, mainly completed by Jiaxing Port, Ningbo Zhoushan Port, Wenzhou Port and other ports, details can be seen in Table 1.

Table 1 Coal throughput and growth rate of major ports in 2021

Port	Coal throughput (tons)	Two-year average growth rate (%)
NingboZhoushan Port	2500	8.0
Jiaxing Port	3524	1.5
Wenzhou Port	1145	6.4

4.1.2 Prediction of transfer volume from MCRW

Combined with the historical trend and existing planning requirements, the future change trend of coal consumption in the province is estimated, and as the basis for predicting the change trend of the total coal transportation of major ports in the future, it is assumed that the total coal transportation of all ports in 2025 and 2030 will decrease simultaneously according to the declining trend of coal consumption in the province (see Table 2 for details).

Table 2 Forecast of coal consumption in the whole province

Year	Forecast of coal consumption in the province (10,000 tons)	Decrease from 2021 (%)
2020	13100	-
2021	12920	-
2022	12760	-
2023	12620	2.32
2024	12500	3.25
2025	12400	4.02
2026	12310	4.72
2027	12230	5.34
2028	12160	5.88
2029	12090	6.42
2030	12020	6.97

Based on the overall trend of decreasing coal consumption year by year in Zhejiang province, the potential of coal MCRW in the future is mainly predicted based on the expected growth of the waterway traffic of Ningbo Zhoushan Port, Jiaxing Port and Wenzhou Port in the target years (2025, 2030), combined with the natural decline rate of waterway traffic. The specific algorithm is as follows:

Coal MCRW potential of each port = predicted value of coal waterway transportation in target year - coal waterway transportation in base year + natural decline of coal waterway transportation in base year = coal throughput in target year × water transportation ratio in target year - coal throughput in base year × water transportation ratio in base year + coal throughput in base year × water transportation ratio in base year × natural decline rate = coal throughput in target year × (water transportation in target year) Proportion - Water transport proportion in base year)

Among them, the coal traffic volume in the target year (2025, 2030) is calculated in combination with the total coal transportation volume in the base year of each port (2021) and the decreasing trend of coal consumption in the province in the future (The predicted results are shown in Table 3).

Table 3 Prediction result of coal MCRW in Zhejiang province

Port	Prediction result of coal MCRW (10,000 tons)	
	2025	2030
Ningbo Zhoushan Port	240	465
Jiaxing Port	227	330
Wenzhou Port	0	11
Total	467	806

4.2 Analysis of cement transport logistics

4.2.1 Analysis of the potential of MCRW of cement transport

Cement production enterprises are distributed in bands and highly clustered, and cement output is high in the north and low in the south. There are nearly 150 cement enterprises in Zhejiang province, mainly concentrated in Huzhou Changxing, Hangzhou Tonglu, Fuyang, Jiande, Jinhua Lanxi, Quzhou Jiangshan and Changshan and other places. Jiaxing, Ningbo, Quzhou, Jinhua, Huzhou, Hangzhou top six cement output in the province, except Hangzhou, Ningbo due to strong demand for infrastructure construction, local cement is still in short supply, the rest of the cities are the main cities of cement output. Zhoushan, Wenzhou, Taizhou, Lishui and other places have no clinker production, and there are only some grinding mills in the region, so the cement output is relatively low, and it is the main city of cement import.

The cement transportation in Zhejiang province is mainly to serve the local market, the waterway is the main mode of transportation in northern Zhejiang, and the road transport is mainly in western and central Zhejiang. In Zhejiang province, cement mainly flows to Quzhou, Huzhou, Jinhua and other northern and western Zhejiang areas, and to Wenzhou, Taizhou, Lishui and other southern and eastern Zhejiang areas. Road transport conventional transport distance of 50~200 kilometers, the longest up to 300 kilometers, the main cross-regional flow flow including Changxing into Jiaxing, Hangzhou north market; Tonglu, Fuyang, Zhuji and other places into Hangzhou, Shaoxing market; Jinhua, Quzhou into Taizhou, Wenzhou, Ningbo and other coastal markets, the rest for local, regional short distance road transport. Water transport is mainly sold in Huzhou, Jiaxing and other northern Zhejiang areas to southern Jiangsu, Shanghai and local, water transport accounts for about 70%-80%; Clinker mainly involves the purchase of clinker by sea in coastal grinding stations, Hangjiashao grinding stations and other places through the inland waterway purchase of clinker, some cement plants also need to pass the road short barge. Railway transportation is mainly for Changxing,

Jiangshan, Xuancheng and other places of enterprises sent to Ningbo, Wenzhou area of cement and a small amount of clinker.

Based on the current spatial distribution of cement and clinker production and sales in the province, combined with the characteristics of narrow cement transport radiation, small scale and short transport distance, the potential of MCRW is mainly based on the adjustment of the transport structure of large cement enterprises in the main cement supply places such as Quzhou, Jinhua and Hangzhou, and the increase of the proportion of cement clinker waterway transport at the port. In terms of cement enterprises, Southern Cement and Red Lion cement have gathered most of the production capacity in northwest Zhejiang, and are the key direction of MCRW. At present, the five large cement factories of Southern Cement have drafted the "14th Five-Year" logistics optimization and upgrading plan for Zhejiang Southern Cement according to the logistics conditions of each enterprise and its customers, planning to accelerate the construction of the enterprise terminal of Shanya South and Changshan South and the construction of the client's water transport transfer warehouse, give full play to the water transport conditions of Qiantang water system, and increase the MCRW Increase the sales volume of waterborne cement in Hangzhou city north, Xiaoshan, Shaoxing, Yuyao and other places. In terms of Hongshi cement, half of Tonglu Hongshi cement has been transported to Hangshao direction through the Fuchun River channel, and the future development potential is limited. In terms of port clinker transport, at present, there is still about 30 million tons of gap in our province to be transferred from outside the province, most of which come from Anhui, Liaoning and other provinces, and less of which are imported from Vietnam and other countries. Due to the small radiation range, the port is basically transported by road after receiving and unloading, and some clinker is temporarily transferred to small boat transport or transported to the hinterland through the transit station. With the opening of the waterway, the improvement of facilities and equipment and the development of sea-river combined transport business, there is a certain potential of MCRW.

4.2.2 Prediction of transfer volume from MCRW

From the perspective of cement production and consumption in the province, the existing market fluctuates and shows a trend of gradual saturation, and the natural growth of related transportation volume can be ignored. The direct downstream of cement is infrastructure and real estate, the current growth of the two has entered a flat period, accounting for the proportion of fixed asset investment is basically stable at about 60%, from the recent three years of cement production in our province (respectively 1.34 million, 132 million, 135 million tons), the cement production capacity in our province is basically 130 million in the middle and low fluctuations. Under the background that the country and our province continue to resolve the problem of overcapacity and strictly restrict new production capacity, it is expected that the cement production and consumption in the province will stabilize in the future and enter the structural optimization stage (the productivity layout will gather in the Changxing - Jiande - Qujiang - Changshan - Northwest Zhejiang area). Therefore, when estimating the potential of MCRW of cement, the impact of natural growth on the forecast results is temporarily ignored (The predicted results are shown in Table 4).

Table 4 Prediction result of cement MCRW in Zhejiang province

Core enterprises/ports	Prediction result of coal MCRW (10,000 tons)	
	2025	2030
Jiaxing Port	67	112
Wenzhou Port	100	200
Longyou Port	20	50
Changshan South cement	0	200
Shan Ya South cement	20	60
Lanxi Red Lion cement	20	30
Tonglu Red Lion Cement	20	30
Other cement enterprises	26	137
Toal	273	819

4.3 Analysis of grain transport logistics

4.3.1 Analysis of the potential of MCRW of grain transport

Grain logistics presents the characteristics of large-scale centralized inflow outside Zhejiang Province and small batch scattered transportation within Zhejiang province. In 2021, the total amount of grain imported and imported from outside Zhejiang Province is about 21.86 million tons. In recent years, the quantity and proportion of imported grain in Zhejiang Province have increased significantly, and the net purchase of grain from outside the province has increased slightly year by year. Among them, the transfer from outside the province is the main way to make up for the gap in grain production and demand in our province, which is generally transferred from Jiangsu, Anhui, Heilongjiang, Liaoning, Jilin, Shandong and other neighboring provinces and major grant-producing provinces, about 16.59 million tons, accounting for about 75.8%. Specifically, Jiangsu, Anhui, Jiangxi and other neighboring provinces bought a total of 8.26 million tons, accounting for 49.8%; The purchase volume of major grain-producing provinces such as the three Eastern provinces was 5.13 million tons, accounting for 30.9%; Shandong, Henan and other provinces bought 3.2 million tons, accounting for 19.3%. Foreign imports mainly come from Brazil, the United States, Ukraine, Argentina, Australia, Southeast Asia and other food-producing countries, about 5.27 million tons, accounting for 24.2%, an increase of 1.37 million tons over last year.

Grain transfer from outside the province is mainly by direct road, combined transport of public water and railway container. The transportation of grain imported from abroad is mainly water-water combined transport, supplemented by public water combined transport. The potential of "rotating water" is mainly to guide the inland transportation of grain in neighboring provinces, and to increase the proportion of grain shipping in distant regions. The demand for grain imported from outside Zhejiang Province is strong, especially the purchase of Jiangsu, Anhui, Shanghai, Jiangxi and other neighboring provinces accounts for half of the imported grain imported from outside Zhejiang Province. At present, this part of grain transportation is still dominated by direct road transportation. Through unimpeded inter-provincial water transportation channels such as Zhejiang-Shanghai, Zhejiang-Jiangsu, Zhejiang-Jiangxi, Zhejiang-Anhui and other inland ports, the function of grain receiving and unloading operation can be improved. It has great potential to promote the transfer of grain transport from highway to waterway or public water transport. In addition, with the unimpeded Hangyong Canal and the upgrading and transformation of Zhoushan International Grain and Oil Industry Park, it will be conducive to the northeast, overseas and other distant areas to use water transportation through Ningbo Zhoushan Port to complete the distribution and transfer.

4.3.2 Prediction of transfer volume from MCRW

Based on the projected permanent population and per capita grain consumption of Zhejiang Province in the target years (2025 and 2030) and remaining unchanged (0.35 tons/(person year)), the total annual grain consumption of the province in the future is estimated, and the net grain purchase of the province is obtained by subtracting the annual grain consumption of the province from the annual grain production of the province (stable at about 6 million tons in the past four years). The volume of purchases from other provinces, the volume of imports from abroad and the volume of sales to other provinces are estimated by trend extrapolation according to the proportion data over the years. Since the grain data involves the safety of reserves and it is difficult to obtain detailed flow direction data, it is therefore set that the future sea-river combined grain transport target of Jiaxing Port, the proportion of MCRW in each target year and the waterway level of northern Zhejiang as a reference to complete 7% of the grain MCRW in 2025 and 11% of the grain MCRW in 2030 in the region with four-tier waterway conditions.

According to the estimates, by 2025, Zhejiang's grain purchase volume is expected to be 19.92 million tons, the export volume is expected to be 5.86 million tons, and the grain import volume is 6.43 million tons. Among them, a total of 9.92 million tons were purchased from neighboring provinces such as Jiangsu, Anhui and Jiangxi, and 6.16 million tons were purchased from major grain-producing provinces such as the three eastern provinces. By 2030, China will purchase 22.98 million tons of grain from other provinces, sell 6.76 million tons of grain from other provinces, and import 7.41 million tons of grain. Among them, the neighboring provinces bought 11.44 million tons, and the three eastern provinces bought 7.11 million tons. See Table 5 for details. In summary, the potential of grain MCRW in Zhejiang Province in 2025 is about 1.34 million tons, and the potential of MCRW in 2030 is about 2.43 million tons. The specific forecast results are shown in Table 6.

Table 5 Prediction results of grain consumption and purchase from 2025 to 2030 (unit: 10,000 tons)

Items	2021	2025	2030
Resident population	6540	7420	8300
Grain consumption	2314	2650	2964
Foreign purchase volume	1659	1992	2298
Neighboring provinces such as Jiangsu, Anhui and Jiangxi purchased quantity	826	992	1144
The purchase volume of the three eastern provinces	513	616	711
Overseas import	527	643	741
Sales outside the province	493	586	676

Table 6 Prediction result of grain MCRW in Zhejiang province (unit: 10,000 tons)

Potential direction	2025	2030
Shanghai, Jiangsu direction	49	89
Three eastern provinces, overseas direction	85	154
total	134	243

4.4 Analysis of container transport logistics

4.4.1 Analysis of the potential of MCRW of container transport

In terms of coastal areas, the container throughput of coastal ports in Zhejiang Province in 2021 is 34.89 million TEU, of which 31.08 million TEU is handled by Ningbo Zhoushan Port, accounting for 89%. From the perspective of collection and distribution structure, the proportion of containers transported by road, rail and water is 70:4:26, and most of them are completed by road transportation, only 8.2 million TEU are passed by water, and the collection and distribution structure has not been substantially improved compared with 2020 (70:3:27) and 2017 (41:1:58). In terms of inland waterways, more than 30 inland container routes have been developed in Zhejiang province, and the container throughput of inland ports in the province reached 1.22 million TEU in 2021, more than doubling in the past five years (600,000 TEU in 2017), with an average annual growth rate of 19.4%. In terms of sea-river combined transport, Jiaxing, as the main source of inland container transport in Zhejiang province, has strengthened policy support and financial subsidies for container sea-river combined transport in recent years. At present, there are 19 inland ports and 21 routes docking with Jiaxing Port, covering northern Zhejiang, central Zhejiang and western Zhejiang, and expanding to southern Jiangsu, central Anhui and other places, and the cargo source hinterland is constantly deepening inward.

The potential of container MCRW is mainly in northern Zhejiang, supplemented by eastern Zhejiang, western Zhejiang and southern Zhejiang. The container development foundation in northern Zhejiang is the best. According to the survey, it is expected that by 2025, the container sea-river combined transport throughput target will exceed 1.6 million TEU. In eastern Zhejiang, with the renovation of the Hangyong Canal, the construction of a number of wharf operation areas such as Shaoxing, Zhuji, Xiaoshan and Shangyu, and the exploration and expansion of business from Ningbo to Hangshao and Jinhua in western Zhejiang, the potential of inland container MCRW is expected to be further explored. In western Zhejiang Province, relying on the improvement of the three-level waterway in the middle and upper reaches of Qiantang River and the construction of Qujiang, Longyou, Fangxiadian, Nwbu and other wharf operation areas, some containers from Jinhua, Quzhou to Hangzhou, Ningbo and other places are expected to abandon land and water. However, due to the mature development of railway container logistics on this line and better timeliness, the volume of MCRW in this area is not large. South Zhejiang Province relies on the Oujiang River channel to open up the southwest Zhejiang region directly to the sea, with the further development of Wenzhou to Ningbo Zhoushan branch line, some goods can be through the MCRW through the Oujiang River container river-sea combined transport.

4.4.2 Prediction of transfer volume from MCRW

Jiaxing Port, Ningbo Zhoushan Port, Quzhou Port and Wenzhou Port, the core ports of northern, eastern, western and southern Zhejiang channels, were focused on to predict the potential of container MCRW in the direction of major channels. Due to the lack of specific container throughput collection and distribution structure data, we refer to the container sea-river combined transport and river-river transfer business increment of ports, exclude the natural growth rate of container water transport (take cargo throughput growth rate), and comprehensively predict the MCRW potential of the container in Zhejiang province in the target years (2025, 2030).

Among them, the growth trend of container water transport business volume such as sea-river combined transport and river-river transfer in the northern and eastern Zhejiang channels is extruded to the target year, and the natural growth of container water transport business volume is subtracted on this basis to obtain the container MCRW potential of each channel in the target year. The size of the inland container business in the southern and western Zhejiang channels is small, so all the incremental container business in the future target year can be included in the MCRW potential (assuming that all the business growth is water transport business). The prediction results are shown in Table 7

Table 7 Prediction result of container MCRW in Zhejiang province (unit: 10,000 TEU)

Channel direction	2025	2030
Northern Zhejiang passage	100	174
Eastern Zhejiang passage	0.1	0.2
West Zhejiang passage	5	13
Southern Zhejiang passage	15	25
Total	120.1	212.2

5. Conclusion

The analysis results of MCRW potential of coal, cement, grain and container transportation logistics are sorted out, and the container volume is converted to 8 tons per TEU (the average container weight of Zhejiang province is about 10 tons, the empty container weight is about 2 tons, equivalent to about 8 tons of cargo), and the MCRW potential of the four major cargo types in the province is summarized in the table 8.

It can be seen that by 2025, containers have the greatest potential to contribute to the development of MCRW in Zhejiang Province, followed by coal, cement and grain; By 2030, with the significant improvement of waterway and terminal conditions in western Zhejiang, cement water transport business will be further expanded, although container is still the largest cargo type of "revolution water" contribution potential in the province, but the contribution potential of cement is more than coal, ranking second.

Table 8 The potential of MCRW in four major goods in Zhejiang Province (unit: 10,000 ton)

Cargo name	2025	2030
coal	467	806
cement	273	819
grain	134	243
container	120.1/961	212.2/1698
total	1835	3566

The MCRW potential of each channel is sorted out, and the MCRW potential of the four major water transport channels in Zhejiang Province is summarized in table 9.

Table 9 Potential of MCRW in four major channels in Zhejiang Province (unit: 10,000 tons)

Channel direction	2025	2030
Eastern Zhejiang passage	326	621
Northern Zhejiang passage	1143	1923
Southern Zhejiang passage	220	411
West Zhejiang passage	146	611
Total	1835	3566

It can be seen that by 2025, the development potential of MCRW in the northern Zhejiang Channel is the greatest, and it should be regarded as a key area of attention; The east Zhejiang Passage has great potential of coal MCRW. The potential of MCRW in south Zhejiang passage is mainly cement; The West Zhejiang Channel is still in the stage of construction and development by 2025, and the potential of the relevant cargo MCRW has not been fully stimulated.

By 2030, the scale of container and coal MCRW in northern Zhejiang Channel is expected to continue to expand, further consolidating the first position of MCRW potential in Zhejiang Province; The development of coal and grain MCRW in the eastern Zhejiang Passage continued to advance; The MCRW of the mine construction and cement in the west Zhejiang Passage is obvious, the MCRW of the container is further promoted, and the comprehensive potential scale continues to narrow with that of the East Zhejiang Passage; Due to the limited demand for coal and grain in the inland hinterland of the southern

Zhejiang Channel, it is difficult to effectively promote the MCRW of related cargo transportation logistics, and the increment of container MCRW is small, and the comprehensive potential is small.

Acknowledgements

With special thanks to the Zhejiang Provincial Department of Transportation and Zhejiang Port and Channel Management Center for the data sources and their valuable information sharing.

This research was supported by the independent research and development project of Zhejiang Science Research Institute of Transport [grant number ZK202322; ZK202323]; Science and technology plan project of Zhejiang Provincial Department of Transportation [grant number 202205]; Zhejiang Provincial Science and Technology Plan Project [grant number 2022C35068].

Disclosure statement

No potential competing interest was reported by the authors.

References

1. Dong, J., Li, Y., Li, W., Liu, S., 2022. CO2 Emission Reduction Potential of Road Transport to Achieve Carbon Neutrality in China. Sustainability 14, 5454.
2. Du, H., Li, Q., Liu, X., Peng, B., Southworth, F., 2021. Costs and potentials of reducing CO2 emissions in China's transport sector: Findings from an energy system analysis. Energy 234: 121163.
3. Lu, Q., Chai, J., Wang, S., Zhang, Z.G., Sun, X.C., 2020. Potential energy conservation and CO2 emissions reduction related to China's road transportation. Journal of Cleaner Production 245, 118892.
4. State Council of the PRC, 2022. Work Plan for Promoting the development of multimodal Transport, Optimizing and Adjusting the transport structure (2021-2015). http://www.gov.cn/xinwen/2022-01/07/content_5666976.htm
5. Tran, N.K., and Lam, J.S.L., 2022. Effects of container ship speed on CO2 emission, cargo lead time and supply chain costs. Research in Transportation Business & Management 43, 100723.
6. Ulrich, G., and Karl, W., 2022. Estimating transportation-related CO2 emissions of European city tourism, Journal of Sustainable Tourism, 30:1, 145-168, DOI: 10.1080/09669582.2021.1939708.
7. Ülker, D., Bayırhan, i., Mersin, K., Gazioğlu, C., 2021. A comparative CO2 emissions analysis and mitigation strategies of short-sea shipping and road transport in the Marmara Region. Carbon Management 12(1), 1-12.
8. Wei, X., Chang, Y., Kwon, O., Zhang, N., 2021. Potential gains of trading CO2 emissions in the Chinese transportation sector. Transportation Research Part D: Transport and Environment 90: 102639.
9. Xing, H., Spence, S., Chen, H., 2020. A comprehensive review on countermeasures for CO2 emissions from ships. Renewable and Sustainable Energy Reviews, 134: 110222.
10. Zhou, D., Huang, F., Wang, Q., Liu, X., 2021. The role of structure change in driving CO2 emissions from China's waterway transport sector. Resources, Conservation and Recycling. 171: 105627.

Achieving energy transition through spatial planning - A case of onshore wind energy planning procedures optimization in China

Jinjin Guan[1,*], Jun Xiong[2]

[1]School of Art Design and Media, East China University of Science and Technology, Shanghai, China
[2]China Securities Depository and Clearing Corporation Limited Shanghai Branch, Shanghai, China

Abstract. Wind energy is expected to be an effective solution to mitigate climate change and environmental pollution with broad resource distribution, decarbonization characteristics, and mature technologies. However, the rapid expansion of wind energy may also bring about new issues in space allocation and land use conflicts. Integration research of wind energy into spatial planning framework is imperative. Based on the dynamic change theories and conventional energy planning process, this paper proposed a planning framework with four main phases: 1) preparation, analysis, and orientation; 2) model design; 3) optimization and systematization; 4) implementation and supervision. Under the new framework, the connotation of wind energy planning and existing issues have been re-examined, and the planning procedures have been reorganized to correspond to the Chinese spatial planning system and energy development targets. Principles, priority, and systematization measures were discussed to optimize the planning system from the organizational perspective. The newly constructed framework connected the wind energy development plan and spatial planning into well-organized procedures with definite principles, targets, implementation guidelines, and supervision mechanisms at each planning level. The outcome of the research can be utilized as a guideline strategically positioning of planning goals and content.

1 Introduction

With the progressive energy transition globally, wind energy is recognized as a reliable solution to reduce carbon emissions, mitigate climate change, and increase regional energy security [1][2]. After decades of development, onshore wind energy has developed mature technologies, manufacturing chains, and market mechanisms. Promoted by the Paris Agreement that promised the decarbonization targets, various countries set roadmaps for energy transition.

The Chinese government has also set corresponding targets to improve the proportion of non-fossil energy to 42% by 2035 and 65% by 2050. It also committed a total reduction of CO2 intensity of 66% relative to 2005 [3]. By 2022, the national wind capacity exceeded 329 GW, accounting for 40 % of the global share.

In the energy transition process, the spatial planning regimes have direct influences on energy efficiency and environmental issues of wind energy. With the initial high-quality wind resource sites being occupied, searching available areas for wind farms is getting harder. Especially in intensive land use regions, land use compliance, environmental impact, and local resistance constitute the main obstacles to wind energy expansion [4]. Additionally, the wind turbines and infrastructure are recognized as the visual impairment of the landscape, which causes aesthetic value decline and the loss of local identity [5]. With the increasingly prominent contradiction between spatial planning and energy development, integration research of wind energy into spatial planning framework is imperative.

This paper aims to construct a proper framework to integrate wind energy into the spatial planning regimes of China. The wind energy planning system is optimized from connotation re-examination, existing issues analysis, planning targets assignment, standardization of planning procedures, and sustainability in space utilization to cope with the evolution of spatial planning system and decarbonization demand of China.

2 Methods

This section discusses which phases are needed in the theoretical framework to integrate wind energy and spatial planning, the critical problems in each phase of the framework, and which content needs to be integrated. Drawing on Hill's theory of dynamic change, policy implementation is the externalization performance of the development of public projects, which provides the perspective of dynamic change to abstract the framework from the policy implementation process [6]. Based on the theoretical model for energy planning proposed by Mirakyan and Guio [7], this paper constructed a framework to gradually deconstruct the integration of wind energy and spatial planning with the following four

* Corresponding author: guanjinjin@ecust.edu.cn

phases: 1) Preparation, analysis, and orientations; 2) Model design; 3) Prioritization and systematization; 4) Implementation and supervision (Fig. 1).

This paper discusses the detailed planning regimes and procedures of wind energy planning under the planning framework. Phase Ⅰ laid the foundation for the framework construction by re-examining the connotation of planning system, analyzing the existing issues and evolution targets of the spatial planning system. Phase Ⅱ designed the theoretical model for wind energy planning by identifying planning targets, principles, implementation paths, and detailed procedures of wind energy at each planning level. In phase Ⅲ, recommendations for optimization planning regimes were proposed, including the principles, priority, and systematization for the wind energy planning procedures. Phase Ⅳ focuses on planning implementation and supervision, which tries to learn foreign planning experience and explores a suitable roadmap for Chinese wind energy planning.

3 Results

3.1 Chinese spatial planning regimes evolution

This stage is primarily backgrounded interpretation and policy analysis to clarify the connotation of spatial planning, existing issues, and developing orientation during wind energy and spatial planning integration in China.

Wind energy planning consists of development plan and spatial planning, with comprehensive targets from strategic, industrial, economic, and spatial dimensions. The development plan emphasizes coordinating social, industrial, economic, and administrative systems by technical and policy instruments, while spatial planning focuses more on spatial distribution, resource allocation, and their coupling relationships. However, there is a significant gap between these two plans: the planning principles, planning objectives, data processing methods, and standards are not well connected.

Even though the central government issued the experimental pilot Multiple Planning Integration, there is no precise regulation to coordinate comprehensive planning and sectoral planning. The vast number of departments dealing with sectoral planning cannot coexist in an overlapping spatial range and similar planning contents, causing difficulties in planning implementation at the local level. Repeated planning contents waste social resources and prevent each type of sectoral planning development.

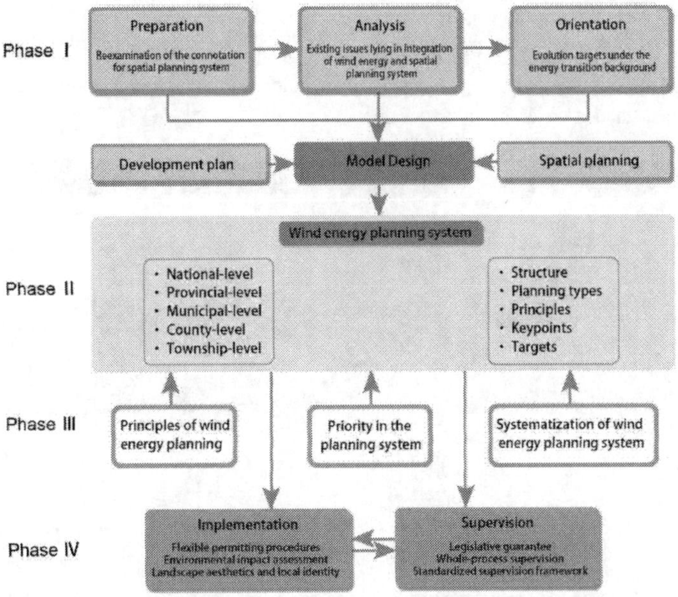

Figure 1. Framework of wind energy planning system

In the last few decades, wind farms have been planned in an extensive model for rapid development. There is no appreciable differentiation of planning methods, objectives, and technological standards among different planning levels, representing a hidden danger to newly planned wind farms under a more complicated social background [8].

The synchronization of sectoral planning in China faces a number of challenges: 1) There are overlapping and misplaced issues among several sectoral planning; 2) Planning durations are not correspondent; 3) Planning principles and tools are not standardized; 4) There is a lack of communication and coordination channel. Learning from Germany, through cross-regional cooperation between different authorities, renewable energy can obtain more efficient and precise development positioning and inter-regional coordination. The Spatial Planning Act (ROG) and wind energy decree in each state set clear laws, procedures, and leading authorities for wind power planning [9].

China's Multiple Planning Integration policy is based on the coupling and convergence of various types of sectoral planning in the national space. It is a seamless fit in the area and an integration of planning objectives, durations, scales, and methods. The new planning system is expected to eliminate the losses caused by internal systems such as cross-regional and cross-departmental repetitive work and resource competition.

Referenced European countries' experiences, a semi-open planning framework provides the interface between national development goals and the implementation of wind farms at regional and local levels [10]. The flexible land use ratio considers local governments' development goals and industrial structure. Furthermore, setting flexible land ratios and buffer distances for wind farms also considers the industrial structure and development priorities of local governments [11].

3.2 Model design

From the macro-perspective, wind energy planning can be classified into two sections: development plan and spatial planning [12]. The Wind Energy Development Plan encompasses detecting, evaluating, and utilizing wind resources and then setting development targets for wind facilities. It should also coordinate the policies, technologies, and open market to design a reasonable roadmap over a midterm or long-term period. The Medium and Long-term Development Plan for Renewable Energy (2050) and the Renewable Energy Five-year Plan (short-term) constitute the Wind Energy Development Plan by setting the national strategic target of installing 1 TW of wind energy by 2050, meeting 17% of domestic electricity consumption [13]. The development plan aims to coordinate general social-economic development with the specific industry development of political, economic, and spatial control policies. However, the achievement of the targets is finally conducted through the spatial planning system. The Wind Energy Development Plan must be combined with spatial-related implementation to locate the strategic objectives at the spatial level.

Among the spatial planning system, wind energy belongs to sectoral planning, developed by the Energy Ministry and other related authorities. The wind energy spatial planning system aims to disintegrate strategic targets of development plans into a vertically structured spatial planning system. It is released by planning authorities at the provincial, municipal, county and township levels, aiming to transfer the abstract targets of the Wind Energy Development Plan to spatial distribution. Other sectoral planning can be divided into assistant and constraint planning according to their relationships with the wind energy industry chain. For instance, the transportation plan, infrastructure facility plan, and electricity grid plan are associated with wind energy. Constraint plans, such as landscape planning, ecological protection planning, and cultural heritage planning, usually compete with wind energy in land use.

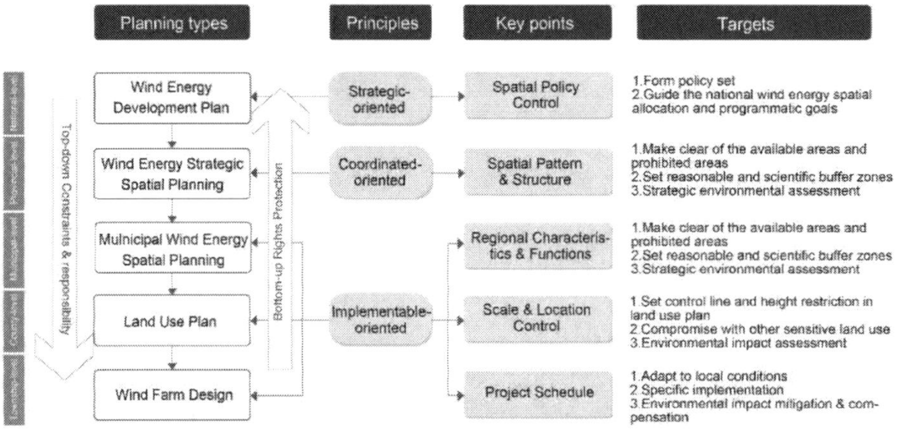

Figure 2. Structure, principle, and targets of wind energy planning system (edited by the author)

There is a great necessity to integrate the development plan and spatial planning of wind energy in the same planning system with a hierarchical structure, as illustrated in Figure 2. The system includes multiple planning levels corresponding to comprehensive planning. Specific principles, key points, and targets are issued at each level. The hierarchical structure features the mechanism of top-down constraints and bottom-up rights protection.

3.3 Optimization and systematization

3.3.1 Principles of wind energy planning

The prosperity of renewable energy development is not independent but is influenced by a complex context of global energy markets, technologies, geopolitics, etc. The dynamic evolution of spatial planning is closely connected to the energy transition. As a basic plan, spatial planning cannot set permanent standards, but it can set principles for wind energy development: environmental protection and sustainability.

- Environmental protection principle

Due to the various environmental impacts (noise, visual impact, flicker shadow, soil erosion, etc.) emitted by wind turbines [14][15], the social opposition to wind energy has become a critical obstacle to wind farm planning and construction. There is a growing consensus among the public that environmental protection should be considered as the priority in wind energy expansion.

- Sustainability principle

It is unrealistic to expect purely quantitative growth in wind energy capacity. Instead, we call for a sustainable roadmap following the principle of fair resource utilization. With a considerable energy consumption market in China, the total primary energy consumption of 4.64 billion tones of standard coal equivalent (tce) in 2018 [16], the energy transition should move from a coal resource-based economy to a sustainable economy based on affordable, stable, and green energy. In another aspect, wind energy development should also be gradual and proceed in parallel with other energy-supporting facilities to reduce the inner waste due to improper planning.

- Intensive land use principle

With the expected population growth and built-up land, the land reserved for renewable energy is limited and unavoidably mixed with ecological space, agricultural land, countryside, forest, etc. An intensive energy transition model combines land for various functions such as photovoltaic, wind power, agriculture, and fishery to maximize economic efficiency. Additionally, energy-efficient spatial structures can improve the density and output of hybrid energy facilities, and promote the spatial allocation between the energy production- and consumption side [17].

3.3.2 Priority in the planning system

Renewable energy can reduce carbon emissions and mitigate environmental pollution. On the other hand, renewable energy has side effects that impact the environment, although far less than conventional energy [18]. The priority of spatial planning is the sustainable use of natural resources, which means irreversible damage is not allowed in the natural environment. Even if many benefits can be obtained in the short term, the necessary conditions for long-term development, namely an excellent ecological environment, cannot be ignored. Wind energy planning should make concessions when it conflicts with ecological resources such as basic farmland, ecological reserves, and national parks.

The priority of environmental protection depends on a comprehensive legislation and regulation system. Guaranteed by the Renewable Energy Law (amended in 2010) and Environmental Impact Assessment Law (amended in 2016), environmental and ecological conservation awareness has been raised. This is reflected in the promulgation of more related regulations, such as " Interim Procedures for the Administration of Examination and Approval of Construction Facilities in National Nature Reserves" (March 2018) and "Notice on Further Strengthening the Management of National Forest Parks" (December 2017).

3.3.3 Systematization of wind energy planning system

Based on the initial framework proposed in phase II (Fig.2), it is necessary to optimize the wind energy planning system with differentiated targets corresponding to comprehensive planning at each level. Under the consensus principles of environmental protection, sustainability, and intensive land use (see 3.3.1), detailed planning procedures are further discussed in Figure 3. It optimizes the framework of wind farm planning from the organizational perspective in China and emphasizes the primary tasks at each step. Wind energy planning has spatial and temporal scopes, targets, planning instruments, and technical indicators corresponding to respective planning stages. For different participants, their responsible and assisting roles are also assigned clearly.

3.4 Implementation and supervision

The wind energy planning implementation contains extensive prework involving the investigation of geological conditions, microclimate, soil, ecological environment, and local acceptance of wind energy. During the planning process, permitting is challenging, which decides whether the project can be implemented. In China, the permitting process contains not only an Environmental Impact Assessment Permit but also Land Use Permit, Planning Permit, and Construction Permit [19]. It is recommended to refer to the case in Germany, where wind turbines with different heights and numbers correspond to different permitting procedures. Wind turbines below 50 meters are subject to the Building Law. In view of the construction of a large number of GW-level wind farms in northern China, the disturbance of wind farms to microclimate, soil erosion, and ecological environment also requires continuous observation and research [18]. Additionally, the change in landscape aesthetics and local identity should be included in the Environmental Impact Assessment from a cultural perspective.

Supervision is a critical part of the spatial planning regime, ensuring the quality of the implementation and bottom-up feedback channel. The complex legislative framework challenges the enforcement of supervision. For high-polluting projects, supervision plays a significant role in permitting process, which decides whether the project can be approved and constructed [20]. The optimization of the supervision system reflects in two aspects. Firstly, the supervision should shift from a single, conclusion-oriented target to whole-process supervision. It enhances supervision from source and process and ensures the legality and rationality of the entire Plan. Secondly, the supervision should shift from single-department supervision to a strict, standardized

 https://doi.org/10.1051/e3sconf/202344103008

supervision framework to not facilitate any problems that may exploit loopholes.

4 Conclusions

This paper proposed a four-phase framework based on the dynamic theory and conventional energy planning system, which aimed to integrate wind energy into spatial planning system under the evolution of Chinese urbanization and Multiple Planning Integration background. The study sequentially reorganized the wind energy planning system with literature research on spatial planning regimes, existing issues, and evolution orientation. Based on the model dealing with multi-level principles, key points, and targets, the proposed framework was optimized by discussing the principles, priority, and standardization of the regimes. And finally, the planning implementation and supervision recommendations were put forward for further research.

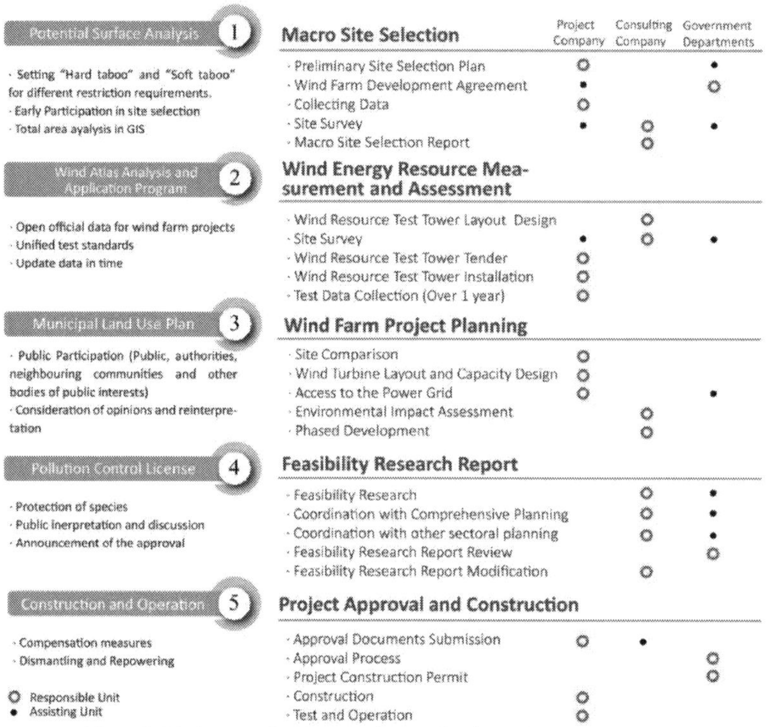

Figure 3. Wind energy planning procedures and optimization recommendations (edited by the author)

References

1. REN21 (Renewable Energy Policy Network for the 21st Century), *Renewables 2018 Global Status Report. A comprehensive annual overview of the state of renewable energy*. 2018.
2. L. C. Lau, K. T. Lee, and A. R. Mohamed, "Global warming mitigation and renewable energy policy development from the Kyoto Protocol to the Copenhagen Accord - A comment," *Renew. Sustain. Energy Rev.*, vol. 16, no. 7, pp. 5280–5284, 2012.
3. CNDRC (China National Development & Reform Commission) and CNREC(China National Renewable Energy Centre), "China Renewable Energy Outlook 2019 Executive Summary," 2019.
4. J. R. F. Diógenes, J. Claro, J. C. Rodrigues, and M. V. Loureiro, "Barriers to onshore wind energy implementation: A systematic review," *Energy Res. Soc. Sci.*, vol. 60, no. October 2018, pp. 1–33, 2020.
5. S. Linke, "Ästhetik, Werte und Landschaft," in *Landschaftsästhetik und Landschaftswandel*, O. Kühne, H. Megerle, and F. Weber, Eds. Wiesbaden: Springer VS., 2017, pp. 23–40.
6. M. Hill, *The Public Policy Process*, Fourth edi. England: Pearson Education, 2005.
7. A. Mirakyan and R. De Guio, "Integrated energy planning in cities and territories: A review of methods and tools," *Renew. Sustain. Energy Rev.*, vol. 22, pp. 289–297, 2013.
8. J. Guan and H. Zepp, "Factors Affecting the Community Acceptance of Onshore Wind Farms: A Case Study of the Zhongying Wind Farm in Eastern China," *Sustainability*, vol. 12, no. 7, p. 6894, 2020.

9. Fachagentur Windenergie an Land, "Überblick zu den Abstandsempfehlungen zur Ausweisung von Windenergiegebieten in den Bundesländern," 2019.
10. J. Guan, "Lessons from German On-shore wind farm planning," *J. Phys. Conf. Ser. 1102 012029*, pp. 1–7, 2018.
11. F. Nkomo, "WWEA Policy Paper Series (PP-02-18-b)," Bonn, Germany, 2018.
12. Z. Dong, "Ten relations of territorial planning in the new era," *Resour. Sci.*, vol. 41, no. 9, pp. 1589–1599, 2019.
13. IEA (International Energy Agency) and ERI (Energy Research Institute), "China Wind Energy Development Roadmap 2050," Beijing, China, 2011.
14. P. Roddis, S. Carver, M. Dallimer, P. Norman, and G. Ziv, "The role of community acceptance in planning outcomes for onshore wind and solar farms: An energy justice analysis," *Appl. Energy*, vol. 226, no. October 2017, pp. 353–364, 2018.
15. S. M. Hammami, S. Chtourou, and A. Triki, "Identifying the determinants of community acceptance of renewable energy technologies: The case study of a wind energy project from Tunisia," *Renew. Sustain. Energy Rev.*, vol. 54, pp. 151–160, 2016.
16. CNBS (Chinese National Bureau of Statistics), "Statistical Bulletin of National Economic and Social Development in 2018," 2019. [Online]. Available: http://www.stats.gov.cn/tjsj/zxfb/201902/t20190228_1651265.html.
17. G. Stoeglehner, "Integrated spatial and energy planning : a means to reach sustainable development goals," *Evol. Institutional Econ. Rev.*, no. 0123456789, 2020.
18. X. Guo *et al.*, "The impact of onshore wind power projects on ecological corridors and landscape connectivity in Shanxi, China," *J. Clean. Prod.*, vol. 254, p. 120075, 2020.
19. J. Guan, "Westerly breezes and easterly gales: A comparison of legal, policy and planning regimes governing onshore wind in Germany and China," *Energy Res. Soc. Sci.*, vol. 9, no. 67, p. 101506, Sep. 2020.
20. L. Li, X. Ren, Y. Yang, P. Zhang, and X. Chen, "Analysis and recommendations for onshore wind power policies in China," *Renew. Sustain. Energy Rev.*, vol. 82, no. December 2016, pp. 156–167, 2018.

 https://doi.org/10.1051/e3sconf/202344103009

Mechanism and Method of Auxiliary Decision-Making for Dispatching Real-time Operation in Electric Spot Market

Jianguo SHI[1*], Zhijie DU[1], Qi GUO[1], Qianpeng HAO[1], Fancheng MENG[1], Qiang LI[1], Shun YAO[1], Wenze LI[1], Rongli LI[2]

[1] Inner Mongolia Power Dispatching Control Center, Hohhot 010010, China
[2] Inner Mongolia Power Science & Research Institute, Hohhot 010010, China

Abstract: Starting from the market rules of the pilot provinces of the domestic spot market, this paper analyzes the mechanism of market intervention and suspension, the mechanism of the power supply period, and the mechanism of market clearing anomaly. At the same time, combined with the real-time operation of the Inner Mongolia power grid, the traditional dispatching mode and the power spot market operation dispatching mode are compared. Given the abnormal frequency, section overload, and insufficient spinning reserve in real-time operation, an auxiliary decision mechanism of manual intervention in the power spot market is proposed. The mechanism includes load bias setting, section limit bias setting, deviation monitoring, and other auxiliary functions, which provide a method of quick intervention under the electric spot market to ensure the safe and stable operation of the power grid.

1. Introduction

On August 28, 2017, the State Council and the National Energy Administration issued the "Notice on the Pilot Construction of the Power Spot Market", establishing eight spot pilot areas, including the Inner Mongolia power grid. Initially, they began the construction of the power spot market [1]. The construction of a power spot market plays a decisive role in giving full play to the guiding role of price signal, ensuring the real-time balance of power and the safe operation of the power grid through the market mechanism. In addition, it is significant for the country to promote the realization of the goal of "carbon peak and carbon neutrality" [2-5]. To the requirements of relevant documents, each pilot province carried out a day-ahead market and a real-time market to explore the road for constructing the electric spot market. With the continuous improvement and update of market trading rules, the trading mechanism of the spot market is gradually maturing. However, many problems must be considered and solved in real-time operation.

There may be various problems during the real-time operation of the power grid, such as the frequency being constantly high or low and the equipment power flow needing to be constantly full or exceeding the limit. So, the dispatcher needs to carry out necessary manual intervention to prevent the situation from deteriorating further. In the traditional dispatching mode, the dispatcher usually adjusts the output of the thermal power unit to control the power flow quickly. However, many interests are involved in the electric spot market, so balancing the relationship between manual intervention and market operation is necessary to ensure a fair market environment. Therefore, it is particularly important to study the mechanism and method of auxiliary decision-making in the real-time dispatching operation to ensure the coordination of necessary manual intervention in market operation and dispatching. This paper first analyzes and summarizes the electric spot market intervention and suspension mechanism in various provinces. Combined with the situation of the Inner Mongolia power grid, an auxiliary decision-making tool for the electric spot market is proposed for the first time in China, which improves the executable ability of quick dispatch intervention in power grid emergencies. In addition, this tool also guarantees the long-term, uninterrupted operation of the electric spot market as well as the safety and stability of the power grid.

2. Electric market intervention mechanism

2.1. Coordination mechanism of manual intervention and market operation

Due to the emergency caused by extreme weather, supply and demand relationship, and emergency events, each electric market has corresponding treatment measures. In the context of a high proportion of new energy access and low energy carbonization, the possibility of an emergency in the power market is relatively increased. The uncertainty and intermittency of new energy generation are the main factors affecting the power balance [6].

*1483434843@qq.com

2.2. Typical spot market manual intervention processing mechanism

Implementing the principle of power grid safety first, the operation of the power spot market should be suspended immediately if there is a threat to the safety of the power grid. Given a short electricity supply, Guangdong Province prioritizes the market demand-side response. If the intraday power balance still cannot meet the requirements, the orderly power consumption plan is started according to government regulations, and the load forecast data is adjusted [7-9]. Zhejiang Province makes thermal power units quit AGC to adjust output in view of the insufficient reserve [10]. The capacity compensation mechanism was established in Shanxi Province to decouple the capacity cost recovery of the unit from the power generation operation, ensuring sufficient system reserve capacity during peak hours [11-12]. Gansu Province adopts the method of emergency intervention of thermal power unit output or call of thermal power unit start-up to deal with power grid safety constraints and accidents. The intervention period is counted separately. All the on-grid electricity of the intervention thermal power unit does not participate in the spot market settlement. The electricity and electricity charges in the day-ahead and real-time spot market are set to zero, and all on-grid electricity is recorded as monthly medium and long-term electricity.

In the case of the day-ahead market suspension, Shanxi Province, Guangdong Province, Shandong Province, and Zhejiang Province prepare the day-ahead power generation plan based on the principle of power grid safety. Day-ahead market clearing is not carried out on the day. The actual execution results and real-time market prices are used for settlement. In the case of the suspension of the real-time market, the settlement period corresponding to the market suspension period shall be based on the actual execution result and the real-time market price of the same period of the previous trading day. As for the suspension of the auxiliary service market, Guangdong Province and Shandong Province shall call the auxiliary service according to the principle of system needs and take the actual execution result and the market price of the auxiliary service in the same period of the previous trading day as the basis for the settlement. Among them, Shandong Province intervenes in the power market using price control when the price is abnormal and declares the corresponding trading period as the price control period [13]. When the electric spot market is suspended in Sichuan Province and Fujian Province, the power dispatching agency will prepare the day-ahead power generation plan and the real-time rolling power generation plan based on the principle of power grid safety and not carry out market transactions and clearing. Besides, the actual execution results will be settled by market trading rules or the method specified by the competent government department.

2.3. Conclusion

After analyzing the relevant rules and measures formulated by the provincial spot market of domestic typical pilot units for insufficient power supply period, insufficient spinning reserve period, price abnormal period, power grid security constraint or sudden failure, market suspension, and other issues, the following conclusions are drawn:

First, in the case of affecting the safe operation of the power grid, each typical pilot provincial spot market takes timely disposal measures according to the principle of safety to prevent the expansion of risks. What's more, to avoid the total suspension of the market as far as possible, the power market of each province takes measures to deal with classification. In other words, certain market intervention measures are taken to prevent the extension of the market suspension.

Second, suspending the spot market will lead to no operating results, resulting in a certain amount of suspension costs. Therefore, all regions have stipulated the settlement basis and settlement method when the market is suspended, and some regions put forward the principle of common cost sharing. In some provinces, the power trading institution adopts the method of price control together with the power dispatching institution and declare the corresponding trading period as the price control period.

Third, when it is in short supply in some periods of the load forecast, the market demand response resources are preferentially invoked to meet the power balance.

3. Research on manual intervention mechanism of Inner Mongolia power grid spot market

3.1. Inner Mongolia power spot market manual intervention mechanism

During the real-time operation of the spot power market, there are many problems, such as abnormal frequency, insufficient spinning reserve, and tie lie over-load, etc. To avoid market risks, the dispatcher should try not to manually adjust the unit generation plan and use a standardized process to deal with such problems.

3.2. Frequency anomaly

The main frequency of the grid will fluctuate near the power frequency under the action of the frequency modulation unit and the power balance constraint is given in equation 1.

$$\sum_{i=1}^{I} p_{i,t} - \sum_{j=1}^{J} P_{j,t} - s_t^+ + s_t^- = 0 \qquad (1)$$

In this equation, $p_{i,t}$ represents the total units output, $P_{j,t}$ represents the total load demand, s_t^+ and s_t^- represents the slack variable. The total units output is equal to total load demand, then the grid frequency remains stable. However, if there is a large continuous deviation of load forecast or a serious blockage of thermal power units, it may lead to insufficient capacity of frequency modulation units, contributing to the continuous abnormal frequency of the system. Under this circumstance, the dispatch agencies in many provinces will immediately call the thermal power unit by the market rules to restore the frequency. Inner Mongolia's power grid innovatively corrects ultra-short-term load forecasts to indirectly change the spot system clearing results to solve the problem of insufficient spinning reserves.

3.3. Section overload

Section power flow is a constraint in the process of market clearing calculation given in equation 2.

$$F_{m,t}^{min} - s_{m,t}^{-} \leq \sum_{i=1}^{I} SF_{m,i,t} p_{i,t} - \sum_{j=1}^{J} SF_{m,j,t} P_{j,t} \leq F_{m,t}^{max} + s_{m,t}^{+} \quad (2)$$

In this equation, $F_{m,t}^{min}$ and $F_{m,t}^{max}$ represents the lower and upper limits of the section, $SF_{m,i,t}$ and $SF_{m,j,t}$ represents the flow transfer factor of the section, $p_{i,t}$ represents the total units output, $P_{j,t}$ represents the total load demand, $s_{m,t}^{+}$and $s_{m,t}^{-}$ represents the slack variable. The section power flow needs to be kept within the constraint range from $F_{m,t}^{min}$ to $F_{m,t}^{max}$ for the grid to be safe and stable. In real-time operation, there may be a large deviation between the predicted value of the bus load and the actual value, which may lead to the continuous overload of the section power flow. This may cause serious grid accidents and should be controlled in time. In traditional dispatching mode, the dispatcher can adjust the output of relevant thermal power units according to sensitivity. However, to avoid market risks during real-time operation [14-15], Inner Mongolia Power Grid changed the section limit value to make the spot system recalculate to solve the problem of section overload for the first time in China.

3.4. Emergency plan of suspension of electric spot market

When there is a special situation that endangers the stable operation of the power grid and affects the order of the power market, the dispatcher shall intervene or suspend the power spot market in time. In addition, dispatchers should deal with accidents and arrange power system operations according to relevant regulations, and timely report to the competent government departments and energy regulatory agencies [16].

When intervening or suspending the power spot market, Inner Mongolia Power Grid takes the following measures:

First, when the day-ahead market is suspended, the corresponding intra-day and real-time markets are suspended simultaneously, restoring to the no-spot trading mode.

Second, when the real-time market is suspended, the power dispatching agency adjusts the output of thermal power units based on the real-time operation and ultra-short-term load forecast information to ensure a balanced power supply. During the suspension of the spot market, it will be restored to no spot trading mode, and the result of day-ahead market clearing will not be settled.

Third, the dispatching agency shall record the reasons for the suspension of the spot market and the measures taken, promptly filing with the competent government departments and energy regulators and announcing to the relevant market members. When the abnormal situation is resolved, and the grid resumes safe operation, the electric spot market will be resumed accordingly.

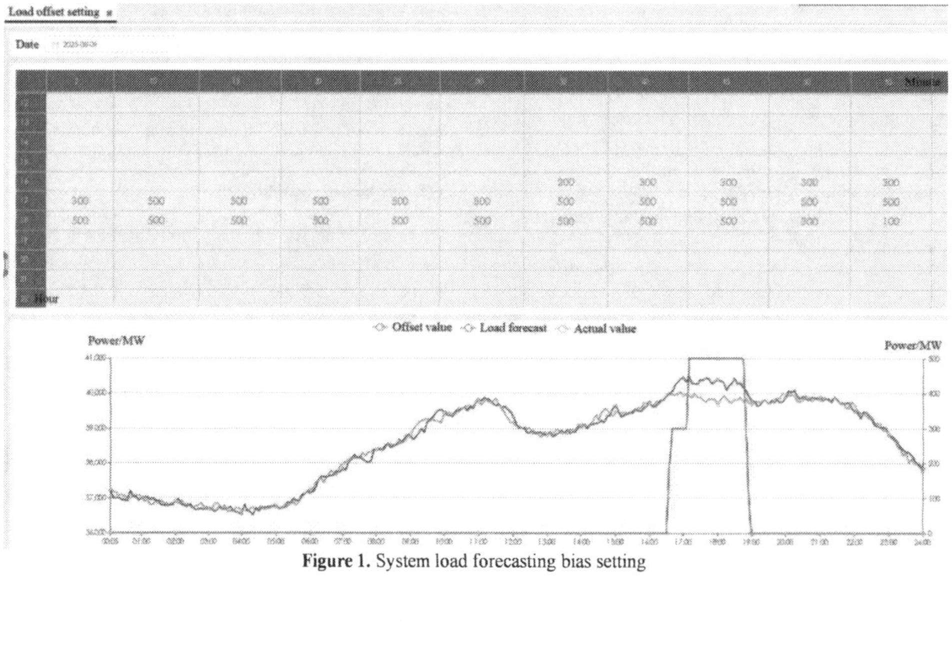

Figure 1. System load forecasting bias setting

4. Power dispatching real-time operation decision aid software

4.1. Load forecasting bias setting

When there is a large continuous deviation of load forecast or a serious blockage of thermal power units, the system frequency may be persistently high or low. To solve this problem effectively, the dispatcher can superimpose a bias quantity based on the forecast load through the software. The Inner Mongolia Power spot system periodically accesses the ultra-short-term system load forecast file, and the data range is the next 4 hours. Meanwhile, the spot system will directly modify the system load forecast data according to the load offset set by the dispatcher, which is used as the reference data for the later clearing calculation. Thus, the accuracy of spot system clearing results will be improved.

The front-end interface is shown in Figure 1, showing three curves, including the predicted load value, the actual load value, and the deviation value. The dispatcher can set the load bias through different periods on this page. However, the historical period is gray and cannot be modified. For example, a maximum load bias of 500 MW was set from 17:30 to 19:00 in Figure 1 so that the ultra-short-term prediction curve would be higher than the actual curve during the modified period. The spot market will calculate according to the adjusted load forecast curve and call the thermal power unit to compensate for the blocked output or unexpected actual load.

4.2. Section limit offset setting

In real-time operation, some sections may be continuously full load due to the large deviation of local bus load prediction. The dispatcher can set a bias quantity based on the normal limit of the section, and the market clearing calculation will control the suction power flow according to the superimposed section limit to solve the situation that the actual section power flow continues to be full.

The front page is shown in Figure 2, through which the dispatcher can set the limitation of each section according to different periods, supporting batch and box selection modifications. For example, the limit of the selected section was modified from 540 MW to 700 MW in Figure 2. Besides, the grey display of historical periods cannot be changed.

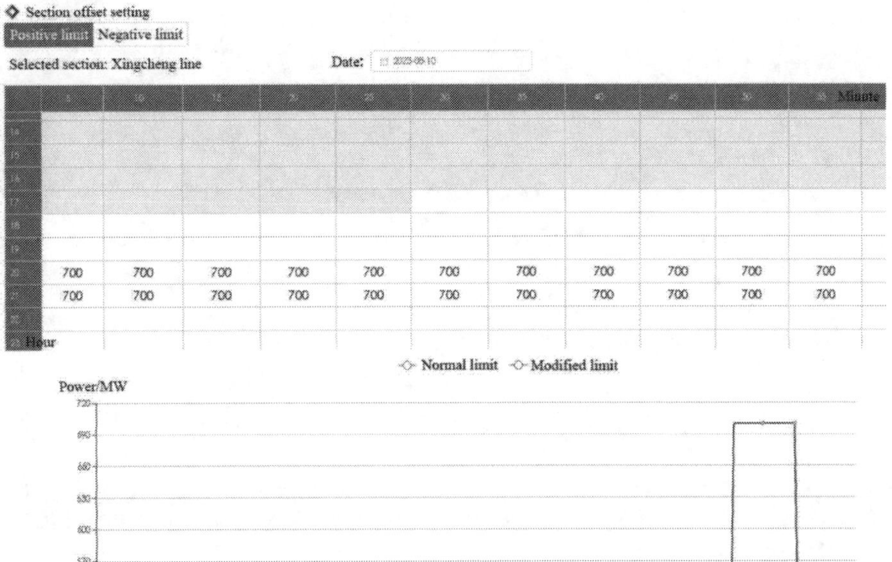

Figure 2. Section limit offset setting

4.3. Deviation monitoring

The reason for the deviation is that the planned power flow needs to be consistent with the actual power flow. Therefore, the active power deviation monitoring function is mainly used to assist the dispatcher in finding and adjusting the deviation between the plan and the actual value in time. The active power deviation monitoring function is mainly divided into four aspects: tie line monitoring, generation monitoring, bus load monitoring, and sectional power flow monitoring.

Tie line monitoring

Name of tie line Find

Name of tie line	Planned value	Actual value	Deviation value	Average value
North China tie line	-3150.00	-3061.78	88.22	-3412.00
Hangu tie line	-965.94	-889.32	76.62	-1015.09
Fengwan tie line	-2184.06	-2172.46	11.60	-2396.91

Figure 3. Tie line monitoring

The tie line monitoring function is mainly responsible for monitoring the power flow information of the Inner Mongolia power grid to the North China Power Grid, including the main pass, the South passage, and the north passage. The deviation amount and percentage are obtained by detecting the actual power flow and comparing it with the planned value, as shown in Figure 3. This feature supports object retrieval, sorting, and filtering by criteria. We click on a specific object to display the historical information, and it supports the curve export function.

Generation monitoring

Name of power plant Find

Name of power plant	Planned value	Actual value	Deviation value	Average value
Fenghuangling 2#G	241.49	296.65	55.16	0.00
Wusitai 2#G	221.00	274.00	53.00	0.00
Kangbashi 1#G	268.25	320.45	52.20	0.00
Huitengliang wind power	3.05	51.94	48.89	11.61
Songjiaqu 4#G	509.66	557.47	47.81	0.00
Fenghuangling 1#G	274.35	314.07	39.72	0.00

Figure 4. Generation Monitoring

Power generation monitoring periodically detects the planned output and actual output of thermal power units and new energy stations, as shown in Figure 4. The deviation value and percentage of deviation are obtained by detecting the actual output and comparing it with the planned output. This function supports object retrieval and arrangement. The historical planned output and actual output can be displayed by clicking on the specific power plant, and the curve export function is supported.

Bus load monitoring

Name of bus load Find

Name of bus load	Planned value	Actual value	Deviation value	Average value
Tailiang line	-0.73	37.77	38.50	-0.73
Jingao 1#T	0.00	37.28	37.28	0.00
Gaoqian line	8.21	40.55	32.34	8.51
Jingao 2#T	0.00	31.70	31.70	0.00
Xingang line 1	21.83	53.10	31.27	19.87
Ejina 1#T	1.00	32.13	31.13	1.00

Figure 5. Bus load monitoring

Bus load monitoring mainly displays the deviation between planned and actual values. The bus load prediction is transformed into the pure bus load prediction after the modeling modification of the small unit. Similarly, the monitoring function measures the actual load of the bus according to the small unit coupling model. Thus, the deviation value can be obtained by comparing the planned and actual values, as shown in Figure 5. This feature supports object retrieval, sorting, and filtering by criteria. We click on a specific object to display the historical predicted value, actual value, deviation value, and support curve export function.

Section power flow monitoring

Name of section Find

Name of section	Planned value	Actual value	Deviation value	Average value
Huayun three units	0.00	921.78	921.78	0.00
Load of Huangqihai & Tianpishan	0.00	911.25	911.25	0.00
Buwu line & Dehe line	-743.54	0.00	743.54	-1907.82
Load of Meiligeng	0.00	623.95	623.95	0.00
Dengxiang line	-169.21	397.12	566.33	415.93

Figure 6. Section power flow monitoring

Compared with the planned power flow, the section power flow monitoring function detects the current section power flow, obtaining the deviation value. Through this function, dispatchers can find the section with a large deviation and then analyze the possible reasons for the large deviation in this section to optimize the spot market operation results. The front-end page, as shown in Figure 6, displays the planned, actual, and deviation values of the power flow of each section and supports retrieval, sorting, and filtering according to requirements. We click on a specific object to display historical trend information and support the curve export function.

4.4 Evaluation of decision aid software

The auxiliary decision software was proposed by Inner Mongolia Power Grid for the first time in China, providing a method of manual intervention for the dispatcher. Besides, it provides a solution for the coexistence of the stable operation of the power grid and the uninterruptible operation of the power spot market. In the traditional dispatching mode, the dispatcher directly adjusts the power generation plan of the unit when various problems occur in real-time operation, such as continuous abnormal frequency, insufficient system backup, and exceeding the limit of section power flow. The advantage of direct intervention is that it can be timely and quickly intervened. However, the downside is that the dispatcher decides the intervention subjectively and may not be the best measure. Under the power spot market, there are many interests involved. So, the dispatcher should try to make decisions with the cooperation of the system, reducing the influence of subjective factors and ensuring the fairness of the market. The intervention through the auxiliary decision software only needs to set the offset amount without adjusting the specific unit plan, avoiding the market risk to the greatest extent and optimizing the spot market results.

5. Conclusion

Based on the real-time operation of the power spot market of the Inner Mongolia power grid, this paper draws the following conclusions:

First, the construction of the power spot market is currently in its infancy, and the pilot provinces are actively exploring and improving the rules of the spot market. When there is a state of emergency or abnormal spot system, the most serious may lead to suspending the spot market. Therefore, emergency measures should be configured in the spot market rules of each province, including power generation plan changes, auxiliary service market adjustments, and electricity settlement mechanisms. When the power spot market cannot persist in operation, it should be suspended in time to ensure the safety of the power grid.

Second, the deviation between the planned value and the actual value in the real-time operation of the power spot market may lead to various abnormal situations. For example, a large deviation of ultra-short-term load prediction or serious power generation obstruction of thermal plants will lead to continuous abnormal frequency. Similarly, a large deviation in the bus load prediction may lead to continuous exceeding of the section power flow. At this time, the dispatchers need to adjust in time to ensure the balance of power supply and demand.

Third, manual intervention will cause a deviation between the plan and actual value in the case of power spot market operation. For example, when the grid spinning reserve capacity is insufficient, some thermal power units may be called to generate beyond the planned value. At this time, on account of the many interests involved in the real-time operation of the spot market, corresponding rules and methods should be considered when manual intervention. In other words, manual intervention should ensure market fairness and avoid market risks as far as possible.

Fourth, to optimize the operation of the power spot market, the Inner Mongolia power grid has established the power flow monitoring system of the spot market for the first time in China. This system allows the unit with blocked output to be found in time and adjusted accordingly to ensure the balance of power supply and demand. Monitoring tie-line power flow helps analyze the power flow transfer to the North China power grid and judge the external power transmission capacity of the Inner Mongolia power grid. Moreover, monitoring the section power flow and bus load helps analyze the reason for the section power flow over-limit and optimize the small unit mounting model.

Fifth, to improve the maneuverability and compliance of manual intervention, Inner Mongolia Power Grid has innovatively developed an auxiliary decision-making tool. The dispatcher adjusts the grid indirectly by setting the load bias and section limit bias via this tool. After that, the spot system calculates the power output of each unit according to the modified data and distributes it. This tool solves the fairness problem in manual intervention from a system perspective.

Acknowledgment

The Young Talents Project of Inner Mongolia Power Company funded this research.

References

1. YANG Chunxiang, ZHANG Xuanrong, YAO Xun, et al. (2022) Construction Analysis Based on Domestic and Foreign Power Spot Market. Journal of Electrical Engineering, 17(4): 290-299.
2. LI Pingjun, FU Zhaoqing, LIU Kangping, et al. (2020) Analysis of Comprehensive Decision-Making of Power Generation Companies Based on Prospect Theory in Electricity Spot Market. Inner Mongolia Electric Power, 38(4): 13-19.
3. M. Khavaninzadeh, M. Rashidinejad and A. Abdollahi. (2015) Benefit of customers competition on power market. In: 2015 20th Conference on Electrical Power Distribution Networks Conference (EPDC). Zahedan, Iran, pp. 60-65.
4. LI Youliang, WANG Zhengfeng, LIANG Xiao, et al. (2022) Analysis of the Impact of Electricity Spot Market on Power Grid Dispatching. Journal of Anhui Electrical Engineering Professional Technique College, 27(2): 19-25.
5. C. Brunner and D. Möst. (2015) The impact of different flexibility options on future electricity spot prices in Germany. In: 2015 12th International Conference on the European Energy Market (EEM). Lisbon, Portugal, pp. 1-6.
6. F. Veselov, I. Erokhina and E. Nikulina. (2020) Modeling of Price Consequences in the Competitive Electricity Market in Russia with the Intensive Development of Non-Carbon Power Plants. In: 2020 13th International Conference "Management of large-scale system development" (MLSD). Moscow, Russia, pp. 1-4.
7. KUANG Minliang. (2023) Reflections on Formulating Medium and Long Term Trading Strategies of Thermal Power Generation Enterprises under Guangdong Electricity Spot Market. Guangdong Electric Power, 36(5): 11-17.
8. MA Hui, CHEN Yuguo, CHEN Ye, et al. (2018) Mechanism design of Southern China (starting from Guangdong Province) electric spot market [J]. Southern Power System Technology, 12(12): 42-48.
9. LIAO Binjie, LI Ruiqi. (2017) Research and design of grid energy dispatch model based on demand response. Electric Power, 50(7): 64-68, 152.
10. ZHOU Ziqing, DENG Hui, FANG Le, et al. (2022) New energy market trading mechanism and benefit analysis of Zhejiang electricity spot market. Zhejiang Electric Power, 41(8): 42-48.
11. WANG Xiaoang, ZOU Peng, REN Yuan, et al. (2022) Problems and Solutions of Medium & Long-term Trading Connected with Electricity Spot Market in Shanxi Province. Power System Technology, 46(1): 21-27.
12. DING Qiang, REN Yuan, HU Xiaojing, et al. (2021) Design and Practice of Joint Optimization Mechanism for Spot Market and Deep Peak Shaving Regulation Market of Shanxi in China. Power System Technology, 45(6): 2220-2227.
13. WANG Yong, YOU Daning, FANG Guanghua, et al. (2020) Mechanism Design and Trial Operation Analysis of Shandong Power Spot Market [J]. Electric Power, 53(9): 38-46.
14. WU Yunliang, LI Bao, LUO Huihong, et al. (2020) Conditional section constraints modeling in spot market clearing. Power System Technology, 44(8): 2819-2829.
15. Ricardo Cartaxo, Ângelo Casaleiro, Ricardo Pastor, et al. (2022) Market Coupling in Europe - Principles and Characteristics. In: 2022 4th International Conference on Power and Energy Technology (ICPET). Beijing, China. pp. 882-887.
16. XIAO Yunpeng, GUAN Yuheng, ZHANG Lan, et al. (2021) Analysis and Construction Path of Risk Hedging Mechanism in Centralized LMP-based Electricity Spot Market. Power System Technology, 45(10): 3982-3991.

Application of New Energy Logistics Vehicles in China: Evidence from Pilot City Surveys

Xiaoyi Li[1,2], Xiaoyu Tan[1,2,*], Meizhen Gao[1,2], Honglei Xu[1,2], Quansheng Huang[1,2], Renjie Wang[1,2], Xuan Ling[3]

[1]Transport Planning and Research Institute, Ministry of Transport, 100028, Beijing, China
[2]Laboratory of Transport Pollution Control and Monitoring Technology, Beijing, China, 100028
[3]Gesellschaft für Internationale Zusammenarbeit, GIZ, Bonn, Germany, 53113

Abstract. In the urban delivery sector, expanding the use of new energy vehicles (NEVs) can have a considerable positive impact on low-carbon sustainable development. To promote the application of new energy vehicle in logistics industry, China's Ministry of Transport launched Green Urban Freight Pilot Program in 2017 and has been achieved with promising results. However, the program has also encountered various restrictions and obstacles. This study carried out local investigations in six pilot cities of the Green Urban Freight Pilot Program. Key challenges of new energy logistics vehicles (NELV) adoption such as insignificant road access privileges, low cost-effectiveness, insufficient charging facilities, etc., were further identified through field research, online interviews and panel discussions. Then, detailed policy suggestions and recommendations targeting to national level and local level are proposed to promoting NELVs development.

1 Introduction

The fast development of China's logistics and delivery sectors has increased demand for short-distance urban freight services greatly. Urban distribution has become an essential component of modern logistics, and the sector desperately needs more logistics vehicles. New Energy Vehicles offer a realistic approach for reducing carbon emissions in China's expanding logistics industry. [1]. Since 2017, Ministry of Transport (MoT) launched the Green Urban Freight Pilot Program for the purpose of promoting new energy logistics vehicles (NELV) in the Chinese urban delivery sector. A total of 46 urban green freight distribution pilot cities were included in the first two batches. The program has yielded excellent results, with over 460 thousand NELVs on the road by 2020, and the number of these cars is growing. [2]. However, the program has experienced a number of constraints and issues, such as insignificant road access privileges, low cost-effectiveness, insufficient charging facilities, etc. Field investigations were conducted in six pilot cities: Chengdu, Foshan, Shenzhen, Yinchuan, Zhengzhou and Zhangjiakou in order to acquire first-hand information on the overall implementation status in the pilot cities of the MoT's Green Urban Freight Pilot Program. This study highlighted effective promotion measures and the key challenges for NELVs promotion, based on a case study of six pilot cities. Particular policy implications for the further promotion of NELVs in China were also proposed.

1.1 NELV application in the urban delivery sector

China is currently the world's largest new energy logistics vehicle production and marketing countries. However, NELV sales have been declining for three years in a row, beginning in 2018 due to the phase-out of purchasing incentives. In 2021, Sales of new energy logistics vehicles in China reached 31,000, growing at a 25% average annual growth rate.[3]. China's NELVs vehicle ownership has now reached 532,000, accounting for 7% of the country's total NEV ownership [4] (Fig. 1). Urban logistics vehicles have favorable circumstances for widespread promotion of new energy vehicles since they are characterized by relatively fixed routes, short- and medium-distance range, high frequency of usage, centralized charging, etc. In terms of charging infrastructure, by the end of 2021, China has built 2.617 million charging piles and 1,298 power stations, forming the world's largest charging and replacing network. The majority of the charging infrastructure is centered in the provinces of Guangdong, Jiangsu, Shaanxi, and Sichuan, as well as Beijing. The infrastructure was built primarily to service public buses and passenger cars, rather than logistics vehicles and others [5].

* Corresponding author: tanxy@126.com

https://doi.org/10.1051/e3sconf/202344103010

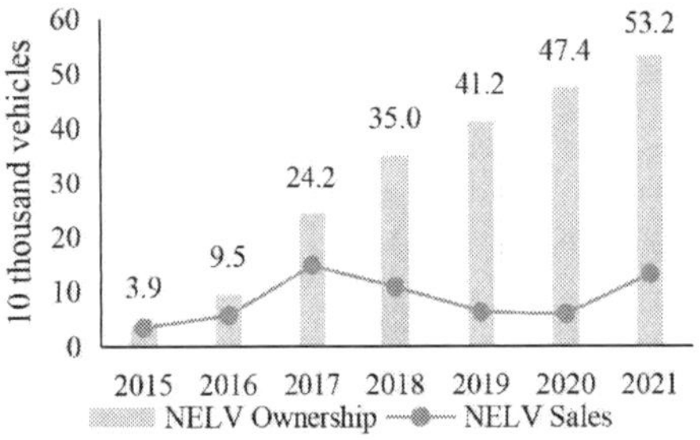

Fig. 1 Car Ownership and Sales of NELVs 2015-2021

1.2 NELV application scenario

NELVs are mostly used in three situations in the urban delivery sector: express deliveries, retailer supplies, and e-commerce sales, with top e-commerce firms typically owning their own fleet of logistics vehicles. However, due to several practical hurdles such as a lack of urban road connectivity, inadequate cost-effectiveness, and a lack of easy charging choices, the use of NELVs for personal delivery services is uncommon.

NELVs are primarily used for point-to-point express delivery within cities, but are rarely used for customised delivery services for individual end-users due to current technical limitations on travel ranges (how far vehicles can go on a single charge), recharging times and a lack of related infrastructure, and challenges with vehicle attendance rates and reliability.

Urban retail supply is also a major scenario for the application of new energy logistics vehicles. As new retail channels emerged, urban convenience stores such as Suning and T-mall began to grow on a big scale. Because urban convenience shops, franchise bakeries, and other urban merchants have set delivery hours, product kinds, and routes, urban retail delivery for these increasing services will remain an essential function that battery-electric logistics trucks may play.

In terms of self-built logistics transportation for e-commerce businesses, The features of logistical needs have changed to include deliveries across a variety of ranges, tiny quantities of items, many batches of units being sent over various periods, and short cycles of delivery traffic as a result of changes in urban inhabitants' consumption habits. Due to present urban traffic regulations and the fact that e-commerce businesses place a high value on time efficiency, it is challenging for conventional gasoline logistics trucks to keep up with the rising logistical needs of e-commerce.

Large-scale e-commerce companies like Alibaba and JD have started using NELVs in their urban delivery transport in response to new energy-saving and emission-reduction policies. JD had accomplished its pledge to replace 100% of its own conventional fuel logistics vehicles with electric vehicles in Beijing by January 2018. [6].

2 Survey

2.1 Survey in six pilot cities

In order to acquire first-hand data, this study conducted field investigation in six pilot cities: Chengdu, Foshan, Shenzhen, Yinchuan, Zhengzhou and Zhangjiakou (Fig. 2). The case cities were chosen for their unique characteristics. Chengdu and Shenzhen are two notable examples of NELV application in the transportation sector. Differences in geographic location, economy and climate of the regions are taken into account (such as Chengdu in the West, Zhengzhou in the central East, Yinchuan in the Northwest, Shenzhen and Foshan in the South). Differences in the types of new energy sources available, related policies and the implementation of supported energy systems were also taken into account (such as Zhangjiakou and Foshan).

Fig. 2 Pilot Cities Investigation Map

Four of the six pilot cities focus on battery electric vehicles promotion, with the other two, Zhangjiakou and Foshan, promoting both battery electric vehicles and fuel-cell electric vehicles (FCEVs). As its hydrogen supplies are produced using green power from nearby wind farms, Zhangjiakou is encouraging FCEVs. Additionally, the city advocated hydrogen as a green transportation choice for the 2022 Winter Olympic Games and offers geographical advantages for green hydrogen. Foshan has been presenting itself as the hydrogen capital of China since 2010, so the technology route for new energy vehicles in Foshan was focused on BELVs and FCEVs. This study analyzed the overall development in the urban delivery sector in the investigated cities by field investigations with local transportation authorities, relevant departments, and transportation enterprises.

The progress in NELV development between cities varies due to different emphasis and effort used within common measures such as through providing financial subsidies, right of way privileges, and support for the new energy infrastructure. There are also individual local measures being implemented, such as formulating relevant standards, building information platforms, optimising the approval process for the new energy infrastructure, and raising public awareness, which impact how the vehicles are being rolled out. A summary

https://doi.org/10.1051/e3sconf/202344103010

of the key characteristics of each case city are presented in Table 1.

Table 1 NELV Promotion Status among Investigated Cities in 2021

City	Number of NELVs	Infrastructure	Authorities	Subsidy Policy	Access Priority
Yinchuan	3000 BEVs	1,067 charging piles planned for 2022	Business Bureau	Vehicle acquisition link with national subsidies, local subsidies	NELVs are completely unlimited, and fuel logistics vehicles are restricted during some hours in the Xinhua business district
Zhengzhou	15000 BEVs	More than 20,000 charging piles	Transportation Bureau	National subsidies for vehicle acquisition, local subsidies for infrastructure construction, and local subsidies for operation	Differential traffic management measures are implemented for NEVs with green number plates, and the classification management system for NELVs is improved to facilitate the road passage of qualified van-type logistics vehicles. In the third ring, NEVs are not restricted except in the morning and evening rush hours. New energy trucks cannot drive into elevated roads.
Chengdu	30000 BEVs 100 FCEVs	23,700 charging piles	Department of Transportation Port and Logistics Office	Vehicle acquisition link with state subsidies	Allows NELVs (BEVs) to pass through urban areas for the full time and not subject to tail number restrictions; In 2020 and subsequent years, there will be no longer issue entry permissions for fuel logistics vehicles (except those that are determined to be necessary to retain)
Shenzhen	25208 BEVs	60,953 charging piles	Transportation Bureau	There are national and local subsidies for the acquisition and use of vehicles	Battery electric light and micro trucks that have completed electronic filing and are subject to supervision are allowed to drive on the rest of the roads in Shenzhen, except for that is prohibited from 7:30 to 21:00 from Monday to Friday; medium BEVS and heavy trucks that are subject to supervision and whose body length does not exceed 6 metres are allowed to drive on the sections and times where ordinary large trucks are restricted the section of Shennan Avenue
Foshan	448 FCEVs	16 hydrogen refuelling stations, 6,109 charging posts	Transportation Bureau	There are national and local subsidies for the acquisition and use of vehicles	Except for the morning rush hour from 7:00 to 9:00, the evening rush hour from 17:00 to 19:00, and the road sections around government agencies, schools and bridges that do not meet the conditions of height and weight restrictions, all other time periods and road sections in Foshan City have been abolished for medium-sized vehicles with an authorised load capacity of less than 5 tons (including) with new energy plates at the provincial level.
Zhangjiakou	5 FCEVs	Over 5,600 charging piles; 4 hydrogen refuelling stations	No	There are national and local subsidies for the acquisition and use of vehicles	No special access priority

3. Findings and discussions

Through field research, online interviews and panel discussions with municipal transport authorities, relevant departments and enterprises, best practice and key challenges of NELV promotion were further identified.

3.1 Effective policies and measures for NELV promotion

Policies and measures for NELV's include priority right of way planning, parking conveniences, industrial management support and related plans are issued locally and nationally to promote the use of NELVs in the urban delivery sector. Some pioneer cities like Shenzhen, Chengdu, and Zhengzhou have earned valuable experience and learnt lessons in these processes, and therefore serve as examples and references for the better promotion and adoption of NELVs in the future.

3.1.1 Implementation of preferential policies for NEVs

To maintain the momentum of new energy development, national purchase subsidies for NEVs, as well as purchase tax exemption policies have been extended to the end of 2022. Locally, subsidies for the purchase of NEVs have been replaced by financial and operational support for vehicle use and operation, in accordance with the national policy, with a view to promote them in an orderly manner. For example, NEVs are provided with special license plates by local governments as an incentive policy, with which they are permitted to access low emission zones. These measures are adopted to cushion the impact caused by cancelling the preferential purchase policy.

3.1.2 Improvement of integrated and specialised planning to enhance NELV's guiding and leading roles

Integrated NEV promotion plans and specialised plans for developing charging facilities have been issued at provincial and municipal levels, supporting the industry's development. Cities such as Shenzhen have integrated plans for the building of charging stations into their city planning, for example, by reserving space for charging piles in NELV drivers' residential compounds. Some cities, like Zhengzhou, have formulated specialised plans for the promotion of NEVs and the construction of charging facilities, and have set goals for supporting the use of NELVs in city planning. These plans have provided the foundation for the development of plans to use NELVs, and tackled problems that operators of the vehicles might face, such as insufficient power capacity without changing facilities, and the lack of standardised charging service fees.

3.1.3 Promotion of preferential right of way for NELVs

The provision of right of way privileges, or priority access to areas of the city at different times for NELVS, hold the key for the vehicles to be seen as a preferred method of transport and distribution for delivery vehicles in urban areas. Therefore to promote the role of NEVs in urban transportation and delivery, Chengdu has clarified their policy relating to the road privileges of different types of vehicles, and imposed no restrictions on the passage of NELVs in urban areas. Shenzhen, Chengdu, and Zhengzhou have completed digital registrations for the right of way of BELVs, with which the normal admittance certificate is not required before they can operate on roads within the cities, except for roads under special regulations.

3.1.4 Innovative operation models to drive NELV application

Cities like Shenzhen and Chengdu are actively exploring innovative operation and management models for promoting and adopting NEVs in their urban delivery sectors. Several cities, such as Shenzhen, have launched the 'Internet + new energy freight transportation' model, helping to reduce information gaps in the freight and logistics sectors by providing information on where and how NELVS can and may operate, such as through programs like the 'Green Truck Taxi', a platform similar to passenger ride-hailing business models and created by Green Wheel Electric Vehicle Co. Ltd. In addition, a considerable number of NELVs are employed in such freight transportation apps as Lalamove and GoGoVaN, providing new opportunities for sharing information about NELVs in operation.

3.2 Main obstacles and challenges for promoting NELVs in China

The study discovered the following constraints and difficulties for the development of NELVs in China through pilot city investigations.

Many cities lack right-of-way policies for new energy logistics vehicles. In comparison to traditional gasoline logistics trucks, most NELVs do not get a priority right of way when it comes to receiving traffic permits to enter urban areas.

Logistics vehicles powered by new energy have no cost advantage. NELVs and its essential components, such battery packs and other equipment, are substantially more expensive to acquire than traditional petroleum logistics Vehicles. Only when leased, NELVsmay match the total cost of ownership of traditional fuel cars. [7].

The existing deployment of public charging facilities, both in terms of their number and geographic distribution, is insufficient to fulfill the needs of NELVs for either immediate or future development.

Inadequate standardisation of NELV manufacture and operation has resulted in a lack of acceptable business management and maintenance procedures, as well as barriers to the successful execution of legislative incentives, such as subsidy allocation.

For NELVs at their after-sales stage, insufficient after-sales services have resulted in inconveniently lengthy maintenance procedures and excessive expenses. The number of after-sales service centers continues to fall short of demand, and there are also shortages of skilled professionals, uniformed quality assurance procedures, and rules for spare component replacement.

With China committing to carbon peaking by 2030 and carbon neutrality by 2060, there will be increased investment in new energy R&D, particularly in BEVs and FCEVs. The analysis demonstrates that for the majority of NELV application scenarios, BELVs can currently satisfy urban transit and delivery needs. However, Cold-chain (refrigerated) delivery is an exception, as existing BELVs cannot match its high-power consumption needs for temperature control during cargo transit. An exception to this is cold-chain (refrigerated) delivery, as existing BELVs cannot match its high-power consumption needs for temperature control during goods transit [8]. Given that cold-chain deliveries often service high-value commodities, somewhat expensive FCEVs may nevertheless be employed more cost-effectively for high-value cold-chain delivery.

4 Policy implications

4.1 Recommendations at the national level

Develop a policy agenda for NELV promotion. With the purpose to accelerate the promotion of NELVs, the purchase cost can be reduced through fiscal policies. Purchase subsidy policies for NELVs should be explored when the existing NEV purchase incentives are phased down. Governments at all levels should develop policies to subsidise the replacement of high emission vehicles of the China III and China IV emission standards with those of China VI and with vehicles powered by natural gas, electricity, and hydrogen. It is important to provide preferential tax policies for replacing and discarding older cars, and to keep in place the NELV vehicle purchase tax exemption.

The operational cost of NELVs could also be reduced through fiscal subsidies. An operation subsidy for NELVs based on their operation mileage should be developed and/or enhanced. The government should also formulate charging subsidy policies (for both the construction and operation of these services). In addition, more R&D funds should be planned to curb technological bottlenecks such as the development of high-performance batteries, the improvement of battery energy density, the upgrading of thermal management systems, the increase of range per charge, and the reduction of fuel cost.

The development and implementation of a New Energy Vehicle Credits (NEVC) mandate for commercial vehicles should be accelerated. Studies should be carried out on conventional commercial vehicles and new energy commercial vehicle manufacturers to understand the characteristics of commercial vehicles under various types, uses, scenarios, users, and regional conditions, as well as the progress and difficulties of the new energy transformation of commercial vehicles. With the consideration of technology and market factors, NEVC regulation for commercial vehicles should be grounded by phases based on different implementation regions, vehicle models, and application scenarios. In addition, the introduction of the NEVC system should consider how to be better at seamlessly transforming from the phasing out progress of subsidy policies to sustaining market momentum.

It is critical to have complete guidelines and standards for NELVs. Authorities should speed up the development of standards for NELVs and its main components and parts. A timetable and roadmap of standardisation for batteries, essential parts and charging facilities should be studied and put forward. Standards for vehicle battery reuse/recycling should be developed as soon as possible. This includes clarifying a unified process and the responsibility and cost allocation among all stakeholders. Through regulations and standards ensuring the vehicle quality of the original OEM, normalising vehicle operation, and controlling the vehicle scrapping process, resource recycling and subsequent effects on environmental protection could be maximised. It is necessary to create a thorough standard for handling NELVs during their whole life cycle. R&D, production, after-sale maintenance, administration of operations, scrapping and recycling are all included in this.

NELV after-sales service systems improvement. After-sale services for NELVs should be regulated, including clarification of responsibilities in maintenance, vehicle recall and replacement, after-sale dispute resolution, battery recycling and safety risk control. Customer convenience should be considered when planning after-sale service networks. The service scope should consider covering businesses along the NEV industrial chain, such as battery recycling, second-hand vehicle trade, and other related services. A standard service procedure needs to be established, including a service hotline for consultation and complaints, and the timely processing of customer requirement should be assured. To satisfy the growing demand for high-skilled workers, a training/education system for highly qualified professionals in after-sale technical services and management on NELVs should be built. Consultation and technical training should be conducted for freight companies to increase their understanding of NELV operation and troubleshooting for higher vehicle utilisation efficiency.

4.2 Recommendations at the local level

Continuous implementation of NELV road privileges. New city admittance permits should only be granted to NELVs. The existing permits currently used by conventional fuel vehicles could thereby only be kept if the vehicle is replaced by a NELV.

Policies on NELV parking privileges for loading/unloading should be implemented while also considering the road share of public bus lanes. Preferential measures on roadside parking should be put into practice. A feasibility study on sharing bus lanes with NELVs during certain periods should also be conducted.

Road priority for NELVs should be maximised. Fossil fuel-related emissions might be further decreased by creating green logistics zones and placing limits on conventional fuel logistics trucks in central areas. Fines should be increased, and road access restriction should be enforced for illegally retrofitted freight delivery purposes passenger vans to avoid harmful competition for freight companies.

Accelerating the construction of the charging facilities. It is important to optimize charging facility layout planning. Based on the analysis of the data from vehicle operation platforms and heat maps of urban logistics vehicles, relevant authorities should better plan the layout of charging stations geographically to match with NELV operation areas and speed up the improvement of charging networks. The service radius of the supporting infrastructure should be reduced.

To speed up the construction of charging infrastructure, the following measures should be adopted. It is important to research and develop fiscal subsidy policies for the development and operation of infrastructure for battery charging, battery swapping, and hydrogen refuelling. It is necessary to give financial support proportionally to the construction of charging piles. Market-oriented development should be encouraged and freed, attracting investments of multiple parties for the construction and operation of charging piles. The construction of the NELV charging network should be prioritised in places like logistics parks, urban hubs and arterial roads.

Fostering of innovative NELV application and organisation models. The viability of battery switching should be investigated further in order to develop a business model that separates car and battery sales (for example, introducing models like Battery-as-a-Service). New vehicle models should be developed based on practical real-world demand. In addition, power packs for battery-swapping vehicles should be standardised, the responsibility of battery safety should be clarified, and pilot programmes of battery-swapping models should be launched as soon as possible. Innovative collaboration strategies between NELV operators and finance institutions should be promoted. The government should play a coordinating role of connecting financing institutions with operators to strengthen their credit and loan support for NELV programmes, and to establish dedicated capital and innovation funds to attract and leverage more investment for industrial development. New technologies, such as Internet+, and cloud computing, should be employed to help establish a smart NELV operation platform for freight efficiency optimisation via big data analysis. The cumulative NELV operation data could then support the selection and customisation of NELV models to better serve various application scenarios, as well as to better improving the administration of urban logistics fleets, and to improve the efficiency of transportation organisations, thus making NELVs stand out in market competition.

5 Conclusion

The study reviewed the development of NELV application in the urban delivery sector. Key challenges and effective measures of NELV adoption in China were identified. The study also presents a complete set of policy suggestions for promoting NELVs to support the growth of a low-carbon urban delivery sector. Recommendations for different types of city can be drawn. Firstly, cities with a developed economy and good financial foundation can learn from Zhengzhou's financial measures to provide financial subsidies for NELVs and the required new energy infrastructure, mainly by means of capital subsidies. Zhengzhou offered additional local subsidies, in addition to national subsidies, for infrastructure construction and operation. Secondly, cities with a less developed economy and weaker financial foundation, but with room for innovation to roll out preferential vehicle right of way policies can learn from the right of way policy of Chengdu, to allow NELVs to maximise their use of right of way privileges and gain a clearer advantage over fuel vehicles. Thirdly, cities with a developed economy and room for innovation in right of way policies, can learn from Shenzhen's example and promote NELVs through the combined efforts of financial subsidies and right of way planning.

Acknowledgement

This research was funded by the National Key Research Development Program of China (Fund NO. 2022YFC3703600), Science and Technology Development Project of Transport Planning and Research Institute Ministry of Transport (Fund NO. 092217-316 and 092117-213).

References

1. Juan, Angel Alejandro, et al. "Electric vehicles in logistics and transportation: A survey on emerging environmental, strategic, and operational challenges." Energies 9.2 (2016): 86.
2. State Post Bureau of The People's Republic of China. (2022). 2021 China Express Development Index Report.
3. The Chinese Central Government's Official Website. (2022). China Governmental Work Report in 2022.
4. Jianjian, Wang, Ye, L., & Jingjing, L. (2020). Market Analysis of Electric Logistics Vehicles in 2019 and Outlook for 2020. Special Purpose Vehicle, 03, 47-52.
5. Jiaqi, W., & Jing, P. (2019). Analysis on Application Status and Trend of New Energy Urban Logistics Vehicles. Internal Combustion Engine & Parts, 12, 196-197.
6. Jiemian News. (2018). JD Logistics will replace its own vehicles in Beijing with new energy vehicles.
7. Qian.,D. (2020). TCO Competitiveness Analysis for New Energy Logistics Vehicles. Special Purpose Vehicle, 03, 65-67.
8. Quak, H., Nesterova, N., & van Rooijen, T. (2016). Possibilities and barriers for using electric-powered vehicles in city logistics practice. Transportation Research Procedia, 12, 157-169.

Clean Energy Contributes to Poverty Alleviation in China

Ping He[a*], Lin Chen[b]

Guilin University of Electronic Technology, Guilin, Guangxi 541004, China

Abstract. Poverty has long plagued humanity. The main factors affecting the gap between the rich and the poor are natural resources, human resources, information resources and science and technology.Using data from various clean energy industries and historical research methods, this paper shows that the clean energy industry is a new way to reduce poverty in developing countries. In the history of agricultural development, land and water resources are the important sources of wealth growth. In the history of industrial development, mineral resources, coal and oil energy have become the material basis for getting rich. In the 21st century with the great development of artificial intelligence, the importance of power resources is becoming more and more prominent, and clean energy such as solar energy and wind energy has become a new natural resource to create social wealth. Chinese successful practice in the development of clean energy industry shows that the clean energy industry is an effective way to get out of poverty. The results of the cooperation between China and African countries in developing clean energy industries show that the energy industry plays an increasingly important role in helping developing countries out of poverty. Although the development of clean energy industry is still restricted by market competition and monopoly organizations, the development road is still tortuous and long, it has a very bright future.

1. Introduction*

Since the 17th century, technological innovation has always led the pace of energy transition. Energy technology is one of the three cutting-edge technologies in the world. Clean energy is an innovative science and technology in the 21st century. As can be seen from Figure 1, in recent years,the clean energy industry has risen rapidly under the promotion of energy technology innovation. Especially after 2007, the research of the clean energy industry is gradually active and continues to heat up, focusing on solar and wind energy, especially solar photovoltaic technology.

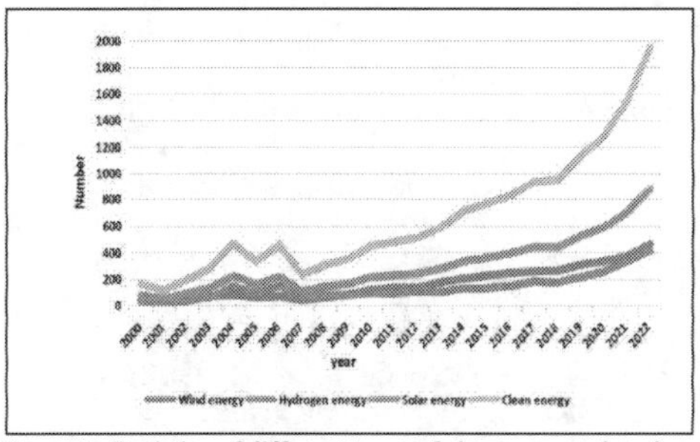

Figure 1. Statistics of different types of clean energy data from 2000 to 2022
Source: Elsevier Science

The existing research on clean energy covers many disciplines, such as energy, new energy materials, environmental science, chemistry, engineering technology and economics, Most of the research results are more than 100 papers. However, in the field of social sciences, clean energy research results are only 67, the number is small, the proportion is low, accounting for less than 5%. At present, most of the academic research focuses on green energy economy[1], the relationship between clean energy and environmental protection [2],the application of a certain material or technology in the clean energy industry [3], and solving energy poverty [4-5], etc., but they neglect clean energy as a new form of energy for human sustainable development. The clean energy is also an important strategic resource for national economic development. This paper breaks through the fields of energy and climate, and focuses on the perspective of clean energy industry as a new way of poverty alleviation in developing countries, which is of vital strategic significance for the development of developing countries. It helps developing countries to get enough attention in clean energy and helps human society realize the fourth energy transition. This study mainly explores a new path of poverty alleviation in developing countries through the clean energy industry, which has very important practical significance.

[a*] Corresponding author: heping@guet.edu.cn, [b] 1159933915@qq.com

2. Materials and Methods

There are two main database resources in this paper: one is government agencies, such as World Bank (IMF), International Energy Agency (IEA), National Bureau of Statistics (NBS), China Renewable Energy Engineering Institute (CREEI); the other is academic electronic library, such as Elsevier Science (SD) and Statista global statistical database.

This paper is a combination of quantitative and qualitative research with a historical and logical systematic review approach through data, journals, books and other publications closely related to energy, economics, technology, etc.

3. Natural Resources are very Important for Wealth Growth

Human life cannot be separated from material goods. With the development of society and the change of production mode, the sources of wealth growth are more and more extensive and rich.

3.1 In the Agricultural Age

Natural resources are important basis for human survival and development. In the age of agricultural civilization, the quantity and quality of land and water had a decisive influence on the growth of wealth. China, India, the Two River basin, Egypt and other areas with rich fertile land and water sources are often able to produce more food and other agricultural products, which are economically developed and have a prosperous society. On the contrary, areas with poor land and poor water sources are often unable to develop agricultural production, leading to social poverty and economic backwardness. In feudal times, China, relying on its vast region and affluent water sources, created rich material wealth and became a country with strong economic power.

3.2 In the Industrial Age

The growth of wealth in the era of industrial civilization mainly depended on industrial production, and energy was an important material basis of industrial production. Firstly, the Industrial Revolution began in Britain in the 18th century and it was based on energy sources and raw materials such as coal, iron and oil, which were the key elements of industrial production. With the industrial revolution, the demand for natural resources was increasing, especially for energy and raw materials such as coal, oil and iron. The large exploitation and utilization of these resources has promoted the rapid growth of industrial production. Secondly, the wealth growth in the industrial age was also closely related to the development and utilization of natural resources. The exploitation and utilization of natural resources have created great wealth for human beings. The exploitation of oil, coal and other energy sources has driven the development of oil and coal industries, thus promotes the economic growth. Kuwait in the Middle East, Saudi Arabia, Iraq, Qatar and other countries get rich by exporting oil. Saudi Arabia relying on high oil prices became the fastest growing country among G20.Saudi Arabia's oil revenue is $326 billion, and its oil export revenue accounted for more than 70% of Saudi Arabia's export revenue in 2022. Australia is the world's largest iron ore resource country, accounting for 29% of the world's economic resources. Australia is also the world ' s largest gold resource country, with an economic resources of 11,000 tons, accounting for one-fifth of the world's total. Mining is an important part of Australia's national economy, accounting for about 7% of its gross national product (GDP). Mining exports accounted for more than 60% of its total exports. Figure 2 shows that the nine of the top 10 countries in the world in total resources (according to statista2021) are in the top 20 in world economic GDP (according to the World Bank)

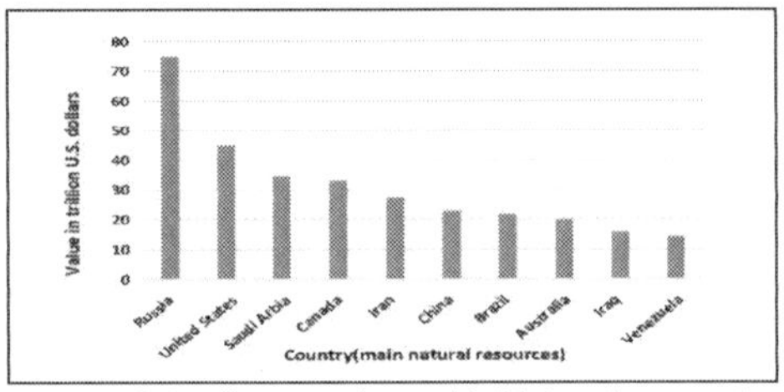

Figure 2. Leading countries worldwide based on natural resource value as of 2021(in trillion U.S.dollars)
Source:statista

3.3 In the Artificial Intelligence Age

In the 21st century of the great development of artificial intelligence, electric power is the core driving force supporting artificial intelligence ' s development. Clean energy is used in the power, information technology and transportation industries, and it has emerged as a new industrial system. Clean energy is seen as a new natural resource,which is one of the important resources such as coal, oil and gas, driving wealth growth[6]. Firstly, developing countries can vigorously develop the solar and wind energy industries in the solar and wind-rich areas by developing clean energy industries. Secondly, clean energy provides a large amount of power for industrial intelligent manufacturing. In the era of artificial intelligence, the large application of industrial intelligent manufacturing enables factories to improve production efficiency and reduce production costs. Finally, with the continuous development of high-speed railway and electric vehicle technology, electric power energy will play an increasingly important role in the future field of transportation[7]. The world ' s mainstream economies mainly generate electricity by coal, oil and natural gas, while developing countries can vigorously promote the new energy vehicle industry and use clean energy in automobile power.The clean energy industry provides a new way for developing countries without or lack of natural energy such as oil and natural gas.

4. The Development Profile of the Clean Energy Industry

In the historical process of the development of clean energy industry, the research and development and application of clean energy undoubtedly occupy a pivotal position. Clean energy refers to those advanced technology systems that have clean, efficient and systematic application, represented by solar energy, wind energy and hydrogen energy. The world is paying increasing attention to accelerating the transition to clean energy, and investment in clean energy is increasing. Figure 3 shows the authoritative statistics of the International Energy Agency, the total global energy investment in 2023 is about $2.8 trillion, of which more than $1.7 trillion will be invested in clean energy, while the investment in fossil energy is only $1.1 trillion, which is far more invested than fossil energy[8]. This undoubtedly marks that clean energy is leading the world into a new era of sustainable development and becoming the main direction of human energy development in the future.

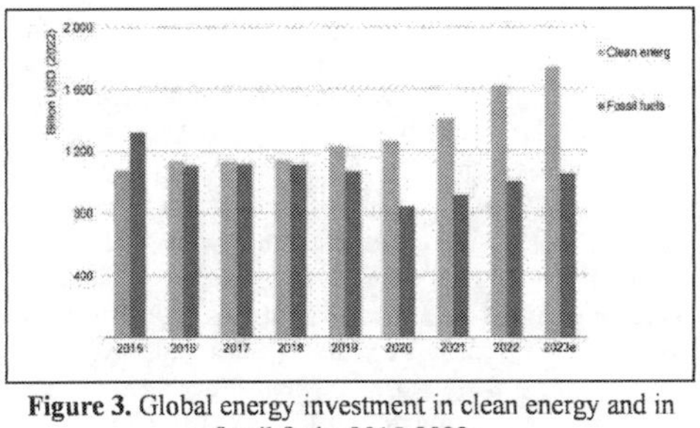

Figure 3. Global energy investment in clean energy and in fossil fuels, 2015-2023e.
Source: The International Energy Agency

The utilization of solar energy has a history of more than 3,000 years, at present, it is mainly used by photovoltaic power generation. In 1954, the American Bell Laboratory invented the mono-crystalline silicon solar cell. In 1996, the United Nations held the World Solar Summit in Zimbabwe. In 2004, under the pressure of a series of energy problems and environmental trends, Europe and the United States began to promote the energy transformation,and this year was called the first year of the world's photovoltaic industry. In 2013, China became the world's largest PV market. Today, solar energy is widely used in the domestic, commercial and industrial fields. Solar panels, solar water heaters, solar cars and other products are gradually popularized, making an important contribution to the alleviation of poverty in developing countries. In the early 20th century, German engineers invented the world's first wind turbine. In the 1960s and 1970s, wind power generation was used on a large scale. Since the 1980s, with the continuous progress of technology and the cost reduction, wind power generation has been widely used worldwide. Since 2000, as the global climate change problem becoming increasingly serious, countries have begun to strengthen international cooperation in the wind energy industry. China, Germany, the United States and other countries are vigorously developing the wind energy industry. In the future, with the continuous progress of technology, the wind energy industry will continue to develop, the global wind power generation is expected to reach more than 400GW by 2030,and the development potential is huge.

5. Chinese Clean Energy Achievements in Poverty Alleviation

Since hydrogen manufacturing technology is not as mature as solar and wind technology, the cost is relatively high, solar and wind are the first choice for poverty alleviation. The global installed cost of solar PV in 2021 fell by about 82% from 2010, the installed cost of onshore wind turbines and offshore wind power fell by about 35% and 41%, respectively, the cost of power generation also fell significantly.

China has now formed a complete clean energy supply chain and led to huge job growth. Global clean energy jobs have surpassed fossil fuels, and the Asia-Pacific region is expected to grow from about 33 million clean energy jobs today to nearly 55 million by 2030[9].

5.1 In the Solar Photovoltaic

In 2009, China launched the Golden Sun Project for poverty alleviation and power generation, and has made significant progress in terms of market, technology and raw materials. In 2013, China became the world's largest PV market. After nearly 30 years of development, China now leads the world in solar energy production and technology.These variations are shown in Figure 4.

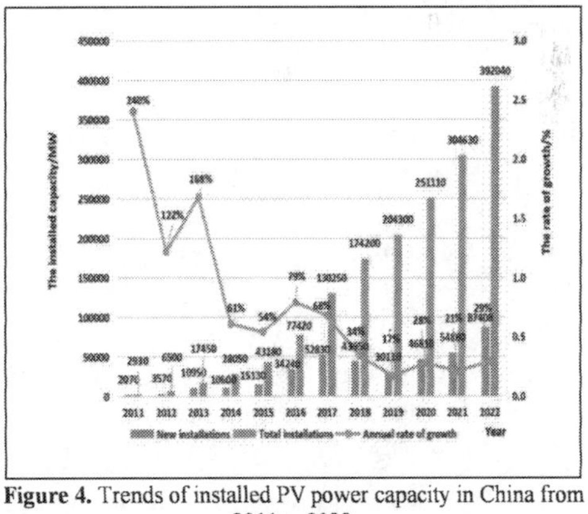

Figure 4. Trends of installed PV power capacity in China from 2011 to 2022.
Source: China Renewable Engineering Institute

Today,as can be seen from Figure 5, China is not only the world leading supplier of clean energy technologies ,but also a net exporter of many clean energy technologies. China has at least 60 percent of the world manufacturing capacity for mass production technologies.Such as solar PV, wind energy systems and batteries, and 40 percent of its electrolytic cell manufacturing capacity[10].In 2022, the total output value of China photovoltaic manufacturing side exceeded 1.4 trillion yuan, the export of photovoltaic products exceeded 51.2 billion US dollars, and the output of poly-silicon and photovoltaic modules ranked first in the world for 12 consecutive years and 16 years respectively[11].

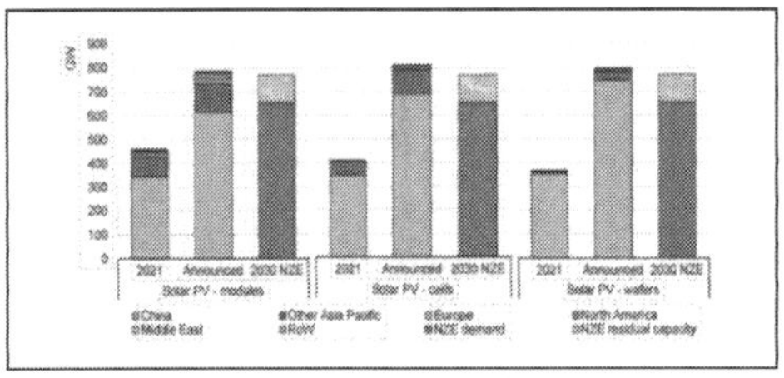

Figure 5. Chinese solar PV manufacturing capacity
Source: The International Energy Agency

China's western region shows significant advantages in the development of solar energy, with its vast territory, low population density and abundant sunlight. The distribution of solar photovoltaic power output in the western region to the eastern region can not only meet the electricity demand in the eastern region, but also achieve economic growth in the western region. In China's poverty alleviation plan, photovoltaic poverty alleviation is one of the top ten targeted poverty alleviation projects. According to the white paper Chinese Energy Development in the New Era released by The State Council of China, from 2014 to 2020, various forms of photovoltaic poverty alleviation projects will be promoted in accordance with government funding. A total of 26.36 million kilowatts of photovoltaic poverty alleviation power stations have been built, benefiting 60,000 poor villages and 4.15 million poor households, generating electricity benefits of about 18 billion yuan per year, and placing 1.25 million public welfare jobs accordingly[12].

5.2 In the Wind Energy

China is one of the world's largest markets for wind power. In 1966, China implemented the Wind Riding Plan and built 19 wind farms. As the technology of flexible grid scheduling becomes more and more mature, battery energy storage facilities can largely solve the problem of wind power instability. By 2023, Chinese installed wind power capacity has exceeded 500 gigawatts, accounting for nearly half of the world's total wind power installed capacity. This shows that China has become a major force in global wind power.These variations are shown in Figure 6.With more than two decades of development, China has a well-established supply chain for the wind power industry. The production of wind turbine nacelles and key components accounts for 60-70% of the global market share ,which makes China a crucial contributor to the global response to climate change[13].

The Chinese government has been vigorously promoting wind power projects in western China to solve local energy problems, boost economic development, and help achieve sustainable development goals. These projects have played a key role in poverty alleviation in the western region. The Chinese government sees the western wind project as an important way to achieve its sustainable development goals.

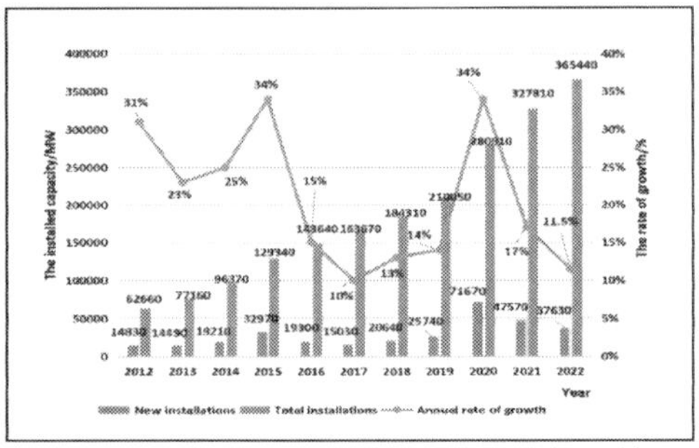

Figure 6: Chinese installed wind power capacity and its change trend from 2011 to 2022
Source: China Renewable Engineering Institute

5.3 In the High Speed Train

China has become the world's largest new energy vehicle production and sales country, As shown in Figure 7, China's 2022 international trade in electric vehicle exports accounted for 35% of global exports. The electric vehicle industry system has greatly increased the GDP growth brought by the automobile industry. While promoting economic growth, Chinese electric vehicle industry system also provides important support for global green mobility. With the continuous advancement of technology and the support of policies, the performance of electric vehicles has been significantly improved in the past few years, making them increasingly competitive in the global market. According to statistics, Chinese new energy vehicle production in 2022 more than 3 million, compared to 2012 more than 10 times[14].

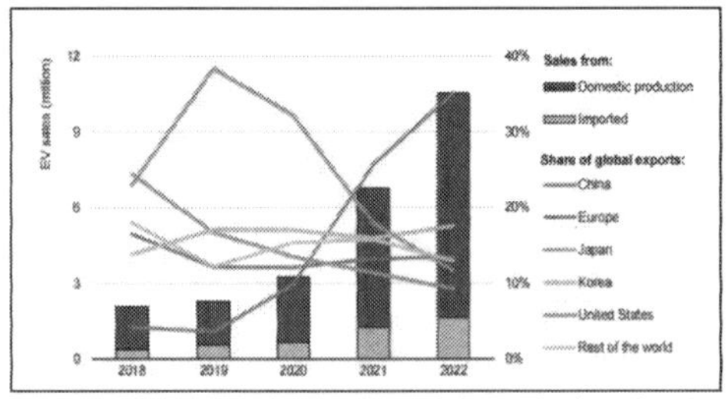

Figure 7. Global electric car sales and share of selected regions in global trade, 2018-2022.
Source: The International Energy Agency

6. Clean Energy in Cooperation between China and Developing Countries

China is a firm supporter and important partner of Africa's green transformation. By 2021, China has helped African countries build hundreds of clean energy and green development projects, and supported African countries to better leverage their advantages in clean energy such as solar and wind power to help Africa achieve poverty alleviation. The Garissa photovoltaic power generation project undertaken by Chinese enterprises in Kenya is currently the largest photovoltaic power station in East Africa. After the Garissa Photovoltaic Power Plant was put into operation in 2019, the annual electricity generation exceeded 76 million KWH, meeting the electricity demand of 70,000 households and more than 380,000 people, effectively alleviating the problem of electricity shortage in Kenya. The Dea Wind Farm in South Africa's Cape Province is China's first wind power project in Africa that integrates investment, construction and operation. The project provides a stable supply of about 760 million KWH of clean electricity to the local area every year, creating local jobs and driving the development of surrounding communities.

According to the latest report Africa Energy Outlook 2022 released by the International Energy Agency, 600 million people in Africa live without electricity, accounting for about 48% of the total population in Africa. The proportion of people without electricity is large. Although Africa has 60% of the world's solar energy resources, it only has 1% of the world's photovoltaic power generation installations, indicating that the development of renewable energy, especially solar energy, has a broad prospect in Africa.

7. Conclusion and Discussion

The development of clean energy industry has been the trend of the period, and now the World Energy Organization has recognized that clean energy is the best choice for the future. On the one hand, human needs a lot of energy for development, and clean energy can cope with the energy crisis and achieve sustainable development. On the other hand, the potential of clean energy is huge.The developing countries can combine their actual situation, and learn from China's advanced clean energy technology, because China's clean energy technology has gradually realized the use of affordable energy.

In the electric power era, The developing countries should catch the opportunity to utilize and vigorously develop the clean energy industry, occupy a dominant position in the energy transformation, and drive the development of related industries to get rid of poverty.

Acknowledgement

This work is supported by the Key Research and Development Program of Guilin University of Electronic Technology [Grant No.YXKSZ05]; Research Program of Guilin University of Electronic Technology[Grant No.YJSZS23]and [Grant No.YJSZSZ1V].

References

1. Fang,W.,Liu, W., Surya Putra, A.R.(2022)Role of research and development in green economic growth through renewable energy development: empirical evidence from South Asia, Renewable Energy .194:1142–1152, https://doi.org/10.1016/j.renene.2022.04.125.
2. Wu,D.S., Xie,Y, Liu D.J.(2023)Rethinking the complex effects of the clean energy transition on air pollution abatement: Evidence from Chinese coal-to-gas policy.energy，283: 128413.https://doi.org/10.1016/j.energy.2023.128413.
3. Yousef, S,Tuckute, S.(2021)Ultra-permeable CNTs/PES membranes with a very low CNTs content and high H2/N2 and CH4/N2 selectivity for clean energy extraction applications,Journal of Materials Research and Technology,15:5114-5127,https://doi.org/10.1016/j.jmrt.2021.10.125.
4. Adusah-Poku,F., Takeuchi,K.(2019) Energy poverty in Ghana: Any progress so far? Renewable and Sustainable Energy Reviews,112:853-864, https://doi.org/10.1016/j. rser.2019.06.038.
5. Paudel, J.(2021)Why Are People Energy Poor? Evidence From Ethnic Fractionalization102:105519, https://doi.org/10.1016/j.eneco.2021.105519.
6. Xu,H.X.,Dong,R.,Cui,Y.,Zan,W.C.,(2022)Does the Photovoltaic Poverty Alleviation Project Promote County Economic Development? : Evidence from 852 Counties in China，Solar Energy,248:5163,https://doi.org/10.1016/j.solener.2022.10.050.
7. Zou,C.N., Ma,F, Pan S.Q., Zhao,Q, Fu G.Y., Zhang,G.S., Yang,Y.C., Yu,H, Liang,Y.B., Lin M.J., Wang,Y, Xiong,B, Liu H.I.(2023)Global Energy Transition Revolution And the Connotation and Pathway of the Green and Intelligent Energy System，Petrol Explor Develop, 50(3): 722-740,https://doi.org/10.1016/S1876-3804(23)60423-9.
8. The International Energy Agency,World Energy Investment 2023pp6-8.
9. The International Energy Agency,World Energy Outlook 2022.
10. The International Energy Agency，The Clean Energy Technology Outlook for 2023.
11. China Renewable Engineering Institute，Chinese Renewable Energy Development Report 2022.P.17.
12. The State Council of China，Chinese energy development in the new era.
13. Global Wind Energy Council,Global wind report 2023.p.53.
14. The International Energy Agency，Global EV Outlook 2023.p.104.

 https://doi.org/10.1051/e3sconf/202344103012

Introduction to Domestic and International Sulfur Hexafluoride (SF_6) Greenhouse Gas Emission Reduction Technologies for Power Grid Enterprises

Wangchao Dong[1,a*], Yue Zhao[1], Fengxiang Ma[1], Feng Zhu[1], Wei Liu[1], Yumei Song[1], Shan Zhu[1], Han Chen[2,b*]

[1] Power Science Research Institute, State Grid Anhui Electric Power Co., Ltd., Hefei, Anhui, China
[2] Center for New Finance Research, International Institute of Finance, University of Science and Technology of China, Hefei, Anhui Province, China

Abstract: Since its introduction in 1930, sulfur hexafluoride (SF_6) has rapidly become the primary insulation material for high-voltage electrical equipment such as transmission lines, switches, transformers, circuit breakers, and reactors, owing to its exceptional electrical properties. However, the flip side of its excellent performance is its extremely high global warming potential (GWP), which is over 23,000 times that of carbon dioxide (CO_2). Managing SF_6 has thus become a key aspect of emission reduction efforts. The primary strategies for reducing SF_6 emissions include minimizing gas leakage during the production, operation, and maintenance of electrical equipment, as well as purifying and recycling emitted SF_6 gas. Various purification and monitoring methods have been developed for routine operations and have already been implemented by power grid companies globally. In addition to ongoing daily monitoring, research is also underway concerning the transformation, degradation, and substitution of SF_6 gases. Degradation and transformation methods primarily focus on high-temperature degradation, photocatalysis, and arc degradation. In the context of substitution, the idea is to use conventional gases or various organic compounds to form SF_6 mixtures or alternative gases, with the aim of reducing effect of leakage while maintaining performance. Finally, recommendations are made for power grid enterprises to reduce SF_6 emissions, emphasizing the importance of source control and the exploration of alternatives. Effective equipment management to prevent gas leakage, comprehensive monitoring and assessment, as well as the proper handling of unusable SF_6 devices, are also crucial steps in emission reduction efforts.

1.Introduction to Sulfur Hexafluoride Gas

Sulfur hexafluoride (SF_6) is an ultra-high-voltage insulation material that also possesses excellent arc extinguishing properties. Since its introduction in 1930, SF_6 quickly became a preferred insulation material for various high-voltage electrical equipment, including high-voltage transmission lines, circuit breakers, transformers, switches, and reactors. It is widely used as an insulating material in these high-voltage electrical devices.

According to reports from the Intergovernmental Panel on Climate Change (IPCC), nearly 80% of globally produced SF_6 gas is applied in the power industry, aside from its usage in magnesium metal industry (4%), electronics industry (8%), related insulation applications (3%), as well as in particle accelerators, fiber optics production, biomedical applications, etc4.

With technological advancements, scientists have discovered other beneficial properties of SF_6, such as its non-flammable nature, non-corrosiveness, and effective insulation against air, which have expanded its applications into areas like metal smelting, integrated circuit manufacturing, and refrigerant production.

However, while SF_6 boasts various excellent electrical properties, it is also recognized as one of the most potent greenhouse gases with severe global warming potential (GWP). Its GWP is over 23,000 times that of carbon dioxide (CO_2), and it can persist in the atmosphere for approximately 3,200 years, making natural degradation within a human lifespan nearly impossible. Consequently, SF_6 is classified as one of the six greenhouse gases that require strict control of emissions by the Kyoto Protocol.

2.Domestic and International SF_6 Gas Emission Reduction Technologies

In order to protect the environment, fulfill emission reduction obligations, and control SF_6 emissions,

[a*] First author: Wangchao Dong, dwchhf@live.com
[b*] Corresponding author: aiguningkesi@mail.ustc.edu.cn

countries around the world have undertaken full lifecycle monitoring of SF_6 electrical equipment. Environmental and power departments in various countries have also implemented numerous technologies to restrict the release of SF_6 gas into the atmosphere, and China is no exception. Currently, the annual usage of SF_6 gas in China's power system alone reaches 5,000 to 6,000 tons and is increasing at a rate of 20% annually. The existing stock of SF_6 gas in active equipment within the power industry is approximately 18,000 tons, with a potential recycling capacity of about 600 tons per year. When converted to CO2 equivalent, this amounts to over 14 million tons1. During the Copenhagen World Climate Conference, China pledged to reduce CO2 emissions per unit of GDP by 40% to 45% by 2020 compared to 2005 levels. With the rapid development of the national power grid, especially high-voltage transmission and distribution lines, and the increasing use of GIS and other switchgear, the consumption of SF_6 gas in the power system has significantly increased. Therefore, the rational and proper recovery, purification, and reuse of SF_6 gas has become an urgent issue to address.

2.1Main Technologies for SF_6 Emission Reduction Already in Use

2.1.1Reducing Gas Leaks from SF_6 Electrical Equipment During Production, Operation, and Maintenance

In general, the reasons for SF_6 gas leaks include:

①Leaks at weld seams;

②Leaks at shell defects;

③Leaks at flange joints;

④Leaks in gasket sealing systems;

⑤Leaks at relay connections;

⑥Leaks in GIS (Gas Insulated Switchgear) enclosures.

These leaks can occur due to improper manufacturing processes or inadequate gas refilling operations. The time distribution of SF_6 gas leaks can be divided into three stages:

①During delivery and acceptance, leaks caused by material, processing, and assembly defects.

②During the first winter and summer cycle, leaks resulting from the impact of temperature changes on sealing materials.

③ Leaks occurring as equipment approaches or exceeds its expected service life due to aging and deterioration of sealing materials.

To address this issue, companies like State Grid Corporation of China have made improvements in processes such as efficient synthesis reactions, crude product purification, and fluoride production processes. They have also strengthened the systematic regulations for operational procedures and equipped themselves with SF_6 leak detectors, online monitoring systems, and recovery devices. For instance, State Grid Corporation of China's Hubei branch has improved the method for detecting SF_6 gas leaks in GIS equipment. They comprehensively monitor the SF_6 emissions from the production, operation, and maintenance processes of SF_6 electrical equipment, ensuring compliance with the national standard for industrial sulfur hexafluoride （GB/T 12022-2014）.

2.1.2Purification and Recovery of Emitted SF_6 Gas

During the use of electrical equipment containing SF_6, there is a potential for SF_6 to undergo decomposition, resulting in the formation of toxic and harmful byproducts, including low fluorides. This can pose risks to both human health and the proper functioning of equipment. To address this concern, it is essential to purify and recover the emitted SF_6 gas to eliminate these impurities. In response, State Grid Corporation of China has established multiple SF_6 gas purification and recovery facilities (as shown in Figure 1).

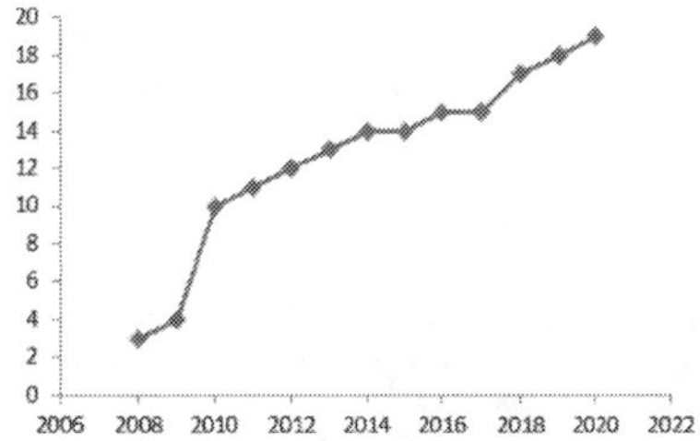

Figure 1 The number of large-scale SF_6 gas purification and recovery facilities in China

The working process of this facility mainly involves collecting the gas from the electrical equipment into the tank through the recovery system of the facility, transporting it to the purification system, removing impurities through the purification system, and then filling the purified gas back into the electrical equipment. This process helps to reduce the emissions of SF_6 and impurities from the electrical equipment to the external environment.

Apart from China, many other countries have also made efforts in the purification and recovery of SF_6 gas. For instance, Solvay Fluor in Germany introduced the "ReUse" program in Europe, proposing on-site treatment of SF_6 waste gas through gas purification vehicles. The Japan Electric Power Company and manufacturers' association have implemented SF_6 recovery plans4; The United States Environmental Protection Agency (USEPA) established SF_6 emission control plans for electrical equipment, magnesium production, die-casting, and other applications as early as 19973.

2.2Partially implemented SF_6 emission reduction technologies

2.2.1Conversion and degradation of SF_6 exhaust gases

SF_6, as a chemically stable non-CO_2 gas, has a long lifecycle with natural degradation mechanisms mainly relying on photolysis and settling, which are inefficient.

Therefore, scientists have proposed various artificial degradation methods to convert SF_6 exhaust gases into non-toxic and harmless products, aiming to reduce the emission of SF_6 gases into the atmosphere. These artificial degradation methods include high-temperature degradation, photocatalytic degradation, arc degradation, etc **错误!未找到引用源。**. In this paper, one of the most advanced degradation methods, the rectifying SiC-Fe_2O_3 heterojunction structure for SF_6 degradation in air, is briefly introduced.

Scientists achieved this by depositing FeOOH onto SiC through self-assembled chemical deposition and designed an in-situ formed SiC-Fe2O3 heterojunction structure during the heating process of SiC-FeOOH composite to catalytically degrade SF_6. The results showed that at 700°C, the initial degradation rate of SF_6 reached 93.6%, with a lifetime of 470 minutes, which is 1.8 times and 1.2 times higher than at 600°C and 650°C, respectively. The degradation amount was 99.0 mL·g^{-1}. Especially for high-concentration SF_6 (35 vol%), the degradation amount reached 615.0 mL·g^{-1}. Through electron coupling within the heterojunction structure, electrons transferred from SiC to SF_6 through Fe_2O_3, thereby reducing the S-F bond energy. This non-uniform electron distribution heterojunction structure exhibits significant potential in the field of SF_6 degradation in the future4 (as shown in Figure 2).

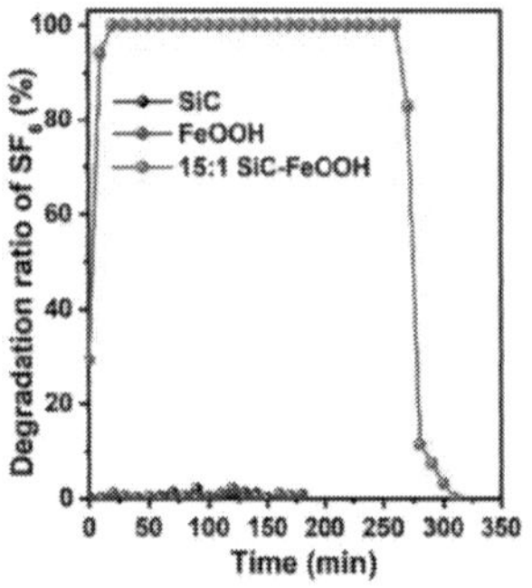

Figure 2 The degradation proportion of SF_6 in SiC, FeOOH, and SiC-FeOOH composite materials.

2.2.2Substituting gases with lower Global Warming Potential (GWP) for SF6

Due to the Kyoto Protocol explicitly listing SF_6 as one of the six greenhouse gases to be restricted in emissions, seeking alternative gases with excellent insulation and ideal arc-extinguishing properties to replace SF_6 has naturally become one of the research directions in the global scientific community. Currently, there are two main categories of SF_6 substitute gases:

① SF_6 mixed gases

Including dry air and its constituent gases: oxygen, nitrogen, carbon dioxide, and methane. The advantages of these gases lie in their environmental friendliness, safety, and low liquefaction temperature (boiling point). However, their insulation performance is only about 0.3 times that of SF_6. As a result, they cannot be used alone but must be mixed with SF_6 or other substitute gases. The addition of other gases reduces the boiling point of SF_6, addressing the issue of gas liquefaction in cold regions. When SF_6 is mixed with SF_6 in a certain proportion, their insulation strength exhibits a strong synergistic effect. Among them, the mixture of SF_6 and nitrogen in a molar ratio of 7:13 shows the strongest synergistic effect5.Therefore, this mixture is often used in the fields of Gas-Insulated Lines (GIL) and Gas Circuit Breakers (GCB) in power equipment (These will be shown in Table 1).

Table 1 The composition and performance of SF_6 mixed gases

Research Institution	SF6 Mixture Composition	Performance or Application
University of Belgrade	$n(SF_6):n(N_2)=7:13$	The strongest synergistic effect occurs in non-uniform fields, and the synergistic effect in non-uniform fields is more significant than in uniform fields.
ALSTOM (France)	$n(SF_6):n(N_2)=1:4$	240kV GIL has been applied at Swiss airports.
SIEMENS (Germany)	$n(SF_6):n(N_2)=1:4$	A GIL with a voltage level of 550kV and a transmission capacity of 300MW has been developed.
Hyosung (Republic of Korea)	$n(SF_6):n(N_2)=1:3$	In the experiment of 170kV/50kA compressed air GCB, its breaking capacity is equivalent to pure SF_6.
The University of Tokyo	$n(SF_6):n(N_2)=1:3$	In the experiment of blowing air GCB, its breaking capacity is close to 80% of pure SF_6.
Manitoba Hydro(Canada)	$n(SF_6):n(N_2)=1:1$	A high-voltage GCB with a voltage level of 115kV and a breaking current of 40kA has been developed.

Furthermore, CF_4 (carbon tetrafluoride) is a compound that exhibits good arc extinguishing performance and a relatively low liquefaction temperature. By adding an appropriate amount of CF_4 to SF_6, the liquefaction temperature of the gas can be reduced, while slightly sacrificing insulation and arc extinguishing performance. This adjustment allows the gas mixture to meet the requirements of usage in cold regions. Various mature SF_6-CF_4 gas mixtures for circuit breakers have been developed, and they can stably operate under severe cold conditions of -40°C.

Saturated halogenated hydrocarbons primarily rely on the strong electronegativity of molecules to adsorb electrons during the discharge process, thereby enhancing the electrical insulation properties. In chemical processes, hydrogen atoms in compounds can be replaced by halogen atoms to increase the electronegativity of gas molecules. Fluorine and chlorine atoms are commonly used as halogen substitutes to ensure low boiling points of gases. Additionally, bromine and iodine atoms offer advantages at the microscopic level, with multiple electron cloud energy levels and larger collision interfaces, enhancing the ability to block discharges **错误!未找到引用源。**. The insulating properties of individual saturated halogenated hydrocarbons are shown in Table 2.

Table 2 Insulation Performance of Saturated Halogenated Hydrocarbons

Compounds	The relative insulation strength (chemical calculation value) compared to SF_6
CH_3F	0.220
CH_2F_2	0.300
CHF_3	0.270
CF_4	0.594
CF_3Cl	0.672
CF_3Cl_2	0.901
CF_3I	1.217
CHF_2CH_3	0.700
C_2F_6	0.775
C_3F_8	1.080

② Fluorinated compounds

Such as perfluoroketones($C_5F_{10}O$,PFK-5110; $C_6F_{12}O$,PFK-6112, the physical and chemical properties of these two perfluorinated ketones are demonstrated in Table 3), perfluoroisobutylene (C_4F_7N), octafluorocyclobutane (c- C_4F_8), trifluoroiodomethane (CF_3I), etc., are being explored as potential substitutes for SF_6. These gases have characteristics including higher insulation strength compared to SF_6 and lower Global Warming Potential (GWP). However, they also have higher boiling points, which necessitates mixing them with gases like N_2 and CO_2 as mentioned earlier. Additionally, fluorinated compounds with carbon can generate carbon deposits after discharge, potentially leading to reduced insulation performance.

Table 3 Comparison of Physicochemical Properties of Perfluoroketones

Compounds	PFK-5110	PFK-6112
Boiling Point/℃	26.9	49
Lifetime in the Atmosphere/Year	0.040	0.014
GWP_{100}	<0.21	~0.29
The relative insulation strength (chemical calculation value) compared to SF_6	2.1	2.7
LC_{50}(4h,Rat)/(g/g)	0.02	>0.1

Among these gases, worth mentioning is trifluoroiodomethane (CF_3I). It has an extremely short atmospheric lifetime (0.005 years) and great development potential. However, it is highly carcinogenic and produces elemental iodine during discharge, which reduces equipment lifespan and poses health and safety risks.

The ideal substitute for SF_6 gas needs to possess high insulation and arc extinguishing strength, low GWP, and a low boiling point. Further research and exploration are required for gases with such characteristics.

3.Conclusion and Suggestions for SF_6 Emission Reduction in Power Grid Enterprises

This paper compiles research findings from both domestic and international reports and provides the following recommendations for SF_6 gas emission reduction in power grid enterprises in China:

Firstly, tackling the source is of paramount importance. This involves two main aspects. On one hand, it is essential to minimize gas leakage from SF_6 electrical equipment during usage, production, operation, and maintenance. On the other hand, there is a need to enhance the degradation and purification recovery of emitted SF_6 gas. Preventing gas leakage can be addressed from three angles:

①For newly acquired SF_6 electrical equipment, a comprehensive documentation system should be established upon receipt. This system should rigorously monitor gas quality, conduct stringent inspections to verify compliance, and closely review essential information such as producer and production date. Simultaneously, strict management measures should be implemented for gas storage areas7.

② For the already operational SF_6 electrical equipment, comprehensive live monitoring is necessary. This encompasses gas composition analysis within the equipment, detection and assessment of gas leakage levels. Maintenance should be carried out for equipment that does not meet application requirements, and regular re-evaluations are essential.

③For unusable SF_6 electrical equipment, prompt retrieval and proper disposal should be conducted. In terms of degradation and purification of gases, the current SF_6 decomposition methods mainly involve arc decomposition, corona spark and point discharge-induced decomposition, and high-temperature decomposition8. It is also possible to employ high temperature, photolysis, discharge, and catalytic methods to purify SF_6 waste gas into completely non-toxic and harmless gases, ultimately achieving the harmless degradation of SF_6 waste gas.

Next, the search for gases with lower GWP to replace SF_6 is crucial. Currently, there are three main categories of substitute gases that have been extensively studied:

①Conventional Gases: This includes compressed air, N_2 (nitrogen), CO_2 (carbon dioxide), etc. Research has indicated that compressed air and N_2 can be used as substitutes for SF_6 in medium-voltage switchgear insulation, with their application expanding to higher voltage levels.

②Perfluorocarbon (PFC) Gases: Various PFC gases have been considered as substitutes for SF_6 in high-voltage equipment. However, their drawback lies in their high liquefaction temperature, which leads to elevated costs. To address this, a solution is to create mixed buffering gases by combining these PFC gases with N_2 or CO_2.

③Trifluoromethyl Iodide (CF_3I) Gas: CF_3I gas is another category9, known for its high dielectric strength.

It has a global warming potential equivalent to CO_2 and decomposes rapidly in the atmosphere.

These alternative gases offer potential solutions for SF_6 replacement, each with its own advantages and challenges that need to be further explored.

Finally, due to varying costs associated with different mitigation strategies, electric grid enterprises should make reasonable choices based on their own budgets, regardless of the selected emission reduction approach. Governments should also implement robust policies and regulations to provide support. Given the urgent need to address global climate change, and considering SF_6's exceptionally high global warming potential and extended lifespan, the academic community, industries, and governments worldwide must unite their efforts. This collective action should stem from a perspective of long-term human interests and the common destiny of humanity, as we collectively confront the challenges posed by SF_6 's greenhouse gas effect10.

Acknowledgments

This paper is supported by the Science and Technology Project of State Grid Corporation of Anhui(no.521205220007).

References

1. Li Z G,Cai W,Li F.(2016) The Full Life Cycle Management of Sulfur Hexafluoride Gas [J]. North China Electric Power,2016,No.439(05):29-34. DOI:10.16308/j.cnki.issn1003-9171.2016.05.005.
2. Xiao S,Shi S Y,Lin J T, et al.(2023) Analysis on the Control Strategy of the Strong Greenhouse Insulating Gas SF_6 in High-voltage Electrical Equipment Under the Goal of"Emission Peak and Carbon Neutrality"[J]. Proceedings of the CSEE ,43(01):339-358. DOI:10.13334/j.0258-8013.pcsee.213248.
3. Dutrow E,Forte R. (1998)Fluorocarbons and sulfur hexafluoride emissions reduction strategy in the United States[J]. Studies in Environmental Science,72. DOI:10.1016/S0166-1116(98)80015-0
4. Xiang M,Jinhai H,Boxu D, et al. (2022)Rectified SiC-Fe_2O_3 heterostructures for high efficient activation and degradation of sulfur hexafluoride in air atmosphere[J]. Chemical Engineering Journal,450(P1). DOI:10.1016/j.cej.2022.137949.
5. Zhang C P,Guo,Q,Quan H. (2022)Past, present and future of SF_6 substitutes for power insulation [J]. Fine Chemicals, 39(05):882-891. DOI:10.13550/j.jxhg.20220009.
6. Wang B S,Yu X J,Hou H, et al.(2020)Review on the Developments of Structure-Activity Relationship and Molecular Design of the Replacement Dielectric Gases for SF_6 [J]. Transactions of China Electrotechnical Society,35(01):21-33. DOI:10.19595/j.cnki.1000-6753.tces.190994.
7. Xiang Z. (2010)SF_6 Emission Reduction to deal with the Global Climate Change [J]. Chinese Journal of Environmental Management, (02):23-27. DOI:10.16868/j.cnki.1674-6252.2010.02.007.
8. Ji Y S,Zhang M,Wang C Y, et al.(2021)Study on Decomposition Products of SF_6/N_2 Gas Mixture Under the Action of Electric Arc [J]. High Voltage Apparatus, 57(03):145-151+156. DOI:10.13296/j.1001-1609.hva.2021.03.022.
9. Lv Z, Nishijima K.(2014)Sulfur Hexafluoride: The Forgotten Greenhouse Gas [J]. China Strategic Emerging Industry,No.17(09):82-84. DOI:10.19474/j.cnki.10-1156/f.2014.09.023.
10. Ren L H.(2020) Sulphur hexafluoride: the most feared greenhouse gas [J]. Ecological Economy, (03):5-8. https://kns.cnki.net/kcms/detail/detail.aspx?FileName=STJJ202003003&DbName=DKFX2020

Numerical simulation study on the heat dissipation characteristics of an air-cooled radiator

Kai Yan*, Feng Li, Jing Ren

CRRC Dalian Locomotive Research Institute Co., Ltd, Dalian, 116000 Liaoning, China

Abstract: This article conducts numerical simulation calculations on the heat dissipation characteristics of an air-cooled radiator. The results show that under natural heat dissipation conditions, the maximum temperature of air-cooled radiators with flat heat dissipation covers decreases by 9.6% compared to radiators without heat dissipation covers, and the maximum temperature of air-cooled radiators with array raised heat dissipation covers decreases by 9.9%; Under forced air cooling conditions, the maximum temperature of an air-cooled radiator with a flat heat dissipation cover decreases by 12.7%, while the maximum temperature of an air-cooled radiator with an array raised heat dissipation cover decreases by 13.3%.

1 Introduction

Air cooled radiators are widely used in various engineering equipment due to their simple structure and high reliability. However, compared to methods such as water cooling and evaporative cooling, the heat transfer coefficient of air cooling is lower. For air-cooled radiators, researchers and engineers improve their heat transfer coefficient and heat transfer performance by improving their fins and other methods [1-5].

Electric motors are an important engineering equipment, and their heat dissipation is mainly achieved through natural cooling or forced air cooling of the motor casing [6-9].

Some researchers studied the chimney effect by using numerical simulation method. The chimney effect on smoke spread behavior was revealed with three-dimensional numerical simulations [10]. The results show that the chimney effect significantly influences smoke movement. Natural convection heat transfer and fluid flow in a thermal chimney with multiple horizontally-aligned cylinders was studied by using RANS/URANS simulations for multiple cylinder columns in one or two rows [11]. It was found that the horizontal pitch influences the balance between the chimney effect and the blockage effect, thus has an optimum with respect to the natural draft velocity. The effect of discrete rings and the chimney effect on the enhancement of the natural convection compound heat transfer in a vertical cylinder was studied numerically [12]. Two dissimilar enhancement mechanisms of boundary layer disturbance and flow acceleration obtained by the use of discrete rings and the chimney effect, respectively, were independently analyzed and then combined as one system.

Some researchers studied the chimney effect by using experimental method. The influence of 'chimney-effect' on fire response of rainscreen façades in high-rise buildings was studied by using experimental method [13]. A new experimental setup is developed and demonstrated for the initial screening of façade assembly at a lab-scale. A critical width of the chimney is established at which maximum vertical fire spread was noticed. A new robust hydrogel composite for efficient photothermal interfacial evaporation was developed [14]. The chimney effect was employed to further enhance the water evaporation rate. The chimney effect-induced airflow in a building drainage system in the spread of severe acute respiratory syndrome coronavirus 2 (SARS-CoV-2) was studied [15]. Results showed that the stack air pressure and temperature distributions suggested that stack aerosols can spread to indoors through pipe leaks which provide direct evidence for the long-range aerosol transmission of SARS-CoV-2 through drainage pipes via the chimney effect.

The relevant research developments both domestically and internationally in recent years reveals that the chimney effect indeed affect the fluid flow in varieties of industry areas [10-15]. However, the enhancement of heat transfer for the motor casing by the chimney effect was not studied by now.

At present, the motor casing is not equipped with a heat sink and only relies on the flat or curved surface of the motor casing for heat dissipation, resulting in limited heat dissipation efficiency. In response to this issue, this article developed a motor end cover heat sink and conducted numerical simulation research on the heat dissipation characteristics of three types of heat sinks with or without the chimney effect.

* Corresponding author email: kaiyan@bjtu.edu.cn

2 Structure of air-cooled radiator

Based on the shape characteristics of the motor end cover, this article designs a heat sink and its improved version, which is installed on the motor end cover.

Figure 1 shows the matching diagram between the radiator and the motor, i.e. the motor end cover.

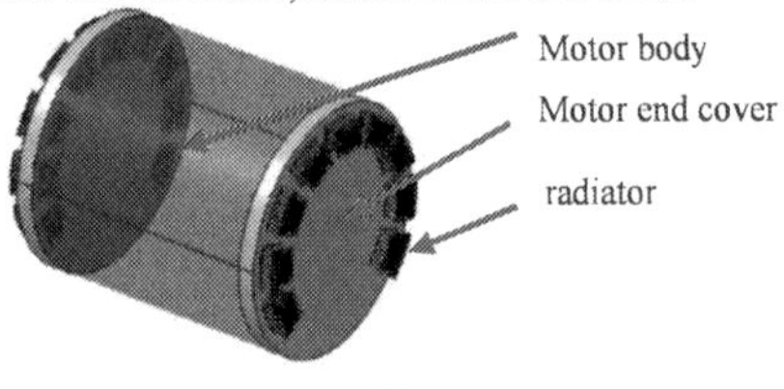

(a)

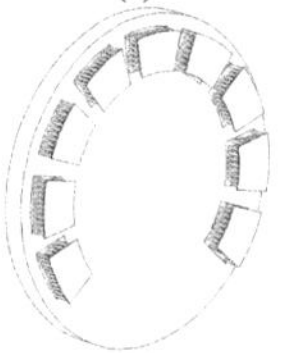

(b)

Fig. 1. Schematic diagram of radiator, motor end cover, and motor.

Figure 2 shows a schematic diagram of the radiator and its improved structure.

From Figure 2, it can be seen that Type 1 radiator includes a heat absorbing column, a radiator body, a heat dissipation column, etc; The Type 2 radiator has added a flat heat dissipation cover plate; The heat dissipation cover plate of the Type 3 heat sink has a stamped array convex structure. A semi enclosed structure is formed between the heat dissipation cover plate and the radiator body, which is beneficial for improving the heat dissipation capacity of the radiator under natural heat dissipation conditions based on the chimney effect [9].

In addition, under the condition of forced cooling by traveling wind, adding a heat dissipation cover plate can gather the traveling wind and allow more air to sweep over the heat dissipation column; Moreover, according to the Bernoulli principle, the faster the air flow rate, the lower the pressure, and the increase in the air flow rate around the heat dissipation column, the lower the air pressure around the heat dissipation column, resulting in more air flowing to the heat dissipation column and increasing the air flow rate around the heat dissipation column. This is also beneficial for improving the heat dissipation capacity of the radiator under forced cooling conditions.

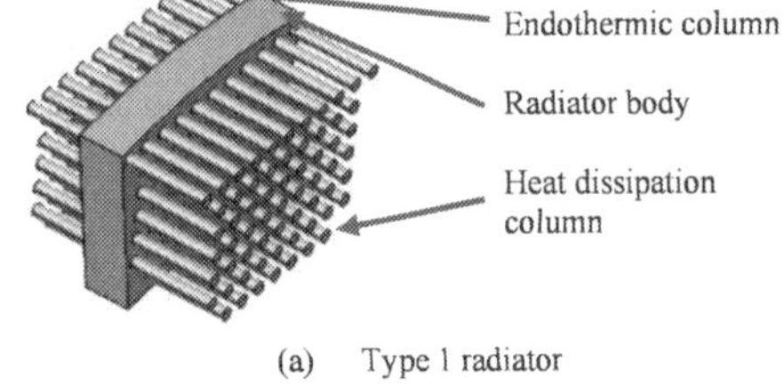

(a) Type 1 radiator

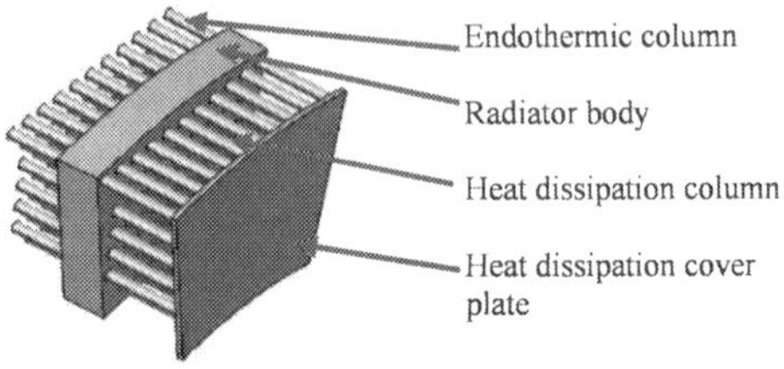

(b) Type 2 radiator

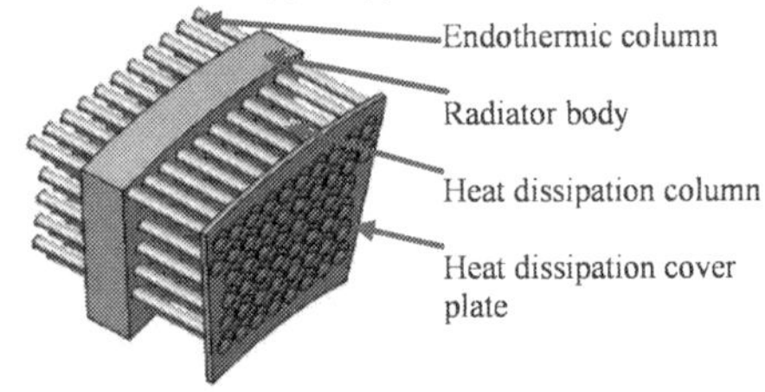

(c) Type 3 radiator

Fig. 2. Structural diagram of radiator and its improved version

3 Numerical simulation calculation method

This article adopts the computational fluid dynamics calculation method to numerically simulate the heat dissipation characteristics of the motor end cover radiator. Air flow usage k- ε Double equation turbulence model.

The following assumptions and simplifications were made in the calculation:

(1) Air flow is an incompressible fluid flow;

(2) Not considering changes in air density caused by temperature changes;

(3) The radiator is a solid of all phases.

Based on this assumption and simplification, establish the fluid continuity equation and momentum equation in vector form in Cartesian coordinate system as follows:

$$\frac{\partial \rho}{\partial t} + div(\rho u) = 0 \tag{1}$$

$$\frac{\partial(\rho u_i)}{\partial t} + div(\rho u_i u) = \frac{\partial(u_{eff} u_i)}{\partial x_i} - \frac{\partial p}{\partial x_i} + S_i \tag{2}$$

$$\frac{\partial(\rho u_j)}{\partial t} + div(\rho u_j u) = \frac{\partial(u_{eff} u_j)}{\partial x_j} - \frac{\partial p}{\partial x_j} + S_j \tag{3}$$

$$\frac{\partial(\rho u_k)}{\partial t} + div(\rho u_k u) = \frac{\partial(u_{eff} u_k)}{\partial x_k} - \frac{\partial p}{\partial x_k} + S_k \tag{4}$$

In equations (1) to (4), ρ is the fluid density, in kg/m^3; t is the time, in seconds; u is the velocity vector, in m/s; p is the pressure, in Pa; x_i、 x_j、 x_k are coordinate components; S_i、 S_j、 S_k are the generalized source term components; μ_{eff} is the effective viscosity coefficient, in Pa·s ($\mu_{eff}=\mu+\mu_t$; μ is the molecular viscosity coefficient, in Pa·s; μ_t is the turbulent viscosity, in N·s/m^2.

The solid structure follows the energy conservation equation:

$$\rho c \frac{\partial T}{\partial t} = \lambda \left(\frac{\partial^2 T}{\partial x^2} + \frac{\partial^2 T}{\partial y^2} + \frac{\partial^2 T}{\partial z^2} \right) \tag{5}$$

In equation (5), x, y, and z are coordinates; ρ is the fluid density in kg/m^3; λ is the thermal conductivity, in W/(m·K); c is the specific heat capacity, in J/(kg·K); t is the time, in seconds; T is the temperature, in K.

Energy equation for air:

$$\frac{\partial(\rho u_i T)}{\partial x_i} = \frac{\partial}{\partial x_i}\left[\left(\frac{\mu}{P_y} + \frac{\mu_t}{\sigma_T}\right)\frac{\partial T}{\partial x_i}\right] + \frac{q_T}{c_p} \qquad (6)$$

In equation (6), μ is the molecular viscosity coefficient, in Pa·s; μ_t is the turbulent viscosity, in N·s/m^2; σ_T is the surface tension, in N/m; q_T is the internal heat source, in W/m^3; c_p is the specific heat capacity at constant pressure, in J/(kg·K).

The initial condition for calculation is that the temperature of each component and air is 40 °C.

The calculation boundary conditions are:

(1) Under natural cooling conditions, the internal heating power of the motor is 10W;

(2) Under forced air cooling conditions, the running wind speed is 10m/s, and the internal heating power of the motor is 100W.

When the maximum temperature change of the radiator is less than 0.001 °C, the calculation stops.

4 Calculation results and analysis

This article conducts simulation calculations on the heat dissipation characteristics of radiators under natural cooling conditions and forced air cooling conditions.

Figure 3 shows the distribution of air flow velocity around the heat dissipation column under natural cooling conditions.

From Figure 3, it can be seen that the maximum air flow velocity around the heat dissipation column is 0.095m/s for Type 1 radiator, 0.099m/s for Type 2 radiator, and 0.0992m/s for Type 3 radiator.

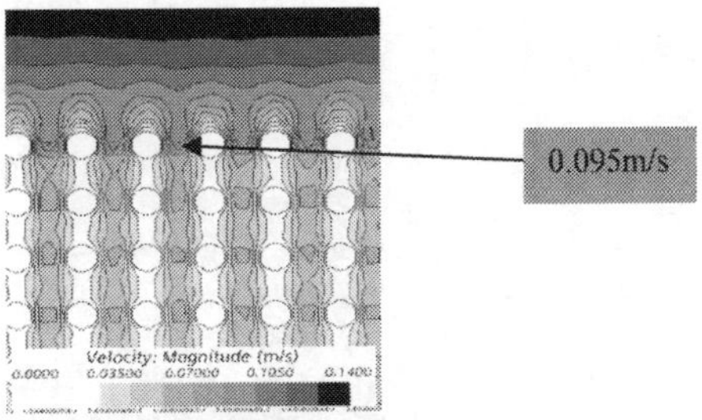

(a) Type 1 radiator

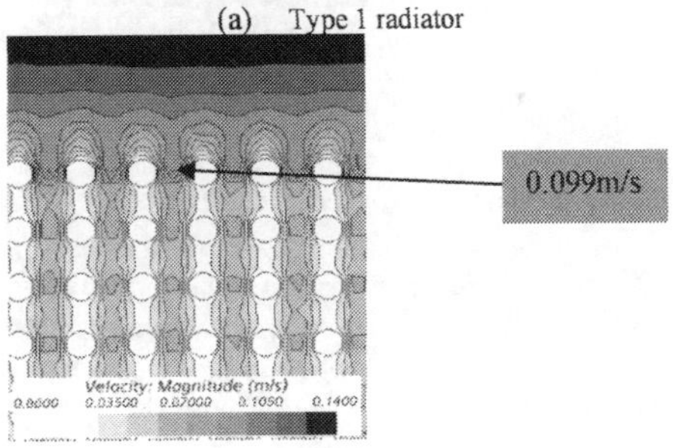

(b) Type 2 radiator

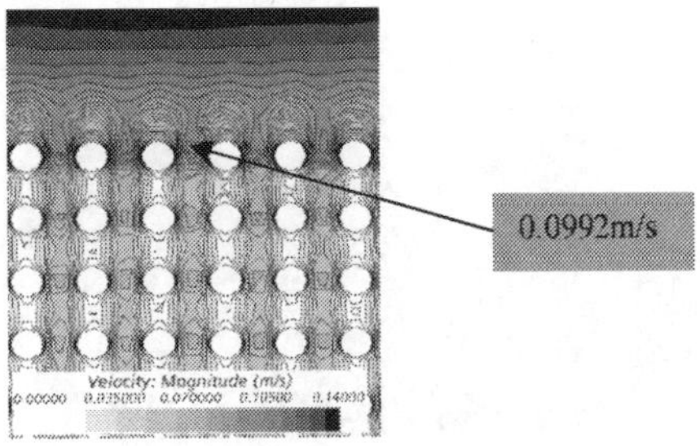

(c) Type 3 radiator

Fig. 3. Distribution of air flow velocity around the heat dissipation column under natural cooling conditions

Figure 4 shows the temperature distribution of the radiator under natural cooling conditions.

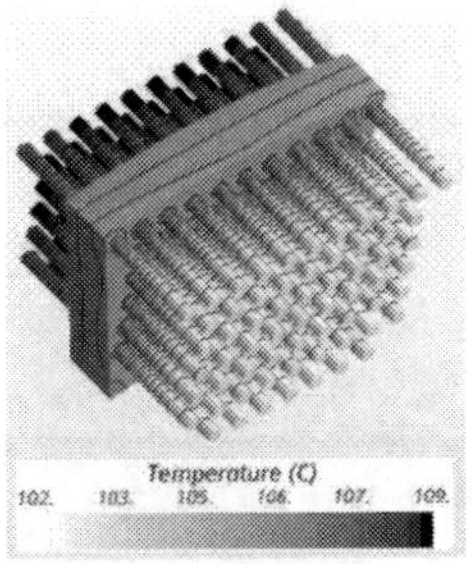

(a) Type 1 radiator

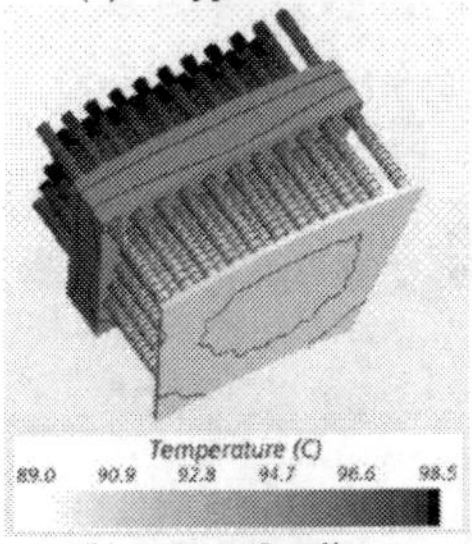

(b) Type 2 radiator

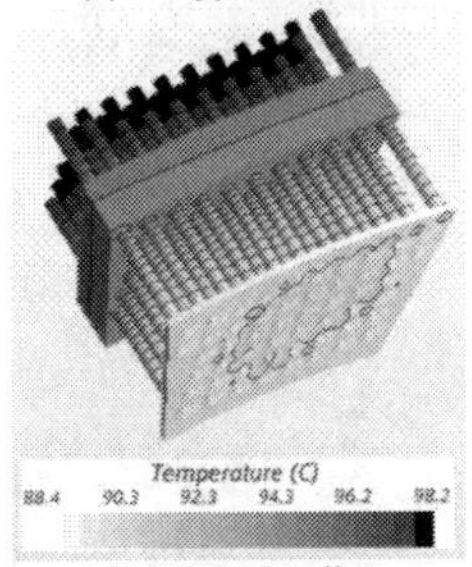

(c) Type 3 radiator

Fig. 4. Temperature distribution diagram of radiator under natural cooling conditions

From Figure 4, it can be seen that the highest temperatures of the heat sink are 109 °C for Type 1 heat sink, 98.5 °C for Type 2 heat sink, and 98.2 °C for Type 3 heat sink.

Figure 5 shows the distribution of air flow velocity and temperature around the radiator under the condition of forced air cooling with running wind.

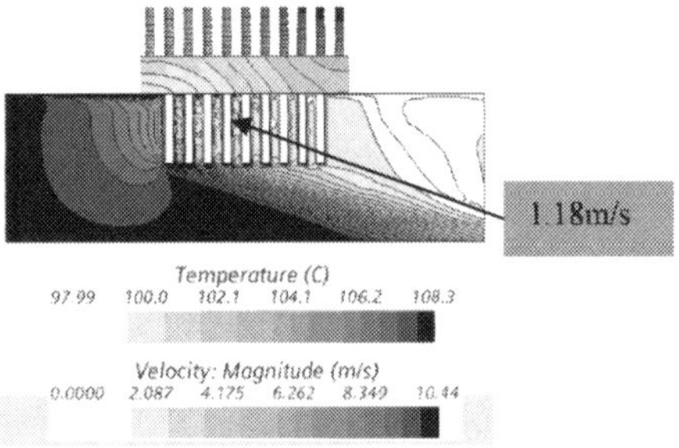

(a) Type 1 radiator

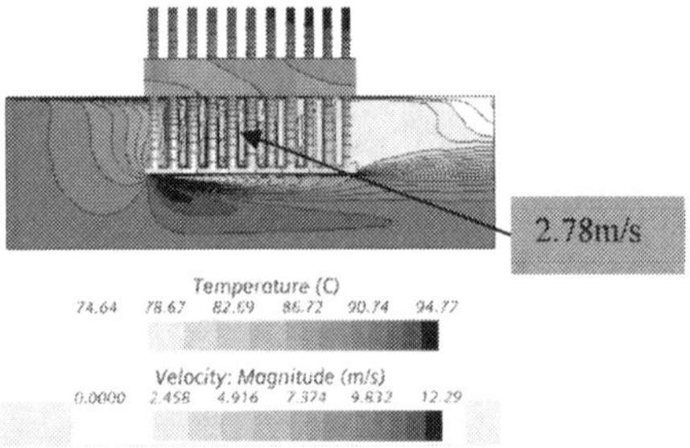

(b) Type 2 radiator

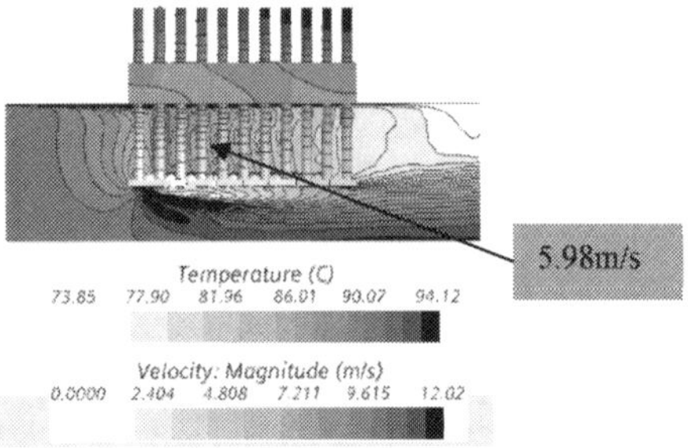

(c) Type 3 radiator

Fig. 5. Distribution of Air Flow Velocity and Radiator Temperature Around the Radiator under Forced Air Cooling Conditions with Traveling Wind

From Figure 5, it can be seen that at the same location, the air flow velocity around the heat dissipation column is 1.18m/s for Type 1 radiator, 2.78m/s for Type 2 radiator, and 5.98m/s for Type 3 radiator, respectively.

From Figure 5, it can also be seen that the highest temperatures of the heat sink are 108.3 ℃ for Type 1 heat sink, 94.7 ℃ for Type 2 heat sink, and 94.1 ℃ for Type 3 heat sink.

5 Conclusion

This article numerically simulates the heat dissipation characteristics of an air-cooled radiator and obtains the following conclusions:

(1) Under natural convection conditions, compared to a heat sink without a heat sink cover (Type 1 heat sink), a heat sink with a flat heat sink cover (Type 2 heat sink) has a 9.6% decrease in maximum temperature, while a heat sink with an array of raised heat sinks cover (Type 3 heat sink) has a 9.9% decrease in maximum temperature.

(2) Under the condition of forced air cooling in the running wind, compared with the heat sink without a heat dissipation cover plate (Type 1 heat sink), the maximum temperature of the heat sink with a flat heat dissipation cover plate (Type 2 heat sink) is reduced by 12.7%, and the maximum temperature of the heat sink with an array raised heat dissipation cover plate (Type 3 heat sink) is reduced by 13.3%.

(3) Under natural cooling conditions, the temperature of the radiator body and heat dissipation column heats the air around the heat dissipation column. Under the action of buoyancy, the air around the heat dissipation column flows upwards, forming a suction effect on the lower temperature air below the radiator, forming a chimney effect and increasing the air flow velocity around the heat dissipation column.

(4) Under forced air cooling conditions, adding a heat dissipation cover plate creates a semi enclosed space between the heat dissipation cover plate and the radiator body, gathering the air flowing around the heat dissipation column and increasing the flow velocity of the air around the cylindrical fins of the radiator. According to Bernoulli's principle, an increase in air flow velocity leads to a decrease in air pressure flowing around the heat dissipation column, which in turn leads to more air flowing towards the heat dissipation column, This increases the amount of forced cooling heat dissipation.

Acknowledgments

The authors acknowledge the support provided by CRRC (2021CYB121, 2022CKB346, 2023CKB343).

References

1. X. Li, *Research on Experiment, Simulation and Application of the Evaporative cooling Heat Exchanger* (Tsinghua University, 2013)
2. Y. Ren, *Dynamic Simulation Study of Pump-Driven Two-Phase Flow Heat Exchange System* (Chongqing University, 2019)
3. X. Sun, *Evaporative Cooling Technology Used for Chip Cooling* (Donghua University, 2011)
4. Z. Li, *Research on Forming and Heat Transfer Performance of Perforated Fin Heat Sink* (Guangdong University of Technology, 2013)
5. B. Chen, G. Gu, Transactions of China Electrotechnical Society **10**, 143-151 (2011)
6. Y. Wei, *Design and optimize on the cooling structure for electric vehicle motor* (Chongqing University of Technology, 2017)
7. X. Lin, *Research on Design and Heat Dissipation for High-Power Servo Motor* (Shenyang University of Technology, 2018)

8. Y. He, et al., *A Natural Heat Dissipation Device Utilizing Chimney Effect* (China patent CN107438349B, 2020
9. A. Muhanmmad, J. China Inst. Water Res. Hydr. Res. **15**, 425-429 (2017)
10. B. Lan, Y. Li, Case Stud. Therm. Eng. **21** (2022)
11. G. Sung, D. Na, S. Yook, Int. J. Heat Mass Tran. **5**, (2022)
12. J. Njoroge, A. Mwaura, Ann. Nucl. Energy, **11** (2022)
13. A. Sharma, K. Mishra, J. Build. Eng. **10** (2021)
14. X. Liang, X. Pei, Desalination, **21** (2022)
15. Q. Wang, Y Li, K. Yuen, J. Hazard. Mater., **2** (2021)

Innovative design and application of the Re-gas Circular Express Box in a double carbon background

Lili Feng[a], Guifei Liu, Xuxiao Shi, Shaochun Mo, Bofeng Chen and Xiaogang Li[b*]

Zhejiang Wanli University, Logistics Department, Ningbo, Zhejiang, China

Abstract. In recent years, with the vigorous development of e-commerce, China's express delivery industry has experienced a peak of development. However, the production of express garbage during this process has led to serious environmental pollution problems. The relevant national policies have also set higher requirements for the reduction, standardization, and recycling of express boxes. Therefore, it is urgent to reduce the size of the express box. It has been observed that many express delivery companies have joined the "green revolution" of promoting express delivery boxes. However, due to factors such as cost and safety, the utilization rate of circular express delivery boxes has not improved. Based on the current situation analysis, this study will address the existing problems and challenges of the current circulating box in practical applications. It will consider designing a green, shared, and circulating express box that can be implemented with existing technology. Therefore, this study is based on RFID technology to conduct innovative research and development, design a circular shared express box named Re-gas, and establish a closed-loop system for the supply chain of express boxes. This will promote the green and efficient transformation of the express industry. However, because the designed and manufactured express box samples have not been tested, and due to limitations imposed by application scenarios, cost control, and industry barriers, the project may have some defects and challenges.

1 Introduction

The Chinese government proposes to increase the national independent contribution, adopt more powerful policies and measures, strive to reach the peak of carbon dioxide emissions by 2030, and achieve carbon neutrality by 2060[1]. According to the statistics and estimation in "Production Characteristics and Management Status of Express Packaging Waste in China," the carbon emission of express boxes in the whole process, from production to garbage disposal, surged from 611,500 tons in 2010 to 23.9584 million tons in 2022. However, the recovery rate of express boxes has not yet reached 20%, and these express wastes are endangering the natural environment's health with a growth rate of over 45%. To solve this problem, the Chinese government has put forward the standards and requirements of green packaging[2]. Under the "Green Revolution" tide, many express delivery companies have submitted their solutions. JD.COM and SF Express have successively developed circulating express boxes such as "Qingliu BOX" and "Feng·BOX," but the existing circulating boxes do not combine portability with protection and safety, and their actual circulation times are small. Therefore, finding a solution to realize the sharing cycle of express boxes, reduce the generation of express garbage, save resources, and protect the environment is an urgent problem to be solved.

In order to address the dilemma more effectively, this study has designed an innovative Re-gas shared express box under the influence of recycling and sharing economy. This study will introduce how to share, recycle, and reuse Re-gas express boxes, focusing on their structure, technology, and application system. The aim is to reduce costs and increase efficiency. Additionally, this study proposes a unique multi-subject joint operation mode for the Re-gas circulation box, which enables "benefit sharing and risk sharing" throughout the entire supply chain of express boxes. The circular express box solution suggested in this study aims to contribute to environmental protection and conservation, promote the clean, low-carbon, green, and sustainable development of the express delivery industry, and aid the country in achieving carbon neutrality as soon as possible[3].

2 Application status and problem analysis of circulation box

2.1 Circulation box application status

At present, China's express delivery industry has reached a peak of development, but it has brought along environmental pollution caused by express garbage. Finding ways to reduce the size of express boxes has

[a]2020013788@zwu.edu.cn, [b*]Corresponding author: showyou@zwu.edu.cn

become a pressing issue that the government is closely monitoring and urgently needs to address. Promoting the use of reusable recycling packaging can effectively reduce the amount of waste generated from express packaging[4]. The Code for Green Packaging of Mail encourages companies to use recyclable express boxes and establish information systems and recycling facilities and equipment for these boxes. It also encourages delivery companies, third-party organizations, and delivery companies to establish a platform for sharing recycling express boxes and gradually expand the use of these boxes. With the support of relevant national policies, many express delivery companies have joined the "green revolution" of express packaging, including Jingdong Logistics, Suning Logistics, SF Express, Shentong, and Yuantong, who have made attempts in the field of green recycling boxes. Since 2016, the Cainiao Alliance launched the "Green Movement Plan", followed by JD.COM's "Green Logistics Plan", Suning's "Green City Plan", and SF Express's "Fengjing Plan".

2.2 Problems Existing in the Application of Circulation Box

The Circulation box has been on the market for 8 years but due to its poor design, high application costs, and low consumer willingness, it is still relatively rare in the market.

2.2.1 The design scheme of circulation box is imperfect

The existing design scheme of the circulation box does not combine portability with protection, and the actual circulation times are small. For example, SF's "Feng · BOX" offers better protection, but it is not light enough. Although the "Qingliu Box" on JD.COM is light in style, its protection is not high.

2.2.2 Application cost is difficult to control

The production cost of circulating boxes is much higher than that of traditional cartons. Additionally, the costs of cleaning, disinfection, maintenance, and reverse logistics of circulating boxes need to be taken into account.

2.2.3 Consumers' willingness to recycle is low

For a long time, consumers have become accustomed to discarding unpacked express boxes as waste and disposing of them as general domestic garbage at will. Judging from the popularization and use of actual recycling boxes, consumers' awareness of recycling is still very weak. Although most people will recycle the recycling box as required, there are also cases where consumers refuse to recycle and dispose of it at will. The uncertainty of consumers' willingness to recycle affects the popularization and use of circular express boxes. Additionally, there are problems such as low product standardization and untimely data link of circulation box logistics[5].

Considering the shortcomings of the existing circulating box in design, economy, convenience, and security, this study carried out innovative research and development and designed a circulating shared express box named Re-gas. The design scheme strictly controls the application cost of the recycling box, standardizes the design of the box, and improves the recovery rate of the recycling box to promote the green and efficient transformation of the express delivery industry[6].

3 Design and development of Re-gas cycle box

3.1 Design concept

Re-gas circulation box is named after four keywords: "Recyclable", "green", "air" and "shared", which embodies its characteristics of recycling, green, air buffer and shared use. The specification of the box body is designed as a cuboid of 175 × 95 × 60 mm, with an opening on one side and a concave broken line between the two transverse sides, which is convenient for folding and transportation after compression. Inflatable membrane interlayer is used to replace filler inside, and inflation valve is connected externally. Square opening is surrounded by three-loop annular zipper, and zipper is fixed and protected by pull ring.

3.2 Development process

First of all, carry out market research to clarify the special needs of transportation of small commodities such as cosmetics and ornaments. Secondly, the preliminary scheme of Re-gas circulation box is designed according to the demand. Then, the technology is realized, including the application of RFID technology, the selection of inflatable film interlayer and the material selection of zipper and pull ring. Finally, the practical application test is carried out to ensure the feasibility and effect of the circulation box in practice.

3.3 Key Design Requirements and Target Recognition

3.3.1 Key Design Requirements

The design requirements of Re-gas recycling box include: sharing, recycling, recycling and green degradation; Cuboid shape, unified specifications; Inflatable film interlayer is adopted inside; The exterior is sealed with zipper and pull ring; Equipped with RFID chip to facilitate information tracking and management.

3.3.2 Target Recognition

The main objectives of Re-gas circulation box include: improving logistics efficiency; Protect user privacy; Reduce logistics costs and improve enterprise profits; Reduce environmental pollution and realize green logistics; Promote the sustainable development of logistics industry.

3.4 Research and development phase

The research and development phase consists of the following three steps.

3.4.1 Design phase

In this stage, the main market research, clear user needs, and based on user needs to design a preliminary plan of Re-gas cycle box.

3.4.2 Technology implementation phase

This stage mainly carries out technical research and implementation, including the selection and application of RFID chips, the selection and production of inflatable film spacers, and the selection of zippers and pull rings.

3.4.3 Test phase

This stage mainly carries out practical application testing, including the convenience of the use of the circulation box, the inflating and deflating operation of the inflating film compartment, the information entry and reading of the RFID chip, the sealing of the zipper and the pull ring.

3.5 Composition and structure of Re-gas cycle box

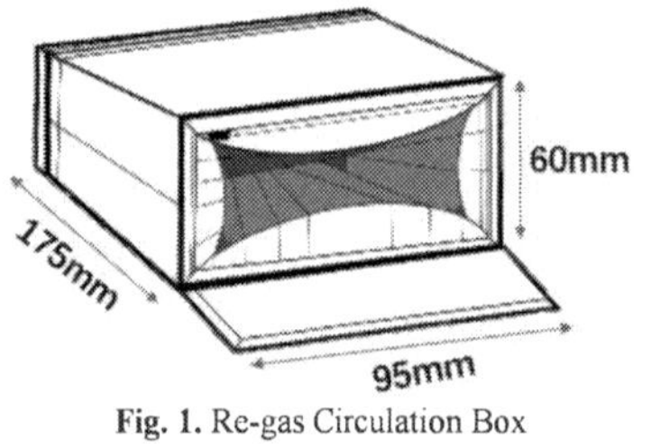

Fig. 1. Re-gas Circulation Box

After investigation, it has been found that small and scattered commodities such as cosmetics and ornaments have more packaging and transportation problems compared to ordinary objects. In light of this phenomenon, this study designed a cuboid-shaped box with dimensions of 175×95×60mm. As shown in Figure 1, the box has an opening on one side and a concave broken line between the two transverse sides, which makes it convenient to fold and transport after compression. An inflatable membrane interlayer is used instead of filling inside the circulating box, and an inflatable valve is connected externally. The square opening is surrounded by a three-loop annular zipper, and the zippers are secured and protected by a pull ring.

3.5.1 Description of the main parts and their functions

(1) RFID electronic chip

An electronic tag is implanted in the outer layer of the Re-gas circulation box, making it convenient to track the box's circulation and record its usage.

(2) Zipper protective cover

To prevent the circulation box from leaking air and losing its protective effect on the transported goods, the extension surface of the box body is used as a protective cover at the sealing place, and the zipper is used to seal it, ensuring that the goods do not fall out of the express box.

(3) Straw fiber pull ring

A straw fiber pull ring is added at the zipper joint outside the box body for fixed protection, ensuring that the circulating box is not opened during the transportation process by the express delivery company and is completely delivered to the receiving consumers. When the consumer picks up the goods, they personally open the special pull ring and place it into the recycling area for recycling and degradation.

(4) Air valve

The circulation box needs to be inflated after placing articles inside, and consumers need to deflate it before removing the articles.

4 Application and operation of Re-gas cycle box

4.1 Application system of circulation box

The owner puts the goods to be transported into the Re-gas circulation box, injects air through the air valve to expand the airbag of the box, the internal film naturally fits the goods for buffer protection, and the external zipper and pull ring of the box are sealed. The logistics information and customer information are entered into the RFID chip by the express delivery company, making it convenient to read the information on the card for sorting, transportation, and distribution. Finally, the consumer removes the pull ring at the pick-up port of the express smart cabinet, opens the zipper, deflates, and takes away the items. The intelligent cabinet automatically demagnetizes the RFID chips and erases customer information[7].

4.2 Circulating box operation system

The Re-gas circulation box can maximize its value only if it is operated reasonably. The operation circulation requires the cooperation of many subjects in the supply chain.

https://doi.org/10.1051/e3sconf/202344103014

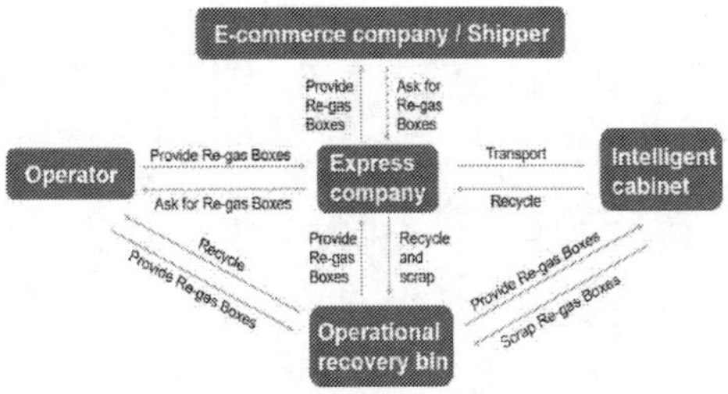

Fig. 2. Circulating box operation system

As shown in Figure 2. From two aspects, forward and reverse logistics, the entire process of Re-gas circular express box circulation is clearly visible under the joint operation of various entities. Specifically, e-commerce companies or shippers have a demand for packaging and shipping goods, so they request express delivery companies to use Re-gas express boxes for packaging and shipping. Express delivery companies initiate Re-gas orders to operators through information platforms. Operators then provide Re-gas express boxes to express delivery companies. Afterwards, the courier company can continue to act as a distributor and lease the Re-gas courier boxes to e-commerce companies or shippers in need. E-commerce companies and shippers use Re-gas express boxes to pack small or scattered goods, which are then shipped and delivered to designated smart cabinets by express delivery companies. Consumers can pick up these goods through small programs.

In comparison to traditional forward logistics, the value of Re-gas express boxes can be realized in reverse logistics. Through the operation and circulation mechanism, Re-gas express boxes can be recycled from consumers to smart cabinets, and retrieved by express delivery companies for recycling. Packaging that needs to be discarded will be recovered by the operator and sent to a recycling warehouse for proper disposal. With the cooperation and joint operation of multiple entities in the express supply chain, a closed-loop system of Re-gas express box supply chain is formed through reverse recycling and recycling.

4.3 Application technology of circulating box

4.3.1 RFID Technology

RFID tags are embedded in fixed positions of the circulation box to record various information such as production, use, recovery, and maintenance. Radio frequency technology allows for repeated reading and writing of electronic tags. Before shipment, the express delivery company inputs logistics information and customer information into the RFID chip. After the consumer picks up the items, the express intelligent cabinet automatically erases the information on the RFID chip, enabling it to be cycled back and forth. This technology offers advantages such as reusability, long service life, the ability to identify multiple targets simultaneously, high efficiency, password protection, and high security.

4.3.2 Modification Technology of PBAT + PLA Alloy

After careful consideration, the material of the circulation box is made through chemical grafting modification. PLA and PBAT are linked and branched to form a new alloy modified material of PBAT and PLA. This material is white and transparent, and can be greatly extended while being completely biodegradable. It is also harmless to the human body and the biological environment.

4.3.3 ERP system

In the operation of the entire system, an ERP system is used. This system is an efficient tool that can manage the logistics, capital flow, and information flow of the supply chain in an all-round and integrated way, share and utilize resources, and fully organize and effectively transmit information.

5 Application value of Re-gas Circulating Box

Today, with the promotion of sustainable development, the pollution problem caused by express boxes is related to their production materials and recovery rate. In order to solve these problems, the Re-gas shared express box designed and developed in this study has achieved innovation and transformation based on the existing express circulation box. The following is its application value.

5.1 Paperless realization of green and low carbon

The Re-gas shared circulation box is made of a two-dimensional code and an RFID chip, and the mobile phone equipped with a reading and writing APP can easily read logistics information. After express delivery, the circulation box can be put into use again only by emptying the data. It not only saves the use of materials and facilitates logistics management but also eliminates information security risks to a great extent[8]. When the box is circulating, only the background database needs to be changed, and the two-dimensional code on the package does not need to be changed.

5.2 Information tracing system improves the recovery rate of circulation box

Relying on information technology, an information traceability system can be established. The RFID tag matched by each Re-gas circulation box is unique, which can help express delivery companies clearly track the logistics status of each express delivery. At the same time, with the RFID chip as the core, combined with the Internet, GPS, online service platform, database, and other support systems, we can accurately grasp the number, distribution, and flow direction of express circulation boxes, and then count the turnover days, times, recovery, and loss of circulation boxes in real-time. Combined with the deposit system, incentive policy, and credit score system, it is possible to improve the recovery rate of recycling boxes[9].

5.3 Repeated sealing with green pull ring reduces environmental pressure

Introduce pull ring accessories to realize repeated sealing and privacy protection. The Re-gas circulation box adopts an environmentally-friendly sealing buckle design independent of the box body, and the pull ring is made of straw fiber as a raw material. Consumers only need to break the pull ring to open the box body. At the same time, the straw fiber pull ring is green and recyclable, and the circulation box needs to be sealed repeatedly, but it will not pollute the environment.

5.4 There is no filling inside the circulation box to reduce the generation of express garbage

The foldable inflatable structure of the circulation box eliminates the need for fillers inside the package, reducing the production of excessive express garbage such as shock-proof bags and bubble films, and minimizing carbon emissions. The interior of the Re-gas circulation box is designed with the foldable inflatable structure of the circulation box eliminates the need for fillers inside the package, reducing the production of excessive express waste such as shock-proof bags and bubble films, as well as carbon emissions. The Re-gas circulation box features an inflatable airbag inside, using air instead of fillers to prevent collisions and falls, providing better protection. When goods are placed in the inflated state, the airbag film naturally conforms to the goods, providing 360-degree protection without any blind spots, resulting in a better protective effect compared to regular express fillers.

The Re-gas recycling box is a reasonable assumption for sharing, reusing, and recycling express boxes in the current situation. By allowing the circulation and sharing of express boxes while maintaining their original functions, it increases the number of times express boxes can be reused, reduces waste of express materials and resources, and contributes to low-carbon environmental protection. This is of great value and significance in promoting the green and sustainable development of the express delivery industry. However, since the designed and manufactured express box samples have not been tested and are limited by application scenarios, cost control, and industry barriers, the project may have defects and face certain challenges.

6 Comparative analysis of Re-gas recycling box and traditional express box

As shown in Table 1, this study compared the advantages and disadvantages of traditional express boxes and new recyclable express boxes from multiple dimensions such as resources, environmental impact and logistics efficiency.

Table 1. Comparative analysis.

	Traditional express box	Re-gas cycle box
Resource utilization efficiency	Low	High
Environmental impact	Pollute	Protect
Logistics efficiency	Low	High
User privacy protection	Strong	Weak

6.1 Resource utilization efficiency

Because of its recyclable characteristics, Re-gas recycling boxes have higher resource utilization efficiency than traditional one-time express boxes[10].

6.2 Environmental impact

Due to its characteristics of one-time use, traditional express boxes produce a lot of packaging waste, which causes greater pressure on the environment[11]. The recyclable and recyclable characteristics of the Re-gas cycle box greatly reduce its impact on the environment[12].

6.3 Logistics efficiency

Re-gas recycling boxes can improve logistics efficiency through the information tracking and management of RFID chips, while traditional express boxes are relatively weak in this respect[8].

6.4 User privacy protection

Re-gas recycling boxes protect user privacy through RFID chips and zippers, pull rings, etc., while the protection measures of traditional express boxes are relatively weak.

7 Summary

By collecting and sorting out relevant literature on the application status and development prospects of green recycling express boxes in the academic frontier, combined with the background, this study designed a more perfect green and shared express box - Re-gas recycling box based on RFID technology, in order to improve the actual utilization rate of recycling express boxes. Through the recycling of express boxes, the generation of express garbage is greatly reduced, and the resulting environmental pollution problem is alleviated. At the same time, a closed-loop system of the express box supply chain, jointly operated by all parties, is constructed to support the operation and circulation of Re-gas circular express boxes. Currently, although RFID technology has proven the feasibility of the above design in various scenarios, there are not enough complete typical cases to support the integrated design of this study. Additionally, due to limitations in application scenarios, cost control, and industry barriers, the project may have defects and face challenges. Future research should focus more on supporting actual cases, so that the designed products can seek more realistic objective data support and feedback verification in specific application scenarios, and promote the design scheme to be more perfect and applicable in real life. It is believed that after overcoming all difficulties, the vision of sharing, recycling, and reusing Re-gas express boxes in the closed-loop system of the supply chain will be realized.

References

1. Wu, L., Wang, Z., Wang, X., Lyu, J. (2021) Impact of High-Proportion Wind Power Integration on Power Angle Stability of Power System. In: The 10th Renewable Power Generation Conference. Online. pp. 52-28. https://doi.org/10.1049/icp.2021.2276
2. He, Y., Liu, P., Zhou, L., Zhang, Y., Liu, Y. (2021) Competitive model of pumped storage power plants participating in electricity spot Market——in case of China. Renewable Energy, 173: 164-176. https://doi.org/10.1016/j.renene.2021.03.087
3. Wang, R., Yang, W., Li, X., Zhao, Z., Zhang, S. (2021) Day-ahead multi-objective optimal operation of Wind–PV–Pumped Storage hybrid system considering carbon emissions. In: 2021 The 2nd International Conference on Power Engineering. Nanning.pp.1270-1279. https://doi.org/10.1016/j.egyr.2022.02.293
4. Sherry. (2020) Green Double 11 starts with rejecting excessive packaging. https://www.toutiao.com/article/6889694583052206595/?&source=m_redirect&wid=1692196451448
5. Hongliang Zhu. (2022) Get more package recycling. http://www.tynews.com.cn/system/2022/11/15/030526036.shtml
6. Zeng, Q., Chen, Z. (2021) Design of recyclable logistics packing box with electronic signature. Internet of things technology,9:82-86. 10.16667/j.issn.2095-1302.2021.09.024.
7. Li, J., Xu, Y. (2020) On the development of recyclable express boxes from the perspective of circular economy. Times economy and trade,16:64-65. 10.19463/j.cnki.sdjm.2020.16.029.
8. Song, B. (2023) Research on package reduction design of express based on green environmental protection concept. Green packaging,8:70-72. 10.19362/j.cnki.cn10-1400/tb.2023.08.014.
9. Sui, L., Han, F. (2023) Research on the development path of shared express packaging under the dual carbon goal. Commercial economy,9:129-131. 10.19905/j.cnki.syjj1982.2023.09.056.
10. Lin, B., Wang, X. (2023) Are Chinese residents willing to pay for green express packaging and to participate in express packaging recycling. International Review of Economics and Finance,88. 10.1016/J.IREF.2023.06.016.
11. Zinkeviciute, Virgilija, Vasiliauskas, et al. (2013) Implementation of the Concept of Green Logistics Reffering to it Applications for Road Freight Transport Enterprises. Business: Theory and Practice, 1 :43-50.
12. José Geraldo Vidal Vieira, Jan C. Fransoo, Carla Deguirmendjian Carvalho. (2015) Freight distribution in megacities: Perspectives of shippers, logistics service providers and carriers. Journal of Transport Geography,46 :46-54.

Research on the implementation path of event carbon neutrality

Peng Hou[1], Qiang Wang[1], Jie Wang[1], Kun Yang[1], Xiyue Du[2] and Siyu Huang[2*]

[1]DATANG CARERA(BEIJING) INVESTMENT CO., LTD, Beijing, 100000, China
[2]Central University of Finance and Economics, Beijing, 100000, China

Abstract. This essay explores the concept of activity-based carbon neutrality as a critical component in achieving carbon emission reduction targets. With the Chinese government's commitment to carbon sequestration by 2030 and carbon neutrality by 2060, activity-based carbon neutrality has gained prominence as a hot topic in environmental discussions. The essay emphasizes the significance of activity-based carbon neutrality in promoting public awareness and participation in carbon reduction efforts. It highlights the existing research gap in this field and stresses the need for exploring methodologies to achieve activity-based carbon neutrality. The essay outlines the accounting standards for activity-based carbon neutrality, focusing on large-scale events and offering practical examples of carbon emission reduction measures during the preparation, execution, and conclusion phases of activities. Additionally, it introduces carbon offsetting methods such as carbon quotas and carbon credits. The article also presents a specific case study of the "Six-Five Environment Day National Event" held in Liaoning Province, which successfully achieved carbon neutrality through various initiatives. Overall, the essay encourages further investigation into activity-based carbon neutrality to enhance overall carbon reduction efforts.

1. Background of Event Carbon Neutrality

Carbon neutrality refers to reducing greenhouse gas emissions to mitigate global warming. With the continuous introduction of national and governmental policies, carbon neutrality has become a hot topic, permeating various aspects of people's lives.

In September 2020, the Chinese government announced the "dual-carbon goal,"[3][8] striving to achieve carbon sequestration by 2030 and carbon neutrality by 2060, actively promoting the effective implementation of the Paris Agreement and participating in global climate governance. In the 2021 National People's Congress, the "dual-carbon goal" was reinforced in the State Council's progress report, with a call for orderly promotion of carbon sinks and carbon neutrality, implementation of carbon sink action plans, and the formal launch of China's first year of implementing the "dual-carbon goal." By the end of 2022, various ministries, the central government, and the State Council had collectively issued nearly a hundred policy statements on the "dual-carbon" goals, including "Comprehensive and Accurate Implementation of the New Development Concept of Carbon Neutrality" and "Action Plan for Carbon Neutrality by 2030."[2] In addition, local governments at all levels have formulated plans to promote carbon neutrality, including the "1+N" policy plan. Some regions and large companies have also proposed specific action plans.

Among these, active carbon neutrality has become a crucial aspect of achieving China's carbon emission targets. As of 2023, the energy-intensive provinces of Shandong and Shanxi have successively released implementation plans for large-scale event carbon neutrality within their respective territories. These plans require event organizers to develop carbon neutrality implementation plans during the preparation stage of large-scale events, carry out emission reduction actions during the events, and calculate greenhouse gas emissions at the conclusion, taking offset measures to achieve carbon neutrality. In the future, carbon neutrality for small and medium-sized events will receive increased attention, with more local governments, industries, and enterprises participating in event carbon neutrality efforts, controlling and limiting carbon emissions during events, making it a trending issue.

Event carbon neutrality refers to reducing carbon emissions during events, including performances, sports events, conferences, forums, exhibitions, etc., excluding activities held within the regular business scope of venues such as theaters, concert halls, parks, and entertainment venues. Based on the number of participants, event carbon neutrality is classified into large-scale and small-medium-scale event carbon neutrality. Large-scale event carbon neutrality encompasses gatherings with a certain number of participants within a specific time and location, with the

*Corresponding author's e-mail: 2022310991@email.cufe.edu.cn

defined number varying by region (e.g., 1000 people in Beijing and 500 people in Shandong).

Event carbon neutrality holds significant educational and informative value. Compared to industrial carbon neutrality, it closely relates to people's daily lives, and advancing event carbon neutrality comprehensively can better raise public awareness of carbon neutrality.

Based on literature research, the existing studies in the field of carbon neutrality mainly focus on industrial production, leaving significant gaps in research on event carbon neutrality. Currently available information related to event carbon neutrality primarily centers on carbon emission accounting methods rather than methodologies for reducing emissions in events. Therefore, research on the implementation path of event carbon neutrality holds substantial research significance.

2. Standards for Calculating Event Carbon Neutrality

Currently, enterprises can follow carbon emission accounting standards such as the "Guidelines for the Implementation of Large-scale Event Carbon Neutrality (Trial)" issued by the Ministry of Ecology and Environment in 2019, as well as local guidelines like the "Beijing Large-scale Event Carbon Neutrality Implementation Guidelines," "Shandong Province Large-scale Event Carbon Neutrality Implementation Plan (Draft for Comments)," and the "Large-scale Event Carbon Neutrality Evaluation Specification" (DB13T5560-2022) by the Hebei Provincial Market Supervision Administration.

According to the guidelines for large-scale event carbon neutrality in various regions, the calculation of greenhouse gas emissions should include all controllable and influenceable emissions during the event's holding phase. When conditions permit, emissions generated during the preparation and conclusion stages of the event should also be accounted for.

The methodology for calculating greenhouse gas emissions for large-scale events should be implemented according to the recommended accounting standards and technical specifications as listed in Table 1.

Table 1. Types of Emission, Emission Sources, Calculation Methods, and Emission Factors for Event Carbon Neutrality

Emission Type	Emission Source		Calculation Method	Emission Factor
Direct Greenhouse Gas Emissions	Fossil fuel combustion in fixed facilities	Includes fixed facilities within large-scale event venues and offices of personnel serving large-scale events, using fossil fuel combustion such as coal boilers, gas boilers, direct-fired machines, gas stoves, etc.	1. "Guidelines for Greenhouse Gas Accounting and Reporting (Trial) for Public Building Operation Units (Enterprises)" issued by the Office of the National Development and Reform Commission (NDRC) (Climate [2015] No. 1722).[1] 2. Refer to Formula (2) for fossil fuel combustion emissions in DB11/T1785-2020.	Emission factors for fossil fuels' lower heating value, carbon content per unit calorific value, and carbon oxidation rate can be obtained from the recommended values in Appendix A, Table A.1 of DB11/T1785-2020.
	Mobile fossil fuel combustion in movable facilities	Includes movable facilities serving large-scale events, such as official vehicles, staff commuting vehicles, etc.	Refer to Formula (2) for fossil fuel combustion emissions in DB11/T1786-2020.	Emission factors for fossil fuels' lower heating value, carbon content per unit calorific value, and carbon oxidation rate can be obtained from the recommended values in Appendix A, Table A.1 of DB11/T1786-2020.
	Emissions from venue and supporting facilities construction	Includes newly constructed facilities, renovated facilities, and temporary constructions serving large-scale events, such as buildings, offices, pavilions, enclosed stadiums, open stadiums, competition venues, open stands, roads, sidewalks, etc.	1. Building material production: Refer to Formula (6.2.1) for carbon emissions during the building material production stage in GB/T51366-2019.	The emission factor for building materials can be obtained from the default values in Table D.0.1 of Appendix D in GB/T51366-2019.[5]
			2. Building material transportation: Refer to Formula (6.3.1) for carbon emissions during the transportation of building materials in GB/T51366-2019.	The emission factor per unit weight transportation distance can be obtained from the default values in Table E.0.1 of Appendix E in GB/T51366-2019.
Indirect Energy-related Greenhouse Gas Emissions	Carbon dioxide emissions from purchased electricity and heat consumption	Includes facilities within large-scale event venues and offices of personnel serving large-scale events, consuming externally purchased electricity and heat, such as ventilation	1. "Guidelines for Greenhouse Gas Accounting and Reporting (Trial) for Public Building Operation Units (Enterprises)" issued by the Office of the National Development and Reform Commission (NDRC) (Climate	The emission factor for annual average grid electricity supply can be obtained from the recommended values in Appendix A, Table A.2 of DB11/T1785-2020.

		systems, air conditioning, lighting, printers, etc.	[2015] No. 1722).	The emission factor for heat supply can be obtained from the recommended values in Appendix A, Table A.2 of DB11/T1785-2020.
		Carbon dioxide emissions from purchased electricity consumption for electric vehicles and other movable facilities. Includes electric vehicles, such as official vehicles, staff commuting vehicles, etc.	1. "Guidelines for Greenhouse Gas Accounting and Reporting (Trial) for Land Transport Enterprises" issued by the Office of the National Development and Reform Commission (NDRC) (Climate [2015] No. 1722). 2. Refer to Formula (5) for emissions from purchased electricity consumption in DB11/T1786-2020.	The emission factor for annual average grid electricity supply can be obtained from the recommended values in Appendix A, Table A.2 of DB11/T1786-2020.
Other Indirect Greenhouse Gas Emissions	Transportation emissions	Transport activities generated by organizers and participants of large-scale events, including flights, high-speed trains, subways, taxis, private cars, etc.	1. "IPCC Guidelines for National Greenhouse Gas Inventories" published by the Intergovernmental Panel on Climate Change (IPCC) in 2006. 2. "Government Greenhouse Gas Conversion Factors for Company Reporting: Methodology" published by the UK Department for Business, Energy, and Industrial Strategy in 2021.	Emission factors for different modes of transportation can be obtained from the parameters in Attachment 5 of "Notice on Key Carbon Emission Unit Management and Carbon Emission Rights Trading Pilot Work in 2020" (Jinghuanfa [2020] No. 6) issued by the Beijing Municipal Ecology and Environment Bureau.
	Accommodation and catering emissions	Activities related to accommodation and catering for participants of large-scale events	1. "ISO 14064-1:2018 Greenhouse gases -- Part 1: Specification with guidance at the organization level for quantification and reporting of greenhouse gas emissions and removals." 2. "Government Greenhouse Gas Conversion Factors for Company Reporting: Methodology" published by the UK Department for Business, Energy, and Industrial Strategy in 2021.	Emission factors for accommodation can be referred to "Defra/DECC (2012) Government GHG Conversion Factors for Company Reporting."
				Emission factors for catering can be calculated from Chapter 3, Table 3-2 of "Carbon Emissions from Residents' Food Consumption: Influencing Factors and Reduction Strategies" in GB/T26157-2010.
	Implicit carbon emissions from large-scale event supplies	Purchased products, raw materials, and materials supplied to large-scale events	1. "ISO 14064-1:2018 Greenhouse gases -- Part 1: Specification with guidance at the organization level for quantification and reporting of greenhouse gas emissions and removals." 2、"Government Greenhouse Gas Conversion Factors for Company Reporting: Methodology" published by the UK Department for Business, Energy, and Industrial Strategy in 2021.	For accommodation emissions, refer to "Defra/DECC (2012) Government GHG Conversion Factors for Company Reporting."
	Emissions from Waste Treatment	Methane emissions from landfilling.	1.Refer to the "Guidelines for Provincial Greenhouse Gas Inventory Compilation (Trial)" issued by the Office of the National Development and Reform Commission (NDRC) in Climate [2011] No. 1041, which includes the "Guidelines for Provincial Greenhouse Gas Inventory Compilation (Trial)."	The solid waste landfill treatment rate, methane generation potential for various types of landfills, methane recovery amount, and oxidation factors can be obtained from the data sources or recommended values in Table 5.1-5.4 of the "Guidelines for Provincial Greenhouse Gas Inventory Compilation

				(Trial)" in Climate [2011] No. 1041.
	Emissions from Carbon Dioxide from Waste Incineration		Carbon content of waste, percentage of mineral carbon in the total carbon content, and combustion efficiency can be obtained from the corresponding emission factor's recommended values in Table 5.5 of the "Guidelines for Provincial Greenhouse Gas Inventory Compilation (Trial)" in Climate [2011] No. 1041.	

3. Implementation Path of Event Carbon Neutrality

3.1. Carbon Emission Reduction in Events

For event-based carbon neutrality, several large-scale events have already adopted various measures to reduce carbon emissions at each stage, providing new insights into achieving event carbon neutrality. According to the accounting standards for event carbon neutrality in various regions, the carbon accounting mainly consists of several stages, including event preparation, event hosting, and event conclusion.

This article will draw from existing cases to introduce the carbon emission reduction process during the stages of event preparation, hosting, and conclusion, providing relevant recommendations to achieve event carbon neutrality.

3.1.1. Event Preparation Stage

During the event preparation stage, companies should first estimate the greenhouse gas emissions generated by the event.

After estimating the greenhouse gas emissions, companies should propose emission reduction measures based on the event plan. According to the relevant carbon neutrality accounting standards, event emissions primarily include transportation emissions, accommodation and catering emissions, implicit emissions from event supplies, and emissions from waste treatment. Specific measures for the event preparation stage for the above four types of emissions are shown in Table 2.

Table 2. Emission Reduction Measures during Event Preparation Stage

Transportation Emissions:	Encourage attendees to use public transportation, new energy vehicles, bicycles, or other low-carbon means of transportation.
	Pre-assess the travel distance of attendees (and facilitate carpooling for nearby participants).
Accommodation and Catering Emissions:	Avoid using single-use plastic cutlery and opt for sustainable alternatives when ordering tableware.
	Choose local, low-carbon, and plant-based catering options whenever possible.
	Encourage VIP guests to bring their own personal items such as toothbrushes.
Emissions from Consumables:	Select energy-efficient and environmentally friendly venues for organizing the event.
	Opt for recyclable consumables and materials during procurement for event preparations.
	Utilize LED lighting systems and maximize natural light during venue setup.
	Promote the use of renewable resources, such as renewable souvenirs and chairs.
Emissions from Waste Disposal:	Set up different types of waste bins in advance for convenient waste sorting.
	Develop a plan for recyclable materials to reduce waste emissions.
	Plant green vegetation around the venue to absorb greenhouse gas emissions
	Establish or renovate wastewater recycling systems around the venue.

3.1.2. Event Implementation Stage

During the event implementation stage, the organizers should diligently execute the emission reduction plan developed during the event preparation stage and further improve the plan based on the actual circumstances of the event. The organizers should carry out actions to control greenhouse gas emissions according to the carbon neutrality implementation plan, ensuring the achievement of the expected emission reduction results.

For the event implementation stage, specific measures for energy conservation and emission reduction should be taken in the areas of transportation emissions, accommodation and catering emissions, emissions from consumables used during the event, and emissions generated from waste disposal. The specific measures to achieve carbon neutrality are shown in Table 3.

Table 3. Emission Reduction Measures during Event Implementation Stage

Transportati on	Set up shuttle buses with designated routes based on participants' addresses to reduce

Emissions	transportation emissions
	Rent new energy vehicles to replace traditional fuel-powered buses.
Accommodation and Catering Emissions	Promote the clean plate campaign and impose penalties for intentional waste behavior
	Use public water dispensers and provide reusable environmentally friendly water cups instead of plastic bottled water and disposable paper cups
	Reduce the provision of disposable amenities (e.g., toothbrushes, slippers)
	Optimize air conditioning temperature and use natural ventilation when unnecessary
Implicit Emissions from Event Supplies	Limit the use of paper banners and pull-up banners, and opt for electronic presentations in PPT format
	Choose electronic versions of conference materials during the event; refrain from using paper forms and documents and switch to electronic formats
	Use LED screens and manage electricity usage effectively, turning off when not in use
	Reasonably schedule meetings during daytime to reduce energy consumption.
	Opt for second-hand items for the event instead of purchasing new ones.
	Utilize electronic business cards and nameplates to save paper.
	Adopt facial recognition and ID verification, eliminating the need for printed badges.
Emissions from Waste Treatment:	Conduct water conservation education, encourage water recycling for toilet flushing, etc.
	Food waste and residue can be used as fertilizer
	Resell event equipment to increase utilization
	Organize paper recycling activities.

3.1.3. Event concluding stage

During the event concluding stage, the organizers consolidate the information on overall carbon emissions reduction during the event, quantifying specific emission reduction values. They calculate the actual greenhouse gas emissions produced based on the real carbon reduction achieved during the event and use it as a basis for carbon offsetting.

At the event concluding stage, the total carbon emission reduction is publicly disclosed to enhance the participants' sense of engagement and accomplishment. This disclosure also serves as a positive example and successful case for other event organizers[9].

3.2. Carbon Emission Offset for Events

In addition to reducing carbon emissions from events, event organizers can achieve carbon neutrality through carbon emission offset methods. The current offset methods include obtaining carbon quotas or carbon credits and establishing carbon sequestration forests.

Organizers determine suitable offset products based on the actual greenhouse gas emissions produced during the event. These offset products mainly fall into two categories: carbon quotas or carbon credit offset and establishment of carbon sequestration forests. Below is a detailed explanation of carbon quotas, carbon credits, and establishing carbon sequestration forests.

3.2.1. Carbon Quotas

"Carbon quotas" refer to the "total amount of greenhouse gases (expressed in carbon dioxide equivalent) that an enterprise is permitted to emit into the atmosphere during a specific period as approved by the competent government department." In simpler terms, to alleviate the pressure of achieving the "peak carbon emissions" target by 2030, the government grants allowances for carbon emissions to key emitting entities to incentivize energy conservation, emission reduction, and promote the green and low-carbon transformation of industries[4].

For example, if the cost of energy conservation and emission reduction is low, Company A may choose to reduce emissions through energy conservation measures or other carbon sequestration methods. After fulfilling its obligations, the remaining "carbon quotas" can be traded in the carbon emission market for economic benefits. On the other hand, if the cost of energy conservation and emission reduction is high[7], and Company A cannot fully meet its carbon emission reduction obligations even after using all its "carbon quotas," it can enter the carbon market to purchase quotas from other companies, government departments, or individuals. Alternatively, Company A can develop its own CCER (China Certified Emission Reduction) projects to obtain "carbon credits" to offset a portion of the quotas (the offset ratio should not exceed 5% of the total quotas). In extreme cases, if Company A fails to offset its carbon emissions or fulfill its obligations within the validity period, it will face significant fines.

3.2.2. Carbon Credits

While "carbon quotas" are based on carbon emissions, "carbon credits" are based on carbon emission reductions.

"Carbon credits" refer to the reduction of carbon emissions achieved through enhanced energy efficiency, pollution reduction, or development reduction measures, which have been verified and confirmed by international organizations, independent third-party entities, or governments. These emission reductions can be entered into the carbon market for trading. Generally, carbon credits are registered and issued in the form of emission reduction projects. The specific differences between carbon quotas and carbon credits are illustrated in Table 4.

Table 4. Comparison of Carbon Quota and Carbon Credit Concepts

	Carbon Quota	Carbon Credit
Differences in Rights	Represents the permissible amount of greenhouse gas emissions.[6]	Represents the reduction in emissions.
Differences in Generation	Issued by the government to companies (either	after the actual emission reduction action takes place,

	paid or free), with a predetermined quantity. Generated	verified by professional institutions.
Differences in Trading	Purpose Meets the need for companies to fulfill their obligations at a lower cost.	Primarily used to fulfill corporate social responsibility requirements.
Differences in Trading Systems	Carbon Emission Trading System (ETS).	Carbon Emission Trading System and voluntary emission reduction system trading.

3.2.3. New Carbon Sink Afforestation (Carbon Sequestration Afforestation)

"New Carbon Sink Afforestation (Carbon Sequestration Afforestation)" refers to a special afforestation activity carried out on land with a determined baseline, with the primary purpose of increasing carbon sequestration. This afforestation and the growth process of the forest (trees) are subject to carbon sequestration measurement and monitoring. Compared to regular afforestation, carbon sequestration afforestation highlights the function of forests as carbon sinks, requiring specific technical requirements for carbon sequestration measurement and monitoring, while emphasizing the multiple benefits of forests. Only afforestation that adheres to the technical requirements of carbon sequestration afforestation and passes inspection and acceptance according to specified procedures, and undergoes carbon sequestration measurement and monitoring by qualified entities as per regulations, can be recognized as a carbon sink forest.

4. Specific Reference Cases

In May 2022, the National Main Venue Event for the 65th Environment Day, successfully organized by the Liaoning Provincial Government, served as an excellent model for activity-based carbon neutrality projects.

The event featured five parallel forums that extensively discussed various ecologically-related issues, showcasing environmental achievements, and summarizing successful experiences. Simultaneously, the organizing committee used public education to raise awareness about environmental protection and carbon neutrality. Through this event, they encouraged adopting low-carbon and eco-friendly lifestyles, fostering a societal atmosphere of collectively building a low-carbon and eco-friendly environment.

Throughout the entire event, the concept of simplicity, greenness, and low carbon was diligently upheld, ultimately achieving carbon neutrality for the National Main Venue Event for the Environment Day[10]. Proper planning was carried out during the preparation, execution, and conclusion stages of the event, employing internet promotion of low-carbon emission reduction concepts and using big data to calculate carbon emissions. Measures for successfully achieving carbon neutrality for this large-scale event are outlined in Table 5.

Table 5. Categorization of Specific Measures for Emission Reduction during the National Environment Day Event at the Hexi Corridor

Phase	Specific Measures	Emission Reduction Direction
Preparation Phase	Plan and budget for emissions related to fossil fuel combustion and transportation during the event, following the "Carbon Neutrality Emission Standards for Events."	Transportation emissions.
	Plan and budget for emissions related to accommodation, catering, and electricity and heat consumption during the event, following the "Carbon Neutrality Emission Standards for Events."	Accommodation and catering emissions.
Event Phase	Encourage participants to bring their own personal items, such as toothbrushes, and promote the practice of clean plate waste reduction.	Accommodation and catering emissions.
	Opt for electronic conference materials instead of printed versions to reduce emissions from conference consumables.	Emissions from disposable conference materials.
	Set up public drinking water facilities and provide reusable eco-friendly water bottles, eliminating plastic bottled water and disposable paper cups.	Emissions from disposable conference materials.
	Purchase "green electricity" to charge new energy vehicles during the event, achieving a carbon reduction of 1.63 tons of carbon dioxide equivalent.	Transportation emissions.
Event Conclusion Phase	Establish carbon-neutral forests in Zhanggutai, Zhanwu County, Fuxin City, Liaoning Province, to offset the carbon emissions from the event.	Creation of new carbon sink forests for offsetting.
	Liaoning Datang International Jinzhou Thermal Power Co., Ltd. voluntarily cancels 60 tons of national carbon emission quotas to offset part of the carbon emissions from this event.	Carbon quota offset.

The various measures implemented during the National Main Venue Event for the 65th Environment Day are highly instructive, and it is anticipated that more and more events will adopt a series of measures to achieve energy conservation and emission reduction in

 https://doi.org/10.1051/e3sconf/202344103015

the future. The initiative of activity-based carbon neutrality is expected to be effectively implemented.

References

1. Schneider, K., Lee, S. H., & Ge, J.(2015) Transforming Rural Communities in China and Beyond.
2. Wang, B., Yu, J., & Wu, R. (2022) Achieving carbon neutrality in China: Legal and policy perspectives. Frontiers in Environmental Science, 10, 2436.
3. Huang, L., Wei, X., & Wang, Q. (2023). Challenges and responses of China's energy legal system under the double carbon target. Frontiers in Energy Research, 11, 1198525.
4. Wang, P., Liu, H., Guo, H., Yu, Y., Guo, Y., Yue, G., ... & Wang, L. (2023). Study on preparation and performance of alkali-activated low carbon recycled concrete: Corn cob biomass aggregate. Journal of Materials Research and Technology, 23, 90-105.
5. Yang, L., Cai, Y., Hong, J., Shi, Y., & Zhang, Z. (2016). Urban distribution mode selection under low carbon economy—A case study of Guangzhou City. Sustainability, 8(7), 673.
6. Luo, R., He, C., Yu, Q., He, L., Zhang, Y., & Ma, W. (2019). Investigating the influence of the implementation of an energy development plan on air quality using WRF-CAMx modeling tools: A case study of Shandong province in China. Atmosphere, 10(11), 660.
7. Li, S., Liu, J., Wu, J., & Hu, X. (2022). Spatial spillover effect of carbon emission trading policy on carbon emission reduction: Empirical data from transport industry in China. Journal of Cleaner Production, 371, 133529.
8. Ministry of Ecology and the Environment.(2019) Beijing Municipal Guidelines for Carbon Neutrality Implementation in Large-scale Events. https://www.gov.cn/gongbao/content/2019/content_5430516.htm
9. Isbat Alam,Shichang Lu,Muddassar Sarfraz & Muhammad Mohsin.(2023).The Interplay of Green Technology and Energy Consumption: A Study of China's Carbon Neutrality and Sustainable Digital Economy. Energies(17). doi:10.3390/EN16176184.
10. Talwar Shalini,Dhir Amandeep,Luqman Adeel,Gupta Jaya & Gugnani Ritika.(2023).Charting the path toward a greener world: A review of facilitating and inhibiting factors for carbon neutrality. Journal of Cleaner Production. doi:10.1016/J.JCLEPRO.2023.138423.

Study on Key Technical Route and Construction Mode of Low-Carbon Park

Zhukui Tan[1], Guangqiu Zhang[2,*], Zongyu Sun[2], Jintang Li[2], Yang Wang[1], Ji Li[2], Wei Xu[2] and Yongxiang Cai[1]

[1] Guizhou Power Grid, Co., Ltd., Guiyang, China
[2] China Academy of Building Research, Beijing, China

Abstract. The potential for energy conservation and emission reduction in parks is enormous, promoting the popularization of low-carbon parks is a necessary means to promote the green and low-carbon transformation of energy consumption and achieve China's 'dual carbon' goals. This article summarizes and extracts four main technical routes for building low-carbon parks through research on low-carbon parks and building cases at home and abroad: energy conservation and low-carbon construction of architectural noumenon, proactive energy conservation and operational carbon reduction, internal development of renewable energy, and external green energy input. Horizontal comparison and qualitative analysis of the four technical routes adopted in each typical case, summarizing and proposing eight models and typical application scenarios for building low-carbon parks, providing a theoretical basis and implementation suggestions for low-carbon development of the park.

1 Introduction

According to the 2022 Research Report of China Building Energy Consumption and Carbon Emissions[1], the total carbon emissions of the whole process of building in China in 2020 will be 5.08 billion tons, accounting for more than 50% of the national carbon emissions, which shows that buildings are the main battlefield for realizing the "Dual Carbon" strategy. It is very important to study and implement the goal of "double carbon" in the construction of the park, which is a building group carrying life and production. According to the "2020 Future smart park white paper"[2], more than 90% of urban residents work and live in the parks, more than 80% of GDP and more than 90% of innovation are generated in the parks, and the urban pattern and planning of our country are basically based on the parks as "units". With a large number of parks in China, highly concentrated elements, vigorous development of innovation activities, and low difficulty in collaboration among participants, it is currently the most suitable scene for low-carbon research and practice. It is precisely because of the practicability and huge potential of research in the park that the application value of energy-saving and carbon-reducing technologies in the park has begun to be valued by the industry, and low-carbon parks have become a focus of attention in implementing the goal of "double carbon".

As the most important part of the park, the building's low-carbon and energy-saving construction is an important part of the realization of low-carbon park. In 2019, the national standard "Technical Standard for Near zero Energy Buildings" (GB/T51350-2019)[3] was officially implemented, which is the first time in the world to clarify the definition, boundary and key technical indicators of near-zero energy buildings and zero-energy buildings in the way of national standards. It provides reference for relevant researchers to study the key technology paths of low carbon, near zero carbon and zero carbon buildings. In addition, the comprehensive energy system of the park is a new energy system that adapts to distributed development and meets the diversified energy needs of the park[4][5]. For the park, optimizing the energy structure and reducing the proportion of energy consumption with high carbon emission is also the most direct path to achieve emission reduction[6]-[8].

On the basis of low-carbon buildings and comprehensive energy, the relevant researchers put forward the overall macro technical measures for the realization of low-carbon parks. Zhou et al.[9] put forward the key points of the optimization strategy for planning and design of green low-carbon parks, including the layout of clean energy, the addition of energy storage facilities, carbon capture devices, carbon utilization or carbon storage logistics warehouses, the expansion of green space carbon sink space, and the construction of digital twin smart workshops to track carbon footprint. Yang et al.[10] proposed to realize the construction and application of low-carbon smart parks through the construction of micro-grid energy management system, all-domain sub-metering system, automatic sensing green buildings and digital management service platform. Zang

* Corresponding author: 1971206558@qq.com

et al.[11] put forward four paths to realize smart low-carbon/zero-carbon parks: using green materials in park construction, low-carbon/zero-carbon energy infrastructure, low-carbon/zero-carbon information infrastructure, and low-carbon/zero-carbon production and life. Chen et al.[12] proposed a zero-carbonization implementation path for business office parks by adopting green energy, green buildings, green transportation, green office and green carbon sink.

Low-carbon park is a green development mode that takes overall consideration of carbon emission and sustainable development, and takes the formation of low-carbon industrial clusters as the ultimate development goal[13]. Low-carbon park is a model of park construction and development, which takes "low carbon" as the goal of park construction and development. However, the above relevant studies only focus on the single building, energy system and other related technologies in the park, and lack the "low-carbon model" study of the park.

Therefore, by investigating typical low-carbon parks and building cases at home and abroad, this paper summarizes and abstractes the main technical paths for the construction of low-carbon parks, makes horizontal comparison and qualitative analysis of the technical routes adopted in each typical case, proposes the typical models and application scenarios for the construction of low-carbon parks, and provides theoretical basis and implementation suggestions for the low-carbon development of parks.

2 Typical low-carbon park project

The research team conducte research and analysis on more than 50 domestic and foreign low-carbon parks and energy-saving building projects, and summarize the main technical routes for the realization of low-carbon parks and buildings. The following seven typical cases are illustrated.

2.1 Low-carbon parks

2.1.1 Apple Park

Apple's new headquarters is located in California, the United States, with a total area of about 175 acres, as shown in Fig. 1, which is a low-carbon technology park dominated by offices. The project officially opened in 2019[14].

Fig.1. Apple Park[14]

The project mainly uses natural ventilation, natural lighting, external shading, plant carbon sink, fresh air heat recovery system, efficient lighting, efficient electrical appliances, radiation terminal, indoor environment monitoring, full fresh air replacement ventilation, solar photovoltaic, biofuel cells and other building energy saving and renewable energy utilization technologies. The building can solve the cold and heat load in the building most of the time through natural ventilation, and the design of natural lighting + task lighting can reduce the lighting power of the station by about 20%.

The peak photovoltaic power generation of the park is about 17kW, which can meet 75% of the daily electricity load, and the California photovoltaic power station supplies the remaining 25% of the energy. At the same time, the building has been connected to the grid, and the surplus electricity can be transmitted by the grid to achieve 100% renewable energy supply[15].

2.1.2 Bay View Campus

Google headquarters Bay View Campus is located in Mountain View, Silicon Valley, California, as shown in Fig. 2, the project covers an area of about 170,000m^2, and officially opened in 2022.

Fig.2. Bay View Campus[16]

The project adopts natural lighting, natural ventilation, automatic shading, green building materials and green construction, indoor and outdoor green landscape, solar photovoltaic, ground source heat pump and other building energy saving and renewable energy utilization technologies.

The 50,000 solar panels in the shape of a dragon scale are installed on the roof of the buildings in the park, which can generate nearly 7MW of electricity, meet about 40% of the energy demand of the park, and reduce nearly 50% of carbon emissions. Meanwhile, green energy companies will provide more than 50% of the electricity to the Google headquarters building, achieving 100% electrification of the park.

2.1.3 EUREF-Campus

Located in the city of Berlin, with an area of 55,000m^2, the EUREF Energy Technology Park is Europe's first zero-carbon smart transformation park, as shown in Fig. 3.

The buildings in the park use high-performance building materials, efficient insulation system, three-layer doors and windows, shading, intelligent lighting, energy management system, energy storage, solar photovoltaic, wind turbines, ground source heat pump and other building energy saving and renewable energy utilization technologies. At the same time, the park will build charging piles and smart grid systems to reduce carbon emissions in the park.

The park deals to purchase biogas and integrates it into the pipeline network to meet 60% of the heating needs of the EUREF-Campus, while relying on ground source heat pump and photovoltaic power generation technology to achieve 80% to 95% renewable energy utilization.

Fig.3. EUREF-Campus[17]

2.1.4 Sino-German Ecopark

The Sino-German Ecopark is located in the West Coast New District of Qingdao, with a start-up area of 34.92km^2, as shown in Fig. 4. The first phase of the project has been completed in 2021.

Fig.4. Passive room technology experience center in Sino-German Ecopark
(Picture source: http://www.sgep.cn/index.htm)

The project adopts natural lighting, natural ventilation, efficient building envelope, intelligent lighting, intelligent building, fresh air heat recovery system, terminal radiation, indoor monitoring system, "double carbon" operating system, lithium battery energy storage, solar photovoltaic, wind turbines, ground source heat pump and other building energy saving and renewable energy utilization technologies.

The installed capacity of distributed PV in the park reaches 16MW, and the energy supply area is nearly 1,000,000 m^2. The construction of passive houses in the first phase of the Sino-German Ecopark totaled 400,000m^2, which is expected to save 1,300,000kWh of primary energy consumption and reduce carbon emissions by 664tons per year.

2.2 Low-carbon buildings

2.2.1 Rocky Mountain Insititute's New Innovation Center

Rocky Mountain Institute's new Innovation Center is located in Colorado, USA, with a construction area of 1450m^2, as shown in Fig. 5. It has been completed and put into operation in 2022, accommodating 50 employees daily.

The building adopts efficient building envelope, natural lighting, natural ventilation, external shading, glare control, indoor sound insulation, thermal storage floor, landscape green plants, fresh air heat recovery system, efficient electrical appliances, efficient lighting, energy control system, solar photovoltaic, lithium battery energy storage and other building energy saving and renewable energy utilization technologies.

The building is equipped with 83kW solar photovoltaic panels to meet the energy needs of 123%~148% of the building. Compared with conventional buildings, heating energy consumption is reduced by 84%, ventilation energy consumption is reduced by 96%, and lighting energy consumption is reduced by 73%.

Fig.5. Rocky Mountain Insititute's New Innovation Center[18]

2.2.2 NREL Research Support Facility

National Renewable Energy Lab (NREL) Research Support Facility is located in Colorado, USA, with an area of 17,000m^2 and a construction area of 203,45m^2, as shown in Fig. 6. The project was completed in 2011 as a green office building.

Fig.6. NREL Research Support Facility
(Picture source: http://www.nrel.gov/)

The project adopts efficient building envelope, electric color glass, natural lighting, natural ventilation, active external shading, green carbon sink, thermal storage floor, fresh air heat recovery system, efficient electrical appliances, radiation terminal, indoor environment monitoring, intelligent lighting, evaporative cooling technology, full fresh air displacement ventilation design, solar photovoltaic, photothermal, ground source heat pump, biofuel cell, biomass boiler and other building energy saving and renewable The energy utilization technology reduces the design target of building energy intensity to 110.57kWh/(m^2·a).

Through solar photovoltaic, photothermal utilization, and biomass boiler hot water system, it can provide 2238MWh of energy for the building every year, accounting for about 30% of the total energy consumption of the building. The actual monitored energy consumption of the project in 2017-2018 is 78.75kWh/(m^2·a), which is 74.4% energy-saving compared with the general office buildings with the same level of use.

2.2.3 CABR Near Zero Energy Demonstration Bbuilding

The near-zero energy demonstration building of China Academy of Building Research (CABR) is located in Beijing, with a construction area of 4025m^2 and its main functions are office and conference, as shown in Fig. 7. The project was officially completed and put into use in 2014.

Fig.7. CABR Near Zero Energy Demonstration Building[19]

The project adopts high efficiency envelope, natural lighting, natural ventilation, adjustable central shading, roof greening, fresh air heat recovery system, efficient electrical appliances, efficient lighting, radiation terminal, indoor environment monitoring, full fresh air displacement ventilation, energy management platform, energy storage, solar photovoltaic, solar collector, ground source heat pump and other building energy saving and renewable energy utilization technologies.

The building energy saving rate is greater than 80%, the building energy consumption level is 25kWh/m^2·a, and the renewable energy utilization rate is 45%.

3 Technology routes of Low carbon park

Through the analysis of typical low-carbon park and building projects at home and abroad, the main technical routes of low-carbon parks can be summarized, including: energy conservation and low-carbon construction of architectural noumenon, proactive energy conservation and operational carbon reduction, internal development of renewable energy, and external green energy input, as shown in Fig. 8. At the same time, the horizontal comparison and qualitative analysis of the four technical routes are carried out, and Table 1 is obtained.

3.1 Energy conservation and low-carbon construction of architectural noumenon

Energy conservation and low-carbon construction of architectural noumenon technology emphasizes the direct use of the natural conditions of the site, through the optimization of planning and architectural design, the building can meet the requirements of heating, cooling and lighting of part of the building under the operation mode of non-mechanical, no energy consumption or less energy consumption, so as to achieve the purpose of reducing building carbon emission. Including the use of efficient building envelope, shading, natural lighting, natural ventilation and green carbon sinks and other means.

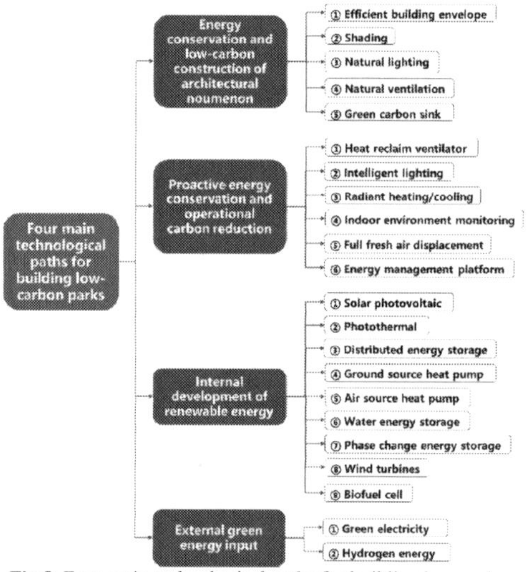

Fig.8. Four main technological paths for building low-carbon parks

Efficient building envelope is one of the key technologies to realize passive house building and ultra-low, near-zero energy building. Efficient building envelope emphasizes high standard insulation design for external walls, roofs, parapet walls, partition walls, floors, underground foundations, and others. Studies have shown that the use of external insulation system is 30% lower than the use of sandwich insulation carbon emissions. At the same time, the efficient building envelope should pay attention to the design of the detailed joints of the cold and hot bridges and the air tightness measures.

Shading has a significant effect on reducing building heat and cold load and improving indoor comfort. When

 https://doi.org/10.1051/e3sconf/202344103016

selecting the performance of the shading device, it is necessary to consider the needs of the building for shading in summer, obtaining heat in winter and natural lighting.

Natural ventilation technology enables indoor cooling through the planning and design of buildings, thereby reducing the energy use of building mechanical ventilation equipment. This kind of reasonable technical design means to deal with the relationship between the building and the natural environment has become the development trend of low-carbon building design.

With the development of passive houses and ultra-low and near-zero energy consumption building systems, energy conservation and low-carbon construction of architectural noumenon technology in buildings will play an increasingly important role.

Table 1. Quantitative analysis of typical low-carbon projects at home and abroad

No.	Name	Energy conservation and low-carbon construction of architectural noumenon	Proactive energy conservation and operational carbon reduction	Internal development of renewable energy	External green energy input	The effect of energy saving and carbon reduction
1	Apple Park	Natural ventilation meets 70% of the building's heating and cooling requirements	Reduced lighting power by 20%	Meet 75% of the building's energy needs	Meet 25% of the building's energy needs	Zero energy consumption
2	Bay View Campus		100% electrification	Meet 40% of the building's energy needs	Meet 50% of the building's energy needs	Neat-zero energy consumption
3	EUREF-Campus			Meet 80%~95% of the building's energy needs	Purchase biogas	Near-zero carbon
4	Sino-German Ecopark	Millions of square meters of passive room, the annual carbon reduction of 46,000 tons, more than 92% energy saving than general office buildings				Low-carbon
5	Rocky Mountain Insititute's new Innovation Center		Compared with conventional buildings, energy consumption for heating is reduced by 84%, air conditioning by 100%, ventilation by 96% and lighting by 73%	Meet 123% ~ 148% of the building's energy needs		Produce enenrgy, negative carbon
6	NREL Research Support Facility	74.4% energy-saving compared with the general office buildings		Meet 30% of the building's energy needs		Neat-zero energy consumption
7	CABR Near Zero Energy Demonstration Bbuilding	The envelope reduces building energy consumption by 47.8%	Reduced lighting power by 20%	Renewable energy utilization rate reached 45%		Neat-zero energy consumption

3.2 Proactive energy conservation and operational carbon reduction

The core of proactive energy conservation and operational carbon reduction technology refers to reducing energy consumption by improving the efficiency of energy use systems, mainly including heat reclaim ventilator, intelligent lighting, radiation terminal, indoor environment monitoring, full fresh air displacement ventilation and energy management and control platforms.

The fresh air heat recovery system uses exhaust air to preheat fresh air in winter and precool fresh air in summer, which can reduce energy consumption caused by treating fresh air, and is an effective means of building energy saving and carbon reduction, as shown in Fig. 9. Studies have shown that the emission reduction of buildings in Beijing, Shanghai and Guangzhou using fresh air heat recovery system only in summer can reach 14%~23%. In addition, the overall energy saving and economic benefits of using fresh air heat recovery system in cold regions are higher than those in other climate regions[20].

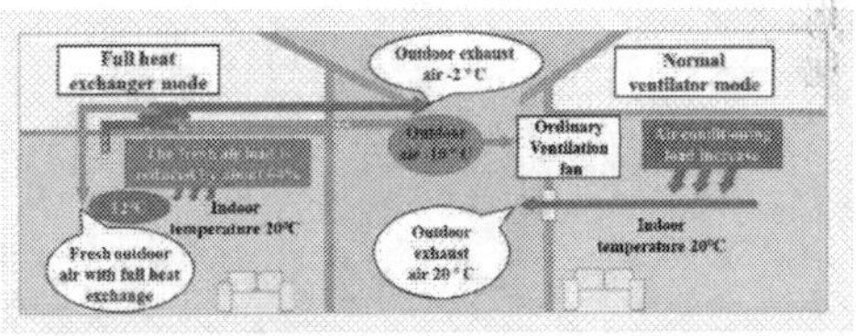

Fig.9. Working principle of fresh air heat recovery system (winter)

Intelligent control and energy-saving lamps should be used for lighting in the park. Common forms of intelligent lighting include light soft start, timing control, brightness intensity adjustment, scene setting, etc., so as to reduce lighting energy consumption and carbon emissions. Intelligent lighting system is usually used with LED lighting, LED lighting has developed rapidly in recent years, is one of the highest luminous efficiency lighting sources, compared with ordinary lighting can save 80%, has been widely used in green buildings, ultra-low energy consumption buildings.

Indoor heating and ventilation terminials of buildings are usually in the form of radiators, fan coils, all-air systems, active cold beams, radiant ceilings/floors, etc. Radiant terminal help reduce energy grade requirements, increase load flexibility, and increase comfort levels.

However, the radiant terminial has high investment and high dehumidification control requirements, so the radiant terminial can be used for partial application demonstration in the low-carbon construction process of the park.

In order to realize intelligent energy monitoring and fine management, improve the operation efficiency and intelligent management level of the park, an integrated energy management system should be built in the park. The system should be fully compatible with heating, air conditioning, lighting, water supply and drainage, power transmission and distribution monitoring, fire alarm, traffic and parking management and other subsystems. It can realize the functions of forecast, supervision, diagnosis, early warning, regulation and optimization of park energy consumption, so as to save park energy consumption and reduce park carbon emissions.

3.3 Internal development of renewable energy

The field of building and park is an important scene for the application of renewable energy, and renewable energy technology is also one of the key technologies to achieve low carbon in building and park. Priority should be given to solar thermal system, ground source heat pump, air source heat pump, etc. The power supply system should give priority to photovoltaic and wind power.

As the most important forms of renewable energy utilization in the park, solar photovoltaic and photothermal utilization are the most important technical ways to reduce carbon emissions in the park. Photovoltaic construction is the most important part of meeting the energy demand of the park, realizing the electrification of the park and building a low-carbon park. The low-carbon park proposes to make full use of the building roof and facade for the installation of photovoltaic modules, encourage self-adoption, and coupled with the energy storage system design to adjust the asynchronism of photovoltaic power generation and load demand, as well as the volatility and uncertainty of solar photovoltaic power generation, so as to improve the effective utilization rate of building photovoltaic power generation. With the continuous improvement of photovoltaic cell power generation efficiency, and the continuous decline of photovoltaic module prices with the advancement of industrialization, photovoltaic construction will become the most direct and effective technical measures to achieve low-carbon parks.

Fossil energy is the main energy for building heating, and the carbon emissions of winter heating in the park account for more than 20% of the total carbon emissions in the park. Therefore, the joint use of clean energy and electric energy according to local conditions is in line with the characteristics of China's resource endowment, and is the only way to re-electrification in the field of heating and cooling. Heat pump is an efficient and low-carbon energy supply method that elevates heat from low temperature to high temperature, and is the best way to efficiently convert electricity into heat, as shown in Fig. 10. As an efficient renewable energy utilization method, heat pump plays an important role in reducing carbon emissions of park heating. The heat source of the heat pump can be divided into air, surface water, groundwater, soil, solar energy, sewage, waste gas, etc.

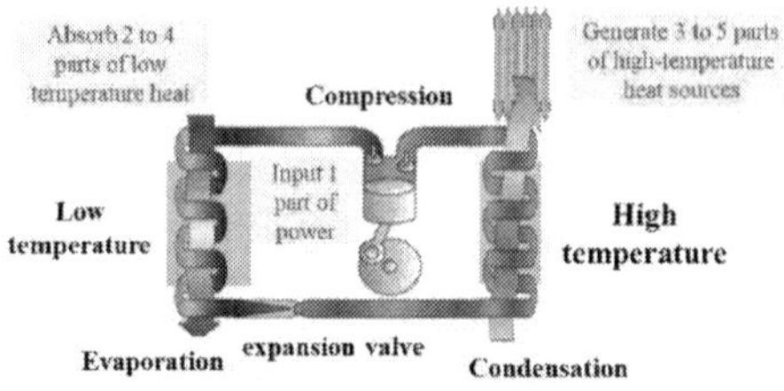

Fig.10. Heat pump system

The initial investment of the heat pump system is not much different from the cost of municipal heating, but the operating cost is much lower than that of municipal heating, showing excellent economy. In addition, a heat pump unit is combined with an energy storage system to form a heat pump energy storage system. By controlling energy storage and discharge, the time transfer of cold and heat source energy production and actual cold and hot energy use can be realized, and the flexible power consumption and peak power load of the heat pump system can be effectively transferred, further improving the economy and utilization efficiency of the system.

Due to the limitation of the park site, the smaller turning radius of the vertical axis wind turbine has better application potential in the park than that of the horizontal axis wind turbine. In areas with good wind energy conditions and far from the grid, it can supply lighting, electrical appliances, communication equipment and other electricity.

3.4 External green energy input

Compared with energy-saving buildings, low-carbon buildings emphasize more on the use of electricity, so it is necessary to consider the cancellation of gas and the realization of gas to electricity, in order to reduce or zero all kinds of direct carbon emissions. Therefore, for parks with sufficient renewable resources, solar energy and wind energy can be fully utilized to generate electricity to meet the demand of the parks. For parks lacking renewable resources, external green energy can be introduced to achieve low carbon goals. The external green energy input is mainly green electricity trading, and the main sources of green electricity are solar energy, wind energy, biomass, geothermal energy, hydrogen energy and so on.

When importing electricity from the external network, the low-carbon park should consider the power composition of the grid at different times, use more electricity and store more electricity when the proportion of wind power is large, use less electricity when the external network is dominated by coal power peaking, and use its own energy storage facilities to release energy storage to reduce the dependence on the external grid.

The development of green electricity market plays an important role in realizing low carbon in parks with limited renewable resources. However, the implementation of the green electricity trading mechanism in the past period of time has also exposed some problems, including the imperfect green electricity pricing mechanism, the popularization of multi-year green electricity trading, and the demand for cross-province and cross-region trading can not be met. Therefore, the future development of the green electricity market is proposed from the above perspectives to better release the potential of green electricity trading and promote the realization of low-carbon parks.

4 Low-carbon park construction

By analyzing the characteristics of the above technical measures, this paper puts forward eight different modes of building low-carbon parks, as shown in Table 2 below.

By further comparing the technical measures applied in the typical low-carbon park construction mode, the typical application scenarios of different modes are summarized as shown in Table 3 below, and carbon emission analysis is carried out for each scenario.

Table 2. Construction modes of typical low-carbon parks

Modes	Names	Energy conservation and low-carbon construction of architectural noumenon	Proactive energy conservation and operational carbon reduction	Internal development of renewable energy	External green energy input
Mode A	Original insulated type	•	•		
Mode B	New energy development type			•	
Mode C	Passive architecture type	•	•		
Mode D	Internal and external combination type	•		•	
Mode E	Comprehensive low-carbon type	•	•	•	
Mode F	Existing building renovation type		•	•	•
Mode G	High density building type	•	•		•
Mode H	Integrated zero carbon type	•	•	•	•

Table 3. Application scenarios of typical low-carbon park modes

Modes	Typical applications scenario	Comments
Mode A	Reconstruction of northern residential community	Focus on energy conservation of architectural noumenon
Mode B	Rural residential building green energy	No focus on net carbon emissions
Mode C	Passive house construction	Focus on building energy efficiency improvement
Mode D	Villa community construction	De-emphasizing operational optimization
Mode E	New construction of comprehensive low-carbon park	Lay emphasis on the overall operation effect
Mode F	Renovation of special climate zones	Energy conservation of architectural noumenon meets the basic requirements, and further upgraded
Mode G	High density commercial building	The use of renewable energy in the park is limited
Mode H	New comprehensive zero-carbon demonstration zone	Can participate in market trading

Mode A: Adopt the technical route of energy conservation and low-carbon construction of architectural noumenon, mainly through strengthening the external wall insulation performance, replacing the external window and other architectural noumenon energy saving means to reduce building energy consumption, and achieve the goal of reducing the cumulative building heating and cooling load by more than 40% in cold areas. This model is mostly suitable for energy saving renovation of residential communities in northern China.

Mode B: Adopt the technical route of internal development of renewable energy, which is mostly suitable for parks or counties and towns that use solar photovoltaic panels to generate power to improve the utilization of renewable energy, and realize the green energy use of rural residents' buildings by not changing the form and function of the building itself. This model usually focuses on regional comprehensive carbon emissions, instead of building energy efficiency and net carbon emissions.

Mode C: Adopt the technical route of energy conservation and low-carbon construction of architectural noumenon+proactive energy conservation and operational carbon reduction, mainly through various energy-saving technologies to construct the best building envelope and indoor environment, improve building energy efficiency, and minimize the building's heating and cooling demand, which is mostly suitable for the construction of passive residential buildings. Relevant researchers have proposed that improving building energy efficiency to ultra-low energy building energy efficiency can reduce the equivalent carbon emissions of all types of residential buildings in various climate zones by 10% ~38% compared with the "General Code for Energy Efficiency and Renewable Energy Application in Buildings" (GB55015-2021), and reduce the carbon emission intensity of public buildings by more than 30%~40% compared with the benchmark buildings[21].

Mode D: The technical route of energy conservation and low-carbon construction of architectural noumenon+internal development of renewable energy is adopted. When the energy conservation and low-carbon construction of architectural noumenon meets the requirements, the optimization effect of building operation is not emphasized, and the villa buildings and residential communities that can install distributed energy such as solar photovoltaic are mainly targeted.

Mode E: Adopt the technical route of energy conservation and low-carbon construction of architectural noumenon+proactive energy conservation and operational carbon reduction+internal development of renewable energy, which is suitable for new comprehensive near-zero carbon parks that pay attention to the overall operation effect of buildings. Compared with mode C, mode E not only improves building energy efficiency, but also further increases the proportion of renewable energy application in the park. At the same time, research shows that if the building reaches the near-zero-energy building energy efficiency index and further increases the building renewable energy application, the carbon reduction rate of the benchmark building can be increased by 15%[22].

Mode F: Adopt the technical route of proactive energy conservation and operational carbon reduction+internal development of renewable energy+external green energy input. It is mainly aimed at the low-carbon transformation of existing public buildings or parks in special climate zones. For example, in mild areas, the building has met the requirements of basic energy conservation, but further improving the performance of the building envelope has no obvious effect on the energy conservation and low-carbon construction of architectural noumenon, so this model can be adopted.

Mode G: Adopt the technical route of energy conservation and low-carbon construction of architectural noumenon+proactive energy conservation and operational carbon reduction+external green energy input, which is mainly suitable for high-density commercial buildings with poor renewable resources in the park, and achieve the low-carbon energy saving goal of the park through green electricity trading.

Mode H: Adopt the technical route of energy conservation and low-carbon construction of architectural noumenon+proactive energy conservation and operational carbon reduction+internal development of renewable energy +external green energy input, so that the net carbon emission of the park meets certain indicator requirements, and the comprehensive carbon emission is less than or equal to zero. This model is suitable for new office industrial parks with good renewable resource conditions, and is the most comprehensive technical model for building a new zero-carbon comprehensive demonstration zone.

5 Conclusions

Through the investigation of low-carbon parks and building projects at home and abroad, this paper summarizes and refines four main technical routes for building low-carbon parks: energy conservation and low-carbon construction of architectural noumenon, proactive energy conservation and operational carbon reduction, internal development of renewable energy, and external green energy input. Horizontal comparison and qualitative analysis were carried out on the four technical routes adopted by each typical project, and eight modes of building low-carbon parks were summarized and proposed. By analyzing the characteristics of different technical measures, the scope of application and typical application scenarios of each mode were obtained, providing theoretical basis and implementation suggestions for low-carbon development of parks.

Acknowledgment

This work is supported by the National Key Research and Development Program of China (No.2022YFE0205300); the Innovation Project of China Southern Power Grid Co., Ltd (No. GZKJXM20210484).

References

1. China Association of Building Energy Efficiency, 2022 research report of China building energy consumption and carbon emissions. Chongqing, 2022.
2. Huawei Technologies Company Limited, 2020 Future smart park white paper. Beijing, 2020.
3. China Academy of Building Research. Technical standard for near zero energy buildings:GB/T 51350-2019. Beijing, China building industry press, 2019.
4. Jordan Jalving, Jaffer Ghouse, Nicole Cortes, Xian Gao, et al. Beyond price taker: Conceptual design and optimization of integrated energy systems using machine learning market surrogates. Applied Energy, 2023, 351: 121767.
5. Xue Yixun, Guo Qinglai, Sun Hongbin, Shen Xinwei, et al. Comprehensive energy utilization rate for park-level integrated energy system. Electric Power Automation Equipment, 2017, 37 (6): 117-123.
6. Jonas Timmerman, Lieven Vandevelde, Greet Van Eetvelde. Towards low carbon business park energy systems: Classification of techno-economic energy models. Energy, 2014, 75 : 68-80.
7. Yin Yongguang, Wang Jinxuan, Zhou Min, Yang Chunfang. Discussion on park energy planning based on "Dual Carbon" policy. Construction science and technology, 2022, (19): 62-64.
8. Lu Liu, Qin Zhijun. Low-carbon planning for park-level integrated energy system considering optimal construction time sequence and hydrogen energy facility. Energy Reports, 2023, 9(7): 554-566.
9. Zhou Youmei, Lei Hao. Optimization strategy of industrial park planning and design from the perspective of double carbon. Architecture & Culture, 2023, (04): 8-10.
10. Yang Zhe, Fang Jiong, Chen Tiantian, Fan Yubo, et al. The construction and application practice of low-carbon smart parks under the dual carbon goal. Innovative Practice in Management of Chinese Power Enterprises. 2023(3): 169-171.
11. Zang Sheng, Zhao Bo. Exploration and practice of smart low-carbon or zero carbon parks. Jiangsu Construction, 2022, (06): 1-3.

12. Chen Xiaolin, Zhang Tiqiang, Zhang Yuning. Research on the implementation path of zero-carbon business office parks under the background of double carbon. Sichuan Chemical Industry,2022,(05):56-58.
13. Liu Lirong, Li Xin, Li Xinyuan, Liu Yong. Planning control path and method on the low-carbon industrial parks based on concept of industrial cluster. Journal of Guangxi University(Natural Science Edition), 2014, 39(04): 933-940.
14. Yue Hengyu, Tang Xiao, Jia Hongkun. Jobs' posthumous work, gathering cutting-edge architectural techniques, Apple Park's inspiration for the development of the construction industry. (2022-06-03).[EB/OL].https://baijiahao.baidu.com /s?id=1734536791386680080&wfr=spider&for=pc.
15. Lv Yanjie, Zhang Shicong, Xu Wei, James Connelly. Best practices and incentive policies for zero energy buildings in the united states. Building Energy Efficiency,2020,48(03):22-30.
16. Official google.org Blog. Reducing our carbon footprint with the direct purchase of renewable energy.(2010-7-20).[EB/OL]http://blog.google.org /2010/07/reducing-our-carbon-footprint-with_35. html.
17. He Jijiang. Germany's energy transformation (4. Berlin EUREF zero carbon energy technology park). China-nengyuan.com, (2018-04-25). [EB/OL] https://www.sohu.com/a/229474663_100049846.
18. Cris McClurg, Iain Campbell. Rocky mountain institute's new innovation center: a net zero energy building. Eco-City and Green Building, 2016, (02): 78-85.
19. Zhang Shicong, Li Huai, Chen Qidong. Tracking research on operation energy consumption of CABR nearly-zero energy demonstration building. Construction science and technology, 2019, (18): 39-44.
20. Zhang Wen, Nie Jinzhe. Study on applicability of heat recovery ventilation system in typical buildings in different climate regions of China. Building Scenice,2022,(08):194-200.
21. Zhang Shicong, Wang Ke, Yang Xinyan, Xu Wei. Research on carbon emission control index of low-carbon, nearly zero-carbon and zero-carbon residential building. Building Science, 2023,39(02): 11-19.
22. Zhang Shicong, Wang Ke, Xu Wei. Research on carbon emission control index of low-carbon, nearly zero-carbon and zero-carbon public building. Building Science,2023,39(02):1-10.

How does carbon emissions trading scheme affect emission reduction decisions of coal-fired power plants? An evolutionary game theoretic perspective

Bo Xu[1], Liucheng Wu[1] and Jiexin Wang[2,*]

[1]School of Business Administration, Northeastern University, Shenyang 110167, China
[2]Institute of Economics, Tsinghua University, Beijing 100084, China

ABSTRACT. Carbon emissions trading scheme (CETS) is widely regarded as a cost-effective market-based regulation for carbon abatement. In the context of CETS, this study develops an evolutionary game model that incorporates two representative coal-fired power plants and a government. Our model captures the interplay of emission reduction strategies between coal-fired power plants and endogenously incorporates government regulatory decisions. We analyze the strategic decisions of coal-fired power plants by discussing the dynamics and equilibrium of the game. Our findings demonstrate that in the absence of government implementation of CETS, coal-fired power plants refrain from investing in carbon abatement. However, with the enforcement of CETS, along with sufficient penalties for excessive carbon emissions, coal-fired power plants become inclined to invest in emission reduction. Furthermore, the willingness of coal-fired power plants to invest in carbon abatement exhibits a negative relationship with both the quota and the cost of emission reduction.

1. Introduction

In recent years, the environmental challenges arising from excessive greenhouse gas emissions have garnered significant attention. Consequently, governments worldwide have placed great importance on the implementation of effective policies aimed at promoting renewable energy consumption and curbing carbon emissions[1-2]. Given that fossil fuel-based energy production contributes substantially to global carbon emissions[3], it becomes imperative to prioritize emissions reduction within this sector as a key strategy for climate change mitigation.

Carbon Emissions Trading Scheme (CETS) is perceived as a viable market-based solution for carbon abatement. Under the CETS framework, coal-fired power plants receive an allocation of carbon emission quotas. They are compelled to purchase carbon permits from other plants if their carbon emissions exceed their quotas. Conversely, these plants have the option to sell any surplus carbon permits at their disposal[4]. At the end of a given compliance period, power plants must hold enough carbon quotas or face severe penalties. Extant research suggests that the effectiveness of the CETS hinges strongly on the design of its core policy indicators, namely the price of carbon permits, the quota allocation, and the penalty scheme[5-7]. These policy indicators may impact the production decisions of coal-fired power plants by influencing their production costs, consequently leading to a reduction in carbon emissions. Therefore, it is crucial to examine how these policy indicators may influence the efficacy of the CETS.

This study aims to investigate how CETS affects the carbon reduction decisions of coal-fired power plants, considering both the strategic decisions of the government and the interplay between power plants. Our study is mainly related to three streams of literature on CETS: (1) the impact of CETS on the reduction of carbon emissions and its effects on social welfare; (2) the implications of CETS on manufacturers; and (3) the influence of CETS on the adoption of green technology and investments in carbon abatement.

The first stream of literature relevant to our work concerns the carbon abatement and social welfare effects of CETS. Governments have implemented a range of policies aimed at mitigating carbon emissions, each yielding diverse outcomes. Holland et al.[8] conducted a comparative analysis of renewable fuel standards, low carbon fuel standards, ethanol subsidies, and cap-and-trade systems (a form of CETS which establishes the aggregate carbon quota based on total carbon emission targets), specifically focusing on their efficacy in reducing greenhouse gas emissions from transportation fuels. They also evaluated the externalities associated with different policy approaches. Their findings suggest that cap-and-trade systems may entail lower societal costs and offer a more effective means of addressing environmental externalities. Du et al.[9] developed a Stackelberg game model for multi-stage carbon

* Corresponding author: wang-jx22@mails.tsinghua.edu.cn

abatement that incorporates firms' considerations regarding quota allocation and the additional cost of abatement. They subsequently identified the optimal grandfathering scheme from a governmental perspective, aiming to maximize social welfare. In the pursuit of Pareto optimal designs for CETS, Feijoo and Das[10] devised a two-layer mathematical-statistical model incorporating the DC optimal power flow (DC-OPF) model and policy optimization. Within the cap-and-trade program context, He et al.[11] explored the optimal production decisions of a self-pricing firm and the optimal cap-setting decisions of a regulator. Their study constructed a Stackelberg game model, wherein the firm seeks to maximize profits as a follower, while the regulator acts as the leader striving to enhance social welfare. Results indicate that as emission intensity increases, the ideal total emissions and the optimal emission cap initially rise but eventually decline. Employing a game-theoretic model, Li et al.[12] analyzed the response of a firm's operational decisions regarding sustainable energy consumption and low carbon production to changes in the cap-and-trade policy. Their findings highlight the importance of policies that elevate consumer preference for low-carbon items as a crucial step towards maximizing total social welfare, particularly if technological upgrades are necessary. In summary, this stream of literature indicates that CETS has the potential to entail lower societal costs, effectively mitigate environmental externalities, and enhance social welfare. Meanwhile, the quota system stands as a fundamental component of CETS, and its reasonable formulation and allocation serve as the bedrock for the effective implementation of CETS.

The second stream of relevant work studies the effects of CETS on manufacturers. Using Stackelberg game theory, Xu et al.[13] and Xu et al.[14] investigated the operational decisions of a multi-product manufacturer under cap-and-trade and carbon tax policies. They compared the effects of these policies on total carbon emissions, the manufacturer's profit, and social welfare. Considering cap-and-trade and consumers' low-carbon preference, Ding et al.[15] developed a Stackelberg game to analyze the equilibrium decisions under different conditions. They found that the carbon price and quota have an impact on the manufacturer and retailer's profit when the manufacturer adopts encroachment decisions. According to Curtis and Lee [16], the NOx Budget Trading Program (NBP), which was put into place in 19 states across the country in 2004, received a heterogeneous response from manufacturing firms of all ages and sizes. Huang et al. [17] utilized a duopoly model to explore the behavior of competing manufacturers under CETS. Their results demonstrate that CETS can promote cooperation among manufacturers and may not necessarily have a negative impact on their profits. Furthermore, the cost advantage and high product substitution can accelerate the low-carbon transition for low-cost manufacturers. Chai et al. [18] studied how to make a monopolistic manufacturer, involved in manufacturing and remanufacturing operations, profitable under a carbon cap-and-trade mechanism in a single period. Their findings indicated that cap-and-trade favors remanufacturing in both the ordinary and green markets, with the carbon trading price playing a more important role in carbon emissions and manufacturer profitability than the carbon cap. Based on a closed-loop supply chain model, Hu et al.[19] analyzed the trade-off between the carbon tax and cap-and-trade system through numerical studies. Their findings reveal that the cap-and-trade system is better suited for the Chinese remanufacturing industry, as it outperforms the carbon tax system in terms of manufacturer profits, social welfare, and consumer surplus. This stream of literature highlights the impact of CETS on manufacturers' market power, profits, and their efforts for low-carbon transition. However, the existing literature lacks emphasis on the participants within the wholesale power market, specifically the coal-fired power plants.

The third stream of literature focuses on the impact of CETS on green technology adoption and carbon reduction investments. Zheng et al.[20] analyzed the decision-making of duopoly manufacturers under a cap-and-trade system. The result shows that the optimal production quantity of duopoly manufacturers and the initial carbon quota are both related to the level of green technology. Liu et al.[21] developed a dual-objective non-linear programming model that considers cap-and-trade and carbon abatement technology investment. This model shows that under a cap-and-trade system, investing in carbon abatement technology not only enables manufacturers to generate profits but also reduces greenhouse gas emissions. Given the significance of promoting the willingness of green technology adoption, Yang et al.[22] studied the difference between the effects of grandfathering and benchmarking on enterprises' green technology investment and product pricing. The findings indicate that grandfathering leads to a greater reduction in carbon emissions, while benchmarking is more effective in encouraging green technology investment. Drawing upon evolutionary game theory, Zhang et al.[23] investigated the influence of government policy on manufacturers' decision-making in the presence of cap-and-trade, demonstrating that imposing a high penalty can encourage manufacturers to adopt green technology. Pan et al.[24] proposed a two-party game model involving a greener and a dirtier manufacturer, highlighting the significant impact of carbon price and green technology costs on the implementation of green technology. Chen et al.[25] investigated firms' decisions and profits under peak-valley price policy and cap-and-trade within the electric power industry via a Stackelberg game. The findings indicate that the peak-valley policy helps enhance firms' willingness to invest in low-carbon technology. Overall, this body of literature suggests that corporate investment in emission reduction, such as the adoption of abatement technologies, may necessitate incentives provided by CETS. Furthermore, corporate willingness to reduce emissions is influenced by the design of policy indicators.

Despite the extensive literature on the impact of CETS, there is a lack of studies specifically investigating how CETS influence the emission reduction decisions of coal-fired power plants, which play a pivotal role in the

 https://doi.org/10.1051/e3sconf/202344103017

electricity market. Wholesale electricity markets typically exhibit an oligopolistic structure, characterized by a small number of dominant power plants that share similar production costs and engage in interdependent decision-making[26-27]. Therefore, it is imperative to consider the interaction among these plants when analyzing abatement decisions. Additionally, most current studies neglect to explore the implications of strategic government decisions on the effectiveness of CETS. In fact, the government's choice to regulate or not is also influenced by cost-benefit considerations. Hence, as a decision-making entity, the government should not be treated as exogenous to the model.

To address the aforementioned issues, we propose an analytical framework within the context of CETS, which comprises two representative coal-fired power plants and a government. Given that wholesale power markets are often oligopolistic, it is reasonable to consider the interactions between these two representative coal-fired power plants. The investment decision of one plant regarding emission reduction will impact the availability of carbon permits in the market, thereby influencing the cost-benefit tradeoff and emission reduction decision of the other plant. Moreover, we incorporate the government as a decision maker into the model, endogenizing its regulatory decision-making process. Then, based on the assumption of bounded rationality, we use evolutionary game theory to model the dynamics of participants in our proposed framework. We obtain evolutionary stable strategies (ESSs) under different scenarios and draw some important conclusions by analyzing these ESSs. Finally, numerical analysis is carried out to validate our findings.

The main contributions of this study can be summarized as follows: (1) We develop an evolutionary game theoretic framework to analyze the influence of CETS on the carbon abatement decisions of coal-fired power plants. Our model incorporates both interacting representative coal-fired power plants and a government, thereby providing an avenue to investigate how government regulatory decisions and the carbon abatement choices made by other power plants influence the investment decisions of a focal power plant in carbon abatement; (2) We analyze the impact of CETS on coal-fired power plants by discussing ESSs under different scenarios. Our findings reveal that without the implementation of CETS by the government, coal-fired power plants will not spontaneously invest in emissions reductions. It is only when the government enforces CETS and imposes sufficient penalties for excessive carbon emission that coal-fired power plants will choose to invest in emission reduction. We also find that the effectiveness of CETS (i.e., its ability to stimulate power plants to invest in carbon abatement) is negatively related to both the quota and carbon abatement cost. These findings provide a solid theoretical basis for policymakers to design reasonable policy indicators for CETS.

The rest of this paper proceeds as follows. In the next section, we introduce the structures and assumptions of our evolutionary game model. In Section 3, we analyze the stability of equilibrium points and some corollaries are proposed. In Section 4, numerical simulations are presented. Section 5 concludes.

2. Model structures and assumptions

2.1. Problem description and symbol definition

In this research, we assume that the governments and coal-fired power plants are players with bounded rationality. After recognizing each other's strategies, they constantly learn and adjust their behavior according to the payoffs of their participants.

There are three ways for coal-fired power plants to fulfill their carbon abatement obligations: (1) invest in carbon abatement equipment; (2) purchase permits from other power plants; (3) pay for the penalty. If a coal-fired power plant chooses to invest in carbon abatement equipment, it will have a surplus of carbon emissions permits, which also brings tremendous carbon abatement costs. Otherwise, they need to buy the permits or pay for the penalty for its excessive carbon emission.

The coal-fired power plant T_1 and T_2 can choose not to invest in carbon abatement (N) or to invest in carbon abatement (R), and the corresponding probabilities are x (or y) and $1-x$ (or $1-y$), respectively. The government can choose not to supervise (implement CETS) or to supervise carbon emissions of coal-fired power plants, with probabilities of z and $1-z$ respectively. Their interactions are shown in Fig. 1.

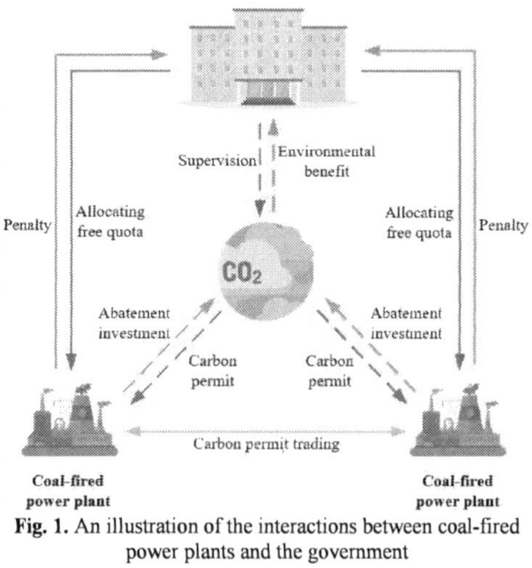

Fig. 1. An illustration of the interactions between coal-fired power plants and the government

The definitions and descriptions of symbols in this paper are shown in Table 1.

Table 1 Explanation of symbols

Symbols	Definition and description
P_e	The price (yuan / kWh) per unit of electricity
P_f	The penalty per unit of carbon emissions

	(kg) that exceed the quotas
P_{CET}	The price of carbon permits needed to offset a unit of carbon emissions (kg)
Q_h	The electricity (kWh) produced by coal-fired power plant in a cycle
Q_r	Carbon emission reduction of coal-fired power plants after carbon abatement investment
C_r	The costs that plants need to pay for carbon abatement
C_e	The reputation loss caused by government inaction
C_p	The costs of government supervision
β	The quotas allocated by the government to coal-fired power plants
ε	Carbon emissions (kg) per unit of electricity (kw) produced
γ	Net loss of social welfare resulting from unit carbon emissions (kg)

2.2. Model assumptions

Assumption 1: The government and coal-fired power plants are boundedly rational participants.

Assumption 2: T_1 and T_2 are two representative coal-fired power plants in power market. Both T_1 and T_2 have the same production capacity.

Assumption 3: The initial carbon emission of coal-fired power plants is greater than the quotas allocated by the government.

Assumption 4: P_{CET} is lower than P_f.

Assumption 5: Quotas are allocated to coal-fired power plants by the government for free.

2.3. Payoff matrix

(1) If the government does not supervise carbon emissions and both T_1 and T_2 do not invest in carbon abatement, then coal-fired power plants don't have to pay for carbon abatement, their payoffs can be expressed as:

$$R_{n1} = P_e \times Q_h \tag{1}$$

The government has to pay for environmental costs and reputation loss, its payoff can be expressed as:

$$R_{g1} = -2\times\gamma\times\varepsilon\times Q_h - C_e \tag{2}$$

(2) If the government choose to supervise and T_1 (or T_2) choose to invest in carbon abatement, then T_1 (or T_2) has to pay for carbon abatement, its payoff can be expressed as:

$$R_{r1} = P_e \times Q_h - C_r \tag{3}$$

T_2 (or T_1) 's payoff can be expressed as Eq. (1); the government's payoff can be expressed as:

$$R_{g2} = -2\times\gamma\times(\varepsilon\times Q_h - Q_r) - C_e \tag{4}$$

(3) If the government chooses not to supervise and both T_1 and T_2 choose to invest in carbon abatement, then the payoffs of coal-fired power plants can be expressed as Eq. (3), the government's payoff can be expressed as:

$$R_{g3} = -2\times\gamma\times(\varepsilon\times Q_h - 2\times Q_r) - C_e \tag{5}$$

(4) If the government chooses to supervise and neither T_1 nor T_2 chooses to invests in carbon abatement, then coal-fired power plants have to pay for penalty, and their payoffs can be expressed as:

$$R_{n2} = P_e\times Q_h - P_f\times(\varepsilon\times Q_h - \beta) \tag{6}$$

The government has to pay for environmental costs and supervision costs, its payoff can be expressed as:

$$R_{g4} = -2\times\gamma\times\varepsilon\times Q_h - C_P \tag{7}$$

(5) If the government chooses to supervise and T_1 (or T_2) chooses to invest in carbon abatement, then T_1 (or T_2) can get additional income by selling permits, its payoff can be expressed as:

$$R_{r2} = P_e\times Q_h - C_r + P_{CET}\times(\varepsilon\times Q_h - \beta) \tag{8}$$

T_2 (or T_1) 's payoff can be expressed as:

$$R_{n3} = P_e\times Q_h - P_{CET}\times(\varepsilon\times Q_h - \beta) \tag{9}$$

The government's payoff can be expressed as:

$$R_{g5} = -2\times\gamma\times(\varepsilon\times Q_h - Q_r) - C_P \tag{10}$$

(6) If the government chooses to supervise and both T_1 and T_2 choose to invest in carbon abatement, then the payoffs of coal-fired power plants can be expressed as Eq. (3), the government's payoff can be expressed as:

$$R_{g6} = -2\times\gamma\times(\varepsilon\times Q_h - 2\times Q_r) - C_P \tag{11}$$

The revenue matrix of tripartite game model are shown in Table 2.

Table 2 Revenue Matrix of Tripartite Game Model

Government	Coal-fired power plants T_1	Coal-fired power plants T_2: N	R
Non-supervision	N	(R_{g1}, R_{n1}, R_{n1})	(R_{g2}, R_{n1}, R_{r1})
	R	(R_{g2}, R_{r1}, R_{n1})	(R_{g3}, R_{r1}, R_{r1})
Supervision	N	(R_{g4}, R_{n2}, R_{n2})	(R_{g5}, R_{n3}, R_{r2})
	R	(R_{g5}, R_{r2}, R_{n3})	(R_{g6}, R_{r1}, R_{r1})

3. Model analysis

3.1. Replicator dynamic equation

The expected revenue of T_1 who chooses not to invest in carbon abatement can be calculated as follows:

$$U_X = yzR_{n1} + y(1-z)R_{n2} + (1-y)zR_{n1} + (1-y)(1-z)R_{n3}$$

(12)

The expected revenue of T_1 who chooses to invest in carbon abatement can be calculated as follows:

$$U_{1-X} = yzR_{r1} + y(1-z)R_{r2} + (1-y)zR_{r1} + (1-y)(1-z)R_{r1} \tag{13}$$

The expected revenue of T_1 can be calculated as follows:

$$\overline{U}_X = xU_X + (1-x)U_{1-X} \tag{14}$$

The replicator dynamic equation of T_1 can be calculated as follows:

$$F(x) = \frac{dx}{dt} = x(U_X - \overline{U}_X) \tag{15}$$

The expected revenue of T_2 who chooses not to invest in carbon abatement can be calculated as follows:

$$U_Y = xzR_{n1} + x(1-z)R_{n2} + (1-x)zR_{n1} + (1-x)(1-z)R_{n3} \tag{16}$$

The expected revenue of T_2 who chooses to invest in carbon abatement can be calculated as follows:

$$U_{1-Y} = xzR_{r1} + x(1-z)R_{r2} + (1-x)zR_{r1} + (1-x)(1-z)R_{r1} \tag{17}$$

The expected revenue of T_2 can be calculated as follows:

$$\overline{U}_Y = yU_Y + (1-y)U_{1-Y} \tag{18}$$

The replicator dynamic equation of T_2 can be calculated as follows:

$$F(y) = \frac{dy}{dt} = y(U_Y - \overline{U}_Y) \tag{19}$$

The expected revenue of government who chooses not to supervise carbon emissions can be calculated as follows:

$$U_Z = xyR_{g1} + x(1-y)R_{g2} + (1-x)yR_{g2} + (1-x)(1-y)R_{g3} \tag{20}$$

The expected revenue of government who chooses to supervise carbon emissions can be calculated as follows:

$$U_{1-Z} = xyR_{g4} + x(1-y)R_{g5} + (1-x)yR_{g5} + (1-x)(1-y)R_{g6} \tag{21}$$

The expected revenue of government can be calculated as follows:

$$\overline{U}_Z = zU_Z + (1-z)U_{1-Z} \tag{22}$$

The replicator dynamic equation of government can be calculated as follows:

$$F(z) = \frac{dz}{dt} = z(U_Z - \overline{U}_Z) \tag{23}$$

The Jacobian matrix of replicator dynamic equations is as follows:

$$J = \begin{bmatrix} (1-2x)(U_X - U_{1-X}) & x(1-x)(z \times P_f \times Q_C - P_f \times Q_C) & x(1-x)(y \times P_f \times Q_C + P_{CET} \times Q_C) \\ y(1-y)(z \times P_f \times Q_C - P_f \times Q_C) & (1-2y)(U_Y - U_{1-Y}) & y(1-y)(x \times P_f \times Q_C + P_{CET} \times Q_C) \\ 0 & 0 & (1-2z) \times (C_P - C_e) \end{bmatrix} \tag{24}$$

where:

$$Q_C = \varepsilon \times Q_h - \beta \tag{25}$$

3.2. Evolutionary stable strategy (ESS)

Proposition 1: The equilibrium points (x, y, z) of replicator dynamic equations are: $(0, 0, 0)$, $(0, 0, 1)$, $(0, 1, 0)$, $(1, 0, 0)$, $(0, 1, 1)$, $(1, 0, 1)$, $(1, 1, 0)$, $(1, 1, 1)$ and $(x_0, y_0, 0)$, where

$$x_0 = y_0 = \frac{C_r - P_{CET} \times Q_C}{P_f \times Q_C} \tag{26}$$

The conditions for each ESS are shown in Table 3.

Table 3 Corresponding conditions of ESSs

ESS	Condition
$(0, 0, 0)$	$C_r - P_{CET} \times Q_C < 0$, $C_P - C_e < 0$
$(0, 1, 0)$	$C_r - P_{CET} \times Q_C > 0$,
$(1, 0, 0)$	$C_r - P_f \times Q_C - P_{CET} \times Q_C < 0$, $C_P - C_e < 0$
$(1, 1, 0)$	$C_r - P_f \times Q_C - P_{CET} \times Q_C > 0$, $C_P - C_e < 0$
$(1, 1, 1)$	$C_P - C_e > 0$

Proof: Let $F(x) = 0$, $F(y) = 0$, $F(z) = 0$, then the equilibrium points $(0, 0, 0)$, $(0, 0, 1)$, $(0, 1, 0)$, $(1, 0, 0)$, $(0, 1, 1)$, $(1, 0, 1)$, $(1, 1, 0)$, $(1, 1, 1)$ can be obtained. When $x_0, y_0 \in [0, 1]$, the equilibrium point $(x_0, y_0, 0)$ can be obtained. According to the Lyapunov stability principle, if the eigenvalues corresponding to the equilibrium point have real negative parts, the equilibrium point is an evolutionary equilibrium point. The eigenvalues corresponding to the equilibrium point are shown in Table 4.

It can be seen from Table 4 that there are eigenvalues with real part greater than 0 in the Jacobian matrix at equilibrium points $(0, 0, 1)$, $(0, 1, 1)$, $(1, 0, 1)$ and $(x_0, y_0, 0)$, therefore, they cannot be asymptotically stable points. According to the Lyapunov stability principle, the general conditions for $(0, 0, 0)$, $(0, 1, 0)$, $(1, 0, 0)$, $(1, 1, 0)$ and $(1, 1, 1)$ to be asymptotically stable points are obtained.

Table 4 Eigenvalues of Jacobi matrix at each equilibrium point

Equilibrium points	Eigenvalues and symbols
$(0, 0, 0)$	$C_r - P_{CET} \times Q_C, C_r - P_{CET} \times Q_C, C_P - C_e$
$(0, 0, 1)$	$C_r, C_r, C_e - C_P$
$(0, 1, 0)$	$C_r - P_{CET} \times Q_C - P_f \times Q_C, P_{CET} \times Q_C - C_r, C_P - C_e$
$(0, 1, 1)$	$C_r, -C_r, C_e - C_P$
$(1, 0, 0)$	$P_{CET} \times Q_C - C_r, C_r - P_{CET} \times Q_C - P_f \times Q_C, C_P - C_e$
$(1, 0, 1)$	$-C_r, C_r, C_e - C_P$
$(1, 1, 0)$	$P_{CET} \times Q_C + P_f \times Q_C - C_r, P_{CET} \times Q_C + P_f \times Q_C - C_r, C_P - C_e$
$(1, 1, 1)$	$-C_r, -C_r, C_e - C_P$
$(x_0, y_0, 0)$	$x_0 \times (1 - x_0) \times P_f \times Q_C, -x_0 \times (1 - x_0) \times P_f \times Q_C, C_P - C_e$

Corollary 1: If the government does not implement CETS, coal-fired power plants will not choose to invest in carbon abatement.

Proof: According to the stability point analysis, if the government doesn't implement regulation, only the equilibrium point $(1, 1, 1)$ is an evolutionary equilibrium point, which means that coal-fired power plants are reluctant to spend additional costs for carbon abatement.

Corollary 2: If the government implements regulation, coal-fired power plants will choose to invest in carbon abatement only if P_f is large enough.

Proof: According to inequalities $C_r - P_f \times Q_C - P_{CET} \times Q_C < 0$ and $C_r - P_f \times Q_C - P_{CET} \times Q_C > 0$, under the premise of $C_P - C_e < 0$, coal-fired power plants don't choose to invest in carbon abatement if (27) is satisfied:

$$P_f < \frac{C_r}{Q_C} - P_{CET} \tag{27}$$

Under the premise of inequality $C_P - C_e < 0$, coal-fired power plants invest in carbon abatement if (28) is satisfied:

$$P_f > \frac{C_r}{Q_C} - P_{CET} \tag{28}$$

Corollary 3: If the government implements regulation, the carbon abatement willingness of coal-fired power plants is negatively related to the quotas.

Proof: According to (25), $C_r - P_{CET} \times Q_C < 0$, $C_r - P_{CET} \times Q_C > 0$, $C_r - P_f \times Q_C - P_{CET} \times Q_C < 0$ and $C_r - P_f \times Q_C - P_{CET} \times Q_C > 0$, under the premise of $C_P - C_e < 0$, the point $(1, 1, 0)$ is an evolutionary equilibrium point if (29) is satisfied:

$$\varepsilon \times Q_h - \frac{C_r}{P_f + P_{CET}} < \beta \tag{29}$$

Under the premise of $C_P - C_e < 0$, the points $(0, 1, 0)$ and $(1, 0, 0)$ are evolutionary equilibrium points if (30) is satisfied:

$$\varepsilon \times Q_h - \frac{C_r}{P_{CET}} < \beta < \varepsilon \times Q_h - \frac{C_r}{P_f + P_{CET}} \tag{30}$$

Under the premise of $C_P - C_e < 0$, the point $(0, 0, 0)$ is an evolutionary equilibrium point if (31) is satisfied:

$$\beta < \varepsilon \times Q_h - \frac{C_r}{P_{CET}} \tag{31}$$

It can be seen from the above that with continuous reduction of quotas, coal-fired power plants gradually tend to invest in carbon abatement, that is, the fewer quotas are, the better the effectiveness of CETS is.

Corollary 4: If the government implements regulation, the carbon abatement willingness of coal-fired power plants is negatively related to the cost of carbon abatement.

Proof: According to (25), $C_r - P_{CET} \times Q_C < 0$, $C_r - P_{CET} \times Q_C > 0$, $C_r - P_f \times Q_C - P_{CET} \times Q_C < 0$ and $C_r - P_f \times Q_C - P_{CET} \times Q_C > 0$, under the premise of $C_P - C_e < 0$, the point $(1, 1, 0)$ is an evolutionary equilibrium point if (32) is satisfied:

$$P_f \times Q_C + P_{CET} \times Q_C < C_r \tag{32}$$

Under the premise of $C_P - C_e < 0$, the points $(0, 1, 0)$ and $(1, 0, 0)$ are evolutionary equilibrium points if (33) is satisfied:

$$P_{CET} \times Q_C < C_r < P_f \times Q_C + P_{CET} \times Q_C \tag{33}$$

Under the premise of $C_P - C_e < 0$, the point $(0, 0, 0)$ is an evolutionary equilibrium point if (34) is satisfied:

$$C_r < P_{CET} \times Q_C \tag{34}$$

It can be seen from the above that with continuous reduction of carbon abatement costs, coal-fired power plants gradually tend to invest in carbon abatement, that is, the lower the carbon abatement cost is, the better the effectiveness of CETS is.

https://doi.org/10.1051/e3sconf/202344103017

4. Numerical simulations

Following Nie et al.[28], we perform numerical simulations in different scenarios to validate our conclusions. For all scenarios, we set $P_e = 0.635$, $Q_h = 97500$, $\varepsilon = 0.835$, $P_{CET} = 0.8$, $\gamma = 0.8$, $Q_r = 20000$. Table 5 presents the corresponding values for other parameters. We assign values to the parameters based on the conditions of each scenario and then conduct numerical simulations, and the results are demonstrated in Table 5 and Figure 2-4. All the ESSs and corollaries were verified through numerical analysis, ensuring the validity of our main conclusions.

Table 5 Influence of changes in relevant indicators on ESS

Corresponding corollary	P_f	β	C_r	C_e	C_P	Corresponding conditions	ESS	Result
Corollary 1 (Scenario 1)	0.9	80000	2000	50000	60000	$C_P - C_e > 0$	$(1,1,1)$	Fig. 4
Corollary 2 (Scenario 2)	0.6	80000	2000	60000	50000	$C_P - C_e < 0$, $P_f < \frac{C_r}{Q_C} - P_{CET}$	$(1,1,0)$	Fig. 4
	0.9	80000	2000	60000	50000	$C_P - C_e < 0$, $P_f > \frac{C_r}{Q_C} - P_{CET}$	$(1,0,0)$ $(0,1,0)$	Fig. 3
Corollary 3 (Scenario 3)	0.9	100000	2000	60000	50000	$C_P - C_e < 0$, $\varepsilon \times Q_h - \frac{C_r}{P_f + P_{CET}} < \beta$	$(1,1,0)$	Fig. 4
	0.9	80000	2000	60000	50000	$C_P - C_e < 0$, $\varepsilon \times Q_h - \frac{C_r}{P_{CET}} < \beta < \varepsilon \times Q_h - \frac{C_r}{P_f + P_{CET}}$	$(1,0,0)$ $(0,1,0)$	Fig. 3
	0.9	60000	2000	60000	50000	$C_P - C_e < 0$, $\beta < \varepsilon \times Q_h - \frac{C_r}{P_{CET}}$	$(0,0,0)$	Fig. 2
Corollary 4 (Scenario 4)	0.9	80000	3000	60000	50000	$C_P - C_e < 0$, $P_f \times Q_C + P_{CET} \times Q_C < C_r$	$(1,1,0)$	Fig. 4
	0.9	80000	2000	60000	50000	$C_P - C_e < 0$, $P_{CET} \times Q_C < C_r < P_f \times Q_C + P_{CET} \times Q_C$	$(1,0,0)$ $(0,1,0)$	Fig. 3
	0.9	80000	1000	60000	50000	$C_P - C_e < 0$, $C_r < P_{CET} \times Q_C$	$(0,0,0)$	Fig. 2

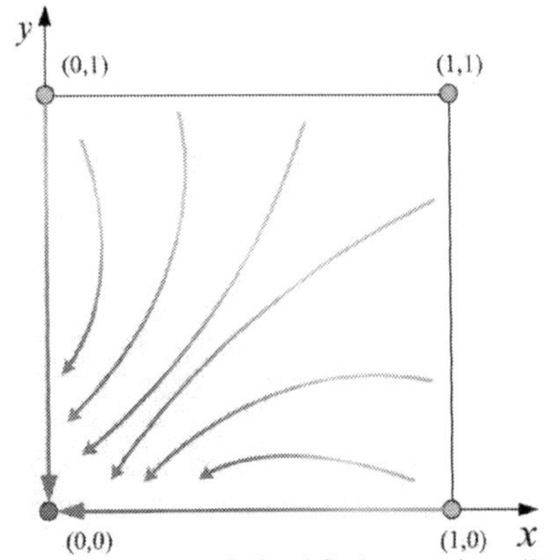

Fig. 2. The evolution trend of coal-fired power plants: (0, 0, 0) is an asymptotically stable point, that is, all coal-fired power plants invest in carbon abatement.

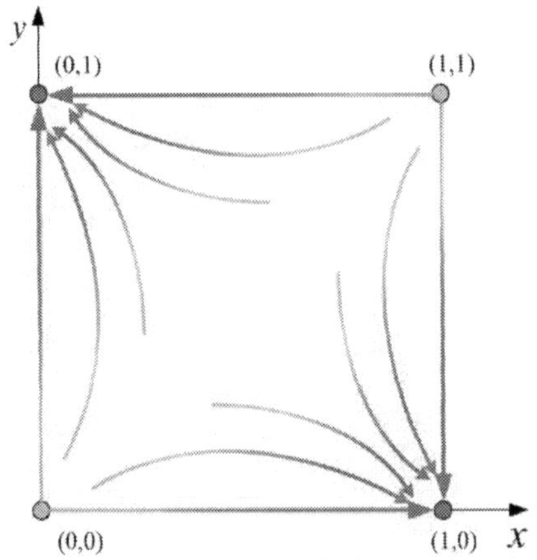

Fig. 3. The evolution trend of coal-fired power plants: (1, 0, 0) and (0, 1, 0) are asymptotically stable points, that is, one of the coal-fired power plants chooses to invest in carbon abatement.

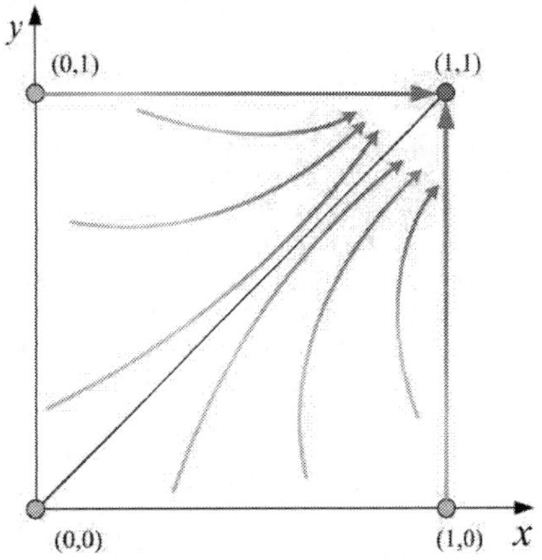

Fig. 4. The evolution trend of coal-fired power plants: (1, 1, 1) and (1, 1, 0) are asymptotically stable points, that is, coal-fired power plants don't invest in carbon abatement

5. Conclusion

In this study, we propose an analytical framework rooted in evolutionary game theory to study the impact of CETS on the strategic decisions of coal-fired power plants. Firstly, in the context of CETS, we construct a tripartite evolutionary game theoretic framework to model the dynamics of representative coal-fired power plants and the government under different scenarios. Secondly, by analyzing the ESS of each scenario, we analyze the impact of CETS on the strategic choices of coal-fired power plants. Finally, we corroborate our findings through numerical simulations.

The conclusions of this study are as follows: First, coal-fired power plants will not invest in carbon abatement if the government chooses not to implement CETS in all scenarios; Second, CETS is ineffective if the penalty for per unit carbon emission is not high enough. Coal-fired power plants will invest in carbon abatement only if the penalty exceeds a certain value; Third, the efficacy of CETS is negatively related to the carbon emission quota and the carbon abatement cost of power plants.

The findings of this paper may provide following managerial implications: Firstly, the government should implement CETS to promote the investment of carbon abatement of coal-fired power plants. Second, the government should establish a dynamical punishment mechanism to make sure that the penalty for excessive carbon emission is strictly higher than the cost of purchasing carbon emission permits; Third, the government should dynamically adjust the carbon emission quota in order to promote carbon abatement while protecting sufficient power supplies of the power market. Finally, the government should encourage innovation in carbon abatement technologies to effectively mitigate the carbon abatement cost for coal-fired power plants.

Author contributions

Bo Xu: Conceptualization, Methodology, Writing—review & editing, Supervision, Project administration. Liucheng Wu: Methodology, Investigation, Writing—original draft, Software, Visualization. Jiexin Wang: Conceptualization, Methodology, Investigation, Writing—original draft, Writing—review & editing.

Data availability

The data used in this study are available from the authors upon reasonable request.

Declarations

Ethical approval: Not applicable.

Consent to participate: All authors of this article consent to participate.

Consent for publication: All authors of the article consent to publish.

Competing interests: The authors declare that they have no known competing financial interests or personal relationships that could have appeared to influence the work reported in this paper.

Acknowledgments

This study was supported by the National Social Science Funds of China (22BGL307), National Training Program of Innovation and Entrepreneurship for Undergraduates (S202210145191) and the Fundamental Research Funds for the Central Universities (N2224001).

References

1. Wang, S., Wang, J.X., & Fan, F. (2021). The hidden mediating role of innovation efficiency in coordinating development of economy and ecological environment: Evidence from 283 Chinese cities. *Environmental Science and Pollution Research.* 28(34), 47668-47684.
2. Xu, B., Wang, J.X., Guo, M.Y., Lu, J.Y., Li, G.H., & Han, L. (2021). A hybrid demand response mechanism based on real-time incentive and real-time pricing. *Energy.* 231.
3. El Khatib, S., Galiana, F.D., 2019. Investigating emission regulation policy in the electricity sector:modeling an oligopolistic electricity market under hourly cap-and-trade. *Energy Economics.* 78: 428-443.
4. Chen, Z.R., Nie, P.Y., 2020. Implications of a cap-and-trade system for emission reductions under an asymmetric duopoly. *Business Strategy and the Environment.* 29(8): 3135-3145.
5. Fang, G. C., Tian, L.X., Liu, M.H., Fu, M., Sun, M., 2018. How to optimize the development of carbon trading in China-Enlightenment from evolution rules

of the EU carbon price. *Applied Energy*. 211: 1039-1049.

6. Lin, B.Q., Jia, Z.J., 2018. Impact of quota decline scheme of emission trading in China: A dynamic recursive CGE model. *Energy*. 149: 190-203.
7. Lin, B.Q., Jia, Z.J., 2019. Energy, economic and environmental impact of government fines in China's carbon trading scheme. *Science of The Total Environment*. 667: 658-670.
8. Holland, S.P., Hughes, J.E., Knittel, C.R., Parker, N.C., 2015. Unintended Consequences of Carbon Policies: Transportation Fuels, Land-Use, Emissions, and Innovation. *The Energy Journal*. 36(3): 35-74.
9. Du, S.F., Zhu, Y.J., Zhu, Y.G., Tang, W.Z., 2020. Allocation policy considering firm's time-varying emission reduction in a cap-and-trade system. *Annals of Operations Research*. 290: 543-565.
10. Feijoo, F., Das, T.K. 2014. Design of Pareto optimal CO2 cap-and-trade policies for deregulated electricity networks. *Applied Energy*. 119: 371-383.
11. He, P., Dou, G.W., Zhang, W., 2017. Optimal production planning and cap setting under cap-and-trade regulation. *Journal of the Operational Research Society*. 68(9): 1094-1105.
12. Li, G., Zheng, H., Ji, X., Li, H.F., 2018. Game theoretical analysis of firms' operational low-carbon strategy under various cap-and-trade mechanisms. *Journal of Cleaner Production*. 197: 124-133.
13. Xu, X.Y., Xu, X.P., He, P., 2016. Joint production and pricing decisions for multiple products with cap-and-trade and carbon tax regulations. *Journal of Cleaner Production*. 112: 4093-4106.
14. Xu, X.P., Yu, Y.G., Dou, G.W., Ruan, X.M., 2022. The choice of cap-and-trade and carbon tax regulations in a cap-dependent carbon trading price setting. *Kybernete*. 51(8): 2554-2577.
15. Ding, F., Lu, Z.P., Jin, M.F., Sun, L.C., 2022. Manufacturer's Encroachment and Carbon Emission Reduction Decisions Considering Cap-and-Trade Regulation and Consumers' Low-Carbon Preference. *International Journal of Environmental Research and Public Health*. 19(16).
16. Curtis, E.M., Lee, J.M., 2018. The reallocative and heterogeneous effects of cap-and-trade. *Economics Letters*. 172: 93-96.
17. Huang, C., Du, S.F., Wang, B., Tang, W.Z., 2022. Accelerate or hinder it? Manufacturer transformation under competition and carbon emission trading. *International Journal of Production Research*. 61(18): 6230-6250.
18. Chai, Q.F., Xiao, Z.D., Lai, K.H., Zhou, G.H., 2018. Can carbon cap and trade mechanism be beneficial for remanufacturing? *International Journal of Production Economics*. 203: 311-321.
19. Hu, X., Yang, Z.J., Sun, J, Zhang, Y.L., 2020. Carbon tax or cap-and-trade: Which is more viable for Chinese remanufacturing industry? *Journal of Cleaner Production*. 243.
20. Zheng, Y., Zhang, G.X., Zhang, W..W., 2018. A Duopoly Manufacturers' Game Model Considering Green Technology Investment under a Cap-and-Trade System. *Sustainability*. 10(3).
21. Liu, L., Zhang, Z.S., Wang, Z., 2021. Two-sided matching and game on investing in carbon emission reduction technology under a cap-and-trade system. *Journal of Cleaner Production*. 282.
22. Yang, W., Pan, Y.C., Ma, J.H. Yang, T.Y., Ke, X., 2020. Effects of allowance allocation rules on green technology investment and product pricing under the cap-and-trade mechanism. *Energy Policy*. 139.
23. Zhang, S.Y., Wang, C.X., Yu, C., 2019. The evolutionary game analysis and simulation with system dynamics of manufacturer's emissions abatement behavior under cap-and-trade regulation. *Applied Mathematics and Computation*. 355: 343-355.
24. Pan, Y.C., Hussain, J., Liang, X.Y., Ma, J.H., 2021. A duopoly game model for pricing and green technology selection under cap-and-trade scheme. *Computers & Industrial Engineering*. 153.
25. Chen, W., Ma, Y.K., Bai, C.G., 2021. The Impact of Carbon Emission Quota Allocation Regulations on the Investment of Low-Carbon Technology in Electric Power Industry Under Peak-Valley Price Policy. *IEEE Transactions on Engineering Management*.
26. Bushnell, J. (2003). A mixed complementarity model of hydrothermal electricity competition in the western United States. *Operations Research*. 51(1): 80-93.
27. Devine, M. T., & Siddiqui, S. (2023). Strategic investment decisions in an oligopoly with a competitive fringe: An equilibrium problem with equilibrium constraints approach. *European Journal of Operational Research*. 306(3): 1473-1494.
28. Nie, Q., Zhang, L., Tong, Z., & Hubacek, K. (2022). Strategies for applying carbon trading to the new energy vehicle market in China: An improved evolutionary game analysis for the bus industry. *Energy*. 259.

E3S Web of Conferences **441**, 03018 (2023) https://doi.org/10.1051/e3sconf/202344103018
CELCT 2023

Influence mechanism of renewable portfolio standard on energy producers

Wan-ting SHEN , Yi-hang JING and Bo Xu*

School of Business Administration, Northeastern University, Shenyang 110167, China

Abstract. Renewable energy Portfolio Standard (RPS) policy plays a significant role in reducing carbon dioxide emissions and promoting the development of renewable energy. Based on bounded rationality, this paper constructed an evolutionary game model to analyze energy power production decision-making, involving both fossil and renewable power producers as participants. Through the discussion of evolutionary game equilibrium, the study revealed the following findings: (1) marketised TGC price benefits the reduction of fossil energy power production; (2) The implementation of RPS policy does not necessarily lead to an increase in total renewable energy production; (3) Under RPS policy, the higher proportion of renewable energy quota is not always advantageous for reducing fossil energy power production.

1 Introduction [a]

Governments worldwide are increasingly focused on environmental protection and sustainable development, thus enforcing strict environmental regulations[1]. According to the International Energy Agency (IEA), 42% of global CO2 emissions come from the electricity sector, with fossil power generation being a major contributor. Thus, it is crucial for China's low-carbon energy transition to reduce fossil energy production in a reasonable and scientific way. RPS policy is a mandatory requirement for the proportion of renewable energy generation in the market, implemented by a country or region through legislation[2]. TGCs are tradable certificates issued by the government, representing a corresponding amount of renewable energy generation. Fossil energy power producers can fulfill their quota obligations under the RPS policy by purchasing TGCs.

Currently, both domestic and international scholars have mainly focused on researching the impact of RPS policy on renewable energy production and the impact of related renewable energy policies on production decisions of power production producers. Nishio, K., & Asano shows that the purchase of certificates from renewable generators will enable effective implementation of the RPS[3]. Wang et al. examined the implementation effects of different policies in the Chinese electricity market and found that RPS policy is more favorable for increasing renewable energy production[4]. ZhuFan and Lin (2020) made a system dynamics model for analyzing the strategic interactions of stakeholders in a three-party evolutionary game[5]. Chen et al. developed an evolutionary game model between manufacturing producers and the government under the RPS policy to analyze their choices regarding technological innovation[6].

This study firstly establishes an evolutionary game analysis framework that considers power producers as game entities under the assumption of bounded rationality. Subsequently, through analyzing the equilibrium outcomes of the evolutionary game, we discuss the influence of market-based TGC prices on the production decisions of different types of producers under RPS policy.

2 Model Description and Construction

Based on the assumption of bounded rationality, this paper constructs an evolutionary game model that includes N fossil energy power producers and N renewable energy power producers as participating agents in the context of market-based TGC prices. The description of parameters are shown in Table 1.

2.1. Symbol description

Table 1 Description of parameters

Items	Descriptiom
α	Quota reqquirement under RPS
$q_{T/R}$	Low production of energy producers (kWh)
$kq_{T/R}$	High production of energy producers (kWh)
P_{TGC}	TGC price (yuan/kWh)
P_e	On grid electricity price (yuan/kWh)
c	Unit cost of fossil energy power producers (yuan/kWh)
Δc	Unit difference cost between renewable and fossil energy power producers

* Corresponding author: xub@mail.neu.edu.cn

https://doi.org/10.1051/e3sconf/202344103018

Table 1 Description of parameters(continued)

Items	Descriptiom
N	Number of fossil or renewable energy power producers
m	Slope of TGC price function
n	Intercept of TGC price function
Q_{supply}	Total renewable energy production
Q_{demand}	Total fossil energy production

2.2 Assumptions of the model

In order to facilitate analysis and improve the rationality of the research without losing generality, this study make the following assumptions:

(1) Both fossil energy and renewable energy power producers have two pure strategies, namely high production strategy (H) and low production strategy (L). (2) If the low production strategy is chosen, the energy power producer generates electricity, while using the high production strategy, the producer generates electricity. We suppose, it is worth pointing out that the value of does not affect the conclusion of this paper. (3) The power producers are the market entities responsible for meeting the RPS obligations. (4) In the TGC trading market, there are homogeneous fossil energy power producers and homogeneous renewable energy power producers, who are bounded rational decision-making agents aiming to maximize their own interests. (5) The demand for TGCs from fossil energy power producers can be fully satisfied, and the TGCs provided by renewable energy power producers can all be sold out completely. (6) The TGC price is expressed as a linear function, which is negatively correlated with supply and positively correlated with demand. (7) The electricity generated by renewable energy power producers can be exchanged for an equivalent quantity of TGCs[7].

2.3 Model establishment

The trading mechanism between fossil energy and renewable energy power producers is illustrated in Figure 1. Power producers sell electricity in the power market to gain revenue. Simultaneously, in the TGC market, fossil energy and renewable energy power producers serve as the demand side and supply side of TGCs, respectively.

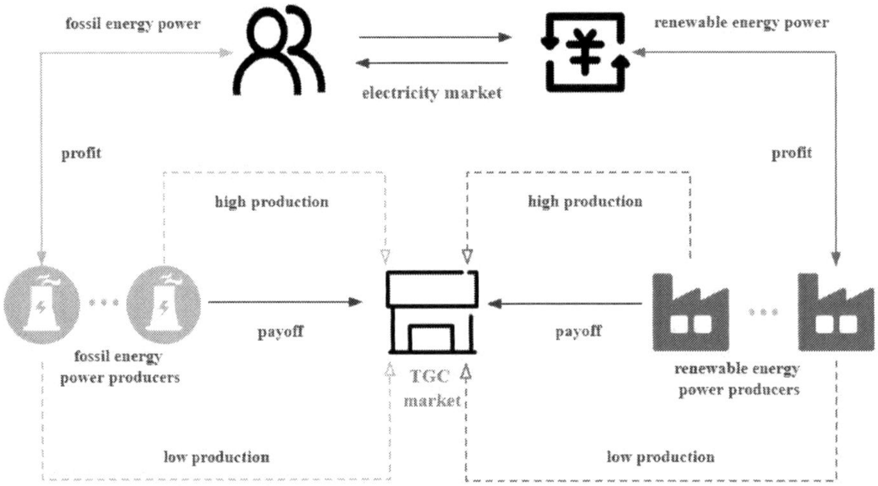

Fig. 1. Market transaction framework.

Under the RPS policy, the government sets a quota α, All power producers are required to fulfill the corresponding quota. Specifically, renewable energy power producers can only sell $(1-\alpha)$ portion of their electricity generation exchanged as TGC to generate revenue, while fossil energy power producers need to purchase α portion of TGC based on their own electricity production. Let $x(0 \le x \le 1)$ be the proportion of renewable energy power producers choosing the low production strategy, and then $(1-x)$ denotes the proportion of renewable energy power producers choosing the high production strategy. Similarly, let $y(0 \le y \le 1)$ be the proportion of fossil energy power producers choosing the low production strategy, and then $(1-y)$ denotes the proportion of fossil energy power producers choosing the high production strategy. Each player has two pure strategies to choose from: the low production strategy (L)($q_{T/R}$) and the high production strategy (H)($kq_{T/R}$, $k>1$). The strategy selection for power generation output will affect the price of TGC and the participants' revenue, each player can make choice which best favor its interest. The payoff matrix for the four different strategy combinations of power production decisions by the two players is shown in Figure 2.

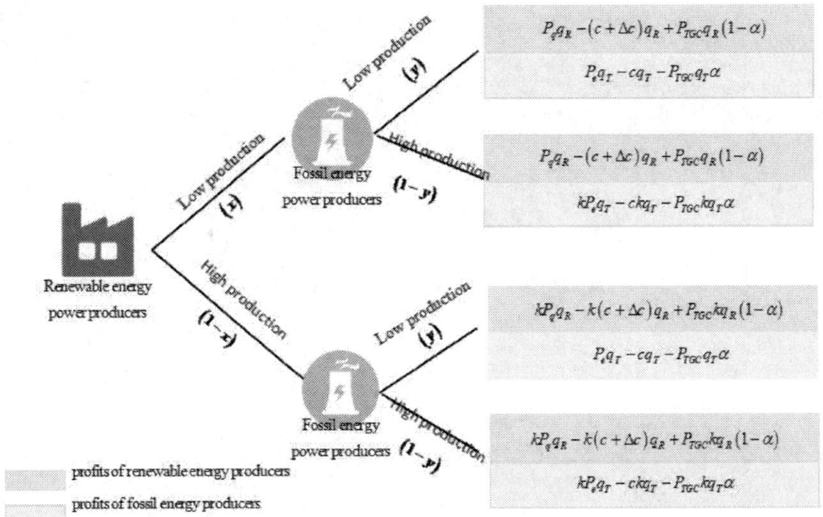

Fig. 2. Game tree and revenue of two kinds of power generation producers.

For analytical convenience, the price of TGC is set as a linear function, that is, the supply and demand of TGCs are vertically related to the energy output of all power generation enterprises [8], that is, $Q_{supply} - Q_{demand} = xNq_R + (1-x)Nkq_R - (1-y)Nkq_T - yNq_T$. Thus, the price of TGC can be expressed as :

$$P_{TGC} = m - n(xNq_R + (1-x)Nkq_R - (1-y)Nkq_T - yNq_T) \quad (1)$$

Where, m and n are the slope and intercept of the TGC price function, respectively.

3 Equilibrium Analysis

Based on the assumptions of bounded rationality and evolutionary stability strategy theory, and represent the rates of change over time in the proportions of power producers choosing the low production strategy for renewable and fossil energy respectively. The replicator dynamic equations for the behaviour of the replicators [10] are given by Equations (2) and (3).

$$\begin{aligned} F(x) &= \frac{dx}{dt} = x(Eg_L - Eg) \\ &= (1-x)xq_R(c+\Delta c - P_e + (\alpha-1)[m + (x-2)nNq_R + (2-y)nNq_T)] \end{aligned} \quad (2)$$

$$\begin{aligned} F(y) &= \frac{dy}{dt} = y(Eb_L - Eb) \\ &= (y-1)yq_T(P_e - c + \alpha[-m + (2-x)nNq_R + (y-2)nNq_T)] \end{aligned} \quad (3)$$

The equations above describe the adjustment rates of the behaviour strategies for renewable energy and fossil energy power producers. The game reaches a relatively stable state only when the dynamic equations of the replicators are equal to zero. When and, eight equilibrium points can be obtained: (0,1), (1,1), (0, y_5), (1, y_6), (x_7, 0), (x_8, 1), where y_5、 y_6、 x_7、 x_8 are expressed as Equations (4), (5), (6), and (7), respectively:

$$y_5 = \frac{c - P_e + (m - 2nNq_R + 2nNq_T)\alpha}{nN\alpha q_T} \quad (4)$$

$$y_6 = \frac{c - P_e + (m - nNq_R + 2nNq_T)\alpha}{nN\alpha q_T} \quad (5)$$

$$x_7 = \frac{P_e - c - \Delta c + (m - 2nNq_R + 2nNq_T)(1-\alpha)}{nN(-1+\alpha)q_R} \quad (6)$$

$$x_8 = \frac{P_e - c - \Delta c + (m - 2nNq_R + nNq_T)(1-\alpha)}{nN(-1+\alpha)q_R} \quad (7)$$

Whereas Equations (8) and (9) represent the average payoff functions for renewable energy power producers in the low and high production strategies, respectively. Among them, $\pi_1 = \pi_2 = P_e q_R - (c+\Delta c)q_R + P_{TGC} q_R (1-\alpha)$, $\pi_3 = \pi_4 = kP_e q_R - k(c+\Delta c)q_R + P_{TGC} k q_R (1-\alpha)$.

$$Eg_L = y\pi_1 + (1-y)\pi_2 = P_e q_R - (c+\Delta c)q_R + P_{TGC} q_R (1-\alpha) \quad (8)$$

$$Eg_H = y\pi_3 + (1-y)\pi_4 = kP_e q_R - k(c+\Delta c)q_R + P_{TGC} k q_R (1-\alpha) \quad (9)$$

Similarly, Equations (10) and (11) provide the average payoff functions for fossil energy power producers in the low and high production strategies, respectively. Among them, $\Pi_1 = \Pi_3 = P_e q_T - c q_T - P_{TGC} q_T \alpha$, $\Pi_2 = \Pi_4 = kP_e q_T - ckq_T - P_{TGC} k q_T \alpha$.

$$Eb_L = x\Pi_1 + (1-x)\Pi_3 = P_e q_T - c q_T - P_{TGC} q_T \alpha \quad (10)$$

$$Eb_H = x\Pi_2 + (1-x)\Pi_4 = kP_e q_T - ckq_T - P_{TGC} k q_T \alpha \quad (11)$$

The average payoffs of renewable and fossil energy producers can be defined in the following way:

$$\begin{aligned} Eg &= xEg_L + (1-x)Eg_H \\ &= (2-x)q_R(P_e - c - \Delta c + (1-\alpha)P_{TGC}) \end{aligned} \quad (12)$$

$$\begin{aligned} Eb &= yEb_{\square} + (1-y)Eb_{\square} \\ &= (2-y)q_T(P_e - c - \alpha P_{TGC}) \end{aligned} \quad (13)$$

The Jacobian matrix can be obtained by differentiating the replicator dynamic Equations (2) and (3).

$$\begin{pmatrix} J_1 & J_2 \\ J_3 & J_4 \end{pmatrix} \quad (14)$$

$$\begin{cases} J_1 = q_R\{(-1+2x)P_e - nN(2-6x+3x^2)(-1+\alpha)q_R - (-1+2x)[c - m + m\alpha + \Delta c - nN(-2+y)(-1+\alpha)q_T]\} \\ J_2 = (1-x)xq_R(nNq_T - nN\alpha q_T) \\ J_3 = -nN(-1+y)y\alpha q_R q_T \\ J_4 = q_T[c - 2cy + m\alpha - 2my\alpha + (-1+2y)P_e + nN(-2+x+4y-2xy)\alpha q_R + 2nN\alpha q_T - 6nNy\alpha q_T + 3nNy^2\alpha q_T] \end{cases} \quad (15)$$

By calculating the determinant and trace of the Jacobian matrix, and satisfying and , the equilibrium point is an evolutionarily stable strategy (ESS); when and , it is an unstable point; when , it is an saddle point[1]. By analysing and discussing the stability of the equilibrium points using the Jacobian matrix, the following theorem and corollary are obtained. The ESS under different conditions can be found in Table 2.

Table 2. ESS and corresponding conditions

Equilibrium point	det(J)	tr(J)	Local stability	conditions
(0,0)	+	-	ESS	$q_T > q_R + \frac{m}{2nN}$ and $0 < \alpha < \frac{P_e - c}{m + 2nN(q_T - q_R)}$
(0,1)	+	-	ESS	$q_T > 2q_R - \frac{m}{nN}$ and $\frac{c - P_e}{nN(2q_R - q_T) - m} < \alpha < 1$
(1,0)	+	-	ESS	$q_R > 2q_T + \frac{m}{nN}$ and $0 < \alpha < 1 - \frac{c + \Delta c - P_e}{nN(2q_T - q_R) + m}$ or $0 < \alpha < \frac{P_e - c}{P_{TGC}}$ and P_{TGC} is fixe-d
(1,1)	+	-	ESS	$1 > \alpha > \frac{P_e - c}{P_{TGC}}$ and P_{TGC} is fixed

Theorem 1: If the price of TGC is fixed and $0 < \alpha < \frac{P_e - c}{P_{TGC}}$, (0,0) is ESS; when the TGC price is fixed and $\frac{P_e - c}{P_{TGC}} < \alpha < 1$, (0,1) is ESS.

Proof: Under the conditions of $0 < \alpha < \frac{P_e - c}{P_{TGC}}$, when the TGC price is fixed, the determinant of the Jacobian matrix at point (0,0) is $k^2 q_R[(P_e - c)(q_T + q_R) + P_{TGC}\alpha(q_R - q_T) + (P_{TGC} - \Delta c)q_R] < 0$, and the trace is , $-k[(P_e - c)(q_T + q_R) + P_{TGC}\alpha(q_R - q_T) + (P_{TGC} - \Delta c)q_R] < 0$ According to the stability criterion for equilibrium points[9], (1,0) is ESS. Similarly, (0,1) is ESS when $0 < \alpha < \frac{P_e - c}{P_{TGC}}$.

Theorem 2: In the case of $q_T > 2q_R - \frac{m}{nN}$, if the quota is met $\frac{P_e - c}{m + nN(q_T - 2q_R)} < \alpha < 1$, (0,1) is ESS.

Corollary 1: Compared to a marketised TGC price, the fixed TGC price will suppress the total production of fossil energy.

Proof: In the case of a fixed TGC price, the profits of renewable and fossil energy power producers are not affected by each other's production levels. Moreover, when the TGC supply and demand can be met, the profits of energy power prodcuers are only affected by their own production decisions. In order to achieve higher profits, producers typically adopt a high production strategy. According to Theorem 1, in the fixed TGC price scenario, (0,0) or (0,1) is the ESS, and renewable energy power procducers will choose the high-production strategy. Fossil energy power producers tend to choose the high production strategy when the quota is low, but choose the low production strategy when the quota is high. Combined with Theorem 2, in a market-driven TGC pricing system, in order to achieve the current market conditions where fossil energy power production far exceeds renewable energy power production, it is preferable for fossil energy power producers to adopt the low production strategy. At this point, the quota requirement $\frac{P_e - c}{m + nN(q_T - 2q_R)}$ is relatively smaller and easier than $\frac{P_e - c}{P_{GCT}}$ to meet, indicating that the marketised TGC price enables the government to suppress the increase in fossil energy power production with a lower quota requirement.

Theorem 3: Under the terms of $q_T > q_R - \frac{m}{2nN}$ and $0 < \alpha < \frac{P_e - c}{m + 2nN(q_T - q_R)}$, (0,0) is ESS.

Corollary 2: When renewable energy production far exceeds fossil energy production $q_T > q_R - \frac{m}{2nN}$, implementing RPS policy increases renewable energy power production.

Proof: When $2q_R - \frac{m}{2nN} > q_T > q_R - \frac{m}{2nN}$, according to Theorem 3, a lower quota satisfies $0 < \alpha < \frac{P_e - c}{m + 2nN(q_T - q_R)}$ makes (0,0) ESS. While $q_T > 2q_R - \frac{m}{nN}$ and a higher quota satisfies $\frac{P_e - c}{m + nN(q_T - 2q_R)} < \alpha < 1$ lets (0,1) ESS, which means, N renewable energy power prodcuers will choose high production, as a result, the renewable energy production quantity will be $2Nq_R > 1.5Nq_R$. Therefore, when fossil energy power production far exceeds renewable energy power production, the RPS policy increases renewable energy production.

Theorem 4: According to the terms of $q_T < q_R - \frac{m}{2nN}$ and $\frac{P_e - c - \Delta c}{m + 2nN(q_T - q_R)} + 1 < \alpha < 1$, (0,0) is ESS.

Theorem 5: Under the terms of $q_T < \frac{1}{2}q_R - \frac{m}{2nN}$ and $0 < \alpha < 1 + \frac{c + \Delta c - P_e}{nN(q_T - 2q_R) - m}$, (1,0) is ESS.

Corollary 3: The ratio of requirement quota does not necessarily have to be higher to have a reduction in fossil energy production.

Proof: From theorem 5, when $q_T < \frac{1}{2}q_R - \frac{m}{2nN}$ and $0 < \alpha < 1 + \frac{c + \Delta c - P_e}{nN(q_T - 2q_R) - m}$, (1,0) is ESS, this implies that N fossil energy power producers will tend to choose the high production strategy. And c ombine Theorem 2, when

$q_T > 2q_R - \frac{m}{nN}$ and $\frac{P_e - c}{m + nN(q_T - 2q_R)} < \alpha < 1$, fossil energy power producers tend to choose the low production strategy. Therefore, a lower quota is beneficial to increase fossil energy power production.

Corollary 4: The implementation of RPS policy may not necessarily lead to an overall increase in renewable energy production.

Proof: According to theorem 5, when $q_T < \frac{1}{2}q_R - \frac{m}{2nN}$ and $0 < \alpha < 1 + \frac{c + \Delta c - P_e}{nN(q_T - 2q_R) - m}$, (1,0) is ESS. This indicates that N renewable energy power producers will eventually choose the low production strategy. As a result, the amount of renewable produced Nq_R will be lower than the initial renewable energy production quantity $1.5Nq_R$. Therefore, the implementation of the RPS policy with lower quota requirements may even decrease renewable energy production.

4 Conclusions

Based on the results of this research, the main conclusions of the above analysis are as follows: Firstly, the results suggest that the marketised price of TGC negatively impacts the overall level of fossil energy production and facilitates a decrease in carbon emissions. Secondly, the implementation of RPS policy may not necessarily lead to an increase in the overall amount of renewable energy production. If the amount of renewable energy power production vastly exceeds the amount of fossil energy power production, then under low quota requirements with a lower quota, the RPS policy will result in a reduction in renewable energy production but an increase in fossil energy power production. Thirdly, under RPS policy, increased quotas are not inherently harmful to the development of fossil energy production, and the policy criteria should be adaptably regulated in accordance with the levels of energy generation.Based on the research findings, we proposed the following policy implications: First, the government ought to embrace a marketised TGC pricing strategy when implementing the RPS policy. Second, when implementing the RPS policy, the government should continuously modify the policy specifics, such as quotas, based on the real power production scenario.

However, there are some limitations to the above study that need to be improved. The study only considers the impact of energy production on the profits of power producers in the electricity and TGC markets, without taking into account factors such as penalties and subsidies. There is ample room for expansion and improvement in future studies.

Acknowledgments

Supported by National Training Program of Innovation and Entrepreneurship for Undergraduates (202210145014).

References

1. Chu, L. K., & Le, N. T. Menviron Sci Pollut R **29**, 2866 (2021)
2. Xin-gang, Z., Ling, W., & Ying, Z. Energy Policy **143**, 111576 (2020)
3. Nishio, K., & Asano, H. Energy Policy **34**, 2387 (2006)
4. Wang, H., Su, B., Mu, H., Li, N., Gui, S., Duan, Y., Kong, X. Energy Policy **139**, 111319 (2020)
5. Zhu, C., Fan, R., & Lin, J. Energy Policy **136**, 111072 (2020)
6. Chen, H., Wang, J., & Miao, Y. AEJ, **60**, 2139–2147 (2021)
7. Song, X., Han, J., Shan, Y., Zhao, C., Liu, J., & Kou, Y. J Clean Prod **264**, 121518 (2020)
8. Xin-gang, Z., Ling-zhi, R., Yu-zhuo, Z., & Guan, W. Energy, **162**, 505–516 (2018).
9. Sandholm, William H.. Theor Eco, **5**, 27–50 (2010)
10. Z. Zhu, K. W. Chan, S. Bu, S. W. Or, X. Gao and S. Xia, T-PWRS **36**, 5975-5978 (2021)

Research on Load Modeling Method for Typical Low Carbon Energy Consumption Scenarios in Border and Cross border Regions Considering Seasonal Migration Characteristics

Shumin Chen[1], Shukui Liang[2*], Hao Zhang[1], Guangzeng You[1], Biao Qiao[2], Yipeng Qin[2], Lu Wang[2]

[1]China Southern Power Grid, Kunming, 650220, China
[2]Institute of Building Environment and Energy Conservation, China Academy of Building Science, Beijing, 100013, China

Abstract. With the process of urbanization and the 'the Belt and Road' initiative, the cross-border energy demand in southwest China has grown rapidly, driving the development of the energy system. The accuracy of load forecasting directly affects the application of energy systems, so it is crucial to conduct research on load forecasting for energy terminals in border and cross-border areas. However, there is a seasonal shift in the diverse energy consumption loads in border and cross-border regions, and currently, research on load forecasting and simulation of typical low-carbon energy consumption scenarios under this feature is basically in a blank state. Based on existing problems, this article conducts research on load modeling methods under the significant 'seasonal migration' characteristics of border and cross-border loads, conducts research on characteristic industries in border and cross-border areas, establishes typical low-carbon energy consumption scenarios and simulation models in border and cross-border areas, and uses sensitivity analysis method of dynamic simulation to analyze the impact of different influencing factors on the load of various building types, The Monte Carlo simulation prediction method is used to obtain the sensitivity probability distribution of various influencing characteristic factors, and the typical energy consumption building load model is modified. Finally, by comparing the energy consumption simulation results with statistical results, the accuracy of simulation energy consumption prediction is verified to be higher than 90%.

1 Introduction

China is in the process of rapid industrialization and urbanization, and the contradiction between energy supply and demand is becoming increasingly prominent. At present, China is facing two major problems: energy shortage and environmental pollution, and energy conservation and clean utilization have become a major strategic demand for the country. In recent years, comprehensive and intelligent energy technologies have received increasing attention, and the research and application of related energy technologies have gradually shifted from centralized heating in northern regions and comprehensive energy system utilization in more developed regions to personalized energy utilization in small and medium-sized cities. Under the 'the Belt and Road' initiative, the demand for energy in the areas bordering Myanmar, Laos and Vietnam in southwest China is growing rapidly, and the demand for various energy technologies is also increasing.

The growth of energy demand in border and cross-border regions has directly driven the development and application of related energy system technologies. Due to the rapid development of urbanization in China, the total carbon emissions in the construction industry are high, accounting for a high proportion. In 2019, the energy consumption of buildings in the operational stage in China was 1.03 billion tce, accounting for 46.1% of the total energy consumption of buildings, with HVAC systems accounting for more than half of the energy consumption in the operational stage[1]. In order to optimize energy scheduling and develop relevant strategies to reduce energy consumption in heating and cooling systems, accurately predicting the cold, heat, electricity, and gas consumption of regional energy terminals is crucial for the application of energy system related technologies, optimization of energy planning and design schemes, and improvement of energy operation efficiency in the region. Therefore, based on the characteristics of border and cross-border regions, it is particularly important to conduct more in-depth research on the different types of hourly cooling, heating, electricity, and gas loads of energy terminals. This article aims to provide a method for modeling the energy load of cross-border regions by studying the hourly load characteristics of typical buildings' cold, hot, and electrical systems in typical low-carbon energy consumption scenarios. This can solve the key problems of unclear demand for characteristic energy

* Corresponding author: Shukui Liang: sk9337@qq.com

use, insufficient research on the characteristics of cold and hot electrical multiple load data, and the lack of modeling methods for typical energy load simulation in border and cross-border regions, providing basic theories and technologies for the healthy development of energy technology in border and cross-border regions.

2 Overseas and Domestic Research Status

The existing load forecasting modeling methods can be divided into three types: physical based white box models, data-driven black box models, and gray box models in between. The white box model is based on heat transfer mechanisms and typically predicts cold and hot loads through simulation. The establishment of a white box model can be based on some commercial software, such as EnergyPlus and TRNSYS. The grey box model simplifies building heat treatment, and due to the dynamic, random, and time-varying cooling and heating loads, the prediction accuracy of the grey box model may be negatively affected[2]. The black box model is widely used in many systems with available input and output measurement data.

Azar[3] et al. conducted a systematic analysis of the factors affecting building energy performance in Kuwait using mixed statistics and energy modeling methods. By collecting local building characteristics and energy information, they analyzed data and driving factors affecting energy use. He used EnergyPlus to establish a building performance simulation model for typical commercial buildings in Kuwait City, which set basic information such as building geometric parameters, design parameters, and operating schedules. Finally, the model was calibrated by verifying the range of different input parameters, in order to apply the model to actual commercial buildings. Ruiqing Du[4] et al. coupled the parameterized model of building effects with the building energy model to establish a weather research forecast model for studying the air conditioning loads of different building categories under local climate conditions in Hong Kong. The establishment of the model takes into account the parameters and characteristics of urban microclimate, natural environment, and different building types. This model is also widely used to simulate high/low rise residential and commercial buildings in cities such as Beijing, Tokyo, Daidai, and Borzano.

At present, research on building load modeling mostly focuses on a single building type, while research on load models for multiple building types mainly focuses on economically developed regions or large cities. Research on the seasonal migration analysis of characteristic multiple loads in border and cross-border regions and the establishment of dynamic simulation models for typical low-carbon energy consumption scenarios is basically in a blank stage. However, due to the special geographical location, there are some energy issues in border and cross-border areas: residents in the southwest border area have frequent trade exchanges, but due to factors such as labor methods and resource conditions, the activity area is unstable and there is a possibility of migration. Moreover, the border tourism industry has the characteristics of spatiotemporal evolution, with significant seasonal differences in passenger flow distribution. The seasonal migration characteristics of industrial and commercial activities directly exacerbate the instability of load in border and cross-border areas. The instability of energy supply restricts the research on flexible and efficient collaborative technologies for new energy systems in border and cross-border regions. Therefore, this study aims to establish a model of typical low-carbon energy consumption scenarios in border and cross-border regions, simulate the multiple loads of cold and hot electricity, clarify the impact of seasonal migration characteristics of border and cross-border regional loads on characteristic energy demand, obtain a method for simulating typical energy consumption loads, and provide basic theories and technologies for the healthy development of energy technology in border and cross-border regions.

3 Research on the Industrial Characteristics of Border and Cross border Regions

Based on the above issues, this study conducts research on the modeling method of multi industry and multi energy loads under the significant "seasonal migration" characteristics of border and cross-border loads. Conduct research on the characteristic industries (tourism, interconnected railway economy, agriculture, industry, etc.) in border and cross-border areas. Based on the research and data collection, the current situation of border areas is mainly focused on tourism and production. The development and evolution of regional tourism usually rely on certain basic support and condition guarantees. Taking Yunnan Province as an example, as the forefront of Southwest China's opening up, the border areas of Yunnan Province are located in the hub area of the four countries and two regions, making them important nodes in the internal and external transportation routes. Good transportation development conditions provide key support for border tourism activities. The number of tourists in border and cross-border areas has periodicity and seasonality. The peak season of tourism in Yunnan is mainly concentrated in May October each year, and the number of tourists is relatively small at other times. The trend of the annual cycle is basically consistent. At the same time, the development of the tourism industry also has transferability. Taking Pu'er, Yunnan as an example, from 2007 to 2017, the development focus of the tourism industry in the border areas of Yunnan was roughly distributed between 100.623 ° E to 100.773 ° E, and 23.611 ° N to 23.728 ° N. It has always been located within the territory of Pu'er City, showing a migration trend from northwest (2007-2008) to southeast (2008-2013) to northwest (2013-2017), overall showing a migration towards the northwest direction. The development of the tourism industry is closely related to the level of openness in border areas, investment in tourism factors, and economic development level. Influenced by various factors, the migration characteristics of the tourism industry are obvious in the development process. For the analysis of industrial and agricultural research results in border and cross-border areas, take the sucrose industry as an example. The sucrose industry is a combination of industry and agriculture, and is a traditional pillar industry in the border areas of Yunnan. As a production factor for sugarcane cultivation, land supply directly affects the planting area of sugarcane. According to the statistics of sugarcane cultivation in Yunnan border areas from 2011 to 2017, the supply and demand gap between land supply area and sugarcane demand area decreased from the 2011/2012 crushing season to the 2013/2014 crushing season, and has since increased year by year. The inability of land production factors to meet market demand has constrained the development of the sugar industry, resulting in sugar making enterprises not being able to achieve saturation production, high equipment idle rate, and thus affecting the energy demand of the industry, increasing the instability and seasonal changes in the load in border areas.

Based on the research situation, the cyclical, seasonal, and migratory characteristics of industrial load demand in border and cross-border regions were analyzed, and combined with data information such as building types, climate conditions, industrial characteristics, resource characteristics, and population information, a typical low-carbon energy consumption scenario with border and cross-border regional characteristics was constructed.

4 Building models for low-carbon energy consumption scenarios in typical border and cross-border regions

Establish building models covering different regions, industries, and building forms based on typical low-carbon energy consumption scenarios in cross-border regions. Establish a building load simulation model for multiple typical cities in Yunnan and Guangxi provinces in the southwestern cross-border region, which is basically the same as the local office, hotel, residential, and hospital buildings. The simulation model includes the basic form of the building, enclosure structure, energy supply equipment, and other related structures and parameters of the target building. Use local historical climate data to conduct dynamic cold and hot electrical load simulation of typical buildings in typical cities.

In order to improve building energy efficiency and sustainability, building energy consumption simulation tools are used to design, debug, analyze parameters, and optimize simulations of one or more building energy systems. This study mainly uses TRNSYS software to simulate and calculate building energy consumption.

4.1 Basic information of typical building terminal simulation models

There are many types and scopes of typical building models, which involve a lot of information. This article presents and explains the architectural forms, energy consumption characteristics, main setting information, and building model simulation of five types of public and residential buildings, including office buildings, shopping malls, hotels, hospitals, and schools. Basic information of typical energy consulting builds can be found in Table 1.

Table 1. Basic information of typical energy consuming buildings.

Typical energy-efficient office buildings	Floor space	36，640m^2
	Functional zoning	Office, interior area
	Energy consumption characteristics	Continuous use on weekdays from 8:00 to 18:00, with concentrated personnel and high peak cooling load
	HVAC system forms under different climate zone conditions	Severe cold areas: air conditioning systems - fan coil units with fresh air Heating System - Radiator Cold regions: air conditioning systems - fan coil units with fresh air Heating System - Radiator Hot summer and cold winter areas: air conditioning systems - fan coil units with fresh air Heating system - fan coil unit with fresh air Hot summer and warm winter areas: air conditioning systems - fan coil units with fresh air Heating system - None
	notes	Zoned dual duct fan coil system
Typical energy consuming hotel buildings	Floor space	38，200m^2
	Functional zoning	Guestrooms, restaurants, conferences, banquets
	Energy consumption characteristics	Operating 24 hours a day, open year-round, with high personnel fluctuations and varying peaks of internal and external disturbances
	HVAC system forms under different climate zone conditions	Severe cold areas: air conditioning systems - fan coil units with fresh air/full air Heating System - Radiator Cold regions: air conditioning systems - fan coil units with fresh air/full air Heating System - Radiator Hot summer and cold winter areas: air conditioning systems - fan coil units with fresh air/full air Heating system - fan coil unit with fresh air/full air Hot summer and warm winter areas: air conditioning system - fan coil unit with fresh air/full air Heating system - fan coil unit with fresh air/full air
	notes	A five-star hotel with 1-3 floors of podium for large-scale conference and catering functions
	Floor space	30，286m^2

Typical Energy Emporium Buildings	Functional zoning	Shopping malls, restaurants, cinemas
	Energy consumption characteristics	Long operation time, high personnel density, high heat dissipation, large fresh air volume, and high cooling and heating load
	HVAC system forms under different climate zone conditions	Severe cold and cold regions: air conditioning systems - all air Heating System - Radiator Hot summer and cold winter areas: air conditioning system - all air Heating System - All Air Hot summer and warm winter areas: air conditioning system - all air Heating system - None
	notes	Including first floor dining and first floor cinema
Typical energy-efficient hospital buildings	Floor space	52，162m^2
	Functional zoning	Outpatient, surgical, testing, ICU
	Energy consumption characteristics	Long usage time, high personnel density, and no holidays. High requirements for indoor air purification and high load of fresh air for air conditioning. Large medical equipment.
	HVAC system forms under different climate zone conditions	Severe cold and cold regions: air conditioning systems - fan coil units with fresh air/full air Heating System - Radiator Hot summer and cold winter areas: air conditioning systems - fan coil units with fresh air/full air Heating system - fan coil unit with fresh air/full air Hot summer and warm winter areas: air conditioning system - fan coil unit with fresh air/full air Heating system - fan coil unit with fresh air/full air
	notes	Comprehensive Medical Technology Building of Third Class Hospital
Typical Energy School Buildings	Floor space	52，162m^2
	Functional zoning	Classrooms, multimedia classrooms, teacher offices
	Energy consumption characteristics	Not used on weekends, during winter and summer vacations, with a long daily usage period
	HVAC system forms under different climate zone conditions	Severe cold and cold regions: Air conditioning system - None Heating System - Radiator Cold regions: air conditioning systems - unit air conditioning units Heating System - Radiator Hot summer and cold winter areas: air conditioning systems - unit air conditioning units Heating System - Unit Air Conditioner (Heat Pump) Hot summer and warm winter areas: air conditioning systems - unit air conditioning units Heating system - None
	notes	High school and secondary university teaching buildings
Typical energy consuming residential buildings	Floor space	12，145m^2
	Functional zoning	residence
	Energy consumption characteristics	The heating energy consumption in the northern region is relatively high, while the air conditioning energy consumption in the southern region is relatively high. Other energy consumption is less affected by the climate, mainly influenced by the living habits of users.
	HVAC system forms under different climate zone conditions	Severe cold areas: Air conditioning system - None Heating System - Radiator Cold regions: air conditioning systems - unit air conditioning units Heating System - Radiator Hot summer and cold winter areas: air conditioning systems - unit air conditioning units Heating System - Unit Air Conditioner Hot summer and warm winter areas: air conditioning systems - unit air conditioning units Heating system - None

4.2 Simulation modeling parameter settings

The establishment of a typical energy consuming building terminal simulation model involves various model parameters such as enclosure structure, internal settings, electromechanical systems, etc. The building shape, size, orientation, internal space division, and usage functions must comply with the actual building situation and be representative. On the basis of a large amount of data collection and organization work, this article compares the forms and characteristics of various types of typical energy consuming building terminals, references relevant literature and standard specifications, and determines the various parameters required for modeling.

1) Enclosure structure

The project has determined the typical urban building model enclosure structure information required for the project according to relevant domestic standards and regulations. The basic information of typical building models is shown in Table 2, and some reference standards are shown in Table 3.

https://doi.org/10.1051/e3sconf/202344103019

Table 2. Information contained in typical building models.

General Information	Building Information	Enclosure structure	Equipment
Position	Number of floors	Exterior wall	Lighting
Total building area	Aspect ratio	Roof	HVAC system form
Socket load	Window to wall ratio	Ground	Cooling equipment
Ventilation requirements	Window position	External window	Equipment efficiency
Personnel density	Sunshade	Inner wall	Control strategy
Indoor temperature and humidity conditions	Floor height	Permeation	
Run Policy	Orientation		

Table 3. Reference standards.

Parameter	Setting basis
Indoor design parameters	Code for design of heating, ventilation, and air conditioning in civil buildings [5]
Indoor lighting, personnel, equipment parameters	Energy efficiency design standard for public buildings [6] Specification for technical requirements for architectural design of subway stations in rail transit Unified standard for industrial building design Energy efficiency standards for buildings in various regions

Due to the vast territory of our country, the climate varies greatly among different regions. The northern region mainly considers the winter cold insulation of buildings, usually using an external wall insulation system[7]. By referring to the relevant design standards for energy efficiency in residential and public buildings in various regions and comparing them in simulation software, the enclosure structure practices of different types of buildings in different cities are obtained. The heat transfer coefficient of the exterior window of the building model is obtained by using the WINDOW6 software component DOE-2 format window parameter database and using Trnsys calculation software for energy consumption simulation calculation.

2) Indoor and outdoor calculation parameters of buildings.

The hourly meteorological data for each city in the simulation of building energy consumption in this study are EPW and TMY data, mainly sourced from data provided by the US Department of Energy[8]. The design temperature for building heating is usually set at 16 ℃~22 ℃, and the relative humidity is controlled within the range of 15%~55%. The design temperature range for air conditioning cooling is 22 ℃~28 ℃, and the relative humidity is between 40%~70%. The maximum allowable indoor wind speed is generally about 0.3m/s. Refer to the 'Code for Design of Heating, Ventilation, and Air Conditioning in Civil Buildings' (GB50736-2012). The energy consumption simulation model of the fresh air system is established in accordance with the provisions of the national standards 'Energy Efficiency Design Standards for Public Buildings' (GB50189-2015) and 'Design Code for Heating, Ventilation, and Air Conditioning in Civil Buildings' (GB50736-2012) for the minimum designed fresh air volume. The indoor personnel density and operating schedule of buildings vary greatly among different types of buildings. According to relevant standards, the per capita occupied area and usage time characteristics of different types and levels of buildings are determined.

3) Building energy system parameters

In this study, office buildings and hotel buildings used fan coil units with fresh air systems, shopping mall buildings used all air systems, hospital buildings used fan coil units with fresh air systems and all air systems, and schools and residential buildings used unit air conditioning.

The lighting system is mainly affected by outdoor illumination and user usage. This study sets up the lighting system in the model building according to the 'Building Lighting Design Standards'[9].

The power equipment in buildings mainly includes elevators, domestic water supply pumps, and other equipment. The configuration of the number of elevators is based on the 'Residential Design Code'[10], 'Office Building Design Code', and relevant literature on elevator electrical equipment, to determine the number of elevators and elevator power.

5 Research on Multidimensional Characteristics of Seasonal Migration Simulation Model Based on Uncertainty Analysis

This study first uses dynamic simulation calculations and a logistic regression machine learning method based on single factor sensitivity analysis to investigate the impact of single factor changes on the multivariate load changes of various types of energy terminal buildings, including the number of tourists received throughout the year, the total number of urban passenger transport, the annual output of the sugar industry, and the total output of grain. Secondly, the Monte Carlo simulation method is used to predict the characteristics of uncertain factors, and the sensitivity probability distribution of each influencing characteristic factor to the influence of multiple loads is obtained. Subsequently, in order to avoid the limitations of the single factor sensitivity analysis method and comprehensively consider the coupling effect of various influencing characteristic factors on the target relationship, the Monte Carlo simulation method is used to obtain the sensitivity weight ratio of the uncertain characteristics prediction results of each influencing characteristic factor to the building load. This completes the quantitative analysis of seasonal migration influencing characteristic factors and is used to correct the multiple loads of various energy consuming building terminals in typical cross-border low-carbon energy consumption scenarios. It has played a good guiding role in the formulation of regional system planning and achieved significant results[11].

5.1 Sensitivity analysis of seasonal migration influencing feature factors based on dynamic simulation and logistic regression learning algorithm

This article takes the hotel building as an example to study the multi-dimensional characteristics of seasonal migration simulation models. The seasonal migration characteristics of other types of energy use terminal buildings are obtained similarly. Conduct dynamic simulation calculations of building multiple loads when the number of tourists received throughout the year, the total number of urban passenger transport, the annual output of the sugar industry, and the total output of grain are subject to single factor changes. Introduce a machine learning algorithm called logistic regression to analyze the sensitivity of feature factor classification based on the single factor impact results.

This paragraph discusses the impact of the number of tourists received throughout the year on hotel buildings in typical cross-border low-carbon energy consumption scenarios. Assuming that other application conditions remain unchanged and the number of tourists received throughout the year varies within the range of 0% -200% of the initial value of 20 million, complete the simulation calculation of building cold, hot, and electrical multiple load under 100 sets of random independent variables. This article uses logistic regression method to conduct data mining on the relevant characteristics of feature influencing factors and complete feature classification analysis of single factor influence.

After analyzing the impact of the number of tourists received throughout the year on the load of hotel buildings in typical cross-border low-carbon energy consumption scenarios, the main impact distribution is shown in the following figure.

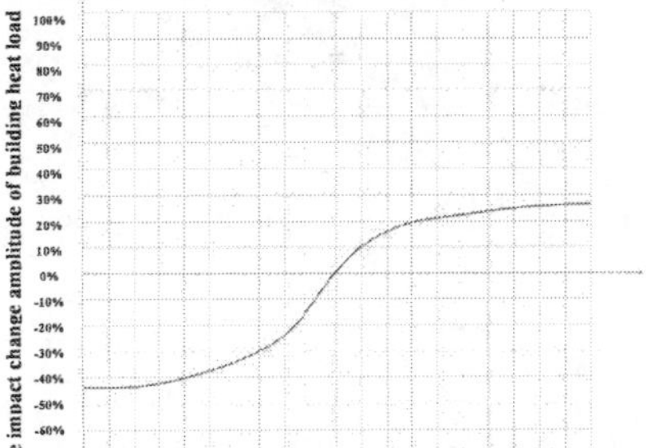

Fig. 1. Sensitivity Analysis of the Average Variation of Building Heat Load under Different Annual Visitor Reception Conditions.

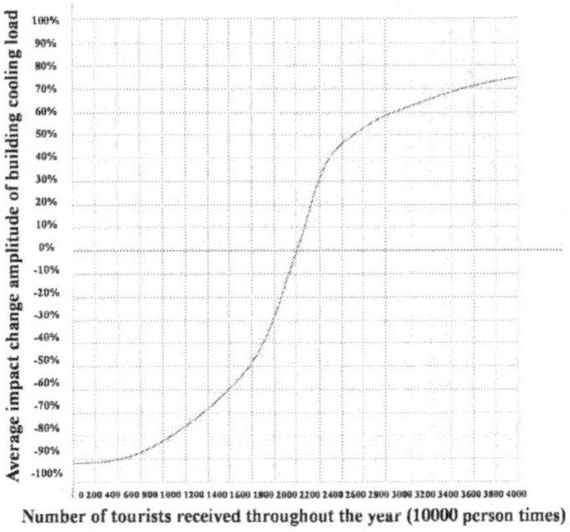

Fig. 2. Sensitivity Analysis of the Average Variation of Building Cooling Load under Different Annual Visitor Reception Conditions.

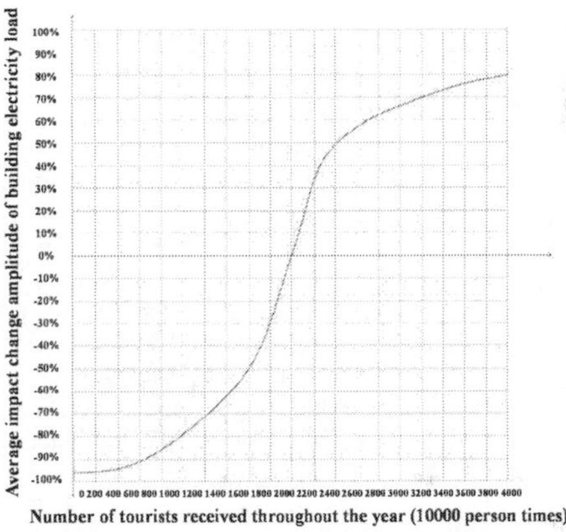

Fig. 3. Sensitivity Analysis of the Average Variation of Building Electricity Load under Different Annual Visitor Reception Conditions.

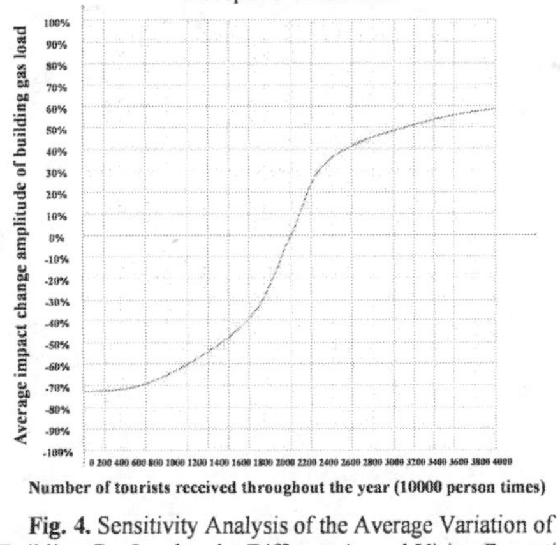

Fig. 4. Sensitivity Analysis of the Average Variation of Building Gas Load under Different Annual Visitor Reception Conditions.

The calculation results of the average range of changes in the cold, hot, and electrical loads of hotel buildings under different conditions of receiving tourists throughout the year are shown in Figures 1-4. The number of tourists has a significant impact on the cooling load, electrical load, and gas load of the hotel, while the impact on the hotel's heating load is not significant.

Simulate sugar production, total grain production, and urban passenger traffic using the same method to obtain sensitivity analysis of various influencing factors on the load changes of hotel buildings in typical cross-border low-carbon energy consumption scenarios. The impact of total passenger volume conditions in different cities on the cold and hot electrical load of hotels is similar to the impact on the number of tourists received throughout the year. The changes in annual production and total grain production of different sugar industries have no significant impact on the cooling load, electrical load, and gas load of hotels, and do not directly affect the heating load and gas load of hotels.

5.2 Comprehensive impact analysis of cross-border load characteristics based on Monte Carlo uncertainty prediction

This article takes hotel buildings in typical cross-border low-carbon energy consumption scenarios as an example, and uses Monte Carlo simulation as an uncertainty prediction method to obtain the sensitivity probability distribution of various influencing factors on the impact of multiple loads, focusing on the characteristics of the number of tourists received throughout the year, the total number of urban passenger transport, the annual output of the sugar industry, and the total output of grain. Using Monte Carlo simulation method to obtain the sensitivity weights of various influencing characteristic factors on building load, quantitative analysis of seasonal migration influencing characteristic factors is completed, and used to correct the multiple loads of various energy consuming building terminals in typical cross-border low-carbon energy consumption scenarios.

Monte Carlo sensitivity probability distribution prediction of the influence of characteristic factors on multivariate loads. Monte Carlo simulation is an uncertainty prediction method that simultaneously processes multiple variables. The simulation steps are shown in the flowchart. For this project problem, the algorithm process is divided into two layers: the inner layer solves the optimization capacity allocation problem of a multi energy system, obtains the comprehensive evaluation objective of the system's capacity allocation, and brings the results into the evaluation index model to calculate the values of each evaluation index of the system; Perform Monte Carlo simulation on the outer layer to obtain the probability distribution of the target variable values.

1) Sensitivity probability distribution of characteristic factors affecting thermal load

Based on the simulation results of the influence of various characteristic factors on building load in the previous text, using Monte Carlo method, the impact of the number of tourists received throughout the year on the magnitude of heat load variation is shown in Table 4.

Table 4. Characteristics of the impact of the number of tourists received throughout the year on the changes in building heat load.

	Minimum impact amplitude	Average impact amplitude	Maximum impact amplitude
Number of tourists received throughout the year	2.17%	22.33%	42.51%

Based on the sensitivity analysis of the impact of building heat load changes on the number of tourists received throughout the year, the Monte Carlo method was introduced to obtain the probability distribution of the impact sensitivity. The results showed a normal distribution, with an average sensitivity of 22.33%. Probability distribution of the impact of the number of tours received through the year on building heat load changes is shown in Fig. 5.

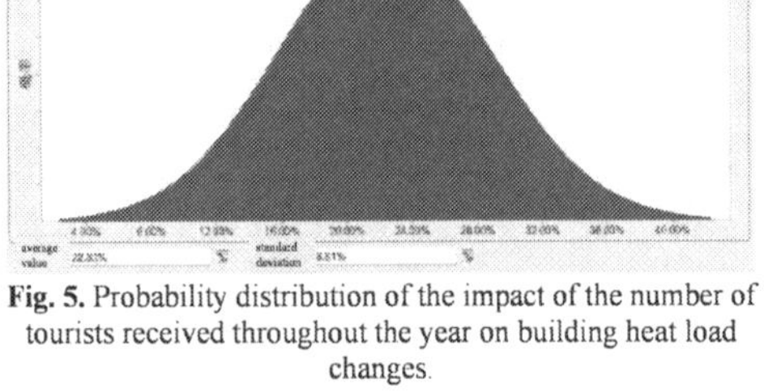

Fig. 5. Probability distribution of the impact of the number of tourists received throughout the year on building heat load changes.

Using the same method, the impact of various factors such as the annual production of the sugar industry, the total number of urban passenger trips, and the total grain production on the magnitude of heat load changes can be obtained. Similarly, the impact characteristics of these four types of seasonal characteristics on the amplitude of changes in cooling load, electrical load, and gas load can be obtained through Monte Carlo method, as shown in the table 5.

Table 5. Characteristics of the impact of changes in building heating, cooling, electricity, and gas loads on the results.

		Minimum impact amplitude	Average impact amplitude	Maximum impact amplitude
Sensitivity probability distribution of characteristic factors affecting heat load	Number of tourists received throughout the year	2.17%	22.33%	42.51%
	Annual production of sugar industry	2.11%	5.26%	13.24%
	Total number of urban passenger trips	2.48%	13.74%	29.82%

	total grain output	0.18%	3.69%	7.26%
Sensitivity probability distribution of characteristic factors affecting cooling load	Number of tourists received throughout the year	9.88%	41.25%	91.97%
	Annual production of sugar industry	0.28%	13.57%	27.59%
	Total number of urban passenger trips	3.45%	32.82%	63.77%
	total grain output	0.31%	8.07%	16.83%
Sensitivity probability distribution of characteristic factors affecting electrical load	Number of tourists received throughout the year	9.92%	43.97%	97.31%
	Annual production of sugar industry	1.13%	16.72%	35.08%
	Total number of urban passenger trips	0.08%	36.11%	72.99%
	total grain output	0.29%	9.43%	18.74%
Sensitivity probability distribution of characteristic factors affecting gas load	Number of tourists received throughout the year	4.39%	33.09%	71.63%
	Annual production of sugar industry	1.17%	2.61%	6.41%
	Total number of urban passenger trips	0.27%	21.35%	43.24%
	total grain output	0.16%	4.27%	8.75%

In this study, four types of seasonal migration characteristics influencing factors, including the number of tourists received throughout the year, the annual output of the sugar industry, the total number of urban passenger transport, and the total output of grain, were studied. Based on sensitivity analysis results, historical data inference method was used to determine the probability distribution of various uncertain parameters in the multi energy system, and an objective function relationship composed of the amplitude of changes in influencing characteristic factors and sensitivity constants was constructed. By utilizing the sensitivity of various influencing factors to load changes, the impact changes of these factors are corrected to reflect the seasonal migration characteristics of typical cross-border low-carbon energy consumption scenarios in the multi factor load simulation model.

Y (Correction relationship for seasonal migration characteristics of load)=Magnitude of load impact on the number of tourists received throughout the year × Sensitivity of the impact of the number of tourists received throughout the year+magnitude of the impact of the annual output change load on the sugar industry × Sensitivity of the annual production impact of the sugar industry+magnitude of the load impact of changes in the total number of urban passenger trips × Sensitivity of the total number of urban passenger trips and the magnitude of the impact of changes in total grain production load × Sensitivity to the impact of total grain production

This study conducted 2000 simple random sampling of random variables to generate random numbers for the objective function of seasonal heat load migration characteristics. From the sampling results of the normal distribution parameters in the figure below, it can be seen that the histogram of the random parameter sampling results is close to the probability density function curve of the parameters, so the sampling frequency is sufficient and the sampling results conform to the distribution characteristics of the parameters is shown in Fig. 6.

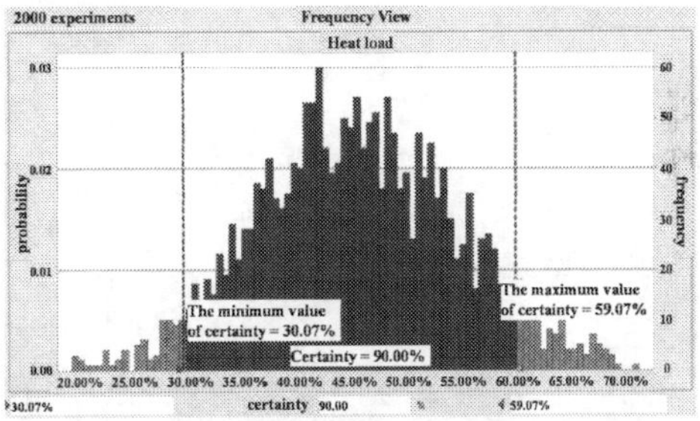

Fig. 6. The overall situation of the normal distribution of the correction relationship function for the seasonal migration characteristics of heat load.

Calculate the probability distribution of the target variable value. The relationship between the distribution of the objective function and the number of simulations is studied, and the results are shown in the figure. From the graph, it can be seen that when the number of simulations reaches 2000, the target variable values converge reliably, indicating good convergence of the algorithm.

Therefore, the sensitivity of the load impact on the number of tourists received throughout the year, the total number of urban passenger transport, the annual production of the sugar industry, and the total grain production is 56.50%, 33.60%, 7.60%, and 2.30%, respectively.

In summary, Y (correction relationship for seasonal migration characteristics of heat load)=magnitude of the impact of heat load on the number of tourists received throughout the year × 56.50%+Heat load impact amplitude of annual production changes in the sugar industry × 33.60%+The magnitude of heat load impact caused by changes in the total number of urban passenger trips × 7.60%+magnitude of heat load impact from changes in total grain production × 2.30%

Using the same method, the correction relationship for seasonal migration characteristics of cooling load can be obtained: Y (correction relationship for seasonal migration characteristics of cooling load)=the magnitude of the impact of cooling load on the number of tourists received throughout the year × 65.30%+The impact of cooling load on the annual production change of the sugar industry × 23.30%+The impact of changes in the total number of urban passenger trips on cooling load × 7.00%+magnitude of impact of changes in total grain production on cooling load × 4.40%

Y (Correction relationship for seasonal migration characteristics of electricity load)=Magnitude of impact of electricity load on the number of tourists received throughout the year × 62.50%+The impact of electricity load on the annual output change of the sugar industry × 28.70%+The magnitude of the impact of changes in the total number of urban passenger trips on electricity load × 6.60%+magnitude of electricity load impact from changes in total grain production × 2.20%

Y (Correction relationship for seasonal migration characteristics of gas load)=Magnitude of impact of gas load on the number of tourists received throughout the year × 72.70%+Annual production changes in the sugar industry, gas load impact amplitude × 25.00%+The impact of changes in the total number of urban passenger trips on gas load × 2.00%+Impact amplitude of gas load on total grain production changes × 0.30%

Based on the method proposed in this study, the sensitivity weights of multiple load seasonal migration characteristics for office, residential, school, hospital, shopping mall, transportation, and industrial building models were studied based on the correction relationship of seasonal migration characteristics for various types of energy consuming building terminal models. The sensitivity weights for the seasonal migration characteristics of multiple loads of various energy consuming building terminals under typical low-carbon energy consumption scenarios in cross-border regions are summarized as Table 6.

Table 6. Sensitivity Weights of Multivariate Load Seasonal Migration Characteristics for Various Typical Energy Use Building Terminal Simulation Models.

		Number of tourists received throughout the year	total number of urban passenger transport	total annual production of sugar industry	total grain production
Office building	Sensitivity weight of heat load impact	17.30%	15.20%	31.80%	35.70%
	Sensitivity weight of cooling load impact	26.40%	18.20%	37.50%	17.90%
	Sensitivity weight of electrical load impact	24.10%	17.50%	35.80%	22.60%
	Sensitivity weight of gas load impact	25.10%	26.30%	22.80%	25.80%
Residential buildings	Sensitivity weight of heat load impact	16.90%	37.20%	19.60%	26.30%
	Sensitivity weight of cooling load impact	17.30%	45.70%	17.50%	19.50%
	Sensitivity weight of electrical load impact	18.50%	44.90%	16.70%	19.90%
	Sensitivity weight of gas load impact	21.40%	41.20%	16.60%	20.80%
School buildings	Sensitivity weight of heat load impact	24.10%	28.60%	23.40%	23.90%
	Sensitivity weight of cooling load impact	34.30%	27.60%	18.30%	19.80%
	Sensitivity weight of electrical load impact	35.20%	26.20%	14.80%	23.80%
	Sensitivity weight of gas load impact	23.80%	25.70%	22.30%	28.20%
Hospital buildings	Sensitivity weight of heat load impact	37.30%	33.50%	21.50%	7.70%
	Sensitivity weight of cooling load impact	42.40%	35.70%	14.10%	7.80%
	Sensitivity weight of electrical load impact	41.80%	36.90%	12.10%	9.20%
	Sensitivity weight of gas load impact	21.70%	24.20%	25.60%	28.50%
Commercial building	Sensitivity weight of heat load impact	47.80%	31.40%	16.50%	4.30%
	Sensitivity weight of cooling load impact	54.10%	23.80%	9.70%	12.40%
	Sensitivity weight of electrical load impact	53.70%	21.50%	14.80%	10.00%
	Sensitivity weight of gas load impact	49.70%	31.50%	8.40%	10.40%
Transportation building	Sensitivity weight of heat load impact	41.60%	42.40%	3.70%	12.30%
	Sensitivity weight of cooling load impact	41.90%	55.70%	1.10%	1.30%
	Sensitivity weight of electrical load impact	42.70%	56.90%	0.08%	0.32%
	Sensitivity weight of gas load impact	31.90%	32.70%	19.40%	16.00%
Industry building	Sensitivity weight of heat load impact	5.10%	6.20%	54.80%	33.90%
	Sensitivity weight of cooling load impact	2.20%	3.70%	72.30%	21.80%

	Sensitivity weight of electrical load impact	1.80%	3.50%	71.50%	23.20%
	Sensitivity weight of gas load impact	3.60%	7.10%	69.80%	19.50%
Hotel building	Sensitivity weight of heat load impact	56.50%	33.60%	7.60%	2.30%
	Sensitivity weight of cooling load impact	65.30%	23.30%	7.00%	4.40%
	Sensitivity weight of electrical load impact	62.50%	28.70%	6.60%	2.20%
	Sensitivity weight of gas load impact	72.70%	25.00%	2.00%	0.30%

In summary, this section completed a single factor sensitivity analysis using dynamic simulation methods combined with logistic regression algorithms. The Monte Carlo simulation uncertainty prediction method was used to obtain the sensitivity weights of each influencing characteristic factor on the impact of multiple loads. The sensitivity weights of typical cross-border low-carbon energy consumption scenarios under the influence of seasonal migration characteristics were corrected for the energy terminal building load model.

5.3 Verification of accuracy of energy consumption calculation results

This article is based on the seasonal migration and multiple load characteristics of typical low-carbon energy consumption scenarios in the southwestern cross-border region. The simulation model of energy consumption building terminals is studied, and the energy consumption simulation results of typical energy consumption building terminal models are compared with statistical energy consumption data. Based on actual cases, a database is applied for load forecasting to verify the accuracy of the model.

Verify the overall accuracy of the building model based on the total energy consumption of the building, and use statistical data of actual operating energy consumption as a reference for comparison. Compare the actual energy consumption statistical information obtained through research with the average of simulated energy consumption calculation results for different types of buildings. The Statistical Table of Building Energy Consumption is shown in Table 7. The comparison results are shown in Fig 7.

Table 7. Statistical Table of Building Energy Consumption [kWh/(m^2. a)].

	Office	hotel	shopping mall	school	hospital	residence
Average energy consumption statistics	89.15	98.73	149.5	60.19	217.51	42.54
Average energy consumption statistics	80	134	200	41	230	43

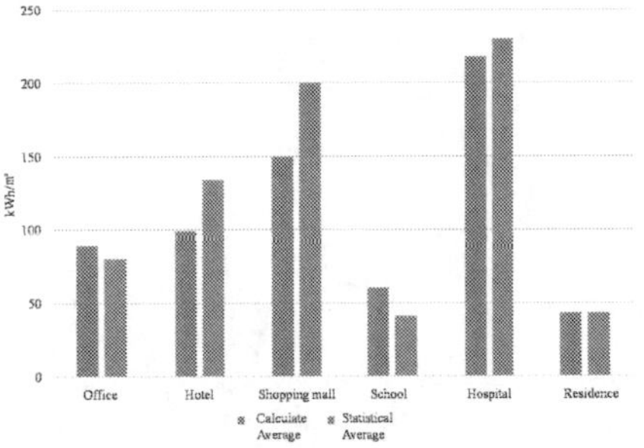

Fig. 7. Comparison between the calculated average energy consumption of buildings except for heating and the actual statistical average of buildings [kW. h/(m^2. a)].

The composition of building electricity consumption is relatively complex. Compare the average simulated energy consumption results of various types of buildings with the actual statistical average of building energy consumption. It can be seen that the average value of energy consumption simulation calculation results is basically at the same level as the actual average value of building energy consumption statistics, with an average difference of less than 10%. The accuracy of the calculation results is more reasonable than 90%. Therefore, it can be considered that the energy consumption simulation calculation model can better reflect the actual energy consumption characteristics of buildings.

The composition of total building energy consumption is relatively complex. The accuracy of the overall building energy consumption calculation results was verified earlier. The calculation results of each sub item building energy consumption were compared with the actual building energy consumption statistics by building type to obtain the accuracy verification of sub item energy consumption. Taking hotel construction as an example.

The proportion of sub item electricity consumption to total electricity consumption in hotel buildings is shown in the following Fig 8.

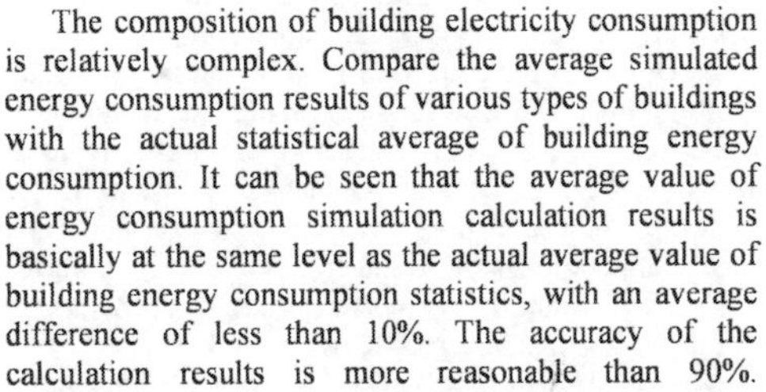

Fig. 8. Comparison of sub item energy consumption ratios in typical hotel buildings.

According to the energy consumption ratio chart, it can be seen that the HVAC energy consumption of hotel buildings accounts for about half of the overall building electricity consumption. The comparison between the sub item energy consumption ratio of hotel buildings and the statistical values of the sub item energy consumption ratio of hotel buildings is shown in the following figure. The simulation results are basically within the accuracy range of 91% compared to the actual values, and the simulation results are relatively reliable. The comparison results are shown in Fig 9.

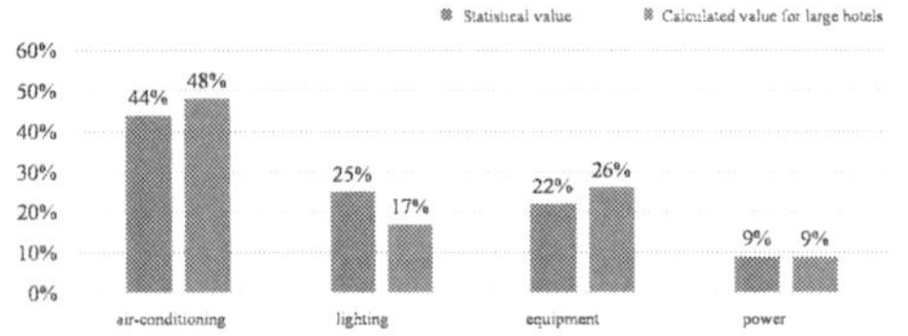

Fig. 9. Comparison between simulated and actual values of sub item energy consumption ratio in typical hotel buildings.

6 Conclusions

This article addresses issues such as unclear energy demand for border and cross-border regions, limited research on the characteristics of cold and hot electrical multiple load data, and missing simulation modeling of typical energy loads in border and cross-border regions. Based on the seasonal migration characteristics of border and cross-border regions, we conducted research on characteristic industries, constructed a key indicator system for typical low-carbon energy consumption scenarios in border and cross-border regions, and conducted research on modeling methods for multi industry and multi energy loads under significant seasonal migration characteristics of border and cross-border loads. At the same time, dynamic simulation, sensitivity analysis, and Monte Carlo methods are combined to analyze the influencing factors of seasonal migration characteristics in typical low-carbon energy consumption scenarios in border and cross-border areas using dynamic simulation and Rocky regression machine learning algorithms. Complete the weight determination of the influencing factors of seasonal and migratory load changes in border and cross-border regions, and use the dynamic characteristics of load migration to perform feature correction and fusion on typical dynamic load simulation data, ensuring the accuracy of typical dynamic load simulation data. Finally, based on the seasonal migration characteristics of border and cross-border energy consumption, the simulation data of typical low-carbon energy consumption scenarios using building terminals were compared with actual operating energy consumption data, and the accuracy of the modeling method was verified to be higher than 90%.

The main innovation points of this article are as follows:

1) This article innovatively introduces the Monte Carlo uncertainty prediction method into the study of seasonal migration load changes in border and cross-border regions, concretizing abstract problems and breaking through the limitations of existing research.

2) The method for modeling the load of typical energy consuming buildings in this study covers six representative types of buildings, including office buildings, hotel buildings, shopping malls, hospital buildings, residential buildings, and schools. It can be replicated and promoted in future research on cross-border regional energy systems.

3) The accuracy of the simulation data of typical low-carbon energy consumption scenarios using building terminals is higher than 90%, and this method is highly reliable.

Acknowledgments

This work is supported by the Research and application demonstration of key technologies for flexible and efficient collaboration of new multi-functional systems in border and cross-border regions (No. YNKJXM20210185).

References

1. R. Lv, et al (2022) Building thermal load prediction using deep learning method considering time-shifting correlation in feature variables. Journal of Building Engineering. 2352-7102.
2. J. Zhao and X. Liu (2018) A hybrid method of dynamic cooling and heating load forecasting for office buildings based on artificial intelligence and regression analysis. Energy & Buildings. 0378-7788.
3. E. Azar, et al (2021) Drivers of energy consumption in Kuwaiti buildings: insights from a hybrid statistical and building performance simulation approach. Energy Policy. 150-1121544.
4. R. Du et al (2023) Effect of local climate zone (LCZ) and building category (BC) classification on the simulation of urban climate and air-conditioning load in Hong Kong. Energy. 271-127004.
5. GB50736-2012, Code for Design of Heating, Ventilation, and Air Conditioning in Civil Buildings [S]. Beijing: China Building Industry Press, 2012.
6. GB50189-2015, Energy Efficiency Design Standards for Public Buildings [S]. Beijing: China Standards Publishing House, 2015.
7. Lu Yaoqing Practical Heating and Air Conditioning Design Manual [M] Beijing: China Construction Industry Press, 2008.
8. Building Energy Consumption Simulation Meteorological Data Website [EB/OL].
9. GB50034-2013, Building Lighting Design Standard [S] Beijing: China Construction Industry Press, 2013.
10. GB50096-2011, Code for Residential Design [S]. Beijing: China Construction Industry Press, 2011.
11. T.T. Chow and Apple L.S. Chan and C.L. Song. Building-mix optimization in district cooling system implementation[J]. Applied Energy, 2004.

https://doi.org/10.1051/e3sconf/202344103020

Status and Prospect of Islanding Detection in Active Distribution Network

Zhonghao SHUANG*, Tao HUANG, Xiujuan HUA, Guangfu XU, Wei HOU

NR Electric Co., Ltd., Nanjing 211102, China

Abstract: A large number of distributed power sources are injected to the distribution network, which can cause island operation in planned or unplanned situations. The islanded operation caused by the fault will cause the inverter to go off the grid in serious fault conditions such as short circuit between phases and disconnection. After the island is formed, if the common connection point before the island exchanges more power, a sudden change in the voltage within the island occurs. The frequency and phase angle changes of the voltage in the island are closely related to the characteristics of the inverter phase-locked loop, and the frequency mutation generally does not cause sudden changes in frequency due to a certain inertia of the phase-locked loop. Due to sudden changes in load power angle, input fault ride-through, etc., sudden changes in the phase angle of voltage in islands may occur. In order to solve the possible problems of power quality deterioration in islands, passive, active, and remote islanding detection schemes can be adopted, and the three schemes have their own advantages and disadvantages. In actual operation, it is necessary to consider the cooperation relationship between island detection, protection, safe automatic control and fault ride through.

1 Introduction

With the rapid development of clean energy, a large number of inverter-type distributed energy sources, represented by photovoltaics (PV), are penetrating into the distribution grid, making the grid more active. After a substantial amount of distributed PV is integrated, the operation of the system can easily go into an islanding mode when it loses external power, a phenomenon that is frequently observed in field operations.

The reasons for islanding vary, and the mechanisms for changes in frequency and phase angle within the island after it's formed are complex[1,2]. Moreover, there's a lack of simple, dead zone-free, low-cost island detection principles and configuration methods that have minimal impact on existing secondary equipment. According to the standard 《GB/T 37408-2019 Technical Requirements for Grid-Connected PV Inverters》, PV based on the grid connection voltage level, can be divided into: Class A inverters connected to the public grid at 35kV and 10kV, and Class B inverters connected on the user side at 380V and through the 10(6) kV voltage level[3,4]. Class A inverters may not have islanding protection capabilities, while Class B inverters should have the ability to quickly detect and immediately disconnect from the grid, with the islanding protection action time not exceeding 2 seconds. This highlights the importance of island detection for inverter grid connection. However, there's currently a lack in the industry of a comprehensive summary and analysis regarding the causes of islanding, its operating mechanisms, detection methods, and their impact on protection systems, which hinders the further development of clean energy.

This paper first analyzes the causes of islanding, discussing both planned and unplanned reasons for its occurrence and their effects on island detection. It examines the changing mechanisms of voltage amplitude, frequency, and phase angle within the island. Building on this, it introduces the mainstream island detection methods and their coordination with existing protection and automatic systems, finally summarizing and looking forward to current island detection technology and configurations.

2 Causes and Issues of Islanding

As shown in Figure 1, in a typical 10kV distribution system with PV, based on different grounding methods of 10kV, it can be categorized into Low Current Grounding Systems (LCGS) and High Current Grounding Systems (HCGS). Taking the most widely used passive islanding protection as an example: islanding detection can be configured as an independent protection device that collects voltage at the Point of Common Coupling (PCC) and trips during islanding. The conventional criteria are line/phase over/under voltage, over/under frequency. It can also be integrated into the PV inverter, collecting AC side voltage. In addition to the conventional criteria, it generally also judges negative sequence over-voltage,

* Corresponding author's email address: shuangzh@nrec.com

and when conditions are met, it sends a pulse before tripping the contactor, commonly referred to as "machine tripping." Due to the presence of negative sequence suppression control, the PV inverter mainly outputs positive sequence components.

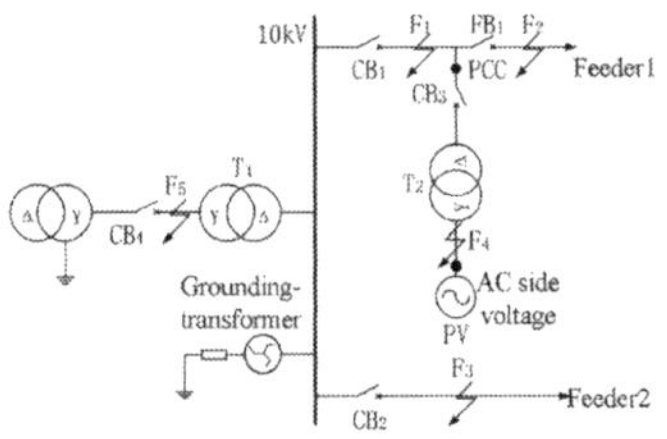

Fig. 1 Typical active distribution network structure

2.1 Causes of Islanding Formation

2.1.1 Planned Islanding

Planned islanding primarily occurs during maintenance when switch CB4 is isolated. Users expect the 10kV busbar to be de-energized. If, after disconnecting CB4, the PV on the feeder continues to back-feed power, then an energized 10kV busbar could endanger the safety of maintenance personnel. Users generally hope that after the switch CB4 is disconnected, all PV would automatically trip and not operate in island mode.

2.1.2 Unplanned Islanding

Unplanned islanding typically occurs due to faults leading to tripping, resulting in the loss of connection with external systems. Possible faults include single-phase grounding, phase-to-phase short circuit, three-phase short circuit, line breakage, etc. The fault may occur on the current line or on neighboring lines, among other locations.

2.2 Changes in Voltage Amplitude, Frequency, and Phase Angle within the Island

Whether it is planned or unplanned islanding, after the PCC is disconnected from the grid and the island is formed, the changes in voltage amplitude, frequency, and phase angle within the island are critical issues related to island detection and power quality. As shown in Figure 2 of a typical photovoltaic power generation system, after the formation of the island, the photovoltaic inverter, being the only power source within the island, dictates the changes in voltage amplitude, frequency, and phase angle.

During islanding, according to Figure 2, the voltage at the PCC $\dot{U}_{PCC}$ is equivalent to the voltage formed by the inverter's output current $\dot{I}_{inv}$ on the equivalent load impedance G(jω).

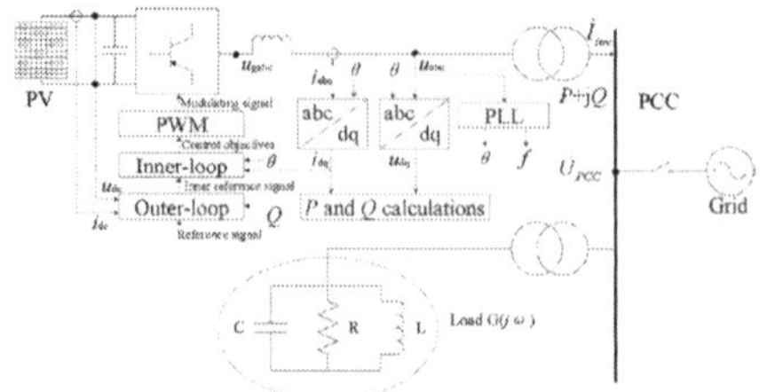

Fig.2 Typical PV power generation system

Ignoring the load branch circuit, transformer impedance, and only considering the fundamental frequency, we have the following formula:

$$\dot{U}_{PCC} = \dot{I}_{inv} \cdot G(j\omega) \tag{1}$$

$$\dot{I}_{inv} = \frac{u_{gabc}}{Z_g + G(j\omega)} \tag{2}$$

In the formula, $\boldsymbol{u_{gabc}}$ represents the fundamental frequency voltage output of the inverter, and $\boldsymbol{Z_g}$ represents the impedance from the inverter output to the PCC.

2.2.1 Voltage Amplitude Variation

Based on equation (1), The amplitude variation of $\boldsymbol{U_{pcc}}$ is closely related to the load characteristics. If there is a significant power exchange at the PCC before the islanding, a sudden change in $\dot{U}_{PCC}$ is very likely after islanding occurs. Even if the power exchange at the PCC is minimal or even zero before islanding, due to the fluctuations of photovoltaic and load, there might still be a possibility of sudden increase or decrease in voltage amplitude.

2.2.2 Frequency Variation

Based on formulas (1) and (2), the steady-state frequency of $\boldsymbol{U_{pcc}}$ is determined by the frequency of $\boldsymbol{u_{gabc}}$, and its phase angle is jointly determined by the phase angle of $\boldsymbol{u_{gabc}}$ and the load angle of G(jω). To determine the frequency and phase angle variations of $\boldsymbol{U_{pcc}}$, it's essential first to discuss the variations in the inverter output voltage $\boldsymbol{u_{gabc}}$.

The frequency and phase angle of the photovoltaic inverter's output voltage and current are closely related to the Phase-Locked Loop (PLL). PLL utilizes synchronization technology to ensure that its phase is in sync with the voltage at the sampling point, which is typically either the PCC or the low-voltage side of the isolation transformer. A commonly employed synchronization method is the Synchronous Reference Frame PLL (SRF-PLL). Its fundamental principle involves controlling uq to be zero in the dq coordinate system to achieve phase-locking. As illustrated in Figure 3, $\boldsymbol{R_g}$, $\boldsymbol{L_g}$, and $\boldsymbol{X_g}$ respectively represent the resistance, inductance, and reactance from the inverter outlet to the PCC, $\boldsymbol{I_{invd}}$ and $\boldsymbol{I_{invq}}$ are the direct-axis and quadrature-axis components of the inverter's output current, ω and θ are the angular speed and phase angle generated by the SRF-PLL, and k_p and ki denote the proportional and integral coefficients, respectively.

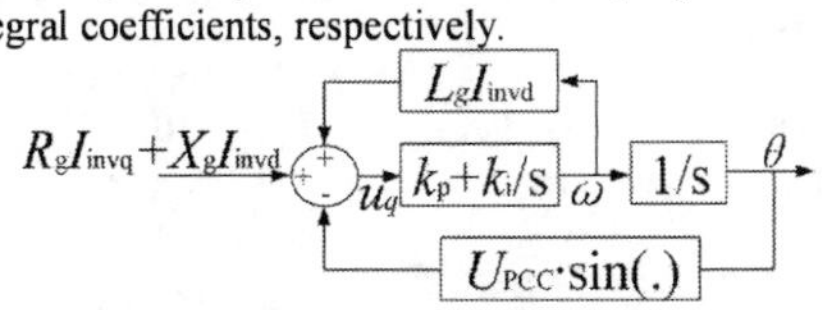

Fig. 3 Principle of SRF-PLL

Based on the SRF-PLL principle, when the tracking error of the reactive angle u_q enters the PI control, it inherently exhibits inertia characteristics. After the occurrence of an islanding event, an abrupt change in the external voltage UPCC will not immediately lead to a significant shift in the phase angle θ of u_{gabc}. This suggests that the synchronization frequency and phase angle of the SRF-PLL change gradually, analogous to the trend in frequency and phase angle changes observed when a synchronous machine encounters an external fault. As illustrated in Figure 4, within the current inner-loop control, i_{dref} and i_{qref} represent the control target values generated by the outer loop, with T_d and T_q as the filtering constants, and u_{gabc} denotes the output voltage of the inverter. The phase angle θ produced by the SRF-PLL serves as the reference phase angle for generating the voltage u_{gabc} within the current inner loop. Based on Equations (1) and (2), it can be inferred that the steady-state frequency of U_{pcc} will not undergo abrupt changes. It's worth noting that, since the impedance angle of G(jω) correlates with the frequency, when the power factor of the inverter's output deviates significantly from the impedance angle of G(jω) at the fundamental frequency, a relatively large frequency offset from the fundamental frequency is required within the islanding area to achieve the control target.

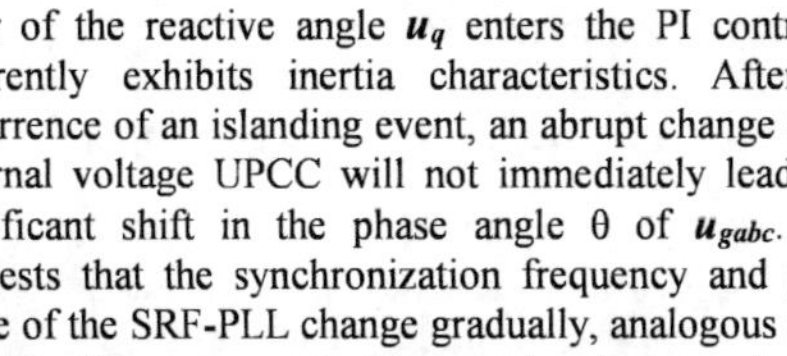

Fig. 4 Current inner loop control

2.2.3 Phase Angle Variation

Before the occurrence of islanding, the inverter operates according to the rated power factor, which implies that G(jω) in Equations (1) and (2) is close to being purely resistive. Once islanding takes place, G(jω) transforms into a local load. This could potentially result in a sudden change in the load power angle, subsequently leading to a sudden change in the phase angle of U_{pcc}. Figure 5 illustrates the simulated waveforms of U_{pcc} and I_{inv} after islanding occurs at 0.8s. The figure contrasts the cases where the load power factor remains unchanged (as represented by U_{pcc} and I_{inv}) and where it varies (as represented by U_{pcc}-PFChg and I_{inv}-PFChg). As evident, if there's a change in the load power factor following the onset of islanding, a noticeable sudden change in the voltage phase angle will occur.

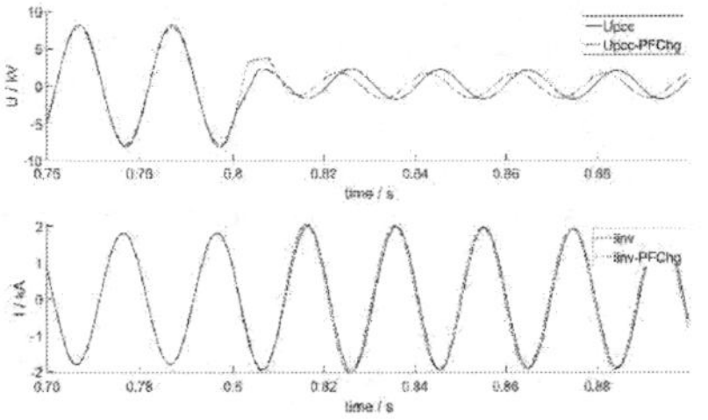

Fig.5 Effect of power factor change on the phase of PCC voltage and inverter output current

Moreover, the phase angle of I_{inv} is influenced not only by θ but also by the proportion of i_{dref} and i_{qref} (i.e., the power factor). If the inverter's output power factor remains unchanged after islanding, the phase angle of the output current won't undergo a sudden change. Otherwise, in scenarios such as low voltage ride through (LVRT), high voltage ride through (HVRT), and changes in the load power angle, alterations in the power factor can result in sudden shifts in the phase angle of the output current.

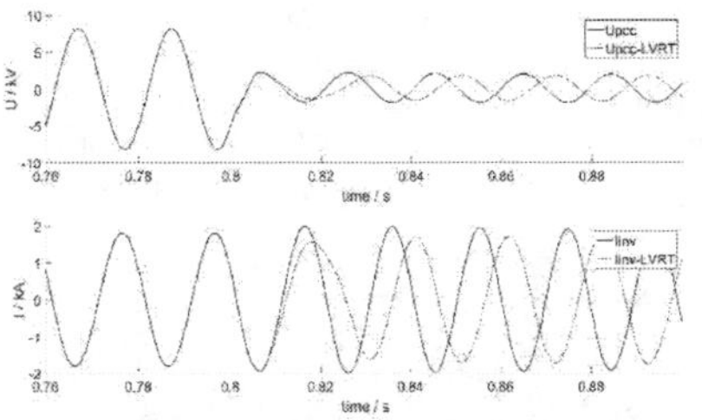

Fig.6 Effect of LVRT on the phase of PCC voltage and inverter output current

For inverters equipped with fault ride-through capabilities, upon entering fault ride-through mode, the target for voltage outer loop control gets locked by the LVRT/HVRT characteristic, and the current inner loop has minimal inertia[5]. As a result, the output current phase angle might undergo a sudden shift due to abrupt changes in the power factor. When the frequency remains constant, as per formula (1), it's easy to deduce that U_{pcc} will also exhibit a phase angle shift. Figure 6 illustrates the simulated waveforms of U_{pcc} and I_{inv} after islanding occurs at 0.8s, contrasting between scenarios without LVRT activation (as seen with U_{pcc} and I_{inv}) and with LVRT activation (as represented by U_{pcc}-LVRT and I_{inv}-LVRT). It's evident that upon activating LVRT after islanding, there's a marked shift in both the voltage and current phase angles.

2.2.4 Conclusions on Voltage Amplitude, Frequency, and Phase Angle Changes within the Island

From the analysis, it is clear that after the onset of an island, there might be abrupt shifts in the voltage amplitude and phase angle within the island. The phase angle and frequency might either oscillate or diverge, leading to a deterioration of power quality within the island and posing threats to the safety of the power supply. Even under conditions like stable sunlight and fewer sensitive loads for consumers, there's a possibility that the island could operate stably for a certain duration. In such instances, while there might not be significant variations in voltage, frequency, or phase, there remains a risk of damaging consumers' electrical appliances.

3 Islanding Detection Scheme

From the previous analysis, it's clear that islanding operation can pose several challenges to operation, maintenance, and electrical safety. Therefore, it's essential to implement a reliable islanding detection scheme. The three primary islanding detection methods in the mainstream are passive, active, and remote.

3.1 Passive Detection Method

During an islanding event, electrical parameters such as frequency, phase angle, voltage, and harmonics within the island might undergo sudden changes. Passive islanding detection identifies islanding based on these observed shifts in electrical parameters. Common criteria for detection include over/under-voltage, over/under-frequency, Rate of Change of Frequency (ROCOF), Rate of Change of Power (ROCOP), sudden voltage phase shifts, and harmonic distortion rate.

3.1.1 Detection Principle

The principle behind passive islanding detection is straightforward and cost-effective. It has been widely employed in distribution networks integrated with Distributed Generation (DG). However, a significant drawback of this method is the potential for a non-detection zone. If, during islanding, the load is in equilibrium with the DG power output, the frequency, voltage, and phase angle within the island could remain stable for a certain duration. This stability leads to a failure in detection. The non-detection zone of passive islanding detection is associated with the load power factor[6-9]. Passive islanding detection becomes ineffective when conditions described by formulas (3) and (4) are met:

$$(U_{PCC0}/U_{\max})^2 - 1 \le \Delta P / P_{inv0} \le (U_{PCC0}/U_{\min})^2 - 1 \quad (3)$$

$$M[1-(f_0/f_{\min})^2] \le \Delta Q / P_{inv0} \le M[1-(f_0/f_{\max})^2] \quad (4)$$

$$M = R\sqrt{\frac{C}{L}} \quad (5)$$

In the formula, ΔP and ΔQ represent the active and reactive power differences after the PCC is disconnected, and they can be considered as the active and reactive power transmitted by the PCC before the islanding. P_{inv0} is the active power output of the PV before islanding. U_{min} refers to the under-voltage protection threshold; U_{max} refers to the over-voltage protection threshold; f_{min} indicates the under-frequency protection threshold; f_{max} indicates the over-frequency protection threshold. U_{PCC0} is the PCC voltage before the islanding, and f_0 is the PCC frequency prior to islanding. M denotes the power quality factor of the load, with the load impedance being represented by the parallel model $G(j\omega)=R//j\omega L//1/j\omega C$. The IEEE expert group believes that the actual power distribution network load's power quality factor will not exceed 2.5. According to the IEEE Std.929-2000, the power quality factor of the islanding protection test circuit is set to 2.5, while both IEC62116-2008 and Q/GDW618-2011 require the power quality factor of the islanding protection test circuit to be 1.

3.1.2 Impact of Various Faults on Detection

Faults of different types and locations can have varied effects on islanding detection. The general process of islanding occurrence is illustrated in Figure 7.

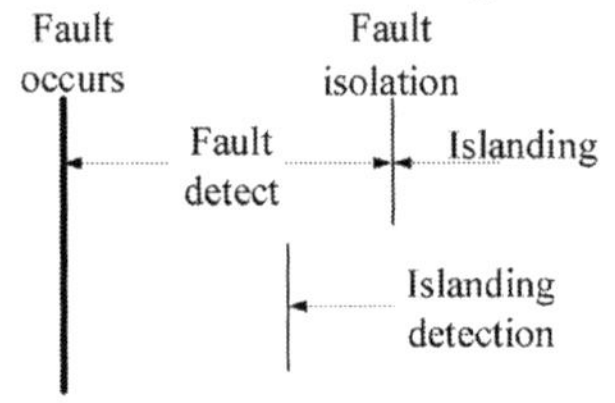

Fig. 7 Islanding process

Based on the various fault scenarios occurring at different locations in Figure 1, the islanding detection situation before fault isolation is shown in Table 1[10].

Table 1 Island detection before fault isolation

Fault Scenario	Islanding Detection Situation before Fault Isolation
F1, F2, F3 Single Phase Grounding	**LCGS**: Due to the isolation transformer T2, PV is unaffected by zero-sequence voltage; PCC might detect over-voltage in the non-faulty phase and trip. **HCGS**: Both PV and PCC might detect low line or phase voltage, resulting in trip-off and circuit breaker tripping.
F1, F2 Phase-to-Phase Short Circuit	PV might detect negative sequence voltage and trip; PCC might detect low voltage in the faulted line or phase and trip.
F1, F2 Three Phase Short Circuit	Both PV and PCC might detect considerably low line or phase voltage, eventually leading to trip-off and circuit breaker tripping.
F1 Line Break	During single or double phase line breaks, PCC might trip due to low line or phase voltage. For Δ/Y isolation transformer, PV might detect negative sequence voltage and trip. Three-phase line break equates to islanding; whether to trip or not depends on voltage and frequency changes within the island.
F2 Line Break	During single or double phase line breaks, the system might experience certain over phase voltage. Due to the existence of the isolation transformer, PV operates normally unless the PCC downstream line is too long or heavily loaded; otherwise, PCC is generally unaffected. A three-phase line break means

	PV loses contact with the load, having no impact on PV or PCC, but it might reverse the current flow of the main line.
F3, F4 Phase-to-Phase or Three Phase Short Circuit	Both PV and PCC might detect negative sequence voltage or low line or phase voltage, leading to erroneous circuit breaker tripping and machine trip-off.
F3 Line Break	Regardless of single, double, or three-phase line breaks, PV remains unaffected. If feeder 2 is lengthy or heavily loaded, it might cause certain over line or phase voltage. However, the possibility of PCC line or phase over-voltage is relatively low.
F4 Single Phase Grounding	**LCGS**: After single-phase grounding, the non-faulty phase rises to line voltage, leading to phase over-voltage trip-off; PCC remains unaffected. **HCGS**: Both PV and PCC might detect low line or phase voltage, resulting in trip-off and circuit breaker tripping.
F4 Line Break	Regardless of single, double, or three-phase line breaks, they all lead to trip-off; PCC remains unaffected.

It can be observed that, in the event of a severe fault on this line (two-phase, three-phase short circuit, line break, etc.), the PV will disconnect in a short time. In case of severe faults on other lines or within the PV itself, the PV might disconnect erroneously. Under LCGS, when a single phase is grounded, the PV does not disconnect.

3.2 Active Detection Method

To address the blind spots of passive islanding detection, some researchers have proposed active islanding detection by injecting small disturbance signals into the distribution grid. During grid-tied operation, these disturbances are absorbed by the system, with minimal impact on power quality. However, during islanding, the effect of these disturbances is amplified, and the passive islanding detection can then determine the islanding situation based on the magnitude of the disturbances. There are various active islanding detection methods, including Active Frequency Drift (AFD), Slide-mode Frequency Shift (SMS), and Reactive Power Disturbance method, etc[11,12]. The following introduces the principle of active islanding protection using the most typical Reactive Power Disturbance method, illustrated using a typical system in Figure 2.

The Reactive Power Disturbance method detects islanding by injecting imbalanced reactive power to disturb the system frequency. In Figure 2, based on the dq-axis transformation theory, the inverter voltage $\dot{U}_{PCC}$ and current $\dot{I}_{inv}$ are transformed to the *dq-axis* coordinate system to obtain U_d, U_q, I_d, and I_q, with $\dot{U}_{PCC}$ set in the direction of the *d-axis*. The relationship between I_d, I_q, and the PCC voltage frequency f during islanding is as follows[13, 14]:

$$\frac{I_q}{I_d}\cdot\frac{1}{2M}=\frac{f-f_n}{f_n} \tag{6}$$

In the formula, M represents the quality factor of the load impedance, and f_n stands for the resonance frequency of the load impedance. It is evident that when Id remains constant, the change in PCC voltage frequency (f) linearly correlates with I_q. Under normal operation of the inverter, the power factor is typically required to be close to 1.0, implying that I_q is set to zero. However, if I_q is not zero, indicating a demand for the inverter to produce reactive power, an islanding situation may lead to a system frequency offset due to unbalanced reactive power, thus detecting the island. In regular grid-connected conditions, the grid absorbs the imbalanced reactive power, preventing any frequency shift. The extent of frequency disturbance relates to the ratio of I_q/I_d and the quality factor M. When M is larger, a significant reactive power is needed to cause a noticeable frequency deviation, analogous to the meaning of formula (4).

Additionally, differing control strategies might lead to injecting disturbances into the grid connection point in opposing directions, causing a dilution effect where disturbances cancel each other out, resulting in a larger detection blind spot.

3.3 Remote Detection Method

The remote method for islanding protection involves islanding detection from the grid side. This primarily uses methods like the power line carrier communication method and topology identification method[15, 16].

The power line carrier communication method entails the installation of a signal generator on the grid side that continuously sends a weak signal. On the PV side, a local receiver captures this signal. If islanding occurs and the signal disappears, it is judged as an islanding event. Given the condition of achieving a certain signal strength, only one signal generator is required to detect islanding in high-density photovoltaics without any blind spots. Just like general carrier communication, this method has challenges, such as the signal being largely affected by the characteristics of load impedance and the continuity of power line impedance. There are adaptations where the power line carrier communication is transitioned to wireless communication, effectively sidestepping the limitations of carrier communication, but this comes at a significantly increased cost.

The topology identification method constructs a wide-area network using communication methods like optical fibers and wireless to collect information about switch positions, current, and voltage. A remote central processing unit, based on the topology identification algorithm, determines the occurrence of islanding. Common algorithms for topology identification include depth or breadth-first searches and the self-multiplication of full connection matrices. The topology identification method does not have dead zones or power quality issues. However, the response time and reliability are contingent on establishing a high-reliability and low-latency communication network, which can be quite expensive.

3.4 Comparison of Approaches

Table 2 compares the advantages and disadvantages of three islanding detection methods:

Table 2 Comparison of three islanding detection schemes

Method	Response Speed	Cost	Reliability
Passive	Average, depends on the rate of voltage frequency deterioration	Low, protection functions can be integrated into new energy interface devices or inverter controllers	Average, with dead zones present
Active	Fast, depends on load quality factor, etc.	Low, only requires changes to inverter control strategy, but significant impacts on power quality with large numbers of PVs	High, with smaller dead zones; multiple PVs might affect each other
Remote	Fast, depends on communication delay	High, requires communication channels	High, completely without dead zones

If PV inverters integrate both active and passive islanding detection methods, they can meet the islanding detection requirements in most cases. However, when the load quality is high or the load has complex nonlinear characteristics, there might be issues like prolonged islanding detection time and low sensitivity. Additionally, with the massive integration of PV and inconsistent controller control strategies, problems such as interference offsetting each other and decreased power quality may arise.

4 Integration of Islanding Detection with Other Secondary Functions

In an active distribution network equipped with islanding detection, it is essential to consider the coordination with existing line protection, transformer protection, reclosing, automatic-standby-switching, and other secondary devices. The general principle is that the integration of islanding detection should not impact the existing protection logic and actions. It is generally believed that inverters have a smaller current loop control inertia, meaning that after a fault, the inverter quickly enters a steady state, and the transient process only lasts a few milliseconds.

4.1 Coordination with Line and Component Protection

Based on the various fault scenarios that occur at different positions in Figure 1, it can be observed that during faults, PV might induce issues like external extraction, augmentation, and reverse fault currents. This requires validation of protection configurations and adjustment of set values. For the commonly used three-segment overcurrent protection in feeders: For overcurrent segments I and II, settings are typically based on short-circuit current in the minimum operational mode. The fault current provided by PV is limited and doesn't have a substantial impact; Overcurrent segment III might be set based on the load current under the maximum operational mode and a reliability coefficient. This setting might be affected by PV. However, considering that the overcurrent segment III generally acts as backup protection with a significant delay, augmentation generally wouldn't trigger overstepping actions. External extraction might cause non-operation under I and II segment failure scenarios, but the probability is low. Hence, island detection and overcurrent protection don't need special coordination.

In weak grid systems, if the load side of the line differential protection contains a high proportion of PV, the initiation of a phase-to-phase short circuit might lead to a lag in the phase angle of the fault current at both ends. This can affect the sensitivity of the differential protection and might even cause it to refuse operation. Given that fiber-optic differential protection serves as the primary protection and its action time is usually within 30ms, the low voltage ride-through time of PV generally exceeds this duration. Therefore, relying solely on island detection cannot address this issue.

For lines where reverse power protection is installed at the starting point, upon islanding, there's a possibility of power from within the island being fed back to the external load. In this situation, the island detection time should be shorter than the action time of reverse power protection to avoid expanding the power outage scope due to the operation of reverse power protection.

In small current grounding systems, single-phase grounding is the main type of fault. Grounding line selection devices based on transient principle are usually configured. Since the effect of inverter-type distributed power generation on transient processes is akin to that of large-capacity loads, PV won't influence grounding line selectors based on the transient principle.

For the issue where PV might cause over-voltage at the neutral point of the Y-side of the transformer, it's generally more severe in distribution networks with higher penetration rates. The island detection time should be less than the action time of transformer gap protection.

4.2 Coordination between PCC and Inverter's Built-in Islanding Detection

If an islanding detection device is installed at the PCC, there needs to be a coordination mechanism with the inverter's built-in islanding detection. Common coordination methods include segmented setting coordination, disabling the inverter's islanding detection, and no protection coordination. Here, not having protection coordination is recommended: in most actual sites, PCC islanding detection devices and inverter islanding detection do not employ coordinated measures. The rationale for not having protection coordination is that, during islanding, whether the inverter is disconnected first or the PCC switch is not crucial. Either disconnection doesn't affect the other islanding detection's judgment: if the PCC switch disconnects first, an island forms, and then the inverter's islanding detection acts, cutting off each inverter; if the inverter trips first, the PCC loses voltage, activating the low-voltage protection of the PCC islanding detection device.

4.3 Coordination with Low/High Voltage Ride-Through

According to the standard 《GB/T 37408-2019 Technical Requirements for Grid-Connected Photovoltaic Inverters》, Class A inverters must be equipped with a low/high voltage ride-through function (hereinafter referred to as **fault ride-through**). Inverter fault ride-through can maintain reliable grid-connected operation during system low/high voltage conditions without disconnecting from the grid (usually for the duration of fault reclosure), providing reactive power support for grid voltage stability. Since the electrical quantities of fault ride-through and islanding operation are hard to distinguish (fault ride-through might also experience frequency changes), inverters usually activate only one of either fault ride-through or islanding detection.

When the inverter engages in fault ride-through and the PCC has islanding detection, the islanding detection duration should exceed the fault ride-through duration. Otherwise, when islanding detection trips, the inability to output reactive power renders the fault ride-through ineffective. In this scenario, reclosure and backup auto-switching should adopt a no-check method. If a no-voltage check method is used, it becomes impossible to simultaneously satisfy the relationship of islanding detection being shorter than reclosure time, islanding detection duration being longer than fault ride-through duration, and fault ride-through duration exceeding reclosure time. It's worth noting that fault ride-through only makes sense before fault isolation. In islanding scenarios, injecting reactive power is similar to reactive power disturbance-based anti-islanding protection, only increasing frequency deviation. However, this indirectly boosts the sensitivity of islanding detection.

5 Conclusion and Outlook

The large-scale integration of distributed energy resources can lead to planned or unplanned islanding operations. Within these unplanned islanding scenarios, aside from the single-phase grounding in low current grounding systems, other severe faults could result in islanding detection actions or false triggers. After islanding occurs, the internal voltage, frequency, and phase angle may experience significant distortion. Specifically, if there's a considerable power exchange voltage magnitude at the Point of Common Coupling (PCC) before the fault, there will be a sudden change. The voltage frequency might gradually shift due to the inertia of the phase-locked loop, and the voltage phase angle could abruptly change because of variations in the load angle, fault ride-through, and other reasons. This underscores the potential deterioration in the power quality within the island. To address the above challenges, there are three islanding detection methods: Passive detection methods, which are straightforward in principle but have issues like detection dead zones and setting adjustments. The size of these dead zones relates to the power deficit at the PCC before islanding and load power factor; Active detection methods can reduce the dead zone to some extent, but they can also be influenced by different control strategies and interactions between inverters; Remote detection methods can overcome the aforementioned challenges, but due to their high costs, they are less commonly implemented in practice.When utilizing passive or active islanding detection principles, coordination with protection and automation systems is essential. This minimizes the impact on distributed energy sources related to transformer neutral over-voltage, reclosing synchronization, and fault self-healing. Additionally, the coordination with fault ride-through needs consideration. In such cases, if reclosing is configured, it should be set to unchecked, and the islanding detection time should be slower than fault ride-through.

Given the existing challenges in islanding detection, the following areas warrant further exploration: Adaptive passive islanding detection methods that can dynamically adjust settings based on the power deficit at the PCC before islanding, thus reducing the protection dead zone; Coordinated control strategies for active anti-islanding after a vast integration of distributed energy resources, aiming to synchronize interference injection across a wide area of distributed sources, resolving the problem of one diminishing as another increases; Research on islanding detection schemes based on weak communication, which can account for the impact on communication channels under fault conditions and reduce the construction and maintenance costs of remote islanding detection; Studies on the coexistence mechanism between fault ride-through and islanding detection on the inverter side. This involves considering the longest fault isolation time from the external grid and transitioning from fault ride-through to islanding detection, achieving an integrated local configuration of both fault ride-through and islanding detection.

References

1. H Yang, B Zhang, Y Song, et al. Simulation and Discussion on Mechanism of Dynamic Interaction Between Frequency and Voltage in Islanded Power Grid[J]. 2013, 37(12): 3503-3508.
2. T Bi, S Liu, A Xue, et al. Fault Characteristics of Inverter-interfaced Renewable Energy Sources[J]. 2013, 33(13): 165-171.
3. Shanghai Electrical Equipment Testing Institute Co., Ltd., Beijing Institute of Electrical Technology and Economics for Mechanical Industry. NB/T 32004-2018 Technical specification of PV

grid-connected inverter[S]. National Energy Administration, 2018.

4. China Electricity Council. GB/T 37408-2019 Technical requirements for photovoltaic gird-connected inverter[S]. 2019.
5. X Kong, Z Zhang, X Yin, et al. Study on Fault Current Characteristics and Fault Analysis Method of Power Grid With Inverter Interfaced Distributed Generation[J]. 2013, 33(34): 65-74.
6. S Wu, Q Xu, X Yuan, et al. Anti-Islanding Detection Standards for Distributed PV Power Generations and Analysis on Factors Influencing Non-Detection Zone of Islanded PV Generation[J]. 2015, 39(04): 924-931.
7. G San, Q Zhao, X Guo, et al. Islanding Detection for Photovoltaic Grid-Connected Inverters Based on Intermittent Frequency Disturbance[J]. 2009, 33(11): 83-86.
8. Zeineldin H H, Saif A, Salama M M A, et al. Three-dimensional Non-detection Zone for Assessing Anti-islanding Detection Schemes[J]. Electric Power Components and Systems. 2010, 38(6): 621-636.
9. Zhihong Y, A. K, Yu Z, et al. Evaluation of anti-islanding schemes based on nondetection zone concept[J]. IEEE Transactions on Power Electronics. 2004, 19(5): 1171-1176.
10. J Miao, X U Guangfu, L I Zhenghong, et al. Analysis on Neutral Point Overvoltage during Grounded Fault in Distribution Network with Inverter-interfaced DGs and Countermeasures[J]. Electric Power Construction. 2018, 39(5): 70-76.
11. Y Yu, H Wang, W Fan, et al. Mixed Islanding Detection Method for Grid - connected Photovoltaic System Based on Slip - mode Frequency Shift Method[J]. 2018, 30(08): 32-37.
12. J Zhang, X Zhang, K Peng, et al. Multi-Distributed Generation Island Detection Method Based on Zero-Sequence Voltage Positive Feedback Control[J]. 2022: 1-16.
13. Z Tang, R Zhu, H Zhang, et al. Identification research of the island and the disturbance based on reactive current disturbance[J]. 2016, 44(22): 159-165.
14. C Zhong, T Jing, M Yang. A New Islanding Detection Method With Periodic Reactive Current Disturbance[J]. 2014, 29(03): 270-276.
15. Velasco D, Trujillo C L, Garcerá G, et al. Review of anti-islanding techniques in distributed generators[J]. Renewable and Sustainable Energy Reviews. 2010, 14(6): 1608-1614.
16. X Hua, M Jiang, G Xu, et al. A Station Domain Islanding Judgment Methond Based on Voltage Class to Topology Analysis[J]. 2018, 37(04): 97-101.

Exploration and practice of zero-carbon intra-park in the context of Carbon peaking and Carbon neutrality Goals： A case study on an intra-park

Jun Zhang [1], Yipeng Qin [2,*], Chao Feng [3], Biao Qiao [2], Jialiang Song [1], Shukui Liang [2], Lichun Mu [1] and Lu Wang [2]

[1] Guoshun Green Construction Technology Co., LTD, 251411 Jinan, China
[2] China Academy of Building Research, 100013 Beijing, China
[3] Shandong Guoshun Construction Group Co. LTD, 250306 Jinan, China

Abstract. Aiming at the planning of zero carbon park under the background of double carbon, this paper takes an industrial park as an example, and analyzes and evaluates the industrial planning situation, energy resources situation and energy planning situation of the park based on the concept of "passive priority, active optimization". Taking the project as the blueprint, integrating the building energy-saving measures and the advanced nature and carbon reduction ability of the integrated smart energy technology, starting from the overall optimization layout of the park, covering the industrial structure, buildings, transportation, energy and other key areas of the park, the comprehensive application of low-carbon technology forms such as building micro-environment, ultra-low energy consumption building system, efficient integrated energy system and renewable energy. Total carbon emissions as the core goal, to achieve the goal zero carbon zone. This study for the project and project time zero carbon building and future zero carbon park planning to provide the reference of the same kind of park.

1 Introduction

As the basic unit of a city, the park gathers various resource elements such as industry, function, innovation and human resources [1], and is the national and regional engines of economic growth [2], as well as highly aggregated units of social production and life [3]. The zero-carbon upgrading of the park determines the effectiveness and speed of the implementation of the "double carbon" strategy [4]. The current study focused on measuring the sources of carbon emissions in industrial parks, with direct and indirect emissions. Electricity consumption and renewable energy offsets were all taken into account [5]. From an energy use point of view, energy saving and emissions reduction are mainly from the energy, production, transportation, building, and living aspects of parks [6]. Therefore, adjusting the industrial structure, optimizing the energy structure, improving the overall energy efficiency and accelerating the construction of zero carbon parks will become the focus of implementing precise emission reduction and achieving carbon peak and carbon neutrality. Literature [7] the analysis of typical low carbon development both at home and abroad as well as domestic and international low-carbon city development zone model of development, from the policy, the park low-carbon industry circulation loop, low carbon cycle economy and low carbon propaganda Angle are summarized, but applies only to the national development industrial park and other large industrial park, concrete measures and rough granularity. Literature [8] analyzes the typical forms of the integrated energy system of the future industrial park, and discusses the characteristics and connotation of the integrated energy system of the park. Literature [9] summarizes the low-carbon development model of national pilot industrial parks, puts forward the low-carbon development focus of different types of industrial parks, and points out the focus point of low-carbon development of industrial parks. Literature [10] summarizes the research progress of low-carbon park and the key problems to be solved in the future by combing relevant research literature.

At present, there have been a considerable amount of research and practice on how to build low-carbon parks in China. Based on a industrial park as an example, this paper based on the concept of "passive priority, active optimization" to carry out the zero carbon park planning and construction, from the industrial layout, construction and a single building energy saving, comprehensive low carbon energy system, production system and ecological landscape optimization for further work, from the Angle of the park carbon emissions aspects has carried on the quantitative analysis, To provide the reference and help for the zero carbon construction.

* Corresponding author: 750686335@qq.com

2 Project overview

The project is located in shandong province, the industrial park covers an area of 114700 m2, planning a total construction area of 130000 m2, including the floor area of 120000 m2, underground is about 10000 m2. The park contains about 90000 m2, passive digital processing center residential experiment and demonstration base of about 10000 m2, research and development office center of 15000 m2.

Adhering to the "passive priority to reduce demand, active optimization efficiency" concept, from the layout optimization, construction monomer design, low carbon energy system coordination, to low carbon production technology introduced. Overall layout, carry out the top-level design of zero carbon park, consider the direct or indirect carbon emissions of enterprise production, buildings, park transportation and other aspects in an overall way, and comprehensively promote the transformation of zero carbon production, zero carbon buildings, zero carbon transportation and other application scenarios.

3 Technical route

Zero carbon park refers to the systematic integration of the concept of carbon neutrality in the planning, construction, management and operation of the park. Relying on the zero carbon operating system, carbon neutrality measures such as energy saving, emission reduction, carbon sequestration and carbon sink are integrated by digital means, and low-carbon industrial development, green energy transformation, facility agglomeration and sharing, and resource recycling utilization are realized by intelligent management. Park internal carbon emission and absorption balance itself, production ecological life depth integration of new industrial park.

When creating zero carbon park, need to consider the factors including the energy structure, energy efficiency and unit energy consumption intensity, carbon emissions and carbon intensity, the influence of these factors mainly came from the park building, can use in the field of industrial and traffic demand. Therefore, the construction of zero exploration park mainly starts from energy, construction and transportation, and digitalization runs through the whole process.

The zero-carbon smart park is achieved through the smart energy system of demand-side energy saving + supply-side efficient utilization, supply-side renewable energy use + supply-demand matching.

4 Technical solution

4.1 Layout design optimization

Based on the climatic environment characteristics of Shandong, where the project is located, combined with the actual function of the building and the innovative living concept, the computer physical environment simulation technology is used to simulate and analyze the sunshine, wind environment, lighting system and daylighting and shading, so as to form the overall layout in line with the evaluation requirements of the green park.

(1) The layout of the building group is dense, and the height of the building is considered to be low in the south and high in the north, so as to facilitate ventilation in summer and shelter northwest or northeast wind in winter. The group layout of the building has a great impact on the microclimate environment of the plot and the quality of the public space, so the group layout of the building should be tested and simulated. Through deduction, the daylighting and ventilation in the building should be ensured, the wind speed and wind pressure of the site meet the requirements, and the power consumption of the building due to daylighting and ventilation should be reduced.

(2) to the north and south direction is as a benchmark, monomer building toward the east and by west Angle should not exceed 30 °. Maximize the effective use of sunlight throughout the year, while avoiding cold winter invasion, increase indoor lighting and ventilation, improve indoor comfort.

(3) the reasonable development and utilization of underground space, in the case of guarantee rate and meet the local natural lighting and ventilation of underground car park.

4.2 Single building design optimization

The project plant in accordance with the green factory, smart factory, happiness factory and tourism factory as the goal. According to the "Green Industrial Building Evaluation Standard" GB/T58378-2013, the application of green technology, construction of three-star green industrial building, clean production, to achieve sustainable development.

The research and development office center and the passive residential experimental demonstration base are planned to build near-zero energy consumption buildings. The scheme comprehensively considers the use of optimized building scheme, window-to-wall ratio, fresh air heat recovery, natural ventilation, high-performance building envelope, high-performance doors and Windows, external shading design, high air tightness and intelligent lighting to minimize the demand for cooling and heating of the building. At the same time, the use of efficient cold and heat sources and other renewable energy sources to improve the building's comprehensive energy saving rate and renewable energy utilization rate. The following is explained by taking the R&D office center as an example.

4.2.1 Envelope

(1) High-performance envelope structure is one of the key basic technologies for building energy saving. To realize building energy conservation and emissions reduction, USES the high performance of retaining structure is indispensable, through efficient thermal

insulation technology can avoid building palisade structure heat bridge, reduce construction cold heat loss, protect the main body structure, reduce the temperature stress, increase the life of structure. The difference between the near-zero energy consumption building with high energy saving level and the general building is that not only the external wall and roof are insulated, but also the parapet, partition wall, floor, underground foundation and floor are required to carry out thermal insulation design according to higher standards.

(2) High-performance Windows and doors, enhance thermal insulation and air tightness, reduce the heat transfer load and fresh air load of the building, and minimize the heat gain from radiation in summer by setting adjustable external shading; This project increase hovel in complex building hall design, reduce the cold winter. At the same time, the project uses high-performance exterior Windows and exterior doors.

(3) There are many factors for building heat gain in summer, among which solar radiation is an important factor affecting the cooling demand of buildings in summer, and solar illumination will bring large building heat gain in summer.

(4) In the design of high-performance buildings, the generation of thermal Bridges is more strictly controlled, and the non-thermal Bridges are designed for the outer protective structure of the building to reduce the local heat loss of the building, lower than the indoor dew point temperature, and avoid the damage of building mildew and condensation caused by moisture intrusion. During the design and construction, it should be ensured that there is no thermal bridge on the roof, exterior wall, ground and exterior Windows of each building.

(5) In order to reduce indoor seepage load, need to take a series of technical measures of dealing with the envelope connection nodes, gas-tight can meet specified requirements. Such as the connection between doors and Windows and external walls; Sealing treatment of pipe through wall; Install the seal of the switch socket hole and so on.

4.2.2 Efficient heat recovery system

The complex building of this project is an office building with dense personnel and large demand for fresh air. This plan requires the building to set up fresh air heat recovery. In the system plan, heat recovery technology is used to utilize the surplus cold and surplus heat of exhaust air.

4.2.3 Efficient cooling and heat sources

Based on renewable energy and energy efficiency project system, the engineering complex building adopts high efficient heat recovery fresh air units and fan coil air conditioning system, cold source using ultra-low temperature air-cooled heat pump units + water storage; The heat source adopts ultra-low temperature air-cooled heat pump + water heat storage, which fully improves the energy utilization rate of the building and reduces the operating cost.

4.2.4 Energy efficient lighting

The energy consumption of lighting system generally accounts for 20%-40% of building energy consumption. For buildings with stable personnel and lighting equipment such as office buildings, the energy saving potential is great. The project lighting system adopts LED lighting, and the project will use intelligent lighting control technology to control the lighting system according to personnel activities and outdoor sunlight.

4.3 Smart low-carbon energy design

Smart low carbon energy system includes cold and heat source system, domestic hot water system, solar photovoltaic system, smart micro grid system and smart low carbon control system.

(1) Based on the project is nearly zero energy consumption target and demonstration benefits, fully considering the project basic conditions and the owner, the cold and heat source system configuration scheme of the project. According to the project energy resources conditions and load characteristics, priority is given to the selection of air source heat pump and energy storage system and other energy-saving technologies.

In research and development office center, for example, at the end of project adopt fan-coil unit plus fresh air system, heat pump units in summer a separate design for the return water temperature on 7/12 ℃, cool storage system to participate in the design for the return water temperature on 7/15 ℃, winter design for 45/40 ℃ by monitoring the return water temperature.

Heat pump heating, decrease the design conditions of water temperature not only can improve the COP value of heat pump units, reduce the operation cost, and meet with the project of energy conservation and environmental protection concept. According to load conditions during operation, to set heat pump units out of the water temperature in stages. At the beginning and end of the heating season, set a lower water supply temperature; At the beginning and end of the cooling supply, set a higher water supply temperature to improve system efficiency and reduce system operating costs.

This project is preferred by end of the heat pump system for building foundation of cold and hot load, peak period of electricity during the day can release.

(2) using solar energy light heat as living hot water heat source, solar thermal system is the efficient use of solar energy is a renewable energy, concentrated solar thermal system has the advantages of high integration, stability of water supply, and beautiful. The roof of the project is set up with a full glass vacuum tube collector to form a solar heat collection system. The full glass vacuum tube solar collector is used, and the collector is set in series and parallel.

(3) This project adopts the solar photovoltaic (pv) + battery as regional renewable energy power generation of local power supply. The PV modules of this power station project are installed on the roof of the joint workshop, using BAPV installation with slope, the

azimuth Angle of the module installation is 14°, and the total installed capacity is 2391.12kW.

(4) This project is based on the park energy system to establish a smart micro-grid system as the energy basis for the application of smart energy in near-zero energy consumption buildings. By setting up photovoltaic power generation, power storage system, flexible load absorption equipment, energy equipment close to the demand side, make full use of new energy and renewable energy power generation, realize the miniaturization system of local power generation and demand matching, which has the characteristics of clean and environmental protection, high flexibility and high efficiency. When controlled reasonably, it can significantly improve the reliability of energy consumption and reduce the investment in transmission and distribution network.

(5)The project is equipped with a composite energy system composed of air source heat pump, water storage, solar hot water and solar photovoltaic. At the same time, the building adopts high-performance active sunshade, intelligent lighting and other equipment, and combines the actual use function and building energy monitoring needs to set up relevant energy management platform. Energy system informatization, automation, digital features is an important support of intelligent energy, the project is based on the actual use requirements to build high intelligent and strong demonstration, the wisdom of the energy management cloud platform. The multi-energy complementary system in smart energy involves massive data, and there are requirements for data processing and calculation, real-time data analysis and data sharing. The application of big data and cloud computing technology in smart energy is to solve the complexity problem of data type, computing theory and system architecture. Cloud computing is an intelligent foundation based on big data, data computing technology as the means, and cloud platform as the infrastructure. Big data and cloud computing technology Through the establishment of cloud computing platform, the massive dynamic data in the intelligent energy processing analysis and intelligent application, so as to realize the scheduling decision.

4.4 Optimization of landscape ecological programs

The project design and implement "people-oriented" the guiding ideology, through elaborate design, the environment design in the factory the factory and dormitory, complex well reflect integrity and practicability of reasonable division, on the design of the scenic spot the pursuit of nature and harmony. Based on the characteristics of large factory floor area, high comprehensive runoff system, large net rainwater flow, small green roof space, low green land rate and less pavement area, the technology library of sponge city in the park is constructed based on the principles of economic feasibility and technical rationality, and the guidance of function strength and suitability is put forward.

4.5 Low carbon production technology

(1) On the basis of product production logistics, determine the workshop process layout, so that the workshop logistics and factory logistics are fully integrated, so that the workpiece logistics is smooth, convenient and orderly, and unnecessary parts are reduced: create conditions for lean production, economic batch distribution, on-time processing and distribution, and improve the production efficiency of workpiece processing and distribution.

(2) Adopt the production organization mode combining process specialization and productization, set up the cutting section, welding processing section and painting section, and improve the processing and production efficiency.

(3) New automatic welding production equipment to reduce labor intensity and improve production efficiency.

(4) The sorting area is set to reduce the cutting production time occupied by the material picking after cutting. The cleaning and sorting storage area of the cutting plate is set at the back end of the cutting machine to realize the fixed-point material picking

5 Calculation of carbon emissions

Through the application of the overall low-carbon technology in the park, the comprehensive energy consumption per unit building area of HVAC system, lighting, domestic hot water and elevator of the complex building is 92.25kWh/(m2.a), and the annual carbon emission is about 425 tons, under the condition of air tight measures and without considering the external photovoltaic access of the building. The comprehensive energy consumption of rental housing unit building area is 67.52 kWh/ (m2.a), and the annual carbon emission is about 205 tons. Park plant construction is carried out in accordance with the green building three star, without considering the production under the condition of energy consumption, the HVAC, lighting, living hot water in carbon emissions about 1200 tons. Joint workshop roof adopts BAPV with slope to install pv single crystal components, with a total installed capacity of 2391.12 kW, an annual capacity of 2.377 million kWh, 1862 tons of carbon dioxide.

This project aims at zero/low carbon park. Through the efficient utilization of renewable energy, near zero energy buildings, and the application of smart low carbon energy related technologies, the total annual carbon emission of buildings in the park is about 1822 tons of HVAC, lighting, domestic hot water and elevator system. The carbon emission reduction of photovoltaic renewable energy generation in this project can meet its carbon emission. Achieve the goal of zero carbon building and energy field.

6 Conclusion

The project in a park as the goal, zero carbon/low carbon energy investment and energy technologies by integrating park, based on passive and active optimization design principles, from the source to reduce carbon emissions in the industrial park. The top-level layout design park layout, construction, transportation, production and other aspects, build give priority to with renewable energy of zero carbon energy system, and supporting infrastructure, such as intelligent control platform effectively integrated comprehensive energy planning. Through the application of near-zero energy buildings, smart low-carbon energy and other related technologies to reduce the basic carbon emissions of the park, the carbon reduction of photovoltaic renewable energy power generation can meet its carbon emissions. From the park, construction level to the indoor outdoor green, low carbon design, through the establishment of low carbon environment and improve living behavior.

(1) the complex building of office and apartments rental housing 2 buildings designed and built according to nearly zero energy consumption building standards. Among them, the comprehensive energy saving rate of the complex building is 60.56%, the energy saving rate of the building body is 48.71%, and the utilization rate of renewable energy is 30%. The overall energy saving rate of rental housing is 61.13%, the energy saving rate of the building itself is 53.17%, and the utilization rate of renewable energy is 30%.

(2) Based on the low level of energy consumption in the zero-carbon park, the concept of smart energy is innovatively penetrated and demonstrated, the construction and control of smart energy system is combined with zero-carbon technology, and a micro-grid and intelligent management platform are planned to be built in the future. This project intends to use the smart energy management system of the park to establish the first smart energy demonstration with zero carbon characteristics in Shandong Province through the informatization of energy systems, the intelligence of operation links, and the synergy and complementarity of multiple energy sources.

(3) Through the planning and design of zero carbon park, the annual carbon emission of the near-zero energy consumption building complex is about 425 tons; The annual carbon emissions of rental housing are about 205 tons. The plant in the park is constructed according to the three-star green building. Without considering the production energy consumption, the annual carbon emission of HVAC, lighting and domestic hot water is about 1200 tons. The roof of the joint factory adopts BAPV to install photovoltaic single crystal modules along the slope, which can reduce carbon dioxide by 1,862 tons per year, and the carbon emission of the park reaches the standard of approaching zero.

References

1. UNIDO. (2018) International guidelines for industrial parks. Available online: http://www.unido.org/sites/default/files/2020-05/International_Guidelines_for_Industrial_Parks_EN.pdf
2. Van Beers D., Tyrkko K., Flammini A., Barahona C., Susan C. (2020) Results and lessons learned from assessing 50 industrial parks in eight countries against the international framework for eco-industrial parks. Sustainability, 12, 10611.
3. Dai Y., Day S., Masi D., Golgeci I. (2022) A synthesised framework of eco-industrial park transformation and stakeholder interaction. Bus. Strategy Environ. 31, 3122-3151.
4. Zhu T., Feng L., Ji Y.J. (2014) Exploration on the promotion path of low-carbon industrial park construction and development in China. ENVIRONMENTAL PROTECTION, 42: 43-45
5. Ji F., Dong Y., Sun X., Liu L., Du J. (2022) Industrial park heat integration considering centralized and distributed waste heat recovery cycle systems. Appl. Energy, 318, 119207
6. Tritto A., Camba A. (2022) State-facilitated industrial parks in the belt and road initiative: towards a framework for understanding the localization of the Chinese development model. World Dev. Perspect, 28, 100465.
7. Yin H.C., Zhou X.Y. (2022) Research on low-carbon development model of domestic and overseas typical cities and parks. SHANGHAI ENERGY CONSERVATION, 04: 363-369
8. Tang X.Y., Zhao Z.L., Li Q.S., Ma X.Y., Liu J.S., Zhao W.X., Zhang Y. (2018) Morphological characteristics and evolutionary route of integrated energy system in industrial parks. SOUTHERN POWER SYSTEM TECHNOLOGY, 12: 9-17
9. U X. (2018) An analysis on classified low-carbon development pathways of national pilot industrial parks. China population, resources and environment, 28: 32-39
10. YAN P.X. (2019) The research progress and future development of low-carbon industrial park in China. Ecological Economy, 35: 26-30+87

Dual-carbon development opportunities and challenges in the property industry: a key role in China's low-carbon transition

Haiyang Du[1], Litao Su[2], Zilin Yang[3], Dongkui Zhang[1], Dan Wang[1], Liping Geng[1], Fang Yu[1], Peng Hou[1], Xiaojuan Wang[3,*]

[1]DATANG CARERA (BEIJING) INVESTMENT CO., LTD, Beijing, 100010, China
[2]China Information and Communication Technology Group Co., Ltd, Wuhan, 430000, China
[3]Central University of Finance and Economics, Beijing, 100010, China

Abstract. This paper aims to explore the key role of the property industry in China's low-carbon transition, and analyze the opportunities and challenges of its dual-carbon development. Firstly, the urgency of global greenhouse gas emissions and the importance of property industry in carbon emissions are introduced, then the definition and significance of carbon neutral and dual carbon goals are analyzed, and the international and domestic dual carbon policies and commitments are reviewed. Then it lists the main sources and components of carbon emissions of the property industry in detail, and evaluates the carbon footprint of the property industry. Next, it discusses the opportunities of new technology and innovation, renewable energy, carbon finance and carbon market for the dual-carbon development of the property industry. At the same time, it also discusses the challenges of technical constraints and cost pressures, the uncertainty and landing problems of policies and regulations, and the difficulty and constraints of landlord and tenant participation. Finally, the paper puts forward policy suggestions to promote the dual-carbon development of the property industry, including setting clear carbon reduction targets and schedules, providing policy support and financial incentives, and establishing industry cooperation and sharing mechanism.

1. Introduction

1.1 Research background

Climate change is one of the major challenges facing the world at present, and its serious impact has already involved the global ecological environment security and the sustainable development of human society. The increase in global greenhouse gas emissions is one of the main causes of climate change. As greenhouse gases play a key role in the absorption and radiation of heat in the Earth's atmosphere, global temperatures continue to rise, extreme weather phenomena occur frequently, ecological balance is damaged, and human society is faced with great risks.

In this context, low-carbon transition has become an important issue for global action. As an important support for urban development, the real estate industry has huge opportunities and challenges for dual-carbon development. Property management involves the operation and maintenance of a large number of buildings and facilities, and the construction sector is one of the major sources of greenhouse gas emissions. Data show that the carbon emissions of the whole process of construction in China exceed 5 billion tons, accounting for 50.9% of the total carbon emissions in the country[1]. Therefore, to strengthen the double carbon development of property industry is of great significance. By promoting the low-carbon transformation of property management, reducing energy consumption and carbon emissions, it can contribute to the global response to climate change. However, the real estate industry is facing various challenges in the process of realizing the dual-carbon development, including technological innovation, investment cost, management ability and so on.

1.2 Research purpose

The purpose of this paper is to explore the opportunities and challenges of dual-carbon development in the property industry. Through the introduction of climate change caused by the increase of global greenhouse gas emissions, the importance and urgency of the property industry in carbon emissions are introduced. We will then explore the opportunities for dual-carbon development in the property industry, including the application of low-carbon technologies, green building design and operational management innovation. At the same time, we will also analyze the challenges facing the property industry, such as economic pressure and insufficient policy support. Ultimately, we aim to provide theoretical guidance to promote the dual-carbon development of the property industry and promote the key role of China's low-carbon transition.

* Corresponding author: 1546994122@qq.com

1.3 Overview of the paper structure

This paper will be divided into the following parts to elaborate. First, the second part will detail the issue of climate change caused by the increase in global greenhouse gas emissions, including the types of greenhouse gases, the sources of emissions, and the impact on the climate, with the aim of explaining why action is needed to address climate change. Secondly, the third part will focus on the importance and urgency of the property industry in carbon emissions, explore the reasons why the construction industry is a major source of greenhouse gas emissions, and analyze the potential and impact of property management on carbon emission reduction. The fourth part will focus on the dual-carbon development opportunities of the property industry, including the application of low-carbon technology, green building design and operation management innovation, in order to demonstrate the important role of the property industry in sustainable development. Then, the fifth part will discuss the challenges faced by the dual-carbon development of the property industry, such as the pressure of economic efficiency, insufficient policy support, etc., through in-depth analysis of the nature and reasons of the challenges, the corresponding solutions are proposed. Finally, the sixth part will give a summary and conclusion, review the main viewpoints and research findings of this paper, and propose possible research directions in the future, with a view to promoting the dual-carbon development of the property industry and contributing to China's low-carbon transition. Through the discussion and analysis of the above parts, this paper will provide readers with a comprehensive understanding of the opportunities and challenges of the dual-carbon development of the property industry and the basis for in-depth thinking.

2. Analysis of carbon neutrality and dual carbon targets

2.1 Definition and significance of carbon neutral and dual carbon targets

Carbon neutrality is the process of bringing net emissions to zero by reducing greenhouse gas emissions and balancing remaining emissions with absorption. The two-carbon target refers to further reducing greenhouse gas emissions on the basis of carbon neutrality and achieving a higher level of carbon reduction. Carbon neutrality and dual carbon targets are of great significance in the prevention and control of climate change. Through carbon neutrality, it can effectively slow down the climate change caused by greenhouse gases, protect the ecological environment and the sustainable development of human society. The dual carbon target will further strengthen the reduction of emissions and promote the global development of a lower carbon future.

2.2 International and domestic dual carbon policies and commitments

Countries have developed a series of dual-carbon policies and commitments. For example, in 2020, the European Union proposed the European Green Deal, which aims to achieve carbon neutrality by 2050. The United States has also resumed its participation in the Paris Agreement and has proposed a "carbon neutral 2050" plan. As one of the world's largest greenhouse gas emitters, China has also actively responded and proposed a "2060 carbon neutral target". These international and domestic dual carbon policies and commitments have laid the foundation for promoting global cooperation on greenhouse gas emission reduction.

2.3 The role and responsibility of the property industry in carbon neutrality

As a large energy consumer and one of the main sources of carbon emissions, the property industry plays an important role and responsibility in the process of carbon neutrality. First, the property industry can reduce energy consumption and carbon emissions by adopting low-carbon technologies and green building design. For example, the use of solar energy and geothermal energy as clean energy supplies, the use of energy-saving lamps and intelligent energy management systems can effectively reduce energy consumption and carbon emissions. Secondly, property managers can innovate operation management models to improve energy efficiency and reduce waste. For example, the introduction of information technology and big data analysis to accurately monitor energy use, timely detection and resolution of energy waste problems. In addition, the property industry can promote sustainable lifestyles, encourage residents and tenants to reduce energy consumption and carbon emissions, and participate in carbon neutral actions.[2]

3. Carbon emission status of the property industry

3.1 Sources and composition of carbon emissions from the property industry

The main sources of carbon emissions in our life can be divided into three categories, namely direct carbon emissions, indirect carbon emissions and implicit carbon emissions. Direct carbon emissions include the burning of fossil fuels such as coal; Indirect carbon emissions include the use of electricity, water and other energy sources; Implicit carbon emissions mainly refer to the carbon emissions implied by the daily procurement of equipment, such as tables, chairs, computers, etc., during the processing of these equipment.[3]

In the property industry, direct carbon emissions come from heating systems, and indirect carbon emissions mainly come from building energy consumption, air conditioning and water treatment. Here is a breakdown of the sources of these emissions:

a. Heating system: Heating in winter requires a large amount of energy, such as the direct combustion of fossil fuels such as coal or the large consumption of electric heating systems, which will lead to high carbon emissions directly generated by the property industry.

b. Means of transportation: there are generally parking lots in the area under property management, and the carbon emissions generated by cars and connecting traffic in the area are also the main sources of carbon emissions in the process of property management.

c. Building energy consumption: The energy consumption of buildings is the most important source of carbon emissions in the property industry. Mainly including electricity use, lighting, elevator operation, etc., is an important form of indirect carbon emissions.

d. Ventilation and air conditioning: The operation of air conditioning systems consumes a lot of electricity and often uses refrigerants, which will lead to increased carbon emissions.

e. Water treatment: The water treatment process in property management also requires energy input, such as pump operation, water heating, etc. In this process, certain carbon emissions will be generated indirectly.

f. Equipment procurement: In the process of property management, it is not only necessary to invest a lot of human resources, but also to purchase tables, chairs, computers and other equipment, which are the main sources of implied carbon emissions.

3.2 Assessing the carbon footprint of the property sector

In order to quantify the contribution of the property industry in terms of carbon emissions, this paper will evaluate the carbon footprint of the property industry, analyze the energy use in the process of property management and the circumstances that lead to carbon emissions. This paper will quantify the main sources of carbon emissions discussed above, in which C_n is the carbon emissions of each emission source, and A_n is the carbon emission coefficient of indirect carbon emission sources. By summing and calculating the total carbon emissions of each emission source, the energy consumption and total carbon emissions in property management are calculated.

$$C = \sum_{n=1}^{N} C_n \quad (1)$$

$$C_n = A_n * E_n \quad (2)$$

$$C = \sum_{n=1}^{N} (A_n * E_n) \quad (3)$$

Meanwhile, energy use in property management is closely related to carbon emissions. For example, the power consumption habits of property managers and residents, the efficiency of energy equipment, the implementation of energy management measures and other factors will affect energy consumption and carbon emission levels. For example, property management also involves daily cleaning work, greening maintenance and other activities, which may also generate certain energy consumption and carbon emissions.[4]

Through the above method, it can help to quantitatively assess the carbon emission contribution of the property industry. The following is the process for assessing the carbon footprint of the property sector:

a. Collect data: Collect data related to the property industry, including energy consumption, water use, waste discharge, etc. This data can be obtained from property management companies, energy suppliers, water authorities and others.

b. Carbon emission calculation: Convert the collected data into carbon emission data. By using a reliable carbon emission factor, it is possible to convert energy consumption, water use and waste emissions into corresponding carbon emissions.

c. Boundary determination: The boundary of the assessment is determined, i.e. the scope covered by the assessment. This may include the overall operation of the property, or the operation of specific buildings and equipment.

d. Data analysis: Analyze and organize the collected data to calculate the total carbon emissions of the property. Total emissions can be further broken down into emissions from different sources, such as energy emissions, water supply emissions, etc.

e. Comparison and evaluation: The carbon emission of the property is compared with the standard or similar properties to assess its performance in terms of carbon emissions. The results of the assessment can be compared with past data to observe trends of improvement or deterioration.

f. Formulate emission reduction strategies: According to the evaluation results, formulate strategies and plans to reduce carbon emissions, such as increasing carbon sink reserves in the community, improving energy efficiency, using clean energy, promoting water-saving measures, increasing waste recycling, etc.

g. Monitoring and reporting: After the implementation of the emission reduction strategy, the carbon emissions of the property are continuously monitored and the results are regularly reported. This helps track progress and adjust strategies in a timely manner.

By conducting a property industry carbon footprint assessment, property management companies can better understand their contribution to carbon emissions and take appropriate measures to reduce their carbon footprint and promote sustainable development.[5]

4. The dual-carbon development opportunities of the property industry

4.1 The application of new technologies and innovations in carbon emission reduction

With the continuous progress of science and technology, new technologies and innovations have great potential for carbon emission reduction in the property industry. At present, they have been applied to the following fields and have some potential impacts on carbon emission reduction in the property industry:

a. Intelligent buildings: Intelligent buildings can achieve efficient use of energy and automated management through the use of advanced sensors, adaptive control systems and other technologies, thereby reducing carbon emissions.

b. Energy management system: The intelligent energy management system can monitor and control the energy consumption of buildings, provide accurate data analysis, and help property managers optimize energy waste to achieve the purpose of emission reduction.

c. Green building materials: The development and application of more environmentally friendly building materials, such as renewable materials and low-carbon materials, can reduce energy consumption and carbon emissions in the construction process.

4.2 The promotion and application of renewable energy in the property industry

Renewable energy is one of the important ways to reduce carbon emissions, and it has a wide application prospect in the promotion and application of property industry. For example: solar photovoltaic power generation system, the use of solar energy, convert it into electricity and other energy supply to the property building

a. Solar photovoltaic power generation: Installing solar photovoltaic systems to convert light energy into electricity to supply property buildings can not only reduce the demand for traditional energy, but also reduce carbon emissions.

b. Wind energy utilization: Under suitable conditions, use wind energy to generate electricity, provide clean energy for property sites, and reduce dependence on traditional energy.

c. Biomass energy utilization: The use of waste and crop residues and other biomass resources for energy conversion, such as biomass power generation, biofuels, etc., effectively reduce carbon emissions.

4.3 The impact of carbon finance and carbon market on the property industry

The development of carbon finance and carbon market has brought new opportunities to the property industry, which can be summarized in the following aspects:

a. Carbon trading and carbon quota: With the development of the carbon market, the property industry can participate in carbon trading, offset carbon emissions by purchasing carbon quota, and sell excess carbon quota through emission reduction measures to obtain economic benefits.

b. Carbon credit and carbon certification: Through the implementation of carbon emission reduction projects, the property industry can obtain carbon credits, participate in carbon credit trading, and demonstrate its low-carbon and environmental protection image through carbon certification to improve brand value and market competitiveness.

c. Carbon finance support: Carbon finance provides financing support and investment opportunities for the property industry, such as the financing of carbon emission reduction projects, green bonds, etc., which promotes the application of low-carbon technologies and innovations. [6]

The application of new technologies and innovations, the promotion and application of renewable energy, and the impact of carbon finance and carbon markets provide broad development opportunities for the property industry to achieve the dual-carbon goal. Property managers and stakeholders need to take active action to take advantage of these opportunities to drive the property industry towards low-carbon development and contribute to sustainable urban development. [7]

5. The dual-carbon development challenge of the property industry

5.1 Technical limitations and cost pressures

Technical limitations: The adoption of new technologies and innovative solutions requires significant R&D and testing efforts and may present technical feasibility and stability challenges. In addition, the energy upgrading of older buildings also faces technical limitations.

Economic cost pressure: The introduction of low-carbon technologies, the implementation of energy management systems, and the transformation of building energy all require capital investment, which may bring economic cost pressure to property enterprises. [8]

5.2 Uncertainty and landing problems of policies and regulations

Uncertainty of policies and regulations: There may be uncertainty in the formulation and adjustment of relevant policies and regulations such as carbon emission allowances and carbon trading mechanisms, which brings certain troubles to the long-term planning and strategic decision-making of the property industry.

Implementation problems: The implementation of policies and regulations requires the support and cooperation of relevant supervision, implementation and evaluation institutions, but sometimes the construction and operation of these institutions may lag, resulting in increased difficulty in the implementation of relevant policies.

Solutions to these challenges can include:

a. Technology research and development and innovation: Increase support for the research and development and application of low-carbon technology, encourage enterprises to carry out technological innovation, reduce the cost of low-carbon technology and improve the feasibility.

b. Financial support and policy incentives: the government can provide financial support, tax relief and other incentive measures to reduce the economic cost pressure of property enterprises to participate in carbon emission reduction.

c. Strengthen policy communication and guidance: Government departments need to communicate with the property industry in a timely and effective manner, clarify the direction and timetable for the formulation of policies and regulations, and provide better policy guidance and business support for property enterprises.

5.3 The difficulty and constraints of landlord and tenant participation

The participation of owners and tenants is the key to achieving the dual-carbon goal of the property industry, but it also faces some difficulties and constraints:

a. Awareness and cognition problems: Some owners and tenants may lack sufficient awareness and understanding of the importance and specific measures of carbon emission reduction.

b. Conflict of interest and constraints: Owners have their own investment return and interest considerations, and tenants also pay more attention to comfort and price when choosing properties, which may exert certain constraints on the promotion of carbon emission reduction actions.

To address the above challenges, the following improvements can be proposed:

a. Publicity and education: raise the carbon awareness of owners and tenants and enhance their enthusiasm to participate in carbon emission reduction by organizing publicity activities and conducting training courses.

b. Incentive mechanism: Establish incentive mechanism, such as certification and reward of low-carbon real estate, to encourage owners and tenants to choose carbon emission reduction properties.

c. Win-win cooperation: Property managers can cooperate with tenants to jointly promote energy conservation and emission reduction projects, such as sharing energy information, implementing energy-saving renovation, etc., to achieve the common interests of both parties.

Through the implementation of the above measures, it can promote owners and tenants to better participate in carbon emission reduction, and promote the property industry to move towards the dual-carbon development goal.[9]

6. Policy Suggestions for promoting the dual-carbon development of the property industry

6.1 Provide policy support and financial incentives

The Government may consider exploring the provision of policy support and financial incentives to promote the dual-carbon development of the property industry. Specific measures could include:

a. Provide tax relief policies related to carbon emission reduction to reduce the economic cost pressure of property enterprises.

b. Set up a special fund to support property enterprises to carry out carbon emission reduction technology transformation and facility upgrading projects.

c. Establish a carbon trading mechanism to encourage property enterprises to actively participate in the carbon market and promote carbon emission reduction actions.

d. Increase support for green building certification and incentives, and encourage property enterprises to develop and manage low-carbon and sustainable buildings.

6.2 Establish an industry cooperation and sharing mechanism

In order to strengthen the dual-carbon development of the property industry, the government can promote the establishment of industry cooperation and sharing mechanisms. This could include:

a. Organize exchange seminars, training courses and other activities within the industry to promote experience sharing and technical cooperation among property enterprises.

b. Establish an energy data sharing platform to encourage property enterprises to share information on energy use and management to improve overall energy efficiency.

c. Support property enterprises to participate in joint energy procurement and reduce energy costs and carbon emissions through centralized procurement.

Through the above policy recommendations, we can further promote the dual-carbon development of the property industry and help achieve carbon emission reduction targets. The government, property enterprises and relevant institutions should work closely together to form a virtuous circle of policy support, technological innovation and industry cooperation, and jointly promote the property industry to a low-carbon and sustainable future.

References

1. Shengjie Zhang. Green building materials will become the "main battlefield" of zero-carbon transformation of buildings. N. China Energy News, 017,(2023)
2. Nie Yao, Li Duyuan, Lin Borong, Xiao Wei. Exploration and practice of park comprehensive energy planning under the background of dual carbon: a case study of a science and Technology innovation city in Xiongan .J. Building Science :1-9, (2023)
3. Hu Jiabing. The idea of ecological building innovation path under dual-carbon situation .J. Guangdong Building Materials, **07**, 39,(2023)
4. Ivalin P ,Alicia L ,Georgios M , et al. Decarbonizing real estate portfolios considering optimal retrofit investment and policy conditions to 2050 .J. iScience, 26,(2023)

5. Zhang Ting. Thinking on the development path of construction industry based on the goal of "double carbon" .J. Green Building, **04**, 49-52, (2023)
6. Chi Z ,Jianliang Z ,Yanan L . Financial inclusion and low-carbon architectural design strategies: solutions for architectural climate conditions and architectural temperature on new buildings. .J. Environmental science and pollution research international,30,(2023)
7. NI Haiqiong,YU Yaxin,MAO Jianlong. The role of carbon footprint and building carbon emissions on dual carbon J. Green Building, **04**,11-14,(2022)
8. Fu C ,Zhou C . Examining the Impact of Real Estate Development on Carbon Emissions Using Differential Generalized Method of Moments and Dynamic Panel Threshold Model .J. Sustainability,15,(2023)
9. Yang Meng. Low Carbon and Property Management .J. China Property Management, **01**,16-19,(2010)

Innovative Paths and Strategies for Property Management in the Context of Dual-Carbon Strategy

Litao Su[1], Qiang Wang[2], Shuo Shi[3], Yingang Feng[4], Liping Geng[4], Haiyang Du[4], Dongkui Zhang[4], Dan Wang[4], Xiaojuan Wang[3,*]

[1]China Information and Communication Technology Group Co., Ltd, Wuhan, 430000, China
[2]Information Science & Intelligence (Beijing) Intelligent Technology Co. Ltd, Beijing, 100000, China
[3]Central University of Finance and Economics, Beijing 100000, China
[4]DATANG CARERA (BEIJING) INVESTMENT CO., LTD, Beijing, 100000, China

Abstract. In the context of China's strategy to promote carbon peaking and carbon neutrality, the construction industry has come under the spotlight for its higher carbon emissions. Property management plays an important role in the operation phase of buildings. Therefore, it is of great significance to innovate property management paths and strategies for energy saving and emission reduction in the property sector. In this paper, we will first discuss the concept and connotation of "dual-carbon" strategy, as well as the relationship between property management and carbon emissions. Secondly, it discusses the challenges faced by property management under the "dual-carbon" strategy and the benefits of accomplishing the transformation, in order to explain the necessity and value of innovative property management. Then, the three innovative paths of property management, namely intelligent building and energy management, application and promotion of renewable energy, and carbon footprint monitoring and assessment, are discussed and the key strategies are proposed from certain perspectives. Then, successful case of property management is introduced, and the experiences for their success are summarized. Finally, based on the above innovation paths, key strategies and successful cases, suggestions are made from three perspectives: policies and regulations, enterprises and owners, and technology and talents.

1 Introduction

Climate change is a major issue facing all humankind today, and climate change, characterized by global warming and an increase in extreme weather, is affecting the sustainable development of the planet. The burning of fossil fuels and the release of carbon dioxide from human production and consumption activities are contributing to continued global warming, posing a serious threat to the ecological security of the planet and to the survival of humankind itself. Environmental risks account for five of the top ten critical global risks in the Global Risks Report 2022, with the failure of climate actions topping the list.

In response to global warming, in December 2015, nearly 200 parties from around the world signed the Paris Agreement, which sets out to limit the increase in global average temperature to no more than 2 degrees Celsius from the pre-industrial temperature rise and to work towards limiting it to no more than 1.5 degrees Celsius, as well as to provide for a unified arrangement for global action to combat climate change after 2020.

China has actively responded to the Paris Climate Change Agreement and formulated the strategic goal of "double carbon". 2021, the CPC Central Committee and the State Council issued the "Opinions on the Complete and Accurate Implementation of the New Development Philosophy and Doing a Good Job in Carbon Peak and Carbon Neutral Work", which is aimed at China's carbon peak and carbon neutral goals for the time planning and strategic deployment. General Secretary Xi attaches great importance to carbon peak and carbon neutral work, emphasizing that carbon peak and carbon neutral is a broad and profound economic and social systemic change, and incorporating carbon peak and carbon neutral into ecological civilization construction.

The construction industry is the key target of China's "dual-carbon" work. According to China Building Energy Consumption and Carbon Emission Research Report (2022),[1] the total carbon emission of the whole process of building in 2020 will be 5.08 billion tons of carbon dioxide, accounting for 50.9% of the total national carbon emission. Among them, the carbon emission in the operation stage of the building accounts for 42.52% of the total carbon emission in the whole process of the building. Property management should realize green and low-carbon transformation, thus making positive contributions to energy saving and emission reduction in the operation phase of buildings.

* Corresponding author: 1546994122@qq.com

This paper analyses the carbon emissions from 3 stages of the whole building process, summarizes the role of property management modes on the carbon emission reduction of the building operation stage and even the whole process of the property. On this basis, this paper explores the strategy of low-carbon transformation of property management, and provides suggestions for systematic upgrading of the overall property management system, so as to help the national "double-carbon" goal.

2 Dual-carbon strategy and property management

2.1 Concept and implications of a dual-carbon strategy

On September 22, 2020, President Xi Jinping announced at the 75th session of the UN General Assembly that China would strive to peak carbon dioxide emissions by 2030, and work towards the goal of carbon neutrality by 2060.1 China's carbon neutrality goal is to achieve carbon dioxide emissions by 2030, and to achieve carbon neutrality by 2060.

In October 2021, the Central Committee of the Communist Party of China and the State Council issued the "Opinions on the Complete and Accurate Implementation of the New Development Concept and Doing a Good Job in Carbon Peaking and Carbon Neutrality" and the "Action Plan for Peak Carbon by 2030", which jointly constructed the top-level design of China's carbon peaking and carbon neutrality "1+N" policy system, and the relevant policies on key areas of carbon emissions and at the provincial and municipal levels have been introduced one after another. Relevant policies in key areas of carbon emission and at provincial and municipal levels have been introduced one after another.

In August 2022, the Ministry of Science and Technology (MOST), the National Development and Reform Commission (NDRC), the Ministry of Industry and Information Technology (MIIT), and other nine departments issued the "Implementation Plan for Scientific and Technological Support for Carbon Peak Achievement and Carbon Neutrality (2022-2030)", which puts forward the scientific and technological support initiatives for realizing the goal of carbon peak achievement by 2030.

2.2 The relationship between property management and carbon emissions

The whole building process includes the building materials production stage, the building construction stage and the building operation stage. Among them, the carbon emissions from the building operation stage accounted for 42.52% of the carbon emissions from the whole building process in 2020, while the national carbon emissions from the whole building process accounted for more than 50% of the national carbon emissions. Property management is the operation and maintenance for the building operation stage, so its innovative path and strategy will have a great impact on the carbon emissions in the building operation stage, and if its low-carbon transformation process links the building materials production, building design and building construction stages, it will realize the carbon emission reduction in the whole process of building.[2]

According to Zhao Linfu (2021), green property management includes five elements: energy saving management, water saving management, waste classification management, environmental greening management and pollution prevention management.[3] Among them, energy saving management and pollution prevention management are closely related to carbon emissions. Energy conservation management targets the intelligent upgrading of three high energy-consumption systems, namely the central air-conditioning system, the water supply equipment system and the public lighting system, and the improvement of related energy-saving technologies. The energy-saving technologies involved in the Green Building Evaluation Standards (GB/T50378-2019) that are related to direct carbon emissions mainly include air conditioning temperature control technologies, lighting technologies, greening technologies for building materials, drainage technologies and landscape greening technologies. Pollution prevention and control management in reducing carbon emissions is mainly to optimize the energy structure, increase the proportion of clean energy use, improve the efficiency of energy use, and improve the recycling rate of materials.

3 The need and value of innovation in property management

3.1 The Challenges of Property Management in a Dual-Carbon Context

Property management underperforms on both energy consumption and carbon emissions. Property service enterprises should pay attention to the reduction of both energy consumption and carbon emissions in property management, and focus on lower carbon emissions per unit of energy consumption. This requires enterprises to optimize their energy structure, increase the proportion of clean energy and improve energy utilization.

3.2 The value and benefits of innovative property management

3.2.1 Innovative property management can bring positive impact on carbon emission reduction

Property service enterprises should focus on green concepts when carrying out property management innovation, and complete green and low-carbon transformation in addition to improving property safety and quality. Currently, China's two green building rating standards, namely the Green Building Evaluation

Standard (GB/T50378-2019) and the Green Retrofit Standard for Existing Buildings (GB/T51141-2015), establish a five-level indicator system based on safety and durability, health and comfort, convenience of life, resource conservation and environmental liveability, where the secondary indicators contain entries that are closely related to carbon emission reduction The secondary indicators contain entries closely related to carbon reduction.[4]

Fig. 1. Carbon emissions from each stage of the building process.
Data resource: Building Energy and Emissions Database

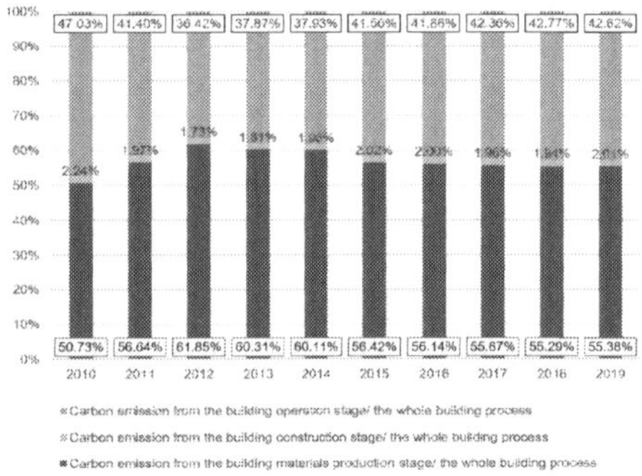

Fig. 2. Percentage of carbon emission at each stage.
Data resource: Building Energy and Emissions Database

As shown in Figure 1, from the viewpoint of carbon emissions, carbon emissions in the construction phase have been at a low level from 2010 to 2019, and carbon emissions in the production phase of building materials and the operation phase of buildings are both at a high level and showing an upward trend. Among them, carbon emissions from the materials production stage declined significantly in 2015, decreasing from 2895,688,200 tons in 2014 to 2552,484,200 tons in 2015, a year-on-year decrease of 11.85%, and the upward trend has levelled off since then. And the carbon emissions from the building operation stage basically maintained a growth rate of about 4%. As shown in Figure 2, from the point of view of the proportion of carbon emissions at each stage, the building operation stage of the building alone has been increasing since 2012, growing from 38% in 2012 to 43% in 2019. This shows that reducing carbon emissions in the operation phase is the key to reducing carbon emissions in the whole life cycle of buildings.

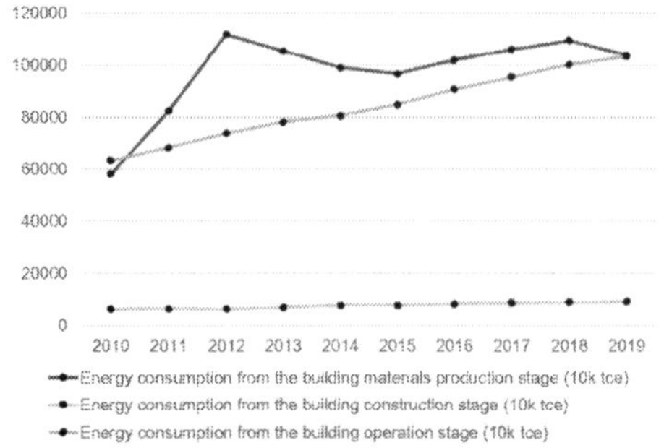

Fig. 3. Energy consumption from each stage of the building process.
Data resource: Building Energy and Emissions Database

As shown in Figure 3, from the point of view of energy consumption, the energy consumption of the building materials production stage was at the highest level among the three stages from 2011 to 2018, and a reduction in energy consumption was realized since the peak of energy consumption in 2013, while the energy consumption of the building operation stage has been growing at a fast rate and exceeded the energy consumption of the building materials production stage in 2019. This shows that reducing energy consumption in the building operation phase is the key to reducing total building energy consumption.

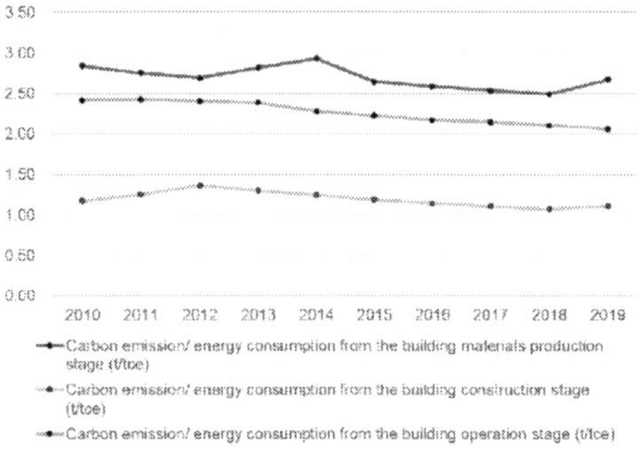

Fig. 4. Carbon emission/ energy consumption from each stage of the building process.
Data resource: Building Energy and Emissions Database

As shown in Figure 4, from the viewpoint of the ratio of energy consumption to carbon emissions, the ratio has been decreasing during the building operation stage of the building, which indicates that the energy structure has been optimized and the use of clean energy has increased during this stage. Combined with the high and continuous growth of carbon emissions and energy consumption in this stage, only working on energy and increasing the use of clean energy cannot completely solve the problem, but requires systematic innovation and change in property management.

In addition, according to the China Building Energy Consumption and Carbon Emission Research Report (2022), in 2020, carbon emissions from the building materials production stage of the whole building process decreased by 7.1% year-on-year, and those from the construction stage decreased by 1.5%, while those from the operation stage increase by 1.5% year-on-year, which is the only stage among the three stages with year-on-year carbon emissions growth. This shows that to realize the low-carbon development of the whole process of construction, it is inevitable to focus on the low-carbon development of the building operation stage, and the green low-carbonization of the building operation stage requires innovation in the property management mode. In conclusion, it is of great significance for property service enterprises to innovate property operation and management pathways to realize "double-carbon" goal.

3.2.2 Innovative property management can bring economic benefits

The whole life cycle includes the processes of building materials selection, building design, building construction, operations management. CHEN Si-qin (2009) pointed out that the construction cost and operation and maintenance cost is the relationship between one and the other, and the operation and maintenance cost is 7.7 times of the construction cost. Therefore, property service enterprises can reduce the whole life cycle cost and bring economic benefits by docking with the project designers in advance, presetting the construction and operation programs, and investing costs in systematic upgrading of properties and improvement of green building technology. [5]

In terms of building materials, CHEN Si-qin (2009) pointed out that 60% of ecological building materials can be recycled, while many existing building materials cannot be recycled, which not only results in a waste of resources, but also increases the cost of cleaning up old materials and the cost of purchasing new materials. In terms of energy, the new property management can be digitized and intelligently transformed, using Internet of Things, artificial intelligence and other technologies to save electricity and energy consumption, reducing the cost of energy consumption.

Property service enterprises innovate in property management, apply energy-saving technologies, and utilize intelligent systems to help owners save the consumption of water and electricity and other resources, thus reducing the expenditure on water and electricity and other living expenses, and making owners the beneficiaries. Property service enterprises are recognized by owners, customer stickiness increases, market competitiveness increases, and enterprises are inspired to further green and low-carbon transformation in property management, thus forming a virtuous cycle.

Overall, for property service enterprises, the short-term operating costs of innovating property operation and management modes will increase due to research innovations, design innovations, system upgrades, and operation and maintenance. However, in the long run, the green and low-carbon transformation of property operation and management will bring innovations in technology and products and services, which will increase the core competitiveness of the enterprise, make it the choice of more customers, and bring performance to the enterprise. For the owners, innovative property operation and management can reduce the expenditure of water, electricity and energy; for the society, innovative property operation and management can reduce the waste of building materials, energy and other resources, and enhance the recyclability of resources and the sustainability of properties, with obvious increase in social and economic benefits.

4 Property management innovation path

4.1 Intelligent Buildings and Energy Management

Wang Rongming (2023) summarizes intelligent building technologies as BIM technology, Internet of Things (IoT) technology, 3D printing technology, intelligent weak power technology, cloud computing and big data technology. [6] Among them, BIM technology is the prerequisite for the application of other intelligent building technologies, which is of great significance in the architectural design and construction phases, and has an impact on the logistics operation and management industry in the later stages; 3D printing technology mainly plays a role in the architectural design phase; and IoT technology plays a fundamental role in the building operation and maintenance phase, which is complemented by other intelligent building technologies, and jointly completes the optimization of the technologies in the operation of properties.

In addition, the intelligent building energy management system based on IoT technology can play the role of rational allocation of energy. Intelligent building energy management system is generally composed of three parts, i.e., perception layer, network layer and management layer. In these three parts, the system can be integrated with intelligent building technologies such as Internet of Things (IoT) technology, intelligent weak power technology, cloud computing and big data technology. The perception layer connects the intelligent equipment on site with the system through the Internet of Things, and transmits the data collected by the equipment, etc. to the network layer, which realizes the incoming, transforming and sending of the data, and carries out data storage and management in the application layer through the Internet of Things cloud computing.

4.2 Promotion and application of renewable energy

4.2.1 Electricity generation

Renewable energy power generation includes solar power, hydroelectric power, wind power and bioenergy power. The development and utilization of solar energy is the main focus, and solar energy can be collected through high-efficiency thermal collection devices.

Renewable energy can be applied to power generation for public lighting systems, such as lawn lamps and street lamps.

4.2.2 Heat energy generation

Heat can be generated through solar, geothermal, etc. According to Liu Hanzhang (2019) solar water heating system can save more than 10 kWh of thermal energy per day per household on average. [7] Ground source heat can overcome the obstacles to heat pumping from air sources and is therefore more efficient. It can provide heat in the winter through the heat pump for intelligent building, and at the same time accumulate part of the coldness preparing for the summer; and summer can make the building's heat transfer back to the earth, while the earth to accumulate a certain amount of heat and then for winter use. This helps to reduce the energy consumption of the air conditioning system to control the temperature.

4.2.3 Water consumption

Efficient utilization of water resources can be accomplished through water recycling and utilization techniques. Utilizing the rainwater collection system, rainwater is efficiently collected and then utilized. Domestic wastewater is collected, filtered and decontaminated for irrigation or recycled as green water for intelligent buildings. Water recycling can reduce the sewage load and sewage costs, making efficient use of water resources.

4.3 Carbon footprint monitoring and assessment

Carbon footprint assessment and detection can be accomplished by constructing a BIM-LCA system.

BIM technology can be used to record the carbon footprint. At present, BIM technology can store all data in the whole process of construction into the database and realize the sharing and transmission of information. The government can connect the BIM software of each logistics service enterprise with the management platform, and save the data of the whole process of the building, including the data related to carbon emissions, into the database of the platform. Based on the data related to carbon emissions, a carbon footprint assessment system can be established to record and assess the carbon footprint in real time.

The LCA (Life Cycle Assessment) method can be used for carbon footprint assessment. LCA is a method used to assess the environmental impact of a product, service or process throughout its life cycle, and is considered by the European Commission to be the most appropriate method for measuring environmental impact. [8] LCA often consists of four phases: goal and scope definition, life cycle invention, environmental impact assessment and interpretation. [9] Companies usually start by sorting out and counting data on the manufacturing paths of their products, verifying the energy used by their suppliers for the production of raw materials, calculating the emissions from transportation processes, and then defining and calculating the energy used within their own factories.

5 Innovative Strategies for Property Management

5.1 Policy and regulatory support

CHEN Si-qin (2009), pointed out that there are economic externality problems in green building, which need to be regulated by policy incentives.[6] In the case of economic externalities, there is a contradiction between the needs and interests of various subjects, and the market is not able to effectively allocate resources, which requires certain policy interventions to adjust.

Requiring property service enterprises to disclose information on energy consumption and carbon emissions on time, and evaluating or rewarding or penalizing the effects of emission reduction can help promote low-carbon transformation of enterprises. At the same time, it can also improve the enterprise carbon trading policy, after the disclosure of relevant information for carbon quota trading, help to incentivize enterprises to save energy and reduce emissions.

In the process of exploring the green and low-carbon upgrading of property management modes, property service enterprises need to invest a large amount of capital and thus have financing needs. The government can formulate green incentive policies to give financing support to enterprises that meet certain green operation standards. At present, there are building green bonds, REITs, green building loans and other financial policies, which play a role in stimulating and helping property service enterprises in green transformation.

The government can formulate tax reduction policies and provide tax reduction programs for enterprises that meet green and low-carbon standards. The government can also provide financial subsidies to enterprises that pass the operational scheme to support the implementation of operational management optimization.

The government can subsidize owners who purchase green operation buildings to stimulate the general public's desire to purchase green buildings, which can increase the market competitiveness of green and low-carbon property service enterprises, and at the same time achieve the purpose of promoting the green and low-carbon transformation of more property service enterprises.

5.2 Key strategies for enterprises and owner engagement

Mechanisms for cooperation between enterprises and developers. Coordination between preliminary building design and later property operation. The property service enterprises will interface with the relevant persons in charge at the stages of building materials selection, architectural design and building construction, and optimize the system levels and system-to-system connections as far as possible, so as to maximize the coordination with the later operations such as energy saving, intelligent regulation and control, and reduction of the later costs, and to avoid limitations of the later innovative property management caused by the poor design in the early stage.

Cultivation of green logistics operations management awareness for enterprises and owners. Publicize and guide enterprises and property owners on the concept of green property operation. Property service enterprises generally have the problem of not recognizing the concept of green property management. Owners do not pay much attention to property management, and only care about issues such as rent and house price, but lack awareness of property operation, not to mention the association between property operation and green low-carbon thinking. Therefore, we should help enterprises to understand the concept of green property operation and upgrade the system and technology instead of stopping at the surface of greening; we should publicize the concepts of green building and green operation to the owners, so that the owners can better cooperate with the enterprises and consider the green and low-carbon factors when they are renting housing.

Businesses communicate with owners to connect. Through the owners' committee, the property service enterprise and the owners can communicate with each other, break the information barrier, and discuss the decision-making together. Both play the role of owners of enterprises to supervise, but also make owners more understanding, cooperate with the work of enterprises, so that the property to realize the green operation.

5.3 Technical and human resources support

Apply big data and artificial intelligence technology to enhance the intelligent decision-making capability of the property management hub, and develop from the Informa ionization construction stage to the intelligent operation stage.

Utilize intelligent building technologies such as BIM technology, Internet of Things technology, intelligent weak power technology, cloud computing technology, etc., to build intelligent regulation and control system. Give full play to the visualization attributes of BIM technology and the ability to intelligently manage the whole cycle of the building; give full play to the advantages of Internet of Things technology in connecting multiple devices and timely transmission of various data; give full play to the management and control performance of intelligent weak power technology on multiple operating systems; and give full play to the advantages of cloud computing and big data technology in sharing and analysing the data of building resources.

Establish an organizational structure for green operation management and set up dedicated posts. Improve salary and remuneration to attract talents.

6 Case study: successful property management innovation practices

6.1 Typical Cases

Under the guidance of the national "dual-carbon" goal, Country Garden integrates the concept of sustainable development into the whole process of construction, and actively explores green and low-carbon transformation in the processes of green construction, green building, green construction, green operation and green leasing. Country Garden practices energy saving, emission reduction and consumption reduction, makes breakthroughs in key technologies of zero-energy and zero-carbon buildings, and upgrades platforms and systems through digitalization and intelligence to promote dual-control of energy consumption and carbon emissions and sustainable development.

Below is the introduction of Country Garden passive house representative project - Taizhou Nine Zhangs House: Taizhou Nine Zhangs House is the first large-scale passive low-energy green building district in Taizhou, and won the "13th Five-Year" National Key Research and Development Program of near-zero energy consumption residential building demonstration project. The project introduces low-carbon, energy-saving and green concepts from Germany, and has passed the Passive House Institute (PHI) Passive House Certification, China Passive Ultra-low Energy Green Building Certification, and China Green Building 2-star Certification.

The project adopts a variety of passive technologies: advanced thermal insulation of the external wall, through the thicker outer insulation material package, so that the heat transfer of the external wall is in a uniform and constant state; the use of doors and windows with good airtightness and heat insulation. The heat transfer coefficient is $\leqslant 0.85W/m^2K$, and the airtightness level should not be lower than the requirement of grade 8, so as to prevent heat loss as much as possible; adopt highly efficient new air heat recovery system, which integrates heating, cooling, heat recovery, dehumidification, haze removal function, and air-source heat pump, with 24-hour constant temperature, humidity, and oxygen; adopt the technology of no cold and heat bridge; improve the overall airtightness of the building, such as adopting high airtightness and high waterproof external windows + external hanging installation + tarpaulin method.

Due to the characteristics of building materials and operation system, passive building can ensure fresh air, low noise, no humidity, maintain the appropriate temperature for human body, reduce the frequency of air conditioning and heating, and reduce the use of energy by 90%.

6.2 Analysis of lessons learned from successful cases

Selected as one of China's top 100 property service companies for many consecutive years, Country

Garden's various initiatives in green property management are worthy of reference.

6.2.1 Saving energy and improving energy utilization efficiency

a. Country Garden improves high-energy-consuming equipment. The hotel lighting equipment saves 5,000MWh of electricity in a whole year.

b. The application of new magnetic levitation central air-conditioning technology saves 150kWh of electricity in the central air-conditioning system and 1,300kWh of electricity in the whole lifecycle.

c. Photovoltaic power generating panels are laid on the roofs of the hotels to make full use of solar energy.

6.2.2 Saving water resources and avoiding water pollution

a. Use sanitary ware with high water efficiency and equipped with water-saving pipes and fittings;

b. Select corrosion-resistant and durable materials for pipes and valves with good tightness.

c. Develop rainwater collection and utilization system and use the recycled rainwater for public water use in construction sites.

d. Adopt water-saving system combining sprinkler irrigation and drip irrigation, and improve the utilization rate of water resources for greening and irrigation in the public areas.

e. Residents' domestic sewage is disinfected and treated by a three-stage septic tank and reaches the tertiary discharge standard before being incorporated into municipal sewage pipelines.

6.2.3 Establishing digital management platforms

a. Construct IT platforms, such as the development and operation integration platform, automated testing platform and other platforms to manage the whole process of the project and improve the integration level of the project life cycle.

b. Build RPA robots to realize the integrated and digital management of multiple business scenarios, such as land resource information and business data statistics.

c. Build the one-stop operation platform which enables one-stop operation of project management and improve management efficiency.

6.2.4 Establishing an open, transparent and sustainable supply chain

a. Work closely with different types of suppliers, such as architects, designers, service providers, contractors, etc., to ensure the integration of projects from development and design to operation and maintenance;

b. Formulates a series of systems, such as "Supplier Resource Management Process", "Supplier Penalty Mechanism", etc., to continually inspect and assess the hardware facilities and operational capabilities of suppliers;

c. Sets up the internal control group to continuously improve the process of supply chain responsibility fulfilment.

7 Suggestions for promoting innovation in property management in a dual-carbon context

7.1 Policy recommendations

Improve the green operation management system by improving the green operation management indexes and building the evaluation system of green operation management; appropriate economic subsidies are supposed to be given to consumers who purchase green buildings; green channels can be opened for enterprises that carry out green operation and management, and preferential tax and credit policies will be adopted, and risks will be shared with enterprises. [10] compulsory information disclosure to improve the transparency of operation and management information and open the reward and punishment mechanism; designate specialized departments or encourage third-party enterprises to conduct green operation and management performance evaluation and monitoring.

7.2 Advice for enterprises and owners

Solid multi-party partnership by establishing a strategic alliance and reach a consensus on cooperation with architects, designers, service providers, contractors, etc. for front-end operation and management design; Outsource non-core business of an enterprise to devote limited resources to the iterative upgrading of core business; experience sharing by building a green operation and management information sharing platform to share technology and management experience; deepen conceptual understanding of the green operation concept and internal transformation; strengthen communication between enterprises and owners that Enterprises publicize green operation concepts and proactively disclose operation data to owners, and owners monitor enterprises.

7.3 Technical and research recommendations

Improve the theoretical framework of green management, explore the correlation and improvement space within the operation management system, and overcome the systematic problems of green property management; Upgrade energy-saving and emission reduction technologies and materials and model the use of them; apply artificial intelligence to improve intelligent decision-making capabilities and seamlessly connect platforms with real-world scenarios; encourage industry-university-research linkage that enterprise practice and university research complement each other's strengths, create talent training bases, and encourage colleges to send technical talents to enterprises.

References

1. Research Report on Energy Consumption and Carbon Emissions of Buildings in China, Construction and Architecture 2, 57-69(2023)
2. Y. Han, K. Kikuta, M. Hayashi, Research on the Whole Lifecycle Emission Reduction Effect of Buildings with Different Structures in Severely Cold Regions—A Case Study in China. Energies **16**, 5285(2023)
3. L. Zhao, Green property management development accelerates under "dual-carbon" targets. Housing & Real Estate **28**,38-39(2021)
4. Z. Chen, Practicing a green and low-carbon property management model that contribute to carbon neutrality. Property Aspect **7**, 94-95(2022)
5. S. Chen, Life Cycle Cost-benefit Analysis of Green Building Based on Economic & Development Countermeasures. Ecology & Intelligent buildings **37**, 53-56(2009)
6. R. Wang, Research on the application of intelligent building technology in modern construction engineering. Intelligent City. Intelligent City **9**, 26-28(2023)
7. H. Zhang, Application and discussion of renewable energy in intelligent buildings in Hunan province. Green Environmental Protection Building **10**, 32—35(2019)
8. M. Lavagna, A. Campioli, A.D. Valle, S. Giorgi, Design Strategies Toward Low-Carbon Buildings and Neighborhoods. The Use of LCA to Support a Project Proposal for Reinventing Cities. Lecture Notes in Networks and Systems, vol **482**(2022)
9. S.M. Jordaan, LCA Framework, Methods, and Application. Wells to Wire, 13-29(2021)
10. G. Liu, Y. Jie, Green building operations management in the context of carbon neutrality innovation path and strategy research. Construction Economy **43**, 98-104(2022)

Effect of Green Finance on Industrial Carbon Emissions in China ——Empirical Analysis Based on Provincial Panel Data

Yingying Zhou[1*], Zhuoran Zhou[1], Youxin Zhang[1], Yilin Chang[2]

[1]China University of Mining and Technolog, School of Economics and Management, 221116 Xuzhou, China
[2]Zhongnan University of Economics and Law, Wenlan School of Business, 430073 Wuhan, China

Abstract. Based on the Environmental Kuznets Curve (EKC), this paper empirically analyzes the impact of green finance development on industrial carbon emissions in China by using the panel data of Chinese mainland province. It is found that the development of green finance has significantly suppressed the industrial carbon emissions in China. Heterogeneity test shows that the inhibition effect on carbon emission in central China is the most obvious, and the inhibition effect on carbon emission in eastern and western regions decreases in turn. Technological progress significantly inhibits carbon emissions, especially in central China, followed by the western region and finally the eastern region. It is suggested to improve the green and low-carbon financing system, support the optimization of energy consumption structure and guide substantive technological progress, so as to promote the realization of carbon emission reduction targets.

1 Introduction

With the deepening of industrialization, a large number of greenhouse gases, mainly carbon dioxide, are emitted, and it is an indisputable fact that the global temperature is rising. Green finance guides all economic entities to pay attention to the natural ecological balance through their own activities, which is a financial service for improving the environment, saving resources and coping with climate change, and helps to achieve the goal of "double carbon". Since the issuance of "Guiding Opinions on Building a Green Financial System", local governments have made concerted efforts to jointly build a green financial framework. It is widely believed that combining green finance with environmental policies can effectively control the level of environmental pollution, promote the greening and sustainability of economic development. On one hand, policies related to green finance guide financial institutions to provide loans to environmentally friendly businesses, thereby promoting the development of the green economy and reducing carbon emissions. On the other hand, green finance can stimulate innovation in green technologies, improve energy equipment at the provincial level, and reduce carbon emissions.With so many inputs and outputs, does the development of green finance actively and effectively curb carbon emissions? Based on the data of 30 provinces in mainland China from 2005 to 2020, this paper explores the actual impact of green financial development and technological progress on industrial carbon emissions in China by building a panel data regression model.

2 Literature review

At present, there are many research results on the influencing factors of carbon emissions, including economic growth, industrial structure and financial development. Among them, scholars from all walks of life hold different opinions on financial development.

Green finance and carbon emission reduction are the relationship between ways and purposes, so the impact of green finance development on carbon emission reduction has become a highly valued issue in economics. Bai et al. (2022), Liu Feng et al. (2022) and Li Yunyan et al. (2023) believe that there are regional differences in the role of green finance in carbon emissions. Bai et al. (2022) think that green finance has become an important means to reduce carbon emissions in the eastern region, and the impact of green finance on carbon emissions in the central region is inverted U-shaped, while the driving force of green finance on carbon emissions reduction in the western region is weak ([1]).

Liu Feng et al. (2022) think that green finance has the most obvious effect on carbon emissions in the central region, but not in the eastern region. Li Yunyan and Zhang Shuo (2023) think that economic development has a positive regulatory effect on the carbon reduction effect of green finance, and the carbon reduction effect of green finance in developed regions is stronger than that in underdeveloped regions.Yasir et al. (2023) utilizing the Delphi and fuzzy Analytical Hierarchy Process(FAHP) method specify that green finance development is a very crucial strategy to minimize carbon emissions. Muhammad

* Corresponding author: zhouyingying@cumt.edu.cn

Umar et al. (2023) pointed out that focusing on green finance and green innovation can help achieve sustainable environmental goals ([2]). Md Al Mamun et al. (2023) used a large sample of 46 countries to verify that green finance significantly reduces carbon emissions in the short and long term ([3]).

Mao Yanjun, Qu Yingbo (2023) and Chen X et al. (2021) think that there is spatial spillover effect in the development of green finance. Mao Yanjun and Qu Yingbo (2023) pointed out that the development of local green credit can not only significantly inhibit local carbon dioxide emissions, but also have a significant negative impact on neighboring carbon dioxide emissions through spatial spillover effect. Chen X et al. (2021) proposed that the development of green finance can not only reduce the carbon emissions in this region, but also restrain the carbon emissions in neighboring regions.

To sum up, it is of theoretical and practical significance to study the relationship between green finance development and carbon emission level.

3 Model construction, data source and variable measurement

3.1. Model Building

Grossman and Krueger first pointed out the "inverted U" relationship between pollutants (sulfur dioxide and dust) and per capita income level. Panayotou (1993) thought that the relationship between economic development and environmental pollution showed an "inverted U" curve, and named it Environmental Kuznets Curve (EKC). Based on the analysis framework of environmental Kuznets curve (EKC) and referring to the practice of scholars such as Liu Feng and Huang Ping (2022), this paper sets the following model, simplifies the logarithmic quadratic function, eliminates the strong fluctuation of data, and tests the impact of green finance on carbon emissions. Equation 1 is the model constructed in this study.

$$\ln co_{2it} = c + \alpha_1 \ln gf_{it} + \alpha_2 \ln gdp_{it} + \alpha_3 \ln rd_{it} + \alpha_4 \sum Z_{it} + Year_t + \mathrm{Re}gion_i + \varepsilon_1 \quad (1)$$

In Equation 1, it represents the carbon emission intensity of the region in the first year and represents the development level of green finance. It means the level of economic development, which means that the R&D expenditure in the year accounts for the local GDP in that year, and it is a set of control variables that affect the explained variables. It is a constant term and a random disturbance term. In addition, year ($year_t$) fixed effect and city ($Region_i$) fixed effect are added to the model to control the influence of unobservable factors in the year and city. The control variables mainly include urbanization level (city), environmental regulation (reg) and industrial structure (ind), proportion of local fiscal expenditure (cz) and proportion of foreign investment (FDI).

3.2. Data sources and variable measurement

Carbon emission data comes from China Carbon Accounting Database (CEADs), and other data come from China National Bureau of Statistics, Securities Times, Times Finance, China Insurance Yearbook, and statistical yearbooks of 32 statistical bureaus such as Beijing Municipal Bureau of Statistics, Tianjin Municipal Bureau of Statistics, Hebei Provincial Bureau of Statistics and Shanxi Provincial Bureau of Statistics. The implementation effect of national policies is often influenced by the strict constraints and effective implementation of provinces, so this paper uses the samples of 30 provinces in Chinese mainland (Tibet Autonomous Region, Hong Kong Special Administrative Region, Macao Special Administrative Region and Taiwan Province are not selected for research). This paper selects the data of 30 provinces in China from 2005 to 2020 for empirical test.

The development level of different provinces is different, so the comparability of directly comparing carbon dioxide emissions is relatively low. Therefore, this article characterizes carbon emissions through carbon emission intensity (carbon emission efficiency), and measures the proportion of carbon dioxide emissions in each province to per capita GDP in the region. It represents the carbon dioxide (CO_2) released per unit of GDP per capita produced, with a unit of 100000 tons per yuan. The Green Finance Index (gf) considers data availability and constructs measurement indicators from four dimensions: green credit, green insurance, green investment, and government support for green industries. The urbanization level (city) is measured using the proportion of urban population in each province to the total urban population; The level of economic development is characterized by provincial per capita GDP unit: 10000 yuan (gdp); The environmental regulation variable (reg) is characterized by industrial sulfur dioxide emissions at the provincial level. The industrial structure (ind) uses the proportion of the secondary industry at the provincial level to GDP. R&D expenses (rd) are represented by the proportion of provincial-level R&D expenses to local GDP for the current year. The proportion of local fiscal expenditure (cz) is represented by the proportion of provincial level fiscal expenditure to local GDP in the current year. The proportion of foreign investment (FDI) is measured by the ratio of the total output value of foreign-invested industrial enterprises to the total industrial output value of the city. Table 1 is a statistical description of each variable.

Table 1. Statistical description of variables

variable	Variable name	average number	Standard deviation	minimum value	maximum
CO_2	*Carbon emission (carbon intensity)*	*1.1072*	*0.8224*	*0.0700*	*4.5348*
gf	*Development level of green finance*	*0.1607*	*0.1031*	*0.0441*	*0.8390*
city	*Urbanization level*	*0.5516*	*0.1401*	*0.2687*	*0.8960*

gdp	*Level of economic development*	*4.4429*	*2.8224*	*0.5052*	*16.4890*
FDI	*Proportion of foreign investment*	*0.0217*	*0.0198*	*0.0001*	*0.1210*
reg	*Environmental regulation*	*56.8539*	*45.2342*	*0.1800*	*200.2000*
ind	*industrial structure*	*0.4319*	*0.0770*	*0.2862*	*0.7315*
rd	*R&D expenses*	*0.0147*	*0.0111*	*0.0018*	*0.0644*
cz	*Proportion of local fiscal expenditure*	*0.2253*	*0.0995*	*0.0792*	*0.6430*

4 Empirical analysis

Import the data into STATA, log the data with gen program, perform regression fitting on the log data, and control the time effect of fixed effect. Table 2 shows the benchmark regression results of the effect of green finance on carbon emission reduction in industrial sectors.

4.1. The Benchmark Model Regression

The template is designed so that author affiliations are not repeated each time for multiple authors of the same affiliation. Please keep your affiliations as succinct as possible (for example, do not differentiate among departments of the same organization). This template was designed for two affiliations.

Table 2. Benchmaek regression of carbon emission reduction effect of green finance

	(1)	(2)	(3)	(4)
	basic model	*Explanatory variable Lag one period*	*Urbanization, industrial structure and R&D expenses lag behind by two periods.*	*Final model*
lngf	-0.2529** (0.09)	-0.1885* (0.09)	-0.2642** (0.10)	-0.3521** (0.11)
lncity	0.1362 (0.15)	-0.0770 (0.16)	0.3309* (0.15)	0.3869* (0.17)
lngdp	-0.5953*** (0.05)	-0.5392*** (0.06)	-0.6720*** (0.06)	-0.6735*** (0.07)
lnFDI	-0.05** (0.01)	0.0535*** (0.02)	-0.0385** (0.01)	-0.0348* (0.01)
lnreg	0.0588*** (0.02)	0.0694*** (0.02)	-0.0593*** (0.02)	0.0576*** (0.01)
lnind	-0.0013 (0.09)	0.1839* (0.09)	0.1399 (0.09)	0.2293* (0.09)
lnrd	0.1644** (0.051)	0.0810 (0.05)	0.1673*** (0.05)	0.1731*** (0.05)
lncz	0.2761*** (0.07)	0.1170 (0.08)	0.2273** (0.07)	0.2650*** (0.08)
C	0.9046** (0.28)	0.3352 (0.31)	1.2405*** (0.32)	1.3104*** (0.36)

Note: *, * * and * * mean significant at the levels of 5%, 1% and 0.1% respectively.

As can be seen from Table 2, after fixed-effect regression of years and regions, the regression results are significant except for industrial structure (ind) and urbanization level (city), and the F test is significant, rejecting the original hypothesis and thinking that the fixed-effect regression effect is better than the mixed regression result. After random effect regression, the explanatory variable t test decreased significantly, and BP test was carried out, and the test result p value was 0. In this paper, hausman test is used to choose fixed effect model or random effect model. The test results show that the P value is 0, far less than 0.001, so the original hypothesis is rejected and the fixed effect model is selected for regression analysis. The t-test for industrial structure (ind) and urbanization level (city) is not significant. Considering that the transformation of industrial structure and urbanization process may have a lag effect on carbon dioxide emissions, all explanatory variables are taken as lagged and fixed effect regression is performed again. The proportion of urbanization level, R&D expenses, and local fiscal expenditure in the explanatory variables with a lag of one period is still not significant. Therefore, the analysis of urbanization level, industrial structure, and R&D expenses with a lag of two periods is conducted.

4.2. Model determination and interpretation

Model interpretation is based on the benchmark regression results in Table 2, and column (4) in Table 2 is selected as the final model interpretation variable coefficient. This model eliminates the interference of urban and time-invariant factors by taking urbanization level (city) as two periods, industrial structure (ind) as three periods, R&D expenditure (rd) as two periods and local fiscal expenditure (cz) as one period. At the same time, this paper adds urban fixed effect and time fixed effect. Column (4) of Table 2 shows that the influence coefficient of green finance on carbon emission reduction is -0.3521, which shows that green finance contributes to carbon emission reduction through 1% significance test. Economically significant, the probability of carbon emission reduction will increase by 0.3521 percentage points for every 1 percentage point increase in the green financial development index. Green finance supports the development of energy conservation and environmental protection, clean coal, carbon emission reduction and other fields through the role of capital orientation and environmental information disclosure, and promotes the green transformation of market players to help save energy and reduce carbon. The influence coefficient of urbanization level (city) is 0.3869, and the significance test of 5% shows that the improvement of urbanization level will increase carbon emissions.

The influence coefficient of GDP per capita is -0.6735, and the significance test of 0.1% shows that GDP growth contributes to carbon emission reduction and provides more financial support for green finance. It is worth mentioning that the fifth explanatory variable, that is, the development of industrial structure lags behind by three periods, and the reason why this variable needs to be selected by one period has been discussed above. The effect of carbon emission reduction is negatively affected by the fact that the industrial structure is three stages behind, and its influence coefficient is 0.2293, which is positive and passes the significance level test of 5%, indicating that the higher the level of industrial structure development, the lower the possibility of carbon emission reduction. From the economic point of view, every 1 percentage point increase in the development index of green industrial structure will reduce the probability of carbon emission reduction by 0.2293 percentage points. The reason may be that with the development of industrial structure, the demand for carbon emission will further increase, but at the same time, the carbon emission reduction technology has not been able to develop with it, resulting in the carbon emission reduction effect being lower than the demand for carbon emission, which is finally reflected in the decrease of carbon emission reduction effect with the increase of the proportion of industrial structure, that is, the secondary industry.

4.3. Heterogeneity analysis

The economic development of provinces in China is unbalanced, and the impact of green finance on carbon emissions may have regional differences. This paper will be divided into eastern, central and western regions to test the impact of green finance on regional heterogeneity of carbon emissions. Specifically, the virtual variable (virtual 1) is set, and the western region is set to 2, the central region is set to 1 and the eastern region is set to 0, thus generating three interactive variables, namely 0 (the eastern province of China), 1 (the central province of China) and 2 (the western province of China). At the same time, the interaction model is used to verify and observe three interaction variables (Table 3).

Table 3. Heterogeneity Analysis of the Impact of Green Finance on Carbon Emissions in Eastern, Central and Western Provinces of China

variable	coefficient	Standard error
city	*-1.7123****	*0.24*
gdp	*0.2289****	*0.11*
FDI	*-0.0769***	*0.03*
reg	*0.4148****	*0.03*
ind	*1.0752****	*0.19*
rd	*0.2105****	*0.06*
cz	*-0.2311**	*0.11*
0	*-0.2940**	*0.12*
1	*-0.3416***	*0.11*
2	*-0.1213*	*0.12*
c	*-2.4102****	*0.44*

Note: *, * * and * * mean significant at the levels of 5%, 1% and 0.1% respectively

As can be seen from Table 3, the influence coefficient of green finance in the eastern region is -0.2940, which has passed the 5% level significance test; the influence coefficient of interaction item in 1 (central province of China) is -0.3416, with a p value of 0.002, which has passed the 1% level significance test; the influence coefficient of interaction item in 2 (western province of China) is -0.1213, and the difference between the influence coefficient of interaction item in 1 (central province of China) and that in 2 (western province of China) is the largest. There are great regional differences in the carbon emission reduction effect of green finance development between the central provinces of China and the western provinces of China, and the carbon emission reduction effect of green finance development is more obvious in the central provinces of China. Combined with the eastern provinces of China, compared with the western provinces of China and the central provinces of China, the green financial index coefficient of the eastern provinces of China is slightly lower than that of the central provinces of China, but it is still much higher than that of the western provinces of China, indicating that the central provinces of China also have better carbon emission reduction effects.

5 Conclusions and suggestions

Based on the panel data of 30 provinces in Chinese mainland from 2005 to 2020 (excluding Tibet, Hong Kong, Macao and Taiwan due to lack of data), this paper discusses the relationship between green finance, industrial restructuring and carbon emissions, and draws the following conclusion: the development of green finance in China has a significant inhibitory effect on carbon emissions. Heterogeneity test shows that there are regional differences in the above inhibition effects, and the central region has the most obvious inhibition effect on carbon emissions, followed by the eastern and western regions. It is suggested to improve the green and low-carbon financing system; Improve the supervision system of green finance, standardize the green financial market, strengthen risk management and prudent supervision, and guard against risks in the field of green finance. Formulate policies according to local conditions and regions; Provide relevant policy and financial support for technologies that reduce carbon emissions, and promote the development of carbon emission reduction technologies ([4-5]).

6 Discussion

The effect of green finance on industrial carbon emissions in China based on provincial panel data has been analyzed in this paper. Meanwhile, other contents, such as indirect impact channels and spatial spillover effects, are worth further in-depth research. It is also should be paid attention to how to increase medium and long-term financial support, and promote substantive green innovation for enterprises. Financial institutions should continue expanding the scale of green credit, set targets for green loan growth, and innovate green financial products

Acknowledgments

This research was funded by the Fundamental Research Funds for the Central Universities, grant number 2023SK04.

References

1. Bai J , Chen Z , Yan X. "Research on the impact of green finance on carbon emissions: evidence from

China,"Economic Research, 2022, vol.35, pp.6965-6984.

2. Muhammad Umar and Adnan Safi, "Do green finance and innovation matter for environmental protection? A case of OECD economies," Energy Economics, 2023, vol.119, pp.1-11.
3. Md Al Mamun, Sabri Boubaker, Duc Khuong Nguyen, "Green finance and decarbonization: Evidence from around the world," Finance Research Letters,2022, vol.46, part B, pp.1-7.
4. Li, C., Yasir, A. S., & Sharafat, A. "Evaluating the factors of green finance to achieve carbon peak and carbon neutrality targets in China,".Sustainability, 2023, vol.15, pp.1-22.
5. Gao Y, Shen Z. "Carbon emission reduction effect of green financial reform policy," China Environmental Science, 2022, vol. 42, pp. 4849-4859.

Transition towards carbon neutrality: A structured review on current policies and measures in international shipping

Xiaofang Wu*, Yunqi Hu, Mengying Chen and Zeyu Jiang

College of Harbour and Coastal Engineering, Jimei University, Xiamen, 361021, PR China

Abstract. Current studies lack a comprehensive understanding of the status of carbon emission reduction in international shipping from the perspectives of policies and measures, which tends to increase uncertainty in policy-making. The study aims to use a structured review approach to systematically collect and analyse carbon reduction policies as well as technical, regulatory, and economic measures in international shipping in recent years to support international shipping carbon neutrality decisions. The results show that most regions are committed to achieving zero emissions by 2050, but the reduction targets of international maritime organizations are lagging; while hydrogen, ammonia and green methanol fuels and ship wind power have a high potential to advance emission reduction, they are still limited by technology, regulation, cost and support; market-based trading measures can reflect emission reduction targets, but carbon prices are not yet mature; other measures also suffer from emissions shifting, low emission reduction potential or unsustainability. Tracking the progress of policies and measures to reduce carbon emissions in international shipping benefits to "know yourself as well as the enemy" for supporting decisions.

1 Introduction

International shipping, which is responsible for more than 80% of international trade freight, accounts for 2%-3% of global greenhouse gas emissions [1]; according to the International Maritime Organization (IMO) Fourth Greenhouse Gas Study, carbon dioxide (CO_2) emissions from international shipping will still increase by 50% in 2050 compared to 2018 and by 90-130% compared to 2008 while maintaining current policy levels [1].

To achieve the United Nations 2030 Sustainable Development Goals and the temperature control targets from the Paris Agreement, various initiatives have been put forward to reduce carbon emissions from international shipping in recent years, such as the IMO published the Initial IMO Strategy on Reduction of GHG (greenhouse gas) Emissions from Ships in 2018; the Ministry of Transport of the People's Republic of China proposed to reduce carbon emissions per unit of transport turnover of operating ships by 3.5% in 2025 compared to 2020 [2]; the European Union proposed the "FuelEU" maritime initiative to set maximum limits on the GHG content of ship fuel [3]. However, has the policy pace of carbon reduction in shipping been consistent across regions in recent years and what are the differences? How about the corresponding measures against the backdrop of continuous updating of shipping carbon reduction policies?

Currently, only a few scholars kept eyes on relevant policies and measures to reduce carbon emissions from shipping, e.g., Bouman, Lindstad [4] and Balcombe, Brierley [5] attempted to assess the potential of emission reduction measures; Serra and Fancello [6] provided a preliminary discussion of possible scenarios and challenges for reducing emissions from shipping by 2020; Xing, Spence [7] overviewed the concepts, approaches, characteristics, barriers, and prospects of abatement measures for ships. However, they are fragmented or outdated in understanding the abatement policies and measures for shipping carbon emissions. This study, therefore, uses a structured review method to systematically collect relevant information and analyze the abatement progress across regions. It will help decision-makers and scholars to gain a comprehensive understanding of carbon emissions reduction strategies in international shipping, and provide a knowledge base for carbon neutrality in shipping.

2 Methods and materials

A literature review has been regarded as an important tool for knowledge management, helping to collect and assess existing knowledge and identify research questions to develop and expand relevant knowledge to support decision-making [8]. A structured, procedural, replicable literature review method contributes to exploring knowledge scientifically and rigorously [9]. This study therefore uses the structured review method to systematically collect and analyse the current status of carbon reduction policies and measures in international

* Corresponding author: xiaofang.wu@jmu.edu.cn

shipping for knowledge management. The first step is to collect and refine the information from literature databases and relevant official websites by using keyword searching and then extracting the key content for further analysis [10].

In this study, the databases of the Web of Science (SCI) and China National Knowledge Infrastructure (CNKI), which cover a wide range of literature in both English and Chinese, are selected. The keywords "maritime OR shipping OR marine transportation AND carbon" and "海运+航运+海洋运输 AND 碳排放+碳减排+低碳+零碳+脱碳" are respectively used for the SCI and CNKI abstract searching. The time range was set in 2018 because it is an important phase following the release of the IMO's initial reduction strategy and it can help to verify the process and effectiveness of the strategy's implementation. After excluding the News and Meeting Abstract and those papers belonging to research areas of physical sciences, life sciences, and arts and humanities and merely focusing on aviation, inland waterways, naval vessels, or submarine cables without mentioning shipping policies and measures based on their titles and abstracts, there were 83 articles from SCI remained for further studies on May 31, 2022; the CNKI refinement, which excluded the News and abstracts and those papers focusing on aviation, inland waterways, Sulphur emission, or black carbon, without mentioning shipping policies and measures according to their titles and abstracts, kept 102 articles on the same day. On the other hand, related international and regional organizations are also selected and searched based on experts' consultations (shown in Table 1).

Table 1. Relevant organizations for data collection.

International	IMO
	International Transport Forum (ITF)
	Global Maritime Forum (GMF)
	International Council On Clean Transportation (ICCT)
	International Chamber of Shipping (ICS)
Regional	European Federation for Transport and Environment (T&E)
	European Commission
	United Kingdom government
	Ministry of Transport of the People's Republic of China
	United States Environmental Protection Agency
	Maritime and Port Authority of Singapore
	Ministry of Land, Infrastructure, Transport and Tourism, Japan
	Ministry of Oceans and Fisheries, South Korea
	Norwegian Shipowners' Association
	Lloyd's Register of Shipping
	American Bureau of Shipping
	China Classification Society
Consultancies	University Maritime Advisory Services (UMAS)
	CE Delft
Companies	Maersk
	Wärtsilä
	ExxonMobil

As initial results, we can find the literature from databases paid much attention to a variety of specific measures on carbon emission reduction for shipping such as alternative fuel use [11, 12], carbon capture technologies [13, 14], energy efficiency management [15], and carbon tax [16]. Both topics from the two databases are very similar. As for the organizations' knowledge, we can see the IMO guides the carbon emission reduction from shipping by proposing not only specific abatement targets but also short, medium and long-term measures [17]; the ITF, an international inter-governmental organization in the transport sector, also published a pathway towards zero-carbon shipping [18]; other international organizations including GMF, ICCT, and ICS discussed the policies and potential options for approaching the carbon reduction ambition [19-21]. Additionally, regional organizations aim to issue detailed policies or measures as well as abatement routes [22] and consultancy organizations represented by CE Delft and UMAS have carried out much research in recent years to support them [3, 23, 24]. Companies mainly integrated shipping carbon reduction into their corporate social responsibility reports.

3 Results

Based on the above data collected, this study summarizes the carbon emission reduction policies in international shipping since 2018 as well as specific measures from technical, management, and economic aspects based on their characteristics. The technical measures focus on improving the abatement technology, process, equipment, or materials; the management ones tend to help achieve the carbon emission reduction targets through planning, organizing, coordinating, and controlling; and the economic measures are characterized by market or economic pathways.

3.1. Current policies for carbon abatement in international shipping

3.1.1 International policies

In 2018, the IMO set an initial target to reduce the greenhouse gas emissions from ships, namely to reduce the global carbon intensity of shipping by at least 40% by 2030 compared to 2008, and to strive for a 70% reduction by 2050; and to reduce total greenhouse gas emissions from shipping by at least 50% by 2050 compared to 2008, and to strive for zero greenhouse gas emissions in this century [25].

3.1.2 Regional policies

Regional policies on carbon abatement in international shipping can be shown as follows:

- Europe: In recent years, through the European Green Deal, the European Commission has proposed the EU's greenhouse gas emissions will be at least 55% of the 1990 level by 2030 and carbon neutral by 2050 [26].

- UK: The UK government encouraged investment in maritime infrastructure to promote the use of carbon-neutral fuels and renewable energy to ensure that the IMO target of a 50% reduction in maritime GHG emissions by 2050 can be achieved [27].

- China: In September 2020, China announced its "Carbon Peaking, Carbon Neutrality" target, i.e., peaking carbon emissions by 2030 and achieving carbon neutrality by 2060.

- USA: The United States government rejoined the Paris Climate Agreement in January 2021 and proposed to reduce the GHG emissions by 50-52% of 2005 levels by 2030; achieve a 100% carbon-free electricity by 2035 and a net-zero economy by 2050 [28].

- Singapore: In March 2022, the Singapore Maritime Decarbonization Blueprint: Towards 2050, published by the MPA, proposed a transition to a low-carbon future through the use of cleaner energy, automation and digitalization by 2030; and a net-zero emission economy by 2050 [29].

- Japan: In March 2020, through the release of the Roadmap to Zero Emissions from International Shipping, Japan proposed that by 2030, the carbon intensity of international shipping will be reduced by at least 40% compared to 2008; by 2050, total annual GHG emissions from international shipping will be reduced by at least 50% compared to 2008; and as soon as possible in this century, GHG emissions from international shipping will cease [30].

- South Korea: In 2020, the South Korean government released the 2030 Green Ship-K Initiative to enable to gradual reduction of 70% of GHG emissions in 2030 in ship design, future fuels, renewable energy, and equipment by exploring advanced emission-free technologies emissions by 2030, and towards net zero emissions by 2050 [31].

- Norway: Norway has adopted a national plan to reduce GHG emissions by 40% in 2030 compared to 1990 and to become a low-emission society by 2050 [32]. In addition, members of the Norwegian Shipowners' Association will reduce GHG emissions from ships by 50% by 2030 compared to 2008, order only ships with zero-emissions technology from 2030 onwards, and have a zero-emissions fleet by 2050 [33].

3.2 Current measures for carbon abatement in international shipping

3.2.1 Technological measures

The existing technical measures to reduce carbon emissions from international shipping can be presented as follows:

(1) Alternative fuel technologies

- Introduction: Current alternative fuel technologies cover LNG, biofuels, methanol, hydrogen and ammonia [23].

- Implementation: By 2018, over $500 million had been invested in EU LNG offshore bunkering projects [34]; biofuels and methanol have been tried since around 2009 and 2015 respectively, while hydrogen and ammonia fuels from around 2019 and 2020 respectively [23].

- Advantage and disadvantage: On a life-cycle basis, the use of LNG can reduce emissions by 18% more than conventional fuels [11], but it's difficult to achieve the IMO's emission reduction targets [34, 35] and prone to leakage problems [36]; Conventional biofuels also do not reduce carbon emissions due to the destruction of forests and grasslands, but advanced biofuels can reduce carbon emissions significantly [37]. Grey methanol, produced from natural gas, produces more greenhouse gases, but green methanol, produced from the fusion of hydrogen and CO_2, is considered an important carbon-neutral option [37]. Hydrogen and ammonia are zero-emission fuels, but are flammable and toxic, posing safety risks [37].

- Barriers: Uncertainties in regulatory regimes, low technology maturity, high production costs, restricted supporting infrastructure and high retrofitting costs have hindered the progress of alternative fuel use [12, 23].

- Future pathways: The emission reduction effect of using LNG is considered to be temporary [34]. Although technologies for biofuel and methanol are developing rapidly, they do not provide sufficient emission reductions and are considered to be a "bridge" fuel [37]. It is expected that hydrogen and ammonia marine engines will be commercially available on a large scale by the mid-21st century [37].

(2) Engine technology

- Introduction: Reducing carbon emissions by limiting engine power [38] or increasing the thermal efficiency and intelligence of the engine [7].

- Implementation: Advanced and intelligent engine management, common rail or electronically controlled fuel injection, variable geometry turbochargers, fuel slide valve upgrades, and cylinder fuel consumption optimization are currently being used to improve engine efficiency [7]. Limiting engine power is also considered a simple way to meet EEXI requirements for older ships with few modifications to the ship and no impact on basic performance [38], and amendments to the engine power limitation guidelines have been approved by IMO [39].

- Advantage and disadvantage: The use of advanced and intelligent engine technology has now almost reached the limit of energy efficiency, with less than 1% potential for further fuel savings [7]; as current ship operating speeds are already well below maximum speed, CO_2 reductions are hardly proportional to engine power limits, and only a power limit of more than 50% will reduce CO_2 emissions from existing ships to below the expected 2030 level [38].

- Barriers: There is less scope for emission reductions with advanced and intelligent engines [7] and mandatory engine power limits [38].

- Future pathways: Flexible engine power options, such as hybrid and electric propulsion, may be considered in the future [7].

(3) Ship propulsion technology

- Introduction: Improve ship propulsion design [40] or use wind propulsion technology [7] to reduce energy consumption.

- Implementation: Shipping companies are focusing on improving the design of ship propulsion units [40]; wind propulsion is still in the early stages of development or testing [18].

- Advantage and disadvantage: Optimized design of ship propulsion units may improve energy efficiency by up to 15% [40]; wind energy has a high potential for fuel savings as a renewable energy source.

- Barriers: Wind propulsion technology is still immature and has high installation and operating costs [18].

- Future pathways: Studies predicted that wind propulsion systems will gradually gain acceptance from 2025 onwards, despite the relatively low maturity of the market [18].

(4) Carbon capture technology

- Introduction: The captured CO2 is liquefied and stored temporarily on board a ship and then transported to a storage site for sequestration or reuse [13].

- Implementation: This technology has now been tried on LNG vessels [13]. Post-combustion carbon capture processes have been explored [41], as well as membrane and solvent-based carbon capture technologies [14, 42].

- Advantage and disadvantage: The technology can be implemented in the short term and has the potential to reduce CO2 emissions from the maritime industry by 65% [13, 43].

- Barriers: The technology faces high investment costs, limited space on board, and a lack of full-scale industrialization [43].

- Future pathways: It is still in the development and demonstration phase but will play an important role in reducing emissions due to its significant emission reduction and ship suitability [44].

(5) Hull antifouling technology

- Introduction: Reducing hull roughness and improving energy efficiency in shipping [5].

- Implementation: Significant funding is currently being invested in antifouling paints to prevent bacteria from adhering to the hull to reduce hull drag [5].

- Advantages and disadvantages: Modern paints are divided into biocides, which tend to affect the environment, and biocide-free paints, which suffer from non-stick, relatively expensive and insufficient durability [5].

- Barriers: Existing coatings face challenges in terms of performance, cost, ease of application and ecological impact [5].

- Future pathways: More suitable coatings, such as fish mucus-based excavations, will be explored in the future [45].

(6) Shore power technology

- Introduction: Ships are powered by shore-based electrical systems while in port, shutting down auxiliary engines and reducing energy consumption [46].

- Implementation: Some dozens of ports currently have shore power facilities. 80% of the energy used by certain types of ships in port in 2020 will come from shore power or similar environmentally friendly sources [46].

- Advantage and disadvantage: The measure has the advantage of being implementable in the short term [46]. Studies have shown that the implementation of shore power can reduce carbon emissions by between 3% and 60% [46].

- Barriers: The implementation of shore power facilities is influenced by their compatibility, financial cost of construction, and public acceptance [46].

- Future pathways: Core EU ports are required to establish shore power or LNG fuel supply facilities by 2025, and other countries are following this measure [18], but the economic incentives and regulatory framework are still under discussion [7, 47].

Of these, the use of hydrogen ammonia fuels and green methanol, ship wind propulsion technology and use of non-hazardous paints have high potential for emission reductions, but are currently difficult to implement; engine technology, ship propulsion design and carbon capture technology have less potential for emission reductions and are short-term in nature; from a whole life cycle perspective, Liquefied Natural Gas (LNG) technology, biofuels, grey methanol and shore power technologies may be susceptible to other GHG emissions or emissions displacement and are characterized as transitional measures.

3.2.2 Management measures

The existing management measures to reduce carbon emissions from international shipping can be presented as follows:

(1) Scheduling optimization,

- Introduction: Optimizing the scheduling of ships, containers, berths, shore bridges, and yard bridges can reduce energy consumption by increasing efficiency [48].

- Implementation: Scheduling optimization algorithms incorporating carbon emission targets are still under development [7, 43]; there is a strong appetite for optimized scheduling among shipping companies and terminal operators [7].

- Advantage and disadvantage: It is beneficial to reduce ship turnaround times and improve service efficiency and competitiveness [7].

- Barriers: How to carry out multi-objective optimization is an obstacle to optimal scheduling in the context of carbon reduction [48, 49].

- Future pathways: Further improving the quality of the algorithm under uncertainty and optimizing scheduling from a multi-objective perspective [48].

(2) Speed optimization

- Introduction: Reducing carbon emissions by reducing the speed at which ships sail [7].

- Implementation: In a sluggish shipping market with excess capacity, slow sailing has been used as an important option to save fuel [7].

- Advantage and disadvantage: Since speed is a cubic function of power and fuel consumption, slow sailing can reduce fuel consumption [15], but long-term slow sailing can form carbon deposits to affect diesel engine efficiency and increase CO2 emissions [50]. Slow sailing can increase voyage time and reduce vessel transport workload, resulting in lost revenue [51]; it also has the potential to lead to an expansion of the world fleet and increased competition; and is less likely to promote technological innovations that reduce emissions [7].

- Barriers: Slow sailing is constrained by factors like voyage frequency, charter time, customer demand, and additional operating costs [7].

- Future pathways: Studies advised that slow sailing can be implemented within a regulatory framework based on fair market and voluntary action, but need not become mandatory [7].

(3) Fleet deployment and route optimization

- Introduction: Fleet deployment is about minimizing the time and cost of transport within environmental constraints [52]; route optimization considers the allocation of ships in the context of carbon emissions [53], and the selection of routes with minimal ship resistance and calm weather conditions [7].

- Implementation: Research has begun to explore multi-objective fleet deployment and route allocation planning models that incorporate carbon reduction [52, 53]; weather forecasting systems have also been widely developed and applied to shipping companies [7].

- Advantage and disadvantage: Fleet deployment and route allocation optimization models that take into account carbon emission reductions can contribute to low-carbon decision-making in shipping [52, 53]; reducing vessel resistance to navigation and integrating sea and weather conditions into routes can help save energy [7].

- Barriers: Optimization models and weather route forecasts that consider carbon emissions still have limitations [52, 54].

- Future pathways: The model and the accuracy of weather route prediction will be improved and practical applications will be attempted [52, 54].

(4) Energy efficiency management

- Introduction: Improving the energy efficiency of ships and saving energy to reduce carbon emissions [55].

- Implementation: This is carried out based on rules, including EEDI, Ship Energy Efficiency Management Plan (SEEMP), Energy Efficiency Operational Indicator (EEOI), Energy Efficiency Performance Indicator (EEPI), EEXI and CII [7, 55-57]. At this stage only EEDI and SEEMP are mandatory [7]; EEXI and CII will be implemented from 2023.

- Advantage and disadvantage: The rules-based energy efficiency management can contribute to global maritime governance [7]. However, the existing rules still do not maximize the efficiency of the global fleet [57]. For example, due to the prevalence of deceleration, the engine load of most ships is much lower than the engine load allowed by EEXI, so EEXI has little effect in reducing emissions [56]; due to the slow renewal of ships and the small proportion of new ships, EEDI implementation will only reduce emissions by about 2% by 2030 [7].

- Barriers: If EEDI is implemented on existing ships, it is limited by financial support, although it may bring 1%-6% emission reductions by 2030 [7]. While SEEMP is mandatory for new or existing ships on board, it currently relies on the voluntary setting of energy efficiency targets and self-monitoring [55]. Other targets have not been agreed due to a lack of commercial data or scientific value [55].

- Future pathways: In line with the IMO emissions reduction process, attention will be given in the short term to implement the existing ship EEXI and ship CII rating mechanisms [17]; and further refinement of EEDI and SEEMP [17].

(5) Monitoring and forecasting

- Introduction: Predicting and monitoring carbon emissions from shipping is used to support carbon reduction decisions.

- Implementation: Scenario analysis is used to predict carbon emissions from ships based on automated ship identification systems [58], multivariate interactive grey models [59], and energy consumption [60]. Besides relying on energy statistics, ship fuel consumption data collection systems and emission monitoring reports near real-time shipping emission monitoring are also being explored [61].

- Advantage and disadvantage: Predicting and monitoring shipping carbon emissions can help to quantitatively support international shipping carbon reduction decisions.

- Barriers: Data limitation and sensitivity, as well as the complexity, uncertainty, and accuracy of results may affect the reliability and effectiveness [61].

- Future pathways: The accuracy and reliability of forecasting and monitoring are continuously improved [62].

(6) Assessment and evaluation

- Introduction: Assessing the international shipping carbon reduction measures can help shipping carbon reduction decisions.

- Implementation: The reduction measures were assessed through techno-economics [63, 64], cost-effectiveness [65], life cycle [66], climate impact [23], and multiple objectives [67].

- Advantage and disadvantage: Assessing the shipping carbon reduction measures helps to judge their availability and adaptability, but the accuracy needs to be improved, and the activity-based ship emission inventories are underestimated [68].

- Barriers: Lack of basic data [69], uncertain parameters [66] and case study support [67] may affect the assessment and evaluation.

- Future pathways: Overcoming the barriers to the assessment and evaluation.

To sum up, most of these measures optimize the allocation of resources and improve energy efficiency. In contrast, the optimization of scheduling, speed, fleet, and route is more likely to be supported by stakeholders, but is limited by the accuracy and reliability of the model; the implementation of the commanding and mandatory energy efficiency management measures has pressure, but they are conducive to achieving phased progress in emission reduction; the monitoring and forecasting as well as assessment and evaluation are as supportive decisions tools.

 https://doi.org/10.1051/e3sconf/202344103025

3.2.3 Economic measures

The existing management measures to reduce carbon emissions from international shipping can be presented as follows:

(1) Emissions taxes and levies

- Introduction: A corresponding fee that is pre-set on fossil fuel consumption or CO2 emissions and requires market actors to pay for their emissions [20].

- Implementation: By May 2022, 36 carbon tax policies have been implemented globally [70]. Studies estimated that to achieve a 50% reduction in GHG emissions by 2050, the carbon price should be US$173 per ton. To achieve the 2050 target, the carbon price should be US$191 per ton, while the current carbon price is much lower [20].

- Advantages and disadvantages: The value of carbon emission limits in container shipping and total transport costs are negatively correlated, and higher carbon taxes can have an impact on the choice of maritime transhipment ports [71]; maritime carbon tariffs can hurt economic development, trade exports and welfare levels [72].

- Barriers: The emissions taxes and levies are largely determined by policymakers, making it difficult to reflect emission reductions and secure emission reduction targets [20]; uncertain taxes and levies also tend to influence business decisions [20].

- Future pathways: Setting prices that are commensurate with emission reduction targets [37]; to reduce price uncertainty, price bands can be set [20].

(2) Feebates

- Introduction: A variant of a tax or levy that sets benchmarks for emissions and carbon intensity, with a rebate if emissions are below the benchmark and a levy if they are above [20].

- Implementation: IMO short-term measures of carbon reduction can provide a basis for benchmarking the rebate, for example by setting an operational efficiency benchmark based on annual operational CII or using EEXI for benchmarking technical efficiency [20].

- Advantage and disadvantage: The tax and fee mechanism could easily lead to or exacerbate a two-tier market as low-emitting ships would be rewarded and high-emitting ships would be financially burdened by penalties [20].

- Barriers: It is difficult to ensure that environmental objectives are met; the feasibility and environmental benefits of a charging mechanism depend largely on the ability of the regulator to set benchmarks and rebates [20].

- Future pathways: Benchmarking tax and fee mechanisms is an important issue to address in the future [20].

(3) Emissions trading system (ETS)

- Introduction: A quantitative instrument with an emissions target or baseline set by the regulator, but not including a carbon price [20]. The measure is divided into (i) cap-and-trade systems, which fix an emissions cap and regulate emissions by auctioning or allocating emission allowances; and (ii) baseline and credit systems, which set a baseline for emission levels and issue credits to entities below the baseline level to bank or sell [20].

- Implementation: By May 2022, a total of 32 ETS policies had been implemented globally [70]. The EU began discussions in 2007 to extend its ETS to the maritime sector [73], but draft legislation to include shipping in the EU ETS had not been adopted by May 2022.

- Advantage and disadvantage: Cap-and-trade systems can achieve emission reductions quickly and cheaply, with a high degree of certainty about the effect of reductions [20]. A baseline and credit system can provide participants with an incentive to keep emissions below the baseline and avoid buying permits for excess emissions, and further reduce emissions and generate marketable credits [20]. However, the implementation of an ETS will increase costs and a market-determined carbon price may also lead to price volatility, increasing market uncertainty and business risk [74].

- Barriers: Geographical scope restrictions and the method of allocating emission allowances can easily hinder ETS implementation [73].

- Future pathways: Setting baselines related to environmental targets [37]; introducing price corridors to reduce uncertainty in carbon prices or allowances, and establishing accurate monitoring and enforcement systems to avoid non-compliance and cheating [20].

(4) Subsidies

- Introduction: Reducing the cost of zero-emission fuels by subsidizing fuel production and research and development costs [20]. Subsidies can be implemented through cash payments, tax breaks or direct financial support [20].

- Implementation: Most members of the Organization for Economic Co-operation and Development currently provide subsidies through tax breaks and fiscal advances, including tonnage taxes, special fiscal treatment for seafarers and ship fuel exemptions [20].

- Advantage and disadvantage: It helps to reduce the price of zero-emission technologies and fuels, stimulates research development and innovation, and closes the competitive gap [20], but tends to create overproduction of goods; manufacturers also risk making profits through subsidies rather than reducing the price of their products [20].

- Barriers: The lack of systematic data makes it difficult to ensure that subsidies are comprehensive, fair and transparent, thus affecting the effectiveness of subsidy policy implementation [47].

- Future pathways: Improving the transparency of subsidies [47] and, where possible, paralleling them with other policies and regulations [37].

By contrast, emissions taxes, levies and feebates are mostly price-setting measures by policymakers, which have a direct effect on emission reduction, but are less likely to correspond to emission reduction targets and to lead to a situation where "the strong are stronger and the weak are weaker", affecting market competition; subsidies can avoid distortions in market competition, but they are not sustainable and can easily lead to overproduction of commodities or diversion of subsidies; ETS incorporates the market and helps stimulate the initiative to reduce emissions, and the certainty of ETS's emission reduction effect is stronger, but the current low

https://doi.org/10.1051/e3sconf/202344103025

carbon price makes it difficult to guarantee the achievement of emission reduction targets.

4 Discussion

As for the policy objective of carbon emission reduction in international shipping, most regions are committed to achieving zero emissions and carbon neutrality by 2050, while the IMO's emission reduction target is relatively lagging, which may hinder the Paris Agreement's goal achievement of temperature control of 1.5°C. For this reason, the ICS and countries such as the UK, the US, China, Japan and Norway have proposed to adjust the IMO's initial strategic goal of emission reduction to achieve zero emissions from international shipping by 2050 [75]. At this stage, the IMO is mainly promoting the emission reduction process by improving energy efficiency and carbon intensity targets, while the EU, the UK, Singapore, Korea, and other regions have started to explore renewable, clean energy or alternative fuels to promote zero emissions from shipping; in addition, the ETS has been extended to shipping by the EU first. Overall, policy objectives and initiatives to reduce carbon emissions from international shipping are being strengthened and advanced.

Measures that have relatively minor changes to existing ships and operational mechanisms, such as engine technology, ship propulsion plant design technology, hull anti-fouling technology, and the optimization of scheduling, speed, fleet, and route, are relatively easy to implement for achieving emission reductions through fuel savings and energy efficiency improvement in shipping, but are currently limited by advancing technology and optimization models to affect their emission reduction potentials; if carbon capture devices are retrofitted, the additional business operating costs will need to be considered; the LNG technology and grey methanol are less effective in reducing emissions due to incomplete combustion or leakage of greenhouse gases or pollutants; biofuel and shore power technologies are essentially equivalent to emissions displacement, and there is uncertainty about their effectiveness from a life-cycle perspective. It can also be observed that in recent years there has been a growing interest in hydrogen, ammonia and green methanol renewable energy use and market-based measures (e.g., ETS), in addition to the active response and promotion of IMO mandatory energy efficiency management. Among these, truly clean or renewable energy sources are bound to be an important solution to decarbonising shipping and effectively achieve zero emissions but are still limited by technology, regulation, cost, and support [12, 18, 23]. While ETS can draw on market forces as opposed to taxes and subsidies and correspond to emission reduction targets, uncertainty about the price of carbon, geographical scope limitations and the allocation of emission allowances [73] have hindered their implementation, and draft legislation to including shipping in the ETS in the EU region has not yet been adopted and remains resistant. We take nature-based solutions that are inspired by or learn from nature [76] and market-based tactics as priorities for reducing added negative interventions to ecosystems and society.

As a shipping industry serving the international trade economy, carbon emission reduction from shipping has been a hot topic in shipping management in recent years in the context of the Sustainable Development Goals and the Paris Agreement temperature control targets. While this study provides a comprehensive collection of information from literature databases and relevant organizations, it focuses only on progress and the status of representative regions from 2018 to May 2022 due to time and resource constraints. The 78th IMO Marine Environment Protection Committee meeting in June 2022 also discussed carbon reduction from shipping, including support for EEXI, SEEMP and CII implementation, advancing impact assessments and mid-term mitigation measures, and revising the initial strategy for carbon reduction in shipping, which still require further follow-up. Although international shipping involves multiple administrative jurisdictions, geographic regions and disciplines, and carbon reduction policy measures are difficult to advance, the overview in this paper shows that efforts are being made by many parties, especially for the research and development of renewable energy technologies, the mandatory implementation of energy efficiency targets and the promotion of ship emissions trading systems.

5 Conclusion

Carbon emissions from international shipping threaten global climate change and ecosystem health, yet the current pace of their reduction remains slow. Through a structured review of literature databases and relevant organizations' websites, this study collects and compares the international and regional policies as well as technical, management, and economic measures of carbon emission reduction in shipping to support decision-making and academic research.

The study found that most of the regional emission reduction targets tend to be zero-emission or carbon-neutral by 2050, which is more stringent than the initial strategy targets for emission reduction released by IMO in 2018. To meet the Paris Agreement's 1.5°C temperature control target, many regions are putting pressure on IMO to revise their emission reduction strategies in mid-2023. Regions have explored renewable and low-carbon fuels and incentives to reduce emissions, but are still at the initiative and planning stage. Specifically, hydrogen, ammonia and green methanol fuels and ship wind propulsion technology measures have high emission reduction potential, but are still limited by technology development, regulatory regimes, cost investment and the level of supporting facilities; engine technology, ship propulsion technology, hull anti-fouling paint technology and shipping optimization management are relatively easy to implement, but have limited emission reduction potential; LNG technology, biofuels, grey methanol and shore power technology are also relatively easy to implement but their potentials in emissions reductions are limited from a full life cycle perspective; carbon capture technologies are constrained by cost; and the tax system tends to lead to distorted market competition and is not conducive to sustainable economic incentives; the ETS, an entirely market-dependent trading system, is also underdeveloped due to the impact of low and uncertain carbon prices. Tracing the nature of shipping carbon neutrality that aims to sustain the ecosystem and society, it is recommended to take nature-based and market-based solutions as priorities for avoiding unpredictable impacts.

Although this study focuses only on the last five years of relevant information and policies of shipping representative regions due to time and resource constraints, it is a critical period since IMO's breakthrough release of its initial emission reduction strategy and it can also reflect important progress in carbon reduction in international shipping. Nevertheless, we will further track the latest progress of international shipping carbon reduction policies and measures, and try to assess the specific impacts caused by the implementation of international shipping carbon reduction measures, including the economic costs of inputs and the external ecological benefits they generate, to support scientific decision-making on the economic incentives for international shipping carbon reduction and improve the efficiency and effectiveness of policies and measures.

References

1. Faber, J., et al., Fourth IMO GHG Study 2020 (2021)
2. Ministry of Transportation of China. Circular of the Ministry of Transport on the issuance of the "14th Five-Year Plan" for the development of green transport (2021)
3. Baresic, D., et al. Closing the Gap: An Overview of the Policy Options to Close the Competitiveness Gap and Enable an Equitable Zero-Emission Fuel Transition in Shipping (2022)
4. Bouman, E.A., et al., Transportation Research Part D-Transport and Environment, **52**, 408-421 (2017)
5. Balcombe, P., et al., Energy Conversion and Management, **182**,72-88 (2019)
6. Serra, P. and G. Fancello, Sustainability,12(8) (2020)
7. Xing, H., S. Spence, and H. Chen, Renewable & Sustainable Energy Reviews, **134** (2020)
8. Snyder, H., Journal of Business Research, **104**, 333-339 (2019_
9. Massaro, M., J. Dumay, and J. Guthrie, Accounting, Auditing & Accountability Journal, **29**(5), 767-801 (2016)
10. Wu, X., L. Zhang, and M. Luo, Environment, Development and Sustainability, **22**(3),1729-1747 (2018)
11. Al-Douri, A., et al., Canadian Journal of Chemical Engineering, **100**(6), 1178-1186 (2022)
12. Mallouppas, G., C. Ioannou, and E.A. Yfantis, Energies, **15**(4) (2022)
13. Ros, J.A., et al., International Journal of Greenhouse Gas Control, **114** (2022)
14. Oh, J., et al., Separation and Purification Technology, **282** (2022)
15. Goicoechea, N. and L.M. Abadie, Energies, **14**(22) (2021)
16. Lin, S., et al., Maritime Policy & Management, (2022)
17. IMO. IMO's work to cut GHG emissions from ships (2022)
18. ITF. Decarbonising Maritime Transport: Pathways to zero-carbon shipping by 2035 (2018)
19. Smith, T., et al. A Strategy for the Transition to Zero-Emission Shipping (2021)
20. Baresic, D., et al. Closing the Gap: An Overview of the Policy Options to Close the Competitiveness Gap and Enable an Equitable Zero-Emission Fuel Transition in Shipping (2022)
21. Hughes, E. FuelEU Maritime – Avoiding Unintended Consequences (2021)
22. European Federation for Transport and Environment. Roadmap to decarbonising European shipping (2018)
23. CE Delft and Ecorys. Assessment of impacts from accelerating the uptake of sustainable alternative fuels in maritime transport (2021)
24. Nelissen, D., A. Kleijn, and J. Faber. FuelEU Maritime and EU ETS: Sound incentives for the fuel choice? (2022)
25. IMO. The Initial IMO Strategy on Reduction of GHG Emissions from Ships (2018)
26. European Commission. A European Green Deal (2021)
27. UK Department for Transport. Maritime 2050: navigating the future (2019)
28. U.S. EPA. Climate Change (2022)
29. Maritime and Port Authority of Singapore. Maritime Singapore Decarbonisation Blueprint: Working Towards 2050 (2022)
30. MLITT of Japan. Roadmap to Zero Emission from International Shipping (2020)
31. Ministry of Oceans and Fisheries of the Republic of Korea. 2030 Greenship-K Promotion Strategy (2020)
32. Norwegian Government. The Government's action plan for green shipping (2019)
33. Norwegian Shipowners' Association. Zero emissions in 2050 (2020)
34. Baresic, D., et al. LNG as a marine fuel in the EU (2018)
35. Pavlenko, N., et al. The climate implications of using LNG as a marine fuel (2020)
36. Lv, L., How did Japan seize the opportunity to develop new energy vessels? Guangdong Shipbuilding, **40**(6),2 (2021)
37. Taylor, J., et al. Future Maritime Fuels in the USA – the options and potential pathways (2022)

38. Rutherford, D., et al. Limiting engine power to reduce CO2 emissions from existing ships (2020)
39. IMO. Resolution MEPC.335(76) (2021)
40. Wärtsilä. Our journey to date - decarbonising shipping (2020)
41. Stec, M., et al., International Journal of Greenhouse Gas Control, **108** (2021)
42. Guler, E. and S. Ergin, International Journal of Greenhouse Gas Control, **110** (2021)
43. Romano, A. and Z. Yang, Ocean & Coastal Management, **214,** 105936 (2021)
44. Wang, S., Decarbonisation towers are coming? China Ship Inspection, (05), 82-85 (2021)
45. Yan, M., et al., Applied Bionics and Biomechanics, (2022)
46. Daniel, H., J.P.F. Trovao, and D. Williams, Etransportation, **11** (2022)
47. ITF. Maritime Subsidies: Do They Provide Value for Money? (2019)
48. Kenan, N., A. Jebali, and A. Diabat, Computers & Industrial Engineering, **165** (2022)
49. Wang, T., et al., Transportation Science, **54**(5), 1307-1331 (2020)
50. Dere, C., et al., Journal of Engineering for the Maritime Environment (2022)
51. Faber, J., J. Király, and A. Kleijn. Fleet-level compliance with the CII Regulation (2021)
52. Rodriguez, M.H., et al., International Journal of Production Economics, **243** (2022)
53. Li, H., Research on fleet deployment for container liner considering carbon emission (2019)
54. Boren, C., M. Castells-Sanabra, and M. Grifoll, Journal of Engineering for the Maritime Environment (2022)
55. Zhang, S., Methods to rate the operational energy efficiency of ocean-going ships engaged in international voyages (2020)
56. Rutherford, D., X. Mao, and B. Comer. Potential CO2 reductions under the Energy Efficiency Existing Ship Index (2020)
57. IMO. Draft life cycle GHG and carbon intensity guidelines for maritime fuels (2020)
58. Lu, W., Research on ship carbon emission prediction based on AIS data mining (2019)
59. Cao, Y., et al., Applied Soft Computing, **104** (2021)
60. Czermanski, E., et al., Energies, **14**(2) (2021)
61. Dou, X., et al., The Innovation, **3**(1), 100182-100182 (2022)
62. Ma, X., Study on driving factors and mitigation policies of CO_2 emissions from China's international seaborne freight transport (2020)
63. Register, L.s. and UMAS. Techno-economic assessment of zero-carbon fuels (2020)
64. Stolz, B., et al., Nature Energy, **7**(2), 203-212 (2022)
65. Czermanski, E., et al., Frontiers in Energy Research, **8** (2020)
66. Aseel, S., et al., Energies, **14**(19) (2021)
67. Hu, H., J. Yuan, and V. Nian, Transport Policy, **82**, 148-157 (2019)
68. Percic, M., et al., Applied Energy, **309** (2022)
69. Tai, H.-H. and Y.-H. Chang, Reducing pollutant emissions from vessel manoeuvring in port areas. Maritime Economics & Logistics (2022)
70. World Bank. Carbon Pricing Dashboard (2022)
71. Xu, L. and M. Yanping, Journal of Shanghai Maritime University, **40**(02), 12-17 (2019)
72. Hu, J., Research on the impact of maritime carbon tariff on China's economy (2020)
73. Christodoulou, A., et al., Energies, **14**(13) (2021)
74. Parker, S., et al. Harnessing the EU ETS to reduce international shipping emissions (2022)
75. Marine Environment Protection Committee. MEPC Session 78 (2022)
76. Cohen-Shacham, E., et al. Nature-based Solutions to address global societal challenges (2016)

AUTHOR INDEX